VOLUME ONE HUNDRED AND SIX

VITAMINS AND HORMONES

Thyroid Hormone

VOLUME ONE HUNDRED AND SIX

VITAMINS AND HORMONES

Thyroid Hormone

Series Editor

GERALD LITWACK, PhD

Toluca Lake, North Hollywood, California

ELSEVIER

ACADEMIC PRESS

An imprint of Elsevier

Cover photo credit:
Moitreyi Das, Mausam Ghosh, Kusumika Gharami, Sumantra Das
Thyroid Hormone and Astrocyte Differentiation
Vitamins and Hormones (2018) **106**, pp. 283–312.

Academic Press is an imprint of Elsevier
50 Hampshire Street, 5th Floor, Cambridge, MA 02139, United States
525 B Street, Suite 1800, San Diego, CA 92101-4495, United States
The Boulevard, Langford Lane, Kidlington, Oxford OX5 1GB, United Kingdom
125 London Wall, London, EC2Y 5AS, United Kingdom

First edition 2018

ISBN: 978-0-12-814116-8
ISSN: 0083-6729

Publisher: Zoe Kruze
Acquisition Editor: Ashlie Jackman
Editorial Project Manager: Joanna Collett
Production Project Manager: Vignesh Tamil
Cover Designer: Victoria Pearson

Typeset by SPi Global, India

CONTENTS

Contributors *xv*
Preface *xix*

1. Local Regulation of Thyroid Hormone Signaling **1**
Alexander G. Little

1. Introduction 2
2. Bioavailability 3
3. Bioactivity 7
4. Conclusions 11

References 12
Further Reading 17

2. Thyroid Hormone Transport and Transporters **19**
Doreen Braun and Ulrich Schweizer

1. Introduction 20
2. Thyroid hormone Transport 22
3. Monocarboxylate Transporter 8 23
4. Mouse Models of Allan–Herndon–Dudley Syndrome 33
5. Thyroid Hormone Transport and Physiology 35
6. Further Clinical Implications for Thyroid Hormone Transporters 37
7. Conclusions and Future Directions 38

Acknowledgments 39
References 39

3. Nuclear Import and Export of the Thyroid Hormone Receptor **45**
Jibo Zhang, Vincent R. Roggero, and Lizabeth A. Allison

1. Introduction 46
2. Thyroid Hormone Receptor Nucleocytoplasmic Shuttling 47
3. Nuclear Import Pathways of TR 51
4. Nuclear Export Pathways of TRα1 54
5. Mislocalization of TR and Pathogenesis 58
6. Conclusions and Future Directions 61

Acknowledgments 62
References 62

4. Molecular Basis of Nongenomic Actions of Thyroid Hormone **67**

Paul J. Davis, Jack L. Leonard, Hung-Yun Lin, Matthew Leinung, and Shaker A. Mousa

1. Introduction 68
2. Regulation of the State of the Cytoskeleton by Thyroid Hormone 70
3. Regulation of Mitochondrial Respiration by Thyroid Hormone 71
4. Nongenomic Actions of Truncated and Intact TRs in Cytoplasm on Bone Cells, Neurons, and Tumor Suppression 72
5. Cytoplasmic Proteins in Addition to TRs That Mediate Nongenomic Actions of Thyroid Hormone 75
6. Nongenomic Actions of Thyroid Hormone Initiated at Plasma Membrane 75
7. Thyroid Hormone Receptor on Plasma Membrane Integrin αvβ3 76
8. Thyroid Hormone αvβ3 Receptor and Angiogenesis 77
9. Thyroid Hormone αvβ3 Receptor and Tumor Cell Proliferation 78
10. Thyroid Hormone αvβ3 Receptor and Regulation of Gene Expression 81
11. Thyroid Hormone αvβ3 Receptor and Cell Physiology 85
12. Discussion 86

Acknowledgments 90

References 90

5. The Mechanism of Negative Transcriptional Regulation by Thyroid Hormone: Lessons From the Thyrotropin β Subunit Gene **97**

Shigekazu Sasaki, Akio Matsushita, Go Kuroda, Hiroko M. Nakamura, Yutaka Oki, and Takafumi Suda

1. Introduction 98
2. Structure and Function of Thyroid Hormone Receptors 99
3. Mechanism of T3-Dependent Positive Regulation 101
4. Does a nTRE Exist in the TSHβ Gene? 103
5. Negative Regulation of the TSHβ Gene Is Not the Mirror Image of Positive Regulation 105
6. Basal Expression of the TSHβ Gene Is Maintained by Thyrotroph-Specific Transcription Factors, GATA2 and Pit1 106
7. A CV1 Cell Reconstitution System That Mimics Thyrotrophs 108
8. TR Is Tethered by GATA2 to Mediate Negative Regulation of the TSHβ Gene by T3 108
9. NHRs Tethered by DNA-Binding Transcription Factor Mediate Ligand-Dependent Negative Regulation 109
10. Negative Regulation of Genes Other Than the TSHβ Gene May Not Be the Mirror Image of Positive Regulation 110
11. Negative Regulation of Genes Other Than the TSHβ Gene May Be Mediated via a Tethering Mechanism 111

12. Target Molecule of T3-Bound TRs Tethered by GATA2 112
13. Artifactual Negative Regulation via Plasmid Backbones 116
14. Conclusion and Future Directions 117
Acknowledgments 117
References 118

6. Thyroid Autoimmunity: An Interplay of Factors 129

Stephen J. Merrill and Sarah B. Minucci

1. Introduction 130
2. Interacting Factors 132
3. Results and Discussion 138
4. Conclusion 139
Acknowledgment 139
References 139

7. Thyroid Hormone Receptor Antagonists: From Environmental Pollution to Novel Small Molecules 147

Louise S. Mackenzie

1. Introduction 148
2. TR Antagonists 149
3. Synthetic TR Antagonists 155
4. Environmental TR Antagonists 158
5. Concluding Remarks 159
References 159

8. Thyroid Hormone Stimulation of Adult Brain Fatty Acid Oxidation 163

Mikaela M. Sifuentes and James D. Lechleiter

1. Introduction 164
2. Nuclear and Mitochondrial Thyroid Hormone Receptor Isoforms 165
3. Thyroid Hormone and Brain Metabolism 167
4. Thyroid Hormone Regulation of Fatty Acid Oxidation 169
5. Clinical Applications 180
6. Conclusion 184
References 185

9. The NF-κB Family of Transcription Factors and Its Role in Thyroid Physiology 195

Carla Reale, Tiziana Zotti, Ivan Scudiero, Pasquale Vito, and Romania Stilo

1. Introduction 196
2. The NF-κB Family 196

3. Canonical Activation of NF-κB 201
4. NEMO/IKKγ-Dependent IKK Complex Activation 202
5. NF-κB in Thyroid Physiology 204
6. Conclusion 206
References 207

10. Thyroid Hormone Regulation of Adult Neurogenesis 211

Sashaina E. Fanibunda, Lynette A. Desouza, Richa Kapoor, Rama A. Vaidya, and Vidita A. Vaidya

1. Thyroid Hormone and the Adult Brain 213
2. Making New Neurons in the Adult Brain: A Role for Thyroid Hormone 215
3. Thyroid Hormone and Adult Hippocampal Neurogenesis 216
4. Thyroid Hormone Influence on Adult SVZ Neurogenesis 227
5. Underlying Molecular Mediators for the Regulation of Neurogenesis by Thyroid Hormone 232
6. Functional Implications of the Neurogenic Effects of Thyroid Hormone 235
7. Conclusion 239
Acknowledgments 240
References 240

11. Thyroid Hormone and the White Matter of the Central Nervous System: From Development to Repair 253

Laura Calzà, Vito A. Baldassarro, Mercedes Fernandez, Alessandro Giuliani, Luca Lorenzini, and Luciana Giardino

1. Introduction 254
2. TH Metabolism in the CNS 255
3. Tissue Hypothyroidism in the CNS 261
4. TH, Developmental Myelination and Neonatal Hypoxic/Ischemic Insults 265
5. TH and Myelin Turnover/Repair in Adulthood 267
6. Therapeutic Implications 271
Acknowledgments 272
References 272
Further Reading 280

12. Thyroid Hormone and Astrocyte Differentiation 283

Moitreyi Das, Mausam Ghosh, Kusumika Gharami, and Sumantra Das

1. Introduction 283
2. Morphology of Astrocytes During Brain Development 285
3. Morphology of Astrocytes in Diseases 287
4. Regulators of Astrocyte Morphology 288
5. THs on Astrocyte Function 291

6. Effect of THs on Astrocyte Morphogenesis 293
7. Mechanism of TH Action on Astrocyte Morphogenesis 297
8. Conclusion 305
References 306

13. Thyroid Hormone in the CNS: Contribution of Neuron–Glia Interaction 313

Mami Noda

1. Introduction 314
2. Receptors and Function of TH in the CNS 314
3. Abnormal Level of TH in the CNS 315
4. Transportation of THs to the Brain and Metabolism in Astrocytes 317
5. Functional Role of TH in Microglia 317
6. Astrocytes and Oligodendrocytes Differentiation by T3 322
7. Pathological Effects of TH on Glial Cells 323
8. Sex- and Age-Dependent Effects of Hyperthyroidism 323
9. Possible Influence of TH-Induced Neuron–Glia Interaction on Psychiatric Symptoms 324
10. Conclusion and Future Directions 325
Acknowledgment 326
References 326

14. Thyroid Hormone Signaling in Retinal Development, Survival, and Disease 333

Fan Yang, Hongwei Ma, and Xi-Qin Ding

1. Introduction 334
2. Retinal Structure and Function 335
3. TH and Retinal Development 337
4. TH and Photoreceptor Survival 342
5. TH and Retinal Diseases 344
6. Conclusion and Future Perspectives 345
Acknowledgments 346
References 346

15. Thyroid Hormone Signaling in the Development of the Endochondral Skeleton 351

Richard C. Lindsey, Patrick Aghajanian, and Subburaman Mohan

1. Introduction 352
2. TH Physiology 353
3. TH in the Developing Skeleton 357
4. TH in the Adult Skeleton 364

5. Skeletal Effects of TR Signaling 365
6. Summary and Conclusions 368
Acknowledgments 370
References 370

16. Thyroid Hormone and Skeletal Development 383
Cecilia H.A. Gouveia, Manuela Miranda-Rodrigues, Gisele M. Martins, and Bianca Neofiti-Papi
1. Introduction 384
2. Overview of Skeletal Development 386
3. Overview of TH Physiology 389
4. Expression of TH Physiology Components in the Skeleton 393
5. The Effects of TH 398
6. Mechanisms of Action of TH to Promote Skeletal Development 410
7. The Skeletal Phenotype and Development in TH-Physiology-Linked-Mutant Mice 429
8. The Skeletal Phenotype and Development in Patients With RTH 444
9. Concluding Remarks 449
10. Final Conclusions and Future Directions 452
Acknowledgments 452
References 453

17. Thyroid Hormone Role and Economy in the Developing Testis 473
Arturo Hernandez
1. Introduction 474
2. Thyroid Hormone Status and the Developing Testis: Studies on Animal Models of Developmental Hypo- and Hyperthyroidism 475
3. Endogenous Determinants of TH Action in the Developing Testis 482
4. Conclusion 491
Acknowledgments 492
References 492

18. Signaling Pathways in Thyroid Cancer 501
Shan Jin, Yun-Tian Yang, and Wuyuntu Bao
1. Introduction 501
2. Thyroid Cancer-Related Signaling Pathways 502
3. Residual Tumor-Related Signaling Pathways 504
4. "Reclaim Therapy"-Related Signaling Pathways 505
5. Improvement in Anticancer Ability and Signaling Pathways 507

6. Thyroid Cancer-Targeted Therapeutic Drugs 508
7. Conclusion and Future Directions 510
Acknowledgments 511
References 511

19. Roles of Thyroid Transcription Factor 1 in Lung Cancer Biology 517

Cody A. Phelps, Shao-Chiang Lai, and David Mu

1. Introduction 518
2. Genetic Alterations of TTF-1 in Cancer 518
3. Dynamics of Gene Content of a Core Amplicon 519
4. TTF-1$^-$ Lung Squamous Cell Carcinomas and TTF-1$^-$ Lung ADs 520
5. TTF-1 and Small Cell Lung Cancer 521
6. Pro- and Antioncogenic Activities of TTF-1 522
7. The Link Between TTF-1 and Lipid/Cholesterol Metabolism in the Lung 530
8. The Connection Between TTF-1 and Secretome 532
9. Conclusion and Future Directions 533
Acknowledgment 533
References 535

CONTRIBUTORS

Patrick Aghajanian
Musculoskeletal Disease Center, Loma Linda VA Healthcare System, Loma Linda, CA, United States

Lizabeth A. Allison
College of William and Mary, Williamsburg, VA, United States

Vito A. Baldassarro
CIRI-SDV; FaBit, University of Bologna, Bologna; IRET Foundation, Ozzano Emilia, Italy

Wuyuntu Bao
Affiliated Hospital of Inner Mongolia Medical University, Hohhot, China

Doreen Braun
Institut für Biochemie und Molekularbiologie, Rheinische Friedrich-Wilhelms-Universität Bonn, Bonn, Germany

Laura Calzà
CIRI-SDV; FaBit, University of Bologna, Bologna; IRET Foundation, Ozzano Emilia, Italy

Moitreyi Das
Indian Institute of Chemical Biology, Kolkata, India

Sumantra Das
Indian Institute of Chemical Biology, Kolkata, India

Paul J. Davis
Albany Medical College, Albany; Pharmaceutical Research Institute, Albany College of Pharmacy and Health Sciences, Rensselaer, NY, United States

Lynette A. Desouza
Tata Institute of Fundamental Research, Mumbai, India

Xi-Qin Ding
University of Oklahoma Health Sciences Center, Oklahoma City, OK, United States

Sashaina E. Fanibunda
Tata Institute of Fundamental Research, Mumbai, India

Mercedes Fernandez
FaBit, University of Bologna, Bologna, Italy

Kusumika Gharami
Indian Institute of Chemical Biology, Kolkata, India

Mausam Ghosh
Indian Institute of Chemical Biology, Kolkata, India

Luciana Giardino
CIRI-SDV; DIMEVET, University of Bologna, Bologna; IRET Foundation, Ozzano Emilia, Italy

Alessandro Giuliani
DIMEVET, University of Bologna, Bologna, Italy

Cecilia H.A. Gouveia
Institute of Biomedical Sciences; Experimental Pathophysiology Program, School of Medicine, University of São Paulo, São Paulo, SP, Brazil

Arturo Hernandez
Center for Molecular Medicine, Maine Medical Center Research Institute, Scarborough, ME, United States

Shan Jin
Affiliated Hospital of Inner Mongolia Medical University, Hohhot, China

Richa Kapoor
Tata Institute of Fundamental Research, Mumbai, India

Go Kuroda
Hamamatsu University School of Medicine, Shizuoka, Japan

Shao-Chiang Lai
Leroy T. Canoles Jr. Cancer Research Center, Eastern Virginia Medical School, Norfolk, VA, United States

James D. Lechleiter
UT Health San Antonio, San Antonio, TX, United States

Matthew Leinung
Albany Medical College, Albany, NY, United States

Jack L. Leonard
University of Massachusetts Medical School, Worcester, MA, United States

Hung-Yun Lin
PhD Program for Cancer Biology and Drug Discovery, College of Medical Science and Technology, Taipei Medical University, Taipei, Taiwan

Richard C. Lindsey
Musculoskeletal Disease Center, Loma Linda VA Healthcare System; Loma Linda University School of Medicine, Loma Linda, CA, United States

Alexander G. Little
Rosenstiel School for Marine and Atmospheric Science, University of Miami, Miami, FL, United States

Luca Lorenzini
IRET Foundation, Ozzano Emilia, Italy

Hongwei Ma
University of Oklahoma Health Sciences Center, Oklahoma City, OK, United States

Louise S. Mackenzie
School of Life and Medical Sciences, Hertfordshire University, Hatfield, United Kingdom

Gisele M. Martins
Institute of Biomedical Sciences; Experimental Pathophysiology Program, School of Medicine, University of São Paulo, São Paulo, SP; Federal University of Espírito Santo, Vitória, ES, Brazil

Akio Matsushita
Hamamatsu University School of Medicine, Shizuoka, Japan

Stephen J. Merrill
Marquette University, Milwaukee, WI, United States

Sarah B. Minucci
Virginia Commonwealth University, Richmond, VA, United States

Manuela Miranda-Rodrigues
Institute of Biomedical Sciences, University of São Paulo, São Paulo, SP, Brazil

Subburaman Mohan
Musculoskeletal Disease Center, Loma Linda VA Healthcare System; Loma Linda University School of Medicine, Loma Linda, CA, United States

Shaker A. Mousa
Pharmaceutical Research Institute, Albany College of Pharmacy and Health Sciences, Rensselaer, NY, United States

David Mu
Leroy T. Canoles Jr. Cancer Research Center, Eastern Virginia Medical School, Norfolk, VA, United States

Hiroko M. Nakamura
Hamamatsu University School of Medicine, Shizuoka, Japan

Mami Noda
Laboratory of Pathophysiology, Graduate School of Pharmaceutical Sciences, Kyushu University, Fukuoka, Japan

Yutaka Oki
Hamamatsu University School of Medicine, Shizuoka, Japan

Bianca Neofiti-Papi
Institute of Biomedical Sciences; Experimental Pathophysiology Program, School of Medicine, University of São Paulo, São Paulo, SP, Brazil

Cody A. Phelps
Leroy T. Canoles Jr. Cancer Research Center, Eastern Virginia Medical School, Norfolk, VA, United States

Carla Reale
Biogem Consortium, Ariano Irpino, Italy

Vincent R. Roggero
College of William and Mary, Williamsburg, VA, United States

Shigekazu Sasaki
Hamamatsu University School of Medicine, Shizuoka, Japan

Ulrich Schweizer
Institut für Biochemie und Molekularbiologie, Rheinische Friedrich-Wilhelms-Universität Bonn, Bonn, Germany

Ivan Scudiero
Biogem Consortium, Ariano Irpino, Italy

Mikaela M. Sifuentes
UT Health San Antonio, San Antonio, TX, United States

Romania Stilo
Università del Sannio, Benevento, Italy

Takafumi Suda
Hamamatsu University School of Medicine, Shizuoka, Japan

Rama A. Vaidya
Medical Research Centre, Kasturba Health Society, Mumbai, India

Vidita A. Vaidya
Tata Institute of Fundamental Research, Mumbai, India

Pasquale Vito
Biogem Consortium, Ariano Irpino; Genus Biotechnology; Università del Sannio, Benevento, Italy

Fan Yang
University of Oklahoma Health Sciences Center, Oklahoma City, OK, United States

Yun-Tian Yang
Affiliated Hospital of Inner Mongolia Medical University, Hohhot, China

Jibo Zhang
College of William and Mary, Williamsburg, VA, United States

Tiziana Zotti
Genus Biotechnology, Università del Sannio, Benevento, Italy

PREFACE

The thyroid gland has been referred to as "the master gland" because it secretes the thyroid hormones that control the metabolic—and protein synthetic—rates of most of the cells of the body. Thyroid follicle cells respond positively to the thyroid-stimulating hormone (TSH) secreted by the anterior pituitary. In turn, TSH is stimulated by the thyroid-releasing hormone (TRH) that is secreted by the paraventricular nucleus in the hypothalamus. The secretion of TRH is regulated by neuropeptide Y, leptin from the adipocyte and two melanocyte-related peptides, alpha-melanocyte-stimulating hormone, and the agouti-related peptide. TSH binds and activates its receptor on the plasma membrane of the thyroid follicle cell and stimulates the passage of iodide through the sodium-iodide cotransporter (symporter [NIS]) to the cell interior. Iodide is subsequently oxidized to iodine, which is, in turn, transported into the follicle where thyroglobulin is stored. Thyroglobulin is iodinated and subsequently cleaved, after return to the cell, to yield the two thyroid hormones, thyroxine (T4) and triiodothyronine (T3) and other smaller iodinated products that can be recycled for further synthesis of hormones. Although the amount of T4 secreted into the circulation greatly exceeds the amount of T3, it is the latter which is the preferred ligand for the nuclear thyroid hormone receptor of target cells. T4 also can be converted to T3 by deiodinases inside the target cell. Finally, elevation of thyroid hormones in the circulation feed backs negatively on the secretion of TRH and TSH, providing a homeostatic mechanism.

Much of the information in this volume deals with the basic aspects of thyroid hormone actions. There is emphasis on signaling pathways, transport, nongenomic effects, autoimmunity, antagonism, physiology, and activities in specific tissues and cancers.

As in the past, the information in this book proceeds from basic research aspects to organ systems to disease connections. Basic science opens with a chapter on: "Local regulation of thyroid hormone signaling" by A.G. Little. The next chapter is entitled: "Thyroid hormone transport and transporters" by D. Braun and U. Schweizer. J. Zhang, V.R. Roggero, and L.A. Allison focus on: "Nuclear import and export of the thyroid hormone receptor." The "Molecular basis of nongenomic actions of thyroid hormone" is the topic of P.J. Davis, J.L. Leonard, H.-Y. Lin, M. Leinung, and S.A. Mousa. This is followed by "The mechanism of negative transcriptional regulation by thyroid hormone: lessons from the thyrotropin β subunit gene" by

S. Sasaki, A. Matsushita, G. Kuroda, H.M. Nakamura, Y. Oki, and T. Suda. S.J. Merrill and S.B. Minucci describe: "Thyroid autoimmunity: an interplay of factors." "Thyroid hormone receptor antagonists: from environmental pollution to novel small molecules" is the subject of a chapter by L.S. Mackenzie. "Thyroid hormone stimulation of adult brain fatty acid oxidation" is discussed by M.M. Sifuentes and J.D. Lechleiter. The last chapter in the basic science section is: "The NF-κB family of transcription factors and its role in thyroid physiology" by C. Reale, T. Zotti, I. Scudiero, P. Vito, and R. Stilo.

The section on tissues starts with "Thyroid hormone regulation of adult neurogenesis" by S.F. Fanibunda, L.A. Desouza, R. Kapoor, R.A. Vaidya, and V.A. Vaidya. L. Calzà, V.A. Baldassarro, M. Fernandez, A. Giuliani, L. Lorenzini, and L. Giardino contribute: "Thyroid hormone and the white matter of the central nervous system: from development to repair" and M. Das, M. Ghosh, K. Gharami, and S. Das report on "Thyroid hormone and astrocyte differentiation." Furthermore, M. Noda focused on: "Thyroid hormone in the CNS: contribution of neuron–glia interaction." Next, "Thyroid hormone signaling in retinal development, survival, and disease" is provided by F. Yang, H. Ma, and X.-Q. Ding. Continuing in this section, R.C. Lindsey, P. Aghajanian, and S. Mohan report on: "Thyroid hormone signaling in the development of the endochondral skeleton" and C.H. Gouveia, M. Miranda-Rodrigues, G.M. Martins, and B.N. Papi describe: "Thyroid hormone and skeletal development." This section is completed by A. Hernandez whose chapter is on: "Thyroid hormone role and economy in the developing testis."

The last section consists of two chapters, one by S. Jin, Y.-T. Yang, and W. Bao who contribute: "Signaling pathways in thyroid cancer" and by C.A. Phelps, S.-C. Lai, and D. Mu who describe: "Roles of thyroid transcription factor 1 in lung cancer biology."

The illustration on the cover is Fig. 9 from chapter 12 by M. Das, M. Ghosh, K. Gharami, and S. Das entitled: "Thyroid hormone and astrocyte differentiation." The legend to this figure is: "Model of the TH-mediated signaling leading to differentiation of astrocytes."

The cooperation of Helene Kabes of Elsevier, Oxford, United Kingdom, and of Vignesh Tamilselvvan of Elsevier/S and T Book Production, Chennai, India, in the production of this book is most appreciated.

Gerald Litwack
North Hollywood, CA
July 6, 2017

CHAPTER ONE

Local Regulation of Thyroid Hormone Signaling

Alexander G. Little[1]
Rosenstiel School for Marine and Atmospheric Science, University of Miami, Miami, FL, United States
[1]Corresponding author: e-mail address: alexander.little@utoronto.ca

Contents

1. Introduction 2
2. Bioavailability 3
   2.1 TH Transporters 3
   2.2 Deiodinases 4
   2.3 Intracellular Reservoir Proteins 6
3. Bioactivity 7
   3.1 Thyroid Receptors 7
   3.2 Integrin αvβ3 9
   3.3 Thyroid Response Elements 10
4. Conclusions 11
References 12
Further Reading 17

Abstract

Thyroid hormone (TH) plays important signaling roles in mammalian growth, development, and thermogenesis. Traditionally its actions were thought to be regulated predominately through modulation of free plasma concentrations, intracellular breakdown by deiodinase enzymes and nuclear processing by thyroid receptors. In the past decade, however, there has been some departure from this classical model, whereby regulatory changes at different levels of organization simultaneously modulate its bioavailability and bioactivity. Here I review the recent literature to emphasize how subtle changes at these various levels of regulation can underlie tissue-specific and temporal (development and disease) changes in mammalian systems. For instance, it has become increasingly clear that TH acts both centrally to control sympathetic output and peripherally to regulate metabolism in target tissues. An interesting caveat is that central actions by TH may simultaneously alter its effects at peripheral tissues. In addition to T3, other TH derivatives (e.g., rT_3 and T_2) have also been shown to possess physiological activity, meaning intracellular processing by deiodinase enzymes is likely more complex than previously thought. Importantly, these TH derivatives can act both through the traditional genomic thyroid receptor pathways for transcriptional control, or more direct, fast-acting, nongenomic mechanisms. This fine-tuning means

Vitamins and Hormones, Volume 106
ISSN 0083-6729
http://dx.doi.org/10.1016/bs.vh.2017.06.004

tissue-specific, ontogenic, and pathological changes in TH processing at any one level can have different consequences for TH bioavailability and bioactivity depending on parallel changes in upstream and/or downstream pathways.

1. INTRODUCTION

The classical model for TH signaling involves centrally mediated release of TH into the bloodstream, where it transcriptionally regulates energy metabolism through its interaction with thyroid receptors (TRs) in the nuclei of target cells. Recently, however, there has been some departure from this classical model. For instance, it is now known that other TH derivatives (e.g., rT_3, T_2, TETRAC, and TRIAC) also possess physiological activity and that THs and their derivatives can act outside the classical "genomic" pathway (Cioffi, Senese, Lanni, & Goglia, 2013). Accordingly, these responses have been termed "nongenomic" (Lösel & Wehling, 2003), meaning their actions are initiated outside the nucleus, but not necessarily implying the absence of downstream changes in transcriptional programs. Importantly, the genomic and nongenomic mechanisms of TH action do not always represent mutually exclusive processes, but can occur sequentially or simultaneously (see Hammes & Davis, 2015 for review).

TH actions can be regulated peripherally, where TH acts directly at peripheral tissues, or centrally, where its actions are mediated via the CNS (typically the hypothalamus; see Mullur, Liu, & Brent, 2014; Warner & Mittag, 2012). Importantly, ontogenetic differences exist in the way cells and tissues process circulating concentrations of free-plasma TH (Bianco, 2011). These local levels of regulation are important because they represent the fundamental basis for tissue specificity of TH action. Additionally, many mechanisms for the local regulation of TH are plastic. This means cells and tissues are not passive in TH signaling, but instead can dampen, amplify, or redirect signaling pathways independently from fluctuations in circulating levels of free-TH to fine-tune its actions.

My aim here is to provide an overview for local regulation of TH signaling in mammals, emphasizing how tissue-specific differences or temporal changes at these local levels can have major effects on TH action. As such, this review is does not represent a comprehensive summary for each level of regulation independently. For this purpose, I direct readers to a number of great reviews. For simplicity, I categorize these levels of regulation based on whether they modulate local TH bioavailability or bioactivity. While I take a systematic approach to these local levels of TH regulation and their implications for downstream physiology, it should be noted that developmental requirements, stressor responses, and various disease states are likely driven by changes at multiple levels of regulation, rather than any one level exclusively.

2. BIOAVAILABILITY

Bioavailability of TH is regulated locally by intracellular transporters at the cell surface, deiodinase enzymes, and cytosolic reservoir proteins. These pathways represent mechanisms whereby target cells and tissues can uncouple intracellular concentrations of TH from circulating levels to dampen or amplify local actions.

2.1 TH Transporters

Owing to its relatively large size and uncharged nature, TH has only low rates of passive diffusion across the plasma membrane (Bernal, Guadaño-Ferraz, & Morte, 2015). Thus, target cells, with some exceptions (see integrin *αvβ3* later), must express membrane transporters to facilitate TH uptake (Friesema, Docter, Moerings, et al., 1999). There are a growing list of proteins known to transport TH, but the majority are promiscuous in terms of their ligand specificity, where TH uptake likely represents a secondary or tertiary function (see Bernal et al., 2015 for review). A handful of TH transporters, including monocarboxylate transporters 8 and 10 (MCT8 and MCT10), L-type amino acid transporter 1 and 2 (LAT1 and LAT2), and the organic anion-transporting polypeptide 1c1 (OATP1C1), however, have exclusive or high specificity for TH and its derivatives (see Visser & Friesema, 2011 for review).

MCT8 is expressed predominately in brain, heart, liver, kidney, and the adrenal and thyroid glands, whereas MCT10 is expressed in intestine, liver, kidney, muscle, and placenta (see Visser & Friesema, 2011). High expression levels of these transporters in liver, kidney, and the thyroid gland likely promote the release of newly derived T_3 into the bloodstream (Müller, Mayerl, Visser, et al., 1999). MCT8 facilitates transport of T_4, T_3, and rT_3, whereas MCT10 transports T_4 and T_3. Importantly, however, transport by MCT8 and MCT10 is bidirectional, where it drives TH influx and efflux simultaneously to achieve rapid equilibrium (Friesema, Jansen, Jachtenberg, et al., 2008). Recent work shows that increasing expression levels of MCT8 and MCT10 enhances intracellular T_3 accumulation, but does not significantly alter nuclear T_3 activity in human choriocarcinoma cells (van Mullem et al., 2016). Parallel increases in T_3 metabolism by deiodinase enzyme 3 suggest that the bidirectionality of MCT8 and MCT10 transport acts to concentrate T_3 at the cell periphery, where it is metabolized to T_2 by deiodinase enzyme 3 (see later; van Mullem et al., 2016). This means that MCT8 and MCT10 not only mediate intracellular concentrations of T_3, but their specific expression profiles determine T_3 sublocalization in ways that promote either its nuclear activity or its cytoplasmic activity (see later) and metabolism into rT_3 and T_2 (i.e., at high expression levels). It is important to note, however, that the expression of cytosolic TH-binding proteins provides another layer of regulatory control for these effects (see later). This provides a scenario whereby D3 activity, and therefore rT_3 and T_2 production, depends not only on D3 expression levels but also on TH transport dynamics that sublocalize T_3 to the cell periphery.

In mammals, LAT1 and 2 are expressed almost ubiquitously (see Kinne, Schülein, & Krause, 2011), whereas OATP1C1 is expressed in the Leydig cells of the testis and in brain barrier cells (Visser & Friesema, 2011; Visser, Friesema, Jansen, & Visser, 2008). LAT1 and 2 have relatively high affinity for T_2 (3,3′), whereas OATP1C1 has highest affinity for rT_3 and T_4 (Kinne et al., 2011). Relatively little work has been carried out on TH transport by LAT1 and 2, presumably because THs represent only a subset of their target ligands (del Amo, Urtti, & Yliperttula, 2008). While OATP1C1 appears to be a major transporter of THs in mice, it is questionable whether it plays a similar role in primates and other mammals (Mayerl, Müller, Bauer, et al., 2014). On the other hand, OATP1C1 polymorphisms have been associated with fatigue and depression in humans, indicating some sort of physiological role for TH transport (Van Der Deure, Appelhof, Peeters, et al., 2008). OATP1C1 is expressed inversely proportional to TH levels during hypo- and hyperthyroidism (Roberts, Woodford, Zhou, et al., 2008), which may suggest a compensation strategy to maintain intracellular TH concentrations.

2.2 Deiodinases

Deiodinases are a group of selenoproteins that have long been known to regulate the bioavailability of active THs in target tissues (Bianco, 2011). They thereby represent one of the main mechanisms underlying tissue specificity in the classical TH signaling model (Christensen, Cammack, & Wray, 1995; Köhrle, 1995). These enzymes function to increase TH activity by metabolizing inactive TH precursors to their active form(s) or reduce activity by deactivating active THs (Bianco & Kim, 2006). Mammals have three deiodinase paralogs, dio1, dio2, and dio3, which encode isozymes D1, D2, and D3, respectively (see Bianco & Kim, 2006). These membrane-anchored proteins form homodimers to catalyze the removal of iodo groups from precursor iodothyronines, with different isozymes targeting outer- vs inner-ring iodo groups. For instance, D3 removes an iodo group from the inner-ring, thereby metabolizing T_4 into rT_3, and T_3 into T_2 (Bianco, 2011). D2 removes iodo groups from the outer-ring, thereby metabolizing T_4 into T_3, and reverse rT_3 into T_2 (Bianco, 2011). D1 on the other hand removes iodo groups from both the outer- and inner-rings, and may therefore function as a scavenger enzyme, liberating iodine from inactive metabolites (Darras, Houbrechts, & Van Herck, 2015). It is also suggested, however, that D1 may function to produce T_3 in the liver, kidney, and thyroid gland for release into circulation (Galton, 2005).

In the classical sense, D2 would promote TH activity by increasing relative concentrations of T_3, while D3 would reduce activity by decreasing intracellular T_3. There is increasing evidence, however, that downstream metabolites, such as rT_3 and T_2, possess biological activity (see Senese, Cioffi, De Lange, Goglia, & Lanni, 2014). Thus, the role of intracellular deiodinase metabolism is likely more complex than originally thought, meaning deiodinases may not merely serve as a switch for T_3 activity, but may also tailor TH actions by modulating the relative concentrations of its active derivatives. In mammals, deiodinases are expressed ubiquitously throughout development, and in a wide range of tissues (see Galton, 2005 for review). D2 is sublocalized in the endoplasmic reticulum, where it metabolizes T_4 to T_3 for nuclear import (Arrojo e Drigo et al., 2011). D3 on the other hand is sublocalized in the plasma membrane, where it generally functions to protect cells and tissues from excessive levels of T_3 (Arrojo e Drigo et al., 2011). Thus, barring the recently discovered activities of rT_3 and T_2, D2 and D3 work almost antagonistically to modulate intracellular levels of T_3 (Galton et al., 2014). While D2 has a relatively short half-life of approximately 2h, the half-life of T_3 is closer to 12h (Baqui, Botero, Gereben, et al., 2003). These differences in half-lives suggest that rapid changes in intracellular T_3 levels are likely more effective via regulatory changes in expression of D2, rather than D3. Accordingly, D2 plays a primary role in rapid thermogenic response to cold exposure in mammals, where an approximately 26-fold increase in brown adipose D2 activity enhances intracellular concentrations of T_3 to promote oxidative metabolism and mitochondrial uncoupling for heat production (Arrojo e Drigo et al., 2011).

D2 is also critical in the T_4-mediated negative feedback loop, regulated through the hypothalamic–pituitary axis (Arrojo e Drigo et al., 2011), where increased D2 activity in the pituitary promotes the negative feedback response (Rosene, Wittmann, Drigo, et al., 2013). This is interesting because pituitary D2 is itself downregulated by rising levels of T_4 (Abend, Fang, Alex, et al., 1991), suggesting that rising levels of T_4 depress TH secretion at the thyroid gland, while simultaneously dampening its own feedback response (Little, 2016). In mice and presumably other mammals, hypothalamic–pituitary D2 expression is under the control of the pituitary adenylate cyclase-activating polypeptide (PACAP). Because PACAP is a master regulator of homeostatic response, both in terms of psychogenic and physiological stress (Rudecki & Gray, 2016), its potential role(s) in the dysfunctional TH feedback control characteristic of different disease states (Fliers, Bianco, Langouche, & Boelen, 2015) represents a valuable undertaking for future study. Another interesting caveat is that emerging evidence supports a PACAP role in circadian responses to photoperiod (Hannibal, 2016). The possibility therefore exists that diurnal or seasonal changes in circulating levels of free-TH may be driven through PACAP/D2-dependent regulation of negative feedback control via the pituitary.

2.3 Intracellular Reservoir Proteins

Because the primary TH transporter, MCT8, is bidirectional, a number of strategies exist to concentrate intracellular TH above its otherwise expected equilibrium point. μ-Crystallin (CRYM) is an NADPH-dependent cytoplasmic TH-binding protein that serves as an intracellular TH reservoir, while simultaneously reducing free-TH in the cytoplasm to promote further influx (van Mullem et al., 2016). CRYM is predominately expressed in the skin, lens, and skeletal muscle but is also found in retina, brain, heart, kidneys, and liver (Aoki, Ito, & Ito, 2000; Seko, Ogawa, Li, Taimura, & Ono, 2016). While some work suggests that CRYM expression sequesters T_3 from the nucleus to reduce its genomic actions (Mori et al., 2002; Seko et al., 2016), recent studies show that CRYM may actually facilitate shuttling of T_3 from the cytoplasm to the nucleus to sustain or promote its genomic actions (Choi, Moskalik, Ng, et al., 2015; Takeshige, Sekido, Kitahara, et al., 2014; van Mullem et al., 2016). However, the ambiguity in these finding may have to do with relative expression levels of D3. For instance, CRYM expression enhances cellular uptake and transcriptional activity of T_3 in the absence of D3, but this activity is ablated when D3 is expressed at high levels (van Mullem et al., 2016). CRYM homodimerizes in the cytoplasm, where it has high affinity for T_3 while bound by NADPH, but subsequently releases T_3 upon dissociation from NADPH (Borel, Hachi, Palencia, et al., 2014). It is worth noting that CRYM also binds free-T_2, with highest affinity in the absence of NADPH (Moreno, Silvestri, Lombardi, et al., 2003). The biological importance of NADPH-dependent T_3 binding by CRYM is not currently known, and may be difficult to determine considering CRYM also possesses NADH-dependent ketamine reductase activity (Borel et al., 2014; Hallen, Cooper, Jamie, Haynes, & Willows, 2011). Interestingly though, NADPH-dependent binding of T_3 is itself regulated by the presence of THs, vitamin D, and insulin (Suzuki, Suzuki, Mori, et al., 2007). In theory, this suggests that the sequestration and reservoir functions of CRYM may be regulated independently from changes in its overall expression profiles.

3. BIOACTIVITY

The bioactivity of TH is determined through its interactions with classical nuclear receptor TRs, newly identified TR variants, the integrin $\alpha v\beta 3$ receptor at the cell surface, and changes in local genetic architecture that alter the accessibility of thyroid response elements (TREs). In terms of bioactivity, it is important to note that differences in TH actions can be driven both by changes in expression profiles of TRs but also by the availability and activity of coregulator partners and other interacting proteins.

3.1 Thyroid Receptors

TRs represent the downstream targets in the classical TH signaling model. They belong to a class of ligand-activated transcription factors called nuclear receptors, which are characterized by a modular structure consisting of an N-terminal regulatory domain (A/B), a DNA-binding domain (DBD), a hinge region, a ligand-binding domain (LBD), and a C-terminal domain (Robinson-Rechavi, Garcia, & Laudet, 2003). While TRs most often form heterodimers with retinoid X receptor (Paquette, Atlas, Wade, & Yauk, 2014), they are also active as monomers, homodimers, and even heterotrimers

(Habibi, Nelson, & Allan, 2012; Mengeling, Lee, & Privalsky, 2008). They bind short nucleotide sequences representing TREs in the promoters of target genes to regulate transcription in a T_3-dependent, and potentially T_2-dependent, manner (Mendoza, Navarrete-Ramírez, Hernández-Puga, et al., 2003). Binding T_3 triggers conformational changes that generally promote shedding of corepressors (such as SMRT and NCoR1) for the recruitment of coactivator complexes to induce transcription (Pascual & Aranda, 2013). Importantly, TRs are expressed in a broad range of tissues (Flamant & Gauthier, 2013).

There are two TR isoforms in mammals, *THRA* and *THRB*, where each paralog encodes four distinct TR variants (TRα1; TRα2; TR$\Delta\alpha$1; TR$\Delta\alpha$2; TRβ1; TRβ2; TRβ3; TR$\Delta\beta$3; see Sirakov & Plateroti, 2011 for review). Of the four isoforms encoded by *THRA*, only TRα1 binds both DNA and T_3 (Yen et al., 2006). It therefore represents the main TRα isoform in terms of the classical TH signaling pathway (Pessemesse et al., 2014). However, two TR$\Delta\alpha$1 variants, p43 and p28, translocate to the mitochondrial matrix and inner membrane, respectively, to regulate mitochondrial function in response to T_3 (Lombardi et al., 2015). p43 appears to play key roles in glucose homeostasis (Bertrand et al., 2013; Blanchet et al., 2012) and skeletal muscle development (Pessemesse, Schlernitzauer, Sar, et al., 2012), while the physiological role of p28 is not currently known. Of the four isoforms encoded by *THRB*, TR$\Delta\beta$3 is the only one missing a characteristic NR domain, in the form of the LBD (González-Sancho, García, Bonilla, & Muñoz, 2003). Importantly, local expression patterns of these isoforms have different outcomes for TH action (see Amma, Campos-Barros, Wang, Vennström, & Forrest, 2001; Harvey & Williams, 2002; Lazar, 1993; Ribeiro, Bianco, Kaneshige, et al., 2010).

While differences in the relative expression profiles of TR isoforms can drive different signaling pathways across tissues and throughout development (Chiamolera, Sidhaye, Matsumoto, et al., 2012; Ribeiro et al., 2010), a single TR isoform can also signal different transcriptional programs depending on the specific nuclear environment (Flamant & Gauthier, 2013; Fozzatti, Lu, Kim, & Cheng, 2011). TR selectivity in target genes may result from differences in posttranslational modifications (Liu et al., 2015) or differences in the coregulator and other TR-interacting protein repertoires characteristic of different cells and tissues (Astapova, 2016; Flamant & Gauthier, 2013). In respect to the former, sumoylation (addition of small ubiquitin-like modifiers) of TRα and TRβ is necessary for the activation of preadipocyte proliferation and differentiation pathways

(Liu et al., 2015). There is also increasing evidence for cross talk between TRs and other transcription factors, where TRs can disrupt the activities of other transcription factors through tethering interactions via its hinge region (Matsunaga, Sasaki, Suzuki, et al., 2015; Pascual & Aranda, 2013; Weitzel, 2008).

3.2 Integrin αvβ3

It was long proposed that a cell surface receptor was responsible for the relatively fast-acting effects of TH, but only relatively recently was integrin *αvβ3* implicated in this role (Davis, Davis, & Cody, 2005). Integrins are a family of transmembrane receptors that coordinate cell–cell and cell–extracellular matrix protein interactions (Ruoslahti & Pierschbacher, 1987). Integrin *αvβ3* was originally identified as a vitronectin receptor (see Horton, 1997), but has since been shown to bind a wide range of protein and nonprotein ligands (Xiong, Stehle, Diefenbach, et al., 2001), including TH (Bergh et al., 2005) and several of its derivatives (Davis, Glinsky, Lin, et al., 2015; Davis, Leonard, Lin, & Incerpi, 2015). Activation by TH triggers signal transduction cascades that promote different cellular responses involved in angiogenesis (Davis et al., 2005), cell proliferation (Scarlett et al., 2008; Stenzel, Wilsch-Bräuninger, Wong, Heuer, & Huttner, 2014), and glucose uptake (Incerpi, Hsieh, Lin, et al., 2014). As a transmembrane receptor, integrin *αvβ3* is inherently more dependent on circulating levels of TH for activity, which means cell- or tissue-specific sensitivities may therefore correlate well with overall expression levels. It has therefore received much attention for its role(s) in tumorgenesis due to its high expression in many types of cancer (Cai & Chen, 2006; Varner & Cheresh, 1996).

Integrin *αvβ3* interacts with both T_4 and T_3 via at two distinct sites (S1 and S2) that activate different signal transduction cascades (Lin, Sun, Tang, et al., 2009). The S1 domain of integrin αvβ3 binds T_3 to promote Src-phosphatidylinositol 3-kinase (PI3K)-mediated transport of TRα from the cytoplasm to the nucleus (Davis, Glinsky, et al., 2015; Davis, Leonard, et al., 2015; Lin et al., 2009). Alternatively, the S2 domain binds T_3 and T_4, but has higher affinity for T_4, to promote extracellular signal-regulated kinases (MAPK/ERK1/2) signaling (Davis, Glinsky, et al., 2015; Davis, Leonard, et al., 2015). T_4 activation at the S2 domain results in the phosphorylation and translocation of TRβ from the cytoplasm to the nucleus (Hammes & Davis, 2015; Lin et al., 2009), as well as the internalization of the *αvβ3* subunits (Hammes & Davis, 2015) and phosphorylation of estrogen receptor α (ERα). Phosphorylation of TRβ Ser-142 promotes dissociation of transcriptional corepressors for coactivators, while internalization of *αvβ3* results in the nuclear translocation of *αv*, where it complexes with MAPK and E1A-binding protein p300, a coactivator, to enhance the expression of target genes, including TRβ1, ERα, and HIF-1α (see later; Hammes & Davis, 2015). Phosphorylation at Ser-167 has been shown to activate of ERα in a ligand-independent manner (Joel et al., 1998). Thus, activation of integrin *αvβ3* by either T_3 at S1, or T_3/T_4 at S2 each promote distinct signaling pathways that prime TRα and TRβ1, respectively, for genomic action by T_3. Interestingly, TETRAC has recently emerged as a potential therapeutic agent for various cancers, where it blocks T_4-induced expression of cancer-related genes (e.g., ERα; HIF-1α; VEGF) via integrin *αvβ3* to slow tumor cell proliferation (Davis, Glinsky, et al., 2015; Davis, Leonard, et al., 2015; Schmohl, Müller, Wechselberger, et al., 2015).

3.3 Thyroid Response Elements

TRs bind to the regulatory regions of target genes via TREs, which are short nucleotide-binding motifs (Harvey & Williams, 2002). The classic-binding motif for TRs was originally identified as two AGGTCA half-sites oriented as a direct repeat (DR) separated by a 4-nucleotide spacer (DR4; Lazar, 2013). The numbers of half-sites, spacer lengths, and orientations, however, have been shown to vary considerably (Paquette et al., 2014). In some cases, TRs even associate with sequence motifs that have no resemblance to the canonical TRE half-site sequences, but whether this represents direct binding to DNA is not known (Sinha & Yen, 2016). There is increasing evidence that the nucleotide sequences of specific TREs influence both the configuration (i.e., heterodimer; homodimer or monomer), binding efficiency, and transcriptional actions of TR binding (Grøntved, Waterfall, Kim, et al., 2015; Velasco, Togashi, Walfish, et al., 2007). TRs have bimodal activity, where they bind TREs to repress or activate gene transcription in response to the presence of T_3 (Fondell, Ge, & Roeder, 1996; Glass, Holloway, Devary, & Rosenfeld, 1998; Xu, Glass, & Rosenfeld, 1999). In their unbound state, TRs generally act as transcriptional repressors, but shed corepressor complexes for the recruitment of coactivators in the presence of T_3 (Zhang & Lazar, 2000). However, negative TREs (nTREs) have also been described (see Santos, Fairall, & Schwabe, 2011 for review). nTREs reverse the regulatory control of TRs, meaning TRs promote transcriptional activity in their unbound state and recruit corepressor complexes to depress transcription when bound by T_3 (Santos et al., 2011). This means that differences in chromatin ultrastructure between tissues may underlie tissue-specific changes in TH sensitivity by altering TR accessibility to specific response elements. In other words, TREs that are accessible in some tissues are not necessarily accessible in others, either due to changes in chromatin ultrastructure (Alenghat, Meyers, Mullican, et al., 2008) or occupation by other transcription factors or DNA-binding proteins (Liu & Brent, 2010). Chromatin remodeling or shifts in occupation frequencies by other DNA-binding proteins may therefore be regulated in ways that alter TH sensitivity within a tissue temporally. Importantly, however, recent work shows some departure from this biomodal model for transcriptional regulation by TRs (Grøntved et al., 2015). While some unliganded TRs do in fact interact with DNA sequence motifs in certain regions, many TRs only interact with TREs when bound by T_3 (Grøntved et al., 2015).

4. CONCLUSIONS

Taken together this work suggests that there is considerable overlap between central and peripheral regulation by TH. One possibility is that TH signaling via the CNS can remodel how peripheral tissues process circulating TH, thereby regulating its local actions. For instance, it is suggested that TH acting at the level of the hypothalamus stimulates gluconeogenesis in the liver by adjusting local TH sensitivities of hepatocytes (Fliers, Klieverik, & Kalsbeek, 2010). A second possibility for overlap is that local changes in TH sensitivity at peripheral tissues may alter their sensitivity to TH-mediated central inputs. In rodents for instance, increased expression of TRs in brown adipose tissue acts to amplify adrenergic stimulation of thermogenesis (Ribeiro, Carvalho, Schultz, et al., 2001), which is mediated, at least in part, by the central actions of TH (López, Varela, & Vázquez, 2010). Another example is that TRα-mediated signaling in the heart can modulate autonomic inputs that regulate cardiac function in response to environmental stressors (Mittag, Davis, Vujovic, Arner, & Vennström, 2010).

It is becoming increasingly clear that individual cells do not merely respond to circulating levels of TH, but harness it using a number of mechanisms to fine-tune its local actions. Because the thyroid axis plays a common role in human disease, including metabolic syndromes and various cancers, an increased understanding of these local regulatory pathways is uncovering ever new therapeutic possibilities (Davis, Glinsky, et al., 2015; Gu, Gelsomino,

Covington, Beyer, et al., 2015). This is especially true in the sense that local TH actions are likely balanced across multiple levels of regulation, rather than driven from any one level specifically. This means that disruption at any one level may be compensated for via actions at another. For instance, increased D2 expression in BAT of MCT8-deficient mice maintains local T_3 concentrations despite a reduction in overall TH uptake (Di Cosmo et al., 2013). A caveat is that seemingly subtle changes can be additive across levels of regulation or can counter balance each other to maintain local TH action. It is therefore important to consider disease-related changes in TH signaling pathways in the context of the other levels of regulation that may directly, or indirectly, modify their ultimate effects.

REFERENCES

Abend, S. L., Fang, S. L., Alex, S., et al. (1991). Rapid alteration in circulating free thyroxine modulates pituitary type II 5′ deiodinase and basal thyrotropin secretion in the rat. *The Journal of Clinical Investigation*, *88*, 898–903. http://dx.doi.org/10.1172/JCI115392.

Alenghat, T., Meyers, K., Mullican, S. E., et al. (2008). Nuclear receptor corepressor and histone deacetylase 3 govern circadian metabolic physiology. *Nature*, *456*, 997–1000. http://dx.doi.org/10.1038/nature07541.

Amma, L. L., Campos-Barros, A., Wang, Z., Vennström, B., & Forrest, D. (2001). Distinct tissue-specific roles for thyroid hormone receptors β and α1 in regulation of type 1 deiodinase expression. *Molecular Endocrinology*, *15*, 467–475.

Aoki, N., Ito, K., & Ito, M. (2000). mu-Crystallin, thyroid hormone-binding protein, is expressed abundantly in the murine inner root sheath cells. *Journal of Investigative Dermatology*, *115*, 402–405. http://dx.doi.org/10.1046/j.1523-1747.2000.00070.x.

Arrojo e Drigo, R., Fonseca, T. L., Castillo, M., Salathe, M., Simovic, G., Mohácsik, P., et al. (2011). Endoplasmic reticulum stress decreases intracellular thyroid hormone activation via an eIF2a-mediated decrease in type 2 deiodinase synthesis. *Molecular Endocrinology*, *25*, 2065–2075.

Astapova, I. (2016). Role of co-regulators in metabolic and transcriptional actions of thyroid hormone. *Journal of Molecular Endocrinology*, *56*, R73–R97.

Baqui, M., Botero, D., Gereben, B., et al. (2003). Human type 3 iodothyronine selenodeiodinase is located in the plasma membrane and undergoes rapid internalization to endosomes. *The Journal of Biological Chemistry*, *278*, 1206–1211.

Bergh, J. J., Lin, H. Y., Lansing, L., Mohamed, S. N., Davis, F. B., Mousa, S., et al. (2005). Integrin αVβ3 contains a cell surface receptor site for thyroid hormone that is linked to activation of mitogen-activated protein kinase and induction of angiogenesis. *Endocrinology*, *146*, 2864–2871.

Bernal, J., Guadaño-Ferraz, A., & Morte, B. (2015). Thyroid hormone transporters—Functions and clinical implications. *Nature Reviews. Endocrinology*, *11*, 406–417. http://dx.doi.org/10.1038/nrendo.2015.66.

Bertrand, C., et al. (2013). Mice lacking the p43 mitochondrial T3 receptor become glucose intolerant and insulin resistant during aging. *PLoS One*, *8*(9), e75111.

Bianco, A. C. (2011). Minireview: Cracking the metabolic code for thyroid hormone signaling. *Endocrinology*, *152*, 3306–3311. http://dx.doi.org/10.1210/en.2011-1104.

Bianco, A. C., & Kim, B. W. (2006). Deiodinases: Implications of the local control of thyroid hormone action. *The Journal of Clinical Investigation*, *116*, 2571–2579.

Blanchet, E., et al. (2012). Mitochondrial T3 receptor p43 regulates insulin secretion and glucose homeostasis. *The FASEB Journal*, *26*(1), 40–50.

Borel, F., Hachi, I., Palencia, A., et al. (2014). Crystal structure of mouse mu-crystallin complexed with NADPH and the T3 thyroid hormone. *FEBS Journal*, *281*, 1598–1612. http://dx.doi.org/10.1111/febs.12726.

Cai, W., & Chen, X. (2006). Anti-angiogenic cancer therapy based on integrin αvβ3 antagonism. *Anti-Cancer Agents in Medicinal Chemistry*, *6*(5), 407–428.

Chiamolera, M. I., Sidhaye, A. R., Matsumoto, S., et al. (2012). Fundamentally distinct roles of thyroid hormone receptor isoforms in a thyrotroph cell line are due to differential DNA binding. *Molecular Endocrinology*, *26*, 926–939. http://dx.doi.org/10.1210/me.2011-1290.

Choi, J., Moskalik, C., Ng, A., et al. (2015). Regulation of thyroid hormone-induced development in vivo by thyroid hormone transporters and cytosolic binding proteins. *General and Comparative Endocrinology*, *222*, 69–80. http://dx.doi.org/10.1016/j.ygcen.2015.07.006.

Christensen, M. J., Cammack, P. M., & Wray, C. D. (1995). Tissue specificity of selenoprotein gene expression in rats. *The Journal of Nutritional Biochemistry*, *6*, 367–372.

Cioffi, F., Senese, R., Lanni, A., & Goglia, F. (2013). Thyroid hormones and mitochondria: With a brief look at derivatives and analogues. *Molecular and Cellular Endocrinology*, *379*, 51–61.

Darras, V. M., Houbrechts, A. M., & Van Herck, S. L. J. (2015). Intracellular thyroid hormone metabolism as a local regulator of nuclear thyroid hormone receptor-mediated impact on vertebrate development. *Biochimica et Biophysica Acta*, *1849*, 130–141. http://dx.doi.org/10.1016/j.bbagrm.2014.05.004.

Davis, P. J., Davis, F. B., & Cody, V. (2005). Membrane receptors mediating thyroid hormone action. *Trends in Endocrinology and Metabolism*, *16*, 429–435. http://dx.doi.org/10.1016/j.tem.2005.09.007.

Davis, P. J., Glinsky, G. V., Lin, H. Y., et al. (2015). Cancer cell gene expression modulated from plasma membrane integrin αvβ3 by thyroid hormone and nanoparticulate tetrac. *Frontiers in Endocrinology*, *5*, 240.

Davis, P. J., Leonard, J. L., Lin, H. Y., & Incerpi, S. (2015). Receptor proteins for nongenomic actions of thyroid hormone. *Immunology, Endocrine & Metabolic Agents in Medicinal Chemistry*, *15*, 94–97.

del Amo, E. M., Urtti, A., & Yliperttula, M. (2008). Pharmacokinetic role of L-type amino acid transporters LAT1 and LAT2. *European Journal of Pharmaceutical Sciences*, *35*, 161–174.

Di Cosmo, C., Liao, X. H., Ye, H., Ferrara, A. M., Weiss, R. E., Refetoff, S., et al. (2013). Mct8-deficient mice have increased energy expenditure and reduced fat mass that is abrogated by normalization of serum T3 levels. *Endocrinology*, *154*, 4885–4895.

Flamant, F., & Gauthier, K. (2013). Thyroid hormone receptors: The challenge of elucidating isotype-specific functions and cell-specific response. *Biochimica et Biophysica Acta*, *1830*(7), 3900–3907.

Fliers, E., Bianco, A. C., Langouche, L., & Boelen, A. (2015). Thyroid function in critically ill patients. *The Lancet Diabetes and Endocrinology*, *3*, 816–825.

Fliers, E., Klieverik, L. P., & Kalsbeek, A. (2010). Novel neural pathways for metabolic effects of thyroid hormone. *Trends in Endocrinology and Metabolism*, *21*, 230–236.

Fondell, J. D., Ge, H., & Roeder, R. G. (1996). Ligand induction of a transcriptionally active thyroid hormone receptor coactivator complex. *PNAS*, *93*, 8329–8333.

Fozzatti, L., Lu, C., Kim, D. W., & Cheng, S. Y. (2011). Differential recruitment of nuclear coregulators directs the isoform-dependent action of mutant thyroid hormone receptors. *Molecular Endocrinology*, *25*, 908–921.

Friesema, E. C., Docter, R., Moerings, E. P., et al. (1999). Identification of thyroid hormone transporters. *Biochemical and Biophysical Research Communications*, *254*, 497–501.

Friesema, E. C. H., Jansen, J., Jachtenberg, J.-W., et al. (2008). Effective cellular uptake and efflux of thyroid hormone by human monocarboxylate transporter 10. *Molecular Endocrinology, 22*, 1357–1369. http://dx.doi.org/10.1210/me.2007-0112.
Galton, V. A. (2005). The roles of the iodothyronine deiodinases in mammalian development. *Thyroid, 15*, 823–834.
Galton, V. A., et al. (2014). Life without the iodothyronine deiodinases. *Endocrinology, 155*(10), 4081–4087.
Glass, C. K., Holloway, J. M., Devary, O. V., & Rosenfeld, M. G. (1998). The thyroid hormone receptor binds with opposite transcriptional effects to a common sequence motif in thyroid hormone and estrogen response elements. *Cell, 54*, 313–323.
González-Sancho, J. M., García, V., Bonilla, F., & Muñoz, A. (2003). Thyroid hormone receptors/THR genes in human cancer. *Cancer Letters, 192*, 121–132.
Grøntved, L., Waterfall, J. J., Kim, D. W., et al. (2015). Transcriptional activation by the thyroid hormone receptor through ligand-dependent receptor recruitment and chromatin remodelling. *Nature Communications, 28*, 6.
Gu, G., Gelsomino, L., Covington, K. R., Beyer, A. R., et al. (2015). Targeting thyroid hormone receptor beta in triple-negative breast cancer. *Breast Cancer Research and Treatment, 150*, 535–545.
Habibi, H. R., Nelson, E. R., & Allan, E. R. (2012). New insights into thyroid hormone function and modulation of reproduction in goldfish. *General and Comparative Endocrinology, 175*, 19–26.
Hallen, A., Cooper, A. J., Jamie, J. F., Haynes, P. A., & Willows, R. D. (2011). Mammalian forebrain ketimine reductase identified as μ-crystallin; potential regulation by thyroid hormones. *Journal of Neurochemistry, 118*, 379–387.
Hammes, S. R., & Davis, P. J. (2015). Overlapping nongenomic and genomic actions of thyroid hormone and steroids. *Best Practice & Research. Clinical Endocrinology & Metabolism, 29*, 581–593.
Hannibal, J. (2016). PACAP in the circadian timing system: Learning from knockout models. In D. Reglodi & A. Tamas (Eds.), *Pituitary adenylate cyclase activating polypeptide—PACAP*. Switzerland: Springer International Publishing, pp. 227–237.
Harvey, C. B., & Williams, G. R. (2002). Mechanism of thyroid hormone action. *Thyroid, 12*, 441–446.
Horton, M. A. (1997). The αvβ3 integrin "vitronectin receptor". *The International Journal of Biochemistry & Cell Biology, 29*, 721–725.
Incerpi, S., Hsieh, M. T., Lin, H. Y., et al. (2014). Thyroid hormone inhibition in L6 myoblasts of IGF-I-mediated glucose uptake and proliferation: New roles for integrin αvβ3. *American Journal of Physiology. Cell Physiology, 307*, C150–C161.
Joel, P. B., Smith, J., Sturgill, T. W., Fisher, T. L., Blenis, J., & Lannigan, D. A. (1998). pp90rsk1 regulates estrogen receptor-mediated transcription through phosphorylation of Ser-167. *Molecular and Cellular Biology, 18*, 1978–1984.
Kinne, A., Schülein, R., & Krause, G. (2011). Primary and secondary thyroid hormone transporters. *Thyroid Research, 4*(1), S7.
Köhrle, J. (1995). Thyroid hormone deiodinases—A selenoenzyme family acting as gate keepers to thyroid hormone action. *Acta Medica Austriaca, 23*, 17–30.
Lazar, M. A. (1993). Thyroid hormone receptors: Multiple forms, multiple possibilities. *Endocrine Reviews, 14*(2), 184–193.
Lazar, M. A. (2013). Thyroid hormone receptors: Multiple forms, multiple possibilities. *Endocrine Reviews, 14*, 184–193. http://dx.doi.org/10.1210/edrv-14-2-184.
Lin, H.-Y., Sun, M., Tang, H. Y., et al. (2009). l-Thyroxine vs. 3,5,3′-triiodo-l-thyronine and cell proliferation: Activation of mitogen-activated protein kinase and phosphatidylinositol 3-kinase. *American Journal of Physiology. Cell Physiology, 296*, C980–C991.

Little, A. G. (2016). A review of the peripheral levels of regulation by thyroid hormone. *Journal of Comparative Physiology. B, 9*, 1–2.
Liu, Y. Y., Ayers, S., Milanesi, A., Teng, X., Rabi, S., Akiba, Y., et al. (2015). Thyroid hormone receptor sumoylation is required for preadipocyte differentiation and proliferation. *The Journal of Biological Chemistry, 290*, 7402–7415.
Liu, Y. Y., & Brent, G. A. (2010). Thyroid hormone crosstalk with nuclear receptor signaling in metabolic regulation. *Trends in Endocrinology and Metabolism, 21*, 166–173.
Lombardi, A., et al. (2015). Regulation of skeletal muscle mitochondrial activity by thyroid hormones: Focus on the "old" triiodothyronine and the "emerging" 3,5-diiodothyronine. *Frontiersin Physiology, 6*, 237.
López, M., Varela, L., & Vázquez, M. J. (2010). Hypothalamic AMPK and fatty acid metabolism mediate thyroid regulation of energy balance. *Nature Medicine, 16*, 1001–1008.
Lösel, R., & Wehling, M. (2003). Nongenomic actions of steroid hormones. *Nature Reviews. Molecular Cell Biology, 4*, 46–55.
Matsunaga, H., Sasaki, S., Suzuki, S., et al. (2015). Essential role of GATA2 in the negative regulation of type 2 deiodinase gene by liganded thyroid hormone receptor β2 in thyrotroph. *PLoS One, 10*, e0142400.
Mayerl, S., Müller, J., Bauer, R., et al. (2014). Transporters MCT8 and OATP1C1 maintain murine brain thyroid hormone homeostasis. *The Journal of Clinical Investigation, 124*, 1987–1999. http://dx.doi.org/10.1172/JCI70324.
Mendoza, A., Navarrete-Ramírez, P., Hernández-Puga, G., et al. (2003). 3,5-T2 is an alternative ligand for the thyroid hormone receptor β1. *Endocrinology, 154*, 2948–2958.
Mengeling, B. J., Lee, S., & Privalsky, M. L. (2008). Coactivator recruitment is enhanced by thyroid hormone receptor trimers. *Molecular and Cellular Endocrinology, 280*, 47–62. http://dx.doi.org/10.1016/j.mce.2007.09.011.
Mittag, J., Davis, B., Vujovic, M., Arner, A., & Vennström, B. (2010). Adaptations of the autonomous nervous system controlling heart rate are impaired by a mutant thyroid hormone receptor-α1. *Endocrinology, 151*, 2388–2395.
Moreno, M., Silvestri, E., Lombardi, A., et al. (2003). Identification of 3,5-diiodo-L-thyronine-binding proteins in rat liver cytosol by photoaffinity labeling. *Endocrinology, 144*, 2297–2303.
Mori, J.-i., et al. (2002). Nicotinamide adenine dinucleotide phosphate-dependent cytosolic T3 binding protein as a regulator for T3-mediated transactivation. *Endocrinology, 143*(4), 1538–1544.
Müller, J., Mayerl, S., Visser, T. J., et al. (1999). Tissue-specific alterations in thyroid hormone homeostasis in combined Mct10 and Mct8 deficiency. *Endocrinology, 155*, 315–325.
Mullur, R., Liu, Y. Y., & Brent, G. A. (2014). Thyroid hormone regulation of metabolism. *Physiological Reviews, 94*, 355–382.
Paquette, M. A., Atlas, E., Wade, M. G., & Yauk, C. L. (2014). Thyroid hormone response element half-site organization and its effect on thyroid hormone mediated transcription. *PLoS One, 9*, e101155. http://dx.doi.org/10.1371/journal.pone.0101155.
Pascual, A., & Aranda, A. (2013). Thyroid hormone receptors, cell growth and differentiation. *Biochimica et Biophysica Acta, 1830*, 3908–3916.
Pessemesse, L., Lepourry, L., Bouton, K., Levin, J., Cabello, G., Wrutniak-Cabello, C., et al. (2014). p28, a truncated form of TRα1 regulates mitochondrial physiology. *FEBS Letters, 588*, 4037–4043.
Pessemesse, L., Schlernitzauer, A., Sar, C., et al. (2012). Depletion of the p43 mitochondrial T3 receptor in mice affects skeletal muscle development and activity. *FASEB Journal: Official Publication of the Federation of American Societies for Experimental Biology, 26*, 748–756.

Ribeiro, M. O., Bianco, S. D. C., Kaneshige, M., et al. (2010). Expression of uncoupling protein 1 in mouse brown adipose tissue is thyroid hormone receptor-beta isoform specific and required for adaptive thermogenesis. *Endocrinology*, *151*, 432–440. http://dx.doi.org/10.1210/en.2009-0667.

Ribeiro, M. O., Carvalho, S. D., Schultz, J. J., et al. (2001). Thyroid hormone-sympathetic interaction and adaptive thermogenesis are thyroid hormone receptor isoform-specific. *The Journal of Clinical Investigation*, *108*, 97–105. http://dx.doi.org/10.1172/JCI12584.

Roberts, L. M., Woodford, K., Zhou, M., et al. (2008). Expression of the thyroid hormone transporters monocarboxylate transporter-8 (SLC16A2) and organic ion transporter-14 (SLCO1C1) at the blood-brain barrier. *Endocrinology*, *149*, 6251–6261.

Robinson-Rechavi, M., Garcia, H. E., & Laudet, V. (2003). The nuclear receptor superfamily. *Journal of Cell Science*, *116*, 585–586.

Rosene, M. L., Wittmann, G., Drigo, R. A. E., et al. (2013). Inhibition of the type 2 iodothyronine deiodinase underlies the elevated plasma TSH associated with amiodarone treatment. *Endocrinology*, *151*, 5961–5970. http://dx.doi.org/10.1210/en.2010-0553.

Rudecki, A. P., & Gray, S. L. (2016). PACAP in the defense of energy homeostasis. *Trends in Endocrinology and Metabolism*, *27*, 620–632.

Ruoslahti, E., & Pierschbacher, M. D. (1987). New perspectives in cell adhesion: RGD and integrins. *Science*, *238*, 491–497. http://dx.doi.org/10.1126/science.2821619.

Santos, G. M., Fairall, L., & Schwabe, J. W. R. (2011). Negative regulation by nuclear receptors: A plethora of mechanisms. *Trends in Endocrinology and Metabolism*, *22*, 87–93. http://dx.doi.org/10.1016/j.tem.2010.11.004.

Scarlett, A., Parsons, M. P., Hanson, P. L., Sidhu, K. K., Milligan, T. P., & Burrin, J. M. (2008). Thyroid hormone stimulation of extracellular signal-regulated kinase and cell proliferation in human osteoblast-like cells is initiated at integrin αVβ3. *The Journal of Endocrinology*, *196*, 509–517.

Schmohl, K. A., Müller, A. M., Wechselberger, A., et al. (2015). Thyroid hormones and tetrac: New regulators of tumour stroma formation via integrin αvβ3. *Endocrine-Related Cancer*, *22*, 941–952.

Seko, D., Ogawa, S., Li, T. S., Taimura, A., & Ono, Y. (2016). μ-Crystallin controls muscle function through thyroid hormone action. *FASEB Journal: Official Publication of the Federation of American Societies for Experimental Biology*, *30*, 1733–1740.

Senese, R., Cioffi, F., De Lange, P., Goglia, F., & Lanni, A. (2014). Thyroid: Biological actions of "nonclassical" thyroid hormones. *The Journal of Endocrinology*, *221*, R1–R2.

Sinha, R. A., & Yen, P. M. (2016). Mechanisms for thyroid hormone action in the CNS. In N. Koibuchi & P. M. Yen (Eds.), *Thyroid hormone disruption and neurodevelopment*. New York: Springer, pp. 3–21.

Sirakov, M., & Plateroti, M. (2011). The thyroid hormones and their nuclear receptors in the gut: From developmental biology to cancer. *Biochimica et Biophysica Acta-Molecular Basis of Disease*, *1812*, 938–946. http://dx.doi.org/10.1016/j.bbadis.2010.12.020.

Stenzel, D., Wilsch-Bräuninger, M., Wong, F. K., Heuer, H., & Huttner, W. B. (2014). Integrin αvβ3 and thyroid hormones promote expansion of progenitors in embryonic neocortex. *Development*, *141*, 795–806.

Suzuki, S., Suzuki, N., Mori, J.-I., et al. (2007). Micro-crystallin as an intracellular 3,5,3′-triiodothyronine holder in vivo. *Molecular Endocrinology*, *21*, 885–894. http://dx.doi.org/10.1210/me.2006-0403.

Takeshige, K., Sekido, T., Kitahara, J. I., et al. (2014). Cytosolic T3-binding protein modulates dynamic alteration of T3-mediated gene expression in cells. *Endocrine Journal*, *61*, 561–570.

Van Der Deure, W. M., Appelhof, B. C., Peeters, R. P., et al. (2008). Polymorphisms in the brain-specific thyroid hormone transporter OATP1C1 are associated with fatigue and depression in hypothyroid patients. *Clinical Endocrinology*, *69*, 804–811.

van Mullem, A. A., van Gucht, A. L., Visser, W. E., Meima, M. E., Peeters, R. P., & Visser, T. J. (2016). Effects of thyroid hormone transporters MCT8 and MCT10 on nuclear activity of T3. *Molecular and Cellular Endocrinology*, *437*, 252–260.
Varner, J. A., & Cheresh, D. A. (1996). Integrins and cancer. *Current Opinion in Cell Biology*, *8*, 724–730.
Velasco, L. F., Togashi, M., Walfish, P. G., et al. (2007). Thyroid hormone response element organization dictates the composition of active receptor. *The Journal of Biological Chemistry*, *282*, 12458–12466.
Visser, W., & Friesema, E. (2011). Minireview: Thyroid hormone transporters: The knowns and the unknowns. *Molecular Endocrinology*, *25*, 1–4.
Visser, W. E., Friesema, E. C. H., Jansen, J., & Visser, T. J. (2008). Thyroid hormone transport in and out of cells. *Trends in Endocrinology and Metabolism*, *19*, 50–56. http://dx.doi.org/10.1016/j.tem.2007.11.003.
Warner, A., & Mittag, J. (2012). Thyroid hormone and the central control of homeostasis. *Journal of Molecular Endocrinology*, *49*, R29–R35.
Weitzel, J. M. (2008). To bind or not to bind—How to down-regulate target genes by liganded thyroid hormone receptor? *Thyroid Research*, *1*, 1.
Xiong, J. P., Stehle, T., Diefenbach, B., et al. (2001). Crystal structure of the extracellular segment of integrin alpha Vbeta3. *Science*, *294*, 339–345. http://dx.doi.org/10.1126/science.1064535.
Xu, L., Glass, C. K., & Rosenfeld, M. G. (1999). Coactivator and corepressor complexes in nuclear receptor function. *Current Opinion in Genetics & Development*, *9*, 140–147.
Yen, P. M., Ando, S., Feng, X., Liu, Y., Maruvada, P., & Xia, X. (2006). Thyroid hormone action at the cellular, genomic and target gene levels. *Molecular and Cellular Endocrinology*, *246*(1), 121–127.
Zhang, J., & Lazar, M. A. (2000). The mechanism of action of thyroid hormones. *Annual Review of Physiology*, *62*, 439–466.

FURTHER READING

Aranda, A., & Pascual, A. (2001). Nuclear hormone receptors and gene expression. *Physiological Reviews*, *81*, 1269–1304.
Lin, H.-Y., Tang, H. Y., Davis, F. B., et al. (2012). Nongenomic regulation by thyroid hormone of plasma membrane ion and small molecule pumps. *Discovery Medicine*, *14*, 199–206.

CHAPTER TWO

Thyroid Hormone Transport and Transporters

Doreen Braun, Ulrich Schweizer[1]
Institut für Biochemie und Molekularbiologie, Rheinische Friedrich-Wilhelms-Universität Bonn, Bonn, Germany
[1]Corresponding author: e-mail address: uschweiz@uni-bonn.de

Contents

1. Introduction 20
2. Thyroid hormone Transport 22
3. Monocarboxylate Transporter 8 23
   3.1 The Syndrome of Inherited MCT8 Deficiency 24
   3.2 Mutations in MCT8 Impair Thyroid Hormone Transport 25
   3.3 Concepts for Treatment of MCT8 Deficiency 32
4. Mouse Models of Allan–Herndon–Dudley Syndrome 33
5. Thyroid Hormone Transport and Physiology 35
6. Further Clinical Implications for Thyroid Hormone Transporters 37
   6.1 Polymorphisms in Thyroid Hormone Transporters May Affect Thyroid Hormone Levels 37
   6.2 Inhibition of Thyroid Hormone Transport by Drugs 37
   6.3 Intestinal Uptake of T_4 and Thyroid Hormone Transport 38
7. Conclusions and Future Directions 38
Acknowledgments 39
References 39

Abstract

Thyroid hormones orchestrate developmental processes and are among the most important regulators of energy metabolism. Thyroid hormone actions are mostly, but not exclusively, mediated by nuclear hormone receptors. As amino acid derivatives, thyroid hormones need plasma membrane transporters in order to reach their nuclear receptors. Several transporters from different gene families mediate thyroid hormone uptake into cells. Monocarboxylate transporter 8 is a specific thyroid hormone transporter found mutated in patients with severe psychomotor retardation and strangely abnormal thyroid hormone constellations. These patients display a syndrome in which some organs are exposed to increased thyroid hormone signaling, while other organs are lacking thyroid hormone signaling due to complete lack of thyroid hormone uptake. Investigations in many organ systems using mouse models of thyroid hormone transmembrane transporter deficiency have helped complete our picture of thyroid hormone metabolism and action in the body during development and under different

Vitamins and Hormones, Volume 106
ISSN 0083-6729
http://dx.doi.org/10.1016/bs.vh.2017.04.005

physiological conditions. Incorporating the concept of thyroid hormone transmembrane transport has helped understand previously enigmatic drug interactions and may explain how the hormonal set points in the hypothalamus–pituitary–thyroid axis are established.

1. INTRODUCTION

Thyroid hormones (THs) orchestrate fundamental processes in vertebrate physiology. Energy expenditure is profoundly regulated by the availability of thyroid hormone as shown by pioneering studies of Magnus-Levy (Magnus-Levy, 1895; Mullur, Liu, & Brent, 2014). Thyroid hormones are also essential for developmental processes. For example, metamorphosis of amphibians, i.e., the involution of the tadpole's tail, involution of gills, and development of lungs, and also CNS development and maturation of the intestinal tract from vegetarian to predatory life style all depend on thyroid hormone (Brown et al., 1995). Parts of this mechanism are conserved in mammals. Children lacking adequate thyroid hormone production develop the syndrome of cretinism, which prominently involves neurodevelopmental delay, impaired hearing, and stunted growth. Hypothyroidism may also develop later in life due to thyroid disease. These patients can be effectively treated with oral thyroid hormone replacement therapy. Hyperthyroidism is another frequent disease, often due to somatic mutations in the TSH receptor, TSH receptor agonistic antibodies as in Graves' disease, or during the course of autoimmune thyroid destruction. Clinical practice shows that deviations from a person's normal thyroid hormone levels massively impair the patients' well-being. The hypothalamus–pituitary–thyroid axis (HPT axis or thyroid hormone axis) is therefore regulated on many levels. Thyroxine (T_4, 3,3′,5,5′-tetraiodothyronine), the main product of the thyroid gland, is essentially a prohormone. The active, thyroid hormone receptor-binding hormone is T_3 (3,3′,5-triiodothyronine), a lesser product of the thyroid gland and mostly derived from deiodination of T_4 (Fig. 1). Deiodinases (Dio) are selenocysteine-containing enzymes capable of removing iodide from iodothyronines (Bianco, Salvatore, Gereben, Berry, & Larsen, 2002). While Dio1 and Dio2 exhibit 5′-deiodinase activity through which T_4 can be activated to T_3, Dio1 and Dio3 exhibit 5-deiodinase activity which invariably inactivates T_4 and T_3. Since Dio1–3 are expressed in a developmentally regulated and cell type-specific manner, thyroid hormone levels can be adjusted locally to yield physiologically adequate thyroid hormone levels—largely

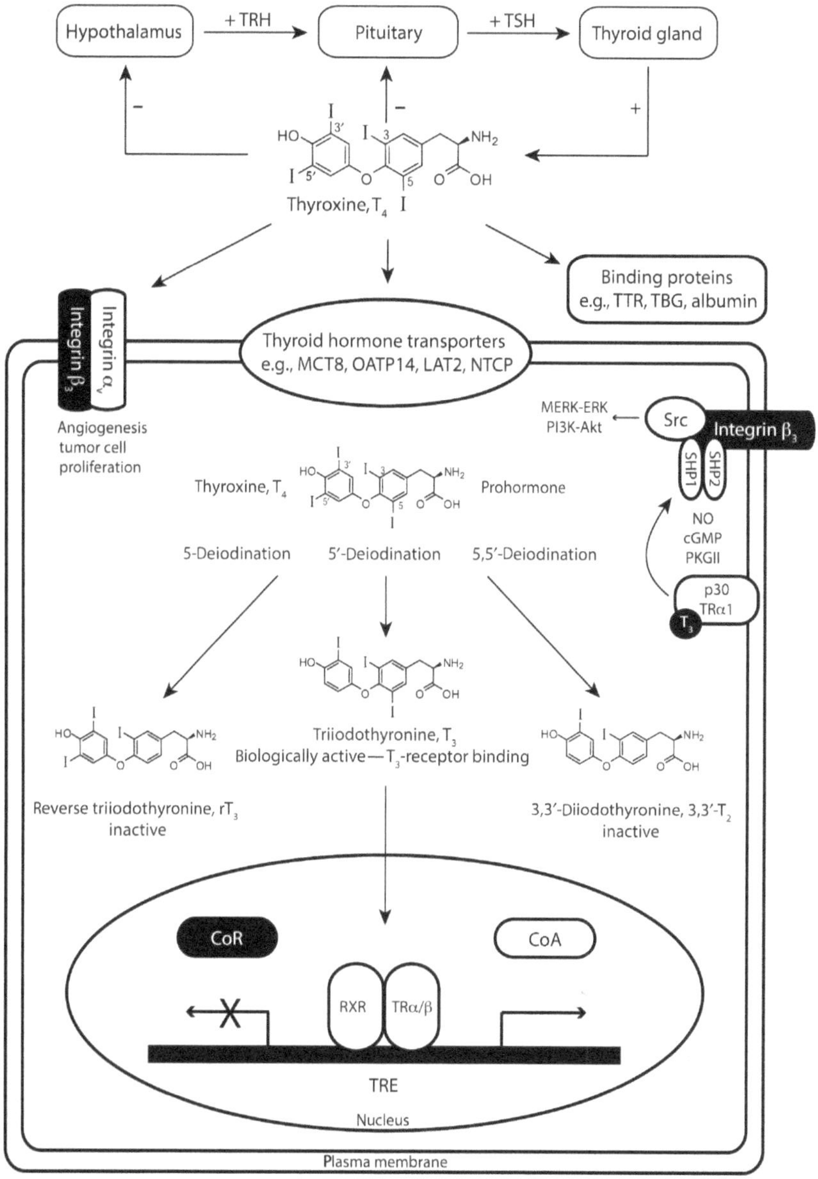

Fig. 1 Thyroid hormone signaling. Thyroxine (T_4) is released by the thyroid gland in response to TSH from the pituitary which is under control by TRH of the hypothalamus. T_4 can signal from the extracellular aspect via $\alpha_v\beta_3$ integrin. After cellular uptake T_4 is converted to T_3 which binds cytoplasmic TRα1 and triggers noncanonical cGMP-dependent signaling. Canonical signaling involves T_3 binding to nuclear TRα and TRβ which heterodimerize with retinoid-X-receptor (RXR) and control transcription of target genes. Deiodinases not only activate T_4 by T_3 formation but also inactivate T_4 and T_3 to yield rT_3 and 3,3′-T_2, respectively.

independent from circulating thyroid hormone levels (Schweizer, Weitzel, & Schomburg, 2008). T_4 can signal via a noncanonical, integrin-dependent mechanism (Davis, Leonard, & Davis, 2007) and a recently discovered cGMP-dependent mechanism involving a cytoplasmic isoform of TRα (Kalyanaraman et al., 2014) (Fig. 1). Most thyroid hormone effects, however, are mediated via the nuclear thyroid hormone receptors, TRα and TRβ. Accordingly, mutations in the genes encoding TRβ cause the syndrome of resistance to thyroid hormone (Dumitrescu & Refetoff, 2013). Recently, mutations in TRα have been discovered (Bochukova et al., 2012; van Mullem et al., 2012). Together, both syndromes together affect almost all known thyroid hormone effects. Therefore, thyroid hormone needs to cross the plasma membrane to reach their nuclear receptor targets.

2. THYROID HORMONE TRANSPORT

Thyroid hormones are notorious for their tendency to stick to laboratory vessel walls and were therefore regarded hydrophobic molecules. Accordingly, the majority of thyroid hormones are bound to plasma proteins, thyroid hormone-binding globulin (TBG), transthyretin (prealbumin), and serum albumin (Schweizer & Köhrle, 2012). The free hormone hypothesis posits that only the free fractions of T_4 and T_3 are available for cellular uptake, while the protein-bound fraction serves as a buffered pool which secures ubiquitous availability of hormones. As amino acid metabolites containing an aminopropionic acid moiety, thyroid hormones are expected to need protein transporters help them cross the plasma membrane. From the 1970s several groups studied thyroid hormone uptake into cells (Krenning, Docter, Bernard, Visser, & Hennemann, 1978; Rao, Eckel, Rao, & Breuer, 1976). Several thyroid hormone transport systems were described, e.g., a Na^+-dependent, high-affinity, low-capacity system and an energy-independent, low-affinity, high-capacity system both present in liver and other cell types. Some researchers found that thyroid hormone uptake occurred through amino acid uptake systems L and T, while later also organic anion transporters were found to transport thyroid hormone (Hennemann et al., 2001; Schweizer & Köhrle, 2012). Today we know that the sodium/taurocholate cotransporting polypeptide (NTCP, SLC10A1) is the Na^+-dependent high-affinity system in liver which also transports sulfated thyroid hormone (Friesema et al., 1999). Several transporters belong to the organic anion-transporting polypeptide (OATP) family: Oatp1–3 in rodents, SLCO4A1 and SLCO1A2 in humans, as well as the T_4-specific

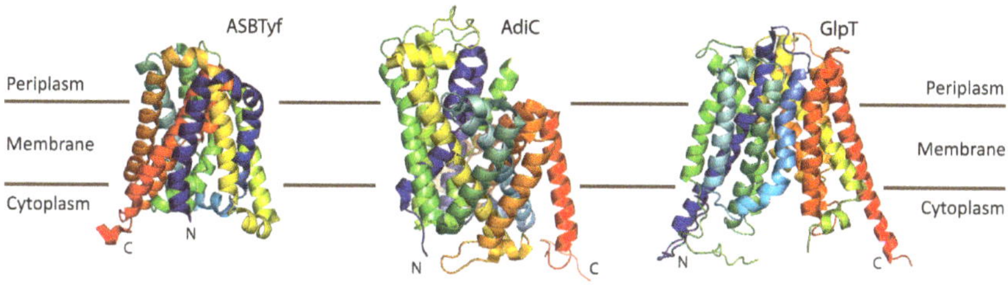

Fig. 2 Crystal structures of prokaryotic transmembrane transport proteins which are homologous to mammalian thyroid hormone transporters. The apical bile acid transporter from *Yersinia* (ASBTYf) belongs to the sodium-dependent bile acid transporters and is homologous to hepatic sodium–taurocholate cotransporting polypeptide NTCP. The arginine:agmatine antiporter AdiC is homologous to L-type amino acid transporters LAT1 and LAT2. The glycerol-phosphate transporter GlpT from *Escherichia coli* belongs to the MFS family of transporters and is homologous to MCTs and OATPs.

SLCO1C1/Slco1c1 in humans and rodents (Abe et al., 1998; Sugiyama et al., 2003). System L amino acid transporters comprise LAT1 (SLC7A5) and LAT2 (SLC7A8) and mediate energy-independent thyroid hormone uptake into neurons, astrocytes, and possibly other cell types (Blondeau, Beslin, Chantoux, & Francon, 1993; Chantoux, Blondeau, & Francon, 1995). System T amino acid transporter TAT1 is also known as monocarboxylate transporter 10 (MCT10, SLC16A10) which transports T_3, but not T_4 (Friesema et al., 2008). Thyroid hormone transporters represent an interesting example of convergent evolution, since several transporter families evolved members capable of thyroid hormone transport despite their structural differences (Fig. 2). Structures of thyroid hormone transporters have been recently reviewed by us (Schweizer, Johannes, Bayer, & Braun, 2014).

3. MONOCARBOXYLATE TRANSPORTER 8

MCT8/SLC16A2 was first identified as a transporter gene with unknown function on the X chromosome (Lafreniere, Carrel, & Willard, 1994). Based on its similarity to the aromatic amino acid transporter TAT1, Friesema et al. analyzed whether MCT8 is capable of thyroid hormone transport (Friesema et al., 2003). They found that MCT8 is not an amino acid transporter, but a very specific thyroid hormone transporter expressed in many tissue and organs, e.g., liver, kidney, brain, and pituitary (Friesema et al., 2003). Shortly after, the same group reported a new syndrome of X-linked psychomotor retardation with very peculiar thyroid hormone levels, i.e., high T_3, low T_4, and normal TSH, which was associated with mutations in the *MCT8* gene (Friesema et al., 2004). Independently, Dumitrescu et al. reported mutations in the *MCT8* gene in patients with unusual thyroid function tests and neurological abnormalities (Dumitrescu, Liao, Best, Brockmann, & Refetoff, 2004). Together these two publications established the profound physiological role of MCT8 in the thyroid hormone axis and neurological function and, therefore, of thyroid hormone plasma membrane transport. Only then it was realized that families carrying an X-linked mental retardation syndrome which was described in 1944 as the Allan–Herndon–Dudley syndrome (Allan, Herndon, & Dudley, 1944) also carry mutations in the *MCT8* gene (Schwartz et al., 2005). Apparently, the abnormal thyroid function tests in Allan–Herndon–Dudley syndrome patients were not reported over a period of six decades while under genetic study, and the genetics of a neurologic condition were finally solved by endocrinologists.

3.1 The Syndrome of Inherited MCT8 Deficiency

The diagnostic hallmarks of MCT8 deficiency are high T_3, low T_4, and normal TSH in plasma in male patients with severe mental retardation (Fu & Dumitrescu, 2014). Patients usually come to medical attention within the first year after birth with hypotonia and feeding problems. While normal during pregnancy and after birth, affected children fail to attain head control and show central hypotonia and quadriplegia, often irritability, and communication skills remain rudimentary with most patients never attaining speech (Dumitrescu et al., 2004; Friesema et al., 2004; Schwartz et al., 2005). Later, dystonia and involuntary movements may develop, sometimes seizures. Because of the high plasma T_3 levels, sex hormone-binding globulin (SHBG) is elevated and plasma cholesterol is low (Biebermann et al., 2005; Dumitrescu & Refetoff, 2013; Herzovich et al., 2007; Wemeau et al., 2008). Bone age was reportedly advanced (Herzovich et al., 2007). Remarkably, cardiac function is normal at young age, possibly because MCT8 is also expressed in the heart and the tissue is thus unresponsive to the elevated T_3 in MCT8 patients (Biebermann et al., 2005). Accordingly, patients show classical signs of hypothyroidism, i.e., cold skin and constipation (Schwartz et al., 2005), while other symptoms are rather consistent with hyperthyroidism, e.g., increased plasma ammonium and lactic acid, as well as muscle wasting (Herzovich et al., 2007). Apparently, depending on whether a tissue or organ expresses MCT8 and possibly other thyroid hormone transporters or not, the tissue shows signs of local hypo- or hyperthyroidism. Myelination depends on thyroid hormone signaling and was reported delayed in MCT8 patients (Holden et al., 2005), but seems to normalize after 5 years of age. Microcephaly is often described and some authors suggested cerebral atrophy (Kakinuma, Itoh, & Takahashi, 2005). Life expectancy may be normal; at least several patients have reached ages beyond 70 years (Schwartz et al., 2005).

3.2 Mutations in MCT8 Impair Thyroid Hormone Transport

Pathologic mutations are most likely loss of function, because there is no apparent difference between patients with large deletions in the *MCT8* gene or nonsense or missense mutations (Dumitrescu et al., 2004; Friesema et al., 2004; Friesema, Visser, & Visser, 2010; Refetoff & Dumitrescu, 2007). Due to the X-chromosomal location of MCT8, hemizygous mutations precipitate the full phenotype in affected males. Accordingly, heterozygous female carriers are not affected, while biased X-chromosome inactivation may lead to disease in female carriers (Frints et al., 2008). Uptake of thyroid hormone into cells can be easily studied in cells exposed to ^{125}I-labeled thyroid hormone. Fibroblasts from *MCT8* patients show reduced or no uptake of ^{125}I-T_3 (Visser et al., 2009). Interestingly, patients carrying the mutations S194F, L434W, L568P, and possibly ΔF501 exhibit reportedly a milder phenotype with some capacity to walk and speak (Jansen, Friesema, Kester, Schwartz, & Visser, 2008; Visser et al., 2009). If expressed in cells and analyzed for thyroid hormone uptake, the respective mutant proteins retained some reduced functionality. Thus, the severity of the MCT8 syndrome may correlate with the thyroid hormone transport capacity of mutant proteins. While the effects of large gene deletions and nonsense mutations are obvious, much can be learned from missense mutations (Table 1). These may point at critical positions in the protein and help explain MCT8 function (Kinne et al., 2009).

MCT8 belongs to the major facilitator superfamily of transporters and accordingly contains 12 transmembrane helices (TMHs) arranged as a tandem repeat of a bundle of 6 TMHs (Schweizer et al., 2014). N- and C-termini are cytosolic and the two bundles of six TMHs are linked by a long cytoplasmic linker (Fig. 2). A homology model of MCT8 has been presented and allowed to map pathological mutations onto the structure model (Kinne et al., 2010). Mutations may directly interfere with transport activity or affect proper protein folding or transmembrane insertion (Schweizer et al., 2014).

Table 1 Pathogenic Mutations in *MCT8*

Mutation		Type	Effects on MCT8	References
	delEx1	Deletion	No T_3 uptake activity in JEG3 cells	Jansen et al. (2007) Friesema et al. (2004) Visser et al. (2009)
	delEx2-4	Deletion		Vaurs-Barriere et al. (2009)
	delEx2-6	Deletion		Capri et al. (2013)
	delEx3	Deletion	No T_3 uptake activity (JEG3 cells)	Jansen et al. (2007) Yamamoto et al. (2013)
	delEx3-4	Deletion	No T_3 uptake activity (JEG3 cells)	Jansen et al. (2007) Friesema et al. (2004)
	delEx6	Deletion		Friesema et al. (2010)
c.1A > T	M1L	Missense		Frints et al. (2008)
c.127insGGCAGC	insGS43	Insertion		Friesema et al. (2010)
c.565insATC	insI189	Insertion	No T_3 and T_4 uptake activity (JEG3 & COS1 cells) Cytoplasmatic localization (JEG3 cells)	Jansen et al. (2008) Holden et al. (2005) Kinne et al. (2009)
c.575A > G	H192R	Missense		Friesema et al. (2010) Capri et al. (2013)

Table 1 Pathogenic Mutations in *MCT8*—cont'd

Mutation		Type	Effects on MCT8	References
c.578A > T (?)	N193I	Missense	Female patient: no expression of mutant MCT8-N193I in fibroblasts due to unfavorable nonrandom x-inactivation	Frints et al. (2008)
c.581C > T	S194F	Missense	Reduced T_3 and T_4 uptake activity (JEG3 & COS1 cells) Plasma membrane localization (JEG3 cells)	Jansen et al. (2008) Schwartz et al. (2005) Kinne et al. (2009)
c.629insA	K209fsX241	Nonsense		Frints et al. (2008)
c.630insA	N210fsX30	Insertion		Friesema et al. (2010)
c.630insG	N210fsX30	Insertion		Friesema et al. (2010) Filho et al. (2011)
c.del631-44	R211fsX25	Nonsense	No T_3 uptake activity (JEG3 cells)	Jansen et al. (2007)
c.656G > A	W219X	Nonsense		Tonduti et al. (2013)
c.661G > A	G221R	Missense	No T_3 and T_4 uptake activity (JEG3 & COS1 cells)	Schwartz and Stevenson (2007) Vaurs-Barriere et al. (2009) Kersseboom et al. (2013) Capri et al. (2013)
c.670G > A	A224T	Missense		Friesema et al. (2010) Boccone, Dessì, Meloni, and Loudianos (2013)

Continued

Table 1 Pathogenic Mutations in *MCT8*—cont'd

Mutation		Type	Effects on MCT8	References
c.671C>T	A224V	Missense	No T_3 uptake activity (JEG3 & CHO-K1 cells) No plasma membrane localization (FTC-238 cells)	Jansen et al. (2007) Biebermann et al. (2005) Friesema et al. (2004) Kinne et al. (2009)
c.688-690delTCT	delF230	Deletion	No T_3 and T_4 uptake activity (JEG3 & COS1 cells) Cytoplasmatic localization (JEG3 cells)	Jansen et al. (2008) Schwartz et al. (2005) Kinne et al. (2009)
c.703G>A	V235M	Missense	No T_3 and T_4 uptake activity (JEG3 & COS1 cells) Cytoplasmatic localization (JEG3 cells)	Jansen et al. (2008) Schwartz et al. (2005) Kinne et al. (2009)
c.706insGTG	insV236	Insertion	No T_3 and T_4 uptake activity (JEG3 & COS1 cells)	Kakinuma et al. (2005) Friesema et al. (2010) Kinne et al. (2009) Kersseboom et al. (2013)
c.733C>T	R245X	Nonsense	No T_3 uptake activity (JEG3 cells)	Jansen (2007) Friesema et al. (2004)
c.812G>A	R271H	Missense	Modest T_3 uptake (JEG3 cells)	Jansen et al. (2007) Fuchs, Pfarr, Pohlenz, and Schmidt (2009) Kinne et al. (2009)

Table 1 Pathogenic Mutations in *MCT8*—cont'd

Mutation		Type	Effects on MCT8	References
c. 844G > C	G282C	Missense	Residual uptake activity in Flp-in 293 and COS1 cells but not in JEG3 cells	Friesema et al. (2010) Kersseboom et al. (2013)
c.962C > T	P321L	Missense	No T_3 and T_4 uptake activity (JEG3 & COS1 cells)	Vaurs-Barriere et al. (2009) Kersseboom et al. (2013)
c.1003C > T	Q335X	Nonsense		Herzovich et al. (2007) Vaurs-Barriere et al. (2009)
c.1018delC	L340X	Nonsense		Friesema et al. (2010)
c.1102A > T	R368X	Nonsense		Tsurusaki et al. (2011)
c.1201G > A	G401R	Missense		Namba et al. (2008)
c.1212delT	A405fsX12	Deletion		Dumitrescu et al. (2004)
c.1301T > G	L434W	Missense	Reduced T_3 and T_4 uptake activity (JEG3 & COS1 cells) Plasma membrane localization (JEG3 cells)	Jansen et al. (2008) Schwartz et al. (2005) Kinne et al. (2009)
c.1333C > T	R445C	Missense		Friesema et al. (2010) Capri et al. (2013) Groeneweg et al. (2013)
c.1343C > A	S448X	Nonsense	No T_3 and T_4 uptake activity (JEG3 & COS1 cells) No protein expression (JEG3 cells)	Jansen et al. (2008) Schwartz et al. (2005)

Continued

Table 1 Pathogenic Mutations in *MCT8*—cont'd

Mutation		Type	Effects on MCT8	References
c.1343-1344insGCCC	S448fsX5	Insertion		Friesema et al. (2010)
c.1358A > T	D453V	Missense	No T_3 and T_4 uptake activity (JEG3 & COS1 cells)	Friesema et al. (2010) Kersseboom et al. (2013)
c.1412T > C	L471P	Missense	No T_3 uptake activity (JEG3 & MDCK1 cells) Reduced T_3 uptake activity in JEG1 cells	Jansen et al. (2007) Friesema et al. (2004) Kinne et al. (2009) Tonduti et al. (2013)
c.1428G > ?	M476I	Missense		Schwartz et al. (2007)
c.1492G > A	D498N	Missense		Groeneweg et al. (2013)
c.1500-1502delCTT	delF501	Deletion	Reduced T_3 and T_4 transport activity (uptake & efflux) in fibroblasts and JEG3 cells Plasma membrane localization	Visser et al. (2009)
c.1535T > C	L512P	Missense	Residual T_3 uptake in JEG1 cells but not in MDCK1 cells	Dumitrescu et al. (2004) Kinne et al. (2009)
c.1558C > T	Q520X	Nonsense		Vaurs-Barriere et al. (2009) Capri et al. (2013)
c.1610C > T	P537L	Missense	Residual uptake activity in COS1 but not in JEG3 cells	Papadimitriou et al. (2008) Friesema et al. (2010) Kersseboom et al. (2013)

Table 1 Pathogenic Mutations in *MCT8*—cont'd

Mutation		Type	Effects on MCT8	References
c.1649delA	Y550fsX17	Deletion		Namba et al. (2008)
c.1673G > A	G558D	Missense	Residual uptake activity in COS1 & MDCK1 cells but not in JEG1 & JEG3 cells	Schwartz et al. (2007) Frints et al. (2008) Friesema et al. (2010) Kinne et al. (2009) Kersseboom et al. (2013)
c.1690G > A	G564R	Missense	No T_3 and T_4 transport activity (uptake & efflux) in fibroblasts & JEG3 cells	Visser et al. (2009)) Tsurusaki et al. (2011)
c.1703T > C	L568P	Missense	Reduced T_3 and T_4 uptake activity (JEG3 & COS1 cells) Plasma membrane localization (JEG3 cells)	Jansen et al. (2008) Schwartz et al. (2005) Kinne et al. (2009)
c.1826delC	P609fsX70	Deletion		Vaurs-Barriere et al. (2009) Capri et al. (2013)
c.1835delC	P612fsX68	Deletion		Maranduba et al. (2006)

3.3 Concepts for Treatment of MCT8 Deficiency

3.3.1 Raising Low T_4

The paradoxical thyroid function tests (high T_3, low T_4, normal TSH) in *MCT8* patients raised the questions whether and how these should be normalized. With the aim to normalize T_4 levels (because T_4 is expected to enter the brain), T_4 was supplemented in one of the index patients (Biebermann et al., 2005). While T_4 could be normalized, T_3 increased further. Both pituitary and liver responded adequately to circulating thyroid hormones. However, sweating and tachycardia occurred, while the neurological phenotype was not ameliorated. The tachycardia implies that the heart is not entirely unresponsive to raised thyroid hormone levels.

3.3.2 Lowering Elevated T_3

High plasma T_3 may represent a clinical problem in the management of *MCT8* deficiency, as most patients have low body weight and signs of protein catabolism (Herzovich et al., 2007). The reason is that some tissues express additional thyroid hormone transporters which, in the absence of MCT8 function, allow access of high plasma T_3 to the organ, causing local signs of hyperthyroidism. Thus, high T_3 may cause a catabolic situation in adipose tissue and muscle, resulting in muscle weakness. Another early attempt to normalize thyroid function tests in MCT8 patients employed the antithyroid drug propylthiouracil combined with T_4 replacement therapy (Wemeau et al., 2008). While this therapy finally resulted in significant weight gain and normalized tachycardia, no improvement of the psychomotor retardation was observed either.

What remained unclear is why T_4 is paradoxically low in the face of elevated T_3. As MCT8 contributes to thyroid hormone release from the gland, it was speculated that T_3 would be more efficiently released than T_4 in the absence of MCT8 function (Di Cosmo et al., 2010). However, due to concerns whether the patient's thyroid gland changed its structure and in the light of pathological findings in the mouse model, thyroidectomy was performed on an *MCT8* patient (Wirth et al., 2011). The patient received L-T_4 replacement therapy after thyroidectomy and L-T_4T_3 remained elevated in the face of low T_4, indicating that differential release from the thyroid gland cannot be the sole reason for the abnormal T_3/T_4 ratio. Possibly a dysregulation of the hypothalamic–pituitary system occurs because MCT10 is able to compensate for MCT8 by allowing T_3 enter thyroid hormone-sensitive cells, but not T_4 (Alkemade et al., 2005, 2011).

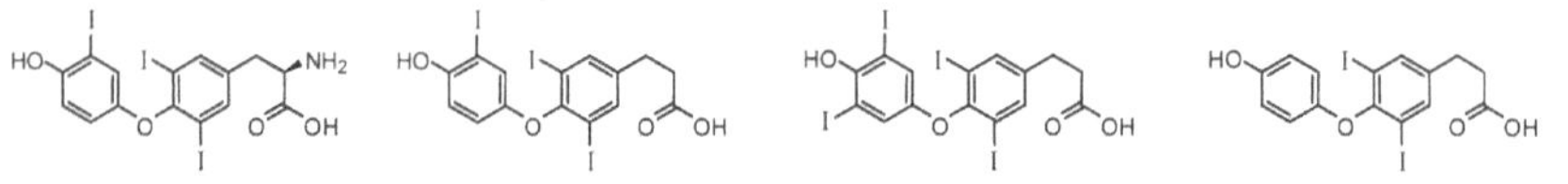

Triiodothyronine, T_3 Triiodothyroacetic acid, TRIAC Tetraiodothyroacetic acid, TETRAC Diiodothyropropionic acid, DITPA

Fig. 3 Molecular structures of T_3 and thyromimetic compounds which enter cells independently of MCT8. TETRAC is activated by deiodination for TRIAC, the compound which binds the nuclear TR.

3.3.3 Replacing Thyroid Hormone With an MCT8-Independent Agonist

Another rationale is to find a thyroid hormone receptor-activating molecule which can exert its effect independently of cellular uptake by MCT8 and treat the patients with this compound. One such compound is 3,5-diiodothyropropionic acid (DITPA) (Fig. 3). DITPA treatment brought the patients' thyroid function tests close to normal and the decline of T_3 correlated with significant weight gain (Verge et al., 2012). While SHBG levels were normalized, neurological measures were not improved.

TRIAC and TETRAC are thyroid hormone metabolites which are generated through sequential decarboxylation and oxidation of the resulting amine to the corresponding carboxylic acid. Yet, these compounds are known to activate nuclear TR and are capable of cellular entry independent of MCT8. A pertinent question now is whether MCT8 patients may benefit from treatment with a thyromimetic compound (DITPA, TETRAC, or TRIAC) when initiated very early, possibly in utero before the neurodevelopmental defect occurs (Fig. 3).

4. MOUSE MODELS OF ALLAN–HERNDON–DUDLEY SYNDROME

Rodent models have been important tools for increasing our understanding of thyroid hormone transport in the mammalian body. After the realization that mutations in *MCT8* lead to psychomotor retardation in patients, it was shown that *Mct8* is widely expressed in neurons throughout the central nervous system (Heuer et al., 2005). Then mice were engineered lacking functional Mct8 (Dumitrescu, Liao, Weiss, Millen, & Refetoff, 2006; Trajkovic et al., 2007). The mice replicated the abnormal hormone constellation of *MCT8* patients. T_3 content and T_3-dependent gene activity were decreased in brain and increased in liver. Uptake of T_4 from plasma into brain was not affected, while uptake of T_3 into brain was virtually absent in the *Mct8*-deficient mouse model (Trajkovic et al., 2007). A mild neurological phenotype was revealed in *Mct8*-deficient mice which suggested that also in the brain some cells depend on Mct8, while others are independent of Mct8 because of other transporter protein expression (Wirth et al., 2009). Using inhibitors directed against thyroid hormone transporters other than Mct8, thyroid hormone uptake into cultured cortical neurons was found to contain a component similar to -type amino acid transporters (Wirth et al., 2009). A comparative neurodevelopmental study showed that Mct8 and Lat2 are (co)expressed from early development in mouse brain, while in human brain MCT8 is expressed in embryonic neurons and LAT2, in contrast, only in microglial cells (Wirth et al., 2009). Murine primary astrocytes also coexpress several thyroid hormone transporters including Mct8 and Lat2 and a probenecid-sensitive transporter (Braun, Kinne, et al., 2011; Braun, Wirth, et al., 2011). Recently, inhibition by probenecid of hormone transport into primary astrocytes demonstrated functional coexpression of Slco1c1 with Mct8 (Schnell et al., 2013).

Compensation of lack of Mct8 function occurs not only in neurons and astrocytes but also along the blood–brain barrier. The uptake of radiolabeled T_4, but not T_3, in *Mct8*-deficient mice suggested that a compensating T_4 transporter may be present in mouse brain. Indeed, Slco1c1/Oatp14 is specifically expressed in rodent microcapillary brain endothelial cells (Sugiyama et al., 2003), but much less so in humans (Roberts et al., 2008).

When Slco1c1 was genetically inactivated along with Mct8, T_4 uptake into the brain was virtually abolished in the double deficient mice, leading to a hypothyroid situation in their brains (Mayerl et al., 2014). T_3 and T_4 content in the forebrain was reduced below the already low levels in *Mct8*-deficient mice. Maturation of parvalbumin-positive GABAergic interneurons was significantly decreased in the cortex of double knockout mice consistent with their developmental dependence on thyroid hormone. Moreover, myelination was clearly reduced (Mayerl et al., 2014).

The *Mct8*-deficient mouse model was used to investigate the possible treatment with DITPA (Di Cosmo, Liao, Dumitrescu, Weiss, & Refetoff, 2009). DITPA normalized thyroid function tests in $Mct8^{-/y}$ mice, including thyroid hormone-dependent gene expression in the brain (Di Cosmo et al., 2009).

The *Mct8*-deficient mouse model was also treated with TETRAC (Horn et al., 2013). TETRAC can be metabolized to TRIAC by Dio2 and replace T_3 in an *in vitro* assay of Purkinje cell development. TETRAC cellular uptake is independent of Mct8 and Sclo1c1, and injection of TETRAC into athyroid Pax8-deficient mice rescued cerebral thyroid hormone-dependent gene expression and parvalbumin-positive

GABAergic interneuron development. Pituitary TSH was also repressed by TETRAC in *Mct8*-deficient mice, showing that this compound can replace thyroid hormone and its uptake is independent of Mct8 (Horn et al., 2013).

5. THYROID HORMONE TRANSPORT AND PHYSIOLOGY

Thyroid hormone transport is of importance throughout the body. Every cell type responsive to nuclear thyroid hormone receptors needs thyroid hormone transmembrane transport proteins (Fig. 4). The interplay of thyroid hormone transporters Mct8, Lat2, Slco1c1, and possibly Mct10 in cerebral thyroid hormone transport was already discussed above and explains the coexistence of hyper- and hypothyroid states of cell populations within one organ.

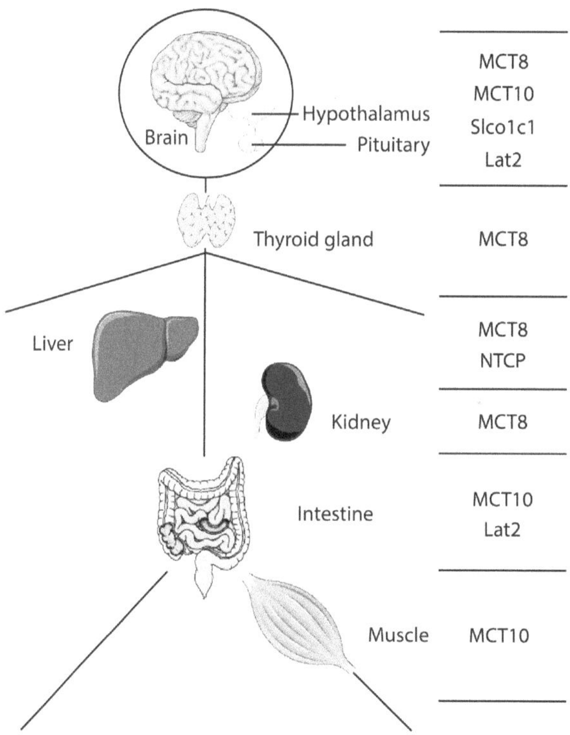

Fig. 4 Tissues and organs depending on thyroid hormone transmembrane transport. Transporters involved in thyroid hormone homeostasis in each organ are indicated.

Similarly, the coexpression of Mct8 along with other transporters in the liver causes a hyperthyroid state in liver, because elevated plasma T_3 is available to hepatocytes even in the absence of Mct8. Accordingly, plasma cholesterol is reduced, plasma SHBG is elevated, and type I deiodinase (Dio1) is induced in *Mct8*-deficient mice and patients (Dumitrescu et al., 2006; Wemeau et al., 2008).

The situation in the heart is not as clear. MCT8 seems the most important thyroid hormone transporter in the heart and patients usually do not exhibit increased heart rate when diagnosed with *MCT8* (Biebermann et al., 2005). In older patients, however, increased heart rate has been reported (Wemeau et al., 2008).

The kidney is involved in maintenance of adequate thyroid hormone levels by resorption of filtrated thyroid hormones from primary urine. Free thyroid hormones or hormones bound to plasma transfer proteins are taken up along the tubular epithelium—by direct transport or via endocytosis of plasma proteins. In the absence of Mct8 in mice, renal thyroid hormone levels are increased along with T_3-dependent gene expression, possibly suggesting that release of thyroid hormones from the cells into the plasma is impaired (Trajkovic-Arsic, Visser, et al., 2010). In contrast, T_3 and T_4 are lost with the urine. Mct10 is coexpressed with Mct8 in kidney tubular epithelium. Combined inactivation of *Mct8* and *Mct10* exaggerated the thyroid hormone accumulation in the kidney, while paradoxically normalizing plasma T_4 levels (Muller et al., 2014). Lat2 is also expressed in the kidney tubular epithelium. Mice deficient in Lat2 exhibited aminoaciduria, but no changes of plasma thyroid hormone levels (Braun, Kinne, et al., 2011; Braun, Wirth, et al., 2011). These findings suggest that Mct8 and Mct10 both contribute to thyroid hormone efflux from kidney tubular epithelial cells, while Lat2 is more important for amino acid resorption in the same cells.

Thyroid hormone transmembrane trafficking is also important for the function of the thyroid gland. Thyroid hormones are made in the follicular lumen on the apical side of thyroid epithelial cells. Endocytosis of thyroglobulin (fragments) is followed by degradation to free hormones by lysosomal proteases (Friedrichs et al., 2003). The next step, release from lysosomal vesicles, is likely mediated by membrane transporters, possibly including Lat2 (U.S. unpublished). Cytoplasmic thyroid hormones are then thought to leave the gland by transmembrane transporters including MCT8 (Di Cosmo et al., 2010; Trajkovic-Arsic, Muller, et al., 2010; Wirth et al., 2011).

Inner ear development depends on thyroid hormone signaling (Forrest, 1996). Lack of TRβ or Dio2 expression leads to hearing impairment in mice (Ng et al., 2004). Several thyroid hormone transporters are expressed in the inner ear in a cell type-specific and developmentally regulated pattern (Sharlin, Visser, & Forrest, 2011).

6. FURTHER CLINICAL IMPLICATIONS FOR THYROID HORMONE TRANSPORTERS

6.1 Polymorphisms in Thyroid Hormone Transporters May Affect Thyroid Hormone Levels

An association of polymorphisms in SLCO1C1 with fatigue and depression has been observed in hypothyroid patients under T_4 replacement therapy (van der Deure, Appelhof, et al., 2008), but none of the polymorphisms in Slco1c1 changed transport activity *in vitro* (van der Deure, Hansen, et al., 2008). Correlations of thyroid hormone levels in probands with polymorphisms in thyroid hormone transporter genes have been investigated also for other transporters, but observed effects were weak, not replicated in other studies or sex specific (van der Deure et al., 2009).

6.2 Inhibition of Thyroid Hormone Transport by Drugs

Tyrosine kinase inhibitors (TKIs) are a relatively new class of drugs effective against many types of cancer. TKIs interact with the ATP-binding site of tyrosine kinases (e.g., vascular endothelial growth factor receptor, VEGFR; platelet-derived growth factor receptor, PDGFR; stem cell factor receptor c-KIT; RET kinase), avoiding the transfer of a phosphate moiety from ATP to tyrosine in effector proteins. Abnormalities in thyroid hormone metabolism were noted in TKI-treated patients including hypothyroidism, destructive thyroiditis, and thyroid atrophy (Desai & Stadler, 2006; Mannavola et al., 2007). Furthermore, athyroid patients under levothyroxine ($\text{L-}T_4$) replacement therapy suddenly needed an increased amount of $\text{L-}T_4$ during treatment with TKIs to suppress their TSH (Wong et al., 2007).

The molecular mechanisms of TKI-induced defects in thyroid hormone metabolism are not yet well understood. One mode of action, in particular of sunitinib and axitinib, is destruction of the thyroid gland possibly as a consequence of inhibition of VEGFR and subsequent capillary dysfunction and insufficient blood supply of the gland observed in mouse models treated with axitinib (Kamba et al., 2006). Negative effects on thyroid hormone biosynthesis were suggested with inhibition of peroxidases and the growth of rat thyroid cells (Salem, Fenton, Marion, & Hershman, 2008; Wong et al., 2007). Another possible mode of action of TKI may be inhibition of thyroid hormone transporters. The increased requirement of L-T_4 in athyroid patients during TKI treatment excludes negative effects on the thyroid gland itself, but *in vitro* studies demonstrated impaired thyroid hormone transport due to inhibition of thyroid hormone transporters, e.g., MCT8 (Braun & Schweizer, 2014; Braun et al., 2012; Illouz, Braun, Briet, Schweizer, & Rodien, 2014). This example shows that thyroid hormone transport should be considered in the future when discussing unwanted drug effects or tissue-specific action of TR-activating drugs.

6.3 Intestinal Uptake of T_4 and Thyroid Hormone Transport

Levothyroxine is one of the most frequently prescribed medications worldwide. Oral administration is common, but we do not currently know how and where and through which transporters T_4 is taken up in the intestinal tract. Interestingly, some athyroid patients suffer from T_4 malabsorption, leading to a syndrome of oral thyroxine resistance (Tönjes et al., 2006). Several reasons for oral thyroid hormone resistance have been proposed: inadequate resorption due to interaction in the gut with nutrients (e.g., coffee), coeliac or hepatic diseases, interference with comedication, and psychosocial misbehavior (Münchausen syndrome) (Ain, Refetoff, Fein, & Weintraub, 1991). In a small number of oral thyroid hormone resistance patients all these reasons could be excluded. Rather, these patients appear deficient in an unidentified intestinal thyroid hormone transporter.

7. CONCLUSIONS AND FUTURE DIRECTIONS

The last decade has seen a rapid development of the concept of thyroid hormone transport. While in three decades before the concept was followed only by few visionary researchers who realized that thyroid hormones are in essence amino acids needing transporters to cross the plasma membrane, the identification of MCT8 has finally established the importance of thyroid hormone transport in the clinical disciplines. Investigations into the functions of thyroid hormone transporters in mouse models and with respect to many organs have broadened our appreciation of this novel concept. Thyroid hormone transporters are affected by seemingly unrelated drugs possibly explaining so far enigmatic clinical findings. There may yet remain clinical implications of thyroid hormone transporters unexplored.

ACKNOWLEDGMENTS

The authors thank the Deutsche Forschungsgemeinschaft (DFG) for financial support and our colleagues, national and international, for many stimulating discussions.

REFERENCES

Abe, T., Kakyo, M., Sakagami, H., Tokui, T., Nishio, T., Tanemoto, M., et al. (1998). Molecular characterization and tissue distribution of a new organic anion transporter subtype (oatp3) that transports thyroid hormones and taurocholate and comparison with oatp2. *The Journal of Biological Chemistry*, *273*(35), 22395–22401.

Ain, K. B., Refetoff, S., Fein, H. G., & Weintraub, B. D. (1991). Pseudomalabsorption of levothyroxine. *JAMA*, *266*(15), 2118–2120.

Alkemade, A., Friesema, E. C., Kalsbeek, A., Swaab, D. F., Visser, T. J., & Fliers, E. (2011). Expression of thyroid hormone transporters in the human hypothalamus. *The Journal of Clinical Endocrinology and Metabolism*, *96*(6), E967–E971. http://dx.doi.org/10.1210/jc.2010-2750.

Alkemade, A., Friesema, E. C., Unmehopa, U. A., Fabriek, B. O., Kuiper, G. G., Leonard, J. L., et al. (2005). Neuroanatomical pathways for thyroid hormone feedback in the human hypothalamus. *The Journal of Clinical Endocrinology and Metabolism*, *90*(7), 4322–4334. http://dx.doi.org/10.1210/jc.2004-2567.

Allan, W., Herndon, C. N., & Dudley, F. C. (1944). Some examples of the inheritance of mental deficiency: Apparently sex-linked idiocy and microcephaly. *American Journal of Mental Deficiency*, *48*, 325–334.

Bianco, A. C., Salvatore, D., Gereben, B., Berry, M. J., & Larsen, P. R. (2002). Biochemistry, cellular and molecular biology, and physiological roles of the iodothyronine selenodeiodinases. *Endocrine Reviews*, *23*(1), 38–89.

Biebermann, H., Ambrugger, P., Tarnow, P., von Moers, A., Schweizer, U., & Grueters, A. (2005). Extended clinical phenotype, endocrine investigations and functional studies of a loss-of-function mutation A150V in the thyroid hormone specific transporter MCT8. *European Journal of Endocrinology*, *153*(3), 359–366. 153/3/359 [pii]. http://dx.doi.org/10.1530/eje.1.01980.

Blondeau, J. P., Beslin, A., Chantoux, F., & Francon, J. (1993). Triiodothyronine is a high-affinity inhibitor of amino acid transport system L1 in cultured astrocytes. *Journal of Neurochemistry*, *60*(4), 1407–1413.

Boccone, L., Dessì, V., Meloni, A., & Loudianos, G. (2013). Allan-Herndon-Dudley syndrome (AHDS) in two consecutive generations caused by a missense MCT8gene mutation. Phenotypic variability with the presence of normal serum T3 levels. *European Journal of Medical Genetics*, *56*(4), 207–210. http://dx.doi.org/10.1016/j.ejmg.2013.02.001. [Epub 2013 Feb 16].

Bochukova, E., Schoenmakers, N., Agostini, M., Schoenmakers, E., Rajanayagam, O., Keogh, J. M., et al. (2012). A mutation in the thyroid hormone receptor alpha gene. *The New England Journal of Medicine*, *366*(3), 243–249. http://dx.doi.org/10.1056/NEJMoa1110296.

Braun, D., Kim, T. D., le Coutre, P., Köhrle, J., Hershman, J. M., & Schweizer, U. (2012). Tyrosine kinase inhibitors noncompetitively inhibit MCT8-mediated iodothyronine transport. *The Journal of Clinical Endocrinology and Metabolism*, *97*(1), E100–E105. http://dx.doi.org/10.1210/jc.2011-1837.

Braun, D., Kinne, A., Brauer, A. U., Sapin, R., Klein, M. O., Köhrle, J., et al. (2011a). Developmental and cell type-specific expression of thyroid hormone transporters in the mouse brain and in primary brain cells. *Glia*, *59*(3), 463–471. http://dx.doi.org/10.1002/glia.21116.

Braun, D., & Schweizer, U. (2014). Authentic bosutinib inhibits triiodothyronine transport by monocarboxylate transporter 8. *Thyroid, 24*(5), 926–927. http://dx.doi.org/10.1089/thy.2013.0660.

Braun, D., Wirth, E. K., Wohlgemuth, F., Reix, N., Klein, M. O., Grüters, A., et al. (2011b). Aminoaciduria, but normal thyroid hormone levels and signalling, in mice lacking the amino acid and thyroid hormone transporter Slc7a8. *The Biochemical Journal, 439*(2), 249–255. http://dx.doi.org/10.1042/BJ20110759.

Brown, D. D., Wang, Z., Kanamori, A., Eliceiri, B., Furlow, J. D., & Schwartzman, R. (1995). Amphibian metamorphosis: A complex program of gene expression changes controlled by the thyroid hormone. *Recent Progress in Hormone Research, 50*, 309–315.

Capri, Y., Friesema, E. C., Kersseboom, S., Touraine, R., Monnier, A., Eymard-Pierre, E., et al. (2013). Relevance of different cellular models in determining the effects of mutations on SLC16A2/MCT8 thyroid hormone transporter function and genotype-phenotype correlation. *Human Mutation, 34*(7), 1018–1025. http://dx.doi.org/10.1002/humu.22331. [Epub 2013 May 1].

Chantoux, F., Blondeau, J. P., & Francon, J. (1995). Characterization of the thyroid hormone transport system of cerebrocortical rat neurons in primary culture. *Journal of Neurochemistry, 65*(6), 2549–2554.

Davis, P. J., Leonard, J. L., & Davis, F. B. (2007). Mechanisms of nongenomic actions of thyroid hormone. *Frontiers in Neuroendocrinology, 29*(2), 211–218.

Desai, A. A., & Stadler, W. M. (2006). Novel kinase inhibitors in renal cell carcinoma: Progressive development of static agents. *Current Urology Reports*, 7(1), 16–22.

Di Cosmo, C., Liao, X. H., Dumitrescu, A. M., Philp, N. J., Weiss, R. E., & Refetoff, S. (2010). Mice deficient in MCT8 reveal a mechanism regulating thyroid hormone secretion. *The Journal of Clinical Investigation, 120*(9), 3377–3388.

Di Cosmo, C., Liao, X. H., Dumitrescu, A. M., Weiss, R. E., & Refetoff, S. (2009). A thyroid hormone analog with reduced dependence on the monocarboxylate transporter 8 for tissue transport. *Endocrinology, 150*(9), 4450–4458.

Dumitrescu, A. M., Liao, X. H., Best, T. B., Brockmann, K., & Refetoff, S. (2004). A novel syndrome combining thyroid and neurological abnormalities is associated with mutations in a monocarboxylate transporter gene. *American Journal of Human Genetics, 74*(1), 168–175.

Dumitrescu, A. M., Liao, X. H., Weiss, R. E., Millen, K., & Refetoff, S. (2006). Tissue-specific thyroid hormone deprivation and excess in monocarboxylate transporter (mct) 8-deficient mice. *Endocrinology, 147*(9), 4036–4043.

Dumitrescu, A. M., & Refetoff, S. (2013). The syndromes of reduced sensitivity to thyroid hormone. *Biochimica et Biophysica Acta, 1830*(7), 3987–4003. http://dx.doi.org/10.1016/j.bbagen.2012.08.005.

Filho, H. C., Marui, S., Manna, T. D., Brust, E. S., Radonsky, V., Kuperman, H., et al. (2011). Novel mutation in MCT8 gene in a Brazilian boy with thyroid hormone resistance and severe neurologic abnormalities. *Arquivos Brasileiros de Endocrinologia e Metabologia, 55*(1), 60–66.

Forrest, D. (1996). Deafness and goiter: Molecular genetic considerations. *The Journal of Clinical Endocrinology and Metabolism, 81*(8), 2764–2767.

Friedrichs, B., Tepel, C., Reinheckel, T., Deussing, J., von Figura, K., Herzog, V., et al. (2003). Thyroid functions of mouse cathepsins B, K, and L. *The Journal of Clinical Investigation, 111*(11), 1733–1745.

Friesema, E. C., Docter, R., Moerings, E. P., Stieger, B., Hagenbuch, B., Meier, P. J., et al. (1999). Identification of thyroid hormone transporters. *Biochemical and Biophysical Research Communications, 254*(2), 497–501. http://dx.doi.org/10.1006/bbrc.1998.9974.

Friesema, E. C., Ganguly, S., Abdalla, A., Manning Fox, J. E., Halestrap, A. P., & Visser, T. J. (2003). Identification of monocarboxylate transporter 8 as a specific thyroid hormone transporter. *The Journal of Biological Chemistry*, *278*(41), 40128–40135.

Friesema, E. C., Grueters, A., Biebermann, H., Krude, H., von Moers, A., Reeser, M., et al. (2004). Association between mutations in a thyroid hormone transporter and severe X-linked psychomotor retardation. *Lancet*, *364*(9443), 1435–1437.

Friesema, E. C., Jansen, J., Jachtenberg, J. W., Visser, W. E., Kester, M. H., & Visser, T. J. (2008). Effective cellular uptake and efflux of thyroid hormone by human monocarboxylate transporter 10. *Molecular Endocrinology*, *22*(6), 1357–1369.

Friesema, E. C., Visser, W. E., & Visser, T. J. (2010). Genetics and phenomics of thyroid hormone transport by MCT8. *Molecular and Cellular Endocrinology*, *322*(1–2), 107–113.

Frints, S. G., Lenzner, S., Bauters, M., Jensen, L. R., Van Esch, H., des, P., et al. (2008). MCT8 mutation analysis and identification of the first female with Allan-Herndon-Dudley syndrome due to loss of MCT8 expression. *European Journal of Human Genetics: EJHG*, *16*(9), 1029–1037.

Fu, J., & Dumitrescu, A. M. (2014). Inherited defects in thyroid hormone cell-membrane transport and metabolism. *Best Practice & Research Clinical Endocrinology & Metabolism*, *28*(2), 189–201. http://dx.doi.org/10.1016/j.beem.2013.05.014.

Fuchs, O., Pfarr, N., Pohlenz, J., & Schmidt, H. (2009). Elevated serum triiodothyronine and intellectual and motor disability with paroxysmal dyskinesia caused by a monocarboxylate transporter 8 gene mutation. *Developmental Medicine and Child Neurology*, *51*(3), 240–244. http://dx.doi.org/10.1111/j.1469-8749.2008.03125.x. [Epub 2008 Oct 17].

Hennemann, G., Docter, R., Friesema, E. C., de Jong, M., Krenning, E. P., & Visser, T. J. (2001). Plasma membrane transport of thyroid hormones and its role in thyroid hormone metabolism and bioavailability. *Endocrine Reviews*, *22*(4), 451–476.

Herzovich, V., Vaiani, E., Marino, R., Dratler, G., Lazzati, J. M., Tilitzky, S., et al. (2007). Unexpected peripheral markers of thyroid function in a patient with a novel mutation of the MCT8 thyroid hormone transporter gene. *Hormone Research*, *67*(1), 1–6.

Heuer, H., Maier, M. K., Iden, S., Mittag, J., Friesema, E. C., Visser, T. J., et al. (2005). The monocarboxylate transporter 8 linked to human psychomotor retardation is highly expressed in thyroid hormone-sensitive neuron populations. *Endocrinology*, *146*(4), 1701–1706.

Holden, K. R., Zuniga, O. F., May, M. M., Su, H., Molinero, M. R., Rogers, R. C., et al. (2005). X-linked MCT8 gene mutations: Characterization of the pediatric neurologic phenotype. *Journal of Child Neurology*, *20*(10), 852–857.

Horn, S., Kersseboom, S., Mayerl, S., Muller, J., Groba, C., Trajkovic-Arsic, M., et al. (2013). Tetrac can replace thyroid hormone during brain development in mouse mutants deficient in the thyroid hormone transporter mct8. *Endocrinology*, *154*(2), 968–979. http://dx.doi.org/10.1210/en.2012–1628.

Illouz, F., Braun, D., Briet, C., Schweizer, U., & Rodien, P. (2014). Thyroid effects of tyrosine kinase inhibitors. *European Journal of Endocrinology*. http://dx.doi.org/10.1530/EJE-14-0198.

Jansen, J., Friesema, E. C., Kester, M. H., Milici, C., Reeser, M., Grüters, A., et al. (2007). Functional analysis of monocarboxylate transporter 8 mutations identified in patients with X-linked psychomotor retardation and elevated serum triiodothyronine. *The Journal of Clinical Endocrinology and Metabolism*, *92*(6), 2378–2381. [Epub 2007 Mar 13].

Jansen, J., Friesema, E. C., Kester, M. H., Schwartz, C. E., & Visser, T. J. (2008). Genotype-phenotype relationship in patients with mutations in thyroid hormone transporter MCT8. *Endocrinology*, *149*(5), 2184–2190.

Kakinuma, H., Itoh, M., & Takahashi, H. (2005). A novel mutation in the monocarboxylate transporter 8 gene in a boy with putamen lesions and low free T4 levels in cerebrospinal fluid. *The Journal of Pediatrics*, *147*(4), 552–554.

Kalyanaraman, H., Schwappacher, R., Joshua, J., Zhuang, S., Scott, B. T., Klos, M., et al. (2014). Nongenomic thyroid hormone signaling occurs through a plasma membrane-localized receptor. *Science Signaling*, 7(326) ra48. http://dx.doi.org/10.1126/scisignal.2004911.

Kamba, T., Tam, B. Y., Hashizume, H., Haskell, A., Sennino, B., Mancuso, M. R., et al. (2006). VEGF-dependent plasticity of fenestrated capillaries in the normal adult microvasculature. *American Journal of Physiology Heart and Circulatory Physiology*, *290*(2), H560–H576. http://dx.doi.org/10.1152/ajpheart.00133.2005.

Kersseboom, S., Kremers, G. J., Friesema, E. C., Visser, W. E., Klootwijk, W., Peeters, R. P., et al. (2013). Mutations in MCT8 in patients with Allan-Herndon-Dudley-syndrome affecting its cellular distribution. *Molecular Endocrinology*, *27*(5), 801–813. http://dx.doi.org/10.1210/me.2012-1356. [Epub 2013 Apr 2].

Kinne, A., Kleinau, G., Hoefig, C. S., Grüters, A., Köhrle, J., Krause, G., et al. (2010). Essential molecular determinants for thyroid hormone transport and first structural implications for monocarboxylate transporter 8. *The Journal of Biological Chemistry*, *285*(36), 28054–28063. http://dx.doi.org/10.1074/jbc.M110.129577.

Kinne, A., Roth, S., Biebermann, H., Köhrle, J., Grüters, A., & Schweizer, U. (2009). Surface translocation and tri-iodothyronine uptake of mutant MCT8 proteins are cell type-dependent. *Journal of Molecular Endocrinology*, *43*(6), 263–271. http://dx.doi.org/10.1677/JME-09-0043.

Krenning, E. P., Docter, R., Bernard, H. F., Visser, T. J., & Hennemann, G. (1978). Active transport of triiodothyronine (T3) into isolated rat liver cells. *FEBS Letters*, *91*(1), 113–116.

Lafreniere, R. G., Carrel, L., & Willard, H. F. (1994). A novel transmembrane transporter encoded by the XPCT gene in Xq13.2. *Human Molecular Genetics*, *3*(7), 1133–1139.

Magnus-Levy, A. (1895). Über den respiratorischen Gaswechsel unter dem Einfluss der Thyroiden sowie unter verschiedenen physiologischen Zuständen. *Berliner. Klinische Wochenschrift*, *32*, 650–652.

Mannavola, D., Coco, P., Vannucchi, G., Bertuelli, R., Carletto, M., Casali, P. G., et al. (2007). A novel tyrosine-kinase selective inhibitor, sunitinib, induces transient hypothyroidism by blocking iodine uptake. *The Journal of Clinical Endocrinology and Metabolism*, *92*(9), 3531–3534. http://dx.doi.org/10.1210/jc.2007-0586.

Mayerl, S., Muller, J., Bauer, R., Richert, S., Kassmann, C. M., Darras, V. M., et al. (2014). Transporters MCT8 and OATP1C1 maintain murine brain thyroid hormone homeostasis. *The Journal of Clinical Investigation*, *124*(5), 1987–1999. http://dx.doi.org/10.1172/JCI70324.

Muller, J., Mayerl, S., Visser, T. J., Darras, V. M., Boelen, A., Frappart, L., et al. (2014). Tissue-specific alterations in thyroid hormone homeostasis in combined Mct10 and Mct8 deficiency. *Endocrinology*, *155*(1), 315–325. http://dx.doi.org/10.1210/en.2013-1800.

Mullur, R., Liu, Y. Y., & Brent, G. A. (2014). Thyroid hormone regulation of metabolism. *Physiological Reviews*, *94*(2), 355–382. http://dx.doi.org/10.1152/physrev.00030.2013.

Ng, L., Goodyear, R. J., Woods, C. A., Schneider, M. J., Diamond, E., Richardson, G. P., et al. (2004). Hearing loss and retarded cochlear development in mice lacking type 2 iodothyronine deiodinase. *Proceedings of the National Academy of Sciences of the United States of America*, *101*(10), 3474–3479.

Rao, G. S., Eckel, J., Rao, M. L., & Breuer, H. (1976). Uptake of thyroid hormone by isolated rat liver cells. *Biochemical and Biophysical Research Communications*, *73*(1), 98–104.

Refetoff, S., & Dumitrescu, A. M. (2007). Syndromes of reduced sensitivity to thyroid hormone: Genetic defects in hormone receptors, cell transporters and deiodination. *Best Practice & Research Clinical Endocrinology & Metabolism*, *21*(2), 277–305.

Roberts, L. M., Woodford, K., Zhou, M., Black, D. S., Haggerty, J. E., Tate, E. H., et al. (2008). Expression of the thyroid hormone transporters MCT8 (SLC16A2) and OATP14 (SLCO1C1) at the blood-brain barrier. *Endocrinology*, *149*(12), 6251–6261.

Salem, A. K., Fenton, M. S., Marion, K. M., & Hershman, J. M. (2008). Effect of sunitinib on growth and function of FRTL-5 thyroid cells. *Thyroid*, *18*(6), 631–635. http://dx.doi.org/10.1089/thy.2007.0336.

Schnell, C., Shahmoradi, A., Wichert, S. P., Mayerl, S., Hagos, Y., Heuer, H., et al. (2013). The multispecific thyroid hormone transporter OATP1C1 mediates cell-specific sulforhodamine 101-labeling of hippocampal astrocytes. *Brain Structure & Function*. http://dx.doi.org/10.1007/s00429-013-0645-0.

Schwartz, C. E., May, M. M., Carpenter, N. J., Rogers, R. C., Martin, J., Bialer, M. G., et al. (2005). Allan-Herndon-Dudley syndrome and the monocarboxylate transporter 8 (MCT8) gene. *American Journal of Human Genetics*, 77(1), 41–53.

Schwartz, C. E., & Stevenson, R. E. (2007). The MCT8 thyroid hormone transporter and Allan-Herndon-Dudley syndrome. *Best Practice & Research Clinical Endocrinology & Metabolism*, *21*(2), 307–321. Review.

Schweizer, U., Johannes, J., Bayer, D., & Braun, D. (2014). Structure and function of thyroid hormone plasma membrane transporters. *Eur Thyroid J*, *3*(3), 143–153.

Schweizer, U., & Köhrle, J. (2012). Function of thyroid hormone transporters in the central nervous system. *Biochimica et Biophysica Acta*, *1830*(7), 3965–3973. http://dx.doi.org/10.1016/j.bbagen.2012.07.015.

Schweizer, U., Weitzel, J. M., & Schomburg, L. (2008). Think globally: Act locally. New insights into the local regulation of thyroid hormone availability challenge long accepted dogmas. *Molecular and Cellular Endocrinology*, *289*(1–2), 1–9. S0303-7207(08)00128-7 [pii]. http://dx.doi.org/10.1016/j.mce.2008.04.007.

Sharlin, D. S., Visser, T. J., & Forrest, D. (2011). Developmental and cell-specific expression of thyroid hormone transporters in the mouse cochlea. *Endocrinology*, *152*(12), 5053–5064. http://dx.doi.org/10.1210/en.2011-1372.

Sugiyama, D., Kusuhara, H., Taniguchi, H., Ishikawa, S., Nozaki, Y., Aburatani, H., et al. (2003). Functional characterization of rat brain-specific organic anion transporter (Oatp14) at the blood-brain barrier: High affinity transporter for thyroxine. *The Journal of Biological Chemistry*, *278*(44), 43489–43495.

Tonduti, D., Vanderver, A., Berardinelli, A., Schmidt, J. L., Collins, C. D., Novara, F., et al. (2013). MCT8 deficiency: Extrapyramidal symptoms and delayed myelination as prominent features. *Journal of Child Neurology*, *28*(6), 795–800. http://dx.doi.org/10.1177/0883073812450944. [Epub 2012 Jul 17].

Tönjes, A., Karger, S., Koch, C. A., Paschke, R., Tannapfel, A., Stumvoll, M., et al. (2006). Impaired enteral levothyroxine absorption in hypothyroidism refractory to oral therapy after thyroid ablation for papillary thyroid cancer: Case report and kinetic studies. *Thyroid*, *16*(10), 1047–1051. http://dx.doi.org/10.1089/thy.2006.16.1047.

Trajkovic, M., Visser, T. J., Mittag, J., Horn, S., Lukas, J., Darras, V. M., et al. (2007). Abnormal thyroid hormone metabolism in mice lacking the monocarboxylate transporter 8. *The Journal of Clinical Investigation*, *117*(3), 627–635. http://dx.doi.org/10.1172/JCI28253.

Trajkovic-Arsic, M., Muller, J., Darras, V. M., Groba, C., Lee, S., Weih, D., et al. (2010). Impact of monocarboxylate transporter-8 deficiency on the hypothalamus-pituitary-thyroid axis in mice. *Endocrinology*, *151*(10), 5053–5062. http://dx.doi.org/10.1210/en.2010-0593.

Trajkovic-Arsic, M., Visser, T. J., Darras, V. M., Friesema, E. C., Schlott, B., Mittag, J., et al. (2010). Consequences of monocarboxylate transporter 8 deficiency for renal transport and metabolism of thyroid hormones in mice. *Endocrinology*, *151*(2), 802–809. http://dx.doi.org/10.1210/en.2009-1053.

van der Deure, W. M., Appelhof, B. C., Peeters, R. P., Wiersinga, W. M., Wekking, E. M., Huyser, J., et al. (2008). Polymorphisms in the brain-specific thyroid hormone transporter OATP1C1 are associated with fatigue and depression in hypothyroid patients. *Clinical Endocrinology*, *69*(5), 804–811. http://dx.doi.org/10.1111/j.1365-2265.2008.03267.x.

van der Deure, W. M., Hansen, P. S., Peeters, R. P., Kyvik, K. O., Friesema, E. C., Hegedus, L., et al. (2008). Thyroid hormone transport and metabolism by OATP1C1 and consequences of genetic variation. *Endocrinology*, *149*(10), 5307–5314.

van der Deure, W. M., Peeters, R. P., Uitterlinden, A. G., Hofman, A., Breteler, M. M., Witteman, J., et al. (2009). Impact of thyroid function and polymorphisms in the type 2 deiodinase on blood pressure: The Rotterdam Study and the Rotterdam Scan Study. *Clinical Endocrinology*, *71*(1), 137–144. http://dx.doi.org/10.1111/j.1365-2265.2008.03447.x.

van Mullem, A., van Heerebeek, R., Chrysis, D., Visser, E., Medici, M., Andrikoula, M., et al. (2012). Clinical phenotype and mutant TRalpha1. *The New England Journal of Medicine*, *366*(15), 1451–1453. http://dx.doi.org/10.1056/NEJMc1113940.

Vaurs-Barriere, C., Deville, M., Sarret, C., Giraud, G., Des Portes, V., Prats-Vinas, J. M., et al. (2009). Pelizaeus-Merzbacher-like disease presentation of MCT8 mutated male subjects. *Annals of Neurology*, *65*, 114–118.

Verge, C. F., Konrad, D., Cohen, M., Di Cosmo, C., Dumitrescu, A. M., Marcinkowski, T., et al. (2012). Diiodothyropropionic acid (DITPA) in the treatment of MCT8 deficiency. *The Journal of Clinical Endocrinology and Metabolism*, *97*(12), 4515–4523. http://dx.doi.org/10.1210/jc.2012-2556.

Visser, W. E., Jansen, J., Friesema, E. C., Kester, M. H., Mancilla, E., Lundgren, J., et al. (2009). Novel pathogenic mechanism suggested by ex vivo analysis of MCT8 (SLC16A2) mutations. *Human Mutation*, *30*(1), 29–38.

Wemeau, J. L., Pigeyre, M., Proust-Lemoine, E., d'Herbomez, M., Gottrand, F., Jansen, J., et al. (2008). Beneficial effects of propylthiouracil plus L-thyroxine treatment in a patient with a mutation in MCT8. *The Journal of Clinical Endocrinology and Metabolism*, *93*(6), 2084–2088.

Wirth, E. K., Roth, S., Blechschmidt, C., Hölter, S. M., Becker, L., Racz, I., et al. (2009). Neuronal 3′,3,5-triiodothyronine (T3) uptake and behavioral phenotype of mice deficient in Mct8, the neuronal T3 transporter mutated in Allan-Herndon-Dudley syndrome. *Journal of Neuroscience*, *29*(30), 9439–9449. 29/30/9439 [pii]. http://dx.doi.org/10.1523/JNEUROSCI.6055-08.2009.

Wirth, E. K., Sheu, S. Y., Chiu-Ugalde, J., Sapin, R., Klein, M. O., Mossbrugger, I., et al. (2011). Monocarboxylate transporter 8 deficiency: Altered thyroid morphology and persistent high triiodothyronine/thyroxine ratio after thyroidectomy. *European Journal of Endocrinology*, *165*(4), 555–561. http://dx.doi.org/10.1530/EJE-11-0369.

Wong, E., Rosen, L. S., Mulay, M., Vanvugt, A., Dinolfo, M., Tomoda, C., et al. (2007). Sunitinib induces hypothyroidism in advanced cancer patients and may inhibit thyroid peroxidase activity. *Thyroid*, *17*(4), 351–355. http://dx.doi.org/10.1089/thy.2006.0308.

Yamamoto, S., Okuhara, K., Tonoki, H., Iizuka, S., Nihei, N., & Tajima, T. (2013). A novel deletion mutation of SLC16A2 encoding monocarboxylate transporter (MCT) 8 in a 26-year-old Japanese patient with Allan-Herndon-Dudley syndrome. *Clinical Pediatric Endocrinology*, *22*(4), 83–86. http://dx.doi.org/10.1292/cpe.22.83. [Epub 2013 Oct 26].

CHAPTER THREE

Nuclear Import and Export of the Thyroid Hormone Receptor

Jibo Zhang, Vincent R. Roggero, Lizabeth A. Allison[1]
College of William and Mary, Williamsburg, VA, United States
[1]Corresponding author: e-mail address: laalli@wm.edu

Contents

1. Introduction 46
2. Thyroid Hormone Receptor Nucleocytoplasmic Shuttling 47
   2.1 Translocation Through the Nuclear Pore Complexes 48
   2.2 Analysis of Nucleocytoplasmic Shuttling in Live Cells 50
3. Nuclear Import Pathways of TR 51
   3.1 NLSs in TRα1 52
   3.2 Importins That Mediate TRα1 Nuclear Import 53
4. Nuclear Export Pathways of TRα1 54
   4.1 NESs in TRα1 55
   4.2 Exportins That Mediate TRα1 Nuclear Export 57
5. Mislocalization of TR and Pathogenesis 58
   5.1 TR Mislocalization and Cancer 59
   5.2 TR Mislocalization and RTH Syndrome 60
6. Conclusions and Future Directions 61
Acknowledgments 62
References 62

Abstract

The thyroid hormone receptors, TRα1 and TRβ1, are members of the nuclear receptor superfamily that forms one of the most abundant classes of transcription factors in multicellular organisms. Although primarily localized to the nucleus, TRα1 and TRβ1 shuttle rapidly between the nucleus and cytoplasm. The fine balance between nuclear import and export of TRs has emerged as a critical control point for modulating thyroid hormone-responsive gene expression. Mutagenesis studies have defined two nuclear localization signal (NLS) motifs that direct nuclear import of TRα1: NLS-1 in the hinge domain and NLS-2 in the N-terminal A/B domain. Three nuclear export signal (NES) motifs reside in the ligand-binding domain. A combined approach of shRNA-mediated knockdown and coimmunoprecipitation assays revealed that nuclear entry of TRα1 is facilitated by importin 7, likely through interactions with NLS-2, and importin β1 and the adapter importin α1 interacting with both NLS-1 and NLS-2. Interestingly, TRβ1 lacks NLS-2 and nuclear import depends solely on the importin α1/β1 heterodimer.

Vitamins and Hormones, Volume 106
ISSN 0083-6729
http://dx.doi.org/10.1016/bs.vh.2017.04.002

Heterokaryon and fluorescence recovery after photobleaching shuttling assays identified multiple exportins that play a role in nuclear export of TRα1, including CRM1 (exportin 1), and exportins 4, 5, and 7. Even single amino acid changes in TRs dramatically alter their intracellular distribution patterns. We conclude that mutations within NLS and NES motifs affect nuclear shuttling activity, and propose that TR mislocalization contributes to the development of some types of cancer and Resistance to Thyroid Hormone syndrome.

1. INTRODUCTION

The thyroid hormone receptors (TRs) are members of the nuclear receptor superfamily that exhibit a dual role as activators or repressors of gene transcription in response to thyroid hormone (T_3) and provide a model system for investigating complex networks of cellular trafficking and gene expression. Thyroid hormone enters the cell via uptake by plasma membrane transporters, and then can diffuse through the cytoplasm into the nucleus (Mondal, Raja, Schweizer, & Mugesh, 2016). Although there is evidence for nongenomic mechanisms of action for thyroid hormone (Davis, Goglia, & Leonard, 2016), the cellular response is mediated primarily by TRs. By modulating target gene transcription in response to T_3, TRs play key physiological roles in the regulation of many aspects of development, growth, and metabolism (Mondal et al., 2016; Mullur, Liu, & Brent, 2014). Here, we describe the journey of the intranuclear TRs from the cytoplasm into the nucleus to carry out their genomic functions.

The thyroid hormone receptors, TRα and TRβ, are encoded by separate genes on chromosome 17 and chromosome 3, respectively. Several protein variants that are differentially expressed in specific tissues are encoded by alternative processing of gene transcripts, including the main hormone-binding variants, TRα1, TRβ1, and TRβ2 (Mullur et al., 2014; Pascual & Aranda, 2013). Of these three major variants, TRα1 is predominant in bone, the gastrointestinal tract, cardiac and skeletal muscle, and the central nervous system; TRβ1 is most abundant in the liver and kidney; and TRβ2 is expressed in the hypothalamus, pituitary, cochlea, and retina (Flamant & Gauthier, 2013). In addition, v-*erb*A, a mutated derivative of the cellular locus encoding TRα1 (*c-erb*Aα) carried by the avian erythroblastosis virus, has acquired oncogenic properties. The v-ErbA oncoprotein is unable to bind ligand and acts as a constitutive dominant repressor of transcription regulated by TRα1 in mammalian and avian cells (Wolffe et al., 2000).

In this review, we focus primarily on TRα1 intracellular trafficking, but comparisons are made with TRβ1 and the oncoprotein v-ErbA.

After localizing to the nucleus, TRα1 and TRβ1 (referred to collectively as "TR" for simplicity) function involves a multifaceted cascade of events that results in binding of TR, often as a heterodimer with the retinoid X receptor (RXR), to thyroid hormone response elements (TREs), and culminates in the modulation of target gene expression (Diallo, Wilhelm, Thompson, & Koenig, 2007). In the absence of T_3, TR binds "positive" TREs and represses gene expression in conjunction with corepressors; in the presence of T_3 and coactivators, gene expression is activated. In contrast, at the less common "negative" TREs the opposite occurs; in the absence of T_3, TR promotes transcription, whereas T_3-bound TR is repressive (Weitzel, 2008). Two questions thus arise, how does TR make its way to the nucleus after synthesis in the cytosol, and how is this process regulated? To address these questions, we discuss an intricate picture that is emerging from integrating understanding of gene regulation at the level of DNA–protein interactions with the additional levels of control made possible by cell compartments. A dynamic balance between nuclear import, nuclear retention, and nuclear export of transcription factors is central to coordinating cell signaling and gene expression.

2. THYROID HORMONE RECEPTOR NUCLEOCYTOPLASMIC SHUTTLING

With the discovery that TRα1 and TRβ1 shuttle rapidly between the nucleus and cytoplasm, the long-standing dogma that TRs reside solely in the nucleus tightly bound to DNA was overturned (Baumann, Maruvada, Hager, & Yen, 2001; Bunn et al., 2001). Studies early on used heterokaryon assays, which are an elegant and sensitive way to detect rapid nucleocytoplasmic shuttling of proteins that by other techniques, would appear to be localized in the nucleus at steady state (Bonamy, Guiochon-Mantel, & Allison, 2005; Bunn et al., 2001; Grespin et al., 2008). In this assay, HeLa (human) cells are first transfected with expression vectors for green fluorescent protein (GFP)-tagged TRα1. Then, the GFP-TRα1-expressing human cells are fused with untransfected NIH–3T3 (mouse) cells to form heterokaryons that contain multiple nuclei from both species in a common cytoplasm (Fig. 1). To distinguish the human and mouse nuclei, the cells are stained with the dye Hoechst 33258, which gives a characteristic pattern of "speckles" in the mouse nuclei. After incubation for as little as 1 h,

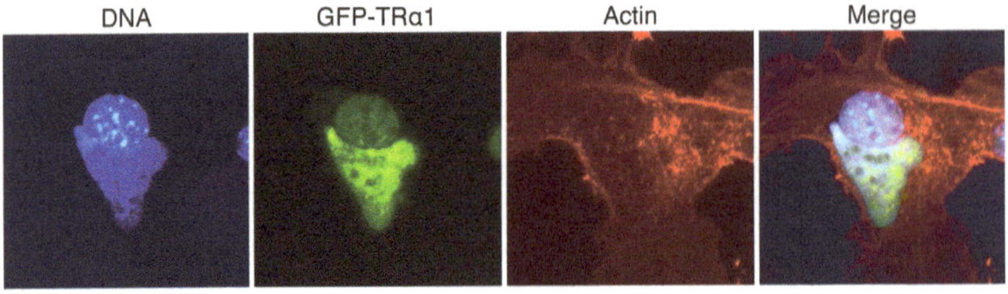

Fig. 1 Heterokaryon assay showing nucleocytoplasmic shuttling of TRα1. For the preparation of heterokaryons, HeLa (human) cells were transfected with GFP-TRα1 expression vectors and then fused with untransfected NIH/3T3 (mouse) cells using 50% polyethylene glycol (PEG) 1500. For the shuttling assays, fused cells were incubated for 2 h at 37°C in culture medium containing cycloheximide (to prevent de novo protein synthesis). The human nucleus (diffuse *blue* staining) was distinguished from the mouse nucleus (speckled) by differential coloration with Hoechst 33258 DNA stain. To visualize heterokaryons, cells were fixed and stained for actin with Vectashield containing TRITC-phalloidin (*red*). GFP-TRα1 (*green*) shuttling was viewed by epifluorescence microscopy.

the results of shuttling can be visualized in fixed cells; the accumulation of green fluorescence in the mouse nucleus in heterokaryons demonstrates export of GFP-TRα1 from the human nucleus and import into the mouse nucleus.

Subsequent studies using both *Xenopus oocytes* and mammalian cells provided further insight into some of the factors modulating shuttling. For example, phosphorylation of TRα1 was shown to play a role in its nuclear retention (Nicoll et al., 2003); and in other studies, dominant negative TR mutants were shown to localize to both nuclear and cytosolic compartments and display altered transport activity (Bonamy et al., 2005; Bondzi et al., 2011; Bunn et al., 2001; DeLong, Bonamy, Fink, & Allison, 2004). These findings suggest that multiple factors, including posttranslational modification and receptor mutations, contribute to TR shuttling dynamics. However, there exists another larger obstacle for TR localization; the nuclear and cytoplasmic compartments are physically separated by the nuclear envelope (Cautain, Hill, de Pedro, & Link, 2015). The presence of this double-membrane barrier surrounding the nucleus creates a requirement for a mechanism that regulates exchange of macromolecules with the cytoplasm.

2.1 Translocation Through the Nuclear Pore Complexes

Macromolecules cross the nuclear envelope through elaborate protein assemblies, approximately 120 MDa in size, called nuclear pore complexes (Kosinski et al., 2016; Lin et al., 2016). The nuclear pore complex has an eightfold symmetry, and translocation occurs through a central channel; extending from the ends of this channel is the nuclear basket and cytoplasmic filaments, which serve as docking sites for transport factors. The 100-nm-wide channel is a passive "soft" barrier that allows diffusion of small molecules and ions, but obstructs diffusion of macromolecules greater than 3 nm in diameter and ~40 kDa in size (Timney et al., 2016). The selective barrier is generated by proteins called nucleoporins that contain intrinsically disordered phenylalanine–glycine (FG) repeats. Selected cargoes destined for the nucleus, even those as large as 40 nm in diameter, can be rapidly transported through this mesh-like barrier.

Proteins such as TRα1 (46 kDa) are translocated into the nucleus by a temperature- and energy-dependent mechanism that requires soluble factors in the cytosol (Roggero et al., 2016). These soluble factors are a family of evolutionarily conserved karyopherin β-like transport factors, with each family member performing a distinct nuclear import, export, or bidirectional transport function. Karyopherins in charge of importing cargo into the nucleus are called importins, whereas those in charge of exporting cargo back to the cytoplasm are called exportins (Cautain et al., 2015; Tran, King, & Corbett, 2014). Most nuclear proteins contain nuclear localization signals (NLSs) that direct binding with importins, and they also may contain nuclear export signals (NESs) harbored within the protein's amino acid sequence that direct nuclear export by interacting with exportins (Chook & Suel, 2011; Kimura & Imamoto, 2014; Marfori et al., 2011; Soniat & Chook, 2015). Besides interacting with protein cargo, karyopherins also weakly bind to the FG repeats in the cytoplasmic filaments to bring the cargo to the nuclear pore complex. Signal-mediated transport also relies on an asymmetrical cellular distribution of the small GTPase Ran in either its GTP or its GDP-bound state. A high nuclear RanGTP concentration is required for dissociation of import complexes that have successfully passed through the nuclear pore complex, and for assembly of export complexes that will exit the nucleus (Wente & Rout, 2010).

Twenty β-family karyopherins have been identified so far in humans. They have similar molecular masses and all contain multiple HEAT repeats (named for the repeat structure found in Huntingtin, Elongation factor 3, protein phosphatase 2A, and the yeast kinase TOR1), but their amino acid sequences are 90% different from each other (O'Reilly, Dacks, & Field, 2011). Each of them transports a distinct group of cargo, but most proteins appear to utilize multiple karyopherins for transport (Chook & Suel, 2011). Karyopherins mostly interact with their cargo directly via a nonclassical

NLS. Notably, importin β1 requires interaction with one of the seven adaptor proteins from the importin α family to assist in the cargo-recognition process; importin α binds directly to a classical NLS within the cargo, while importin β1 interacts with the nucleoporins (Pumroy & Cingolani, 2015). In addition, adding to the diversity of pathways, there are other non-conventional nuclear transport pathways (Wagstaff & Jans, 2009); for example, an importin-independent nuclear import pathway can be accessed by proteins with conserved ankyrin repeats (Lu, Zak, et al., 2014).

Bidirectional transport occurs through the nuclear pore complexes, and it is now clear that many nuclear proteins, including the TRs, undergo shuttling. To visualize shuttling of a protein that is primarily nuclear at steady state requires specialized techniques, such as the heterokaryon assay shown in Fig. 1. The results of heterokaryons assays are typically analyzed in fixed cells, but it is often desirable to monitor the kinetics of shuttling in live cells.

2.2 Analysis of Nucleocytoplasmic Shuttling in Live Cells

The development of fluorescent protein tags (Rodriguez et al., 2017) and fluorescence recovery after photobleaching (FRAP) (Brazda et al., 2014) has led to major advances in quantifying the dynamics of protein shuttling kinetics in live cells. After photobleaching of a region of interest in the nucleus, many transcription factors exhibit complete recoveries within seconds (Mueller, Morisaki, Mazza, & McNally, 2012), indicating that they can diffuse throughout the entire nucleus and are immobilized to nuclear structures only transiently (Groeneweg et al., 2014). For example, approximately half of ligand-bound glucocorticoid receptor is freely diffusing, while the remaining population is bound to DNA, either for short periods of time (~0.7 s) or for longer time periods (~2.3 s) (Groeneweg et al., 2014).

FRAP also has been instrumental in our examination of TR nucleocytoplasmic shuttling (Grespin et al., 2008; Mavinakere, Powers, Subramanian, Roggero, & Allison, 2012; Subramanian et al., 2015). In this case, we use transfected cells with two (or more) nuclei that are expressing GFP-tagged TRα1. One nucleus within these multinucleate cells is exposed to intense laser illumination. This exposure results in the loss of fluorescence within the selected nucleus due to photobleaching of the GFP fluorophore. A series of images are then taken in which fluorescence recovery to the bleached nucleus is measured and compared with the concomitant decrease in intensity within unbleached nuclei. As for heterokaryon assays, near complete equilibration is seen over the course of 1 h (Grespin et al., 2008). Since TRα1 is primarily nuclear at steady state, but shuttles between the nucleus and cytosol, shRNA-mediated knockdown of an essential export factor would be predicted to result in decreased nuclear export of TRα1. This effect is visualized as greater retention of fluorescence in the unbleached nucleus, along with a significantly slower recovery of fluorescence to the bleached nucleus during the FRAP assay (Subramanian et al., 2015).

3. NUCLEAR IMPORT PATHWAYS OF TR

Although an NLS in the hinge domain had been partially characterized for TRs early on (Baumann et al., 2001; Casas et al., 2006; Lee & Mahdavi, 1993; Maruvada, Baumann, Hager, & Yen, 2003), the corresponding complete NLS in the hinge domain of TRα1 was not fully defined until recently (Mavinakere et al., 2012). Another layer of complexity is added to TR shuttling by modular, often overlapping, functional domains; thus, it is imperative to show that a domain is both necessary and sufficient (i.e., transferrable). TRα, TRβ, and other variants share a common protein structure (Fig. 2). This common structure includes

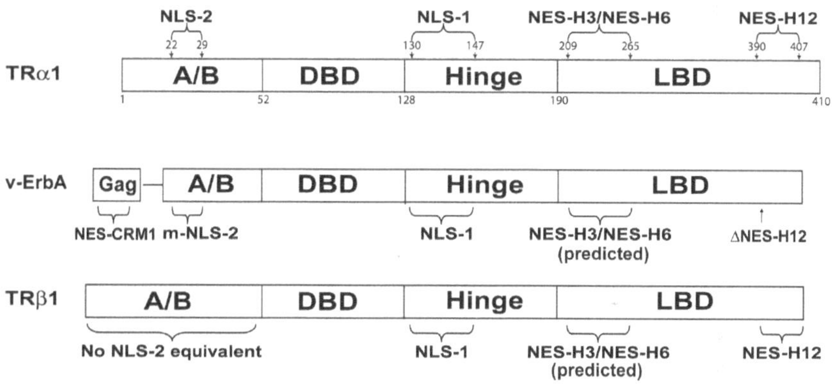

Fig. 2 TRα1 has multiple conserved signals for nuclear import and export. The diagram (not to scale) shows the location of NLS and NES motifs of TRα1, TRβ1, and v-ErbA, with amino acid residue numbers indicated. v-ErbA is a highly mutated variant of TRα1 carried by the avian erythroblastosis virus (AEV). Amino acid sequence changes which contribute to its oncogenic properties include fusion of a portion of AEV Gag to its N-terminus, N- and C-terminus deletions, and 13-amino acid substitutions. The positions of NLS, NES, and putative and predicted NES sequences are indicated in relation to the respective individual domains of TRα1: N-terminal A/B domain (A/B), DNA-binding domain (DBD), hinge domain, and ligand-binding domain (LBD). NES-CRM1, CRM1-dependent NES in the Gag region of v-ErbA; m-NLS-2, inactive NLS-2 equivalent in v-ErbA; NES-H3/NES-H6 (Predicted), based on the sequence identity with NES-H3/NES-H6 in TRα1; ΔNES-H12, deletion of the NES-H12 region in v-ErbA.

multiple functional regions: a less-conserved N-terminal A/B region, a highly conserved DNA-binding domain (DBD), a short hinge region, and a C-terminal ligand-binding domain (LBD). The N-terminal A/B domain contains a constitutive autonomous activation function 1 domain (AF-1), which is used to regulate transcription. By coupling mutagenesis and localization studies, we first tested each domain for NLS activity by constructing cytosolic-localized fluorescent fusion proteins.

3.1 NLSs in TRα1

Through a comprehensive, systematic analysis, two NLSs were identified in TRα1, one in the hinge region and one in the A/B domain that are sufficient to target a cytosolic protein to the nucleus (Fig. 2). No NLS activity was present in the DBD, which in most nuclear receptors is highly conserved in structure and sequence; at least 40% of the sequence is identical among all nuclear receptors (Chen & Young, 2010). There also was no NLS activity present in the LBD, the domain which harbors NES activity (see Section 4.1).

3.1.1 Hinge Domain NLS

A classical bipartite NLS, named NLS-1, resides in the hinge region of TRα1 (Bunn et al., 2001; Mavinakere et al., 2012; Fig. 2). The overall basic amino acid sequence of NLS-1 (^{130}KRVAKRKLIEQNRERRRK^{147}) is sufficient to import a cytosolic-localized fusion protein (GFP-GST-GFP) into the nucleus when it is fused with the hinge region. This bipartite NLS is well conserved in both the oncoprotein v-ErbA and TRβ1. TRα1's hinge domain is a short bridge between the DBD and the LBD. Originally, the hinge domain was considered no more than a flexible rotation "joint" between the DBD and LBD that helps different nuclear receptors to alter their conformations to adapt to different TREs; however, there is ample evidence for multiple functional elements within the hinge domain that contribute to efficient DNA binding, activation function and repression, T_3 binding, and corepressor interactions (Nascimento et al., 2006).

3.1.2 A/B Domain NLS

Compared to the other domains in TR, the N-terminal A/B domain is the most divergent and least well characterized (Thuestad, Kraus, Apriletti, & Saatcioglu, 2000). The A/B domain, however, contributes to transcriptional activation significantly through AF-1 (Tian, Mahajan, Wong, Habeos,

& Samuels, 2006). A second NLS (NLS-2) was identified in the A/B domain of TRα1, comprised of a nonclassical, novel monopartite eight-amino acid sequence (^{22}PDGKRKRK^{29}). NLS-2 successfully targeted a cytosolic GFP-GST-GFP fusion protein into the nucleus and was shown to be necessary for efficient import in the context of the full-length receptor (Mavinakere et al., 2012). We also showed using site-directed mutagenesis that sequence context of NLS-2 is critical for function; flanking amino acid sequences may have a strong negative or positive impact. In addition, at the residue equivalent to arginine 26 in TRα1, v-ErbA has a histidine. Illustrating the importance of this substitution, the mutated A/B domain of v-ErbA failed to import cytosolic GFP-GST-GFP into the nucleus (Mavinakere et al., 2012). Further, a TRα1 A/B domain fusion protein with the v-ErbA-equivalent histidine substitution (R26H) failed to import GFP-GST-GFP into the nucleus, demonstrating that the naturally occurring R26H substitution abrogates the activity of NLS-2, suggesting that this defect in NLS activity contributes to the oncogenic conversion of TRα1 into v-ErbA.

Notably, the novel NLS-2 that we identified in the A/B domain of TRα1 is absent in TRβ1. We and others have noted that TRβ1 has a small cytosolic population (Baumann et al., 2001; Bunn et al., 2001; Maruvada et al., 2003); thus, this cytosolic subpopulation may reflect an altered balance of NLS and NES activity. Interestingly, the NLS-2 sequence is well conserved among other vertebrate species with some variations in chicken and fishes; however, TR from the tunicate *Ciona*, which is proposed to be the ancestral protein for TRα1 and TRβ1, shows no equivalent to rat TRα1 NLS-2 when sequences are aligned, suggesting that appearance of a monopartite NLS in the TRα1 A/B domain is a later evolutionary development (Mavinakere et al., 2012).

3.2 Importins That Mediate TRα1 Nuclear Import

The finding of two NLSs pointed to the possibility of TRα1 using more than one import pathway. We investigated which importins mediate nuclear import of TRα1 using a combined approach of shRNA-mediated knockdown, treatment with the importin β1-specific inhibitor importazole, and coimmunoprecipitation assays in HeLa (human) cells expressing GFP-tagged TRα1 (Roggero et al., 2016). Among all the importins tested in transient transfection assays (importins 4, 5, 6, 7, 8, 9, 13, importin β1, and adaptor importin α variants), only importin 7, importin β1, and adaptor importin α1 knockdown experiments, or treatment with importazole resulted in a significant localization pattern change from primarily nuclear to a more cytosolic distribution of TRα1. To demonstrate direct interaction between TRα1 or TRβ1 and these importins, coimmunoprecipitation assays were performed. Importin 7, importin β1, and adaptor importin α1 were shown to interact with TRα1, while importin 4 as a negative control did not.

3.2.1 Importin α1/β1 Heterodimer

Importin α1 (also known as karyopherin α2) is the general adaptor importin for classical NLS-containing cargo. As an adaptor protein, importin α1 interacts with importin β1 and recognizes protein cargo with a classical NLS to form the complete classical import complex (Cautain et al., 2015). Given that NLS-1 has the hallmark characteristics of a classical NLS, it was not surprising that TR can follow an importin α1/β1-mediated import pathway. Coimmunoprecipitation assays indicate that importin β1 and the adapter importin α1 interact with both NLS-1 and NLS-2 in TRα1. In contrast, TRβ1 nuclear import is facilitated only by importin α1/β1 interacting with NLS-1 (Roggero et al., 2016).

3.2.2 Importin 7

Our data showed that nuclear entry of TRα1 in HeLa cells also is facilitated by importin 7, likely through interactions with NLS-2. TRβ1, which lacks NLS-2, was not shown to interact directly with importin 7 in coimmunoprecipitation assays (Roggero et al., 2016). Importin 7 recognizes a diversity of cargos and is so far the only known β-family karyopherin that can import cargos by interacting with importin β1 (Chook & Suel, 2011). It has not yet been determined whether importin 7 binds to TRα1 as a heterodimer or on its own. Taken together, these findings indicate that TRα1 trafficking into the nucleus is mediated by two distinct pathways, although how multiple pathways serve to regulate nuclear entry in response to cell-specific signals remains a question of interest. Fig. 3 summarizes what is currently known about the pathways followed by TRα1 through the nuclear pore complex.

4. NUCLEAR EXPORT PATHWAYS OF TRα1

Once it was clear that TRα1 and TRβ1 not only enter the nucleus to carry out their functions but also shuttle rapidly between the nucleus and cytoplasm, we turned our attention to the other half of shuttling, nuclear

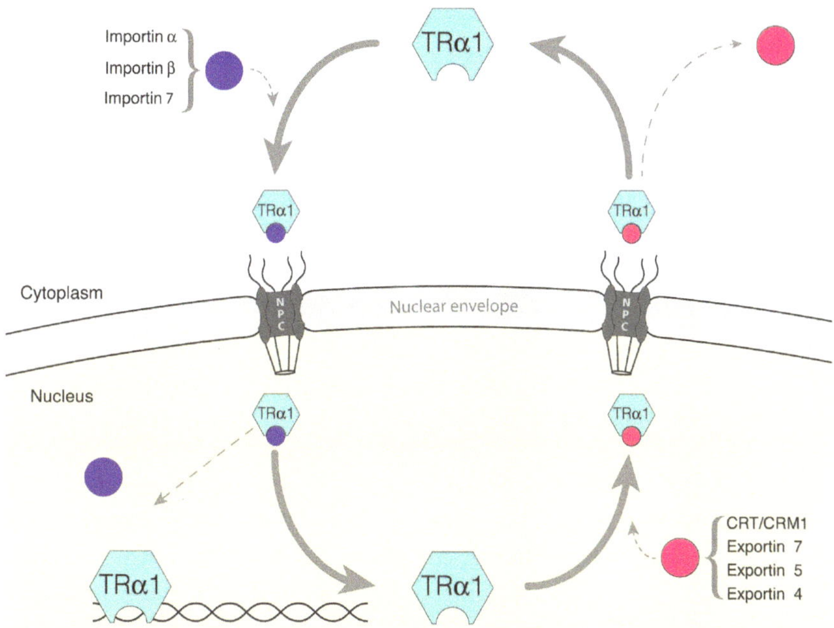

Fig. 3 TRα1 nucleocytoplasmic shuttling pathway. TRα1 binds to specific importins in the cytoplasm. The TRα1-importin complex passes through a nuclear pore complex (NPC) embedded in the nuclear envelope into the nucleus, where the complex is disassembled and TRα1 binds to target genes. TRα1 exits the nucleus through the NPC in association with specific exportins. *CRT*, calreticulin.

export. Our initial studies showed that TRα1 exits the nucleus through two distinct pathways, one CRM1 (exportin 1) dependent, and the other CRM1-independent (Grespin et al., 2008; Mavinakere et al., 2012; Subramanian et al., 2015). Intriguingly, the retroviral oncoprotein v-ErbA, a dominant negative mutant of TRα1, displays altered transport activity due to acquisition of a viral CRM1-dependent NES. As a consequence, v-ErbA localizes to both cellular compartments at steady state (Bonamy et al., 2005; Bondzi et al., 2011; Bunn et al., 2001; DeLong et al., 2004).

4.1 NESs in TRα1

Although we know the viral CRM1-dependent NES in v-ErbA is leucine rich and hydrophobic (DeLong et al., 2004), the TR CRM1-dependent NES remains elusive. As part of our published work, however, we identified novel NESs that are insensitive to leptomycin B, a CRM1 inhibitor, suggesting that they mediate the CRM1-independent export pathway: NESs in helix 3, helix 6, and helix 12 of the LBD were shown to be sufficient to target a nuclear protein to the cytosol. NES activity was not observed in other domains of TR (Mavinakere et al., 2012; Fig. 2). TR's LBD is the most complex domain because of its various activities, including ligand binding, receptor dimerization, and hormone-induced transcriptional activation or repression (Wu & Koenig, 2000). The LBD has been suggested to have an essential role in mediating TR–RXR heterodimer or TR homodimer formation because of the conserved ninth heptad close to the C-terminus.

4.1.1 NES-H12

Twelve α-helices in the LBD form a hollow pocket lined with hydrophobic residues; helix 12 (H12) contains a ligand-dependent activation domain, designated AF-2, that is highly conserved across the nuclear receptor superfamily (Barettino, Vivanco Ruiz, & Stunnenberg, 1994). In the absence of T_3, helix 12 is in an extended position. Upon ligand binding, TR undergoes a conformational change; helix 12 rotates to swing shut and close off the pocket around T_3, releasing corepressors and forming a docking surface for transcriptional coactivators (Dasgupta & O'Malley, 2014; Figueira et al., 2011; Rosen & Privalsky, 2009, 2011; Soriano et al., 2011). SRC1 and p160, for example, are coactivators possessing histone acetyltransferase activity, which "open" up chromatin for transcription.

One of the three NES motifs in TRα1, NES-H12, overlaps with the AF-2 domain, suggesting that competition for binding may modulate TR nuclear retention vs nuclear export. NES-H12 (^{390}VECPTELFPPLFLEVFED^{407}) was shown to be sufficient to partially target a nuclear-localized fusion protein (GFP-GST-GFP-Hinge) to the cytosol and was shown to be necessary for efficient export in the context of the full-length receptor (Mavinakere et al., 2012). Further, mutagenesis studies showed that multiple hydrophobic amino acids are required for efficient NES-H12 export activity, and mutations predicted to disrupt the α-helical structure result in a significant decrease in NES-H12 activity.

4.1.2 NES-H3 and NES-H6

Among the 12 helices, helices 3 and 5 of TRα1 recruit corepressor proteins to suppress transcription (Hu & Lazar, 1999). The two major corepressors are N-CoR1 (nuclear receptor corepressor 1) and SMRT (silencing mediator of retinoic acid and thyroid hormone receptors), also known as

N-CoR2. Besides their own inhibitory effects, N-CoR1 and SMRT also recruit histone deacetylases, which repress transcription by promoting chromatin condensation. The TR–DNA–corepressor complex causes a nonpermissive chromatin structure and inevitably inhibits the basal transcriptional machinery (Mullur et al., 2014). We determined that the region spanning helices H3 to H6 in TRα1 has two regions of transferable NES activity: NES-H3 (^{209}KVDLEAFSEFTKIITPAITRVVDFAKKLPM ^{238}) and NES-H6 (^{237}PMFSELPCEDQIILLKGCCMEIMSLRAAV^{265}). NES-H3 is well conserved among TR variants with a single amino acid change in v-ErbA and TRβ1, respectively. Similarly, NES-H6 is fully conserved among TR variants, except for TRβ1, which has one-amino acid change.

4.2 Exportins That Mediate TRα1 Nuclear Export

There is ample evidence for the existence of multiple, cooperating NES (and NLS) motifs in mediating nucleocytoplasmic shuttling of transcription factors (Dai, Bercury, Jin, & Macklin, 2015; Lu, Antoine, et al., 2014; Mavinakere et al., 2012; Panayiotou et al., 2016; Umemoto & Fujiki, 2012), although the exact role of each in coordinating shuttling is not always well defined. Given that there are multiple NESs facilitating TR export, questions arise regarding their functional equivalence and with which exportins they interact. Our studies have shown that there are multiple pathways for TR nuclear exit, both CRM1 dependent and CRM1 independent.

4.2.1 CRM1

Previously we showed that TRα1 can exit the nucleus by a pathway mediated by the export factor CRM1 (chromosome maintenance factor 1), also known as exportin 1, in cooperation with the Ca^{2+}-binding protein calreticulin (Grespin et al., 2008); however, the exact interaction and mechanism remain unclear. Two main lines of evidence suggested that TR might also follow a CRM1/calreticulin-independent nuclear export pathway. First, under conditions in which TR still shuttles in heterokaryons, shuttling of the oncoprotein v-ErbA is completely blocked in the presence of the CRM1-specific inhibitor leptomycin B (Bunn et al., 2001; DeLong et al., 2004). Also, during FRAP experiments, when one nucleus in a multinucleate HeLa cell was photobleached, recovery of fluorescence in the bleached nucleus in the presence of leptomycin B was reduced by only 60%, relative to recovery in the absence of leptomycin B (Grespin et al., 2008). Second, no CRM1-dependent NES has yet been characterized

(Mavinakere et al., 2012). Indeed, our studies showed that TR interacts directly with calreticulin, but complex formation with CRM1 was not detectable in pull-down assays (Grespin et al., 2008).

4.2.2 Exportins 4, 5, and 7

By coupling RNA interference with FRAP experiments in live HeLa cells, shuttling dynamics of TR were assessed upon shRNA-mediated knockdown of transportins 1 and 2, and exportins 4, 5, 6, and 7 (Subramanian et al., 2015). Only knockdown of exportins 4, 5, and 7 altered TR shuttling dynamics, indicating that multiple exportins influence TR localization. Preliminary coimmunoprecipitation assays show significant protein–protein interactions between TRα1 and exportins 4, 5, and 7 (J. Zhang, unpublished M.S. thesis results).

Although still a small pool, the number of proteins whose nuclear import is mediated by exportin 4 has been expanding since its discovery in 2000, with diverse cargo including eukaryotic translation initiation factor 5A (eIF-5A) and Sox family transcription factors (Gontan et al., 2009; Lipowsky et al., 2000). Exportin 5 is primarily known for its role in exporting small double-stranded pre-microRNAs (Lee, Jiko, Yamashita, & Tsukihara, 2011), but it had previously been shown to mediate androgen receptor export as well (Shank et al., 2008). Exportin 7 (also known as RanBP16) actively exports numerous proteins including 14-3-3σ and p50RhoGAP to the cytoplasm (Mingot, Bohnsack, Jakle, & Gorlich, 2004). Due to its diverse cargo pool, the exportin 7 pathway has been defined as the second most general export pathway, after the CRM1 pathway. Compared to CRM1, whose cargos appear to share a common leucine-rich NES, exportin 7's cargos do not have any distinct structural or functional similarity, although basic residue clusters seem to be important for cargo recognition. Fig. 3 summarizes what is currently known about the pathways followed by TRα1 as it exits the nucleus.

5. MISLOCALIZATION OF TR AND PATHOGENESIS

Prior to our studies on v-ErbA, dominant negative activity was attributed to competition with TR for TREs and/or auxiliary factors involved in transcriptional regulation. Our studies have defined a new mode of action of v-ErbA via subcellular mislocalization, leading to our central hypothesis that mutations targeting subcellular trafficking are pivotal in TR dysregulation (Bonamy & Allison, 2006; Bonamy et al., 2005; Bondzi et al., 2011;

DeLong et al., 2004). We have shown that the oncoprotein is highly mobile and trafficks between the nucleus, cytoplasm, and aggresome (Bonamy et al., 2005; Bondzi et al., 2011; DeLong et al., 2004). The aggresome is one of several types of cytosolic inclusions characterized, in part, by being localized at the microtubule-organizing center next to the nucleus. Misfolded, aggregated proteins are delivered to the aggresome by retrograde transport on microtubules and may eventually be cleared from cells by autophagy (Takalo, Salminen, Soininen, Hiltunen, & Haapasalo, 2013). This latter finding is particularly intriguing in that it points toward a critical role of mislocalization and aggregation of TR in disease.

Aggregation behavior and cytoplasmic sequestration of wild-type TR by mutant TR contribute to dominant negative and gain of function effects. Two synthetic dominant negative mutants of TRβ1, with defects in DNA binding and transactivation, were shown to accumulate in the cytosol at steady state, illustrating that even single amino acid changes in functional domains may lead to a dramatic shift in subcellular distribution of TR (Bunn et al., 2001). Also, a chimeric fusion protein, GFP-tagged v/c/v, has a strikingly altered distribution pattern compared with GFP-tagged v-ErbA. In the chimeric protein, the DBD of v-ErbA was replaced with the DBD of TR that differs by two amino acids. The vast majority of v-ErbA-expressing cells had a whole cell or cytoplasmic distribution, with either a diffuse distribution of GFP-v-ErbA or a localization to bright foci; in contrast, only 38% of cells expressing GFP-v/c/v showed a whole cell or cytoplasmic distribution. In the remaining 62% of cells, GFP-v/c/v was localized to the nucleus, either with a diffuse distribution or in bright foci (DeLong et al., 2004). These findings led us to propose that disease-causing mutations might target subcellular trafficking of TR, in some types of cancer and in Resistance to Thyroid Hormone (RTH) syndrome (Bonamy & Allison, 2006; Mavinakere et al., 2012).

5.1 TR Mislocalization and Cancer

Various forms of cancer, including hepatocellular carcinoma, renal clear cell carcinoma, and cancers of the breast, pituitary, and thyroid, have been linked to TR mutations (Chan & Privalsky, 2010; Conde et al., 2006; Kim & Cheng, 2013; Lin et al., 2013; Wojcicka et al., 2014). It has been proposed that simple dominant negative TR mutants cause endocrine disruption, whereas TR may gain oncogenic function by acquiring an additional ability to recognize a distinct set of target genes (Rosen, Chan, & Privalsky, 2011;

Rosen & Privalsky, 2009, 2011). Other studies suggest that TR functions as a tumor suppressor, because of the correlation between reduced expression of TR, or deletion of TR genes in human cancers (Martinez-Iglesias et al., 2009).

One of our recent findings that links TR mislocalization with cancer stands out as evidence for a pathogenic role of altered NLS activity. As noted earlier (see Section 3.1.2), a R26H substitution in NLS-2 of the oncoprotein v-ErbA abrogates the activity of NLS-2 (Mavinakere et al., 2012). Further, the combination of loss of function with gain of function by dominant negative mutation, and the propensity of TR variants to form cytoplasmic aggregates are reminiscent of the tumor suppressor protein p53 (Ano Bom et al., 2012; Rangel, Costa, Vieira, & Silva, 2014; Silva, De Moura Gallo, Costa, & Rangel, 2014). In fact, aggregates of mutant p53 have prion-like properties, leading to the speculation that cancer may be a "prionoid disease" (Rangel et al., 2014). It will be of interest to assess whether TR mutants in human cancers also display aberrant localization, as has been shown for the oncoprotein v-ErbA.

5.2 TR Mislocalization and RTH Syndrome

In addition to mutations linked to cancer, mutations in TR can give rise to disease, most notably the autosomal dominant genetic syndrome RTH (Bochukova et al., 2012; Dumitrescu & Refetoff, 2013; Moran et al., 2013; Parrilla, Mixson, McPherson, McClaskey, & Weintraub, 1991; Schoenmakers et al., 2013). In the majority of cases, RTH is linked to mutations in the LBD of TRβ1; the highest frequency of RTH mutations occurs in the region corresponding to NES-H12, with another cluster within NES-H3 and NES-H6 (Fig. 2). Many of these TRβ1 RTH mutants have lost T_3 binding and transactivation capacity and some exhibit dominant negative activity (Kim & Cheng, 2013; Rosen & Privalsky, 2011). There is a dearth of information on the contribution of altered nucleocytoplasmic dynamics to the phenotype of RTH. However, our mutagenesis studies on NES-H12 have shown that one RTH-linked mutation in TRβ1, L454S, exhibits moderately reduced export activity, pointing to the intriguing possibility that altered shuttling of TRβ1 may be a contributing factor in RTH (Mavinakere et al., 2012).

A number of cases of RTH have recently been linked to dominant negative mutations in the NES-H12 consensus region of TRα1. The TRα1 mutants have impaired T_3 binding, exhibit negligible hormone-dependent transactivation, and, when coexpressed, inhibit T_3-dependent transactivation by the wild-type TRα1 (Bochukova et al., 2012; Moran et al., 2013; Schoenmakers et al., 2013). The impact of these mutations on the subcellular localization of TRα1 currently is under investigation.

Based on accumulating evidence, we hypothesize that intracellular mislocalization of TR is an important factor to consider in pathogenesis (Bonamy & Allison, 2006; Bonamy et al., 2005; Mavinakere et al., 2012). Of particular interest are those amino acid substitutions located in or near NLS or NES motifs in TR. In support of this hypothesis, mislocalization of other proteins, including oncoproteins and tumor suppressor proteins, as well as defects in the nuclear transport machinery itself has been linked to various human diseases (McLane & Corbett, 2009; Wang & Li, 2014). Having acquired an in-depth understanding of the sequence requirements that are necessary and sufficient for nuclear import and export, we will further explore the correlation between TR trafficking and disease manifestation in the future.

6. CONCLUSIONS AND FUTURE DIRECTIONS

Our studies emphasize the importance of the balance of nuclear import, nuclear retention, and nuclear export in TRα1, TRβ1, and v-ErbA regulation (Bonamy & Allison, 2006; Bonamy et al., 2005; Bondzi et al., 2011; Mavinakere et al., 2012). Conservation of targeting signals among vertebrate species suggests that nucleocytoplasmic shuttling is crucial for the normal functions of TR, and underscores the importance of domain architecture in regulating shuttling.

Future research will continue to push the boundaries of understanding of nuclear receptor function. There is a wealth of literature focusing on the effects of loss of ligand and coregulator binding in TR pathogenesis; however, assays that focus only on one functional characteristic out of context of the suite of interacting partners are limited in their interpretation and scope. For example, when factors for retaining TRβ1 in the nucleus were studied, findings pointed toward a role of corepressor N-CoR1 and RXR in maintaining the receptor in the nucleus (Baumann et al., 2001); however, mutations that were introduced to disrupt protein–protein interactions would have disrupted NES motifs as well, so in retrospect, these studies remain inconclusive. Although the specific role of TR shuttling, whether it is to integrate cell signaling messages in the cytoplasm or mediate cytoplasmic degradation (Bonamy & Allison, 2006), remains to be determined. Receptor localization and trafficking is a central control point in T_3-mediated gene regulation. By taking an integrative approach across multiple levels of analysis—including nuclear import and export activity, intranuclear mobility and retention, protein posttranslational modification, misfolding, and sequestration in cytosolic inclusions—we can link mutations to complex phenotypes and disease, and discern how mutations can lead to loss of function, gain of function, and/or dominant negative behavior. Future research should provide insight into the development of new therapeutic strategies for TR-associated disorders.

ACKNOWLEDGMENTS

This work was supported in part by National Institutes of Health Grant 2R15DK058028 and National Science Foundation Grant MCB 1120513 to L.A.A.

REFERENCES

Ano Bom, A. P., Rangel, L. P., Costa, D. C., de Oliveira, G. A., Sanches, D., Braga, C. A., et al. (2012). Mutant p53 aggregates into prion-like amyloid oligomers and fibrils: Implications for cancer. *The Journal of Biological Chemistry*, *287*, 28152–28162.

Barettino, D., Vivanco Ruiz, M. M., & Stunnenberg, H. G. (1994). Characterization of the ligand-dependent transactivation domain of thyroid hormone receptor. *The EMBO Journal*, *13*, 3039–3049.

Baumann, C. T., Maruvada, P., Hager, G. L., & Yen, P. M. (2001). Nuclear cytoplasmic shuttling by thyroid hormone receptors. *The Journal of Biological Chemistry*, *276*, 11237–11245.

Bochukova, E., Schoenmakers, N., Agostini, M., Schoenmakers, E., Rajanayagam, O., Keogh, J. M., et al. (2012). A mutation in the thyroid hormone receptor α gene. *The New England Journal of Medicine*, *366*, 243–249.

Bonamy, G. M., & Allison, L. A. (2006). Oncogenic conversion of the thyroid hormone receptor by altered nuclear transport. *Nuclear Receptor Signaling*, *4*, e008.

Bonamy, G. M., Guiochon-Mantel, A., & Allison, L. A. (2005). Cancer promoted by the oncoprotein v-ErbA may be due to subcellular mislocalization of nuclear receptors. *Molecular Endocrinology*, *19*, 1213–1230.

Bondzi, C., Brunner, A. M., Munyikwa, M. R., Connor, C. D., Simmons, A. N., Stephens, S. L., et al. (2011). Recruitment of the oncoprotein v-ErbA to aggresomes. *Molecular and Cellular Endocrinology*, *332*, 196–212.

Brazda, P., Krieger, J., Daniel, B., Jonas, D., Szekeres, T., Langowski, J., et al. (2014). Ligand binding shifts highly mobile retinoid X receptor to the chromatin-bound state in a coactivator-dependent manner, as revealed by single-cell imaging. *Molecular and Cellular Biology*, *34*, 1234–1245.

Bunn, C. F., Neidig, J. A., Freidinger, K. E., Stankiewicz, T. A., Weaver, B. S., McGrew, J., et al. (2001). Nucleocytoplasmic shuttling of the thyroid hormone receptor α. *Molecular Endocrinology*, *15*, 512–533.

Casas, F., Busson, M., Grandemange, S., Seyer, P., Carazo, A., Pessemesse, L., et al. (2006). Characterization of a novel thyroid hormone receptor α variant involved in the regulation of myoblast differentiation. *Molecular Endocrinology*, *20*, 749–763.

Cautain, B., Hill, R., de Pedro, N., & Link, W. (2015). Components and regulation of nuclear transport processes. *The FEBS Journal*, *282*, 445–462.

Chan, I. H., & Privalsky, M. L. (2010). A conserved lysine in the thyroid hormone receptor-α1 DNA-binding domain, mutated in hepatocellular carcinoma, serves as a sensor for transcriptional regulation. *Molecular Cancer Research*, *8*, 15–23.

Chen, Y., & Young, M. A. (2010). Structure of a thyroid hormone receptor DNA-binding domain homodimer bound to an inverted palindrome DNA response element. *Molecular Endocrinology*, *24*, 1650–1664.

Chook, Y. M., & Suel, K. E. (2011). Nuclear import by karyopherin-βs: Recognition and inhibition. *Biochimica et Biophysica Acta*, *1813*, 1593–1606.

Conde, I., Paniagua, R., Zamora, J., Blanquez, M. J., Fraile, B., Ruiz, A., et al. (2006). Influence of thyroid hormone receptors on breast cancer cell proliferation. *Annals of Oncology*, *17*, 60–64.

Dai, J., Bercury, K. K., Jin, W., & Macklin, W. B. (2015). Olig1 acetylation and nuclear export mediate oligodendrocyte development. *The Journal of Neuroscience*, *35*, 15875–15893.

Dasgupta, S., & O'Malley, B. W. (2014). Transcriptional coregulators: Emerging roles of SRC family of coactivators in disease pathology. *Journal of Molecular Endocrinology*, *53*, 47–59.

Davis, P. J., Goglia, F., & Leonard, J. L. (2016). Nongenomic actions of thyroid hormone. *Nature Reviews. Endocrinology*, *12*, 111–121.

DeLong, L. J., Bonamy, G. M., Fink, E. N., & Allison, L. A. (2004). Nuclear export of the oncoprotein v-ErbA is mediated by acquisition of a viral nuclear export sequence. *The Journal of Biological Chemistry*, *279*, 15356–15367.

Diallo, E. M., Wilhelm, K. G., Jr., Thompson, D. L., & Koenig, R. J. (2007). Variable RXR requirements for thyroid hormone responsiveness of endogenous genes. *Molecular and Cellular Endocrinology*, *264*, 149–156.

Dumitrescu, A. M., & Refetoff, S. (2013). The syndromes of reduced sensitivity to thyroid hormone. *Biochimica et Biophysica Acta*, *1830*, 3987–4003.

Figueira, A. C., Saidemberg, D. M., Souza, P. C., Martinez, L., Scanlan, T. S., Baxter, J. D., et al. (2011). Analysis of agonist and antagonist effects on thyroid hormone receptor conformation by hydrogen/deuterium exchange. *Molecular Endocrinology*, *25*, 15–31.

Flamant, F., & Gauthier, K. (2013). Thyroid hormone receptors: The challenge of elucidating isotype-specific functions and cell-specific response. *Biochimica et Biophysica Acta*, *1830*, 3900–3907.

Gontan, C., Guttler, T., Engelen, E., Demmers, J., Fornerod, M., Grosveld, F. G., et al. (2009). Exportin 4 mediates a novel nuclear import pathway for Sox family transcription factors. *The Journal of Cell Biology*, *185*, 27–34.

Grespin, M. E., Bonamy, G. M., Roggero, V. R., Cameron, N. G., Adam, L. E., Atchison, A. P., et al. (2008). Thyroid hormone receptor α1 follows a cooperative CRM1/calreticulin-mediated nuclear export pathway. *The Journal of Biological Chemistry*, *283*, 25576–25588.

Groeneweg, F. L., van Royen, M. E., Fenz, S., Keizer, V. I., Geverts, B., Prins, J., et al. (2014). Quantitation of glucocorticoid receptor DNA-binding dynamics by single-molecule microscopy and FRAP. *PLoS One*, *9*, e90532.

Hu, X., & Lazar, M. A. (1999). The CoRNR motif controls the recruitment of corepressors by nuclear hormone receptors. *Nature*, *402*, 93–96.

Kim, W. G., & Cheng, S. Y. (2013). Thyroid hormone receptors and cancer. *Biochimica et Biophysica Acta*, *1830*, 3928–3936.

Kimura, M., & Imamoto, N. (2014). Biological significance of the importin-β family-dependent nucleocytoplasmic transport pathways. *Traffic*, *15*, 727–748.

Kosinski, J., Mosalaganti, S., von Appen, A., Teimer, R., DiGuilio, A. L., Wan, W., et al. (2016). Molecular architecture of the inner ring scaffold of the human nuclear pore complex. *Science*, *352*, 363–365.

Lee, S. J., Jiko, C., Yamashita, E., & Tsukihara, T. (2011). Selective nuclear export mechanism of small RNAs. *Current Opinion in Structural Biology*, *21*, 101–108.

Lee, Y., & Mahdavi, V. (1993). The D domain of the thyroid hormone receptor α1 specifies positive and negative transcriptional regulation functions. *The Journal of Biological Chemistry*, *268*, 2021–2028.

Lin, Y. H., Liao, C. J., Huang, Y. H., Wu, M. H., Chi, H. C., Wu, S. M., et al. (2013). Thyroid hormone receptor represses miR-17 expression to enhance tumor metastasis in human hepatoma cells. *Oncogene*, *32*, 4509–4518.

Lin, D. H., Stuwe, T., Schilbach, S., Rundlet, E. J., Perriches, T., Mobbs, G., et al. (2016). Architecture of the symmetric core of the nuclear pore. *Science*, *352*, aaf1015.

Lipowsky, G., Bischoff, F. R., Schwarzmaier, P., Kraft, R., Kostka, S., Hartmann, E., et al. (2000). Exportin 4: A mediator of a novel nuclear export pathway in higher eukaryotes. *The EMBO Journal*, *19*, 4362–4371.

Lu, B., Antoine, D. J., Kwan, K., Lundback, P., Wahamaa, H., Schierbeck, H., et al. (2014b). JAK/STAT1 signaling promotes HMGB1 hyperacetylation and nuclear translocation. *Proceedings of the National Academy of Sciences of the United States of America*, *111*, 3068–3073.

Lu, M., Zak, J., Chen, S., Sanchez-Pulido, L., Severson, D. T., Endicott, J., et al. (2014a). A code for RanGDP binding in ankyrin repeats defines a nuclear import pathway. *Cell*, *157*, 1130–1145.

Marfori, M., Mynott, A., Ellis, J. J., Mehdi, A. M., Saunders, N. F., Curmi, P. M., et al. (2011). Molecular basis for specificity of nuclear import and prediction of nuclear localization. *Biochimica et Biophysica Acta*, *1813*, 1562–1577.

Martinez-Iglesias, O., Garcia-Silva, S., Tenbaum, S. P., Regadera, J., Larcher, F., Paramio, J. M., et al. (2009). Thyroid hormone receptor β1 acts as a potent suppressor of tumor invasiveness and metastasis. *Cancer Research*, *69*, 501–509.

Maruvada, P., Baumann, C. T., Hager, G. L., & Yen, P. M. (2003). Dynamic shuttling and intranuclear mobility of nuclear hormone receptors. *The Journal of Biological Chemistry*, *278*, 12425–12432.

Mavinakere, M. S., Powers, J. M., Subramanian, K. S., Roggero, V. R., & Allison, L. A. (2012). Multiple novel signals mediate thyroid hormone receptor nuclear import and export. *The Journal of Biological Chemistry*, *287*, 31280–31297.

McLane, L. M., & Corbett, A. H. (2009). Nuclear localization signals and human disease. *IUBMB Life*, *61*, 697–706.

Mingot, J. M., Bohnsack, M. T., Jakle, U., & Gorlich, D. (2004). Exportin 7 defines a novel general nuclear export pathway. *The EMBO Journal*, *23*, 3227–3236.

Mondal, S., Raja, K., Schweizer, U., & Mugesh, G. (2016). Chemistry and biology in the biosynthesis and action of thyroid hormones. *Angewandte Chemie (International ed. in English)*, *55*, 7606–7630.

Moran, C., Schoenmakers, N., Agostini, M., Schoenmakers, E., Offiah, A., Kydd, A., et al. (2013). An adult female with resistance to thyroid hormone mediated by defective thyroid hormone receptor α. *The Journal of Clinical Endocrinology and Metabolism*, *98*, 4254–4261.

Mueller, F., Morisaki, T., Mazza, D., & McNally, J. G. (2012). Minimizing the impact of photoswitching of fluorescent proteins on FRAP analysis. *Biophysical Journal*, *102*, 1656–1665.

Mullur, R., Liu, Y. Y., & Brent, G. A. (2014). Thyroid hormone regulation of metabolism. *Physiological Reviews*, *94*, 355–382.

Nascimento, A. S., Dias, S. M., Nunes, F. M., Aparicio, R., Ambrosio, A. L., Bleicher, L., et al. (2006). Structural rearrangements in the thyroid hormone receptor hinge domain and their putative role in the receptor function. *Journal of Molecular Biology*, *360*, 586–598.

Nicoll, J. B., Gwinn, B. L., Iwig, J. S., Garcia, P. P., Bunn, C. F., & Allison, L. A. (2003). Compartment-specific phosphorylation of rat thyroid hormone receptor α1 regulates nuclear localization and retention. *Molecular and Cellular Endocrinology*, *205*, 65–77.

O'Reilly, A. J., Dacks, J. B., & Field, M. C. (2011). Evolution of the karyopherin-β family of nucleocytoplasmic transport factors; ancient origins and continued specialization. *PLoS One*, *6*, e19308.

Panayiotou, R., Miralles, F., Pawlowski, R., Diring, J., Flynn, H. R., Skehel, M., et al. (2016). Phosphorylation acts positively and negatively to regulate MRTF-A subcellular localisation and activity. *eLife*, *5*, e15460.

Parrilla, R., Mixson, A. J., McPherson, J. A., McClaskey, J. H., & Weintraub, B. D. (1991). Characterization of seven novel mutations of the c-erbA β gene in unrelated kindreds with generalized thyroid hormone resistance. Evidence for two "hot spot" regions of the ligand binding domain. *The Journal of Clinical Investigation*, *88*, 2123–2130.

Pascual, A., & Aranda, A. (2013). Thyroid hormone receptors, cell growth and differentiation. *Biochimica et Biophysica Acta*, *1830*, 3908–3916.

Pumroy, R. A., & Cingolani, G. (2015). Diversification of importin-α isoforms in cellular trafficking and disease states. *The Biochemical Journal*, *466*, 13–28.

Rangel, L. P., Costa, D. C., Vieira, T. C., & Silva, J. L. (2014). The aggregation of mutant p53 produces prion-like properties in cancer. *Prion*, *8*, 75–84.

Rodriguez, E. A., Campbell, R. E., Lin, J. Y., Lin, M. Z., Miyawaki, A., Palmer, A. E., et al. (2017). The growing and glowing toolbox of fluorescent and photoactive proteins. *Trends in Biochemical Sciences*, *42*, 111–129. http://dx.doi.org/10.1016/j.tibs.2016.09.010.

Roggero, V. R., Zhang, J., Parente, L. E., Doshi, Y., Dziedzic, R. C., McGregor, E. L., et al. (2016). Nuclear import of the thyroid hormone receptor α1 is mediated by importin 7, importin β1, and adaptor importin α1. *Molecular and Cellular Endocrinology*, *419*, 185–197.

Rosen, M. D., Chan, I. H., & Privalsky, M. L. (2011). Mutant thyroid hormone receptors (TRs) isolated from distinct cancer types display distinct target gene specificities: A unique regulatory repertoire associated with two renal clear cell carcinomas. *Molecular Endocrinology*, *25*, 1311–1325.

Rosen, M. D., & Privalsky, M. L. (2009). Thyroid hormone receptor mutations found in renal clear cell carcinomas alter corepressor release and reveal helix 12 as key determinant of corepressor specificity. *Molecular Endocrinology*, *23*, 1183–1192.

Rosen, M. D., & Privalsky, M. L. (2011). Thyroid hormone receptor mutations in cancer and resistance to thyroid hormone: Perspective and prognosis. *Journal of Thyroid Research*, *2011*, 361304.

Schoenmakers, N., Moran, C., Peeters, R. P., Visser, T., Gurnell, M., & Chatterjee, K. (2013). Resistance to thyroid hormone mediated by defective thyroid hormone receptor α. *Biochimica et Biophysica Acta*, *1830*, 4004–4008.

Shank, L. C., Kelley, J. B., Gioeli, D., Yang, C. S., Spencer, A., Allison, L. A., et al. (2008). Activation of the DNA-dependent protein kinase stimulates nuclear export of the androgen receptor *in vitro*. *The Journal of Biological Chemistry*, *283*, 10568–10580.

Silva, J. L., De Moura Gallo, C. V., Costa, D. C., & Rangel, L. P. (2014). Prion-like aggregation of mutant p53 in cancer. *Trends in Biochemical Sciences*, *39*, 260–267.

Soniat, M., & Chook, Y. M. (2015). Nuclear localization signals for four distinct karyopherin-β nuclear import systems. *Biochemical Journal*, *468*, 353–362.

Soriano, F. X., Leveille, F., Papadia, S., Bell, K. F., Puddifoot, C., & Hardingham, G. E. (2011). Neuronal activity controls the antagonistic balance between peroxisome proliferator-activated receptor-gamma coactivator-1α and silencing mediator of retinoic acid and thyroid hormone receptors in regulating antioxidant defenses. *Antioxidants & Redox Signaling*, *14*, 1425–1436.

Subramanian, K. S., Dziedzic, R. C., Nelson, H. N., Stern, M. E., Roggero, V. R., Bondzi, C., et al. (2015). Multiple exportins influence thyroid hormone receptor localization. *Molecular and Cellular Endocrinology, 411*, 86–96.

Takalo, M., Salminen, A., Soininen, H., Hiltunen, M., & Haapasalo, A. (2013). Protein aggregation and degradation mechanisms in neurodegenerative diseases. *American Journal of Neurodegenerative Disease, 2*, 1–14.

Thuestad, G., Kraus, I., Apriletti, J., & Saatcioglu, F. (2000). The N-terminal domain of thyroid hormone receptor-α is required for its biological activities. *DNA and Cell Biology, 19*, 389–399.

Tian, H., Mahajan, M. A., Wong, C. T., Habeos, I., & Samuels, H. H. (2006). The N-terminal A/B domain of the thyroid hormone receptor-β2 isoform influences ligand-dependent recruitment of coactivators to the ligand-binding domain. *Molecular Endocrinology, 20*, 2036–2051.

Timney, B. L., Raveh, B., Mironska, R., Trivedi, J. M., Kim, S. J., Russel, D., et al. (2016). Simple rules for passive diffusion through the nuclear pore complex. *The Journal of Cell Biology, 215*, 57–76.

Tran, E. J., King, M. C., & Corbett, A. H. (2014). Macromolecular transport between the nucleus and the cytoplasm: Advances in mechanism and emerging links to disease. *Biochimica et Biophysica Acta, 1843*, 2784–2795.

Umemoto, T., & Fujiki, Y. (2012). Ligand-dependent nucleo-cytoplasmic shuttling of peroxisome proliferator-activated receptors, PPARα and PPARγ. *Genes to Cells: Devoted to Molecular & Cellular Mechanisms, 17*, 576–596.

Wagstaff, K. M., & Jans, D. A. (2009). Importins and beyond: Non-conventional nuclear transport mechanisms. *Traffic, 10*, 1188–1198.

Wang, X., & Li, S. (2014). Protein mislocalization: Mechanisms, functions and clinical applications in cancer. *Biochimica et Biophysica Acta, 1846*, 13–25.

Weitzel, J. M. (2008). To bind or not to bind—How to down-regulate target genes by liganded thyroid hormone receptor? *Thyroid Research, 1*, 4.

Wente, S. R., & Rout, M. P. (2010). The nuclear pore complex and nuclear transport. *Cold Spring Harbor Perspectives in Biology, 2*, a000562.

Wojcicka, A., Piekielko-Witkowska, A., Kedzierska, H., Rybicka, B., Poplawski, P., Boguslawska, J., et al. (2014). Epigenetic regulation of thyroid hormone receptor β in renal cancer. *PLoS One, 9*, e97624.

Wolffe, A. P., Collingwood, T. N., Li, Q., Yee, J., Urnov, F., & Shi, Y. B. (2000). Thyroid hormone receptor, v-ErbA, and chromatin. *Vitamins and Hormones, 58*, 449–492.

Wu, Y., & Koenig, R. J. (2000). Gene regulation by thyroid hormone. *Trends in Endocrinology and Metabolism: TEM, 11*, 207–211.

CHAPTER FOUR

Molecular Basis of Nongenomic Actions of Thyroid Hormone

Paul J. Davis*,†,1, Jack L. Leonard‡, Hung-Yun Lin§, Matthew Leinung*, Shaker A. Mousa†

*Albany Medical College, Albany, NY, United States
†Pharmaceutical Research Institute, Albany College of Pharmacy and Health Sciences, Rensselaer, NY, United States
‡University of Massachusetts Medical School, Worcester, MA, United States
§PhD Program for Cancer Biology and Drug Discovery, College of Medical Science and Technology, Taipei Medical University, Taipei, Taiwan
[1]Corresponding author: e-mail address: pdavis.ordwayst@gmail.com

Contents

1. Introduction 68
2. Regulation of the State of the Cytoskeleton by Thyroid Hormone 70
3. Regulation of Mitochondrial Respiration by Thyroid Hormone 71
4. Nongenomic Actions of Truncated and Intact TRs in Cytoplasm on Bone Cells, Neurons, and Tumor Suppression 72
5. Cytoplasmic Proteins in Addition to TRs That Mediate Nongenomic Actions of Thyroid Hormone 75
6. Nongenomic Actions of Thyroid Hormone Initiated at Plasma Membrane 75
7. Thyroid Hormone Receptor on Plasma Membrane Integrin αvβ3 76
8. Thyroid Hormone αvβ3 Receptor and Angiogenesis 77
9. Thyroid Hormone αvβ3 Receptor and Tumor Cell Proliferation 78
10. Thyroid Hormone αvβ3 Receptor and Regulation of Gene Expression 81
11. Thyroid Hormone αvβ3 Receptor and Cell Physiology 85
12. Discussion 86

Acknowledgments 90

References 90

Abstract

Nongenomic actions of thyroid hormone are initiated by the hormone at receptors in the plasma membrane, in cytoplasm, or in mitochondria and do not require the interaction of nuclear thyroid hormone receptors (TRs) with their primary ligand, 3,5,3′-triiodo-L-thyronine (T_3). Receptors involved in nongenomic actions may or may not have structural homologies with TRs. Certain nongenomic actions that originate at the plasma membrane may modify the state and function of intranuclear TRs. Reviewed here are nongenomic effects of the hormone—T_3 or, in some cases, L-thyroxine

Vitamins and Hormones, Volume 106
ISSN 0083-6729
http://dx.doi.org/10.1016/bs.vh.2017.06.001

(T_4)—that are initiated at (a) truncated TRα isoforms, e.g., p30 TRα1, (b) cytoplasmic proteins, or (c) plasma membrane integrin αvβ3. p30 TRα1 is not transcriptionally competent, binds T_3 at the cell surface, and consequently expresses a number of important functions in bone cells. Nongenomic hormonal control of mitochondrial respiration involves a TRα isoform, and another truncated TRα isoform nongenomically regulates the state of cellular actin. Cytoplasmic hormone-binding proteins involved in nongenomic actions of thyroid hormone include ketimine reductase, pyruvate kinase, and TRβ that shuttle among intracellular compartments. Functions of the receptor for T_4 on integrin αvβ3 include stimulation of proliferation of cancer and endothelial cells (angiogenesis) and regulation of transcription of cancer cell survival pathway genes. T_4 serves as a prohormone for T_3 in genomic actions of thyroid hormone, but T_4 is a hormone at αvβ3 and more important to cancer cell function than is T_3. Thus, characterization of nongenomic actions of the hormone has served to broaden our understanding of the cellular roles of T_3 and T_4.

1. INTRODUCTION

Binding sites for thyroid hormone exist on proteins in plasma, cytoplasm, organelles, and the cell nucleus (Cheng, Leonard, & Davis, 2010). Functions have been assigned experimentally to the binding sites on most of these proteins. Where such sites on proteins underlie acknowledged actions of the hormones, they are designated receptors. But binding sites on proteins also exist that are responsible for availability of the hormone—for example, that retain thyroid hormone in the circulation and in part determine half-life or are involved in cellular uptake of the hormone (Visser, Friesema, & Visser, 2011) or in its presentation to the blood–brain barrier (Richardson, Wijayagunaratne, D'Souza, et al., 2015). Deiodinases (Gereben, Zavacki, Ribich, et al., 2008) have binding sites for their hormonal substrates. Thyroid hormone is a ligand for deiodinases (Gereben et al., 2008), which convert L-thyroxine (T_4) to the principal intracellular bioactive form of the hormone, 3,5,3′-triiodo-L-thyronine (T_3) or to reverse T_3 (rT_3) (see Fig. 1) and thus control duration of action of T_3. In the cell nucleus and in conjunction with specific protein receptors (nuclear thyroid hormone receptors, TRs) and coregulatory proteins, T_3 modulates transcription of a large panel of genes critical to cell function (Brent, 2012). These are *genomic actions* of thyroid hormone. rT_3 is inactive in the cell nucleus and the generation of rT_3 from T_4 outside the thyroid gland restricts the local tissue generation and availability of T_3 and thus limits genomic effects of thyroid hormone. T_4 has

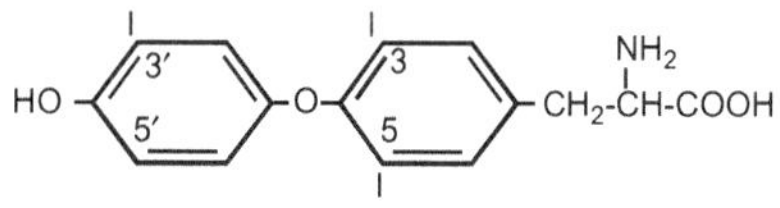

3,5,3′-Triiodothyronine (T_3)

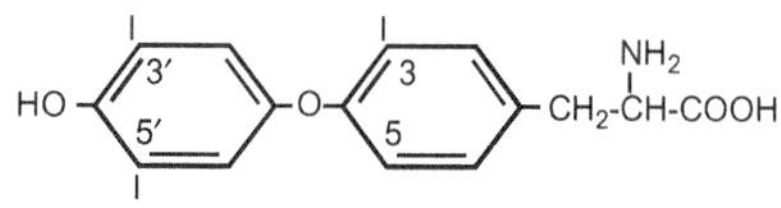

3,3′,5′-Triiodo-L-thyronine (rT_3)

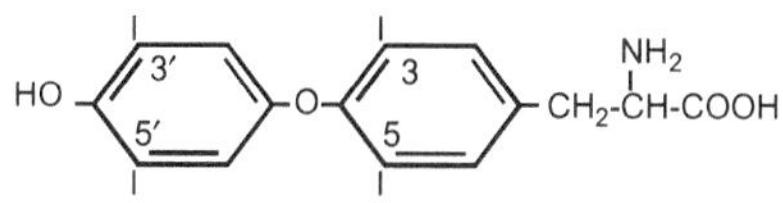

Thyroxine (T_4)

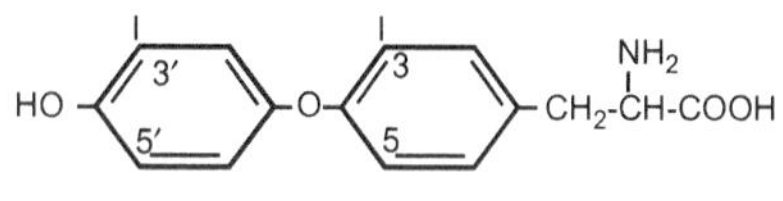

3,3′-Diiodo-L-thyronine (T_2)

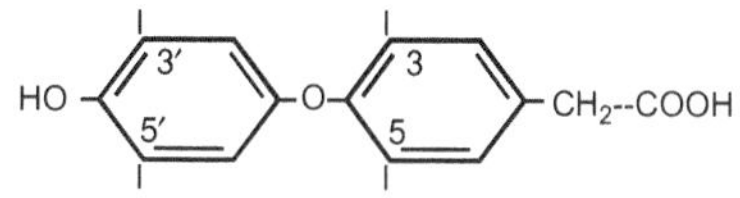

Tetrac

Fig. 1 Chemical structures of thyroid hormone and selected hormone analogs implicated in nongenomic actions of thyroid hormone.

low activity in the cell nucleus at concentrations achieved there and, viewed in the context of genomic functions of thyroid hormone, is a prohormone.

In the past decade, binding sites for thyroid hormone on the plasma membrane of a variety of cells and in cytoplasm have been appreciated to have activities that are highly relevant to cell division—e.g., of endothelial cells/angiogenesis and tumor cells—and to regulation of plasma membrane ion transport activities, to control of nuclear uptake of T_3, to maintenance of the state of the intracellular microfilament system, and to control of respiration in mitochondria. These are *nongenomic actions* of thyroid hormone. That is, they are not initiated by association of T_3 with nuclear TR proteins. Nongenomic actions may be initiated by T_4 (Davis, Davis, Mousa, et al., 2011; Davis, Glinsky, Lin, et al., 2014; Davis, Goglia, & Leonard, 2016) and, in the case of the cytoskeleton, by rT_3 (Davis, Goglia, et al., 2016). The proteins bearing receptors for nongenomic actions of the hormone may be structural components of the plasma membrane, for example, integrin αvβ3. The proteins may be truncated isoforms of nuclear TRs, for example, of TRα, functioning at the plasma membrane, at the cytoskeleton, or in mitochondria to regulate mitochondrial respiration. The nongenomic actions of thyroid hormone on the mitochondrion and on the cellular microfilament systems are reviewed later, as are the nongenomic actions of thyroid hormone that are initiated at the plasma membrane and in cytoplasm.

2. REGULATION OF THE STATE OF THE CYTOSKELETON BY THYROID HORMONE

Studied *in vitro*, in a hypothyroid milieu of nervous system cells, neurons and astrocytes exhibit inadequate development of the actin cytoskeleton (Farwell & Dubord-Tomasetti, 1999; Farwell, Dubord-Tomasetti, Pietrzykowski, et al., 2006). Repair of this dearth of polymerized actin (F-actin) begins within minutes of exposure of the cells to thyroid hormone as T_4 or reverse T_3 (rT_3), but not T_3. The rapid repair without any change in abundance of actin mRNA and the lack of relevance of T_3 to actin chemistry indicate that this hormonal effect is nongenomic in mechanism. The defective actin cytoskeleton limits neuronal migration/motility and impairs the ability of cells to respond to extracellular guidance cues, generate focal contacts, and attach to laminin-coated surfaces.

It has been proposed that a truncated TRα isoform (TRΔα1) functions in cytoplasm as the receptor at which T_4 and rT_3 initiate regulatory steps in actin biochemistry. This isoform does not bind T_3, has no nuclear localization signal, is abundant in brain, and has certain structural features found in proteins that bind the p85α regulatory subunit of actin-relevant phosphatidylinositol 3-kinase (PI3K) (Davis, Goglia, et al., 2016).

In astrocytes, the actin cytoskeleton is attached to the plasma membrane, in part, by interactions between the laminin receptor (α1β2) required to assemble laminin networks on the cell surface (Farwell, Dubord-Tomasetti, Pietrzykowski, et al., 2005; Farwell, Tranter, & Leonard, 1995). Whether these interactions also include the (α1β5) laminin receptor remains to be established. Hoffman et al. showed that a model of thyroid hormone-induced bone resorption was subject to inhibition by an Arg-Gly-Asp (RGD) peptide (Hoffman, Vasko-Moser, Miller, et al., 2002). Such peptides are recognized by a group of integrins, heterodimeric structural proteins of the plasma membrane that regulate cell–cell and cell–extracellular matrix (ECM) protein interactions that are critical to tissue integrity, angiogenesis, and intercellular communication in normal tissues and cancer (Plow, Haas, Zhang, et al., 2000; Seguin, Desgrosellier, Weis, et al., 2015; Weis & Cheresh, 2011).

3. REGULATION OF MITOCHONDRIAL RESPIRATION BY THYROID HORMONE

The existence of the unique calorigenic properties of thyroid hormone has been appreciated for many decades (Magnus-Levy, 1895) and is now recognized to require both nongenomic and genomic mechanisms. Hormone action on the mitochondrion involves both increased cellular respiration and decreased metabolic efficiency. The nongenomic effect is manifested within minutes or several hours of the beginning of hormone exposure and depends upon interaction of T_3 or, possibly, another thyroid hormone analog—3,5-diiodo-L-thyronine (T_2)—with specific receptor proteins in the organelle (mitochondrion) (Lanni, Moreno, & Goglia, 2016). The results of activation of the genomic mechanism are appreciated within a day or more of hormone treatment and involve T_3 actions on nuclear or mitochondrial DNA. The genomic mechanism has recently been reviewed by several groups of authors (Cioffi, Senese, Lanni, et al., 2013; Weitzel & Iwen, 2011; Yehuda-Shnaidman, Kalderon, & Bar-Tana, 2014) and will not be further examined here.

The initial description of binding sites specific for T_3 in the mitochondrial inner membrane was by Sterling in 1975 (Sterling & Milch, 1975), with confirmation by Goglia et al. of the existence of the putative receptor several years later (Goglia, Torresani, Bugli, et al., 1981). More recently, quantitative electron microscopy autoradiography has revealed specific intracellular binding of T_3 in mouse liver mitochondria and four other cellular compartments (Morel, Ricard-Blum, & Ardail, 1996). The study demonstrated similar subcellular binding of hormone both *in vitro* and *in vivo*. Photoaffinity labeling was used by Wrutniak et al. to define the presence in rat liver mitochondrial extracts of two T_3-binding proteins (Wrutniak, Cassar-Malek, Marchal, et al., 1995). These mitochondrial proteins were found to be truncated isoforms of nuclear receptor TRα1 with molecular weights of 43 kDa (p43) and 28 kDa (p28). When T_3 is available, p43 enhances mitochondrial transcription and protein synthesis (Casas, Rochard, Rodier, et al., 1999). p43 is widely expressed (see review by Wrutniak-Cabello, Casas, & Cabello, 2001). The role of p28 in the mitochondrion was not clear until Pessemesse and coworkers recently showed that the protein was directed to mitochondria via a T_3-dependent process and also found the affinity of p28 for T_3 to be higher than that of nuclear TRs (Pessemesse, Lepourry, Bouton, et al., 2014). The affinity equalled that previously reported for

T_3 receptor sites described by others in the inner mitochondrial members (Sterling & Milch, 1975) and may in fact be that protein.

Comprehensive review of the mechanisms involved in regulation of mitochondrial respiration by thyroid hormone has recently been provided by Lanni et al. (2016).

4. NONGENOMIC ACTIONS OF TRUNCATED AND INTACT TRS IN CYTOPLASM ON BONE CELLS, NEURONS, AND TUMOR SUPPRESSION

Discrete truncated, transcriptionally inactive or incompetent cytoplasmic TRα isoforms have been shown to be involved in hormonal regulation of (a) the state of the actin cytoskeleton (see Section 2) and (b) proliferation of bone cells (Kalyanaraman, Schwappacher, Joshua, et al., 2014). Interestingly, the action of thyroid hormone on actin requires T_4 and not T_3, whereas the osteocyte effects reported by Kalyanaraman et al. are T_3-dependent (Kalyanaraman et al., 2014). A lipid raft-associated, truncated TRα (p30) in bone cells mediates the actions of T_3 on mitogen-activated protein kinase (MAPK/ERK; extracellular signal-regulated protein kinases 1/2, ERK1/2) and PI3K that culminate in cell proliferation. p30 TRα1 is produced in bone cell ribosomes from full-length TRα mRNA at an alternative internal translational start site, and its trafficking to the plasma membrane is a function of the protein's posttranslational palmitoylation. We also note that the MAPK and PI3K cell signaling pathways that are activated by these nontraditional TRα gene products also mediate adjunctive inputs to canonical TR-regulated genes from a cell surface receptor for thyroid hormone on an integrin (see Sections 7–9); these inputs affect the expression of signaling pathway-specific target genes that intersect with canonical TR-regulated genes (Davis, Lin, Tang, et al., 2013). This provides for more subtle control of specific gene transcription than is obtained with simple gene activation. Thus, activation by p30 TRα1 of ERKs and PI3K and nitric oxide synthase (NOS) (Kalyanaraman et al., 2014) may also be achieved from the hormone receptor on αvβ3 (Chen, Sjolinder, Wang, et al., 2012; Davis et al., 2011; Lin, Sun, Tang, et al., 2009) as may trafficking of intact nuclear receptors (TRβ1, TRα1) within the cell from cytoplasm to nucleus (Cao, Lin, Luidens, et al., 2009; Davis et al., 2011; Lin, Sun, et al., 2009). Thus, it is conceivable that TR isoform trafficking among compartments, including the plasma membrane, relates to integrin αvβ3. Although the studies of p30 TRα1 were conducted in bone cells, one can assume that this isoform will be found in other cells. This nongenomic mechanism of hormone action is included in Fig. 2.

The intact TRα1 in cytoplasm has been shown by Hiroi and coworkers to bind to the p85 subunit of PI3K and consequently activate endothelial nitric oxide synthase (eNOS) in the course of rescue of ischemic nerve cells (Hiroi, Kim, Ying, et al., 2006). Thus, both intact and specifically truncated isoforms of TRα1 interact with the p85 subunit/PI3K pathway and may activate eNOS, but to different ends in terms of cell function and depending upon the nature of the cell.

Intact TRβ resides in cytoplasm, as noted earlier, and may be directed to the intranuclear compartment by T_4, acting at integrin αvβ3. Moeller et al. have described activation of PI3K by this nuclear receptor protein (Moeller, Cao, Dumitrescu, et al., 2006). This culminates in downstream effects on transcription of genes for the hypoxia-inducible factor-1α (*HIF-1α*), glucose transporter (*GLUT1*), and platelet-type phosphofructokinase (*PFKP*) (Moeller et al., 2006; Moeller, Dumitrescu, & Refetoff, 2005). GLUT1 and PFKP proteins are important contributors to cellular metabolism—cellular glucose uptake and glycolysis, respectively—and HIF-1α has a variety of functions in angiogenesis, energy metabolism, and cancer and non-cancer cell survival. *HIF-1α* gene expression may also be regulated from a cell surface receptor for thyroid hormone (see Sections 7–9). Interaction of TRβ with the p85 subunit of PI3K may also occur in the nervous system where it appears to be essential to the nongenomic contribution of thyroid hormone to the normal maturation of pyramidal neurons in the postnatal brain (Martin, Marron Fernandez de Velasco, Mizuno, et al., 2014).

Kim et al. have shown that TRβ can function nongenomically as a tumor suppressor protein. The model system studied for such a function is the interaction of the TR with simian virus 40 large T antigen (SV40Tag) in an SV40Tag-dependent model of thyroid carcinogenesis (Kim, Zhao, Hanover, et al., 2012). The SV40Tag oncoprotein is an inactivating ligand of p53 and retinoblastoma (Rb) protein, serving to eliminate the tumor suppressing function of p53 and Rb and to permit tumor formation. TRβ binds also to SV40Tag, interfering with the binding by the latter of p53 and Rb and serving to restore their tumor suppressor properties. Finally, the same group has shown that a specific mutation of TRβ (TRβPV) in thyroid cells is carcinogenic (Guigon & Cheng, 2009). Among the interesting properties of TRβPV protein are direct interactions in thyroid cells with the

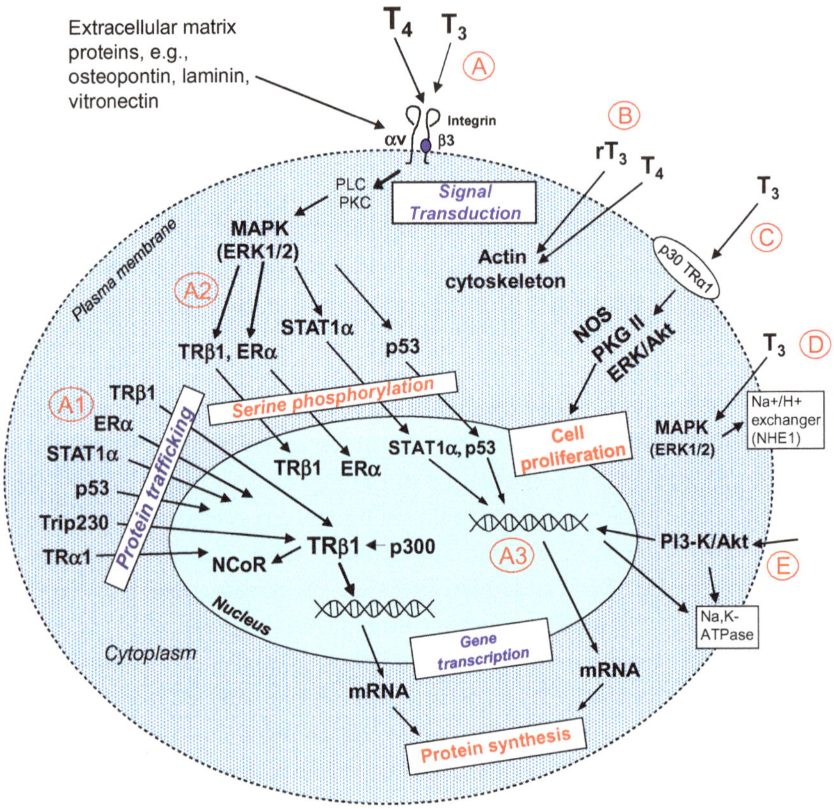

Fig. 2 Selected mechanisms of nongenomic actions of thyroid hormone that begin at the cell surface. (A) Nongenomic actions of the hormone that begin at integrin αvβ3 include regulation of intracellular trafficking of specific proteins (A1) to the nucleus and serine phosphorylation of certain of these proteins (A2) in the course of nuclear entry. Directed to the nucleus from cytoplasm, certain of these proteins may be involved in modulation of transcription of specific genes (A3; see Table 1) and in cell proliferation. The A1, A2, and A3 pathways depend upon activation of phospholipase C (PLC), protein kinase C (PKC), and mitogen-activated protein kinase (MAPK; extracellular signal-regulated kinase 1,2 [ERK1/2]). (B) Nongenomic regulation of the state of the cytoskeleton by T_4 and reverse T_3 (rT_3). (C) Modulation of proliferation of osteocytes initiated at the plasma membrane with binding of T_3 by a truncated TRα1 (p30 TRα1) that is transcriptionally inactive. The TR–T_3 complex activates a series of signal transducing proteins (protein kinase G II [PKG II]; ERK; Akt) and nitric oxide synthase (NOS) to increase cell proliferation. Tumor cell and endothelial cell proliferation may also be stimulated from integrin αvβ3 by T_4 (not shown here; see text). (D) Local activation by T_3 of the plasma membrane sodium–proton (Na^+/H^+) exchanger (NHE1) via MAPK. (E) Activation by T_3 locally of the plasma membrane sodium pump (Na,K-ATPase) via phosphatidylinositol 3-kinase (PI3K)/Akt. The initiation site for this action of the hormone has not been identified. It should be noted that transcription of the gene for the pump may also be affected by this PI3K/Akt mechanism. Details of pathways (A)–(E) are provided in the text. Actions of thyroid hormone on the mitochondrion are not depicted in this figure.

p85α regulatory subunit of PI3K, with β-catenin and with pituitary tumor-transforming gene protein, all of which are relevant to tumor biology.

5. CYTOPLASMIC PROTEINS IN ADDITION TO TRS THAT MEDIATE NONGENOMIC ACTIONS OF THYROID HORMONE

Several thyroid hormone-binding proteins that are structurally unrelated to nuclear TRs have been described in cytoplasm. These cytosolic thyronine-binding proteins (CTBPs) (Nishii, Hashizume, Ichikawa, et al., 1989; Yoshida, Davis, & Schoenl, 1979) have relatively high affinities for T_3 that may regulate nuclear uptake of the hormone. However, certain CTBPs have very specific metabolic functions that are regulated by the binding of iodothyronines. For example, μ-crystallin in cytoplasm has been shown to be ketimine reductase, a member of the pipecolate–lysine degradation pathway in brain; binding of T_3 by this protein reduces its enzyme activity (Hallen, Cooper, Jamie, et al., 2015). Cytoplasmic pyruvate kinase M1 also binds T_3 and kinase activity is reduced in a concentration-dependent manner by the binding of the hormone (Parkison, Ashizawa, McPhie, et al., 1991).

6. NONGENOMIC ACTIONS OF THYROID HORMONE INITIATED AT PLASMA MEMBRANE

The binding of the hormone by the red blood cell membrane *in vitro* was an early example of the existence of a binding site for iodothyronines at the cell surface, and it has been utilized as an index of free hormone and thus of thyroid function (Banos, Tako, Rado, et al., 1967; Goolden, Gartside, Jackson, et al., 1962). That the binding site might be functional in the mature erythrocyte and therefore have a role as a hormone receptor in the cell was not apparent until it was appreciated that membrane ion pumps—such as Ca^{2+}-ATPase (Davis, Davis, & Lawrence, 1989)—could rapidly be activated by thyroid hormone in these enucleate cells. Subsequently, plasma membrane Ca^{2+}-ATPase activity in nucleated cells and activities of the Na,K-ATPase (Lei, Bhargava, & Ingbar, 2009; Lei, Mariash, Bhargava, et al., 2008; Lei, Mariash, & Ingbar, 2004) and the Na^+/H^+ exchanger or antiporter (D'Arezzo, Incerpi, Davis, et al., 2004) were shown to be subject to local, i.e., membrane-based, nongenomic regulation. The nature of the receptor(s) at the plasma membrane that might be linked to these ion pumps was not clear for a number of years, but it was understood that certain kinases had to be rapidly activated in order for thyroid hormone to affect pump activities (Lin, Tang, Davis, et al., 2012). It was also understood that tetraiodothyroacetic acid (tetrac) (Fig. 1), a naturally occurring derivative of T_4 that lacked the amine of the alanine side chain, could block membrane actions of thyroid hormone. A polymeric formulation of thyroid hormone, agarose-T_4, reproduced actions of T_4 at the plasma membrane and was too bulky to gain access to the intracellular space of nucleated cells. Despite scientific awareness of the likely existence of one or more membrane receptors, primary investigative interest in how thyroid hormone works at the molecular level was devoted to the cell nucleus (Cheng et al., 2010; Kahaly & Dillmann, 2005; Zhang & Lazar, 2000) and genomic actions and, to a limited degree, to effects of the hormone on mitochondria (Goglia et al., 1981; Sterling & Milch, 1975).

7. THYROID HORMONE RECEPTOR ON PLASMA MEMBRANE INTEGRIN αVβ3

A specific integrin, αvβ3, in 2005 was shown by us to bear a plasma membrane receptor for thyroid hormone and at which the proangiogenic action of the hormone was initiated (Bergh, Lin, Lansing, et al., 2005). The effects were demonstrated in the chick chorioallantoic membrane (CAM) system and in other angiogenic models (Mousa, Bergh, Dier, et al., 2008; Mousa, Davis, Mohamed, et al., 2006). ERK1/2 mediated the effect of the hormone on angiogenesis, and T_4 was fully active in these nongenomic functions at physiological free hormone levels in cell medium and in cells pretreated with propylthiouracil to prevent cell conversion of T_4 to T_3. Nanoparticulate formulations of T_4 in which the hormone is covalently bound to agarose or poly(lactic-co-glycolic acid) (PLGA) have limited or delayed access to the cell interior, but are biologically active in functions known to be αvβ3 mediated (D'Arezzo et al., 2004; Davis et al., 2004; Mousa, Davis, et al., 2006; Mousa, O'Connor, Bergh, et al., 2005). siRNA knockdown of αvβ3 eliminated nongenomic actions of the hormone on angiogenesis (Bergh et al., 2005), as did addition of monoclonal antibody to αvβ3. Actions of thyroid hormone nongenomically initiated at integrin αvβ3 are included in Fig. 2.

Crystallographic modeling of the thyroid hormone–integrin αvβ3 complex revealed the location of the receptor to be at the head of the integrin, near the RGD recognition site (Cody, Davis, & Davis, 2007; Lin, Cody, Davis, et al., 2011). The αv and β3 monomers both contribute to the structure of the TR site. Modeling also revealed that the site distinguished among T_4, T_3, and tetrac (Lin, Cody, et al., 2011). Mathematical modeling of the kinetics of binding of T_4 and T_3 by the integrin was consistent with the existence of discrete-binding domains at the receptor, one (S1) that bound T_3 exclusively and a second (S2) at which T_4 and T_3 both were ligands (Lin, Cody, et al., 2011; Lin, Sun, et al., 2009). There was also functional specialization, apparent at the two domains, e.g., the T_3-binding (S1) zone activated PI3K, whereas S2 activated MAPK/ERK1/2. S1 regulated the intracellular trafficking of TRα from cytoplasm to nucleus and S2 controlled TRβ1 shuttling. Cancer cell and endothelial cell proliferation were controlled from S2. The existence of the two domains was consistent with mathematical modeling of the kinetics of binding of T_4 and T_3 to the integrin (Lin, Sun, et al., 2009) and modeling of the energetics of hormone binding by quantum mechanics and molecular mechanics (Freindorf, Furlani, Kong, et al., 2012).

8. THYROID HORMONE αVβ3 RECEPTOR AND ANGIOGENESIS

The angiogenesis-relevant properties of thyroid hormone and hormone analogs that are expressed via the hormone receptor of the integrin were studied initially in angiogenic models, including the chick CAM and Matrigel® microtubule assembly systems (Bergh et al., 2005; Davis et al., 2004; Mousa, O'Connor, Davis, et al., 2006). In these systems, T_4 at physiological free hormone concentrations and T_3 at supraphysiological levels have been shown to be as active as conventional vascular growth factors, such as vascular endothelial growth factor (VEGF) (Mousa, Lin, Tang, et al., 2014), basic fibroblast growth factor (bFGF; FGF2) (Davis et al., 2004; Mousa, Lin, et al., 2014), and platelet-derived growth factor (S.A. Mousa, Unpublished) in terms of generation of new vessels. The αvβ3-mediated nongenomic action of the hormone on angiogenesis required ERK1/2 (Bergh et al., 2005; Davis et al., 2004) and was in part dependent upon transcription of the basic fibroblast growth factor (*bFGF; FGF2*) gene and a panel of other genes summarized in Table 1. The action of the hormone on *bFGF/FGF2* gene expression was an early indication that specific gene transcription could be nongenomically controlled by thyroid hormone. Nanoparticulate T_4, as agarose-T_4, has very limited access to the interior of the cell and has been shown to reproduce the effects of T_4 and T_3 in angiogenesis assays. Pharmacologic blockade of the integrin, e.g., with

Table 1 Selected Genes Relevant to Angiogenesis and to Apoptosis Whose Expression Is Subject to Nongenomic Modulation by Thyroid Hormone via the Hormone Receptor on Integrin αvβ3

Gene	Up- or Downregulation of Transcription	References
Proangiogenesis genes		
FGF2	↑	Davis, Mousa, O'Connor, et al. (2004)
HIF-1α	↑	Lin, Sun, et al. (2009)
iNOS	↑	Chen et al. (2012)
MMP-9	↑	Cohen, Flint, Shalev, et al. (2014)
VEGF	↑	Cayrol, Diaz Flaque, Fernando, et al. (2015)
Proapoptosis genes		
APAF-1	↓	Cohen, Ellis, Shinderman, et al. (2015)
CASP3	↓	Cohen et al. (2015)
NOXA	↓	Cohen et al. (2015)
PUMA	↓	Cohen et al. (2015)
Antiapoptosis genes		
COX-2	↑	Lin, Sun, et al. (2009)
SREBP-1	↑	Gnoni, Rochira, Leone, et al. (2012)
XIAP	↑	Lin, Glinsky, Mousa, et al. (2015)

APAF-1, apoptotic protein activating factor 1; *CASP3*, caspase-3; *COX-2*, cyclooxygenase-2; *FGF2*, fibroblast growth factor 2; *HIF-1α*, hypoxia-inducible factor-1α; *iNOS*, inducible nitric oxide synthase; *MMP-9*, matrix metalloproteinase-9; *NOXA* = *PMAIP1*, phorbol-12-myristate-13-acetate-induced protein 1; *PUMA*, p53 upregulated modulator of apoptosis; *SREBP-1*, sterol regulatory element-binding protein 1; *VEGF*, vascular endothelial growth factor; *XIAP*, X-linked inhibitor of apoptosis.

XT-199 (Mousa, Davis, et al., 2006; Mousa, O'Connor, et al., 2006) or tetrac, eliminated the hormonal effect. The spectrum of actions of tetrac or its reformulation covalently bound to a nanoparticle on angiogenesis via αvβ3 is discussed later.

9. THYROID HORMONE αVβ3 RECEPTOR AND TUMOR CELL PROLIFERATION

The molecular basis of stimulation of cancer cell proliferation by thyroid hormone as an integrin-dependent process followed the early studies of angiogenesis. Proliferation of a substantial panel of human and animal tumor cell lines in response to thyroid hormone was demonstrated. Among the cells studied were human breast cancer cells—both estrogen receptor α (ERα)-positive and -negative (Bharali, Yalcin, Davis, et al., 2013; Glinskii, Glinsky, Lin, et al., 2009; Rebbaa, Chu, Davis, et al., 2008)—lung cancer (Meng, Tang, Westfall, et al., 2011; Mousa, Yalcin, Bharali, et al., 2012), glioma (Davis, Tang, Shih, et al., 2006; Lin, Sun, et al., 2009; Lin, Tang, Keating, et al., 2008), thyroid (Lin, Tang, Shih, et al., 2007; Yalcin, Bharali, Dyskin, et al., 2010), kidney (Yalcin, Bharali, Lansing, et al., 2009), pancreatic (Yalcin, Lin, Sudha, et al., 2013), and myeloma (Cohen, Ellis, Khoury, et al., 2011) cancer cell lines. Inhibition of thyroid hormone binding to αvβ3 with tetrac, the nanoparticulate formulation of tetrac (Bharali et al., 2013; Davis et al., 2011, 2014; Yalcin, Dyskin, Lansing, et al., 2010), or other integrin-active agents (e.g., Arg-Gly-Asp [RGD] peptides) (Lin, Sun, et al., 2009) interrupted hormone-induced cell division. MAPK activation was essential to hormone action. There were unique features to hormone action on certain cell lines. For example, in ERα-expressing human breast cancer (MCF-7) cells, T_4 promoted cell proliferation *in vitro* via induction of MAPK-dependent phosphorylation of ER (Tang, Lin, Zhang, et al., 2004). A similar effect was shown in ER-expressing lung cancer cells (Meng et al., 2011) and fulvestrant, an inhibitor of ER activation, blocked the proliferative action of T_4. However, the absence of ERα in human lung cancer cells did not prevent the proliferative effect of thyroid hormone via other mechanisms (Meng et al., 2011). Cross talk between integrin αvβ3 and the estrogen receptor has been reported in ovarian cancer cells to mediate the proliferative effect of thyroid hormone on such cells (Hsieh, Wang, Changou, et al., 2017). This action of T_4 appears to involve cellular internalization of αvβ3, nuclear uptake of integrin αv monomer, and interaction of the monomer with ERα. Elsewhere, αv has been shown to have nuclear coactivator function (Lin, Su, Hsieh, et al., 2013). The implication of such nongenomic effects of iodothyronines in women with ER-expressing cancers is that physiological levels of T_4 in the absence of estrogen may support tumor growth.

The stimulation *in vitro* of cell division of differentiated thyroid cancer (DTC) cell lines by thyroid hormone via αvβ3 was also of note. Such cells clinically are thyrotropin (TSH)-dependent, and L-thyroxine is regularly used to suppress endogenous pituitary secretion of TSH (Davis, Hercbergs, Luidens, et al., 2015). However, in patients with recurrent DTC despite successful suppression of endogenous TSH, we have suggested that the tumor is now TSH-independent and may be responding to exogenous administration of T_4. We have shown that T_4 is antiapoptotic in DTC cells (Lin et al., 2007). The mechanism of this nongenomic effect studied in a model of pharmacologically induced apoptosis is interference by the hormone with specific (Ser-15) phosphorylation of the proapoptotic protein p53 (Lin, Tang, Davis, et al., 2011). The phosphorylation step is MAPK-requiring and is critical to activation of p53 (Loughery, Cox, Smith, et al., 2014). The hormone may be antiapoptotic in a number of cancers and by more than one mechanism (Lin et al., 2015). In contrast, the anti-apoptotic properties of thyroid hormone may be desirable in nonmalignant cells, e.g., myocardiocytes and neurons.

The actions of thyroid hormone *in vitro* on adhesion and migration of myeloma cells have been studied by Cohen et al. (2014). T_4 and T_3 enhance cell adhesion to fibronectin via the hormone receptor on αvβ3, and the integrin, itself, is involved in the adhesion process stimulated by the hormones. The hormones also induce transcription of the matrix metalloproteinase-9 (*MMP-9*) gene involved in cancer remodeling. MAPK activation is involved in the initiation of these actions. Tetrac inhibits these effects.

Thyroid hormone-stimulated hepatocellular carcinoma stroma generation in several mesenchymal stem cell model systems has recently been shown to be αvβ3-dependent (Schmohl, Muller, Wechselberger, et al., 2015). Tetrac was inhibitory in these models. These observations are important because (a) they reveal another dimension of tumor biology that is thyroid hormone-dependent and (b) stem cell function is thyroid hormone- and tetrac-responsive.

Specific chemokine gene transcription can be downregulated by tetrac in breast and thyroid cancer cells by iodothyronines via the receptor on αvβ3 (Davis, Glinsky, Lin, et al., 2016). Other cancers have not been examined. Certain of these cytokine genes code for proinflammatory and angiogenic products that may be important to invasiveness. We have raised the possibility that proinflammatory cytokine production in nonmalignant cells might be reduced by tetrac, but this also remains to be studied.

"Integrin activation" is conventionally defined as regulation from the cell interior and intracellular proteins via signaling pathways of the affinity of these proteins for their ECM ligands (inside-out signaling) (Ginsberg, 2014). Activation is relevant to a panel of cell functions. The shape of the extracellular domain of an integrin may change with activation, thus altering exposure of their binding sites in these highly plastic proteins for ECM proteins. In addition, the relationships of the various domains of integrins—extracellular, transmembrane, intracellular—to one another may also be altered. That "outside origin" or "outside-in" signaling occurs in αvβ3 has been shown in the case of the thyroid hormone-tetrac receptor on the extracellular domain. For example, exposure of prostate cancer cells to radiation *in vitro* activates αvβ3 in minutes (Mousa, Sudha, Yalcin, et al., 2014), apparently enhancing cell–cell interactions and probably decreasing oxygen diffusion and opposing cell division. Acting at its receptor on the extracellular domain of αvβ3, tetrac prevents integrin activation in response to radiation and enhances tumor cell radiosensitivity (Hercbergs, Davis, Davis, et al., 2009). Tetrac formulations may increase radiosensitivity by another mechanism, in that these agents via αvβ3 inhibit tumor cell repair of radiation-induced double-strand DNA breaks (Hercbergs et al., 2009). The molecular basis of neither of these radiosensitization effects has been established.

10. THYROID HORMONE αVβ3 RECEPTOR AND REGULATION OF GENE EXPRESSION

The proliferative actions of thyroid hormone on tumor cells include transcription of a number of genes. The actions include downregulation of expression of tumor suppressor genes (*GDF15*, *IGFBP6*) and of genes whose products block mitochondrial apoptosis (*NIX*, *PUMA*) (Shinderman-Maman, Cohen, Weingarten, et al., 2016). The expression of genes for cell cycle inhibitors, such as *p21* and *p16*, is also downregulated by thyroid hormone. We have indicated earlier that thyroid hormone is antiapoptotic and has been shown to interfere with model proapoptotic drug action *in vitro* (Lin, Davis, Tang, et al., 2009; Lin et al., 2008). It is possible that there is direct extension of hormone action into the cancer chemotherapeutic context, since the *in vitro* cytotoxicity of certain drugs, such as bortezomib vs myeloma cells (Cohen et al., 2015), is inhibited by thyroid hormone. Further, the proangiogenic properties of thyroid hormone have been postulated to interfere with the actions of certain antiangiogenic agents administered to arrest cancer growth (Mousa, Lin, et al., 2014).

Cancer-relevant genes that may be differentially regulated from the integrin by thyroid hormone analogs have been extensively defined with probes such as tetrac and tetrac chemically bound to a 150-nm nanoparticle (named Nanotetrac or Nano-diamino-tetrac) (Davis et al., 2014; Glinskii et al., 2009; Yalcin, Dyskin, et al., 2010). Nanotetrac (Fig. 3) is excluded from the cell interior and acts only at integrin αvβ3 (Bharali et al., 2013). It has been extensively studied in cancer cells and in human cancer cell xenografts. In this compound, the tetrac is covalently bound to a diaminopropane

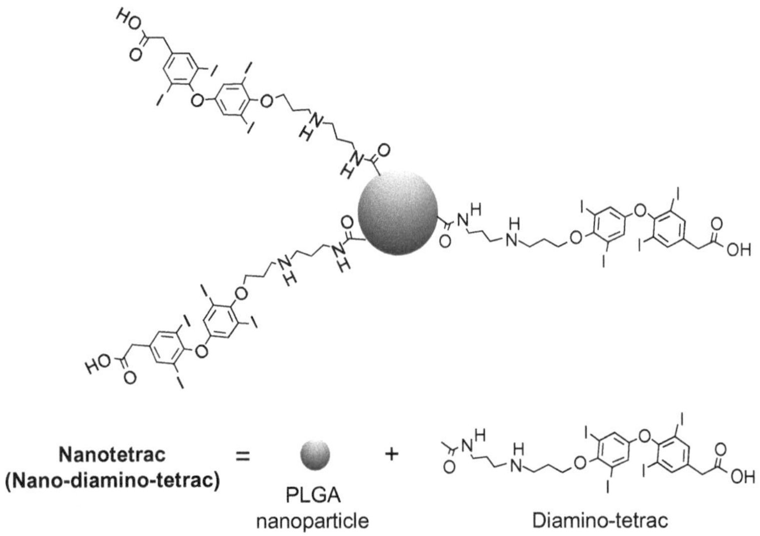

Fig. 3 Schematic diagram of the structure of Nanotetrac (Nano-diamino-tetrac). *PLGA*, poly(lactic-co-glycolic acid).

linker that is attached to a biodegradable PLGA nanoparticle. The nanoparticle retards cellular uptake of the tetrac. The linker is stably ether-bonded to the outer ring hydroxyl group of tetrac. The linker is connected to the PLGA by an amide bond that is imbedded in the nanoparticle and relatively inaccessible to circulating or tissue peptidases. When we studied the nongenomic anticancer activity of Nanotetrac, we found that about 85% of the genes affected by unmodified tetrac were also regulated by Nanotetrac, which, however, had distinctive properties in terms of gene transcription. For example, *EGFR* and *NFκB* expression are downregulated by Nanotetrac, and proapoptotic *BCL2L14* is upregulated by Nanotetrac, but unmodified tetrac does not affect these genes (Davis et al., 2011, 2014). Studies of Nanotetrac and tetrac actions on gene expression have been conducted in media that contained no T_3 or T_4, and thus these actions of Nanotetrac and tetrac on cancer cell gene expression are inherent qualities of the tetrac formulations and do not reflect blockade of T_4- and T_3-binding at αvβ3. However, we have not determined whether T_4 or T_3 in the absence of tetrac may affect transcription of certain of these cancer cell genes. The effects of Nanotetrac on gene transcription in cancer cells and on genes whose products are involved in angiogenesis are summarized in Tables 2A and 2B. It should also be noted that when unmodified tetrac gains access to the cell interior, it is a low-grade thyromimetic at nuclear TRs (Moreno, de

Table 2A Selected Angiogenesis Target Genes Whose Transcription Is Modulated by Nanotetrac (Nano-Diamino-Tetrac), Acting at the Thyroid Hormone-Tetrac Receptor on Integrin αvβ3

Angiogenesis Pathway Target	Action	References
TSP1 (THBS1) expression	↑	Glinskii et al. (2009) and Yalcin, Dyskin, et al. (2010)
bFGF (FGF2) expression	↓	Davis et al. (2004)
VEGFA expression	↓	Davis et al. (2011) and Yalcin, Dyskin, et al. (2010)
EGFR expression	↓	Glinskii et al. (2009)
Cellular Ang-2 content	↓	Mousa et al. (2008)
Cellular MMP-9 content	↓	Cohen et al. (2014)

Ang-2, angiopoietin 2; *bFGF*, basic fibroblast growth factor; *EGFR*, epidermal growth factor receptor; MMP-9, matrix metalloproteinase-9; *TSP1*, thrombospondin 1; *VEGFA*, vascular endothelial growth factor A.

Lange, Lombardi, et al., 2008), rather than the antagonist to thyroid hormone actions that it is at the cell surface integrin.

Various molecular mechanisms are involved in the modulation of gene transcription from αvβ3 by thyroid hormone. Intracellular pools of ERK1/2 and PI3K are activated, and the downstream consequences are expressed on nuclear receptors or on nuclear coactivators or corepressors. These consequences include specific phosphorylation of nuclear receptors, such as TRs (Davis, Lin, et al., 2013; Davis, Shih, Lin, et al., 2000) or ERα (Tang et al., 2004). Another consequence is the transfer of nuclear receptors resident in cytoplasm to the intranuclear compartment (trafficking), usually in conjunction with the importation of MAPK. Other proteins in cytoplasm such as signal transducer and activator of transcription-3 (STAT3) (Lin, Shih, Davis, et al., 1999) and Trip230 (Chen, Chen, Chen, et al., 1999) may be driven to the nucleus by the liganding of thyroid hormone at its receptor on αvβ3. Finally, the binding of T_4 by the integrin may lead to cell internalization of αvβ3, a process in which monomeric β3 is retained in cytoplasm, but the αv monomer is taken up by the nucleus. In the nucleus, αv serves as a coactivator in complexes with p300 or STAT1 for several specific genes, e.g., *ERα*, *TRβ1*, *HIF-1α* and cyclooxygenase-2 (*COX-2*) (Lin et al., 2013). Monomeric αv may also be found in intranuclear complexes with corepressor proteins, such as NCoR and SMRT. It is remarkable that

Table 2B Selected Nongenomic Anticancer Actions of Nanotetrac (Nano-Diamino-Tetrac) Initiated at the Thyroid Hormone-Tetrac Receptor on Integrin αvβ3

Anticancer Effect	Examples of Mechanisms	References
Cell cycle disruption	Downregulation of expression of cyclin genes, cyclin-dependent protein kinase genes	Davis et al. (2011) and Glinskii et al. (2009)
Interruption of cancer cell survival pathway gene expression	Increased expression of proapoptotic *CASP2, CASP3, BCL2L14*; reduced expression of antiapoptotic *MCL1, XIAP*; disordering of catenin pathway via decreased transcription of *CTNNA1, CTNNA2* and increased expression of *CBY1*; enhanced expression of proapoptotic *miR-15A*; decreased expression of multiple *Ras* oncogenes; reduced transcription of stress–defense genes, e.g., *HIF-1α*	Davis et al. (2011), Lin, Cody, et al. (2011), Glinskii et al. (2009), Yalcin, Dyskin, et al. (2010), and Cohen et al. (2014)
Chemosensitization	Reduced plasma membrane P-glycoprotein (P-gp) activity, with increased intracellular residence time in tumor cells of P-gp ligand chemotherapeutic agents	Davis, Lin, et al. (2013)
Disordering of growth factor pathways	Disabling of function of EGFR protein via disrupted cross talk with integrin αvβ3; decreased expression of *EGFR*	Davis et al. (2011) and Glinskii et al. (2009)

BCL2L14, B-cell lymphoma 2-like 14; *CASP2, 3*, caspases 2 and 3; *CBY1*, chibby 1; *CTNNA1, 2*, catenin alpha 1 and 2; *MCL1*, myeloid cell leukemia-1; *XIAP*, X-linked inhibitor of apoptosis.

a component of a cell surface structural protein has been found to have intranuclear functions on gene transcription. It is important to note at this point that these nuclear functions of αv have been defined in lung cancer and ovarian cancer cells and have not been sought as yet in nonmalignant cells.

11. THYROID HORMONE αVβ3 RECEPTOR AND CELL PHYSIOLOGY

We have not detected cytoplasmic thyroid hormone in complexes with monomeric αv or β3 (Lin et al., 2013). This suggests that a plasma membrane transport function for thyroid hormone is not vested in the integrin. It also infers that the internalized integrin monomeric αv—which, as noted earlier, has access to the cell nucleus—is not involved in the importation into the nucleus of the hormone. The β3 monomer is restricted to the cytoplasm, and intracellular functions for this protein have not yet been defined. We have not as yet sought αvβ3 in endosomes in thyroid hormone-treated cells, so that we do not know if the hormone is involved in the cycling of integrin αvβ3 into endosomes and the cell interior and back to the plasma membrane.

In addition to these functions on endothelial cells and cancer cells described earlier, the receptor for thyroid hormone on αvβ3 has been shown to have effects on noncancer cells. The integrin is sufficiently well expressed and/or activated in bone marrow cells to permit certain nongenomic effects of thyroid hormone to be recognized. For example, dividing megakaryocytes pass functional αvβ3 to platelets sufficiently to permit T_4 to induce platelet degranulation and agglutination (Mousa, Davis, Davis, et al., 2010). T_3 does not activate platelets, nor do certain other thyroid hormone analogs, such as diiodothyropropionic acid. T_4 is also involved in glucose uptake in L6 rat myoblast cells; here, the hormone when added alone stimulates cellular glucose uptake but inhibits insulin-like growth factor-1 (IGF-1)-stimulated glucose uptake via cross talk between integrin αvβ3 and the IGF receptor (Incerpi, Hsieh, Lin, et al., 2014). The integrin is also generously expressed by macrophages. Phagocytosis of bacteria by the latter is stimulated by thyroid hormone via integrin αvβ3 (Chen et al., 2012). This hormone-stimulated process is complex, depends upon activation of kinases known to be activated at the integrin, and includes NO generation. It is also important to point out that osteoclasts express large quantities of αvβ3. Thyroid hormone is known to induce bone resorption, and it is likely that the hormone receptor on the integrin plays a major role in this process (Hoffman et al., 2002). If so, then tetrac formulations may have a role in the management of bone demineralization whether it is generalized (osteopenia, osteoporosis) or localized (osteolytic multiple myeloma).

Also mediated by the hormone receptor on the integrin is endothelial cell migration toward a protein cue, e.g., vitronectin (Mousa, Lin, et al., 2014). These studies are conducted in a two-chamber Boyden apparatus. T_4 and T_3 both stimulate cell migration in this model, as does T_4 covalently bound to a nanoparticle that prevents hormonal entry into the cell (S.A. Mousa, Unpublished). The hormonal response is blocked by pharmacologic inhibition of integrin αvβ3 with XT-199 and by inhibition of MAPK activity with PD-98059. This contribution of the hormone is obviously relevant to angiogenesis.

Certain sensory neurons in the embryonic zebrafish brain are touch sensitive during development, and this critical sensory function is regulated by sodium current (I_{Na}). T_4 stimulates I_{Na} in this system (Yonkers & Ribera, 2008, 2009). The hormonal effect on the neuron is acute in onset and is blocked by monoclonal antibody to αvβ3 and by tetrac. T_3 does not reproduce the effect. Thus, nongenomic hormonal actions initiated at the receptor on αvβ3 are contributory to the course of embryogenesis.

As mentioned earlier, NO generation by certain cells can be regulated by integrin αvβ3. Chen and coworkers have reported that thyroid hormone-directed NO production in macrophages is required for killing of engulfed meningococci (Chen et al., 2012). This is a signal tranducing kinase-dependent process.

12. DISCUSSION

It is now clear that the movement of T_3 from the plasma membrane—where it is internalized by one or more specific transport systems, such as MCT8 (Visser et al., 2011)—to the nuclear compartment is not simple diffusion but can involve a repertoire of nongenomic events involving receptor proteins. These events may have downstream consequences on gene transcription or may contribute to processes that are exclusively extranuclear, such as mitochondrial respiration, stabilization of the cytoskeleton, and eNOS activation, thus affecting a number of cell functions.

It is also clear that T_4 functions nongenomically as a hormone, rather than exclusively as a prohormone source of T_3, as it is in the case of genomic actions of thyroid hormone. That T_4 had cellular actions of its own was apparent several decades ago when the effect of this form of the hormone was shown to affect the state of the cytoskeleton or to stimulate cellular transport systems, such as Ca^{2+}-ATPase (Davis et al., 1989), in enucleate cells such as mature human red blood cells that were incapable of deiodinating T_4 or of expressing genomic effects. What has become apparent more recently, however, is that interaction of T_4 with the TR on integrin αvβ3 can nongenomically initiate effects that culminate in expression of specific genes such as *HIF-1α* or *ERα* or that refine gene expression through phosphorylation of TR or modification of transcriptional complexes by adding αv monomer to such complexes.

The overlapping or "intersections" of nongenomic and genomic actions of thyroid hormone have recently been reviewed, and it is apparent that such overlaps occur in the case of steroids, as well (Hammes & Davis, 2015). The significance of the overlaps is substantial. First, as mentioned earlier, the intersections provide for fine-tuning of gene expression, via nongenomically regulated coactivator and corepressor proteins. "Fine-tuning" here refers to the time of onset of effects and modulation of the extent or magnitude of effects. As well, in the case of the integrin αv monomer, this overlapping introduces a previously unappreciated coactivator protein to gene transcription through nongenomic action of T_4 at the plasma membrane. Second, distinctions between nongenomic and genomic actions become blurred. Third, distinctions between actions of nonpeptide hormones are also blurred. That is, thyroid hormone may nongenomically support the transcription of the *ERα* gene in ER-expressing tumors. We have proposed that, in the setting of the tumor-bearing postmenopausal patient, circulating T_4 supports growth or relapse of the cancer. Of interest is that pharmacologic blockade of ERα, e.g., with fulvestrant, may inhibit such actions of thyroid hormone (Meng et al., 2011; Tang et al., 2004). Fourth, the classification of T_4 as a "prohormone" that emerged in the context of genomic actions of T_3 must be expanded to hormonal status in the setting of actions on the cytoskeleton or on a number of cellular functions controlled from the cell surface and integrin αvβ3. These functions include critical regulation of genes important to survival of tumor cells.

Acting at the plasma membrane on αvβ3, thyroid hormone may also nongenomically affect local membrane transport systems (Lin et al., 2012). These include the intracellular pH-regulating Na^+/H^+ exchanger (NHE1), Na,K-ATPase, and P-glycoprotein. That the basal activity of these systems is subject to regulation by the hormone is a critical component of cell homeostasis. Supported by iodothyronines, the Na^+/H^+ exchanger protects against intracellular metabolic acidosis, but at the expense of intracellular $[Na^+]$; the concurrent enhancement of sodium pump activity by T_3 prevents secondary accumulation of $[Na^+]_i$ when NHE1 activity is stimulated. The broad ligand base of P-gp enables this pump desirably to export a large number of toxic substances from normal cells. In cancer cells, however, the stimulated pump undesirably pumps out a variety of chemotherapeutic agents (Davis, Incerpi, Lin, et al., 2015), and this is an important member of the panel of nongenomic tumor support mechanisms activated by the hormone. We emphasize these functions in cancer cells and dividing endothelial cells that express $\alpha v\beta 3$ generously, but the effects may be seen in macrophages (Chen et al., 2012) and in certain other nonmalignant cells, such as osteoclasts (Safer, Crawford, & Holick, 2004; Seguin et al., 2015; Weis & Cheresh, 2011) and myoblasts (Incerpi et al., 2014) that are known to express the integrin.

We have emphasized here that T_4 has important supportive effects on cancer cells that are almost exclusively nongenomic in mechanism. A number of reports have suggested that spontaneous or medically induced hypothyroidism concomitant with cancer may improve survival (Cristofanilli, Yamamura, Kau, et al., 2005; Hercbergs, Goyal, Suh, et al., 2003; Nelson, Hercbergs, Rybicki, et al., 2006; Riesenbeck, Bierer, Hoffmeister, et al., 2011; Schmidinger, Vogl, Bojic, et al., 2011). In an observational study, Hercbergs has recently reported that pharmacologic induction of "euthyroid hypothyroxinemia" importantly improves duration of survival in patients with end-stage cancer of a variety of types (Hercbergs, Johnson, Ashur-Fabian, et al., 2015). The hormonal adjustment achieves a satisfactory quality of life. The treatment strategy is methimazole administration to block endogenous thyroxine production in the thyroid gland combined with the use of exogenous T_3 to maintain clinical euthyroidism. These results are consistent with the impression that tumor cells are more responsive to T_4 than to T_3. It will be important to substantiate these results in other studies.

In the absence of cancer, the nongenomic actions of T_4 and perhaps of T_3 on angiogenesis may be of some importance clinically. We have pointed out in this review and elsewhere (Luidens, Mousa, Davis, et al., 2010; Mousa, Lin, et al., 2014) that the nongenomic proangiogenic properties of thyroid hormone are complex. They involve modulation of transcription of vascular growth factor genes, expression of other genes relevant to blood vessel formation (such as matrix metalloproteinases), the function of plasma membrane receptors for endogenous vascular growth factors, the migration of endothelial cells toward relevant chemical cues, e.g., vitronectin, and potentiation of the angiogenic activities of other host factors, including angiotensin II and bradykinin (Mousa, Lin, et al., 2014). Intravascular administration of T_4 in the ischemic hind limb of the rabbit promotes blood vessel sprouting (El Eter, Rabee, Alkayali, et al., 2007) and in models such as the CAM and endothelial cell microtubule formation in Matrigel®, the proangiogenic activity of physiological concentrations of T_4 and somewhat supraphysiological levels of T_3 have been repeatedly demonstrated (Davis et al., 2004; Mousa, Davis, et al., 2006). Action of the hormone on a wound-healing model has also been reported and attributed to action on keratinocytes (Safer et al., 2004), but it is likely that hormonal proangiogenesis is also an important contributor. Thus, desirable proangiogenesis at sites of ischemia or wound-healing as a function of local thyroid hormone levels is possible.

A second clinical area in which the support of new blood vessel formation by thyroid hormone is of variable desirability is at sites of inflammation. That is, the acute and defensive inflammatory response to local trauma and infection requires angiogenic support. In certain scenarios of chronic inflammation, however, the proangiogenic input from the host's thyroid hormone may be undesirable. Independently of angiogenesis, thyroid hormone appears to contribute to inflammation (Davis, Glinsky, Lin, et al., 2013) and the immune response (De Vito, Balducci, Leone, et al., 2012). Where thyroid hormone contributes integrin-mediated angiogenic support to autoimmune diseases or other unwanted clinical events, interruption of such support needs to be evaluated experimentally as a supplemental intervention. Such an intervention is the use of tetrac formulations that by a complex of actions initiated at $\alpha v\beta 3$ are antiangiogenic. These formulations may also affect pathogenetic cytokine and chemokine production that iodothyronines nongenomically induce (Davis, Glinsky, et al., 2013).

The observation that p30 TRα1 at the plasma membrane in osteocytes can mediate the proliferative response to T_3 (Kalyanaraman et al., 2014) would appear to be important to future systematic approaches to disorders of bone. This may particularly be the case when osteoclast function is increased. We have not emphasized in this review that the osteoclast is, together with dividing endothelial cells, one of a set of unusual nonmalignant cells that generously expresses $\alpha v\beta 3$ and thus is subject to stimulation by thyroid hormone. Focal lytic lesions in multiple myeloma, for example, is a clinical condition of localized bone resorption in which it would at least in theory be desirable concomitantly to block (a) nongenomic action of T_4 on osteoclasts and (b) stimulate normal bone cell division via p30 TRα1 that has been nongenomically activated by the binding of T_3. Similarly, imbalance of bone formation and bone resorption in osteoporosis is a condition in which combined antiosteoclastic, proosteocytic intervention requires experimental consideration.

Pharmacological advantages appear to exist in targeting receptors involved in nongenomic actions of thyroid hormone that are initiated on the cell surface, e.g., on integrin αvβ3 or p30 TRα1. That is, pharmaceutical agents directed to such targets need not be internalized by cells in order to generate actions, and integrin αvβ3 is primarily expressed by and activated in tumor cells, rapidly dividing blood vessel cells and osteoclasts (Davis et al., 2011). Cell- or tissue-selective actions are achievable, leading to development of tetrac covalently coupled to a nanoparticle for tumor-specific delivery, via the nanoparticle, of generic cancer chemotherapeutic agents (Sudha, Bharali, Yalcin, et al., 2017). This approach has reduced systemic toxicity of the chemotherapeutic agent.

ACKNOWLEDGMENTS

Certain experiments described in this review received generous financial support from M. Frank and Margaret Rudy and from Richard Liebich.

Conflicts of interest: Coauthors Davis and Mousa are coinventors of Nano-diamino-tetrac (Nanotetrac), are part-time employees of, and are stockholders in a company that is developing this agent and other tetrac formulations.

REFERENCES

Banos, C. S., Tako, J., Rado, J. P., et al. (1967). Diagnostic value of the 131I-labelled triiodothyronine red blood cell uptake test. *Nuclear-Medizin*, *6*(2), 124–139.

Bergh, J. J., Lin, H. Y., Lansing, L., et al. (2005). Integrin αVβ3 contains a cell surface receptor site for thyroid hormone that is linked to activation of mitogen-activated protein kinase and induction of angiogenesis. *Endocrinology*, *146*(7), 2864–2871.

Bharali, D. J., Yalcin, M., Davis, P. J., et al. (2013). Tetraiodothyroacetic acid-conjugated PLGA nanoparticles: A nanomedicine approach to treat drug-resistant breast cancer. *Nanomedicine*, *8*(12), 1943–1954.

Brent, G. A. (2012). Mechanisms of thyroid hormone action. *The Journal of Clinical Investigation*, *122*(9), 3035–3043.

Cao, H. J., Lin, H. Y., Luidens, M. K., et al. (2009). Cytoplasm-to-nucleus shuttling of thyroid hormone receptor-β1 (Trβ1) is directed from a plasma membrane integrin receptor by thyroid hormone. *Endocrine Research*, *34*(1–2), 31–42.

Casas, F., Rochard, P., Rodier, A., et al. (1999). A variant form of the nuclear triiodothyronine receptor c-ErbAα1 plays a direct role in regulation of mitochondrial RNA synthesis. *Molecular and Cellular Biology*, *19*(12), 7913–7924.

Cayrol, F., Diaz Flaque, M. C., Fernando, T., et al. (2015). Integrin αvβ3 acting as membrane receptor for thyroid hormones mediates angiogenesis in malignant T cells. *Blood*, *125*(5), 841–851.

Chen, Y., Chen, P. L., Chen, C. F., et al. (1999). Thyroid hormone, T3-dependent phosphorylation and translocation of Trip230 from the Golgi complex to the nucleus. *Proceedings of the National Academy of Sciences of the United States of America*, *96*(8), 4443–4448.

Chen, Y., Sjolinder, M., Wang, X., et al. (2012). Thyroid hormone enhances nitric oxide-mediated bacterial clearance and promotes survival after meningococcal infection. *PLoS One*, *7*(7), e41445.

Cheng, S. Y., Leonard, J. L., & Davis, P. J. (2010). Molecular aspects of thyroid hormone actions. *Endocrine Reviews*, *31*(2), 139–170.

Cioffi, F., Senese, R., Lanni, A., et al. (2013). Thyroid hormones and mitochondria: With a brief look at derivatives and analogues. *Molecular and Cellular Endocrinology*, *379*(1–2), 51–61.

Cody, V., Davis, P. J., & Davis, F. B. (2007). Molecular modeling of the thyroid hormone interactions with αvβ3 integrin. *Steroids*, *72*(2), 165–170.

Cohen, K., Ellis, M., Khoury, S., et al. (2011). Thyroid hormone is a MAPK-dependent growth factor for human myeloma cells acting via αvβ3 integrin. *Molecular Cancer Research*, *9*(10), 1385–1394.

Cohen, K., Ellis, M., Shinderman, E., et al. (2015). Relevance of the thyroid hormones-αvβ3 pathway in primary myeloma bone marrow cells and to bortezomib action. *Leukemia & Lymphoma*, *56*(4), 1107–1114.

Cohen, K., Flint, N., Shalev, S., et al. (2014). Thyroid hormone regulates adhesion, migration and matrix metalloproteinase 9 activity via αvβ3 integrin in myeloma cells. *Oncotarget*, *5*(15), 6312–6322.

Cristofanilli, M., Yamamura, Y., Kau, S. W., et al. (2005). Thyroid hormone and breast carcinoma. Primary hypothyroidism is associated with a reduced incidence of primary breast carcinoma. *Cancer*, *103*(6), 1122–1128.

D'Arezzo, S., Incerpi, S., Davis, F. B., et al. (2004). Rapid nongenomic effects of 3,5,3′-triiodo-L-thyronine on the intracellular pH of L-6 myoblasts are mediated by intracellular calcium mobilization and kinase pathways. *Endocrinology*, *145*(12), 5694–5703.

Davis, P. J., Davis, F. B., & Lawrence, W. D. (1989). Thyroid hormone regulation of membrane Ca2(+)-ATPase activity. *Endocrine Research*, *15*(4), 651–682.

Davis, P. J., Davis, F. B., Mousa, S. A., et al. (2011). Membrane receptor for thyroid hormone: Physiologic and pharmacologic implications. *Annual Review of Pharmacology and Toxicology*, *51*, 99–115.

Davis, P. J., Glinsky, G. V., Lin, H. Y., et al. (2013). Molecular mechanisms of actions of formulations of the thyroid hormone analogue, tetrac, on the inflammatory response. *Endocrine Research*, *38*(2), 112–118.

Davis, P. J., Glinsky, G. V., Lin, H. Y., et al. (2014). Cancer cell gene expression modulated from plasma membrane integrin αvβ3 by thyroid hormone and nanoparticulate tetrac. *Frontiers in Endocrinology*, *5*, 240.

Davis, P. J., Glinsky, G. V., Lin, H. Y., et al. (2016). Actions of thyroid hormone analogues on chemokines. *Journal of Immunology Research*, *2016*, 3147671.

Davis, P. J., Goglia, F., & Leonard, J. L. (2016). Nongenomic actions of thyroid hormone. *Nature Reviews Endocrinology*, *12*(2), 111–121.

Davis, P. J., Hercbergs, A., Luidens, M. K., et al. (2015). Recurrence of differentiated thyroid carcinoma during full TSH suppression: Is the tumor now thyroid hormone dependent? *Hormones & Cancer*, *6*(1), 7–12.

Davis, P. J., Incerpi, S., Lin, H. Y., et al. (2015). Thyroid hormone and P-glycoprotein in tumor cells. *BioMed Research International*, *2015*, 168427.

Davis, P. J., Lin, H. Y., Tang, H. Y., et al. (2013). Adjunctive input to the nuclear thyroid hormone receptor from the cell surface receptor for the hormone. *Thyroid*, *23*(12), 1503–1509.

Davis, F. B., Mousa, S. A., O'Connor, L., et al. (2004). Proangiogenic action of thyroid hormone is fibroblast growth factor-dependent and is initiated at the cell surface. *Circulation Research*, *94*(11), 1500–1506.

Davis, P. J., Shih, A., Lin, H. Y., et al. (2000). Thyroxine promotes association of mitogen-activated protein kinase and nuclear thyroid hormone receptor (TR) and causes serine phosphorylation of TR. *The Journal of Biological Chemistry*, *275*(48), 38032–38039.

Davis, F. B., Tang, H. Y., Shih, A., et al. (2006). Acting via a cell surface receptor, thyroid hormone is a growth factor for glioma cells. *Cancer Research*, *66*(14), 7270–7275.

De Vito, P., Balducci, V., Leone, S., et al. (2012). Nongenomic effects of thyroid hormones on the immune system cells: New targets, old players. *Steroids*, *77*(10), 988–995.

El Eter, E., Rabee, H., Alkayali, A., et al. (2007). Role of thyroid hormone analogs in angiogenesis and the development of collaterals in rabbit hind limb ischemia model. *Journal of Thrombosis and Thrombolysis*, *5*(Suppl. 1), 375. in XXIst ISTH Congress.

Farwell, A. P., & Dubord-Tomasetti, S. A. (1999). Thyroid hormone regulates the expression of laminin in the developing rat cerebellum. *Endocrinology*, *140*(9), 4221–4227.

Farwell, A. P., Dubord-Tomasetti, S. A., Pietrzykowski, A. Z., et al. (2005). Regulation of cerebellar neuronal migration and neurite outgrowth by thyroxine and 3,3′,5′-triiodothyronine. *Brain Research. Developmental Brain Research*, *154*(1), 121–135.

Farwell, A. P., Dubord-Tomasetti, S. A., Pietrzykowski, A. Z., et al. (2006). Dynamic nongenomic actions of thyroid hormone in the developing rat brain. *Endocrinology*, *147*(5), 2567–2574.

Farwell, A. P., Tranter, M. P., & Leonard, J. L. (1995). Thyroxine-dependent regulation of integrin-laminin interactions in astrocytes. *Endocrinology*, *136*(9), 3909–3915.

Freindorf, M., Furlani, T. R., Kong, J., et al. (2012). Combined QM/MM study of thyroid and steroid hormone analogue interactions with αvβ3 integrin. *Journal of Biomedicine & Biotechnology*, *2012*, 959057.

Gereben, B., Zavacki, A. M., Ribich, S., et al. (2008). Cellular and molecular basis of deiodinase-regulated thyroid hormone signaling. *Endocrine Reviews*, *29*(7), 898–938.

Ginsberg, M. H. (2014). Integrin activation. *BMB Reports*, *47*(12), 655–659.

Glinskii, A. B., Glinsky, G. V., Lin, H. Y., et al. (2009). Modification of survival pathway gene expression in human breast cancer cells by tetraiodothyroacetic acid (tetrac). *Cell Cycle*, *8*(21), 3562–3570.

Gnoni, G. V., Rochira, A., Leone, A., et al. (2012). 3,5,3'triiodo-L-thyronine induces SREBP-1 expression by non-genomic actions in human HEP G2 cells. *Journal of Cellular Physiology*, *227*(6), 2388–2397.

Goglia, F., Torresani, J., Bugli, P., et al. (1981). In vitro binding of triiodothyronine to rat liver mitochondria. *Pflügers Archiv*, *390*(2), 120–124.

Goolden, A. W., Gartside, J. M., Jackson, D. J., et al. (1962). Uptake of 131I triiodothyronine by red cells. A diagnostic test of thyroid function. *Lancet*, *2*(7249), 218–220.

Guigon, C. J., & Cheng, S. Y. (2009). Novel non-genomic signaling of thyroid hormone receptors in thyroid carcinogenesis. *Molecular and Cellular Endocrinology*, *308*(1–2), 63–69.

Hallen, A., Cooper, A. J., Jamie, J. F., et al. (2015). Insights into enzyme catalysis and thyroid hormone regulation of cerebral ketimine reductase/μ-crystallin under physiological conditions. *Neurochemical Research*, *40*(6), 1252–1266.

Hammes, S. R., & Davis, P. J. (2015). Overlapping nongenomic and genomic actions of thyroid hormone and steroids. *Best Practice & Research Clinical Endocrinology & Metabolism*, *29*(4), 581–593.

Hercbergs, A., Davis, P. J., Davis, F. B., et al. (2009). Radiosensitization of GL261 glioma cells by tetraiodothyroacetic acid (tetrac). *Cell Cycle*, *8*(16), 2586–2591.

Hercbergs, A. A., Goyal, L. K., Suh, J. H., et al. (2003). Propylthiouracil-induced chemical hypothyroidism with high-dose tamoxifen prolongs survival in recurrent high grade glioma: A phase I/II study. *Anticancer Research*, *23*(1B), 617–626.

Hercbergs, A., Johnson, R. E., Ashur-Fabian, O., et al. (2015). Medically induced euthyroid hypothyroxinemia may extend survival in compassionate need cancer patients: An observational study. *The Oncologist*, *20*(1), 72–76.

Hiroi, Y., Kim, H. H., Ying, H., et al. (2006). Rapid nongenomic actions of thyroid hormone. *Proceedings of the National Academy of Sciences of the United States of America*, *103*(38), 14104–14109.

Hoffman, S. J., Vasko-Moser, J., Miller, W. H., et al. (2002). Rapid inhibition of thyroxine-induced bone resorption in the rat by an orally active vitronectin receptor antagonist. *The Journal of Pharmacology and Experimental Therapeutics*, *302*(1), 205–211.
Hsieh, M. T., Wang, L. M., Changou, C. A., et al. (2017). Crosstalk between integrin αvβ3 and ERα contributes to thyroid hormone-induced proliferation of ovarian cancer cells. *Oncotarget*, *8*(15), 24237–24249.
Incerpi, S., Hsieh, M. T., Lin, H. Y., et al. (2014). Thyroid hormone inhibition in L6 myoblasts of IGF-I-mediated glucose uptake and proliferation: New roles for integrin αvβ3. *American Journal of Physiology. Cell Physiology*, *307*(2), C150–C161.
Kahaly, G. J., & Dillmann, W. H. (2005). Thyroid hormone action in the heart. *Endocrine Reviews*, *26*(5), 704–728.
Kalyanaraman, H., Schwappacher, R., Joshua, J., et al. (2014). Nongenomic thyroid hormone signaling occurs through a plasma membrane-localized receptor. *Science Signaling*, 7(326), ra48.
Kim, D. W., Zhao, L., Hanover, J., et al. (2012). Thyroid hormone receptor beta suppresses SV40-mediated tumorigenesis via novel nongenomic actions. *American Journal of Cancer Research*, *2*(5), 606–619.
Lanni, A., Moreno, M., & Goglia, F. (2016). Mitochondrial actions of thyroid hormone. *Comprehensive Physiology*, *6*(4), 1591–1607.
Lei, J., Bhargava, M., & Ingbar, D. H. (2009). Cell-specific signal transduction pathways regulating Na+-K+-ATPase. Focus on "short-term effects of thyroid hormones on the Na+-K+-ATPase activity of chick embryo hepatocytes during development: Focus on signal transduction" *American Journal of Physiology Cell Physiology*, *296*(1), C1–C3.
Lei, J., Mariash, C. N., Bhargava, M., et al. (2008). T3 increases Na-K-ATPase activity via a MAPK/ERK1/2-dependent pathway in rat adult alveolar epithelial cells. *American Journal of Physiology. Lung Cellular and Molecular Physiology*, *294*(4), L749–L754.
Lei, J., Mariash, C. N., & Ingbar, D. H. (2004). 3,3′,5-Triiodo-L-thyronine up-regulation of Na,K-ATPase activity and cell surface expression in alveolar epithelial cells is Src kinase- and phosphoinositide 3-kinase-dependent. *The Journal of Biological Chemistry*, *279*(46), 47589–47600.
Lin, H. Y., Cody, V., Davis, F. B., et al. (2011). Identification and functions of the plasma membrane receptor for thyroid hormone analogues. *Discovery Medicine*, *11*(59), 337–347.
Lin, H. Y., Davis, P. J., Tang, H. Y., et al. (2009). The pro-apoptotic action of stilbene-induced COX-2 in cancer cells: Convergence with the anti-apoptotic effect of thyroid hormone. *Cell Cycle*, *8*(12), 1877–1882.
Lin, H. Y., Glinsky, G. V., Mousa, S. A., et al. (2015). Thyroid hormone and anti-apoptosis in tumor cells. *Oncotarget*, *6*(17), 14735–14743.
Lin, H. Y., Shih, A., Davis, F. B., et al. (1999). Thyroid hormone promotes the phosphorylation of STAT3 and potentiates the action of epidermal growth factor in cultured cells. *The Biochemical Journal*, *338*(Pt. 2), 427–432.
Lin, H. Y., Su, Y. F., Hsieh, M. T., et al. (2013). Nuclear monomeric integrin αv in cancer cells is a coactivator regulated by thyroid hormone. *The FASEB Journal*, *27*(8), 3209–3216.
Lin, H. Y., Sun, M., Tang, H. Y., et al. (2009). L-Thyroxine vs. 3,5,3′-triiodo-L-thyronine and cell proliferation: Activation of mitogen-activated protein kinase and phosphatidylinositol 3-kinase. *American Journal of Physiology. Cell Physiology*, *296*(5), C980–C991.
Lin, H. Y., Tang, H. Y., Davis, F. B., et al. (2011). Resveratrol and apoptosis. *Annals of the New York Academy of Sciences*, *1215*, 79–88.
Lin, H. Y., Tang, H. Y., Davis, F. B., et al. (2012). Nongenomic regulation by thyroid hormone of plasma membrane ion and small molecule pumps. *Discovery Medicine*, *14*(76), 199–206.

Lin, H. Y., Tang, H. Y., Keating, T., et al. (2008). Resveratrol is pro-apoptotic and thyroid hormone is anti-apoptotic in glioma cells: Both actions are integrin and ERK mediated. *Carcinogenesis*, *29*(1), 62–69.

Lin, H. Y., Tang, H. Y., Shih, A., et al. (2007). Thyroid hormone is a MAPK-dependent growth factor for thyroid cancer cells and is anti-apoptotic. *Steroids*, *72*(2), 180–187.

Loughery, J., Cox, M., Smith, L. M., et al. (2014). Critical role for p53-serine 15 phosphorylation in stimulating transactivation at p53-responsive promoters. *Nucleic Acids Research*, *42*(12), 7666–7680.

Luidens, M. K., Mousa, S. A., Davis, F. B., et al. (2010). Thyroid hormone and angiogenesis. *Vascular Pharmacology*, *52*(3–4), 142–145.

Magnus-Levy, A. (1895). Uber den respiratorischen gaswechsel unter dem einfluss der thyroidea sowie unter verschiedenen pathologischen zustanden. *Berliner Klinische Wochenschrift*, *34*, 650–652.

Martin, N. P., Marron Fernandez de Velasco, E., Mizuno, F., et al. (2014). A rapid cytoplasmic mechanism for PI3 kinase regulation by the nuclear thyroid hormone receptor, TRbeta, and genetic evidence for its role in the maturation of mouse hippocampal synapses in vivo. *Endocrinology*, *155*(9), 3713–3724.

Meng, R., Tang, H. Y., Westfall, J., et al. (2011). Crosstalk between integrin αvβ3 and estrogen receptor-α is involved in thyroid hormone-induced proliferation in human lung carcinoma cells. *PLoS One*, *6*(11), e27547.

Moeller, L. C., Cao, X., Dumitrescu, A. M., et al. (2006). Thyroid hormone mediated changes in gene expression can be initiated by cytosolic action of the thyroid hormone receptor β through the phosphatidylinositol 3-kinase pathway. *Nuclear Receptor Signaling*, *4*, e020.

Moeller, L. C., Dumitrescu, A. M., & Refetoff, S. (2005). Cytosolic action of thyroid hormone leads to induction of hypoxia-inducible factor-1α and glycolytic genes. *Molecular Endocrinology*, *19*(12), 2955–2963.

Morel, G., Ricard-Blum, S., & Ardail, D. (1996). Kinetics of internalization and subcellular binding sites for T3 in mouse liver. *Biology of the Cell*, *86*(2–3), 167–174.

Moreno, M., de Lange, P., Lombardi, A., et al. (2008). Metabolic effects of thyroid hormone derivatives. *Thyroid*, *18*(2), 239–253.

Mousa, S. A., Bergh, J. J., Dier, E., et al. (2008). Tetraiodothyroacetic acid, a small molecule integrin ligand, blocks angiogenesis induced by vascular endothelial growth factor and basic fibroblast growth factor. *Angiogenesis*, *11*(2), 183–190.

Mousa, S. S., Davis, F. B., Davis, P. J., et al. (2010). Human platelet aggregation and degranulation is induced in vitro by L-thyroxine, but not by 3,5,3′-triiodo-L-thyronine or diiodothyropropionic acid (DITPA). *Clinical and Applied Thrombosis/Hemostasis*, *16*(3), 288–293.

Mousa, S. A., Davis, F. B., Mohamed, S., et al. (2006). Pro-angiogenesis action of thyroid hormone and analogs in a three-dimensional in vitro microvascular endothelial sprouting model. *International Angiology*, *25*(4), 407–413.

Mousa, S. A., Lin, H. Y., Tang, H. Y., et al. (2014). Modulation of angiogenesis by thyroid hormone and hormone analogues: Implications for cancer management. *Angiogenesis*, *17*(3), 463–469.

Mousa, S. A., O'Connor, L. J., Bergh, J. J., et al. (2005). The proangiogenic action of thyroid hormone analogue GC-1 is initiated at an integrin. *Journal of Cardiovascular Pharmacology*, *46*(3), 356–360.

Mousa, S. A., O'Connor, L., Davis, F. B., et al. (2006). Proangiogenesis action of the thyroid hormone analog 3,5-diiodothyropropionic acid (DITPA) is initiated at the cell surface and is integrin mediated. *Endocrinology*, *147*(4), 1602–1607.

Mousa, S., Sudha, T., Yalcin, M., et al. (2014). In *Actions of nanoparticulate tetraiodothyroacetic acid (nanotetrac) on human prostate carcinoma xenograft growth, vascularity and integrin response to radiation AACR Annual Meeting*. San Diego: CA.

Mousa, S. A., Yalcin, M., Bharali, D. J., et al. (2012). Tetraiodothyroacetic acid and its nanoformulation inhibit thyroid hormone stimulation of non-small cell lung cancer cells in vitro and its growth in xenografts. *Lung Cancer, 76*(1), 39–45.

Nelson, M., Hercbergs, A., Rybicki, L., et al. (2006). Association between development of hypothyroidism and improved survival in patients with head and neck cancer. *Archives of Otolaryngology—Head & Neck Surgery, 132*(10), 1041–1046.

Nishii, Y., Hashizume, K., Ichikawa, K., et al. (1989). Changes in cytosolic 3,5,3′-tri-iodo-L-thyronine (T3) binding activity during administration of L-thyroxine to thyroidectomized rats: Cytosolic T3-binding protein and its activator act as intracellular regulators for nuclear T3 binding. *The Journal of Endocrinology, 123*(1), 99–104.

Parkison, C., Ashizawa, K., McPhie, P., et al. (1991). The monomer of pyruvate kinase, subtype M1, is both a kinase and a cytosolic thyroid hormone binding protein. *Biochemical and Biophysical Research Communications, 179*(1), 668–674.

Pessemesse, L., Lepourry, L., Bouton, K., et al. (2014). p28, a truncated form of TRα1 regulates mitochondrial physiology. *FEBS Letters, 588*(21), 4037–4043.

Plow, E. F., Haas, T. A., Zhang, L., et al. (2000). Ligand binding to integrins. *The Journal of Biological Chemistry, 275*(29), 21785–21788.

Rebbaa, A., Chu, F., Davis, F. B., et al. (2008). Novel function of the thyroid hormone analog tetraiodothyroacetic acid: A cancer chemosensitizing and anti-cancer agent. *Angiogenesis, 11*(3), 269–276.

Richardson, S. J., Wijayagunaratne, R. C., D'Souza, D. G., et al. (2015). Transport of thyroid hormones via the choroid plexus into the brain: The roles of transthyretin and thyroid hormone transmembrane transporters. *Frontiers in Neuroscience, 9*, 66.

Riesenbeck, L. M., Bierer, S., Hoffmeister, I., et al. (2011). Hypothyroidism correlates with a better prognosis in metastatic renal cancer patients treated with sorafenib or sunitinib. *World Journal of Urology, 29*(6), 807–813.

Safer, J. D., Crawford, T. M., & Holick, M. F. (2004). A role for thyroid hormone in wound healing through keratin gene expression. *Endocrinology, 145*(5), 2357–2361.

Schmidinger, M., Vogl, U. M., Bojic, M., et al. (2011). Hypothyroidism in patients with renal cell carcinoma: Blessing or curse? *Cancer, 117*(3), 534–544.

Schmohl, K. A., Muller, A. M., Wechselberger, A., et al. (2015). Thyroid hormones and tetrac: New regulators of tumour stroma formation via integrin αvβ3. *Endocrine-Related Cancer, 22*(6), 941–952.

Seguin, L., Desgrosellier, J. S., Weis, S. M., et al. (2015). Integrins and cancer: Regulators of cancer stemness, metastasis, and drug resistance. *Trends in Cell Biology, 25*(4), 234–240.

Shinderman-Maman, E., Cohen, K., Weingarten, C., et al. (2016). The thyroid hormone-α vβ3 integrin axis in ovarian cancer: Regulation of gene transcription and MAPK-dependent proliferation. *Oncogene, 35*(15), 1977–1987.

Sterling, K., & Milch, P. O. (1975). Thyroid hormone binding by a component of mitochondrial membrane. *Proceedings of the National Academy of Sciences of the United States of America, 72*(8), 3225–3229.

Sudha, T., Bharali, D. J., Yalcin, M., et al. (2017). Targeted delivery of cisplatin to tumor xenografts via the nanoparticle component of nano-diamino-tetrac. *Nanomedicine, 12*(3), 195–205.

Tang, H. Y., Lin, H. Y., Zhang, S., et al. (2004). Thyroid hormone causes mitogen-activated protein kinase-dependent phosphorylation of the nuclear estrogen receptor. *Endocrinology, 145*(7), 3265–3272.

Visser, W. E., Friesema, E. C., & Visser, T. J. (2011). Minireview: Thyroid hormone transporters: The knowns and the unknowns. *Molecular Endocrinology, 25*(1), 1–14.

Weis, S. M., & Cheresh, D. A. (2011). Tumor angiogenesis: Molecular pathways and therapeutic targets. *Nature Medicine, 17*(11), 1359–1370.

Weitzel, J. M., & Iwen, K. A. (2011). Coordination of mitochondrial biogenesis by thyroid hormone. *Molecular and Cellular Endocrinology*, *342*(1–2), 1–7.
Wrutniak, C., Cassar-Malek, I., Marchal, S., et al. (1995). A 43-kDa protein related to c-Erb a α1 is located in the mitochondrial matrix of rat liver. *The Journal of Biological Chemistry*, *270*(27), 16347–16354.
Wrutniak-Cabello, C., Casas, F., & Cabello, G. (2001). Thyroid hormone action in mitochondria. *Journal of Molecular Endocrinology*, *26*(1), 67–77.
Yalcin, M., Bharali, D. J., Dyskin, E., et al. (2010). Tetraiodothyroacetic acid and tetraiodothyroacetic acid nanoparticle effectively inhibit the growth of human follicular thyroid cell carcinoma. *Thyroid*, *20*(3), 281–286.
Yalcin, M., Bharali, D. J., Lansing, L., et al. (2009). Tetraidothyroacetic acid (tetrac) and tetrac nanoparticles inhibit growth of human renal cell carcinoma xenografts. *Anticancer Research*, *29*(10), 3825–3831.
Yalcin, M., Dyskin, E., Lansing, L., et al. (2010). Tetraiodothyroacetic acid (tetrac) and nanoparticulate tetrac arrest growth of medullary carcinoma of the thyroid. *The Journal of Clinical Endocrinology and Metabolism*, *95*(4), 1972–1980.
Yalcin, M., Lin, H. Y., Sudha, T., et al. (2013). Response of human pancreatic cancer cell xenografts to tetraiodothyroacetic acid nanoparticles. *Hormones & Cancer*, *4*(3), 176–185.
Yehuda-Shnaidman, E., Kalderon, B., & Bar-Tana, J. (2014). Thyroid hormone, thyromimetics, and metabolic efficiency. *Endocrine Reviews*, *35*(1), 35–58.
Yonkers, M. A., & Ribera, A. B. (2008). Sensory neuron sodium current requires nongenomic actions of thyroid hormone during development. *Journal of Neurophysiology*, *100*(5), 2719–2725.
Yonkers, M. A., & Ribera, A. B. (2009). Molecular components underlying nongenomic thyroid hormone signaling in embryonic zebrafish neurons. *Neural Development*, *4*, 20.
Yoshida, K., Davis, P. J., & Schoenl, M. (1979). Dissociable and non-dissociable cytoplasmic protein-thyroid hormone interactions. *Biochimica et Biophysica Acta*, *582*(2), 332–345.
Zhang, J., & Lazar, M. A. (2000). The mechanism of action of thyroid hormones. *Annual Review of Physiology*, *62*, 439–466.

CHAPTER FIVE

The Mechanism of Negative Transcriptional Regulation by Thyroid Hormone: Lessons From the Thyrotropin β Subunit Gene

Shigekazu Sasaki[1], Akio Matsushita, Go Kuroda, Hiroko M. Nakamura, Yutaka Oki, Takafumi Suda
Hamamatsu University School of Medicine, Shizuoka, Japan
[1]Corresponding author: e-mail address: sasakis@hama-med.ac.jp

Contents

1. Introduction 98
2. Structure and Function of Thyroid Hormone Receptors 99
3. Mechanism of T3-Dependent Positive Regulation 101
4. Does a nTRE Exist in the TSHβ Gene? 103
5. Negative Regulation of the TSHβ Gene Is Not the Mirror Image of Positive Regulation 105
6. Basal Expression of the TSHβ Gene Is Maintained by Thyrotroph-Specific Transcription Factors, GATA2 and Pit1 106
7. A CV1 Cell Reconstitution System That Mimics Thyrotrophs 108
8. TR Is Tethered by GATA2 to Mediate Negative Regulation of the TSHβ Gene by T3 108
9. NHRs Tethered by DNA-Binding Transcription Factor Mediate Ligand-Dependent Negative Regulation 109
10. Negative Regulation of Genes Other Than the TSHβ Gene May Not Be the Mirror Image of Positive Regulation 110
11. Negative Regulation of Genes Other Than the TSHβ Gene May Be Mediated via a Tethering Mechanism 111
12. Target Molecule of T3-Bound TRs Tethered by GATA2 112
    12.1 Corepressors 112
    12.2 CBP/p300 113
    12.3 p160 Family 113
    12.4 Coregulators for the GATA Family 115
13. Artifactual Negative Regulation via Plasmid Backbones 116
14. Conclusion and Future Directions 117
Acknowledgments 117
References 118

Vitamins and Hormones, Volume 106
ISSN 0083-6729
http://dx.doi.org/10.1016/bs.vh.2017.06.006

Abstract

Thyroid hormone (T3) activates (positive regulation) or represses (negative regulation) target genes at the transcriptional level. The molecular mechanism of the former has been elucidated in detail; however, the mechanism for negative regulation has not been established. The best example of the gene that is negatively regulated by T3 is the thyrotropin (thyroid-stimulating hormone) β subunit (TSHβ) gene. Analogous to the T3-responsive element (TRE) in positive regulation, a negative TRE (nTRE) has been postulated in the TSHβ gene. However, TSHβ promoter analysis, performed in the presence of transcription factors Pit1 and GATA2, which are determinants of thyrotroph differentiation in the pituitary, revealed that the nTRE is dispensable for inhibition by T3. We propose a tethering model in which the T3 receptor is tethered to GATA2 via protein–protein interaction and inhibits GATA2-dependent transactivation of the TSHβ gene in a T3-dependent manner.

1. INTRODUCTION

Microarray analysis revealed that thyroid hormone (T3) activates approximately 30%–70% of its target genes (positive regulation), while it represses the remainder (negative regulation) (Antonini, Sibilio, Dentice, & Missero, 2013; Clement et al., 2002; Feng, Jiang, Meltzer, & Yen, 2000; Hernandez, Morte, Belinchon, Ceballos, & Bernal, 2012; Wood et al., 2002; Yen et al., 2003). Although the molecular mechanisms of positive regulation have been elucidated, those underlying negative regulation are disputed (Lazar, 2003; Santos, Fairall, & Schwabe, 2011; Sasaki, Matsushita, & Nakamura, 2011). Thyrotropin (thyroid-stimulating hormone, TSH) is one of the peptide hormones secreted from pituitary thyrotrophs. It is a heterodimer consisting of an α chain (α-glycoprotein subunit, αGSU) and a β chain (TSHβ) (Shupnik, Ridgway, & Chin, 1989). Although the serum concentration of T3 is less than that of its prohormone, T4, deiodinase activity converts T4–T3 in pituitary thyrotrophs, and T3 binds its receptor (TR) in the nucleus (Gereben et al., 2008). T3 negatively regulates the transcription of the TSHβ and αGSU genes, resulting in drastic (linear-log) decreases of TSH production in thyrotrophs (Fisher et al., 2000). As shown in Fig. 1A, a negative feedback loop is the central mechanism of the hypothalamus–pituitary–thyroid (H–P–T) axis. Because T3 inhibition of TSHβ expression is more potent than that of αGSU (Shupnik et al., 1989), and TSHβ is specific to TSH while αGSU is common to gonadotropins, the transcriptional regulation of the

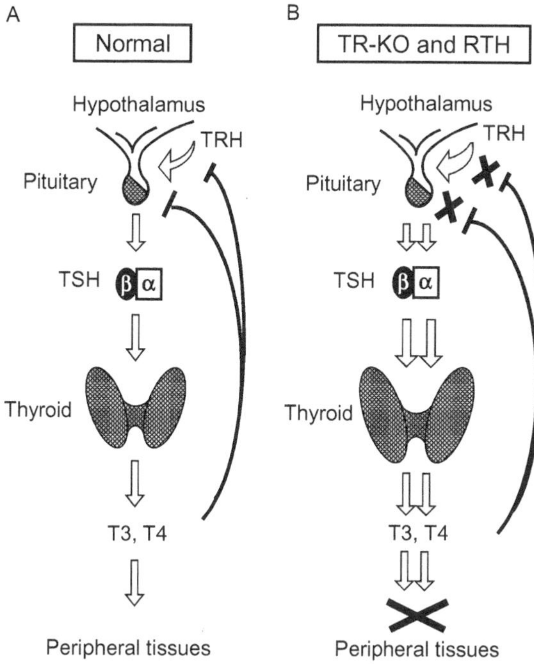

Fig. 1 Hypothalamus–pituitary–thyroid (H–P–T) axis. (A) The secretion of TSH (a heterodimer of TSHβ and αGSU subunits) from pituitary thyrotrophs and TRH from the hypothalamus is negatively regulated by thyroid hormones (T3 and T4). (B) The negative feedback loop is impaired in TR-KO mice (TRβ-KO and TRα plus TRβ-double-KO mice) and in patients with resistance to thyroid hormone (RTH). These findings provide evidence for the involvement of TRβ in the negative regulation of the TSHβ and αGSU genes. Because of increased secretion of TSH, diffuse goiters are often found in RTH patients. *α*, αGSU subunit; *β*, TSHβ subunit; *TRH*, thyrotropin-releasing hormone.

TSHβ gene is the best example of T3-dependent negative regulation. Here, we outline the molecular mechanism of the negative regulation of the TSHβ gene by T3. Lessons from TSHβ gene studies (Sasaki et al., 2011) may provide a general strategy to investigate other genes that are inhibited by liganded nuclear hormone receptors (NHRs).

2. STRUCTURE AND FUNCTION OF THYROID HORMONE RECEPTORS

TRs are ligand-dependent transcription factors belonging to the type II group of NHRs (Cheng, Leonard, & Davis, 2010), while most NHRs for steroid hormones are assigned to the type I group. TRs are encoded by two genes, *Thra* and *Thrb*. As shown in Fig. 2A, the *Thra* gene generates TRα1 and TRα2 via alternative splicing, while the *Thrb* gene generates TRβ1 and TRβ2 using different promoters. TRα1 and TRβ1 are expressed ubiquitously, whereas TRβ2 expression is limited to tissues including the pituitary, hypothalamus, inner ear, and retina (Cheng et al., 2010). Similar to other NHRs, TRs consist of an N-terminal domain (NTD), a central DNA-binding domain (DBD), a hinge region, and a C-terminal ligand-binding domain (LBD). TRα1, TRβ1, and TRβ2 all harbor an intact LBD, while TRα2 does not (Fig. 2A); therefore, TRα1, TRβ1, and TRβ2 are thought to be the functional TRs. Of note, the function of the DBD is not limited

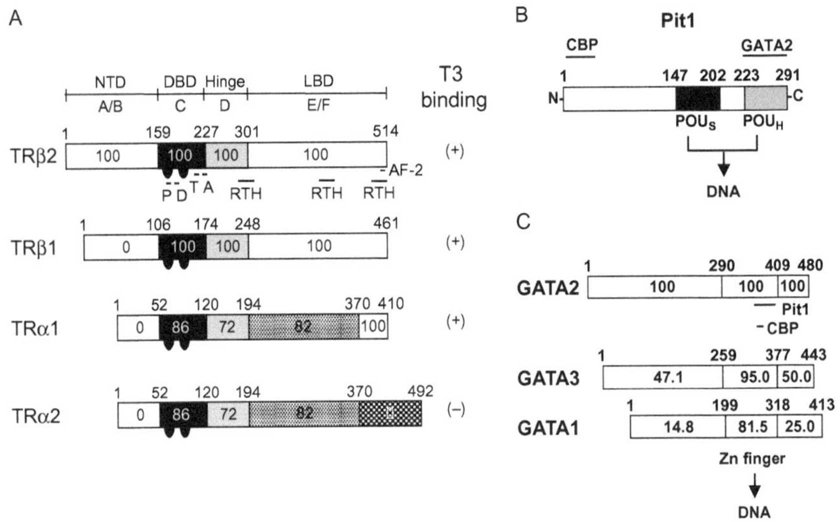

Fig. 2 (A) Schematic representations of TR isoforms. TR consists of an N-terminal region (NTD, A/B domain), a central DNA-binding domain (DBD, C domain), a hinge region (D domain), and a C-terminal ligand-binding domain (LBD, E/F domain). The numbers within the *boxes* represent the amino acid homology (%). While TRα1, TRβ1, and TRβ2 have T3-binding capacity, TRα2 does not bind T3. The P and D *boxes* recognize half-site sequences (typically, AGGTCA) and the number of spacing nucleotides, respectively. T and A *boxes* are involved in dimer formation and polarity of TR–RXR heterodimers on the pTRE. Although the central C domain has been designated as a DBD, its function is not limited to DNA binding (see text). *A*, A box; *AF-2*, activation function 2; *D*, D box; *P*, P box; *T*, T box; *RTH*, three hot spots where mutations are often identified in patients with resistance to thyroid hormone (RTH). (B) Schematic representation of Pit1. *POU$_S$*, POU-specific domain; *POU$_H$*, POU homeodomain; *GATA2*, amino acid sequence interacting with GATA2. (C) Structure of GATA1, 2, and 3. The numbers within the *box* represent the amino acid homology (%). *CBP and Pit*, amino acid sequences interacting with CBP and Pit1, respectively.

to direct DNA binding; it can also function as the interface for protein–protein interaction with other DNA-binding transcription factors, a process termed tethering (Liden, Delaunay, Rafter, Gustafsson, & Okret, 1997; Pfahl, 1993).

3. MECHANISM OF T3-DEPENDENT POSITIVE REGULATION

Before focusing on the negative regulation of the TSHβ gene, we will briefly mention the mechanism of positive regulation because this has profoundly influenced the investigation of negative regulation. For positively regulated genes (Fig. 3A), the TR forms a heterodimer with the retinoid X receptor (RXR) on a short DNA sequence. Although this DNA sequence is usually designated as the T3-responsive element (TRE), we will refer to it

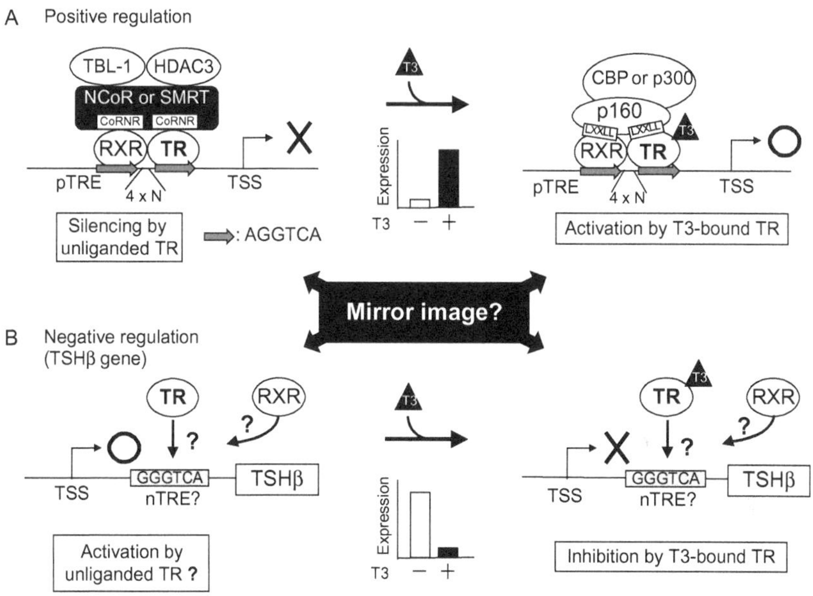

Fig. 3 Schematic representation of T3-dependent positive regulation (A) and negative regulation of the TSHβ gene (B). (A) In positive regulation, TR heterodimerizes with RXR on the pTRE. *Shaded arrows* indicate half-site sequences (typically, AGGTCA); *4xN*, four random nucleotides for spacing. *HDAC*, histone deacetylases; *NCoR*, nuclear receptor corepressor; *RXR*, retinoid X receptor; *SMRT*, silencing mediator for RAR and TR; *TBL-1*, transducin β-like protein 1; *TSS*, transcription start site. (B) Based on the hypothesis that positive and negative regulation may occur as mirror images of each other, the nTRE in the TSHβ gene has been postulated to be immediately downstream of TSS (see text).

in this review as the positive TRE (pTRE) to distinguish it from the putative negative TRE (nTRE) proposed for negatively regulated genes. The canonical pTRE preferred by TRs is composed of a direct repeat of the sequence AGGTCA spaced by four random nucleotides (DR4) (Cheng et al., 2010; Yen, 2001). The specificity of the DNA recognition is determined by not only the DNA sequence of the half-site but also the number of spacer nucleotides. For example, RXR heterodimerizes with peroxisome proliferator-activated receptor, vitamin D3 receptor (VDR), and retinoic acid receptor (RAR) on the direct repeat spaced by one (DR1), three (DR3), and five (DR5) nucleotides, respectively (Cheng et al., 2010). In the absence of T3 (Fig. 3A, left), the TR–RXR heterodimer on pTRE interacts with corepressors, including nuclear receptor corepressor (NCoR) or silencing mediator for RAR and TR (SMRT), which recruits histone deacetylases (HDACs), resulting in active transcriptional repression (silencing). Unexpectedly, however, the silencing by TRβ2 is very weak or undetectable (Privalsky et al., 2009) because its unique NTD neutralizes the inhibitory function of NCoR and SMRT (Yang, Hong, & Privalsky, 1999). Upon T3 binding to TR, the TR–RXR heterodimer releases these corepressors and recruits a member of the p160 family (Fig. 3A, right). The p160 family consists of three members, steroid receptor coactivator 1 (SRC-1), SRC-2 (TIF2/GRIP1), and SRC-3 (AIB1/pCIP/ACTR/TRAM1/RAC3). In the presence of T3, p160 associates with a TR–RXR heterodimer then recruits another class of coactivator, cAMP-responsive element-binding protein-binding protein (CBP) or p300. Both CBP and p300 have endogenous histone acetyltransferase (HAT) activity, which facilitates transcription by acetylating histones. Reporter assays using transient transfection simulate this process at least partly because the transfected plasmid is chromatinized within 24h (Mladenova & Russev, 2009). Liganded type II NHRs also interact with MED1 (TRAP220/DRIP205/ARC205/CRSP200/PBP/MED220), a constituent of the Mediator complex (Fondell, 2013). In the context of positive regulation, MED1 activates the transcription of target genes by controlling the function of RNA polymerase II (Pol-II). The p160 family, CBP/p300 and MED1 possess LXXLL amino acid motifs (L: leucine and X: any amino acid), which is the signature motif for the interaction of these coactivators with liganded NHRs (Coulthard, Matsuda, & Heery, 2003). The monkey kidney-derived cell line, CV1, has often been utilized in studies of positive regulation via NHRs (Misawa et al., 2012; Naar et al., 1991; Tillman, Crone, Kim,

Sprung, & Spindler, 1993; Umesono, Murakami, Thompson, & Evans, 1991).

4. DOES A nTRE EXIST IN THE TSHβ GENE?

In TRβ knockout (TRβ-KO) mice, the negative regulation of the TSHβ gene by T3 is impaired, resulting in elevation of TSHβ and αGSU expression (Fig. 1B; Forrest et al., 1996). This indicates that TRβ mediates the negative regulation of the TSHβ gene. Moreover, in mice that are deficient solely for TRβ2 expression, the negative regulation of the TSHβ gene is impaired to the same extent as in the TRβ-KO mice (Abel et al., 1999). Western blotting revealed that the expression of TRβ2 in the thyrotroph cell line, TαT1 (Yusta, Alarid, Gordon, Ridgway, & Mellon, 1998), is much higher than that of TRα1 or TRβ1 (Nakano et al., 2004). Thus, TRβ2 is the major mediator of the T3 signal in the inhibition of the TSHβ gene. After the essential role of pTRE was established for positive regulation, a nTRE was also postulated in the TSHβ gene (Chin, Carr, Burnside, & Darling, 1993; Cohen & Wondisford, 2013; Shupnik, 2000; Wondisford et al., 1989). Because serum TSH levels drastically increase in hypothyroidism (Fisher et al., 2000), one may assume that negative regulation is the mirror image of positive regulation (Fig. 3A and B; Hollenberg, 2013; Shupnik, 2000; Wondisford et al., 1989). This hypothesis implies that unliganded TR might activate transcription of the TSHβ gene (Fig. 3B). If this is true, deletion analysis of its promoter can identify the nTRE as the DNA sequence required for transactivation by unliganded TR (Fig. 4A). Using a primer extension method in human kidney-derived HEK293 cells transfected with a CAT reporter gene fused with the human TSHβ promoter (Fig. 4A; Wondisford et al., 1989), the transcriptional activity in the absence of T3 was reduced by deletion of a short DNA sequence immediately downstream to the transcription start site (TSS) (Fig. 4A). This sequence, GGGTCA, was reported to be a nTRE because of sequence similarity with a pTRE single half-site (AGGTCA) (Wondisford et al., 1989), and it has been cited as a typical nTRE not only in reviews (Chin et al., 1993; Shupnik, 2000) but also in the textbook (Cohen & Wondisford, 2013; Hollenberg, 2013). However, the existence of this nTRE in the TSHβ promoter (Wondisford et al., 1989) should be interpreted with caution due to the following problems. (1) The original nTRE analysis was performed using HEK293 cells, which do not express Pit1 (Steinfelder, Radovick, &

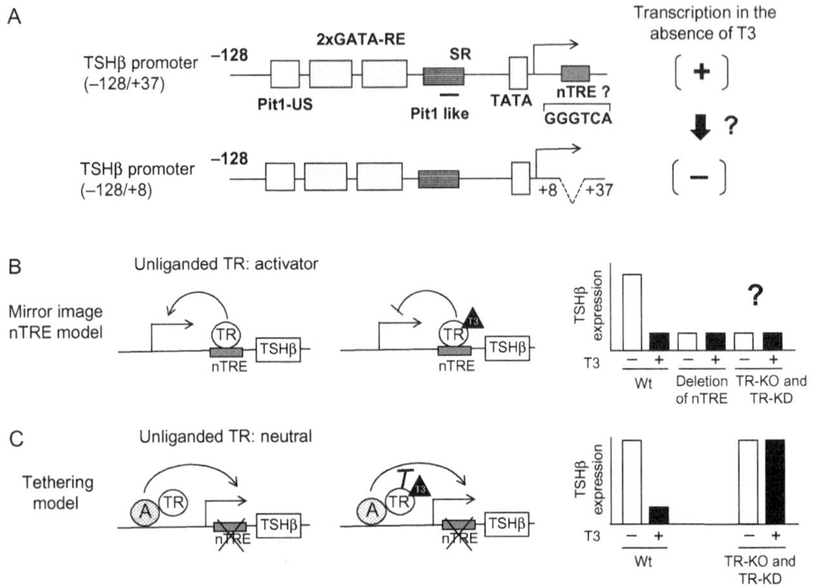

Fig. 4 Two possible definitions of negative regulation by T3. (A) Schematic representation of the TSHβ promoter and its deletion mutant. Using HEK293 cells, the nTRE was reported as a short DNA sequence, deletion of which reduced the activity of the promoter in the absence of T3. *GATA-RE*, GATA-responsive element; *nTRE*, putative negative TRE; *Pit1-like*, sequence similar to the Pit1-binding site in SR; *Pit1-US*, functional Pit1-binding site; *SR*, suppressor region; *TATA*, TATA box. (B) According to the mirror image model, unliganded TR may be a transcriptional activator (*left panel*) and the deletion of the putative nTRE might reduce TSHβ expression (*right panel*). If the mirror image model is true, TSHβ expression in the TR-KO mice and TR-KD cells should be reduced irrespective of T3 concentration; however, this is not the case. (C) In the tethering model, the TSHβ gene is activated by transcription activator "A" (*left panel*) and is negatively regulated by TR tethered to "A" in a T3-dependent manner (*middle panel*). This model is in agreement with TSHβ expression being maintained in TR-KO mice and in TR-KD cells (*right panel*).

Wondisford, 1992) or GATA2 (Imagawa et al., 2002), both of which are essential transcription factors for TSHβ expression (Dasen et al., 1999; Gordon et al., 1997). (2) According to authors (Wondisford et al., 1989), T3-dependent inhibition was detected in HEK293 cells transfected with the CAT reporter gene fused to the human TSHβ promoter; however, they did not state that the cells were cotransfected with a mammalian expression vector for TR in their primer extension assay. Surprisingly, other investigators have been unable to detect endogenous TRs in HEK293 cells by western blotting and reporter assays (Pietrzak & Puzianowska-Kuznicka, 2008; Turowska et al., 2007). (3) As described later, their CAT-based reporter gene (Wondisford et al., 1989) harbors an artificial AP-1 element (binding site for Jun/Fos) derived from pUC/pBR322 plasmids, which is known to induce artifactual inhibition by T3-bound TR (Lopez et al., 1993; Misawa et al., 2012; Sasaki et al., 1999). (4) Because this nTRE sequence (GGGTCA) in the TSHβ gene looks like a single half-site, there was the possibility that RXR might be dispensable for the DNA recognition by TR. However, gel shift assays indicated that binding of the TR monomer with the nTRE, i.e., the single half-site, is much weaker than that of the TR–RXR heterodimer with pTRE (Cohen, Flynn, & Wondisford, 1995; Kurokawa et al., 1993). The negative regulation of the TSHβ gene is specific to T3 and partially to estrogen (E2) and androgen (Cohen & Wondisford, 2013). However, the ligand/receptor selectivity is not easily explained by the nTRE hypothesis (Fig. 3B) because spacer nucleotides cannot be defined (Fig. 3A). Although a genome-wide study for glucocorticoid target genes suggests the existence of glucocorticoid-responsive elements (GREs) that may consist of a single half-site, these GREs do not always mediate the negative regulation of glucocorticoid and its receptor (GR) (Lim et al., 2015). An alternative possibility is that the DNA sequence flanking the nTRE may function as another half-site for RXR. However, this configuration cannot be distinguished from a pTRE (Fig. 3A). The experiments that examined the effect of RXR on the negative regulation by T3 remain inconclusive (Sasaki et al., 2011).

5. NEGATIVE REGULATION OF THE TSHβ GENE IS NOT THE MIRROR IMAGE OF POSITIVE REGULATION

The hypothesis that negative regulation of the TSHβ gene is the mirror image of positive regulation (Fig. 3; Hollenberg, 2013; Shupnik, 2000; Wondisford et al., 1989) was a prerequisite for identification of the nTRE in the TSHβ gene. However, the following observations suggest that it is not the case at molecular level. First, if unliganded TR is a transcriptional activator of the TSHβ gene (Fig. 4B, left panel), TSHβ expression in TRβ-KO mice should be reduced irrespective of serum T3 and T4 levels (Fig. 4B, right panel). However, as mentioned earlier, mice that are deficient for the *Thrb* gene or both *Thra* and *Thrb* genes exhibit elevated TSHβ expression (Abel et al., 1999; Forrest et al., 1996; Gothe et al., 1999; Weiss et al., 1997). Likewise, elevated, not reduced, secretion of TSH is reported in the homozygous patients of a family resistant to thyroid hormone (RTH), where the *Thrb* allele is globally deleted (Sasaki et al., 2011; Takeda, Sakurai,

DeGroot, & Refetoff, 1992). Moreover, TRβ knockdown (TRβ-KD) by RNAi does not decrease TSHβ mRNA levels before T3 treatment in TαT1.1 cells, a derivative of TαT1 cells (Chiamolera et al., 2012). These findings indicate that unliganded TR does not activate the TSHβ gene. Second, silencing function of TRβ2 is extremely weak or almost undetectable compared to TRα1 and TRβ1 (Privalsky et al., 2009; Yang et al., 1999). If the mirror image theory is true, activation of the TSHβ gene by unliganded TRβ2 should be very weak. However, this view is not supported by the fact that TRβ2 is the main mediator of the negative regulation of the TSHβ gene (Abel et al., 1999; Nakano et al., 2004). Thus, it is unlikely that negative regulation of the TSHβ gene is a mirror image of positive regulation.

6. BASAL EXPRESSION OF THE TSHβ GENE IS MAINTAINED BY THYROTROPH-SPECIFIC TRANSCRIPTION FACTORS, GATA2 AND Pit1

Because the cell lines, of which phenotype mimics that of the thyrotrophs, have been limited, early studies for negative regulation often utilized nonthyrotroph cell lines including HEK293, COS, CV1, somatotroph-derived GH3, and choriocarcinoma-derived JEG3 (Sasaki et al., 2011). However, except for JEG3 cells, the inhibition by T3 was usually difficult to detect due to the very low level of basal transcriptional activity before T3 treatment. The fact that TSHβ expression is not maintained by unliganded TR led us to focus on the mechanism maintaining its expression before inhibition by T3 (Fig. 4C, left panel). According to (Wood et al., 1990), the promoter region encompassing nt. −269 to +1 of the TSS was sufficient for the expression of the TSHβ gene in the TSHoma cell line, TtT97. They identified binding sites for the pituitary-specific transcription factor, Pit1 (Cohen & Radovick, 2002; Kashiwabara et al., 2009). Moreover, later analysis using DNA foot printing with TtT97 cell nuclear extracts identified two GATA-responsive elements (GATA-REs) (Fig. 4A) close to the Pit1-binding site (Gordon et al., 1997), which we designated as Pit1-US (Kashiwabara et al., 2009). The GATA family of transcription factors consists of six members, and GATA1, 2, and 3 are classified as hematopoietic GATAs (Bresnick, Katsumura, Lee, Johnson, & Perkins, 2012; Vicente, Conchillo, Garcia-Sanchez, & Odero, 2012). *In vivo* analysis subsequently showed that coexpression of GATA2 with Pit1 determines thyrotroph differentiation in the pituitary (Dasen et al., 1999). In addition, JEG3 cells express GATA2 and 3 (Jorgensen, Quirk, & Nilson, 2004; Steger,

Hecht, & Mellon, 1994). A 30-bp sequence downstream of the GATA-REs is conserved among mouse, rat, and human (Fig. 4A) and is designated as the suppressor region (SR) because its deletion enabled GATA2 to activate this promoter without Pit1 (Kashiwabara et al., 2009). Detailed analysis revealed that, at least in CV1 cells, an unknown protein (SR-binding protein, SRBP) recognizes the SR (Fig. 5) and inhibits transactivation by GATA2 alone. Furthermore, Pit1 binding to a Pit1-like element (Fig. 4A) within the SR interferes with the function of SRBP, resulting in the protection of GATA2 function (derepression). Taken together, these data indicate that GATA2

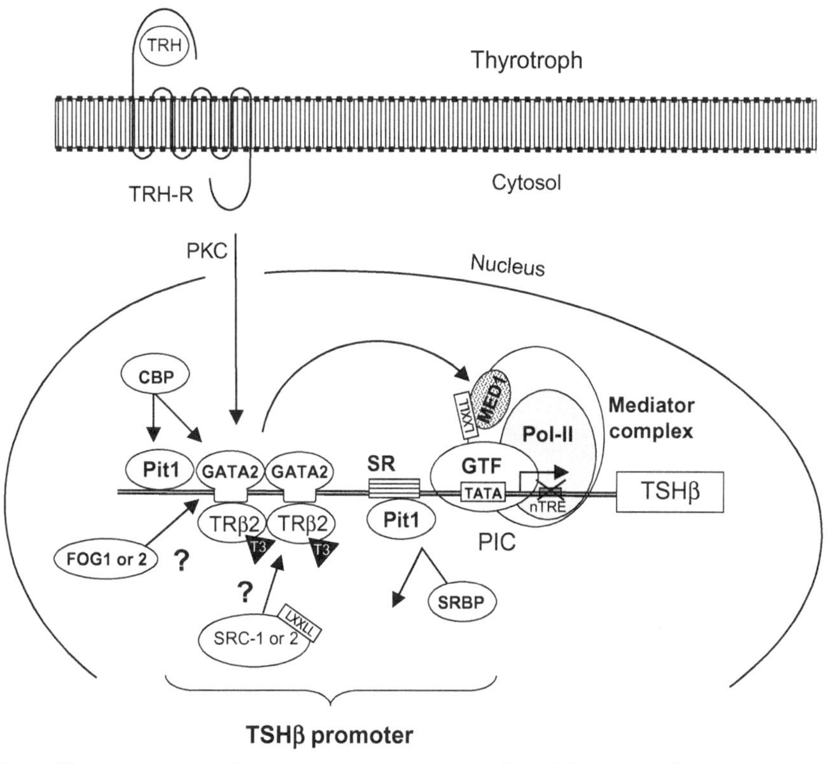

Fig. 5 The mechanism of transcriptional regulation of the TSHβ gene. The configuration of the Pit1-US and GATA-REs may be critical for the recruitment of CBP/p300. In addition to Pit1-US, Pit1 binds with the Pit1-like element in the SR and competes with SR-binding protein (SRBP), resulting in protection of GATA2 function from inhibition by SR (derepression). TRβ2 interacts with the Zn-finger domain of GATA2 via its DBD and interferes with GATA2-dependent transcription in a T3-dependent fashion. *FOG1 or 2*, friend of GATA1 or 2; *GTF*, general transcription factor; *PIC*, preinitiation complex; *Pit1-US*, functional Pit1-binding site; *PKC*, protein kinase C; *Pol-II*, RNA polymerase II; *SR*, suppressor region; *SRC-1 or 2*, steroid receptor coactivator 1 or 2; *TRH-R*, TRH receptor.

but not Pit1 is the true activator driving TSHβ promoter (Kashiwabara et al., 2009).

7. A CV1 CELL RECONSTITUTION SYSTEM THAT MIMICS THYROTROPHS

The TSHβ promoter can be activated in kidney-derived CV1 cells transfected with GATA2 and Pit1 expression plasmids (Gordon et al., 1997). This led us to test whether negative regulation of the TSHβ gene can be simulated in CV1 cells transfected with expression plasmids for GATA2, Pit1, and TRs (Nakano et al., 2004). Using a CAT-based reporter gene, in which the pUC/pBR322-derived AP-1 element (Lopez et al., 1993) was deleted (Sasaki et al., 1999) (described later), we obtained the following results. (1) Negative regulation of the TSHβ gene by T3 was readily observed in CV1 cells as long as GATA2, Pit1, and TRs are coexpressed. CV1 cells are often used to study positive regulation by T3; therefore, this result indicates that thyrotroph-specific factors other than Pit1, GATA2, and TR may not be required for negative regulation of the TSHβ gene. (2) Although all three TRs can mediate negative regulation by T3, repression by TRβ2 is most potent among them. This is in agreement with previous reports showing that TRβ2 is the main mediator of negative regulation of the TSHβ gene (Abel et al., 1999) and that inhibition by T3-bound TRα1 can be detected in TRβ-KO mice. (3) Without GATA2 or Pit1, the TSHβ promoter was not transactivated by unliganded TR alone, of which amount was sufficient for T3-dependent inhibition of GATA2- and Pit1-induced transcription. This *in vitro* observation further demonstrates that negative regulation of the TSHβ gene is not the mirror image of positive regulation.

8. TR IS TETHERED BY GATA2 TO MEDIATE NEGATIVE REGULATION OF THE TSHβ GENE BY T3

We reexamined the function of the nTRE in the TSHβ promoter in the presence of GATA2, Pit1, and TR (Matsushita et al., 2007). Surprisingly, negative regulation of the TSHβ promoter by T3-bound TR is maintained even after the nTRE was completely mutated or deleted. Therefore, it was concluded that this nTRE (Figs. 3 and 4) is not necessary for T3-dependent negative regulation of the TSHβ gene. As noted earlier, GATA2 can transactivate the TSHβ promoter without Pit1 when the SR is deleted (Kashiwabara et al., 2009). We found that the T3-bound TR inhibits this reporter construct driven by GATA2 alone (Matsushita et al., 2007), indicating that the target of T3-bound TR repression is GATA2 and that negative regulation by T3 is independent from the inhibitory effect of the SR. Indeed, coimmunoprecipitation and GST-pull-down assays indicated that the DBD of TR interacts with the Zn-finger region of GATA2 (GATA2-Zf) in a T3-independent fashion (Fig. 2C). These results indicate that negative regulation of the TSHβ gene is mediated by TR tethered by GATA2 and that the TR-DBD does not directly bind to the reported nTRE but instead functions as an interface for protein–protein interaction with GATA2 (Matsushita et al., 2007). This model (Fig. 4C, left and middle panels) is in agreement with the fact that TSHβ expression is maintained in TR-KO mice and TR-KD experiment with TαT1.1 cells (Fig. 4C, right panel). Using CV1 and thyrotroph-derived TαT1 cells, Ohba et al. (Ohba et al., 2011; Sasaki et al., 2011) found that TRH signaling facilitates the GATA2-mediated transcriptional activity of the TSHβ gene induced via the protein kinase C pathway (Fig. 5). Their *in vitro* data correlate well with findings in mice deficient for TR and TRH (Ohba et al., 2011). In addition to GATA2, the orphan NHR, NR4A1 (Nur77), may also be involved in TRH signaling to the TSHβ gene; however, because its binding site is unknown, further investigation is required (Nakajima et al., 2012).

9. NHRs TETHERED BY DNA-BINDING TRANSCRIPTION FACTOR MEDIATE LIGAND-DEPENDENT NEGATIVE REGULATION

The model depicted in Fig. 4C (left and middle panels) is reminiscent of glucocorticoid-dependent inhibition of transcription for genes that are activated by NFκB and/or AP-1 (De Bosscher, Vanden Berghe, & Haegeman, 2003; Herrlich, 2001). In addition, the transcriptional repression by liganded NHR via a tethering mechanism is also reported for genes transactivated by Nur77 (Martens, Bilodeau, Maira, Gauthier, & Drouin, 2005) and GATA family transcription factors (Clabby, Robison, Quigley, Wilson, & Kelly, 2003). ChIP assays using MCF7 cells indicated that ERα can be enriched via tethering as well as direct DNA binding to the classical estrogen-responsive elements (Heldring et al., 2011). In the tethering mechanism by liganded NHR, there is some ligand/receptor selectivity (Caldenhoven et al., 1995; Liden et al., 1997) although it does not directly recognize DNA (Herrlich, 2001; Reichardt et al., 1998). As expected,

GATA2-dependent activation of the TSHβ promoter is inhibited by T3-bound TR and partially liganded ERα (Nagayama et al., 2008), but not by liganded RAR or VDR (Matsushita et al., 2007). Using ChIP assays in TαT1.1 cells with an antibody against TRβ, Chiamolera et al. (2012) reported the enrichment of TRβ at several positions in the TSHβ genome that are different from the previously reported nTRE. However, they did not exclude the possibility of TRβ tethering by DNA-binding transcription factors. In addition, functional analysis by mutating these putative nTREs is necessary because DNA sequences that are directly bound by TRs are not always functional TREs (Ikeda, Wilcox, & Chin, 1996).

10. NEGATIVE REGULATION OF GENES OTHER THAN THE TSHβ GENE MAY NOT BE THE MIRROR IMAGE OF POSITIVE REGULATION

In addition to the TSHβ gene, negative regulation by T3 is reported for the genes encoding αGSU (Chatterjee, Lee, Rentoumis, & Jameson, 1989; Pennathur, Madison, Kay, & Jameson, 1993), type 2 deiodinase (D2) (Christoffolete et al., 2006; Gereben et al., 2008), epidermal growth factor receptor (Thompson, Santon, Shephard, Walton, & Gill, 1992), RSV-LTR (Saatcioglu, Deng, & Karin, 1993), MYH7 (also designated as myosin heavy chain β) (Edwards, Bahl, Flink, Cheng, & Morkin, 1994; Wright, Haddad, Qin, Bodell, & Baldwin, 1999), preproTRH (Hollenberg et al., 1995; Satoh, Yamada, Iwasaki, & Mori, 1996), keratins (Tomic-Canic, Day, Samuels, Freedberg, & Blumenberg, 1996), mouse Na, K-ATPase α3 subunit (Chin, Apriletti, & Gick, 1998), Nm23-H1 (Lin, Shieh, & Hsu, 2000), c-myc (Perez-Juste, Garcia-Silva, & Aranda, 2000), growth hormone (Sanchez-Pacheco & Aranda, 2003), Pit1 (Sanchez-Pacheco, Palomino, & Aranda, 1995), β-amyloid precursor protein (Villa, Santiago, Belandia, & Pascual, 2004), prohormone convertase (PC) 1/3 (Shen, Li, Brent, & Friedman, 2004), PC2 (Shen, Li, Brent, & Friedman, 2005), mouse CD44 (Kim et al., 2005), superoxide dismutase-1 (Santos et al., 2006), phospholamban (Belakavadi, Saunders, Weisleder, Raghava, & Fondell, 2010), type 4 melanocortin receptor (Decherf et al., 2010), CYP27B1 (Kozai et al., 2013), and phospholipase A2 group IIa (Sharma, Thakran, Deng, Elam, & Park, 2013). In some of these genes, a sequence similar to the TSHβ nTRE was reported. However, the definition of negative regulation (Fig. 4B or C) is usually unclear and few studies paid attention to the transactivation mechanism before inhibition by T3 or to artifactual negative regulation (described later). Of note, the expression of αGSU (Forrest et al., 1996; Gothe et al., 1999), D2 (Boelen, Kwakkel, Chassande, & Fliers, 2009; Pelletier, Gauthier, Sideleva, Samarut, & Silva, 2008), preproTRH (Abel, Ahima, Boers, Elmquist, & Wondisford, 2001; Dupre et al., 2004), and MYH7 in the heart (Mansen, Yu, Forrest, Larsson, & Vennstrom, 2001) is maintained in TR-KO mice, suggesting that the negative regulation of these genes is not the mirror image of positive regulation. A similar observation was made by microarray analysis comparing hepatic genes negatively regulated by T3 in wild-type and TRα/TRβ-double-KO mice (Yen et al., 2003).

11. NEGATIVE REGULATION OF GENES OTHER THAN THE TSHβ GENE MAY BE MEDIATED VIA A TETHERING MECHANISM

The findings for the TSHβ gene predict that T3-bound TRs can inhibit genes for which GATA2 (or GATA family members) functions as the main transcriptional activator. The αGSU (Jorgensen et al., 2004; Steger et al., 1994) and endothelin-1 promoters (Dorfman, Wilson, Bruns, & Orkin, 1992) have functional GATA-REs. We found that, when these promoters are stimulated by GATA2 in CV1 cells, T3-bound TR inhibited their activity (Matsushita et al., 2007). Although D2 expression is inhibited by T3 in thyrotrophs (Christoffolete et al., 2006), no nTRE has been identified in its promoter sequence (Gereben et al., 2008). We speculated that GATA2 may play a role in the expression of the D2 gene. Indeed, we found atypical but functional GATA-REs in the D2 gene, the locations of which are different from those predicted by computer (Matsunaga et al., 2015). We showed again that T3-bound TR inhibits GATA2-induced activity of the D2 promoter (Matsunaga et al., 2015). T3-bound TR inhibits transactivation by GATA2 but not by another transcription factor, Nkx2.5, providing evidence that negative regulation is specifically mediated via GATA2 and that cryptic nTREs do not exist in this promoter. Interestingly, multiple functional GATA-REs exist in the GATA2 gene itself (Bresnick et al., 2012). The importance of this autoregulation is established in the GATA-dependent differentiation of the hemopoietic lineage. As expected, we recently found that this positive feedback loop is also repressed by a T3-bound TR (Hirahara et al., 2016). The findings for the TSHβ gene also predict that negative regulation may occur by a T3-bound TR tethered to DNA-binding transcription factors other than GATAs. For example, the cardiac expression of the MYH7 gene is inhibited by T3 (Kahaly & Dillmann, 2005). We found that TR tethered by TEAD family transcription factors on the MCAT (muscle C, A, and T) and A/T-rich elements in the MYH7 promoter mediates T3-dependent negative regulation (Iwaki et al., 2014). Although putative nTREs have been suggested in this gene, we found that inhibition by T3 is maintained even after these sequences are mutated.

12. TARGET MOLECULE OF T3-BOUND TRs TETHERED BY GATA2

Because the protein–protein interaction between GATA2 and TRs is T3 independent (Matsushita et al., 2007), we speculate that T3 binding may modulate the function of the cofactors related to TRs and GATA2. Identification of these molecule(s) is the next important step; however, the situation is similar to the investigation of negative regulation by other NHRs (Chinenov, Gupte, & Rogatsky, 2013; De Bosscher et al., 2003; Rollins, Coppo, & Rogatsky, 2015; Santos et al., 2011).

12.1 Corepressors

It was proposed that unliganded TRs in solution may squelch corepressors from the promoter of the target gene (Tagami, Park, & Jameson, 1999). Upon T3 binding to TRs, corepressors released from TRs might associate with a transcription activator, which maintains basal promoter activity, resulting in transcriptional repression. Yet, this model cannot explain why the expression of the TSHβ gene is not reduced in various kinds of TR-KO mice (Forrest et al., 1996; Gothe et al., 1999). Astapova et al. (Astapova & Hollenberg, 2013; Astapova et al., 2011) generated knock-in mice having a mutant NCoR (NCoRΔID) that cannot associate with unliganded TR. However, in these mice, negative regulation of the TSHβ gene by T3 was maintained although the set point in H–P–T axis was altered. Likewise, mice that harbor a mutant SMRT (SMRTmRID) did not exhibit altered serum TSH, T4 or T3 levels (Astapova & Hollenberg, 2013). Some corepressors interact with NHRs in the presence of cognate ligands (Gurevich, Flores, & Aneskievich, 2007). This class of coregulators includes receptor-interacting protein of 140 kDa (RIP140) and ligand-dependent corepressor (LCoR). One may assume their involvement in negative regulation by liganded NHRs. However, their functions were originally evaluated in the context of the pTRE where NHRs directly bind

DNA (Fig. 3A), and the involvement of these corepressors in ligand-dependent transcriptional repression via a tethering mechanism (Fig. 4C) is unknown.

12.2 CBP/p300

CBP/p300 function as cofactors for certain DNA-binding transcription factors, including NFκB, AP-1, and ligand-dependent NHRs. Kamei et al. (1996) proposed a model where liganded NHRs may interfere with the coactivator function of CBP/p300 for these DNA-binding factors, resulting in reduced transcriptional activity. However, ligand/receptor selectivity is difficult to be explained by this model because multiple NHRs interact with CBP/p300 in the presence of cognate ligands. Moreover, subsequent studies failed to reproduce the relief of ligand-dependent repression by overexpression of CBP or p300 (De Bosscher et al., 1997; De Bosscher, Vanden Berghe, & Haegeman, 2001; Wu, Li, Dietz, & Lala, 2004).

12.3 p160 Family

Negative regulation of TSHβ is relieved in mice deficient for SRC-1 and/or SRC-2 (GRIP1/TIF2) (Takeuchi et al., 2002; Weiss et al., 2002, 1999; Xu et al., 1998). Inhibitions of the genes for proopiomelanocortin (POMC) and corticotropin-releasing hormone (CRH) by liganded GR were also blunted in SRC-1-KO mice (Lachize et al., 2009; Winnay, Xu, O'Malley, & Hammer, 2006). Although there is the possibility of adaptive processes occurring during pituitary development in SRC-1-KO mice (Sadow et al., 2003), cell-based studies using overexpression of SRC-1 (van der Laan, Lachize, Vreugdenhil, de Kloet, & Meijer, 2008) or knockdown of it also reported similar results (Lachize et al., 2009). The ligand-dependent negative regulation is reduced for a mutant TR (Alonso et al., 2009; Ortiga-Carvalho et al., 2005) and VDR (Chen et al., 2000), of which LBDs have defect to interact with p160 without affecting ligand binding. Likewise, overexpression of SRC-2 potentiates negative regulation of the tumor necrosis factor α and IL-6 promoters (An et al., 1999; Cvoro et al., 2006; Nwachukwu et al., 2014) and the AP-1-dependent promoter (Rogatsky, Luecke, Leitman, & Yamamoto, 2002) by liganded ERα and GR, respectively. Involvement of SRC-2 in negative regulation by liganded GR is also demonstrated by genome-wide analyses (Chinenov et al., 2012; Uhlenhaut et al., 2013). It is not easy to imagine that p160s may have a reverse function in negative regulation because they are known to facilitate transcription via

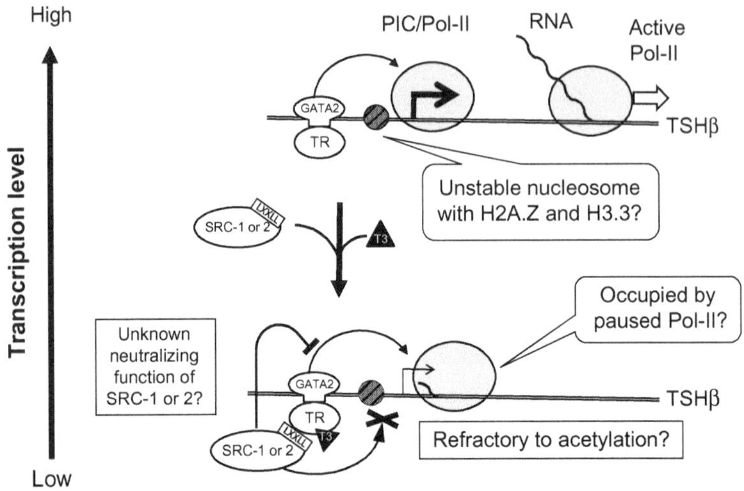

Fig. 6 The model of negative regulation via SRC-1 or 2. In the transcriptionally active gene, the nucleosomes around TSS are unstable because they consist of histone variants, H2A.Z and H3.3. In addition, acetylation level of H3.3 may be already high. In such a status, nucleosome may be refractory to acetylation induced by p160, which is recruited by liganded TR. Moreover, Pol-II may occupy the DNA around TSS even after T3 treatment (Pol-II pausing), resulting in interference with the entry of new Pol-II.

their HAT activity. However, we speculate that a paradoxical situation may occur in the promoters of negatively regulated genes before ligand addition (Fig. 6). First, in the transcriptionally active promoter, the nucleosome immediately upstream of the TSS has few authentic H2A and H3 histones, but harbors the histone variants, H2A.Z and H3.3 (Soboleva, Nekrasov, Ryan, & Tremethick, 2014). The nucleosome-containing H2A.Z and H3.3 is known to be unstable (Schones et al., 2008). Moreover, the "−1" nucleosome may be refractory to additional acetylation by p160 because the acetylation of H3.3 may be already high (Loyola, Bonaldi, Roche, Imhof, & Almouzni, 2006; Subramanian, Fields, & Boyer, 2015). Further investigation is necessary to determine the location of the nucleosome targeted by GATA2-dependent HAT activity in the TSHβ promoter and the sensitivity of histone variants to p160 HAT activity. Second, during active transcription from the gene promoter, the preinitiation complex (PIC) including Pol-II is located immediately downstream of the TSS. This indicates a low density of histones at the TSS. In some cytokine genes (Adelman et al., 2009; Adelman & Lis, 2012; Gupte, Muse, Chinenov, Adelman, & Rogatsky, 2013), Pol-II pauses at immediately downstream of the TSS after synthesizing a short RNA (30–40 mer) when the signaling pathway from a cell-membrane receptor is not activated (Pol-II pausing, Fig. 6). We recently found that expression of αGSU in the gonadotroph-derived cell line, LβT2, is negatively regulated by T3-bound TR. Interestingly, ChIP assays for Pol-II in the αGSU promoter in this cell line indicated that approximately 60% of Pol-II remains close to the TSS after T3 addition. Likewise, genome-wide analysis of genes that are negatively regulated by liganded ERα indicates that, in approximately 70% of these genes, Pol-II stays at the TSS after E2 addition (Hah et al., 2011). Inactive Pol-IIs occupying TSSs may make the DNA refractory to p160-dependent HAT activity and may prevent the reentry of new Pol-II and hinder the transcription cycle. Finally, the repression of the TSHβ gene by T3 is not affected in SRC-3-KO mice (Ying et al., 2005). Similar specificity among p160s is reported in the negative regulation by liganded GR (Rogatsky et al., 2002) and ERα (Nwachukwu et al., 2014). These findings again suggest that involvement of p160 in the negative regulation by liganded NHRs is independent from their HAT activity. We speculate that some p160s have the function that may neutralize the activity of DNA-binding factors and that this function may be determined by the combination between NHRs and DNA-binding transcription factors. It is reported that the elongation of transcription of the IL-8 gene is attenuated by liganded GR (Luecke & Yamamoto, 2005). CyclinT1, a key factor for transcription elongation, not only interacts with SRC-2 (GRIP1) (Kino, Slobodskaya, Pavlakis, & Chrousos, 2002) but also associates with GATA1 (Elagib et al., 2008).

12.4 Coregulators for the GATA Family

T3-bound TR may interfere with the coregulator involved in GATA2-dependent transcription. We found that the TSHβ gene can be transactivated by GATA2 as well as GATA1 and 3 (Fig. 2C) and that T3-bound TR can repress these activities (Matsushita et al., 2007). The amino acid sequence of the GATA2-Zf has high homology with that of GATA1 and 3. As expected, GATA2-Zf is important for the recognition of GATA-RE and also for interactions with multiple transcription factors and cofactors (Vicente et al., 2012). For example, this domain interacts with Pit1 (Dasen et al., 1999; Gordon et al., 2002), TR (Matsushita et al., 2007), MED1 (Gordon et al., 2006), CBP/p300 (Blobel, Nakajima, Eckner, Montminy, & Orkin, 1998), and friend of GATA (FOG) 1 or 2 (Vicente et al., 2012). The expression of the TSHβ gene is reduced in MED1 knockout mice (Ito & Roeder, 2001), suggesting that MED1 is coactivator for

GATA2. Because MED1 also associates with T3-bound TR-LBD via interaction of its LXXLL motifs (Cheng et al., 2010), MED1 may also mediate T3 signal in the context of negative regulation. However, mice with knock-in of a mutant MED1, which cannot interact with liganded NHRs due to mutation of the LXXLL motifs, did not exhibit any apparent gross abnormalities except for defects in pubertal mammary gland development (Jiang et al., 2010). In the myocardium, the MYH7 gene is inhibited by T3 via a tethering mechanism (Iwaki et al., 2014) (described earlier). However, in contrast to the TSHβ gene, MYH7 expression is increased in muscle-specific MED1-KO mice (Chen, Zhang, Birsoy, & Roeder, 2010). FOG1 and 2 interact with GATA2-Zf and function as coregulators (Vicente et al., 2012). They may associate with TRβ (Rouf et al., 2008) and other NHRs (Clabby et al., 2003; Huggins, Bacani, Boltax, Aikawa, & Leiden, 2001). FOGs also associate with the nucleosome remodeling and deacetylase complex, which is known to control GATA-dependent transcription (Lai & Wade, 2011). Endogenous FOG1 and 2 were reported in LβT2 cells (Lo, Zheng, Gong, Crochet, & Halvorson, 2011), in which the αGSU gene is negatively regulated by T3 (data not shown). Matsushita et al. (2007) generated a mutant GATA2 (C295A) in which FOG interaction is disrupted. This mutant GATA2 elicited not only reduced basal transcriptional activity (approximately half in comparison with wild-type GATA2) but also a lack of repression by T3-bound TR.

13. ARTIFACTUAL NEGATIVE REGULATION VIA PLASMID BACKBONES

As mentioned earlier, some liganded NHRs inhibit transcription activity of AP-1 via a tethering mechanism (De Bosscher et al., 2003; Herrlich, 2001; Pfahl, 1993). Unexpectedly, pUC plasmids harbor a functional AP-1 site between nt. 1 and nt. 138, the activity of which is repressed by T3-bound TR (Lopez et al., 1993). Using the BLAST database, we found more than 2000 plasmid constructs bearing this AP-1 site. This site is also included in pBR322 (Yanisch-Perron, Vieira, & Messing, 1985) and pGEM vectors. These plasmids were often used in early molecular biology studies. Unfortunately this site was also present in the plasmids used to analyze the negative regulation of the TSHβ gene (Wondisford et al., 1989). Because of advantages over the CAT assay, the firefly luciferase assay was used to analyze negatively regulated genes. However, at least in CV1 (Tillman et al., 1993), JEG3 (Maia, Harney, & Larsen, 1996), and Hepa1–6 mice liver cancer cells

(Chan, Borowsky, & Privalsky, 2008), the firefly luciferase cDNA functions as a transcriptional enhancer, whose activity is artifactually repressed by T3-bound TR. Computer search of the firefly luciferase cDNA revealed the presence of more than 250 putative DNA elements that can be recognized by transcription factors (Liu & Brent, 2008). Misawa et al. (2012) showed that the firefly luciferase cDNA has an enhancer activity that can be stimulated by a PKC activator, phorbol 12-myristate 13-acetate (PMA), and that this activity is inhibited by T3-bound TR in CV1 and JEG3 cells. Finally, there is the possibility that some reporter plasmids also possess cryptic but functional GATA-REs in their backbone (Ho & Strauss, 2004). The problems of reporter gene plasmids that mediate artifactual negative regulation are described elsewhere (Misawa et al., 2012; Sasaki et al., 2011; Tillman et al., 1993).

14. CONCLUSION AND FUTURE DIRECTIONS

Unliganded TR promotes transcriptional silencing in positive regulation; however, accumulated lines of evidence suggest that it is not a transcriptional activator in the context of negative regulation. This led to the finding that the proposed nTRE in the TSHβ gene is dispensable for repression by T3-bound TR. The identification of a transcription factor maintaining basal transcriptional activity before inhibition by T3 is very important for the analysis of negative regulation. The identification of such a transcription factor allows negative regulation to be studied using cells that endogenously express the factor. Coexpression of it with TR may enable us to study the mechanisms of negative regulation in the cells that have been utilized in the investigation of positive regulation. The information of such factors is also useful to avoid the artifactual negative regulation mediated by plasmid backbones. Moreover, its physical interaction with NHRs should be tested. Investigation of the negative regulation mechanism may provide insight into the critical step in transcription by Pol-II. The lessons from the TSHβ gene are likely to be of universal importance in understanding the mechanism(s) underlying negative regulation by liganded NHRs.

ACKNOWLEDGMENTS

The authors thank Mrs. Mayumi Ito and Miss Mariko Ohishi for their secretarial work. This work was supported in part by a Grant-in-Aid for Scientific Research to S.S. and A.M. from the Ministry of Education, Culture, Sports, Science and Technology of Japan (Grant numbers 15K09425, 24590689, and 26461378).

REFERENCES

Abel, E. D., Ahima, R. S., Boers, M. E., Elmquist, J. K., & Wondisford, F. E. (2001). Critical role for thyroid hormone receptor beta2 in the regulation of paraventricular thyrotropin-releasing hormone neurons. *The Journal of Clinical Investigation, 107*, 1017–1023.

Abel, E. D., Boers, M. E., Pazos-Moura, C., Moura, E., Kaulbach, H., Zakaria, M., et al. (1999). Divergent roles for thyroid hormone receptor beta isoforms in the endocrine axis and auditory system. *The Journal of Clinical Investigation, 104*, 291–300.

Adelman, K., Kennedy, M. A., Nechaev, S., Gilchrist, D. A., Muse, G. W., Chinenov, Y., et al. (2009). Immediate mediators of the inflammatory response are poised for gene activation through RNA polymerase II stalling. *Proceedings of the National Academy of Sciences of the United States of America, 106*, 18207–18212.

Adelman, K., & Lis, J. T. (2012). Promoter-proximal pausing of RNA polymerase II: Emerging roles in metazoans. *Nature Reviews Genetics, 13*, 720–731.

Alonso, M., Goodwin, C., Liao, X., Ortiga-Carvalho, T., Machado, D. S., Wondisford, F. E., et al. (2009). In vivo interaction of steroid receptor coactivator (SRC)-1 and the activation function-2 domain of the thyroid hormone receptor (TR) beta in TRbeta E457A knock-in and SRC-1 knockout mice. *Endocrinology, 150*, 3927–3934.

An, J., Ribeiro, R. C., Webb, P., Gustafsson, J. A., Kushner, P. J., Baxter, J. D., et al. (1999). Estradiol repression of tumor necrosis factor-alpha transcription requires estrogen receptor activation function-2 and is enhanced by coactivators. *Proceedings of the National Academy of Sciences of the United States of America, 96*, 15161–15166.

Antonini, D., Sibilio, A., Dentice, M., & Missero, C. (2013). An intimate relationship between thyroid hormone and skin: Regulation of gene expression. *Frontiers in Endocrinology, 4*, 104.

Astapova, I., & Hollenberg, A. N. (2013). The in vivo role of nuclear receptor corepressors in thyroid hormone action. *Biochimica et Biophysica Acta, 1830*, 3876–3881.

Astapova, I., Vella, K. R., Ramadoss, P., Holtz, K. A., Rodwin, B. A., Liao, X. H., et al. (2011). The nuclear receptor corepressor (NCoR) controls thyroid hormone sensitivity and the set point of the hypothalamic-pituitary-thyroid axis. *Molecular Endocrinology, 25*, 212–224.

Belakavadi, M., Saunders, J., Weisleder, N., Raghava, P. S., & Fondell, J. D. (2010). Repression of cardiac phospholamban gene expression is mediated by thyroid hormone receptor-{alpha}1 and involves targeted covalent histone modifications. *Endocrinology, 151*, 2946–2956.

Blobel, G. A., Nakajima, T., Eckner, R., Montminy, M., & Orkin, S. H. (1998). CREB-binding protein cooperates with transcription factor GATA-1 and is required for erythroid differentiation. *Proceedings of the National Academy of Sciences of the United States of America, 95*, 2061–2066.

Boelen, A., Kwakkel, J., Chassande, O., & Fliers, E. (2009). Thyroid hormone receptor beta mediates acute illness-induced alterations in central thyroid hormone metabolism. *Journal of Neuroendocrinology, 21*, 465–472.

Bresnick, E. H., Katsumura, K. R., Lee, H. Y., Johnson, K. D., & Perkins, A. S. (2012). Master regulatory GATA transcription factors: Mechanistic principles and emerging links to hematologic malignancies. *Nucleic Acids Research, 40*, 5819–5831.

Caldenhoven, E., Liden, J., Wissink, S., Van de Stolpe, A., Raaijmakers, J., Koenderman, L., et al. (1995). Negative cross-talk between RelA and the glucocorticoid receptor: A possible mechanism for the antiinflammatory action of glucocorticoids. *Molecular Endocrinology, 9*, 401–412.

Chan, I. H., Borowsky, A. D., & Privalsky, M. L. (2008). A cautionary note as to the use of pBi-L and related luciferase/transgenic vectors in the study of thyroid endocrinology. *Thyroid, 18*, 665–666.

Chatterjee, V. K., Lee, J. K., Rentoumis, A., & Jameson, J. L. (1989). Negative regulation of the thyroid-stimulating hormone alpha gene by thyroid hormone: Receptor interaction adjacent to the TATA box. *Proceedings of the National Academy of Sciences of the United States of America*, *86*, 9114–9118.

Chen, S., Cui, J., Nakamura, K., Ribeiro, R. C., West, B. L., & Gardner, D. G. (2000). Coactivator-vitamin D receptor interactions mediate inhibition of the atrial natriuretic peptide promoter. *The Journal of Biological Chemistry*, *275*, 15039–15048.

Chen, W., Zhang, X., Birsoy, K., & Roeder, R. G. (2010). A muscle-specific knockout implicates nuclear receptor coactivator MED1 in the regulation of glucose and energy metabolism. *Proceedings of the National Academy of Sciences of the United States of America*, *107*, 10196–10201.

Cheng, S. Y., Leonard, J. L., & Davis, P. J. (2010). Molecular aspects of thyroid hormone actions. *Endocrine Reviews*, *31*, 139–170.

Chiamolera, M. I., Sidhaye, A. R., Matsumoto, S., He, Q., Hashimoto, K., Ortiga-Carvalho, T. M., et al. (2012). Fundamentally distinct roles of thyroid hormone receptor isoforms in a thyrotroph cell line are due to differential DNA binding. *Molecular Endocrinology*, *26*, 926–939.

Chin, S., Apriletti, J., & Gick, G. (1998). Characterization of a negative thyroid hormone response element in the rat sodium, potassium-adenosine triphosphatase alpha3 gene promoter. *Endocrinology*, *139*, 3423–3431.

Chin, W. W., Carr, F. E., Burnside, J., & Darling, D. S. (1993). Thyroid hormone regulation of thyrotropin gene expression. *Recent Progress in Hormone Research*, *48*, 393–414.

Chinenov, Y., Gupte, R., Dobrovolna, J., Flammer, J. R., Liu, B., Michelassi, F. E., et al. (2012). Role of transcriptional coregulator GRIP1 in the anti-inflammatory actions of glucocorticoids. *Proceedings of the National Academy of Sciences of the United States of America*, *109*, 11776–11781.

Chinenov, Y., Gupte, R., & Rogatsky, I. (2013). Nuclear receptors in inflammation control: Repression by GR and beyond. *Molecular and Cellular Endocrinology*, *380*, 55–64.

Christoffolete, M. A., Ribeiro, R., Singru, P., Fekete, C., da Silva, W. S., Gordon, D. F., et al. (2006). Atypical expression of type 2 iodothyronine deiodinase in thyrotrophs explains the thyroxine-mediated pituitary thyrotropin feedback mechanism. *Endocrinology*, *147*, 1735–1743.

Clabby, M. L., Robison, T. A., Quigley, H. F., Wilson, D. B., & Kelly, D. P. (2003). Retinoid X receptor alpha represses GATA-4-mediated transcription via a retinoid-dependent interaction with the cardiac-enriched repressor FOG-2. *The Journal of Biological Chemistry*, *278*, 5760–5767.

Clement, K., Viguerie, N., Diehn, M., Alizadeh, A., Barbe, P., Thalamas, C., et al. (2002). In vivo regulation of human skeletal muscle gene expression by thyroid hormone. *Genome Research*, *12*, 281–291.

Cohen, O., Flynn, T. R., & Wondisford, F. E. (1995). Ligand-dependent antagonism by retinoid X receptors of inhibitory thyroid hormone response elements. *The Journal of Biological Chemistry*, *270*, 13899–13905.

Cohen, L. E., & Radovick, S. (2002). Molecular basis of combined pituitary hormone deficiencies. *Endocrine Reviews*, *23*, 431–442.

Cohen, R. N., & Wondisford, F. E. (2013). Chemistry and biosynthesis of thyrotropin. In L. E. Braverman & D. Cooper (Eds.), *Werner and Ingbar's the thyroid* (10th ed., pp. 149–161). Philadelphia, PA: Wolters Kluwer.

Coulthard, V. H., Matsuda, S., & Heery, D. M. (2003). An extended LXXLL motif sequence determines the nuclear receptor binding specificity of TRAP220. *The Journal of Biological Chemistry*, *278*, 10942–10951.

Cvoro, A., Tzagarakis-Foster, C., Tatomer, D., Paruthiyil, S., Fox, M. S., & Leitman, D. C. (2006). Distinct roles of unliganded and liganded estrogen receptors in transcriptional repression. *Molecular Cell*, *21*, 555–564.

Dasen, J. S., O'Connell, S. M., Flynn, S. E., Treier, M., Gleiberman, A. S., Szeto, D. P., et al. (1999). Reciprocal interactions of Pit1 and GATA2 mediate signaling gradient-induced determination of pituitary cell types. *Cell*, *97*, 587–598.

De Bosscher, K., Schmitz, M. L., Vanden Berghe, W., Plaisance, S., Fiers, W., & Haegeman, G. (1997). Glucocorticoid-mediated repression of nuclear factor-kappaB-dependent transcription involves direct interference with transactivation. *Proceedings of the National Academy of Sciences of the United States of America*, *94*, 13504–13509.

De Bosscher, K., Vanden Berghe, W., & Haegeman, G. (2001). Glucocorticoid repression of AP-1 is not mediated by competition for nuclear coactivators. *Molecular Endocrinology*, *15*, 219–227.

De Bosscher, K., Vanden Berghe, W., & Haegeman, G. (2003). The interplay between the glucocorticoid receptor and nuclear factor-kappaB or activator protein-1: Molecular mechanisms for gene repression. *Endocrine Reviews*, *24*, 488–522.

Decherf, S., Seugnet, I., Kouidhi, S., Lopez-Juarez, A., Clerget-Froidevaux, M. S., & Demeneix, B. A. (2010). Thyroid hormone exerts negative feedback on hypothalamic type 4 melanocortin receptor expression. *Proceedings of the National Academy of Sciences of the United States of America*, *107*, 4471–4476.

Dorfman, D. M., Wilson, D. B., Bruns, G. A., & Orkin, S. H. (1992). Human transcription factor GATA-2. Evidence for regulation of preproendothelin-1 gene expression in endothelial cells. *The Journal of Biological Chemistry*, *267*, 1279–1285.

Dupre, S. M., Guissouma, H., Flamant, F., Seugnet, I., Scanlan, T. S., Baxter, J. D., et al. (2004). Both thyroid hormone receptor (TR)beta 1 and TR beta 2 isoforms contribute to the regulation of hypothalamic thyrotropin-releasing hormone. *Endocrinology*, *145*, 2337–2345.

Edwards, J. G., Bahl, J. J., Flink, I. L., Cheng, S. Y., & Morkin, E. (1994). Thyroid hormone influences beta myosin heavy chain (beta MHC) expression. *Biochemical and Biophysical Research Communications*, *199*, 1482–1488.

Elagib, K. E., Mihaylov, I. S., Delehanty, L. L., Bullock, G. C., Ouma, K. D., Caronia, J. F., et al. (2008). Cross-talk of GATA-1 and P-TEFb in megakaryocyte differentiation. *Blood*, *112*, 4884–4894.

Feng, X., Jiang, Y., Meltzer, P., & Yen, P. M. (2000). Thyroid hormone regulation of hepatic genes in vivo detected by complementary DNA microarray. *Molecular Endocrinology*, *14*, 947–955.

Fisher, D. A., Schoen, E. J., La Franchi, S., Mandel, S. H., Nelson, J. C., Carlton, E. I., et al. (2000). The hypothalamic-pituitary-thyroid negative feedback control axis in children with treated congenital hypothyroidism. *The Journal of Clinical Endocrinology and Metabolism*, *85*, 2722–2727.

Fondell, J. D. (2013). The Mediator complex in thyroid hormone receptor action. *Biochimica et Biophysica Acta*, *1830*, 3867–3875.

Forrest, D., Hanebuth, E., Smeyne, R. J., Everds, N., Stewart, C. L., Wehner, J. M., et al. (1996). Recessive resistance to thyroid hormone in mice lacking thyroid hormone receptor beta: Evidence for tissue-specific modulation of receptor function. *The EMBO Journal*, *15*, 3006–3015.

Gereben, B., Zavacki, A. M., Ribich, S., Kim, B. W., Huang, S. A., Simonides, W. S., et al. (2008). Cellular and molecular basis of deiodinase-regulated thyroid hormone signaling. *Endocrine Reviews*, *29*, 898–938.

Gordon, D. F., Lewis, S. R., Haugen, B. R., James, R. A., McDermott, M. T., Wood, W. M., et al. (1997). Pit-1 and GATA-2 interact and functionally cooperate to activate the thyrotropin beta-subunit promoter. *The Journal of Biological Chemistry*, *272*, 24339–24347.

Gordon, D. F., Tucker, E. A., Tundwal, K., Hall, H., Wood, W. M., & Ridgway, E. C. (2006). MED220/thyroid receptor-associated protein 220 functions as a transcriptional coactivator with Pit-1 and GATA-2 on the thyrotropin-beta promoter in thyrotropes. *Molecular Endocrinology*, *20*, 1073–1089.

Gordon, D. F., Woodmansee, W. W., Black, J. N., Dowding, J. M., Bendrick-Peart, J., Wood, W. M., et al. (2002). Domains of Pit-1 required for transcriptional synergy with GATA-2 on the TSH beta gene. *Molecular and Cellular Endocrinology*, *196*, 53–66.

Gothe, S., Wang, Z., Ng, L., Kindblom, J. M., Barros, A. C., Ohlsson, C., et al. (1999). Mice devoid of all known thyroid hormone receptors are viable but exhibit disorders of the pituitary-thyroid axis, growth, and bone maturation. *Genes & Development*, *13*, 1329–1341.

Gupte, R., Muse, G. W., Chinenov, Y., Adelman, K., & Rogatsky, I. (2013). Glucocorticoid receptor represses proinflammatory genes at distinct steps of the transcription cycle. *Proceedings of the National Academy of Sciences of the United States of America*, *110*, 14616–14621.

Gurevich, I., Flores, A. M., & Aneskievich, B. J. (2007). Corepressors of agonist-bound nuclear receptors. *Toxicology and Applied Pharmacology*, *223*, 288–298.

Hah, N., Danko, C. G., Core, L., Waterfall, J. J., Siepel, A., Lis, J. T., et al. (2011). A rapid, extensive, and transient transcriptional response to estrogen signaling in breast cancer cells. *Cell*, *145*, 622–634.

Heldring, N., Isaacs, G. D., Diehl, A. G., Sun, M., Cheung, E., Ranish, J. A., et al. (2011). Multiple sequence-specific DNA-binding proteins mediate estrogen receptor signaling through a tethering pathway. *Molecular Endocrinology*, *25*, 564–574.

Hernandez, A., Morte, B., Belinchon, M. M., Ceballos, A., & Bernal, J. (2012). Critical role of types 2 and 3 deiodinases in the negative regulation of gene expression by T(3)in the mouse cerebral cortex. *Endocrinology*, *153*, 2919–2928.

Herrlich, P. (2001). Cross-talk between glucocorticoid receptor and AP-1. *Oncogene*, *20*, 2465–2475.

Hirahara, N., Sasaki, S., Matsushita, A., Kuroda, G., Nakamura, M. H., Yamashita, M., et al. (2016). Molecular aspect of the linear-log relationship between thyrotropin and T3 analyzed by the reconstitution system using reporter assay of the GATA2 promoter in CV1 Cells. In *Endocrine society meeting SUN-271*, Boston.

Ho, C. K., & Strauss, J. F., 3rd. (2004). Activation of the control reporter plasmids pRL-TK and pRL-SV40 by multiple GATA transcription factors can lead to aberrant normalization of transfection efficiency. *BMC Biotechnology*, *4*, 10.

Hollenberg, A. N. (2013). Regulation of thyrotropin secretion. In L. E. Braverman & D. Cooper (Eds.), *Werner and Ingbar's the thyroid* (10th ed., pp. 169–182). Philadelphia, PA: Wolters Kluwer.

Hollenberg, A. N., Monden, T., Flynn, T. R., Boers, M. E., Cohen, O., & Wondisford, F. E. (1995). The human thyrotropin-releasing hormone gene is regulated by thyroid hormone through two distinct classes of negative thyroid hormone response elements. *Molecular Endocrinology*, *9*, 540–550.

Huggins, G. S., Bacani, C. J., Boltax, J., Aikawa, R., & Leiden, J. M. (2001). Friend of GATA 2 physically interacts with chicken ovalbumin upstream promoter-TF2 (COUP-TF2) and COUP-TF3 and represses COUP-TF2-dependent activation of the atrial natriuretic factor promoter. *The Journal of Biological Chemistry*, *276*, 28029–28036.

Ikeda, M., Wilcox, E. C., & Chin, W. W. (1996). Different DNA elements can modulate the conformation of thyroid hormone receptor heterodimer and its transcriptional activity. *The Journal of Biological Chemistry*, *271*, 23096–23104.

Imagawa, S., Suzuki, N., Ohmine, K., Obara, N., Mukai, H. Y., Ozawa, K., et al. (2002). GATA suppresses erythropoietin gene expression through GATA site in mouse erythropoietin gene promoter. *International Journal of Hematology*, *75*, 376–381.

Ito, M., & Roeder, R. G. (2001). The TRAP/SMCC/Mediator complex and thyroid hormone receptor function. *Trends in Endocrinology and Metabolism*, *12*, 127–134.

Iwaki, H., Sasaki, S., Matsushita, A., Ohba, K., Matsunaga, H., Misawa, H., et al. (2014). Essential role of TEA domain transcription factors in the negative regulation of the MYH 7 gene by thyroid hormone and its receptors. *PLoS One, 9*, e88610.

Jiang, P., Hu, Q., Ito, M., Meyer, S., Waltz, S., Khan, S., et al. (2010). Key roles for MED1 LxxLL motifs in pubertal mammary gland development and luminal-cell differentiation. *Proceedings of the National Academy of Sciences of the United States of America, 107*, 6765–6770.

Jorgensen, J. S., Quirk, C. C., & Nilson, J. H. (2004). Multiple and overlapping combinatorial codes orchestrate hormonal responsiveness and dictate cell-specific expression of the genes encoding luteinizing hormone. *Endocrine Reviews, 25*, 521–542.

Kahaly, G. J., & Dillmann, W. H. (2005). Thyroid hormone action in the heart. *Endocrine Reviews, 26*, 704–728.

Kamei, Y., Xu, L., Heinzel, T., Torchia, J., Kurokawa, R., Gloss, B., et al. (1996). A CBP integrator complex mediates transcriptional activation and AP-1 inhibition by nuclear receptors. *Cell, 85*, 403–414.

Kashiwabara, Y., Sasaki, S., Matsushita, A., Nagayama, K., Ohba, K., Iwaki, H., et al. (2009). Functions of PIT1 in GATA2-dependent transactivation of the thyrotropin beta promoter. *Journal of Molecular Endocrinology, 42*, 225–237.

Kim, S. W., Ho, S. C., Hong, S. J., Kim, K. M., So, E. C., Christoffolete, M., et al. (2005). A novel mechanism of thyroid hormone-dependent negative regulation by thyroid hormone receptor, nuclear receptor corepressor (NCoR), and GAGA-binding factor on the rat cD44 promoter. *The Journal of Biological Chemistry, 280*, 14545–14555.

Kino, T., Slobodskaya, O., Pavlakis, G. N., & Chrousos, G. P. (2002). Nuclear receptor coactivator p160 proteins enhance the HIV-1 long terminal repeat promoter by bridging promoter-bound factors and the Tat-P-TEFb complex. *The Journal of Biological Chemistry, 277*, 2396–2405.

Kozai, M., Yamamoto, H., Ishiguro, M., Harada, N., Masuda, M., Kagawa, T., et al. (2013). Thyroid hormones decrease plasma 1alpha,25-dihydroxyvitamin D levels through transcriptional repression of the renal 25-hydroxyvitamin D3 1alpha-hydroxylase gene (CYP27B1). *Endocrinology, 154*, 609–622.

Kurokawa, R., Yu, V. C., Naar, A., Kyakumoto, S., Han, Z., Silverman, S., et al. (1993). Differential orientations of the DNA-binding domain and carboxy-terminal dimerization interface regulate binding site selection by nuclear receptor heterodimers. *Genes & Development*, 7, 1423–1435.

Lachize, S., Apostolakis, E. M., van der Laan, S., Tijssen, A. M., Xu, J., de Kloet, E. R., et al. (2009). Steroid receptor coactivator-1 is necessary for regulation of corticotropin-releasing hormone by chronic stress and glucocorticoids. *Proceedings of the National Academy of Sciences of the United States of America, 106*, 8038–8042.

Lai, A. Y., & Wade, P. A. (2011). Cancer biology and NuRD: A multifaceted chromatin remodelling complex. *Nature Reviews Cancer, 11*, 588–596.

Lazar, M. A. (2003). Thyroid hormone action: A binding contract. *The Journal of Clinical Investigation, 112*, 497–499.

Liden, J., Delaunay, F., Rafter, I., Gustafsson, J., & Okret, S. (1997). A new function for the C-terminal zinc finger of the glucocorticoid receptor. Repression of RelA transactivation. *The Journal of Biological Chemistry, 272*, 21467–21472.

Lim, H. W., Uhlenhaut, N. H., Rauch, A., Weiner, J., Hubner, S., Hubner, N., et al. (2015). Genomic redistribution of GR monomers and dimers mediates transcriptional response to exogenous glucocorticoid in vivo. *Genome Research, 25*, 836–844.

Lin, K. H., Shieh, H. Y., & Hsu, H. C. (2000). Negative regulation of the antimetastatic gene Nm23-H1 by thyroid hormone receptors. *Endocrinology, 141*, 2540–2547.

Liu, Y. Y., & Brent, G. A. (2008). Stealth sequences in reporter gene vectors confound studies of T3-regulated negative gene expression. *Thyroid, 18*, 593–595.

Lo, A., Zheng, W., Gong, Y., Crochet, J. R., & Halvorson, L. M. (2011). GATA transcription factors regulate LHbeta gene expression. *Journal of Molecular Endocrinology, 47,* 45–58.

Lopez, G., Schaufele, F., Webb, P., Holloway, J. M., Baxter, J. D., & Kushner, P. J. (1993). Positive and negative modulation of Jun action by thyroid hormone receptor at a unique AP1 site. *Molecular and Cellular Biology, 13,* 3042–3049.

Loyola, A., Bonaldi, T., Roche, D., Imhof, A., & Almouzni, G. (2006). PTMs on H3 variants before chromatin assembly potentiate their final epigenetic state. *Molecular Cell, 24,* 309–316.

Luecke, H. F., & Yamamoto, K. R. (2005). The glucocorticoid receptor blocks P-TEFb recruitment by NFkappaB to effect promoter-specific transcriptional repression. *Genes & Development, 19,* 1116–1127.

Maia, A. L., Harney, J. W., & Larsen, P. R. (1996). Is there a negative TRE in the luciferase reporter cDNA? *Thyroid, 6,* 325–328.

Mansen, A., Yu, F., Forrest, D., Larsson, L., & Vennstrom, B. (2001). TRs have common and isoform-specific functions in regulation of the cardiac myosin heavy chain genes. *Molecular Endocrinology, 15,* 2106–2114.

Martens, C., Bilodeau, S., Maira, M., Gauthier, Y., & Drouin, J. (2005). Protein-protein interactions and transcriptional antagonism between the subfamily of NGFI-B/Nur77 orphan nuclear receptors and glucocorticoid receptor. *Molecular Endocrinology, 19,* 885–897.

Matsunaga, H., Sasaki, S., Suzuki, S., Matsushita, A., Nakamura, H., Nakamura, H. M., et al. (2015). Essential role of GATA2 in the negative regulation of type 2 deiodinase gene by liganded thyroid hormone receptor beta2 in thyrotroph. *PLoS One, 10,* e0142400.

Matsushita, A., Sasaki, S., Kashiwabara, Y., Nagayama, K., Ohba, K., Iwaki, H., et al. (2007). Essential role of GATA2 in the negative regulation of thyrotropin beta gene by thyroid hormone and its receptors. *Molecular Endocrinology, 21,* 865–884.

Misawa, H., Sasaki, S., Matsushita, A., Ohba, K., Iwaki, H., Matsunaga, H., et al. (2012). Liganded thyroid hormone receptor inhibits phorbol 12-O-tetradecanoate-13-acetate-induced enhancer activity via firefly luciferase cDNA. *PLoS One, 7,* e28916.

Mladenova, V., & Russev, G. (2009). Organization of plasmid DNA into nucleosome-like structures after transfection in eukaryotic cells. *Biotechnology & Biotechnological Equipment, 23,* 1044–1047.

Naar, A. M., Boutin, J. M., Lipkin, S. M., Yu, V. C., Holloway, J. M., Glass, C. K., et al. (1991). The orientation and spacing of core DNA-binding motifs dictate selective transcriptional responses to three nuclear receptors. *Cell, 65,* 1267–1279.

Nagayama, K., Sasaki, S., Matsushita, A., Ohba, K., Iwaki, H., Matsunaga, H., et al. (2008). Inhibition of GATA2-dependent transactivation of the TSHbeta gene by ligand-bound estrogen receptor alpha. *The Journal of Endocrinology, 199,* 113–125.

Nakajima, Y., Yamada, M., Taguchi, R., Shibusawa, N., Ozawa, A., Tomaru, T., et al. (2012). NR4A1 (Nur77) mediates thyrotropin-releasing hormone-induced stimulation of transcription of the thyrotropin beta gene: Analysis of TRH knockout mice. *PLoS One, 7,* e40437.

Nakano, K., Matsushita, A., Sasaki, S., Misawa, H., Nishiyama, K., Kashiwabara, Y., et al. (2004). Thyroid-hormone-dependent negative regulation of thyrotropin beta gene by thyroid hormone receptors: Study with a new experimental system using CV1 cells. *The Biochemical Journal, 378,* 549–557.

Nwachukwu, J. C., Srinivasan, S., Bruno, N. E., Parent, A. A., Hughes, T. S., Pollock, J. A., et al. (2014). Resveratrol modulates the inflammatory response via an estrogen receptor-signal integration network. *eLife, 3,* e02057.

Ohba, K., Sasaki, S., Matsushita, A., Iwaki, H., Matsunaga, H., Suzuki, S., et al. (2011). GATA2 mediates thyrotropin-releasing hormone-induced transcriptional activation of the thyrotropin beta gene. *PLoS One, 6,* e18667.

Ortiga-Carvalho, T. M., Shibusawa, N., Nikrodhanond, A., Oliveira, K. J., Machado, D. S., Liao, X. H., et al. (2005). Negative regulation by thyroid hormone receptor requires an intact coactivator-binding surface. *The Journal of Clinical Investigation, 115*, 2517–2523.

Pelletier, P., Gauthier, K., Sideleva, O., Samarut, J., & Silva, J. E. (2008). Mice lacking the thyroid hormone receptor-alpha gene spend more energy in thermogenesis, burn more fat, and are less sensitive to high-fat diet-induced obesity. *Endocrinology, 149*, 6471–6486.

Pennathur, S., Madison, L. D., Kay, T. W., & Jameson, J. L. (1993). Localization of promoter sequences required for thyrotropin-releasing hormone and thyroid hormone responsiveness of the glycoprotein hormone alpha-gene in primary cultures of rat pituitary cells. *Molecular Endocrinology, 7*, 797–805.

Perez-Juste, G., Garcia-Silva, S., & Aranda, A. (2000). An element in the region responsible for premature termination of transcription mediates repression of c-myc gene expression by thyroid hormone in neuroblastoma cells. *The Journal of Biological Chemistry, 275*, 1307–1314.

Pfahl, M. (1993). Nuclear receptor/AP-1 interaction. *Endocrine Reviews, 14*, 651–658.

Pietrzak, M., & Puzianowska-Kuznicka, M. (2008). Triiodothyronine utilizes phosphatidylinositol 3-kinase pathway to activate anti-apoptotic myeloid cell leukemia-1. *Journal of Molecular Endocrinology, 41*, 177–186.

Privalsky, M. L., Lee, S., Hahm, J. B., Young, B. M., Fong, R. N., & Chan, I. H. (2009). The p160 coactivator PAS-B motif stabilizes nuclear receptor binding and contributes to isoform-specific regulation by thyroid hormone receptors. *The Journal of Biological Chemistry, 284*, 19554–19563.

Reichardt, H. M., Kaestner, K. H., Tuckermann, J., Kretz, O., Wessely, O., Bock, R., et al. (1998). DNA binding of the glucocorticoid receptor is not essential for survival. *Cell, 93*, 531–541.

Rogatsky, I., Luecke, H. F., Leitman, D. C., & Yamamoto, K. R. (2002). Alternate surfaces of transcriptional coregulator GRIP1 function in different glucocorticoid receptor activation and repression contexts. *Proceedings of the National Academy of Sciences of the United States of America, 99*, 16701–16706.

Rollins, D. A., Coppo, M., & Rogatsky, I. (2015). Minireview: Nuclear receptor coregulators of the p160 family: Insights into inflammation and metabolism. *Molecular Endocrinology, 29*, 502–517.

Rouf, R., Greytak, S., Wooten, E. C., Wu, J., Boltax, J., Picard, M., et al. (2008). Increased FOG-2 in failing myocardium disrupts thyroid hormone-dependent SERCA2 gene transcription. *Circulation Research, 103*, 493–501.

Saatcioglu, F., Deng, T., & Karin, M. (1993). A novel cis element mediating ligand-independent activation by c-ErbA: Implications for hormonal regulation. *Cell, 75*, 1095–1105.

Sadow, P. M., Koo, E., Chassande, O., Gauthier, K., Samarut, J., Xu, J., et al. (2003). Thyroid hormone receptor-specific interactions with steroid receptor coactivator-1 in the pituitary. *Molecular Endocrinology, 17*, 882–894.

Sanchez-Pacheco, A., & Aranda, A. (2003). Binding of the thyroid hormone receptor to a negative element in the basal growth hormone promoter is associated with histone acetylation. *The Journal of Biological Chemistry, 278*, 39383–39391.

Sanchez-Pacheco, A., Palomino, T., & Aranda, A. (1995). Negative regulation of expression of the pituitary-specific transcription factor GHF-1/Pit-1 by thyroid hormones through interference with promoter enhancer elements. *Molecular and Cellular Biology, 15*, 6322–6330.

Santos, G. M., Afonso, V., Barra, G. B., Togashi, M., Webb, P., Neves, F. A., et al. (2006). Negative regulation of superoxide dismutase-1 promoter by thyroid hormone. *Molecular Pharmacology, 70*, 793–800.

Santos, G. M., Fairall, L., & Schwabe, J. W. (2011). Negative regulation by nuclear receptors: A plethora of mechanisms. *Trends in Endocrinology and Metabolism, 22*, 87–93.

Sasaki, S., Lesoon-Wood, L. A., Dey, A., Kuwata, T., Weintraub, B. D., Humphrey, G., et al. (1999). Ligand-induced recruitment of a histone deacetylase in the negative-feedback regulation of the thyrotropin beta gene. *The EMBO Journal, 18*, 5389–5398.

Sasaki, S., Matsushita, A., & Nakamura, H. (2011). Negative regulation of the thyrotropin beta gene by thyroid hormone. In E. Diamanti-Kandarakis (Ed.), *Contemporary aspects of endocrinology* (pp. 101–138): InTech. http://www.intechopen.com/books/contemporary-aspects-of-*endocrinology*.

Satoh, T., Yamada, M., Iwasaki, T., & Mori, M. (1996). Negative regulation of the gene for the preprothyrotropin-releasing hormone from the mouse by thyroid hormone requires additional factors in conjunction with thyroid hormone receptors. *The Journal of Biological Chemistry, 271*, 27919–27926.

Schones, D. E., Cui, K., Cuddapah, S., Roh, T. Y., Barski, A., Wang, Z., et al. (2008). Dynamic regulation of nucleosome positioning in the human genome. *Cell, 132*, 887–898.

Sharma, P., Thakran, S., Deng, X., Elam, M. B., & Park, E. A. (2013). Nuclear corepressors mediate the repression of phospholipase A2 group IIa gene transcription by thyroid hormone. *The Journal of Biological Chemistry, 288*, 16321–16333.

Shen, X., Li, Q. L., Brent, G. A., & Friedman, T. C. (2004). Thyroid hormone regulation of prohormone convertase 1 (PC1): Regional expression in rat brain and in vitro characterization of negative thyroid hormone response elements. *Journal of Molecular Endocrinology, 33*, 21–33.

Shen, X., Li, Q. L., Brent, G. A., & Friedman, T. C. (2005). Regulation of regional expression in rat brain PC2 by thyroid hormone/characterization of novel negative thyroid hormone response elements in the PC2 promoter. *American Journal of Physiology Endocrinology and Metabolism, 288*, E236–245.

Shupnik, M. A. (2000). Thyroid hormone suppression of pituitary hormone gene expression. *Reviews in Endocrine & Metabolic Disorders, 1*, 35–42.

Shupnik, M. A., Ridgway, E. C., & Chin, W. W. (1989). Molecular biology of thyrotropin. *Endocrine Reviews, 10*, 459–475.

Soboleva, T. A., Nekrasov, M., Ryan, D. P., & Tremethick, D. J. (2014). Histone variants at the transcription start-site. *Trends in Genetics: TIG, 30*, 199–209.

Steger, D. J., Hecht, J. H., & Mellon, P. L. (1994). GATA-binding proteins regulate the human gonadotropin alpha-subunit gene in the placenta and pituitary gland. *Molecular and Cellular Biology, 14*, 5592–5602.

Steinfelder, H. J., Radovick, S., & Wondisford, F. E. (1992). Hormonal regulation of the thyrotropin beta-subunit gene by phosphorylation of the pituitary-specific transcription factor Pit-1. *Proceedings of the National Academy of Sciences of the United States of America, 89*, 5942–5945.

Subramanian, V., Fields, P. A., & Boyer, L. A. (2015). H2A.Z: A molecular rheostat for transcriptional control. *F1000prime Reports, 7*, 01.

Tagami, T., Park, Y., & Jameson, J. L. (1999). Mechanisms that mediate negative regulation of the thyroid-stimulating hormone alpha gene by the thyroid hormone receptor. *The Journal of Biological Chemistry, 274*, 22345–22353.

Takeda, K., Sakurai, A., DeGroot, L. J., & Refetoff, S. (1992). Recessive inheritance of thyroid hormone resistance caused by complete deletion of the protein-coding region of the thyroid hormone receptor-beta gene. *The Journal of Clinical Endocrinology and Metabolism, 74*, 49–55.

Takeuchi, Y., Murata, Y., Sadow, P., Hayashi, Y., Seo, H., Xu, J., et al. (2002). Steroid receptor coactivator-1 deficiency causes variable alterations in the modulation of T(3)-regulated transcription of genes in vivo. *Endocrinology, 143*, 1346–1352.

Thompson, K. L., Santon, J. B., Shephard, L. B., Walton, G. M., & Gill, G. N. (1992). A nuclear protein is required for thyroid hormone receptor binding to an inhibitory half-site in the epidermal growth factor receptor promoter. *Molecular Endocrinology, 6*, 627–635.

Tillman, J. B., Crone, D. E., Kim, H. S., Sprung, C. N., & Spindler, S. R. (1993). Promoter independent down-regulation of the firefly luciferase gene by T3 and T3 receptor in CV1 cells. *Molecular and Cellular Endocrinology*, *95*, 101–109.

Tomic-Canic, M., Day, D., Samuels, H. H., Freedberg, I. M., & Blumenberg, M. (1996). Novel regulation of keratin gene expression by thyroid hormone and retinoid receptors. *The Journal of Biological Chemistry*, *271*, 1416–1423.

Turowska, O., Nauman, A., Pietrzak, M., Poplawski, P., Master, A., Nygard, M., et al. (2007). Overexpression of E2F1 in clear cell renal cell carcinoma: A potential impact of erroneous regulation by thyroid hormone nuclear receptors. *Thyroid*, *17*, 1039–1048.

Uhlenhaut, N. H., Barish, G. D., Yu, R. T., Downes, M., Karunasiri, M., Liddle, C., et al. (2013). Insights into negative regulation by the glucocorticoid receptor from genome-wide profiling of inflammatory cistromes. *Molecular Cell*, *49*, 158–171.

Umesono, K., Murakami, K. K., Thompson, C. C., & Evans, R. M. (1991). Direct repeats as selective response elements for the thyroid hormone, retinoic acid, and vitamin D3 receptors. *Cell*, *65*, 1255–1266.

van der Laan, S., Lachize, S. B., Vreugdenhil, E., de Kloet, E. R., & Meijer, O. C. (2008). Nuclear receptor coregulators differentially modulate induction and glucocorticoid receptor-mediated repression of the corticotropin-releasing hormone gene. *Endocrinology*, *149*, 725–732.

Vicente, C., Conchillo, A., Garcia-Sanchez, M. A., & Odero, M. D. (2012). The role of the GATA2 transcription factor in normal and malignant hematopoiesis. *Critical Reviews in Oncology/Hematology*, *82*, 1–17.

Villa, A., Santiago, J., Belandia, B., & Pascual, A. (2004). A response unit in the first exon of the beta-amyloid precursor protein gene containing thyroid hormone receptor and Sp1 binding sites mediates negative regulation by 3,5,3′-triiodothyronine. *Molecular Endocrinology*, *18*, 863–873.

Weiss, R. E., Forrest, D., Pohlenz, J., Cua, K., Curran, T., & Refetoff, S. (1997). Thyrotropin regulation by thyroid hormone in thyroid hormone receptor beta-deficient mice. *Endocrinology*, *138*, 3624–3629.

Weiss, R. E., Gehin, M., Xu, J., Sadow, P. M., O'Malley, B. W., Chambon, P., et al. (2002). Thyroid function in mice with compound heterozygous and homozygous disruptions of SRC-1 and TIF-2 coactivators: Evidence for haploinsufficiency. *Endocrinology*, *143*, 1554–1557.

Weiss, R. E., Xu, J., Ning, G., Pohlenz, J., O'Malley, B. W., & Refetoff, S. (1999). Mice deficient in the steroid receptor co-activator 1 (SRC-1) are resistant to thyroid hormone. *The EMBO Journal*, *18*, 1900–1904.

Winnay, J. N., Xu, J., O'Malley, B. W., & Hammer, G. D. (2006). Steroid receptor coactivator-1-deficient mice exhibit altered hypothalamic-pituitary-adrenal axis function. *Endocrinology*, *147*, 1322–1332.

Wondisford, F. E., Farr, E. A., Radovick, S., Steinfelder, H. J., Moates, J. M., McClaskey, J. H., et al. (1989). Thyroid hormone inhibition of human thyrotropin beta-subunit gene expression is mediated by a cis-acting element located in the first exon. *The Journal of Biological Chemistry*, *264*, 14601–14604.

Wood, W. M., Ocran, K. W., Kao, M. Y., Gordon, D. F., Alexander, L. M., Gutierrez-Hartmann, A., et al. (1990). Protein factors in thyrotropic tumor nuclear extracts bind to a region of the mouse thyrotropin beta-subunit promoter essential for expression in thyrotropes. *Molecular Endocrinology*, *4*, 1897–1904.

Wood, W. M., Sarapura, V. D., Dowding, J. M., Woodmansee, W. W., Haakinson, D. J., Gordon, D. F., et al. (2002). Early gene expression changes preceding thyroid hormone-induced involution of a thyrotrope tumor. *Endocrinology*, *143*, 347–359.

Wright, C. E., Haddad, F., Qin, A. X., Bodell, P. W., & Baldwin, K. M. (1999). In vivo regulation of beta-MHC gene in rodent heart: Role of T3 and evidence for an upstream enhancer. *The American Journal of Physiology*, *276*, C883–891.

Wu, J., Li, Y., Dietz, J., & Lala, D. S. (2004). Repression of p65 transcriptional activation by the glucocorticoid receptor in the absence of receptor-coactivator interactions. *Molecular Endocrinology*, *18*, 53–62.

Xu, J., Qiu, Y., DeMayo, F. J., Tsai, S. Y., Tsai, M. J., & O'Malley, B. W. (1998). Partial hormone resistance in mice with disruption of the steroid receptor coactivator-1 (SRC-1) gene. *Science*, *279*, 1922–1925.

Yang, Z., Hong, S. H., & Privalsky, M. L. (1999). Transcriptional anti-repression. Thyroid hormone receptor beta-2 recruits SMRT corepressor but interferes with subsequent assembly of a functional corepressor complex. *The Journal of Biological Chemistry*, *274*, 37131–37138.

Yanisch-Perron, C., Vieira, J., & Messing, J. (1985). Improved M13 phage cloning vectors and host strains: Nucleotide sequences of the M13mp18 and pUC19 vectors. *Gene*, *33*, 103–119.

Yen, P. M. (2001). Physiological and molecular basis of thyroid hormone action. *Physiological Reviews*, *81*, 1097–1142.

Yen, P. M., Feng, X., Flamant, F., Chen, Y., Walker, R. L., Weiss, R. E., et al. (2003). Effects of ligand and thyroid hormone receptor isoforms on hepatic gene expression profiles of thyroid hormone receptor knockout mice. *EMBO Reports*, *4*, 581–587.

Ying, H., Furuya, F., Willingham, M. C., Xu, J., O'Malley, B. W., & Cheng, S. Y. (2005). Dual functions of the steroid hormone receptor coactivator 3 in modulating resistance to thyroid hormone. *Molecular and Cellular Biology*, *25*, 7687–7695.

Yusta, B., Alarid, E. T., Gordon, D. F., Ridgway, E. C., & Mellon, P. L. (1998). The thyrotropin beta-subunit gene is repressed by thyroid hormone in a novel thyrotrope cell line, mouse T alphaT1 cells. *Endocrinology*, *139*, 4476–4482.

CHAPTER SIX

Thyroid Autoimmunity: An Interplay of Factors

Stephen J. Merrill[*,1], **Sarah B. Minucci**[†]
[*]Marquette University, Milwaukee, WI, United States
[†]Virginia Commonwealth University, Richmond, VA, United States
[1]Corresponding author: e-mail address: stephen.merrill@marquette.edu

Contents

1. Introduction 130
2. Interacting Factors 132
   2.1 White Adipose Tissue (WAT) 132
   2.2 Leptin and Other Adipokines 133
   2.3 Vitamin D 135
   2.4 Vitamin D Interaction With WAT 136
   2.5 Vitamin D Interaction With the Immune System and Autoimmunity 136
   2.6 Vitamin D and Thyroid Autoimmunity 136
   2.7 The Triggering Event: Breaking of Tolerance 137
3. Results and Discussion 138
4. Conclusion 139
References 139

Abstract

The literature on thyroid autoimmunity has identified many potential factors at play for the initiation and progression of autoimmune thyroid diseases. These factors include genetic susceptibility, environmental factors, some drugs, iodine and selenium, infection, molecular mimics, and immune system defects. The sheer number of feasible factors makes sorting out the necessary agents from the fellow travelers difficult. In addition, many of these factors have the capability to interact—further confusing the picture. Another difficulty in interpreting these data is that most proposed mechanisms are not able to accomplish the triggering event in which the tolerance to self-antigens is actually overcome. In addition, some findings may be describing the conditions present after a disease is diagnosed and may be consequences of the disease rather than a cause. Recent description of the role of adipokines, which include leptin, tumor necrosis factor-alpha, and interleukin-6, in contributing to the inflammatory environment of the thyroid, along with the presence of thyroid Toll-like receptors for pathogen-associated patterns have the potential to deliver that necessary adjuvant signal to break tolerance, seen as necessary in animal autoimmune models. An additional

Vitamins and Hormones, Volume 106
ISSN 0083-6729
http://dx.doi.org/10.1016/bs.vh.2017.07.001

factor, vitamin D_3, due to its interaction both with white adipose tissue (WAT) and the immune system, has a complicated and somewhat controversial story with respect to thyroid autoimmunity. Conflicting results can result when not all factors are considered together.

Aims: To describe the many factors at play in thyroid autoimmunity and how they interact.

Conclusion: Thyroid autoimmunity is the result of an interplay of factors, with adipokines produced by WAT and vitamin D providing immune modulating signals external to the thyroid, while thyrocyte innate responses to environmental conditions provide the necessary adjuvant signal. Shaping the response to be reactive to particular self-antigens and likelihood of a response are due to genetics and molecular mimics.

ABBREVIATIONS

AITD autoimmune thyroid disease
BMI body mass index
CYP27B1 1-alpha-hydroxylase
IL-6 interleukin-6
PAMPs pathogen-associated molecular patterns
PRRs pattern recognition receptors
TG thyroglobulin
TNF-α tumor necrosis factor alpha
TPO thyroid peroxidase
TSH thyroid-stimulating hormone
TSHR thyroid-stimulating hormone receptor
UVB ultraviolet-B radiation
VDR vitamin D receptor
vitamin D 1,25-dihydroxy vitamin D (1,25-dihydroxycholecalciferol)
WAT white adipose tissue

1. INTRODUCTION

There are several autoimmune diseases associated with the thyroid gland, the most common being Graves' disease and Hashimoto's thyroiditis (autoimmune thyroiditis). Graves' disease results in hyperthyroidism with an estimated prevalence of 80/100,000/year in women and 8/100,000/year in men in western countries (McGrogan, Seaman, Wright, & de Vries, 2008). The primary cause of hypothyroidism in the west (where dietary iodine is generally at sufficient levels) is Hashimoto's disease. Hypothyroidism has an estimated incidence of 498.4/100,000/year for females and 2.2/100,000/year for men in western countries (McGrogan et al., 2008).

For this review, we will also be discussing postpartum thyroiditis, one of the several thyroid conditions associated with pregnancy (Pearce, 2015). Taken together, autoimmune thyroid diseases (AITDs) are the most common autoimmune diseases in humans (Hayter & Cook, 2012).

Autoimmunity requires a self-antigen, to which one is normally tolerant, and a process that results in the breaking of that tolerance—resulting in an adaptive immune response to that self-antigen (McLachlan & Rapoport, 2014). Several antigens present in the thyroid have the potential to be a self-antigen involved in autoimmunity. These include thyroglobulin (Tg), thyroid peroxidase (TPO), and the thyroid-stimulating hormone receptor (TSHR). There appears to be several factors potentially involved in the breaking of tolerance to these antigens, including infection (Root-Berstein & Fairweather, 2014; Saeki & Ishihara, 2014), genetic predisposition (Lee, Li, Hammerstad, Stefan, & Tomer, 2015; Medici et al., 2014; Tomer, 2014), environment and environmental chemicals (Bogdanos et al., 2013; Burek & Talor, 2009; Duntas, 2011; Hansen, Brix, Iachine, Kyvik, & Hegedüs, 2006), some drugs (Fabiana et al., 2015) including IFN-α (Tomer, Blackard, & Akeno, 2007), molecular mimics (Benvenga & Guarneri, 2016), dietary iodine (too little or too much) (Burek & Talor, 2009; Hansen et al., 2006; Tomer & Huber, 2009), and selenium (Guastamacchia, Angelo Giagulli, Licchelli, & Triggiani, 2015).

Although the processes involved in Graves' disease and Hashimoto's disease are different and utilize different aspects of the immune response for their immune-mediated destruction (antibody for Graves' and cell-mediated immunity for Hashimoto's), in each there are characteristic antibodies produced that can be monitored and used for diagnosis in each (Damoiseaux, 2013). In Graves' disease, an autoantibody is made to the TSHR. Binding of antibody to that receptor gives a thyroid-stimulating hormone (TSH)—independent signal to the thyroid to increase thyroid hormone production, resulting in hyperthyroidism. The damage to the thyroid in Hashimoto's disease, in addition to cellular mechanisms, may also involve destruction due to the presence of antibodies to Tg and TPO through antibody-dependent cell-mediated cytotoxicity and triggering of the complement cascade (Antonelli, Ferrari, Corrado, Di Domenicantonio, & Fallahi, 2015; Rapoport & McLachlan, 2014). The process by which this immune destruction is initiated in Hashimoto's is a fundamental question of autoimmunity. In postpartum thyroiditis, autoimmunity emerges as a consequence of pregnancy (Pearce, 2015), associated with thyroid dysfunction. Both hyperthyroidism and hypothyroidism can be present during pregnancy, and most women return to normal thyroid hormone levels (euthyroid) after 1 year. However, a fraction develops autoimmune disease (Stagnaro-Green et al., 2010). Of interest in this review is the observation that the placenta is a source of leptin (Masuzaki et al., 1997; Sagawa et al., 2002) which has the potential to affect several aspects of the systems involved.

There are several basic questions one can ask to address thyroid autoimmunity, and autoimmunity in general (Smilek & St. Clair, 2015). One is "why do we have autoimmunity at all" and the second is "why is autoimmune disease relatively rare." Given that all humans have the same antigens present in the thyroid and are exposed to similar environmental factors, why do so few individuals develop AITD? While genetic predisposition can provide a partial answer, interaction with environmental factors are always involved (Guarneri & Benvenga, 2007; Limbach et al., 2016), and understanding the interaction of the factors is needed.

An additional observation also needs to be considered. It has been suggested that the thyroid is one of the most sensitive sites for autoimmunity (Merrill & Mu, 2015), meaning that if autoimmunity is present anywhere, it is likely to also be present in the thyroid. And, it is also likely to have been present in the thyroid first.

2. INTERACTING FACTORS

A suggestion that not all factors involved have been identified above can be seen in the inability to answer several basic questions in the area. These include much higher incidence in women generally thought to be due to immune response, genetic, and hormonal factors (Brunelleschi, 2015; Rubtsov, Rubtsova, Kappler, & Marrack, 2010; Rubtsova, Marrack, & Rubtsov, 2015) and the pattern of autoimmune diseases in multiple autoimmune syndrome (Boccuti et al., 2016; Humbert & Dupond, 1988). The situation appears similar to that of complex multifactorial disorders in genetics studies, where a multitude of potential factors are identified. A complete understanding requires delineating the pathways and the timing at which the factors appear and how they interact (Baron, 2001).

2.1 White Adipose Tissue (WAT)

Fat in the body is stored in two main forms: brown adipose tissue and WAT. Brown adipose tissue is responsible for thermogenesis and displays an inverse relationship with body mass index (BMI) (Cypess et al., 2009). For our purposes, WAT is the most important due to its secretion of adipokines, interaction with vitamin D_3, and its relationship with obesity (Versini, Jeandel, Rosenthal, & Shoenfeld, 2014). Estrogen favors WAT deposition in subcutaneous regions, whereas in men, having more testosterone, WAT is found more in visceral and abdominal regions (Grove, Fried, Greenberg, Xiao, & Clegg, 2010; Ma, Lee, Chisholm, & James, 2015). This sexual dimorphism is important for the sex preference of autoimmune disease (and AITD in particular) due to the fact that subcutaneous adipose tissue secretions are at a higher level (Van Harmelen et al., 1998). In addition, other immunologically active cytokines are positively associated with BMI (Lehr et al., 2012; Sharabiani et al., 2011). The connection between thyroid disruption and diabetes is an important aspect of the discussion below (Hage, Zantout, & Azar, 2011). It is expected that in diabetics, the thyroid disruption, which sets the stage for AITD, would be observed.

2.2 Leptin and Other Adipokines

Leptin, a hormone classified in the cytokine family, was discovered in 1994 through a genetic defect responsible for severe obesity in certain mice (Fantuzzi & Faggiono, 2000; Zhang et al., 1994). Found in humans as well, this hormone is an afferent signal in a feedback loop that controls energy intake and expenditure through appetite. Initially leptin was thought to be the key to control of food intake and feeling full at each meal, but it is now accepted to be part of a long-term system regulating energy storage, as well as an important player in the immune system (Procaccini et al., 2017). Leptin is one of the adipokines, a collection of more than 50 hormones released primarily by WAT (Coelho, Oliveira, & Fernandes, 2013), which include adiponectin, interleukin-6 (IL-6), and tumor necrosis factor alpha (TNF-α).

Secretion of leptin is generally associated with BMI (Klein, Coppack, Mohamed-Ali, & Landt, 1996), but that association shows strong differences between men and women (Kennedy et al., 1997). The higher leptin levels in women (even with the same BMI as men) (Conde et al., 2014) seems to be a result of the higher secretion rate in subcutaneous adipose tissue and may be responsible for the higher level of inflammation in women as indicated by higher levels of adipokines (Maachi et al., 2004).

2.2.1 The Pleiotropic Role of Leptin

A decrease in leptin levels indicates what can be called a "response to starvation," in which the hypothalamus receives signals from the dropping leptin that there is not enough energy stored. This results in stimulating a response by conserving energy already in the body, as well as encouraging the individual to increase nutritional consumption. On the other hand, an increase in leptin indicates a "response to obesity" when an excess amount of energy is stored, signaling to the hypothalamus to utilize more internal energy and reduce intake (Friedman, 2002).

Not only does leptin have energy-regulating properties, but as a cytokine, it also influences behavior of the immune system. Leptin is said to have essentially proinflammatory effects in the body (Conde et al., 2014) and these effects have been found to be both direct and indirect. Because it is an adipokine, secreted by adipocytes, and also a hormone that regulates immune responses, leptin is a connection between metabolism and immune homeostasis (La Cava & Matarese, 2004). In fact, both hyper- and hypoleptinemia are associated with abnormalities of inflammation and the immune system (Fantuzzi & Faggiono, 2000).

With high levels of leptin, inflammation is typically more prevalent (Conde et al., 2014; Procaccini, Pucino, Mantzoros, & Matarese, 2015). In both experimental models and research in patients with inflammatory diseases, including muscular sclerosis and joint inflammation, increased leptin concentration has been found. In fact, studies suggest that leptin plays a central role in the formation of several inflammatory disorders (Procaccini et al., 2015). On the other hand, hypoleptinemia is associated with a low-functioning immune system and increased susceptibility to infection (Fantuzzi & Faggiono, 2000; Friedman, 2002; Loffreda et al., 1998).

Leptin and the associated cytokines are responsible for the communication between other cells in the immune system, including B and T lymphocytes, which perform the direct actions of an adaptive immune response. It has been shown that leptin and the other proinflammatory adipokines suppress the secretion of adiponectin (Deng & Scherer, 2010; Matsubara, Maruoka, & Katayose, 2002) an adipokine involved in insulin sensitivity and considered antiinflammatory. Leptin upregulates secretion of IL-6, IL-12, and TNF-α particularly in WAT (Loffreda et al., 1998). This aspect of leptin alone suggests an important role in several health aspects related to obesity, especially type II diabetes mellitus (Meek & Morton, 2016) and in pregnancy.

2.2.2 Adipokines and Autoimmunity

The association of leptin with autoimmunity has been well recognized (Duntas & Biondi, 2013; Marzullo et al., 2010; Merrill & Mu, 2015), though more often its correlates, obesity or BMI (Biondi, 2010), are identified.

Besides increasing overall inflammation, including in the thyroid, leptin has been shown to disturb thyroid function through disruption of the HPT axis through thyroid-releasing hormone in the hypothalamus (Duntas & Biondi, 2013). The resulting disruption of thyroid functioning is a risk factor for developing overt hypothyroidism (Rosário, Carvalho, & Calsolari, 2016). The adipokines, especially TNF-α, have been identified with increased AITD in a mouse model (Chen, Wei, Sharp, & Braley-Mullen, 2007; Zaccone, Fehérvári, & Cooke, 2003).

2.3 Vitamin D

Vitamin D_3 (cholecalciferol) is primarily produced in the skin when in the presence of ultraviolet-B radiation (UVB). Vitamin D_3 and a related form vitamin D_2 can also be obtained through supplementation and diet. Cholecalciferol is hydroxylated in the liver and enzymatically transformed in the kidney by 1-alpha-hydroxylase (CYP27B1) to form 1,25-dihydroxycholecalciferol (1,25$(OH)_2D_3$), the biologically active form, which has a relatively short half-life. The parathyroid hormone is one of the controllers of CYP27B1 activity, and thus levels of the active form of vitamin D in the blood. CYP27B1 can also be utilized in cells outside the kidney, resulting in active vitamin D in an autocrine or local manner. A seemly recent development is the declining levels of the vitamin in populations (Adams & Hewison, 2010). This decline seems to mirror increases in autoimmune diseases over time (Lerner, Jeremias, & Matthias, 2015) and also increases in certain cancers (Merrill, Ashrafi, Subramanian, & Godar, 2015).

Vitamin D is a fat-soluble vitamin whose primary role is calcium homeostasis, keeping the Ca^{2+} concentration in normal limits. This process involves the parathyroid in a feedback loop, which senses hypocalcemia and increases parathyroid hormone production. This increases 1,25 $(OH)_2D_3$ in the blood. This results in increased absorption of calcium from the gut. The rising Ca^{2+} levels reduce parathyroid activity. Of interest here to note is that autoimmune parathyroid disease can occur (Betterle, Garelli, & Presotto, 2014) but is not seen as related to AITD. Vitamin D is seen as a pleiotropic hormone (Verstuyf, Carmeliet, Bouillon, & Mathieu, 2010) due to its many effects beyond calcium homeostasis described below. Effects of vitamin D are through the vitamin D receptor (VDR), which are found in many cells, including those of the immune system and the thyroid (Haussler et al., 1998).

2.4 Vitamin D Interaction With WAT

One indication of the role of vitamin D in this story of autoimmunity is the interaction of vitamin D with WAT (Abbas, 2017; Mutt, Hyppönen, Saamlo, Järvellin, & Herzig, 2014). Adipose tissue is a primary store of vitamin D (Landrier, Karkeni, Marcotorchino, Bonnet, & Tourniaire, 2016), and the relationship of vitamin D with obesity is controversial (probably due to methodological differences and the store of vitamin D in fat). Supplemental doses of vitamin D in type 2 diabetics have seen significant improvement in A1 (Breslavsky et al., 2013), but a mechanism was not described. It is likely that a reduction in leptin was involved (Menendez, Lage, Peino, Baldelli, & Concheiro, 2001). The relationships between vitamin D levels, adiposity, and parathyroid hormone levels in an older population has been described in Snijder et al. (2005), indicating the close linkage between these quantities.

2.5 Vitamin D Interaction With the Immune System and Autoimmunity

Vitamin D has direct effects on the immune system which relates to the initiation and progression of autoimmune disease (Aranow, 2011; Hayes et al., 2015). In general, vitamin D is seen as related to immune suppression, as part of the body's response to UVB. It has been recognized that vitamin D participates in the innate immune response through the Toll-like receptors (Liu et al., 2006), while reducing inflammatory markers (Ojaimi et al., 2013). This reduction produces a more tolerant immunologic state (Prietl, Treiber, Pieber, & Amrein, 2013). On the other hand, low vitamin D has been associated with increased autoimmune disease in general (Agmon-Levin, Theodor, Segal, & Shoenfeld, 2013; Skaaby, Husemoen, Thuesen, & Linneberg, 2015).

2.6 Vitamin D and Thyroid Autoimmunity

The role of vitamin D in thyroid autoimmunity is controversial (D'Aurizio, Villalta, Metus, Doretto, & Tozzoli, 2015)—probably due to viewing its effect in isolation of other factors. On one hand, vitamin D has been seen to affect the thyroid directly through changes in TSH (Barchetta et al., 2015). This event is often seen as necessary early step in order for thyroid autoimmunity to exist. Other aspects of the association of vitamin D with thyroid autoimmunity include existence of antithyroid antibody, increased thyroid volume, and the association of VDR gene polymorphisms with

AITD (Bizzaro & Shoenfeld, 2015; Kivity et al., 2011). There have been clear associations between low vitamin D and thyroid autoimmunity—especially Hashimoto's (Kim, 2016). A recent metaanalysis of the association between vitamin D and AITD (Wang et al., 2015) made a strong case for a relationship of low vitamin D with AITD. On the other hand, a study by Effraimidis, Badenhoop, Tijssen, and Wiersinga (2012) made the case that vitamin D did not play a role in early thyroid autoimmunity. One should note, however, that the BMI of the participants was suggested that they were lean and not, as a group, affected by high leptin levels. This study points to the need for viewing the factors together—and also seeing that the real role of vitamin D may be to lower the inflammation associated with leptin.

2.7 The Triggering Event: Breaking of Tolerance

Besides the role of leptin in providing a signal which triggers inflammation and affects the HPT axis, the native thyroid cells, thyrocytes, also play a role in the innate immune system. Thyrocytes have the ability to recognize and respond to pathogen-associated molecular patterns (PAMPs) (Kawashima, Tanigawa, Akama, & Yoshihara, 2011) through their pattern recognition receptors (PRRs). The PRRs include Toll-like receptors, Nod-like receptors, and many others. Thyrocytes also have the ability to respond to damage-associated molecular patterns (DAMPs) coming from cellular damage (Kawashima et al., 2013). The response of thyrocytes to these patterns plays two roles in the autoimmune setting. One is to further increase the presence of inflammatory cytokines, TNF-α, IL-1, and IL-6, to promote a specific immune response. Toll-like receptors have been described as adjuvant receptors (Kaisho & Akira, 2002). The second role is the presentation of self-antigens in association with MHC II expression on thyroid cells (Kawashima et al., 2013). Inflammatory molecules have been recognized as an important aspect in the breaking of tolerance in the context of thyroid autoimmunity with like-self (or self) antigens (Akeno et al., 2011). The identified mimics now have the stage set to overcome self-tolerance. Note that if mimics are presented in the system in most animal models without an adjuvant signal, they are not able to generate autoimmunity. Another aspect of this response may be in response to stress, where heat shock protein Hsp60, triggered as a response to the stresses of many conditions, is a mimic. In addition, the stress disruption of the hypothalamic–pituitary–adrenal (HPA) axis is also suggested to play a role in the altering of the immune response characteristics (Klecha, Barreiro Arcos, Frick, Genaro, & Cremaschi, 2008).

2.7.1 Molecular Mimics and the Shape of the Autoimmune Response

The eventual nature and target of the autoimmune response are likely due to which molecular mimic is present. Identified mimics include *Candida albicans* (Vojdani, Rahimian, Kalhor, & Mordechai, 1996), *Bifidobacterium* and *Lactobacillus* species (Kiseleva et al., 2011), *Helicobacter pylori* strains expressing CagA (Figura et al., 1999), *Yersinia enterocolitica* and *Borrelia burgdorferi* (Benvenga, Guarneri, Vaccaro, Santarpia, & Trimarchi, 2004), *Clostridium botulinum* neurotoxin A (Gregoric, Gregoric, Guarneri, & Benvenga, 2011), HCV (Tomer & Villanueva, 2004), and human parvovirus B19 (Mori & Yoshida, 2010). The nature of the response, whether it eventually is seen as involving Th1 or Th2 T helpers is discussed in Rapoport and McLachlan (2014).

Also directing the nature of an autoimmune response is the genetics of the individual. An AITD-susceptible individual can have mutations that are thyroid specific involving the thyroid antigens, polymorphisms in immune activation genes (including HLA-DR3), and polymorphisms for peripheral tolerance (reviewed in Lee et al., 2015).

3. RESULTS AND DISCUSSION

The factors identified in combination are responsible for thyroid autoimmunity. As a test of this statement, let us consider postpartum thyroiditis. As noted, postpartum thyroiditis (not due to Graves' disease) occurs in 5%–8% women in the first year postpartum who were euthyroid prior to pregnancy (Nicholson, Robinson, Smallridge, Ladenson, & Powe, 2006). Women who had a thyroid condition prior to pregnancy often find conditions worsening, probably due to leptin production from the placenta. It would be expected that the incidence of AITD would be increased due to leptin levels. This may be another reason that more women than men have AITD. An interesting observation is that vitamin D is generally low in postpartum thyroiditis (Krysiak, Kowalcze, & Okopien, 2016). It would be expected that vitamin D supplementation may be useful for this population.

A second result from this approach can be seen in the connection between low vitamin D and depressive disorders (Milaneschi et al., 2014). While seasonal affective disorder has one treatment being UVB lamps, it is somewhat surprising that there is a link between anti-TPO antibody with anxiety and mood disorders (Carta et al., 2004). Based on the discussion in this review, this can be understood.

4. CONCLUSION

Thyroid autoimmunity is the result of an interplay of factors, the stage is set with the adipokines providing inflammatory cytokines, and vitamin D providing immune modulating signals external to the thyroid, while thyrocyte innate responses to environmental conditions provide the adjuvant signal. Shaping the response to be reactive to particular self-antigens and likelihood of a response are due to genetics and molecular mimics.

ACKNOWLEDGMENT

S.B.M. was supported in part by an NSF REU site grant, "Computation Across the Disciplines," #ACI-1461264.

Competing interests: Authors have declared that no competing interests exist.

Authors' contributions: S.J.M. was responsible for planning the organization of the review and final editing. S.B.M. contributed to several sections and discussions. Both authors read and approved the final manuscript.

REFERENCES

Abbas, M. A. (2017). Physiological functions of Vitamin D in adipose tissue. *Journal of Steroid Biochemistry and Molecular Biology*, *165*(Part B), 369–381. http://dx.doi.org/10.1016/j.jsbmb.2016.08.004.

Adams, J. S., & Hewison, M. (2010). Update in vitamin D. *Journal of Clinical Endocrinology and Metabolism*, *95*, 471–478.

Agmon-Levin, N., Theodor, E., Segal, R. M., & Shoenfeld, Y. (2013). Vitamin D in systemic and organ-specific autoimmune diseases. *Clinical Reviews in Allergy & Immunology*, *45*(2), 256–266. http://dx.doi.org/10.1007/s12016-012-8342-y.

Akeno, N., Smith, E., Stefan, M., Huber, A., Zhang, W., Keddache, M., et al. (2011). IFN-a mediates the development of autoimmunity both by direct tissue toxicity and through immune cell recruitment mechanisms. *Journal of Immunology*, *186*(8), 4693–4706. http://dx.doi.org/10.4049/jimmunol.1002631.

Antonelli, A., Ferrari, S. M., Corrado, A., Di Domenicantonio, A., & Fallahi, P. (2015). Autoimmune thyroid disorders. *Autoimmunity Reviews*, *14*, 174–180. http://dx.doi.org/10.1016/j.autrev.2014.10.016.

Aranow, C. (2011). Vitamin D and the immune system. *Journal of Investigative Medicine*, *59*(6), 881–886. http://dx.doi.org/10.231/JIM.0b013e31821b8755.

Barchetta, I., Baroni, M. G., Leonetti, F., De Bernardinis, M., Bertoccini, L., Fontana, M., et al. (2015). TSH levels are associated with vitamin D status and seasonality in an adult population of euthyroid adults. *Clinical and Experimental Medicine*, *15*(3), 389–396.

Baron, M. (2001). The search for complex disease genes: Fault by linkage or fault by association? *Molecular Psychiatry*, *6*, 143–149.

Benvenga, S., & Guarneri, F. (2016). Molecular mimicry and autoimmune thyroid disease. *Reviews in Endocrine & Metabolic Disorders*, *17*, 485–498. http://dx.doi.org/10.1007/s11154-016-9363-2.

Benvenga, S., Guarneri, F., Vaccaro, M., Santarpia, L., & Trimarchi, F. (2004). Homologies between proteins of Borrelia burgdorferi and thyroid autoantigens. *Thyroid*, *14*(11), 964–966. http://dx.doi.org/10.1089/thy.2004.14.964.

Betterle, C., Garelli, S., & Presotto, F. (2014). Diagnosis and classification of autoimmune parathyroid disease. *Auotoimmunity Reviews*, *13*, 417–422.

Biondi, B. (2010). Thyroid and obesity: An intriguing relationship. *Journal of Clinical Endocrinology and Metabolism*, *95*(8), 3614–3617. http://dx.doi.org/10.1210/jc.2010-1245.

Bizzaro, G., & Shoenfeld, Y. (2015). Vitamin D and autoimmune thyroid diseases: Facts and unresolved questions. *Immunologic Research*, *61*, 46–52. http://dx.doi.org/10.1007/s12026-014-8579-z.

Boccuti, V., Perrone, A., D'Introno, A., Campobasso, A., Sangineto, M., & Sabbà, C. (2016). An unusual association of three autoimmune disorders: Celiac disease, systemic lupus erythematosus and Hashimoto's thyroiditis. *Autoimmunity Highlights*, 7, 7. http://dx.doi.org/10.1007/s13317-016-0079-9.

Bogdanos, D. P., Smyk, D. S., Invernizzi, P., Rigopoulou, E. I., Blank, M., Pouria, S., et al. (2013). Infectome: A platform to trace infectious triggers of autoimmunity. *Auotoimmunity Reviews*, *12*, 726–740.

Breslavsky, A., Frand, J., Matas, Z., Boaz, M., Barnea, Z., & Shargorodsky, M. (2013). Effect of high doses of vitamin D on arterial properties, adiponectin, leptin and glucose homeostasis in type 2 diabetic patients. *Clinical Nutrition*, *32*, 970–975.

Brunelleschi, S. (2015). Immune response and auto-immune disease: Gender does matter and makes the difference. *Italian Journal of Gender-Specific Medicine*, *2*(1), 5–14.

Burek, C. L., & Talor, M. V. (2009). Environmental triggers of autoimmune thyroiditis. *Journal of Autoimmunity*, *33*, 183–189.

Carta, M. G., Loviselli, A., Hardoy, M. C., Massa, S., Cadeddu, M., Sardu, C., et al. (2004). The link between thyroid autoimmunity (antithyroid peroxidase autoantibodies) with anxiety and mood disorders in the community: A field of interest for public health in the future. *BMC Psychiatry*, *4*, 25. http://dx.doi.org/10.1186/1471-244X-4-25.

Chen, K., Wei, Y., Sharp, G. C., & Braley-Mullen, H. (2007). Decreasing TNF-α results in less fibrosis and earlier resolution of granulomatous experimental autoimmune thyroiditis. *Journal of Leukocyte Biology*, *81*(1), 306–314. http://dx.doi.org/10.1189/jlb.0606402.

Coelho, M., Oliveira, T., & Fernandes, R. (2013). Biochemistry of adipose tissue: An endocrine organ. *Archives of Medical Science*, *9*(2), 191–200. http://dx.doi.org/10.5114/aoms.2013.33181.

Conde, J., Scoece, M., Abella, V., López, V., Pino, J., Gómez-Reimo, J. J., et al. (2014). An update on leptin as immunomodulator. *Expert Review of Clinical Immunology*, *10*(9), 1165–1170. http://dx.doi.org/10.1586/1744666X.2014.942289.

Cypess, A. M., Lehman, S., Williams, G., Tal, I., Rodman, D., Goldfine, A. B., et al. (2009). Identification and importance of brown adipose tissue in adult humans. *New England Journal of Medicine*, *360*, 1509–1517. http://dx.doi.org/10.1056/NEJMoa0810780.

Damoiseaux, J. (2013). Autoantibodies in the grocery shop: Does quantity matter? *Immunologic Research*, *56*(2), 413–419. http://dx.doi.org/10.1007/s12026-013-8419-6.

D'Aurizio, F., Villalta, D., Metus, P., Doretto, P., & Tozzoli, R. (2015). Is vitamin D a player or not in the pathophysiology of autoimmune thyroid diseases? *Autoimmunity Reviews*, *14*, 363–369.

Deng, Y., & Scherer, P. E. (2010). Adipokines as novel biomarkers and regulators of the metabolic syndrome. *Annals of the New York Academy of Sciences*, *1212*, E1–E19. http://dx.doi.org/10.1111/j.1749-6632.2010.05875.x.

Duntas, L. H. (2011). Environmental factors and thyroid immunity. *Annales d'Endocrinologie, 72*, 108–113.

Duntas, L. H., & Biondi, B. (2013). The interconnections between obesity, thyroid function, and autoimmunity: The multifold role of leptin. *Thyroid, 23*(6), 646–653.

Effraimidis, G., Badenhoop, K., Tijssen, J. G. P., & Wiersinga, W. M. (2012). Vitamin D deficiency is not associated with early stages of thyroid autoimmunity. *European Journal of Endocrinology, 167*, 43–48.

Fabiana, P., Francesco, A., Germana, B., Francesco, B., Luciana, T., Teresa, I. M., et al. (2015). Thyroid dysfunction in patients with metastatic carcinoma treated with sunitinib: Is thyroid autoimmunity involved? *Thyroid, 25*(11), 1255–1261. http://dx.doi.org/10.1089/thy.2015.0170.

Fantuzzi, G., & Faggiono, R. (2000). Leptin in the regulation of immunity, inflammation, and hematopoiesis. *Journal of Leukocyte Biology, 68*(4), 437–446.

Figura, N., Di Cairano, G., Lorè, F., Guarino, E., Gragnoli, A., Cataldo, D., et al. (1999). The infection by Helicobacter pylori strains expressing CagA is highly prevalent in women with autoimmune thyroid disorders. *Journal of Physiology and Pharmacology, 50*(5), 817–826.

Friedman, J. M. (2002). The function of leptin in nutrition, weight, and physiology. *Nutrition Reviews, 60*(10), S1–S14.

Gregoric, E., Gregoric, J. A., Guarneri, F., & Benvenga, S. (2011). Injections of Clostridium botulinum neurotoxin A may cause thyroid complications in predisposed persons based on molecular mimicry with thyroid autoantigens. *Endocrine, 39*(1), 41–47. http://dx.doi.org/10.1007/s12020-010-9410-9.

Grove, K. L., Fried, S. K., Greenberg, A. S., Xiao, X. Q., & Clegg, D. J. (2010). A microarray analysis of sexual dimorphism of adipose tissues in high-fat-diet-induced obese mice. *International Journal of Obesity (London), 34*(6), 989–1000. http://dx.doi.org/10.1038/ijo.2010.12.

Guarneri, F., & Benvenga, S. (2007). Environmental factors and genetic background that interact to cause autoimmune thyroid disease. *Current Opinion in Endocrinology, Diabetes, and Obesity, 14*(5), 398–409. http://dx.doi.org/10.1097/MED.0b013e3282ef1c48.

Guastamacchia, E., Angelo Giagulli, V., Licchelli, B., & Triggiani, V. (2015). Selenium and iodine in autoimmune thyroiditis. *Endocrine, Metabolic & Immune Disorders Drug Targets, 15*(4), 288–292.

Hage, M., Zantout, M. S., & Azar, S. T. (2011). Thyroid disorders and diabetes mellitus. *Journal of Thyroid Research, 2011*, 439463. http://dx.doi.org/10.4061/2011/439463.

Hansen, P. S., Brix, T. H., Iachine, I., Kyvik, K. O., & Hegedüs, L. (2006). The relative importance of genetic and environmental effects for the early stages of thyroid autoimmunity: A study of healthy Danish twins. *European Journal of Endocrinology, 154*, 29–38. http://dx.doi.org/10.1530/eje.1.02060.

Haussler, M. R., Whitfield, G. K., Haussler, C. A., Hsieh, J.-C., Thompson, P. D., Selznick, S. H., et al. (1998). The nuclear vitamin D receptor: Biological and molecular regulatory properties revealed. *Journal of Bone and Mineral Research, 13*(3), 325–349.

Hayes, C. E., Hubler, S. L., Moore, J. R., Barta, L. E., Praska, C. E., & Nashold, F. E. (2015). Vitamin D actions on CD4+ T cells in autoimmune disease. *Frontiers in Immunology, 6*, 100. http://dx.doi.org/10.3389/fimmu.2015.00100.

Hayter, S. M., & Cook, M. C. (2012). Updated assessment of the prevalence, spectrum and case definition of autoimmune disease. *Autoimmunity Reviews, 11*, 754–765.

Humbert, P., & Dupond, J. L. (1988). Multiple autoimmune syndromes. *Annales de Médecine Interne, 139*(3), 159–168.

Kaisho, T., & Akira, S. (2002). Toll-like receptors as adjuvant receptors. *Biochemica et Biophysica Acta, 1589*(1), 1–13.

Kawashima, A., Tanigawa, K., Akama, T., & Yoshihara, A. (2011). Innate immune activation and thyroid autoimmunity. *Journal of Clinical Endocrinology and Metabolism, 96*(12), 3661–3671.

Kawashima, A., Yamazaki, K., Hara, T., Akama, T., Yoshihara, A., Sue, M., et al. (2013). Demonstration of innate immune responses in the thyroid gland: Potential to sense danger and a possible trigger for autoimmune reactions. *Thyroid, 23*(4), 477–487.

Kennedy, A., Gettys, T. W., Watson, P., Wallace, P., Ganaway, E., Pan, Q., et al. (1997). The metabolic significance of leptin in humans: Gender-based differences in relationship to adiposity, insulin sensitivity, and energy expenditure. *Journal of Clinical Endocrinology and Metabolism, 82*(4), 1293–1300.

Kim, D. (2016). Low vitamin D status is associated with hypothyroid Hashimoto's thyroiditis. *Hormones, 15*(3), 385–393.

Kiseleva, E., Mikhaiopulo, K., Sviridov, O., Novik, G., Knirel, Y., & Szwajcer Dey, E. (2011). The role of components of Bifidobacterium and Lactobacillus in pathogenesis and serologic diagnosis of autoimmune thyroid diseases. *Beneficial Microbes, 2*(2), 139–154. http://dx.doi.org/10.3920/BM2010.0011.

Kivity, S., Agmin-Levin, N., Zisappl, M., Shapira, Y., Nagy, E. V., Danko, K., et al. (2011). Vitamin D and autoimmune thyroid diseases. *Cellular & Molecular Immunology, 8*, 243–247. http://dx.doi.org/10.1038/cmi.2010.73.

Klecha, A. J., Barreiro Arcos, M. L., Frick, L., Genaro, A. M., & Cremaschi, G. (2008). Immune-endocrine interactions in autoimmune thyroid diseases. *Neuroimmunomodulation, 15*(1), 68–75. http://dx.doi.org/10.1159/000135626.

Klein, S., Coppack, S. W., Mohamed-Ali, V., & Landt, M. (1996). Adipose tissue leptin production and plasma leptin kinetics in humans. *Diabetes, 45*, 984–987.

Krysiak, R., Kowalcze, K., & Okopien, B. (2016). The effect of vitamin D on thyroid autoimmunity in non-lactating women with postpartum thyroiditis. *European Journal of Clinical Nutrition, 70*, 637–639. http://dx.doi.org/10.1038/ejcn.2015.214.

La Cava, A., & Matarese, M. (2004). The weight of leptin in immunity. *Nature Reviews Immunology, 4*(5), 371–379. http://dx.doi.org/10.1038/nri1350.

Landrier, J.-F., Karkeni, E., Marcotorchino, J., Bonnet, L., & Tourniaire, F. (2016). Vitamin D modulates adipose tissue biology: Possible consequences for obesity? *The Proceedings of the Nutrition Society, 75*, 38–46. http://dx.doi.org/10.1017/S0029665115004164.

Lee, H. J., Li, C. W., Hammerstad, S. S., Stefan, M., & Tomer, Y. (2015). Immunogenetics of autoimmune thyroid diseases: A comprehensive review. *Journal of Autoimmunity, 64*, 82–90.

Lehr, S., Hartwig, S., Lamers, D., Famulla, S., Muller, S., Hanish, F. G., et al. (2012). Identification and validation of novel adipokines released from primary human adipocytes. *Molecular & Cellular Proteomics, 11*, 1–13. http://dx.doi.org/10.1074/mcp.M111.010504.

Lerner, A., Jeremias, P., & Matthias, T. (2015). The world incidence and prevalence of autoimmune diseases is increasing. *International Journal of Celiac Disease, 3*(4), 151–155. http://dx.doi.org/10.12691/ijcd-3-4-8.

Limbach, M., Saare, M., Tserel, L., Kisand, K., Eglit, T., Sauer, S., et al. (2016). Epigenetic profiling in CD4+ and CD8+ T cells from Graves' disease patients reveals changes in genes associated with T cell receptor signaling. *Journal of Autoimmunity, 67*, 46–56. http://dx.doi.org/10.1016/j.jaut.2015.09.006.

Liu, P. T., Stenger, S., Li, H., Wenzel, L., Tan, B. H., Krutzik, S. R., et al. (2006). Toll-like receptor triggering of a vitamin D-mediated human antimicrobial response. *Science, 311*, 1770–1773.

Loffreda, S., Yang, S., Lin, H., Karp, C., Brengman, M., Wang, D., et al. (1998). Leptin regulates proinflammatory immune responses. *FASEB Journal, 12*(1), 57–65.

Ma, X., Lee, P., Chisholm, D. J., & James, D. E. (2015). Control of adipocyte differentiation in different fat deposits; implications for pathophysiology or therapy. *Frontiers in Endocrinology, 6*, 1.

Maachi, M., Pieroni, L., Bruckert, J. C., Fellahi, S., Hainque, B., et al. (2004). Systemic low-grade inflammation is related to both circulating and adipose tissue TNFalpha, leptin and IL-6 levels in obese women. *International Journal of Obesity*, *28*, 993–997. http://dx.doi.org/10.1038/sj.ijo.08027.

Marzullo, P., Minocci, A., Tagliaferri, M. A., Guzzaloni, G., Di Blasio, A., De Medici, C., et al. (2010). Investigations of thyroid hormones and antibodies in obesity: Leptin levels are associated with thyroid autoimmunity independent of bioanthropometric, hormonal, and weight-related determinants. *Journal of Clinical Endocrinology and Metabolism*, *95*, 3965–3972. http://dx.doi.org/10.1210/jc.2009-2798.

Masuzaki, H., Ogawa, Y., Sagawa, N., Hosoda, K., Matsumoto, T., Mise, H., et al. (1997). Nonadipose tissue production of leptin: Leptin as a novel placenta-derived hormone in humans. *Nature Medicine*, *3*(9), 1029–1033.

Matsubara, M., Maruoka, S., & Katayose, S. (2002). Inverse relationship between plasma adiponectin and leptin concentrations in normal-weight and obese women. *European Journal of Endocrinology*, *147*, 173–180.

McGrogan, A., Seaman, H. E., Wright, J. W., & de Vries, C. S. (2008). The incidence of autoimmune thyroid disease: A systematic review of the literature. *Clinical Endocrinology*, *69*(5), 687–696.

McLachlan, S. M., & Rapoport, B. (2014). Breaking tolerance to thyroid antigens: Changing concepts in thyroid autoimmunity. *Endocrine Reviews*, *35*(1), 59–105. http://dx.doi.org/10.1210/er.2013-1055.

Medici, M., Porcu, E., Pistis, G., Teumer, A., Brown, S. J., Jensen, R. A., et al. (2014). Identification of novel genetic loci associated with thyroid peroxidase antibodies and clinical thyroid disease. *PLoS Genetics*, *10*(2), e1004123. http://dx.doi.org/10.1371/journal.pgen.1004123.

Meek, T. H., & Morton, G. J. (2016). The role of leptin in diabetes: Metabolic effects. *Diabetologia*, *59*(5), 928–932. http://dx.doi.org/10.1007/s00125-016-3898-3.

Menendez, C., Lage, M., Peino, R., Baldelli, R., & Concheiro, P. (2001). Retinoic acid and vitamin D_3 powerfully inhibit *in vitro* leptin secretion by human adipose tissue. *Journal of Endocrinology*, *170*, 425–431.

Merrill, S. J., Ashrafi, S., Subramanian, M., & Godar, D. E. (2015). Exponentially increasing incidences of cutaneous malignant melanoma in Europe correlate with low personal annual UV doses and suggests two major risk factors. *Dermato-Endocrinology*, 7(1), e1004018.

Merrill, S. J., & Mu, Y. (2015). Thyroid autoimmunity as a window to autoimmunity: An explanation for the sex differences in the prevalence of thyroid autoimmunity. *Journal of Theoretical Biology*, *375*, 95–100. http://dx.doi.org/10.1016/j.jtbi.2014.12.015.

Milaneschi, Y., Hoogendijk, W., Lips, P., Heijboer, A. C., Schoevers, R., van Hemert, A. M., et al. (2014). The association between low vitamin D and depressive disorders. *Molecular Psychiatry*, *19*, 444–451.

Mori, K., & Yoshida, K. (2010). Viral infection in induction of Hashimoto's thyroiditis: A key player or just a bystander? *Current Opinion in Endocrinology, Diabetes, and Obesity*, *17*(5), 418–424. http://dx.doi.org/10.1097/MED.0b013e32833cf518.

Mutt, S. J., Hyppönen, E., Saamlo, J., Järvellin, M.-R., & Herzig, K.-H. (2014). Vitamin D and adipose tissue—More than storage. *Frontiers in Physiology*, *5*, 228. http://dx.doi.org/10.3389/fphys.2014.00228.

Nicholson, W. K., Robinson, K. A., Smallridge, R. C., Ladenson, P. W., & Powe, N. R. (2006). Prevalence of postpartum thyroid dysfunction: A quantitative review. *Thyroid*, *6*(6), 573–582. http://dx.doi.org/10.1089/thy.2006.16.573.

Ojaimi, S., Skinner, N. A., Strauss, B. J. G., Sundararajian, V., Woolley, I., & Visvanathan, K. (2013). Vitamin d deficiency impacts on expression of toll-like receptor-2 and cytokine profile: A pilot study. *Journal of Translational Medicine*, *11*, 176. http://dx.doi.org/10.1186/1479-5876-11-176.

Pearce, E. N. (2015). Thyroid disorders during pregnancy and postpartum. *Best Practice & Research Clinical Obstetrics & Gynaecology*, *29*, 700–706.

Prietl, B., Treiber, G., Pieber, T. R., & Amrein, K. (2013). Vitamin D and immune function. *Nutrients*, *5*, 2502–2521. http://dx.doi.org/10.3390/nu5072502.

Procaccini, C., La Rocca, C., Carbone, F., De Rosa, V., Galgani, M., & Matarese, G. (2017). Leptin as immune mediator: Interaction between neuroendocrine and immune system. *Developmental and Comparative Immunology*, *66*, 120–129.

Procaccini, C., Pucino, V., Mantzoros, C. S., & Matarese, G. (2015). Leptin in autoimmune diseases. *Metabolism*, *64*, 92–104.

Rapoport, B., & McLachlan, S. M. (2014). Graves' hyperthyroidism is antibody-mediated but is predominantly a Th1-type cytokine disease. *Journal of Clinical Endocrinology and Metabolism*, *99*(11), 4060–4061. http://dx.doi.org/10.1210/jc.2014-3011.

Root-Berstein, R., & Fairweather, D. (2014). Complexities in the relationship between infection and autoimmunity. *Current Allergy and Asthma Reports*, *14*, 407. http://dx.doi.org/10.1007/s11882-013-0407-3.

Rosário, P. W. S., Carvalho, M., & Calsolari, M. R. (2016). Natural history of subclinical hypothyroidism with TSH $\leq$10 mIU/l: A prospective study. *Clinical Endocrinology*, *84*(6), 878–881. http://dx.doi.org/10.1111/cen.12939.

Rubtsov, A. V., Rubtsova, K., Kappler, J. W., & Marrack, P. (2010). Genetic and hormonal factors in female-based autoimmunity. *Autoimmunity Reviews*, *9*(7), 494–498. http://dx.doi.org/10.1016/j.autrev.2010.02.008.

Rubtsova, K., Marrack, P., & Rubtsov, A. V. (2015). TLR7, IFN-c, and T-bet: Their roles in the development of ABCs in female biased autoimmunity. *Cellular Immunology*, *294*, 80–83. http://dx.doi.org/10.1016/j.cellimm.2014.12.002.

Saeki, Y., & Ishihara, K. (2014). Infection-immunity liaison: Pathogen-driven autoimmune-mimicry (PDAIM). *Autoimmunity Reviews*, *13*, 1064–1069.

Sagawa, N., Yura, S., Itoh, H., Kakui, K., Takemura, M., Nuamah, M. A., et al. (2002). Possible role of placental leptin in pregnancy. *Endocrine*, *19*, 65–71. http://dx.doi.org/10.1385/ENDO:19:1:65.

Sharabiani, M. T. A., Vermeulen, R., Scoccianti, C., Hosnijeh, F. S., Minelli, L., Sacerdote, C., et al. (2011). Immunologic profile of excessive body weight. *Biomarkers*, *16*(3), 243–251. http://dx.doi.org/10.3109/1354750X.2010.547948.

Skaaby, T., Husemoen, L. L. N., Thuesen, B. H. T., & Linneberg, A. (2015). Prospective population-based study of the association between vitamin D status and incidence of autoimmune disease. *Endocrine*, *50*(1), 231–238. http://dx.doi.org/10.1007/s12020-015-0547-4.

Smilek, D. E., & St. Clair, W. E. (2015). Solving the puzzle of autoimmunity: Critical questions. *F1000 Prime Reports*, 7, 17. http://dx.doi.org/10.12703/P7-17.

Snijder, M. B., van Dam, R. M., Visser, M., Deeg, D. J. H., Dekker, J. M., Bouter, L. M., et al. (2005). Adiposity in relation to vitamin D status and parathyroid hormone levels: A population-based study in older men and women. *Journal of Clinical Endocrinology and Metabolism*, *90*(7), 4119–4123. http://dx.doi.org/10.1210/jc.2005-0216.

Stagnaro-Green, A., Schwartz, A., Gismondi, R., Tinelli, A., Mangieri, T., & Negro, R. (2010). High rate of persistent hypothyroidism in a large-scale prospective study of postpartum thyroiditis in Southern Italy. *Journal of Clinical Endocrinology and Metabolism*, *96*(3), 652–657. http://dx.doi.org/10.1210/jc.2010-1980#sthash.5mZ2YbXT.dpuf.

Tomer, Y. (2014). Mechanisms of autoimmune thyroid diseases: From genetics to epigenetics. *Annual Review of Pathology*, *9*, 147–156. http://dx.doi.org/10.1146/annurev-pathol-012513-104713.

Tomer, Y., Blackard, J. T., & Akeno, N. (2007). Interferon alpha treatment and thyroid dysfunction. *Endocrinology and Metabolism Clinics of North America*, *36*(4), 1051–1066. http://dx.doi.org/10.1016/j.ecl.2007.07.001.

Tomer, Y., & Huber, A. (2009). The etiology of autoimmune thyroid disease: A story of genes and environment. *Journal of Autoimmunity*, *32*(3–4), 231–239. http://dx.doi.org/10.1016/j.jaut.2009.02.007.

Tomer, Y., & Villanueva, R. (2004). Hepatitis C and thyroid autoimmunity: Is there a link? *American Journal of Medicine*, *117*(1), 60–61. http://dx.doi.org/10.1016/j.amjmed.2004.04.004.

Van Harmelen, V., Reynisdottir, S., Eriksson, P., Thorne, A., Hoffstedt, J., Lönnqvist, F., et al. (1998). Leptin secretion from subcutaneous and visceral adipose tissue in women. *Diabetes*, *47*(6), 913–917.

Versini, M., Jeandel, P.-Y., Rosenthal, E., & Shoenfeld, Y. (2014). Obesity in autoimmune diseases: Not a passive bystander. *Autoimmunity Reviews*, *2014*, 981–1000.

Verstuyf, A., Carmeliet, G., Bouillon, R., & Mathieu, C. (2010). Vitamin D: A pleiotropic hormone. *Kidney International*, *78*, 140–145.

Vojdani, A., Rahimian, P., Kalhor, H., & Mordechai, E. (1996). Immunological cross reactivity between Candida albicans and human tissue. *Journal of Clinical & Laboratory Immunology*, *48*(1), 1–15.

Wang, J., Lv, S., Chen, G., Gao, C., He, J., Zhong, H., et al. (2015). Meta-analysis of the association between vitamin D and autoimmune thyroid disease. *Nutrients*, *7*, 2485–2498. http://dx.doi.org/10.3390/nu7042485.

Zaccone, P., Fehérvári, Z., & Cooke, A. (2003). Tumour necrosis factor-α is a fundamental cytokine in autoimmune thyroid disease induced by thyroglobulin and lipopolysaccharide in interleukin-12 p40 deficient C57BL/6 mice. *Immunology*, *108*(1), 50–54. http://dx.doi.org/10.1046/j.1365-2567.2003.01547.x.

Zhang, Y., Proenca, R., Maffei, M., Barone, M., Leopold, L., & Friedman, J. M. (1994). Positional cloning of the mouse *obese* gene and its human homologue. *Nature Medicine*, *372*, 425–432. http://dx.doi.org/10.1038/372425a0.

CHAPTER SEVEN

Thyroid Hormone Receptor Antagonists: From Environmental Pollution to Novel Small Molecules

Louise S. Mackenzie[1]
School of Life and Medical Sciences, Hertfordshire University, Hatfield, United Kingdom
[1]Corresponding author: e-mail address: l.mackenzie2@herts.ac.uk

Contents

1. Introduction 148
   1.1 TR Isotypes 148
   1.2 Hyperthyroidism and Hypothyroidism 149
2. TR Antagonists 149
   2.1 Current Methods to Treat Hyperthyroidism 150
   2.2 Future Methods to Treat Hyperthyroidism 151
   2.3 Amiodarone Side Effects 153
   2.4 Novel Concepts in TH Receptor Antagonism: NSAIDs 153
   2.5 Links to Prostanoid Signaling: Prostacyclin, Beraprost, and GW0742 154
3. Synthetic TR Antagonists 155
   3.1 GC-1 and NH-3 155
   3.2 1-850 156
   3.3 β Aminoketones 156
   3.4 MLS 389544 157
   3.5 TRα- and TRβ-Selective Antagonists 157
4. Environmental TR Antagonists 158
5. Concluding Remarks 159
References 159

Abstract

Thyroid hormone receptors (TRs) are nuclear receptors which control transcription, and thereby have effects in all cells within the body. TRs are an important regulator in many basic physiological processes including development, growth, metabolism, and cardiac function. The hyperthyroid condition results from an over production of thyroid hormones resulting in a continual stimulation of thyroid receptors which is detrimental for the patient. Therapies for hyperthyroidism are available, but there is a need for new small molecules that act as TR antagonists to treat hyperthyroidism.

Many compounds exhibit TR antagonism and are considered detrimental to health. Some drugs in the clinic (most importantly, amiodarone) and environmental pollution

Vitamins and Hormones, Volume 106
ISSN 0083-6729
http://dx.doi.org/10.1016/bs.vh.2017.04.004

exhibit TR antagonist properties and thus have the potential to induce hypothyroidism in some people.

This chapter provides an overview of novel small molecules that have been specifically designed or screened for their TR antagonist activity as novel treatments for hyperthyroidism. While novel compounds have been identified, to date none have been developed sufficiently to enter clinical trials. Furthermore, a discussion on other sources of TR antagonists is discussed in terms of side effects of current drugs in the clinic as well as environmental pollution.

1. INTRODUCTION

Thyroid hormone receptor (TR) is a member of the nuclear receptor superfamily, which alters transcription when activated. The family of receptors also includes other key hormonal nuclear receptors, PPARγ, estrogen, vitamin D, and glucocorticoid receptors, all of which play important roles in homeostasis, growth, and development. In an inactivated state, TRs are bound to DNA and repress transcription; once activated TR acts as a transcriptional activator.

Having such important roles in human health, they have been the focus of pharmaceutical drug development, resulting in thyroid hormone replacement, glucocorticoids (antiinflammatory), thiazolidinedione (diabetes), and estrogens (hormone replacement therapy). However, nuclear receptor signaling is highly complex, and small changes in receptor activation or inactivation will lead to long-term changes in transcription and thus alter the function of individual cells and whole organs. While the pharmacological replacement of these hormones under specific clinical situations can be beneficial, the converse is true and many environmental factors have been found to cause disease through disrupting normal hormonal function. For instance, thyroid hormone (TH) disrupting chemicals can change normal function and TH signaling, and thus cause detrimental effects to the cardiovascular system. One chemical that is frequently released into the environment is Bisphenol A that can be found as a constituent of household dust (Zhang, Li, Gupta, Nam, & Andersson, 2016), is a potent THR antagonist (Sun et al., 2009) that has been linked to a number of serious cardiac effects in infants and adults.

1.1 TR Isotypes

TRs are members of the nuclear receptor superfamily, of which there are two major classes: TRα and TRβ, both of which are the products of separate genes but are expressed in other subtypes: TRα-1, TRα-2, TRβ-1, TRβ-2. All TRs are similar in overall structure with other members of the nuclear receptor such as estrogen receptor. TRs consist of six domains (A–F): domains A (activation) and F (Function) are isoform specific, while domains C (DNA binding), D (hinge), and E (ligand binding) are similar between all isoforms (Apriletti et al., 1998; Ribeiro, Apriletti, Wagner, Feng, et al., 1998; Ribeiro, Apriletti, Wagner, West, et al., 1998; Ribeiro et al., 1995; Wagner et al., 1995). Once ligands bind to the receptor, the domains E and F lead to activation of transcription. Expression of TRs is not ubiquitous, for example, TRα-1 is predominant in heart, whereas TRβ-1 is predominant in the liver, which indicates that there is an important role subtype-selective TR modulator therapeutics.

1.2 Hyperthyroidism and Hypothyroidism

The predominant TR endogenous ligands found in mammals are 3,3′,5-triiodothyonine (T3) and thyroxine (T4), which are produced by the thyroid gland to regulate key homeostatic and physiological processes. Through the regulation of gene transcription, the TRs alter basal and adaptive metabolism, lipid profiles, heart rate, muscle and bone metabolism.

Disruption of the hormones which control these key nuclear receptors may lead to various conditions. A common cause of hyperthyroidism is Graves' disease, which is an autoimmune disease of the thyroid; an increase in circulating TR agonists will lead to hyperthyroidism: tachycardia, heart failure and atrial arrhythmias (Galløe, Rolff, Nordin, Ladefoged, & Mogensen, 1993), excessive weight loss, muscle wasting, attention deficit (Joffe & Sokolov, 2000; Sokolov, Kutcher, & Joffe, 1996; Sokolov, Levitt, & Joffe, 1997), and osteoporosis (Stěpán & Límanová, 1992). Whereas a loss in TR signaling leads to hypothyroidism, with symptoms including metabolic syndrome and atherosclerosis (van Tienhoven-Wind & Dullaart, 2015).

2. TR ANTAGONISTS

Hyperthyroidism is generally treated by the blockade of the thyroid gland in order to reduce T3 and T4 release into the blood stream. However, direct antagonism of the receptor is thought to also result in reversing hyperthyroid symptoms.

The fine balance of TR hormone balance is needed to prevent the wide spread effects of hyperthyroidism without the development of hypothyroidism; while some drugs have been found to have TR antagonist effects (such as amiodarone), many are found as pollutants in the environment, yet some TR-specific synthetic antagonists have been developed with the goal of returning the TR signaling back to a normal medium. However, to date there have been no small molecule TR antagonists developed sufficiently to be taken to clinical trials. Known TR antagonists have been divided by their effects in the environment, therapeutic side effects, and synthetic compounds and are outlined in Tables 1 to 3 and discussed in this chapter.

2.1 Current Methods to Treat Hyperthyroidism

Blockade of TH release form thyroid gland (Utiger, 1995) is achieved using thioamides (carbimazole and propylthiouracil), methimazole, or radioiodine treatment to decrease the activity of the whole organ, or beta blockers and

Table 1 Drugs in Clinic or in Clinical Trials Which Are Potential TR Antagonists

Antagonists	Use/Source	Specificity (Human TRs)	References
Amiodarone	Antiarrhythmic drug	Antagonist of TRα1 and TRβ1	Norman and Lavin (1989)
Beraprost	Prostacyclin (PGI_2) receptor (IP) agonist	TRα1 and TRβ1 antagonist	Perez Diaz, Zloh, Patel, and Mackenzie (2015)
Celecoxib	Nonsteroidal antiinflammatory drug	TRβ1 antagonist	Zloh, Perez-Diaz, Tang, Patel, and Mackenzie (2016)
Desethylamiodarone	Metabolite of amiodarone	TRα1 and TRβ1 antagonist	Bakker, Hudig, Meijssen, and Wiersinga (1998)
Diclofenac	Nonsteroidal antiinflammatory drug	TRβ1 antagonist	Zloh et al. (2016)
Dronedarone and debutyldronedarne	Antiarrhythmic drug and its metabolite (noniodine version of amiodarone)	Inhibits T3 binding to TRα1	Van Beeren et al. (2003)
GW0742	PPARβδ agonist	TRα1 and TRβ1 antagonist	Perez Diaz et al. (2015)

Table 2 Synthetic Molecules Purposely Designed as TR Antagonists

Antagonists	Design Methods	Specificity (Human TRs)	References
1-850	In silico screening for new TR antagonist	Competitively blocks T3 binding to both TRα1 and TRβ1; prevents TR homodimer activity, not TR/RXR heterodimers	Schapira et al. (2003)
GC-1	Designed on the basis of T3 structure	TRβ1 agonist with partial antagonist activity	Chiellini, Nguyen, Apriletti, Baxter, and Scanlan (2002)
MLS 389544	High-throughput screening of small molecule library	Blocks the interaction between TRβ and the steroid receptor coactivator	Hwang et al. (2012, 2011)
NH-3	Designed on the basis of GC-1 structure	Selective TRβ1 antagonist; competes with T3 for ligand binding site	Webb et al. (2002), Nguyen et al. (2002), and Singh et al. (2016)
SJ-AK	High-throughput screening of small molecule library	Blocks coactivator binding to ligand binding site of TRs; selective TRβ1 antagonist	Estébanez-Perpiñá et al. (2007)

calcium antagonists to control some of the symptoms of hyperthyroidism. While methimazole and the thioamides are widely used to disrupt hormone release, their effects take a long period of time to take effect. This is mostly due to the T4 hormone half-life of 8 days, which means it takes 1–2 months before their effects are noticeable. A specific small molecule TR antagonist is therefore a desirable goal of the pharmaceutical industry.

2.2 Future Methods to Treat Hyperthyroidism

While so far novel small molecule TR antagonists have so far not been developed sufficiently to enter clinical trials, the new human monoclonal antibody (TRAb) to thyroid-stimulating hormone receptor (TSHR), K1–70 (Furmaniak et al., 2012), has been trialed in rats and more recently there are plans to conduct clinical trials in man (which at the time of writing was still recruiting). TSHR signaling is linked to the development of hyperthyroidism, and the TRAb monoclonal antibody was isolated form a single sample of a patient with autoimmune hypothyroidism. While this new

Table 3 Molecules Found in the Environment With Proven TR Antagonist Behavior

Antagonists	Use/Source	Specificity (Human TRs)	References
Bisphenol A (BPA)	Monomer of polycarbonate plastics	TRβ1 antagonist	Zoeller (2005) and Moriyama et al. (2002)
Polychlorinated biphenyls (PCBs), polychlorinated dibenzo-*p*-dioxins, and polychlorinated dibenzofurans	Halogenated industrial chemicals	TRβ1 antagonist	Van den Berg, van Raaij, Bragt, and Notten (1991) and Miyazaki, Iwasaki, Takeshita, Kuroda, and Koibuchi (2004)
Diphenylamine derivatives	Synthetic	TRα1 and TRβ1 antagonists	Komatsu, Hirano, Songkram, Kawachi, and Kagechika (2007)
Phthalates: di-*n*-butyl phthalate (DNBP), diisobutyl phthalate (DIBP), di-*n*-octyl phthalate (DNOP), and di-2-ethylhexyl phthalate (DEHP)	Consumer products: building materials, cosmetics, clothing, pharmaceuticals, children's toys, food packaging, cleaning material, insecticides. Identified in drinking water sources in the Yangtze River delta	TRβ antagonists	Shi et al. (2012, 2013) and Shen et al. (2009)
Pyrethroid insecticides: cycloprothrin, cyfluthrin, cyhalothrin, cypermethrin, deltamethrin, etofenprox, fenvalerate, permethrin, and tetramethrin and their metabolites (3-(2,2-dichlorovinyl)-2,2-dimethylcyclopropne carboxylic acid [DCCA] and 3-phenoxybenzoic acid [3-PBA])	Insecticides	TRβ1 antagonists	Du et al. (2010)
Tetrabromobisphenol A (TBBPA)	Flame retardant: halogenated derivative of BPA	TRβ1 antagonist (not TRα antagonist)	Sun et al. (2009) and Oka et al. (2013)
Tetrachlorobisphenol A (TCBPA)	Flame retardant: halogenated derivative of BPA	TRβ1 antagonist	Sun et al. (2009)

biologic treatment would not directly inactivate TRs, it would prevent the creation and secretion of thyroid hormones in the hyperthyroid patient.

2.3 Amiodarone Side Effects

The antiarrhythmic drug amiodarone is known to have TR antagonist properties (Table 1). Structurally similar to T3 and T4, amiodarone is rich in iodine and has been linked to the development of hyperthyroidism. Conversely, the amiodarone metabolite desethylamiodarone has been shown to be a noncompetitive antagonist of TRβ1 (Bakker, van Beeren, & Wiersinga, 1994) and a TRα competitive antagonist (van Beeren, Bakker, & Wiersinga, 1995) and leads to the development of hypothyroidism.

The dual effects of amiodarone have been demonstrated in the Sotalol Amiodarone Atrial Fibrillation Efficacy Trial (SAFE-T) (Batcher et al., 2007). The clinical trial followed patients taking amiodarone over 6 months and measured alterations in circulating thyroid-stimulating hormone (TSH) levels compared to placebo controls. From a group of 247 men on amiodarone, 5.3% patients developed hyperthyroidism (TSH.35 mU/L), 25.8% patients developed subclinical hypothyroidism (TSH > 4.5–10 mU/L), and 5.0% patients developed hypothyroidism (TSH > 10 mU/L) (Batcher et al., 2007).

Indeed, this established a link between patients taking amiodarone and the development of hypothyroidism (Ahmed, Van Gelder, Wiesfeld, Van Veldhuisen, & Links, 2011) and hyperthyroidism has led to the NICE guidelines recommendations that amiodarone usage in patients should undergo 6 monthly clinical reviews along with liver and thyroid function tests and a review of all side effects. While there is interest in developing new TR-selective antagonists for the treatment of hyperthyroidism, it is not recommended that patients use amiodarone due to the effects on heart function.

2.4 Novel Concepts in TH Receptor Antagonism: NSAIDs

Nonsteroidal antiinflammatory drugs (NSAIDs) are taken to treat pain and inflammation, and are the most frequently taken over the counter drug type taken worldwide. They work by inhibiting cyclooxygenase (COX) enzymes, whose products are used by other enzymes to biosynthesise a range of prostanoids. COX is present in two isoforms: COX 1 and COX-2. Generally, COX-1 is expressed constitutively, while COX-2 is induced at sites of inflammation and is the target for NSAIDs. However, NSAIDs such as diclofenac also inhibit COX1, which is linked to increased gastric bleeds.

For this reason, new COX-1 sparing drugs were created such as rofecoxib and celecoxib (COX-2 selective). For people having to take NSAIDs on a daily basis this was an important development for managing chronic pain and inflammation, but unfortunately these drugs were later found to have cardiovascular complications and rofecoxib was removed from the market. Furthermore, the chances of having a myocardial infarction following use of diclofenac were also found to be significantly increased (Kearney et al., 2006). While the majority of side effects of NSAIDs are directly attributed to COX1 inhibition, such as gastric ulcers and increased bleeding times, new evidence has emerged that suggests a novel link between NSAIDs and TRs.

Computational chemistry predictions indicated that diclofenac and celecoxib, but not naproxen could theoretically bind to TRβ (Zloh et al., 2016). While diclofenac is an antagonist of both COX1 and COX2, celecoxib is specific for COX2; both of which demonstrate significant cardiovascular side effects. On the other hand, naproxen has the ability to inhibit both COX1 and COX2 yet is considered relatively cardiosafe (Bhala et al., 2013; Kearney et al., 2006). These findings were confirmed using human cell reporter assays, indicating that diclofenac and celecoxib but not naproxen have direct TRβ antagonistic effects (Zloh et al., 2016).

2.5 Links to Prostanoid Signaling: Prostacyclin, Beraprost, and GW0742

A downstream product of COX1 pathway prostacyclin has also been linked to TR activity (Perez Diaz et al., 2015). Prostacyclin (PGI_2) is a prostaglandin first described by Sir John Vane in 1976 as a potent vasodilator and antithrombotic agent of the vasculature (Moncada, Gryglewski, Bunting, & Vane, 1976). PGI_2 has a very short half-life of 42 s, which makes direct pharmacological analysis difficult, although there are a large number of PGI_2 mimetics available such as iloprost and beraprost which have half-lives of 30–40 min. PGI_2 and its mimetics are ligands for the cell surface prostacyclin receptor (IP) which induces the short-term changes in cell function.

The long-term effects of PGI_2 are known to be important in different organ systems, for example, PGI_2 has important antiinflammatory effects in pulmonary artery and bronchi. PGI_2 has been described as the putative endogenous ligand for the nuclear receptor peroxisome proliferator-activated receptor β/δ (PPARβ/δ) (Forman, Chen, & Evans, 1997). Signaling via PPARβ/δ has a number of effects on cell function and has been shown to be involved in fatty acid metabolism, lipid transport as well as proliferation and differentiation of cells. PPARβ/δ agonists include the synthetic compound GW0742 and GW501516 which completed proof-of-concept clinical trials successfully for dyslipidemia (Ooi, Watts, Sprecher, Chan, & Barrett, 2011) and hypercholesteremia (Olson, Pearce, Jones, & Sprecher, 2012), although further clinical trials were halted due to toxicology issues (Mackenzie & Lione, 2013).

Using computational chemistry methods, a comparison of PGI_2, beraprost, and GW0742 structures revealed an overlap in shapes, and further docking studies predicted an ability for all three compounds to bind to TRs. Further *in vitro* analysis using reporter assays confirmed that GW0742 and beraprost were antagonists of TRα and TRβ. Unfortunately, direct *in vitro* testing of PGI_2 using reporter assays is not possible, although the predicted computational docking of PGI_2 with TR indicated that there is a likelihood that PGI_2 may have TR antagonist properties. However, this is highly unlikely to have any physiological consequences due to the short-lived nature of PGI_2 and the theoretical requirement for very high concentrations of PGI_2 to have this inhibitory effect. These findings do, however, indicate that beraprost and GW0742 have the potential to have long-term effects in cells via TRs, although these results need to be repeated in other *in vitro* and *in vivo* models.

3. SYNTHETIC TR ANTAGONISTS

There are two TR antagonists that are commercially available: MLS000389544 (MLS 389544) (Hwang et al., 2012) and 1-850 (Schapira et al., 2003), both of which have been used experimentally *in vitro* to study TR function. However, there are a number of synthetic TR antagonists that have been reported in the literature (Table 2).

3.1 GC-1 and NH-3

With the aim of creating a TR-selective antagonist, GC-1 was created based on the structure of T3 (Chiellini et al., 2002), which was the first synthetic TR antagonist created although GC-1 had a low binding profile and was later shown to have TR agonist activity, but was useful as a scaffold for further development. Indeed, GC-1 derivatives included NH-3, a TR antagonist which originally showed promise having no TR agonist activity and high binding affinity (Nguyen et al., 2002). Use of highly specific gene reporter assays indicated that NH-3 is a TR antagonist which acts by displacing T3 from the ligand binding site of TR (Hofmann, Schomburg, & Köhrle, 2009). While originally NH-3 was thought to possess partial TR agonist activity, this was found later to be identified as a benzofuran (EBI) side product that was formed during the synthesis of the antagonist

(Singh et al., 2016). NH-3 has been used *in vivo* Xenopus experiments and has shown the importance of TH signaling in early phases of embryonic development (Bronchain et al., 2017).

3.2 1-850

With the aim of identifying a new TR antagonist, virtual screening techniques were employed to screen and select 1-850 as a new candidate drug. This antagonist is easy to synthesize and receptor binding studies indicated that 1-850 competitively blocks T3 binding to both TRα and TRβ ($IC_{50} = 1.5\,\mu M$) with no partial agonist activity. The antagonist structure has a close similarity to T3 and has no binding to the related retinoic acid receptor α (RXR) (Schapira et al., 2003), and further studies have indicated that 1-850 specifically inhibits homodimerized TR nuclear proteins, but has no binding or inhibition of TR/RXR heterodimers (Paquette, Atlas, Wade, & Yauk, 2014). This makes 1-850 a useful pharmacological research tool for deepening our understanding of TR control on transcription and will enable researchers to determine the genes regulated by TR homodimerization as opposed to TR/RXR heterodimerization. For example, a study investigating the regulation of early spontaneous network activity and the development of GABAergic neurons by the thyroid hormone indicated that T3 is important in the maturation of specific cell type interactions (Westerholz, de Lima, & Voigt, 2010). The use of 1-850 inhibited T3 mediated increase in GABAergic soma size which indicated an important role for T3 in rat brain development. Moreover, the use of 1-850 would also indicate that TR homodimerization and not TR/RXR is key to the developmental stage in humans. The specific TR effects described by Paquette et al. (2014) may also explain why 1-850 effects only dampens and not abolishes some T3-mediated responses, for example, in Tilapia fish liver experiments (Navarrete-Ramírez, Luna, Valverde-R, & Orozco, 2014).

3.3 β Aminoketones

The β aminoketones were first identified as inhibitors of TR coactivator interaction through the use of high-throughput screening using a library of 138,000 compounds in an *in vitro* fluorescence polarization assay (Arnold et al., 2005). In this manner, a new series of analogues were derived, such as SJ-AK, which was identified as an irreversible inhibitor which binds to the cysteine groups of TR to prevent binding to the coactivator binding site AF-2 (Estébanez-Perpiñá et al., 2007). However, direct comparison of SJ-AK with NH-3 indicated that neither antagonist achieved a full antagonism of T3-mediated TR signaling when tested in HepG2 cells (Sadana et al., 2011). Moreover, the analogues of SJ-AK possessed other pharmacological properties such as hERG potassium channel activity and related dose-related cardiotoxicity, and new series of these compounds have been developed with improved inhibitory potency and no effects on hERG potassium channels (Hwang et al., 2009).

3.4 MLS 389544

The research group who developed the β aminoketones series of compounds reported that these compounds may lack hERG activity, they still bind to ion channels (Johnson et al., 2011). As an alternative route, they performed quantitative high-throughput screening using a TRβ fluorescence polarization assay against a library of 290,000 small molecules (Johnson et al., 2011). This screen resulted in the identification of methylsulfonylnitrobenzoates (MSNB) as a new class of TR antagonists. Further research identified MLS 389533 (a MSNB) as a selective and potent TRβ antagonist, designed specifically to block the interaction between TRβ and the steroid receptor coactivator (Hwang et al., 2011). While the GC-1 and NH-3 series of TR modulators were synthesized in order to target the ligand-binding pocket of the receptor, MSNB compounds were generated to bind to the four cysteine residues located near the coactivator binding site (Hwang et al., 2011). The MSNB analogues covalently bind to TRβ and are irreversible inhibitors, making MLS 389544 a useful antagonist for *in vitro* studies.

While MLS 389544 has been shown to effectively block TR-mediated gene transcription, this inhibitor has also been used to study the role of nongenomic effects or TRs (Perez Diaz et al., 2015; Zloh et al., 2016). Some nuclear receptors, including TR have nongenomic effects which have been considered to have important homeostatic roles, for example, T3 can induce vasodilation through direct activation of the PI3 kinase pathway (Hiroi et al., 2006). In-depth study as to how the receptor binds to and activates PI3 kinase has not been investigated, although the T3 mediated vasodilation can be inhibited by both NH-3 (Hiroi et al., 2006) the competitive TR antagonist and MLS 389544.

3.5 TRα- and TRβ-Selective Antagonists

Of the TR antagonists listed earlier, the majority were found to have dual TRα and TRβ activities. The different TR subtypes regulate different physiological processed; TRβ expression in the liver linked to the regulation of cholesterol and bile acid metabolism, whereas TRα regulates the heart rate, myocardial cells physiology, and bone development. Both receptor subtypes share ligands that modulate their activity; the ligands themselves are highly diverse and there is a great deal of cross over with other nuclear receptors. For example, environmental plastic-based TR antagonists (see Table 3) also modulate estrogen receptor function. TR subtype-specific effect will correlate with the ratio TRα and TRβ in the organ or cell type as opposed to specific ligands controlling function of the specific TR subtype. Development of selective TRα and TRβ small molecule antagonists may therefore not be clinically relevant or possible. To date, only a couple of the synthetic antagonists have been given to whole animals in *in vivo* experiments, and further exploration of small molecules to identify specific activities of the complex TR activity in humans are required before any drug can be used in clinical trials.

4. ENVIRONMENTAL TR ANTAGONISTS

While it is of great interest that we develop novel drug therapies to treat hyperthyroidism, there is a worldwide concern about the release of TR antagonists into the environment (Table 3). These molecules in the environment must be considered a risk to human health, since they have the potential to induce developmental and neurological effects (Shi et al., 2016, 2013). With concerns focusing on expose to these TR antagonists via drinking water, humans are also exposed on a daily basis to flame retardants, phthalates, halogenated industrial chemicals and insecticides via inhalation of dust particles (Zhang et al., 2016) and consumption of food. How these compounds effect the health of humans and wildlife will have to be monitored on the long-term.

Bisphenol A (BPA), the monomer of polycarbonate plastics (Krishnan, Stathis, Permuth, Tokes, & Feldman, 1993), became labeled as an "Endocrine Disruptor," since having disruptive effects on normal hormonal control over physiological processes. BPA binds to and alters the activity of a number of nuclear hormone receptors, including estrogen receptors (Villalobos et al., 1996, 1995) and TR (Moriyama et al., 2002; Zoeller, 2005). The effects of BPA on TRs are complex; an apparent selective inhibition of TRβ by BPA leads to an increase in thyroxine production, which results in hyperthyroidism in respect to TRα (Zoeller, 2005), which taken together indicates a disruption in normal physiological processes. Exposure during pregnancy is of particular concern, and studies have shown that maternal exposure to BPA leads to an elevation of T4 plasma levels in new born rat pups which is similar to thyroid resistance syndrome in humans (Zoeller, 2005).

5. CONCLUDING REMARKS

There is a need for TR antagonists in the clinic, with current patients with hyperthyroidism lacking therapeutic options. However, the balance of thyroid hormones is crucial for human health, and any new therapy that reaches the clinic will have to have highly specific and potent effects on TR activity. As can be seen from some current drugs in the clinic, most notably amiodarone, disruption of TR activity has consequences on human health. Indeed, we are in constant contact with environmental pollutants that have proven TR antagonistic behavior, which makes some people at risk of hypothyroidism. In order to fully understand the manner in which this multitude of compounds effect TR control of transcription, we need to have further insight into the manner by which these molecules bind to and inactivate TR. It is clear that TR is able to be bound by and inactivated by a wide range of compound types, and our understanding of this phenomenon is still in its infancy.

REFERENCES

Ahmed, S., Van Gelder, I. C., Wiesfeld, A. C., Van Veldhuisen, D. J., & Links, T. P. (2011). Determinants and outcome of amiodarone-associated thyroid dysfunction. *Clinical Endocrinology*, *75*, 388–394.

Apriletti, J. W., Ribeiro, R. C., Wagner, R. L., Feng, W., Webb, P., Kushner, P. J., et al. (1998). Molecular and structural biology of thyroid hormone receptors. *Clinical and Experimental Pharmacology & Physiology. Supplement*, *25*, S2–11.

Arnold, L. A., Estébanez-Perpiñá, E., Togashi, M., Jouravel, N., Shelat, A., McReynolds, A. C., et al. (2005). Discovery of small molecule inhibitors of the interaction of the thyroid hormone receptor with transcriptional coregulators. *The Journal of Biological Chemistry*, *280*, 43048–43055.

Bakker, O., Hudig, F., Meijssen, S., & Wiersinga, W. M. (1998). Effects of triiodothyronine and amiodarone on the promoter of the human LDL receptor gene. *Biochemical and Biophysical Research Communications*, *249*, 517–521.

Bakker, O., van Beeren, H. C., & Wiersinga, W. M. (1994). Desethylamiodarone is a noncompetitive inhibitor of the binding of thyroid hormone to the thyroid hormone beta 1-receptor protein. *Endocrinology*, *134*, 1665–1670.

Batcher, E. L., Tang, X. C., Singh, B. N., Singh, S. N., Reda, D. J., Hershman, J. M., et al. (2007). Thyroid function abnormalities during amiodarone therapy for persistent atrial fibrillation. *The American Journal of Medicine*, *120*, 880–885.

Bhala, N., Emberson, J., Merhi, A., Abramson, S., Arber, N., Baron, J. A., et al. (2013). Vascular and upper gastrointestinal effects of non-steroidal anti-inflammatory drugs: Meta-analyses of individual participant data from randomised trials. *Lancet*, *382*, 769–779.

Bronchain, O. J., Chesneau, A., Monsoro-Burq, A. H., Jolivet, P., Paillard, E., Scanlan, T. S., et al. (2017). Implication of thyroid hormone signaling in neural crest cells migration: Evidence from thyroid hormone receptor beta knockdown and NH3 antagonist studies. *Molecular and Cellular Endocrinology*, *439*, 233–246.

Chiellini, G., Nguyen, N. H., Apriletti, J. W., Baxter, J. D., & Scanlan, T. S. (2002). Synthesis and biological activity of novel thyroid hormone analogues: 5′-aryl substituted GC-1 derivatives. *Bioorganic & Medicinal Chemistry*, *10*, 333–346.

Du, G., Shen, O., Sun, H., Fei, J., Lu, C., Song, L., et al. (2010). Assessing hormone receptor activities of pyrethroid insecticides and their metabolites in reporter gene assays. *Toxicological Sciences*, *116*, 58–66.

Estébanez-Perpiñá, E., Arnold, L. A., Jouravel, N., Togashi, M., Blethrow, J., Mar, E., et al. (2007). Structural insight into the mode of action of a direct inhibitor of coregulator binding to the thyroid hormone receptor. *Molecular Endocrinology*, *21*, 2919–2928.

Forman, B. M., Chen, J., & Evans, R. M. (1997). Hypolipidemic drugs, polyunsaturated fatty acids, and eicosanoids are ligands for peroxisome proliferator-activated receptors alpha and delta. *Proceedings of the National Academy of Sciences of the United States of America*, *94*, 4312–4317.

Furmaniak, J., Sanders, J., Young, S., Kabelis, K., Sanders, P., Evans, M., et al. (2012). In vivo effects of a human thyroid-stimulating monoclonal autoantibody (M22) and a human thyroid-blocking autoantibody (K1-70). *Autoimmunity Highlights*, *3*, 19–25.

Galløe, A. M., Rolff, M., Nordin, H., Ladefoged, S. D., & Mogensen, N. B. (1993). Cardiac performance and thyroid function. The correlation between systolic time intervals, heart rate and thyroid hormone levels. *Danish Medical Bulletin*, *40*, 492–495.

Hiroi, Y., Kim, H. H., Ying, H., Furuya, F., Huang, Z., Simoncini, T., et al. (2006). Rapid nongenomic actions of thyroid hormone. *Proceedings of the National Academy of Sciences of the United States of America*, *103*, 14104–14109.

Hofmann, P. J., Schomburg, L., & Köhrle, J. (2009). Interference of endocrine disrupters with thyroid hormone receptor-dependent transactivation. *Toxicological Sciences*, *110*, 125–137.

Hwang, J. Y., Arnold, L. A., Zhu, F., Kosinski, A., Mangano, T. J., Setola, V., et al. (2009). Improvement of pharmacological properties of irreversible thyroid receptor coactivator binding inhibitors. *Journal of Medicinal Chemistry*, *52*, 3892–3901.

Hwang, J. Y., Attia, R. R., Zhu, F., Yang, L., Lemoff, A., Jeffries, C., et al. (2012). Synthesis and evaluation of sulfonylnitrophenylthiazoles (SNPTs) as thyroid hormone receptor-coactivator interaction inhibitors. *Journal of Medicinal Chemistry*, *55*, 2301–2310.

Hwang, J. Y., Huang, W., Arnold, L. A., Huang, R., Attia, R. R., Connelly, M., et al. (2011). Methylsulfonylnitrobenzoates, a new class of irreversible inhibitors of the interaction of the thyroid hormone receptor and its obligate coactivators that functionally antagonizes thyroid hormone. *The Journal of Biological Chemistry*, *286*, 11895–11908.

Joffe, R. T., & Sokolov, S. T. (2000). Thyroid hormone treatment of primary unipolar depression: A review. *The International Journal of Neuropsychopharmacology*, *3*, 143–147.

Johnson, R. L., Hwang, J. Y., Arnold, L. A., Huang, R., Wichterman, J., Augustinaite, I., et al. (2011). A quantitative high-throughput screen identifies novel inhibitors of the interaction of thyroid receptor beta with a peptide of steroid receptor coactivator 2. *Journal of Biomolecular Screening*, *16*, 618–627.

Kearney, P. M., Baigent, C., Godwin, J., Halls, H., Emberson, J. R., & Patrono, C. (2006). Do selective cyclo-oxygenase-2 inhibitors and traditional non-steroidal anti-inflammatory drugs increase the risk of atherothrombosis? Meta-analysis of randomised trials. *BMJ*, *332*, 1302–1308.

Komatsu, T., Hirano, T., Songkram, C., Kawachi, E., & Kagechika, H. (2007). Novel thyroid hormone receptor antagonists with an N-alkylated diphenylamine skeleton. *Bioorganic & Medicinal Chemistry*, *15*, 3115–3126.

Krishnan, A. V., Stathis, P., Permuth, S. F., Tokes, L., & Feldman, D. (1993). Bisphenol-A: An estrogenic substance is released from polycarbonate flasks during autoclaving. *Endocrinology*, *132*, 2279–2286.

Mackenzie, L. S., & Lione, L. (2013). Harnessing the benefits of PPARbeta/delta agonists. *Life Sciences*, *93*, 963–967.

Miyazaki, W., Iwasaki, T., Takeshita, A., Kuroda, Y., & Koibuchi, N. (2004). Polychlorinated biphenyls suppress thyroid hormone receptor-mediated transcription through a novel mechanism. *The Journal of Biological Chemistry*, *279*, 18195–18202.

Moncada, S., Gryglewski, R. J., Bunting, S., & Vane, J. R. (1976). A lipid peroxide inhibits the enzyme in blood vessel microsomes that generates from prostaglandin endoperoxides the substance (prostaglandin X) which prevents platelet aggregation. *Prostaglandins*, *12*, 715–737.

Moriyama, K., Tagami, T., Akamizu, T., Usui, T., Saijo, M., Kanamoto, N., et al. (2002). Thyroid hormone action is disrupted by bisphenol a as an antagonist. *The Journal of Clinical Endocrinology and Metabolism, 87,* 5185–5190.

Navarrete-Ramírez, P., Luna, M., Valverde-R, C., & Orozco, A. (2014). 3,5-Di-iodothyronine stimulates tilapia growth through an alternate isoform of thyroid hormone receptor β1. *Journal of Molecular Endocrinology, 52,* 1–9.

Nguyen, N. H., Apriletti, J. W., Cunha Lima, S. T., Webb, P., Baxter, J. D., & Scanlan, T. S. (2002). Rational design and synthesis of a novel thyroid hormone antagonist that blocks coactivator recruitment. *Journal of Medicinal Chemistry, 45,* 3310–3320.

Norman, M. F., & Lavin, T. N. (1989). Antagonism of thyroid hormone action by amiodarone in rat pituitary tumor cells. *The Journal of Clinical Investigation, 83,* 306–313.

Oka, T., Mitsui-Watanabe, N., Tatarazako, N., Onishi, Y., Katsu, Y., Miyagawa, S., et al. (2013). Establishment of transactivation assay systems using fish, amphibian, reptilian and human thyroid hormone receptors. *Journal of Applied Toxicology, 33,* 991–1000.

Olson, E. J., Pearce, G. L., Jones, N. P., & Sprecher, D. L. (2012). Lipid effects of peroxisome proliferator-activated receptor-delta agonist GW501516 in subjects with low high-density lipoprotein cholesterol: Characteristics of metabolic syndrome. *Arteriosclerosis, Thrombosis, and Vascular Biology, 32,* 2289–2294.

Ooi, E. M., Watts, G. F., Sprecher, D. L., Chan, D. C., & Barrett, P. H. (2011). Mechanism of action of a peroxisome proliferator-activated receptor (PPAR)-delta agonist on lipoprotein metabolism in dyslipidemic subjects with central obesity. *The Journal of Clinical Endocrinology and Metabolism, 96,* E1568–E1576.

Paquette, M. A., Atlas, E., Wade, M. G., & Yauk, C. L. (2014). Thyroid hormone response element half-site organization and its effect on thyroid hormone mediated transcription. *PLoS One, 9.* e101155.

Perez Diaz, N., Zloh, M., Patel, P., & Mackenzie, L. S. (2015). In silico modelling of prostacyclin and other lipid mediators to nuclear receptors reveal novel thyroid hormone receptor antagonist properties. *Prostaglandins & Other Lipid Mediators, 122,* 18–27.

Ribeiro, R. C., Apriletti, J. W., Wagner, R. L., Feng, W., Kushner, P. J., Nilsson, S., et al. (1998a). X-ray crystallographic and functional studies of thyroid hormone receptor. *The Journal of Steroid Biochemistry and Molecular Biology, 65,* 133–141.

Ribeiro, R. C., Apriletti, J. W., Wagner, R. L., West, B. L., Feng, W., Huber, R., et al. (1998b). Mechanisms of thyroid hormone action: Insights from X-ray crystallographic and functional studies. *Recent Progress in Hormone Research, 53,* 351–392. discussion 392–394.

Ribeiro, R. C., Apriletti, J. W., West, B. L., Wagner, R. L., Fletterick, R. J., Schaufele, F., et al. (1995). The molecular biology of thyroid hormone action. *Annals of the New York Academy of Sciences, 758,* 366–389.

Sadana, P., Hwang, J. Y., Attia, R. R., Arnold, L. A., Neale, G., & Guy, R. K. (2011). Similarities and differences between two modes of antagonism of the thyroid hormone receptor. *ACS Chemical Biology, 6,* 1096–1106.

Schapira, M., Raaka, B. M., Das, S., Fan, L., Totrov, M., Zhou, Z., et al. (2003). Discovery of diverse thyroid hormone receptor antagonists by high-throughput docking. *Proceedings of the National Academy of Sciences of the United States of America, 100,* 7354–7359.

Shen, O., Du, G., Sun, H., Wu, W., Jiang, Y., Song, L., et al. (2009). Comparison of in vitro hormone activities of selected phthalates using reporter gene assays. *Toxicology Letters, 191,* 9–14.

Shi, W., Deng, D., Wang, Y., Hu, G., Guo, J., Zhang, X., et al. (2016). Causes of endocrine disrupting potencies in surface water in East China. *Chemosphere, 144,* 1435–1442.

Shi, W., Wei, S., Hu, X. X., Hu, G. J., Chen, C. L., Wang, X. R., et al. (2013). Identification of thyroid receptor ant/agonists in water sources using mass balance analysis and monte carlo simulation. *PLoS One, 8.* e73883.

Shi, W., Zhang, F. X., Hu, G. J., Hao, Y. Q., Zhang, X. W., Liu, H. L., et al. (2012). Thyroid hormone disrupting activities associated with phthalate esters in water sources from Yangtze River Delta. *Environment International, 42,* 117–123.

Singh, L., Pressly, B., Mengeling, B. J., Fettinger, J. C., Furlow, J. D., Lein, P. J., et al. (2016). Chasing the elusive Benzofuran impurity of the THR antagonist NH-3: Synthesis, isotope labeling, and biological activity. *The Journal of Organic Chemistry, 81,* 1870–1876.

Sokolov, S. T., Kutcher, S. P., & Joffe, R. T. (1996). Changes in thyroid hormone levels associated with desipramine response in adolescent depression. *Progress in Neuro-Psychopharmacology & Biological Psychiatry*, *20*, 1053–1063.
Sokolov, S. T., Levitt, A. J., & Joffe, R. T. (1997). Thyroid hormone levels before unsuccessful antidepressant therapy are associated with later response to T3 augmentation. *Psychiatry Research*, *69*, 203–206.
Stěpán, J. J., & Límanová, Z. (1992). Biochemical assessment of bone loss in patients on long-term thyroid hormone treatment. *Bone and Mineral*, *17*, 377–388.
Sun, H., Shen, O. X., Wang, X. R., Zhou, L., Zhen, S. Q., & Chen, X. D. (2009). Anti-thyroid hormone activity of bisphenol a, tetrabromobisphenol a and tetrachlorobisphenol a in an improved reporter gene assay. *Toxicology In Vitro*, *23*, 950–954.
Utiger, R. D. (1995). Altered thyroid function in nonthyroidal illness and surgery. To treat or not to treat? *The New England Journal of Medicine*, *333*, 1562–1563.
van Beeren, H. C., Bakker, O., & Wiersinga, W. M. (1995). Desethylamiodarone is a competitive inhibitor of the binding of thyroid hormone to the thyroid hormone alpha 1-receptor protein. *Molecular and Cellular Endocrinology*, *112*, 15–19.
Van Beeren, H. C., Jong, W. M., Kaptein, E., Visser, T. J., Bakker, O., & Wiersinga, W. M. (2003). Dronerarone acts as a selective inhibitor of 3,5,3′-triiodothyronine binding to thyroid hormone receptor-alpha1: In vitro and in vivo evidence. *Endocrinology*, *144*, 552–558.
Van den Berg, K. J., van Raaij, J. A., Bragt, P. C., & Notten, W. R. (1991). Interactions of halogenated industrial chemicals with transthyretin and effects on thyroid hormone levels in vivo. *Archives of Toxicology*, *65*, 15–19.
van Tienhoven-Wind, L. J., & Dullaart, R. P. (2015). Low-normal thyroid function and the pathogenesis of common cardio-metabolic disorders. *European Journal of Clinical Investigation*, *45*, 494–503.
Villalobos, M., Becerra, D., Núñez, M. I., Valenzuela, M. T., Siles, E., Olea, N., et al. (1996). Radiosensitivity of human breast cancer cell lines of different hormonal responsiveness. Modulatory effects of oestradiol. *International Journal of Radiation Biology*, *70*, 161–169.
Villalobos, M., Olea, N., Brotons, J. A., Olea-Serrano, M. F., Ruiz de Almodovar, J. M., & Pedraza, V. (1995). The E-screen assay: A comparison of different MCF7 cell stocks. *Environmental Health Perspectives*, *103*, 844–850.
Wagner, R. L., Apriletti, J. W., McGrath, M. E., West, B. L., Baxter, J. D., & Fletterick, R. J. (1995). A structural role for hormone in the thyroid hormone receptor. *Nature*, *378*, 690–697.
Webb, P., Nguyen, N. H., Chiellini, G., Yoshihara, H. A., Cunha Lima, S. T., Apriletti, J. W., et al. (2002). Design of thyroid hormone receptor antagonists from first principles. *The Journal of Steroid Biochemistry and Molecular Biology*, *83*, 59–73.
Westerholz, S., de Lima, A. D., & Voigt, T. (2010). Regulation of early spontaneous network activity and GABAergic neurons development by thyroid hormone. *Neuroscience*, *168*, 573–589.
Zhang, J., Li, Y., Gupta, A. A., Nam, K., & Andersson, P. L. (2016). Identification and molecular interaction studies of thyroid hormone receptor disruptors among household dust contaminants. *Chemical Research in Toxicology*, *29*, 1345–1354.
Zloh, M., Perez-Diaz, N., Tang, L., Patel, P., & Mackenzie, L. S. (2016). Evidence that diclofenac and celecoxib are thyroid hormone receptor beta antagonists. *Life Sciences*, *146*, 66–72.
Zoeller, R. T. (2005). Environmental chemicals as thyroid hormone analogues: New studies indicate that thyroid hormone receptors are targets of industrial chemicals? *Molecular and Cellular Endocrinology*, *242*, 10–15.

CHAPTER EIGHT

Thyroid Hormone Stimulation of Adult Brain Fatty Acid Oxidation

Mikaela M. Sifuentes, James D. Lechleiter[1]
UT Health San Antonio, San Antonio, TX, United States
[1]Corresponding author: e-mail address: lechleiter@uthscsa.edu

Contents

1. Introduction 164
2. Nuclear and Mitochondrial Thyroid Hormone Receptor Isoforms 165
3. Thyroid Hormone and Brain Metabolism 167
4. Thyroid Hormone Regulation of Fatty Acid Oxidation 169
   4.1 The FAO Pathway 170
   4.2 The Carnitine Shuttle 172
   4.3 Mitochondrial Trifunctional Protein 177
   4.4 Cardiolipin 179
5. Clinical Applications 180
   5.1 Injury-Induced Hypothyroidism 180
   5.2 Chronic Alcohol Abuse Neurodegeneration 183
6. Conclusion 184
References 185

Abstract

Thyroid hormone is a critical modulator of brain metabolism, and it is highly controlled in the central nervous system. Recent research has uncovered an important role of thyroid hormone in the regulation of fatty acid oxidation (FAO), an energetic process essential for neurodevelopment that continues to support brain metabolism during adulthood. Thyroid hormone stimulation of FAO has been shown to be protective in astrocytes and mouse models of brain injury, yet a clear mechanism of this relationship has not been elucidated. Thyroid hormone interacts with multiple receptors located in the nucleus and the mitochondria, initiating rapid and long-term effects via both genomic and nongenomic pathways. This has complicated efforts to isolate and study-specific interactions. This chapter presents the primary signaling pathways that have been identified to play a role in the thyroid hormone-mediated increase in FAO. Investigation of the impact of thyroid hormone on FAO in the adult brain has challenged classical models of brain metabolism and widened the window of potential neuroprotective strategies. A detailed understanding of these pathways is essential for any researchers aiming to expand the field of neuroenergetics.

Vitamins and Hormones, Volume 106
ISSN 0083-6729
http://dx.doi.org/10.1016/bs.vh.2017.05.006

1. INTRODUCTION

Mitochondrial function is a principal regulator of essential cell processes, acting as a control center for cell behavior and survival. Changes in metabolism often mediate pathological mechanisms in disease or injury, emphasizing the importance of understanding how mitochondria respond to exposure to various signaling compounds. One such compound is thyroid hormone, a well-established stimulant of mitochondria in nearly all eukaryotic cell types. Secreted by the thyroid gland and distributed systemically via the circulatory system, thyroid hormone plays a major role in the bodily maintenance of energy homeostasis. This is especially important in organs with high energy demand and low tolerance for fluctuations, such as the brain. In the average human adult, the brain consumes 20% of the body's total energetic output, despite making up only 2% of the body weight. This outsized energy demand is tightly regulated: imaging studies of the human brain have revealed that its metabolic activity is strictly maintained at a constant level over time, regardless if one is sitting passively or actively engaged in an activity (Raichle & Gusnard, 2002). This invariableness suggests a close association between metabolic regulation and brain health, a connection that is illustrated by the interruption of energy homeostasis during injury, such as from stroke or trauma.

With its unique metabolic demands and sensitivity to change, the brain is highly responsive to thyroid hormone and interacts with it in a tightly controlled fashion. Thyroid hormone is produced by the thyroid gland, where it is predominantly secreted as T4 (3,5,3′,5′-tetraiodo-L-thyronine, or thyroxine). This is believed to be the primary isoform that is transported across the blood–brain barrier (BBB), which separates the brain from the circulatory system (Crantz, Silva, & Larsen, 1982). In high enough concentrations, T4 may directly interact with thyroid hormone receptors; otherwise, it is converted by iodine removal into its more active form, T3 (3,5,3′-triiodo-L-thyronine). Due to BBB compartmentalization and the higher affinity of brain thyroid hormone transporters for T4 over T3, this activation is a major mechanism for controlling brain sensitivity to thyroid hormone. Across the entire body, circulating T3 is primarily generated by type-1 deiodinase (DIO1), located in the liver and the kidneys. In contrast, DIO1 is undetectable in the adult human central nervous system (Campos-Barros et al., 1996). Instead, thyroid hormone activation is primarily facilitated by type-2 deiodinase (DIO2), preferentially expressed in astrocytes and tanycytes, both of glial lineage (Diano, Leonard, Meli, Esposito, & Schiavo, 2003). Astrocytes serve as the primary site of T3 conversion, accounting for up to 80% of all T3 bound to thyroid hormone receptors in the cerebral cortex (Crantz et al., 1982; van Doom, Roelfsema, & van der Heide, 1982). DIO2 expression and activity are negatively regulated by thyroid hormone, preventing further activation of T4 when sufficient T3 is present (Gereben et al., 2008). Notably, activation of T4 to T3 in the brain is at a much higher percentage (75%) in the brain than in the heart, liver, kidney, and adipose tissue where it ranges from 7% to 20% (Donzelli et al., 2016). Although the significance of this difference is not yet known, it illustrates the brain's unique relationship with thyroid hormone and suggests that the brain is subjected to a high level of T3 activation under normal physiological conditions. As a counterpart to this system, thyroid hormone levels can also be actively "deactivated" by degrading both T4 and T3 into less enzymatically active derivatives. This is accomplished by type-3 deiodinase (DIO3), which in the brain is primarily expressed in neurons. These two deiodinases work as a homeostatic feedback loop to control thyroid hormone levels in the brain (Courtin et al., 2005). DIO2 activity can be downregulated by thyroid hormones as a mechanism of preventing overactivation to T3. This effect occurs rapidly in cultured astrocytes, reaching maximal inhibition at 15 min (Esfandiari, Courtin, Lennon, Gavaret, & Pierre, 1992; Leonard, Siegrist-Kaiser, & Zuckerman, 1990). Thyroid-stimulating hormone, which is released from the pituitary gland to trigger thyroid hormone release, also increases DIO2 activity in cultured astrocytes, demonstrating complimentary mechanisms for regulating T3 signaling (Saunier, Pierre, Jacquemin, & Courtin, 1993). Likewise, brain upregulation of DIO2 activity acts as a protective response against hypothyroidism. Cerebral DIO2 activity is elevated under low levels of T4, maximizing thyroid hormone impact. This is coupled with a corresponding decrease in DIO2 after treatment with T3 or T4 returns thyroid status to normal (Burmeister, Pachucki, & St Germain, 1997).

2. NUCLEAR AND MITOCHONDRIAL THYROID HORMONE RECEPTOR ISOFORMS

Thyroid hormone interacts with the brain by binding to the thyroid hormone receptors, a group of proteins that has grown over recent decades to include a range of products derived from the *THRA* and *THRB* genes. The traditional route of thyroid hormone metabolic regulation is through control of gene transcription. *THRA* and *THRB* were thought to only generate nuclear receptors that bind to hormone response elements in DNA, altering gene expression of metabolic proteins. The *THRA* gene encodes three full-length nuclear receptors, TRα1, TRα2, and TRα3, while *THRB* encodes TRβ1, TRβ2, TRβ3, and TRβ4. TRα1 and its derivatives make up to 70%–80% of all thyroid hormone receptor expression in the adult vertebrate brain (Wallis et al., 2010). This is in contrast with the liver, which primarily expresses TRβ isoforms.

Today, we know that thyroid hormone receptors are also present in the mitochondria, allowing for a more direct modulation of mitochondrial activity independent of nuclear transcription. Speculation about this pathway first arose from experiments dating back to 1951, in which isolated mitochondria exhibited a dose-dependent response to thyroxine within 30–60 min (Martius & Hess, 1951). This was confirmed by other labs reporting responses in as little as 10 min, far too rapid an effect to be mediated by the traditional thyroid hormone pathway (Lehninger & Ray, 1957; Tapley, Cooper, & Lehninger, 1955). Sterling and coworkers were the first to report strong binding of thyroid hormone to mitochondrial fractions, a breakthrough that opened up further investigation into mitochondrial thyroid hormone receptors (Sterling & Milch, 1975). Significantly, they observed a rapid response in isolated mitochondria and found that this was not blocked by the presence of protein synthesis inhibitors; thus, eliminating the possibility that mitochondrial thyroid hormone receptors require transcription of nuclear or mitochondrial DNA (Sterling, Brenner, & Sakurada, 1980). Further investigation led to the report of a high-affinity, low-capacity thyroid hormone receptor in the mitochondria. The presence of multiple isoforms of the mitochondrial thyroid hormone receptor was subsequently confirmed by other studies (Ardail, Lerme, Puymirat, & Morel, 1993; Sterling, 1991). Altogether, four shortened receptors were identified: two isoforms derived from TRα1, one shortened from TRα2, and one from TRβ3.

Of particular interest was the discovery of the varying isoforms resulting from alternative splicing of the *THRA* gene. Sterling's team were again the first to report the identification of a 28-kDa thyroid hormone receptor (p28) localized in the inner mitochondrial membrane and not detected in the nucleus (Sterling, Campbell, & Brenner, 1983). This shortened isoform is homologous to the C-terminal end of TRα1, but it lacks the DNA-binding domain and nuclear localization sequence (NLS), therefore being incapable of independently reaching the nucleus and altering gene expression. Wrutniak and coworkers further identified a 43 kDa isoform (p43) of TRα1 located in the mitochondrial matrix (Wrutniak et al., 1995). Like p28, p43 is not detected in the nucleus, though in contrast it retains the DNA-binding domain and one of the two NLS in TRα1 (Mavinakere, Powers, Subramanian, Roggero, & Allison, 2012). The localization of p43 to the mitochondria is particularly interesting due to its lack of a mitochondrial translocation signal and its import via a nonclassical import mechanism. Wrutniak observed that the removal of the N-terminal end of TRα1, which is shortened in p43, was sufficient to permit mitochondrial transport of the receptor (Wrutniak et al., 1995). This truncation removed one NLS, demonstrating that the second NLS in p43 is insufficient to induce nuclear import. Carazo and colleagues subsequently identified two sequences within the T3-binding domain that direct mitochondrial import of thyroid hormone receptors (Carazo et al., 2012). Although it is unlikely that p43 localizes to the nucleus after binding to T3, Wrutniak determined that p43 binds to multiple sites along mitochondrial DNA, stimulating gene expression. However, later studies suggest that this transcriptional activity is not the only mechanism involved in p43 signaling. Saelim and coworkers reported that T3 stimulation of p43 rapidly modulates calcium release from the endoplasmic reticulum, increasing the mitochondrial membrane potential. This effect was not inhibited by the knockout of the retinoid X receptor, a cofactor required for thyroid hormone receptor transcription. Neither was T3 activity blocked when p43 was modified to remove the DNA-binding domain or inactivate the NLS. Furthermore, T3 treatment increased oxygen consumption within 10 min in cells expressing p43, but not the full-length TRα1, suggesting that mitochondrial localization was required for this rapid, nongenomic effect (Saelim et al., 2004). These data show that even a single thyroid hormone receptor isoform can induce separate signaling pathways over differing time periods. While the exact mechanism of mitochondrial thyroid hormone receptor function remains elusive, these studies have greatly expanded the current understanding of thyroid hormone signaling and brought to light new pathways for rapidly modulating cell metabolism.

3. THYROID HORMONE AND BRAIN METABOLISM

Despite the unique nature of thyroid hormone signaling in the adult brain, it was once thought to be a thyroid hormone-insensitive organ. This perception was influenced by early studies reporting no change in cerebral oxygen consumption in mammals following thyroid hormone treatment (Fazekas, Graves, & Alman, 1951). Subsequent studies also suggested that brain thyroid hormones were so tightly regulated that changes in serum levels had no impact on brain function. However, research over the recent decades has reshaped the predominant viewpoint of the brain's thyroid hormone sensitivity. Animal studies in adult-onset hypothyroidism have found that it significantly alters cerebral substrate use in a reversible, cell type-specific manner and impairs the exchange of metabolic substrates between astrocytes and neurons (Chapa et al., 1995). Similar results have been observed in the cerebellum of hypothyroid adult rats (Constantinou, Chrysanthopoulos, Margarity, & Klapa, 2011). In clinical observations, human patients who have undergone a postadolescence thyroidectomy exhibit disruptive psychological symptoms corresponding to impaired brain mitochondrial metabolism (Park et al., 2014). These effects can be remediated with thyroid hormone replacement therapy, indicating that acute hypothyroidism-induced depression of adult brain metabolism is reversible (Smith & Ain, 1995). Likewise, late-life hypothyroidism is associated with cognitive deficits that can be alleviated with thyroid hormone treatment (Osterweil et al., 1992). Patients with mild adult-onset hypothyroidism also show significantly lower cerebral blood flow in various regions of the brain. Blood flow could not be restored even after 2–7 months of thyroid hormone treatment (Krausz et al., 2004). These studies affirm that even subclinical deficits in thyroid hormone status can significantly impact adult brain function and that not all effects are reversible.

Astrocytes make up approximately half of all the cells in the mammalian brain, and the cells play a major role in the regulation of brain metabolism, controlling cerebral blood flow, exchanging metabolic substrates with neurons, and altering the composition of the extracellular space. Astrocyte endfeet also participates in the interface between the brain and the circulatory system, rendering them particularly sensitive to molecules such as thyroid hormones that enter the brain by crossing the BBB. Indeed, thyroid hormone can rapidly induce changes in astrocyte behavior, influencing differentiation, maturation, and secretion (Gonçalves Trentin, De Aguiar, Castilho Garcez, & Alvarez-Silva, 2003; Manzano, Bernal, & Morte, 2007; Trentin, 2006). This has a profound impact on overall brain function, as astrocytes are a primary source of brain growth factors and actively participate in neuron survival, neurogenesis, axonal growth, and synapse formation. Mutations in the *THRA* gene results in a 57%–72% decrease in cerebral glucose utilization, in which astrocytes are heavily involved

(Itoh et al., 2001). Thyroid hormone treatment of astrocytes has been shown to influence neuron growth and viability. Medium conditioned by secretions of T3-treated astrocytes enhances neurite outgrowth. Correspondingly, astrocyte medium from hypothyroid rats impairs neuritogenesis and neuron survival until treated with T3 (Martinez & Gomes, 2002; Mendes-de-Aguiar et al., 2010). This demonstrates the ability of thyroid hormone to modify neural behavior via astrocyte control.

4. THYROID HORMONE REGULATION OF FATTY ACID OXIDATION

Thyroid hormone regulation of astrocyte metabolism depends in part on its ability to stimulate a metabolic pathway previously thought to be unimportant in the adult brain: fatty acid oxidation (FAO). To summarize briefly, FAO is the process of oxidizing fatty acids to provide substrates for oxidative phosphorylation in ATP production. Although FAO is widely accepted to play a major role in brain development, it has long been thought that fatty acids do not cross the adult BBB or contribute significantly to brain energy production, a viewpoint that has been challenged by a growing body of research. Glucose was previously thought to be the sole energetic substrate utilized in the brain, until observations began to emerge that perfused brains could maintain excitability and oxygen consumption in a glucose-free environment hours after depletion of glucose reserves (Geiger, Magnes, & Geiger, 1952; Grenell & Davies, 1950). Scientists began reporting evidence of FAO in the adult brain in the early 1970s, observing that the process was active in both mature neuron and astrocyte cultures. Whole-animal radiolabeling studies revealed that fatty acids can in fact cross the BBB and become incorporated into the adult brain (Dhopeshwarkar & Mead, 1970; Dhopeshwarkar, Subramanian, & Mead, 1971; Kuge et al., 1995). Furthermore, fatty acid transport proteins can be found in human endothelial cells of the BBB, providing a pathway for long chain fatty acids to be brought into the brain for utilization (Mitchell, On, Del Bigio, Miller, & Hatch, 2011). Astrocytes, which have endfeet that participate in BBB function, were found to preferentially oxidize fatty acids over glucose as fuels for energy production (Edmond, Robbins, Bergstrom, Cole, & de Vellis, 1987). This supports later observations that disrupting FAO in adult *Drosophila* triggers the accumulation of lipid droplets in astrocytes, leading to neurological impairment and a reduced lifespan. The damage can be partially repaired by rescuing FAO in astrocytes alone, reextending the shortened lifespan and highlighting the importance of astrocyte FAO to adult brain function (Schulz et al., 2015). Similar studies of impaired FAO in adult mammals are limited by the fact that the deficiency is lethal during embryogenesis, and even mild impairment would result in irreversible damage attributable to defective development. Nevertheless, advances in imaging and labeling techniques can enable researchers to study brain FAO in healthy animals. Magnetic resonance spectroscopy in live rats has shown that while glucose is the main oxidative fuel for the adult brain, approximately 20% of the brain's total energy demands are satisfied by FAO attributed primarily to astrocytes. Conversely, neurons metabolized glucose and glucose-derived lactate (Ebert, Haller, & Walton, 2003). These and other energetic studies contribute to the growing consensus that astrocytes and neurons use a variety of different substrates in parallel metabolic pathways to meet the high energy demands of the brain. This makes sense from the perspective of survival, as such a multimodal system would conceivably alternate dependence on different substrates based upon availability, or switch off dysfunctional pathways entirely, protecting the cell from damaging byproducts such as reactive oxidative species.

4.1 The FAO Pathway

FAO, also known as β-oxidation, is the process of breaking down fatty acids into substrates utilized in mitochondrial ATP production. Fatty acids inside the cell are "activated" by long chain fatty acyl-CoA ligase, which attaches coenzyme A (CoA), a thiol which functions as an acyl group carrier. Upon activation, the fatty acids then cross the outer and inner mitochondrial membranes into the matrix and enter the oxidative cycle. Each cycle removes two carbons from the fatty acid chain, producing nicotinamide adenine dinucleotide (NADH), flavin adenine dinucleotide (FADH2), and an acetyl-CoA. The shortened fatty acid reenters the cycle until it has been completely reduced to acetyl-CoA, which is transferred to the citric acid cycle. There it is used to create citrate, which undergoes a chain of chemical reactions producing an additional three NADH molecules and one FADH2. Altogether, the oxidation of each two carbons of a fatty acid generates four NADH and two FADH2 molecules, which in turn are consumed by the electron transport chain in the process of producing ATP through oxidative phosphorylation. By regulating FAO, thyroid hormone levels affect the mitochondrial stores of NADH and FADH2, which can serve as reserve fuel for oxidative phosphorylation in scenarios of limited metabolic resources (Fig. 1).

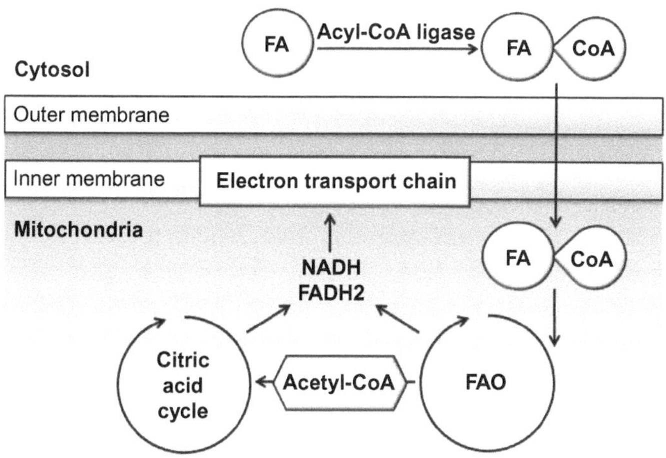

Fig. 1 Overview of fatty acid oxidation. Fatty acids are activated by acyl-CoA ligase into a fatty acyl-CoA and cross into the mitochondria where it enters in the FAO cycle. Acetyl-CoA enters into the citric acid cycle. NADH and FADH2 produced from these processes are used by the electron transport chain to produce ATP.

The ability of FAO to supply acetyl-CoA to the citric acid cycle is especially important in astrocytes, which preferentially utilize fatty acids rather than glucose. In glucose-preferring cell types, such as neurons, glucose is converted into pyruvate via glycolysis. Pyruvate is then transformed into acetyl-CoA for entry into the citric acid cycle by the pyruvate dehydrogenase complex (PDC). While neuronal PDC operates at near maximal levels, in astrocytes, PDC activity is strongly inhibited (Halim et al., 2010). Rather, pyruvate is either converted into lactate via lactate dehydrogenase, or oxaloacetate by the glia-specific pyruvate carboxylase (Cesar & Hamprecht, 1995; Yu, Drejer, Hertz, & Schousboe, 1983). This suggests that FAO is a major contributor to astrocyte acetyl-CoA, compensating for the reduced dependence on glucose.

Since the 1920s, thyroid hormone studies have observed a direct relationship between thyroid state and FAO in multiple cell types (Hoch, 1988). Exposure to T3 initiates a shift in substrate preference from glucose to free fatty acids (Krueger, Ning, Argo, Hyyti, & Portman, 2001) and increases FAO in cultured rat hepatocytes, isolated cardiomyocyte mitochondria, skeletal muscle, mouse fibroblasts, and human astrocytes (Chocron et al., 2012; Lombardi et al., 2009; Sayre et al., 2016; Stakkestad & Lund, 1984). This effect corresponds with observations of human subjects in clinical studies. Hyperthyroid patients exhibit an increased oxidation of fatty acids into carbon dioxide or ketone bodies, while FAO in hypothyroid patients is decreased by more than 50% (Gjedde et al., 2010; Müller & Seitz, 1984). Likewise, mice with a mutation known to cause human generalized thyroid hormone resistance exhibit suppressed FAO (Araki, Ying, Zhu, Willingham, & Cheng, 2009). Given the importance of FAO in the adult brain, its regulation by thyroid hormone likely plays a critical role in the cognitive deficits present in adult-onset hypothyroidism. Indeed, both human and mouse astrocytes exhibit increased ATP production and FAO after only 15 min of T3 treatment, demonstrating rapid responsiveness to thyroid hormone levels in the brain (Sayre et al., 2016). Despite the abundance of research on the impact of thyroid hormone on metabolism, the mechanism by which it alters FAO is not well understood. As will be explained in this chapter, many of the long-term effects of thyroid hormone can be altered by gene transcription via nuclear receptors; however, this is insufficient to explain some rapid responses that occur in isolated mitochondria or in the presence of protein synthesis inhibitors. The investigation into this mechanism is further complicated by the complex nature of thyroid hormone signaling pathways, the existence of compensatory feedback loops, and the lack of appropriate *in vivo* models. This section will discuss the many pathways by which thyroid hormone may alter FAO in the brain.

4.2 The Carnitine Shuttle

Fatty acids can widely range in length and can be divided up into categories of short (<6 carbon atoms), medium (6–12), long (14–20), and very long (>22). Medium and short chain fatty acids are able to passively diffuse across both mitochondrial membranes for FAO, whereas long chain fatty acids require transport via the carnitine shuttle. The carnitine shuttle consists of carnitine palmitoyl transferase I (CPT1) in the outer membrane, with carnitine-acylcarnitine translocase (CAT) and carnitine palmitoyl transferase II (CPT2) both in the inner membrane. First, CPT1 replaces the CoA with a carnitine, enabling it to be transported by CAT into the mitochondrial matrix. The carnitine is then removed by CPT2 and a CoA is reattached, allowing the fatty acid to continue on to the FAO cycle.

For long chain fatty acids such as palmitic acid, the carnitine shuttle represents a major rate-limiting step in FAO. Inhibition of CPT1 rapidly eliminates astrocyte FAO of long chain fatty acids, indicating that there are little to no reserves stored within the mitochondria themselves (Chocron et al., 2012). Therefore, the rate of CPT1 enzymatic activity acts as a primary regulatory mechanism of long chain FAO. In human astrocytes, T3 rapidly increases ATP production only in the presence of functional CPT1, indicating that this response is specifically dependent upon the mitochondrial availability of long chain fatty acids (Sayre et al., 2016). Medium and short chain fatty acids appear insufficient to enable T3 increase of mitochondrial metabolism, possibly due to low availability in astrocytes. Due to their size, very long chain fatty acids are generally metabolized in peroxisomes, making them unlikely substrates for T3-induced FAO in mitochondria.

There are three known isoforms of CPT1, designated CPT1a, CPT1b, and CPT1c. CPT1a makes up the majority of the body's CPT1, as it is the primary isoform expressed in the liver, the major powerhouse of FAO. This same enzyme is also the predominant isoform found in cortical astrocytes (Blázquez, Sánchez, Velasco, & Guzmán, 1998). For this reason, there may be some benefit from the comparison of FAO pathways between hepatocytes and astrocytes. Notably, CPT1a expression and activity is enhanced in hyperthyroid animals, likely having an impact on FAO signaling (Jackson-Hayes et al., 2003; Mynatt, Park, Thorngate, Das, & Cook, 1994). CPT1b, with size and behavior distinct from CPT1a, is located almost exclusively in myocyte mitochondria. The brain-specific CPT1c can be found in both neurons and astrocytes, and it is localized to the endoplasmic reticulum instead of the mitochondria (Sierra et al., 2008; Taïb et al., 2013). Mutation of CPT1c is implicated in spastic paraplegia 73, a late-onset progressive neurological disorder characterized by dysfunctional ceramide metabolism, with impaired conductance in central, but not peripheral nervous tissue (Carrasco et al., 2013). Since de novo ceramide synthesis takes place on the endoplasmic reticulum, the brain specificity of CPT1c suggests that the brain uniquely utilizes a carnitine-mediated transport of long chain fatty acids for this process (Suzuki, Iwasaki, Kato, & Wagai, 1997).

As a rate-limiting step in astrocyte FAO, CPT1 also participates in a nongenomic switching mechanism that allows the cell to alternate between FAO and fatty acid synthesis. CPT1 activity can be acutely regulated by altering the activity of acetyl-CoA carboxylase (ACC), which produces malonyl-CoA, an important substrate for the biosynthesis of fatty acids. The activation of ACC inhibits CPT1 activity, directing cells to alternate between synthesis and oxidation states (McGarry, Leatherman, & Foster, 1978). This function is essential for survival during conditions of nutrient

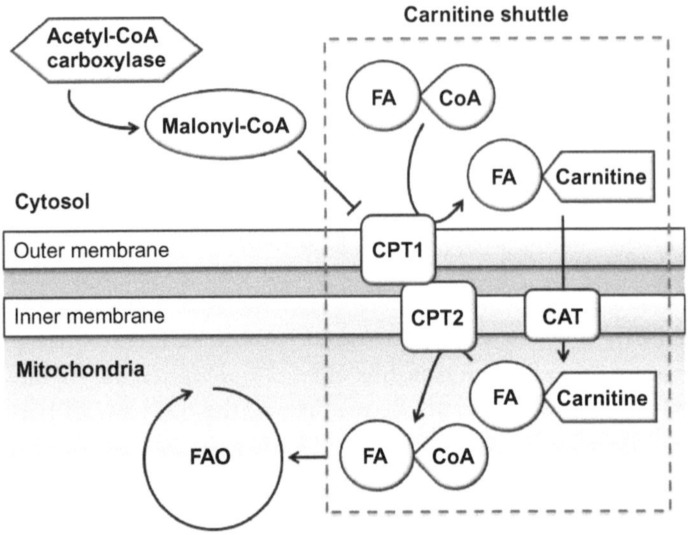

Fig. 2 The carnitine shuttle. CPT1 removes the CoA from the activated fatty acid and adds a carnitine, which allows the fatty acid to be transported into the mitochondria by CAT. CPT2 removes the carnitine and adds a CoA before the fatty acid can enter into the FAO cycle. Acetyl-CoA carboxylase produces malonyl-CoA, which is an endogenous inhibitor of CPT1.

deprivation, which is accompanied by a decrease in malonyl-CoA levels and increased fatty acid transport into mitochondria (Fig. 2).

4.2.1 Adenosine Monophosphate-Activated Protein Kinase

The adenosine monophosphate-activated protein kinase (AMPK) is an ubiquitous signaling enzyme in eukaryotic cells that regulates multiple metabolic signaling pathways in response to cellular stress. Depletion of ATP induces AMPK phosphorylation, thereby activating it and initiating a signaling cascade that switches off ATP-dependent processes and shifts to catabolic mechanisms of generating ATP. Treatment with thyroid hormone has been shown to induce AMPK activation, which in turn phosphorylates ACC, rendering it inactive. This results in lowered levels of malonyl-CoA, derepressing CPT1 transport of fatty acids into the mitochondria and increasing FAO (Lombardi et al., 2009; Yamauchi et al., 2008).

Thyroid hormone may activate AMPK by modulating the activity of upstream regulator calcium/calmodulin-dependent protein kinase kinase-β (CaMKKβ). Intraperitoneal injection of rats with T3 increases CaMKKβ expression in the liver after 8 h (Videla, 2014). This relatively slow response suggests that any CaMKKβ-mediated effects of T3 could be initiated by transcriptional modulation. In myoblasts treated with T3, CaMKKβ phopsphorylation of AMPK was necessary for the thyroid hormone stimulation of FAO, indicating that CaMKKβ can function as a mediator of long-term upregulation of fatty acid consumption (Yamauchi et al., 2008).

The AMPK pathway is especially important in astrocytes due to their role in regulating brain metabolism. AMPK activity in rat cortical astrocytes is threefold higher than in neurons, and further activation of AMPK significantly increases FAO via CPT1 activity in both *in vitro* and *in vivo* studies (Blázquez, Woods, de Ceballos, Carling, & Guzmán, 1999). These observations suggest that thyroid hormone can increase astrocyte FAO simply by increasing availability of long chain fatty acids in the mitochondria. However, not all observations support this hypothesis. A recent study in mouse cortical astrocytes reported that although AMPK activation decreases ACC activity, FAO is not affected (Taïb et al., 2013). Another reported that AMPK phosphorylation is necessary for FAO stimulation, but not sufficient, suggesting that the effect of thyroid hormone may result from a combination of multiple metabolic pathways working in concert (Chocron et al., 2012).

4.2.2 The PI3K/Akt Pathway

Protein kinase B, also known as Akt, is an essential serine/threonine-specific protein kinase at the center of multiple cellular processes essential for cellular growth, metabolism, and survival. Akt interacts with phosphatidylinositol 3-kinase (PI3K) in a signaling pathway that can be activated by a wide range of signals, including thyroid hormone. Exposure of mouse and human endothelial cells to T3 rapidly increases Akt phosphorylation within 20 min via a nongenomic mechanism (Hiroi et al., 2006). In the presence of T3, TRα1 interacts with PI3K, initiating the Akt pathway. This leads to the activation of endothelial nitric oxide synthase (eNOS), which produces the free radical signaling compound nitric oxide. Despite its name, eNOS is present in astrocytes as well as endothelial cells, mediating important functions such as vasodilation (Stobart, Lu, Anderson, Mori, & Anderson, 2013; Wiencken & Casagrande, 1999). In fact, thyroid hormone activation of eNOS mediates T3-induction of increased cerebral blood flow, demonstrating its importance in thyroid hormone-regulated brain metabolism (Hiroi et al., 2006). Importantly, nitric oxide has been shown to increase CPT1a activity by activating soluble guanylyl cyclase, increasing the intracellular levels of cyclic guanosine monophosphate (cGMP) and downregulating ACC (García-Villafranca, Guillén, & Castro, 2003). This provides yet another pathway for derepressing CPT1 by lowering malonyl-CoA. This regulation of FAO is confirmed by experiments in eNOS knockout mice showing a reduction of FAO, increased fatty acid synthesis, and decreased expression of CPT1 (Gouill et al., 2007). However, studies in cultured rat astrocytes suggest that glial CPT1 regulation relies at least in part on a malonyl-CoA-independent pathway (Blázquez et al., 1998) (Fig. 3).

Interestingly, cGMP activity has also been implicated in increasing astrocyte sensitivity to thyroid hormone. A study in cultured rat astrocytes found that stimulation with a cGMP analog increased DIO2 activity in a dose-dependent manner. DIO2 mRNA significantly increased 1 h after exposure, resulting in a detectable increase in DIO2 activity at 2 h. These data suggest that the eNOS pathway may reinforce thyroid hormone signaling in the brain by increasing the conversion of T4 to T3 (Gondou et al., 1998).

In addition to the cGMP-mediated modification of CPT by eNOS, nitric oxide signaling can also modify protein behavior by the posttranslational modification of reducing cysteine residues in a process called *S*-nitrosylation. *S*-nitrosocysteine residues have been detected in many proteins involved in metabolic processes, including CaMKKβ, fatty acid synthase, and long chain fatty acyl-CoA ligase, which activate fatty acids

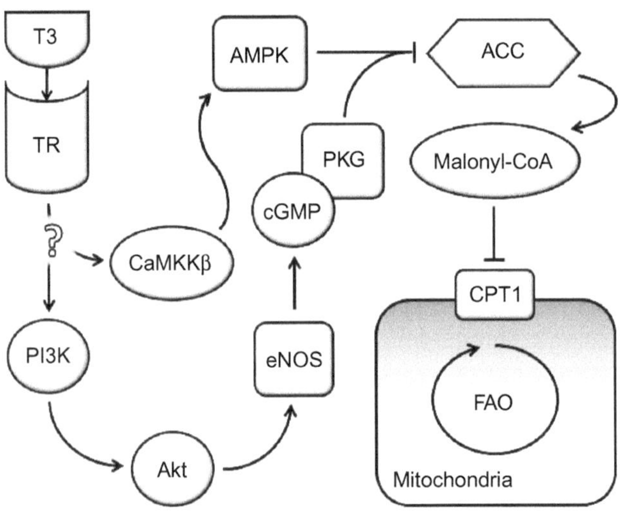

Fig. 3 Nongenomic regulation of CPT by thyroid hormone. Thyroid hormone binds to a thyroid hormone receptor, which may initiate one or both signaling pathways to disinhibit CPT1. Both the AMPK and the PI3K/Akt signaling cascades have been implicated in the T3-mediated increase in CPT1 activity.

before FAO. Nitrosylation of long chain fatty acid-CoA ligase 6 requires neuronal NOS (nNOS), which despite its name is expressed in multiple cell types throughout the body. Modification of fatty acid synthase requires the presence of both eNOS and nNOS (Raju et al., 2015). Like eNOS, nNOS responds to thyroid hormone status, though it appears to have a distinct signaling pathway (Serfozo et al., 2008). Whereas knockout of eNOS results in the formation of lipid droplets in hepatocytes, this is not observed in nNOS null cells, indicating that nNOS is not essential to lipid droplet metabolism and therefore may only be indirectly linked to FAO (Schild et al., 2008).

4.3 Mitochondrial Trifunctional Protein

Upon entering the FAO cycle, a fatty acyl-CoA first interacts with an acyl-CoA dehydrogenase, which introduces a double bond and produces one FADH2. After this initial step, the resulting unsaturated thioester continues on to the mitochondrial trifunctional protein (MTP). MTP is an inner membrane-associated enzyme complex that catalyzes the last three steps of FAO: long chain 2,3-enoyl-CoA hydratase (LCEH), long chain 3-hydroxyacyl-CoA dehydrogenase (LCHAD), and long chain 3-ketoacyl-CoA thiolase (LKAT). The MTP complex consists of four 80 kDa α and four 50 kDa β subunits, which are transcribed from the *HADHA* and *HADHB* genes, respectively. MTPα encodes the LCEH and LCHAD activities, whereas the MTPβ encodes the thiolase activity only. Both genes are linked in a head-to-head orientation on opposite DNA strands that are transcribed from a common bidirectional promoter, suggesting that synthesis of the MTP subunits is coordinated by the same transcription signaling pathway (Orii et al., 1999). This may be due to the fact that the full activity of the MTP complex is heavily dependent upon expression of both subunits and their assembly into a functional unit (Kamijo, Aoyama, Komiyama, & Hashimoto, 1994; Liu et al., 2008). The β subunit is especially reliant on coexpression and complex formation for normal function. Although both subunits contain leader sequences for targeting to the mitochondria, MTPβ cannot localize to the inner membrane in the absence of MTPα (Weinberger, Rinaldo, Strauss, & Bennett, 1995). Certain mutations in the α subunit can also result in the inactivation and rapid degradation in both subunits even when the *HADHB* gene is preserved (Fould et al., 2010; Spiekerkoetter, Khuchua, Yue, Bennett, & Strauss, 2004). Human patients with MTP deficiencies resulting from mutations in either subunit present with severe neonatal metabolic dysfunction and long-term neurological abnormalities. However, MTP deficiency syndrome is rare, since most carriers do not survive gestation. Due to its necessity, MTP is relatively conserved in mammals. Human HADHA shares 98% of identical residues with orangutans, 97% with Rhesus macaques, 86% with sheep, and 84% with mice and rats, with similar levels of homology in HADHB (Madden, 2003) (Fig. 4).

As the primary enzyme involved with FAO and a critical regulator of FAO rate, MTP activity is highly responsive to thyroid hormone. Cells expressing only the mitochondrial p43 increase ATP production and FAO within 15 min of T3 treatment, indicating that thyroid hormone can stimulate FAO via a direct mitochondrial pathway. Furthermore, using an affinity pull-down assay, Chocron and coworkers were able to identify both MTP subunits bound to p43 in isolated mitochondria. Most significantly, T3 treatment rapidly induces the incorporation of MTP subunits into the 520-kDa octameric complex, thereby increasing FAO by allowing MTP to operate a full enzymatic capacity. This effect was not inhibited when fatty acid transport into the mitochondria was blocked, indicating that MTP assembly was not a consequence of increased fatty acid availability (Chocron et al., 2012).

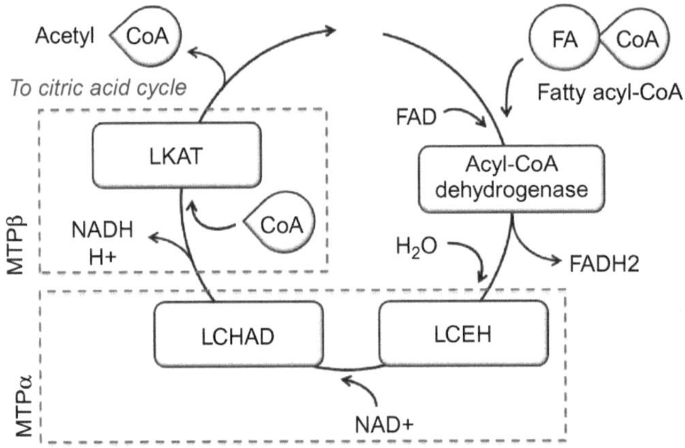

Fig. 4 The fatty acid oxidation cycle. The FAO cycle is facilitated by four enzymes: acyl-CoA dehydrogenase, LCEH, LCHAD, and LKAT. The last three enzymes make up the MTP complex: the α subunit contains LCEH and LCHAD, while the β subunit contains LKAT. One FAO cycle produces an acetyl-CoA and a fatty acyl-CoA that has been reduced by two carbons, which reenters the cycle.

4.4 Cardiolipin

Cardiolipin is an essential phospholipid that makes up to 15%–20% of the inner mitochondrial membrane. It is the only known dimeric phospholipid, and its unique structure plays a major role in maintaining the function of membrane-associated proteins in the mitochondria (Daum, 1985). This is especially important when it comes to the formation of "supercomplexes," large complexes of metabolic proteins that engage in substrate shuttling to increase enzymatic activity (Zhang, Mileykovskaya, & Dowhan, 2002). These supercomplexes may also include the inner membrane-associated MTP, which maintains a close relationship with the complexes of the electron transport chain. During FAO, LCHAD reduces NAD to form NADH, which is taken up by complex I. A study of MTP sensitivity found that concentrations of NAD+/NADH could regulate FAO, but control by this mechanism would require large, nonphysiological fluctuations. Alternatively, the authors proposed a close shuttling pool of NAD+/NADH between MTP and complex I (Eaton, Middleton, & Bartlett, 1998). Evidence supporting this theory has grown. Complex I of the respiratory chain was discovered to have binding capabilities for MTPα, and an acyl-CoA dehydrogenase has also been identified as a complex I assembly factor (Nouws, Te Brinke, Nijtmans, & Houten, 2014; Sumegi & Srere, 1984). Isolated respiratory supercomplexes exhibit FAO substrate channeling, demonstrating that these large units are functionally active (Wang, Mohsen, Mihalik, Goetzman, & Vockley, 2010). Assembly into these supercomplexes is highly dependent upon cardiolipin availability and is associated with increased enzymatic activity (Bazan et al., 2013; Mileykovskaya & Dowhan, 2013).

Thyroid hormone manipulation of cardiolipin in mitochondria has been well documented. Thyroid hormone can rapidly alter the composition of the inner mitochondrial membrane by increasing the activity of cardiolipin synthase, which mediates de novo cardiolipin synthesis (Cao, Cheng, Angel, & Hatch, 1995; Hostetler, 1991). Interestingly, cardiolipin content was also found to be increased via remodeling by monolysocardiolipin acyltransferase (MLCL AT), which has been identified as a shortened isoform of MTPα, with an identical 59 kDa C-terminal end (Taylor & Hatch, 2009). Activity of MLCL AT is also upregulated *in vivo* in rats treated with thyroid hormone (Mutter, Dolinsky, Ma, Taylor, & Hatch, 2000). In fact, Taylor and coworkers also showed that full-length MTPα retained MLCL AT capabilities, leading to the possibility of a dual role of

MTP in thyroid hormone response (Taylor et al., 2012). Interaction of p43 and MTP may initiate concurrent mechanisms of enriching cardiolipin content in the inner membrane, thereby allowing for the incorporation of MTP into higher order supercomplexes. While T3 rapidly increases the association of MTPα into the octameric complex, MTPα can also be observed in even larger complexes at 700 kDa, 800 kDa, and higher (Chocron et al., 2012).

5. CLINICAL APPLICATIONS

The thyroid hormone stimulation of astrocyte FAO has been shown to protect cells against apoptosis in many different cell types under various forms of stress. Investigation of the mitochondrial p43 demonstrated that direct T3 stimulation of mitochondria can increase resistance to apoptotic stimuli. T3 binding to p43 was found to be sufficient to increase the membrane potential of isolated mitochondria, highlighting the lack of nuclear transcription involvement (Saelim, Holstein, Chocron, Camacho, & Lechleiter, 2007). The therapeutic potential of this mechanism has been considered for diseases presenting with metabolic syndrome and FAO deficiencies (Sayre & Lechleiter, 2012). Due to its rapid modulation of mitochondrial function, treatment with thyroid hormone has profound implications for the treatment of brain injuries characterized by glucose deprivation and metabolic dysfunction, such as traumatic brain injury (TBI) or stroke.

5.1 Injury-Induced Hypothyroidism

Brain injuries are often characterized by a drop in circulating thyroid hormone levels. The phenomenon of decreased T3 following stoke has been thoroughly documented over the years in clinical studies all over the world. One of the earliest studies included 65 stroke patients in China, which found that 65% showed low levels of T3 (Liang, 1991). This and two other Chinese studies of 209 total patients reported that low T3 levels are closely correlated with larger lesion size and poor outcome (Chen, Xiao, Liu, & Wang, 2015; Liu et al., 2016). Likewise, a study involving 737 patients with first-time acute stroke in Greece found that 56% of patients exhibited low T3, with a greater neurological impairment and higher mortality rate during the first year. These patients also recovered more slowly, with a smaller percentage independent after 1 year (Alevizaki, Synetou, Xynos,

Pappa, & Vemmos, 2007). This study, along with another from a Chinese cohort of 1072 patients, emphasized the finding that lower T3 was associated with poorer outcome even among patients within a "normal" T3 range (Xu, Li, & Hu, 2016). Similar studies in the United States have reported that T3 levels were predictive of cognitive impairment upon admission and functional recovery up to 1 year after stroke (O'Keefe et al., 2015; Zhang & Meyer, 2010). A study of 88 patients in Lithuania reported that low T3 levels correlated with poor functional and cognitive outcomes, even after controlling for clinical stroke severity (Bunevicius, Kazlauskas, Raskauskiene, Janusonis, & Bunevicius, 2014). Confirmatory results were also published recently from Japan (Suda et al., 2016).

Studies of patients suffering from a TBI have produced similar findings. An early report in the United States noted that patients who died or remained in a vegetative state exhibited T3 levels 30%–50% lower than patients with good recovery (Woolf, Lee, Hamill, & McDonald, 1988). A clinical study of 63 TBI patients in Germany observed an inverse correlation between serum T3 concentration and the number of lesions identified in CT scans (Kleindienst, Brabant, Bock, Maser-Gluth, & Buchfelder, 2009). An Iranian study involving 72 patients documented a decrease in T3 and T4 that correlated with injury severity and mortality rate (Malekpour, Mehrafshan, Saki, Malekmohammadi, & Saki, 2012). Similar studies in Sweden and Russia with a combined 101 patients also reported an association between low T3 and poor outcomes (Olivecrona, Dahlqvist, & Koskinen, 2013; Shamaeva et al., 2012).

Despite these findings, clinicians are skeptical of the value of treating stroke or TBI patients with thyroid hormone. Thyroid hormone dysfunction is seen in up to 70% of hospitalized patients with a wide variety of illnesses. Injury-associated low levels of T3 are termed nonthyroidal illness syndrome (NTIS), and patient outcome is strongly correlated with the severity of T3 deficiency. The cause of NTIS is unknown, though there is some implication of suppression of the pituitary-thyroid axis (Stathatos, Levetan, Burman, & Wartofsky, 2001). The prevailing viewpoint among clinical researchers is that NTIS is a symptom of illness, not a mediator of injury progression. Therefore, studies supporting a relationship between thyroid hormone and recovery point to the prognostic value of T3, not a therapeutic one. The criticism of a lack of evidence supporting T3 supplementation is a valid one, since many human studies lack adequate controls and the effects of thyroid hormone have yet to be clearly defined. Exposure to T3 is not without its own risks of unwanted side effects, and there have yet to be any clinical trials testing the safety of T3 treatment among stroke or TBI patients. Some clinical studies have tested the impact of T3 treatment on various other conditions with NTIS, with mixed outcomes. In the absence of a greater understanding of tissue-specific, genomic, and nongenomic pathways of thyroid hormone, treatment of stroke-associated NTIS remains controversial (Economidou, Douka, Tzanela, Nanas, & Kotanidou, 2011). Such medical dilemmas demonstrate the need for greater understanding of the multiple downstream effects of thyroid hormone, and how it impacts the metabolic status of different tissues.

Despite the lack of human studies supporting thyroid hormone treatment of stroke, much research has been done in animals to understand how T3 affects the poststroke brain. Treatment with T3 has been shown to significantly reduce lesion size in multiple rodent models of ischemia, providing a basis for understanding how this neuroprotection is enacted (Genovese et al., 2013; Mdzinarishvili, Sutariya, Talasila, Geldenhuys, & Sadana, 2013; Rami & Krieglstein, 1992; Sayre et al., 2016). Multiple pathways for T3 neuroprotection have been proposed, much of which implicates FAO. Hiroi and coworkers reported that T3 had no effect on lesion size in eNOS knockout mice, pointing to a PI3K/Akt-eNOS pathway. They also noted that T3 somewhat paradoxically decreased blood pressure, which is consistent with eNOS activation but usually leads to greater severity in stroke, which was not observed (Hiroi et al., 2006). However, this finding is consistent with a FAO-mediated pathway of T3 neuroprotection, given that nitric oxide can induce the derepression of CPT1 and allow for greater trafficking of fatty acids into the mitochondria. This function may especially aid in protecting the penumbra in stroke, where lipid droplets accumulate in multiple cell types surrounding the infarct core (Kudo & Nagayama, 1988). Under conditions of nutrient deprivation, such as in stroke, cultured astrocytes generate large lipid droplets primarily consisting of triacylglycerides, which when consumed by FAO are associated with increased survival (Cabodevilla et al., 2013). In a mouse model of ischemia, blocking fatty acid transport into the mitochondria for FAO results in increased lesion size, demonstrating the importance of FAO in stroke *in vivo*. Decreased FAO capacity in a heterozygous knock down of MTP is also sufficient to block the protective effects of T3 (Sayre et al., 2016). These data suggest an increased reliance on FAO in the brain following injury, and identify MTP as a mediator of T3 neuroprotection.

5.2 Chronic Alcohol Abuse Neurodegeneration

Long-term, heavy alcohol abuse is often associated with the production of a fatty liver. Studies have found that this primarily results from disordered FAO. The abnormal accumulation of lipids, termed alcoholic steatosis, involves inhibition of FAO by accumulation of NADH and the inhibition of fatty acyl-dehydrogenases. In rodents fed ethanol, AMPK activity is also decreased, allowing for high levels of malonyl-CoA. This has the dual effect of suppressing CPT1 transport of fatty acids and leaving fatty acid synthesis unchecked (García-Villafranca et al., 2003). This mechanism of damage is not limited to the liver; alcohol abuse-associated neuropathy and neurocognitive deficits have also been linked to impaired FAO. Chronic alcohol consumption causes brain hypoglycemia, limiting glucose availability in the brain and increasing dependence of brain health on the ability to utilize fatty acids. Alcohol consumption decreases brain expression and activity of CPT1 and CPT2, and restoration of carnitine palmitoyl transferase function improves resistance of isolated brain mitochondria to alcohol insult (Haorah, Rump, & Xiong, 2013). These findings are consistent with observations made decades before: brains from human subjects deceased from alcoholic cirrhosis showed significant signs of neurodegeneration and accumulation of long chain fatty acids (Lesch, Schmidt, & Schmidt, 1973). Impairment of CPT1/CPT2 would selectively inhibit FAO of long chain fatty acids, as medium and short chain fatty acids can diffuse across the mitochondrial membranes without assistance.

Chronic alcohol users also exhibit suppressed T4 and T3 serum levels. Prolonged sobriety is correlated with a return to normal T3, with a reemergence of hypothyroidism upon relapse. Alcohol abuse is well known to directly induce toxicity in thyroid cells and reduce thyroid volume, though this mechanism has yet to be elucidated (Balhara & Deb, 2013). Although T3 supplementation has not yet been investigated as treatment for alcohol abuse-related neuropathies, the current evidence suggests a relationship between thyroid hormone and FAO dysfunction in alcohol dependence disorders and presents a therapeutic opportunity. Recently, agonists of the peroxisome proliferator-activated receptor α (PPARα) have been found to reduce self-administration of alcohol in rats (Karahanian et al., 2015). PPARα ligands increase expression of CPT1a, and overexpression of the receptor in cardiomyocytes induces degradation of malonyl-CoA, providing a dual mechanism for increased fatty acid supply in mitochondria (Campbell et al., 2002; Song et al., 2010). While PPARα is the primary isoform expressed in the liver, it is also expressed throughout the brain in multiple cell types, suggesting that treatment with PPARα agonists will have an impact on brain function (Warden et al., 2016). This provides additional support for the proposal of a FAO-mediated treatment of alcoholic neurodysfunction.

6. CONCLUSION

Contrary to early assumptions that the adult brain is a thyroid hormone-insensitive organ, this critical metabolic hormone initiates multiple brain signaling pathways via genomic and nongenomic mechanisms. Thyroid hormone secretion, activation, and availability are highly regulated throughout the body, and the brain is no exception. Due to the unique feature of the BBB, the brain's local T3 levels can be especially fine-tuned to meet the high energy requirement of synapse maintenance and protect the brain from acute energetic fluctuations during illness or injury. The necessity for responsiveness may account for the existence of the many isoforms of the thyroid hormone receptor and their varying locations within the cell. Not only must thyroid hormone be able to induce long-term changes to bodily metabolism, it must also activate rapid intramitochondrial responses to improve cell survival. This requirement of thyroid hormone can be met by regulating brain fatty acid metabolism, once thought to be of little importance beyond development. Although glucose is the primary energetic substrate for the brain, FAO plays a critical role in astrocyte metabolism, supporting neuron function and providing the protective option of alternative fuel sources in the case of glucose depletion.

Owing to the multifaceted nature of thyroid hormone signaling, it has been challenging to fully explain its mechanism of modulating FAO. This chapter describes several of the most prominent proposals for the primary T3-FAO signaling pathway; however, it is likely that thyroid hormone employs multiple, synergistic mechanisms for regulating cell metabolism. These pathways may include increasing expression of FAO proteins, inducing transport of fatty acids into mitochondria, modifying the assembly of supramolecular complexes, directly interacting with FAO enzymes, or all of the above. Moreover, studies have shown that some of these processes are functionally linked independently of thyroid hormone, which can further complicate efforts to extricate signaling order.

Mitochondrial dysfunction is symptomatic of many brain disorders, making neural metabolism a critical target for treatment. Therefore it is absolutely necessary to consider the impact of thyroid hormone on the brain in disease states, which are often characterized by altered levels of brain T3. For decades, injury-related hypothyroidism has been disregarded as a curious yet inexplicable medical phenomenon, but the extensive documentation of this condition demonstrates its importance to disease progression and recovery. A considerable body of evidence has substantiated the antiapoptotic, neuroprotective effects of thyroid hormone using both *in vitro* and *in vivo* models, indicating that it has more than mere diagnostic value. Thyroid hormones and related mimetics are already approved for human use for treatment for hypothyroidism, making them more accessible than new experimental therapeutics. Despite this, thyroid hormone treatment is hindered by the uncertainty that surrounds its many downstream effects. Although thyroid hormone has been extensively studied in the liver and other tissues, research in the brain has lagged, especially concerning FAO. Many past studies in brain metabolism have failed to consider the contributions of astrocytes or nonglucose substrates, such as fatty acids. Such investigations run the risk of concluding with an incomplete picture of neuroenergetics. Future studies that integrate current knowledge of thyroid hormone and FAO in the adult brain have strong potential to pioneer innovative metabolic strategies of neuroprotection.

REFERENCES

Alevizaki, M., Synetou, M., Xynos, K., Pappa, T., & Vemmos, K. N. (2007). Low triiodothyronine: A strong predictor of outcome in acute stroke patients. *European Journal of Clinical Investigation, 37*(8), 651–657. https://doi.org/10.1111/j.1365-2362.2007.01839.x.

Araki, O., Ying, H., Zhu, X. G., Willingham, M. C., & Cheng, S. Y. (2009). Distinct dysregulation of lipid metabolism by unliganded thyroid hormone receptor isoforms. *Molecular Endocrinology, 23*(3), 308–315. https://doi.org/10.1210/me.2008-0311.

Ardail, D., Lerme, F., Puymirat, J., & Morel, G. (1993). Evidence for the presence of alpha and beta-related T3 receptors in rat liver mitochondria. *European Journal of Cell Biology, 62*(1), 105–113.

Balhara, Y. P. S., & Deb, K. S. (2013). Impact of alcohol use on thyroid function. *Indian Journal of Endocrinology and Metabolism, 17*(4), 580–587. https://doi.org/10.4103/2230-8210.113724.

Bazan, S., Mileykovskaya, E., Mallampalli, V. K. P. S., Heacock, P., Sparagna, G. C., & Dowhan, W. (2013). Cardiolipin-dependent reconstitution of respiratory Supercomplexes from purified *Saccharomyces cerevisiae* complexes III and IV. *Journal of Biological Chemistry, 288*(1), 401–411. https://doi.org/10.1074/jbc.M112.425876.

Blázquez, C., Sánchez, C., Velasco, G., & Guzmán, M. (1998). Role of carnitine palmitoyltransferase I in the control of ketogenesis in primary cultures of rat astrocytes. *Journal of Neurochemistry, 71*(4), 1597–1606.

Blázquez, C., Woods, A., de Ceballos, M. L., Carling, D., & Guzmán, M. (1999). The AMP-activated protein kinase is involved in the regulation of ketone body production by astrocytes. *Journal of Neurochemistry, 73*(4), 1674–1682.

Bunevicius, A., Kazlauskas, H., Raskauskiene, N., Janusonis, V., & Bunevicius, R. (2014). Ischemic stroke functional outcomes are independently associated with C-reactive protein concentrations and cognitive outcomes with triiodothyronine concentrations: A pilot study. *Endocrine*, *45*(2), 213–220. https://doi.org/10.1007/s12020-013-9958-2.

Burmeister, L. A., Pachucki, J., & St Germain, D. L. (1997). Thyroid hormones inhibit type 2 iodothyronine deiodinase in the rat cerebral cortex by both pre- and posttranslational mechanisms. *Endocrinology*, *138*(12), 5231–5237. https://doi.org/10.1210/endo.138.12.5602.

Cabodevilla, A. G., Sánchez-Caballero, L., Nintou, E., Boiadjieva, V. G., Picatoste, F., Gubern, A., et al. (2013). Cell survival during complete nutrient deprivation depends on lipid droplet-fueled β-oxidation of fatty acids. *The Journal of Biological Chemistry*, *288*(39), 27777–27788. https://doi.org/10.1074/jbc.M113.466656.

Campbell, F. M., Kozak, R., Wagner, A., Altarejos, J. Y., Dyck, J. R. B., Belke, D. D., et al. (2002). A role for peroxisome proliferator-activated receptor α (PPARα) in the control of cardiac malonyl-CoA levels: Reduced fatty acid oxidation rates and increased glucose oxidation rates in the hearts of mice lacking PPARα are associated with higher concentrations of malonyl-CoA and reduced expression of malonyl-coa decarboxylase. *Journal of Biological Chemistry*, *277*(6), 4098–4103. https://doi.org/10.1074/jbc.M106054200.

Campos-Barros, A., Hoell, T., Musa, A., Sampaolo, S., Stoltenburg, G., Pinna, G., et al. (1996). Phenolic and tyrosyl ring iodothyronine deiodination and thyroid hormone concentrations in the human central nervous system. *The Journal of Clinical Endocrinology & Metabolism*, *81*(6), 2179–2185. https://doi.org/10.1210/jcem.81.6.8964848.

Cao, S. G., Cheng, P., Angel, A., & Hatch, G. M. (1995). Thyroxine stimulates phosphatidylglycerolphosphate synthase activity in rat heart mitochondria. *Biochimica et Biophysica Acta*, *1256*(2), 241–244. https://doi.org/10.1016/0005-2760(95)00035-B.

Carazo, A., Levin, J., Casas, F., Seyer, P., Grandemange, S., Busson, M., et al. (2012). Protein sequences involved in the mitochondrial import of the 3,5,3′-L-triiodothyronine receptor p43. *Journal of Cellular Physiology*, *227*(12), 3768–3777. https://doi.org/10.1002/jcp.24085.

Carrasco, P., Jacas, J., Sahún, I., Muley, H., Ramírez, S., Puisac, B., et al. (2013). Carnitine palmitoyltransferase 1C deficiency causes motor impairment and hypoactivity. *Behavioural Brain Research*, *256*, 291–297. https://doi.org/10.1016/j.bbr.2013.08.004.

Cesar, M., & Hamprecht, B. (1995). Immunocytochemical examination of neural rat and mouse primary cultures using monoclonal antibodies raised against pyruvate carboxylase. *Journal of Neurochemistry*, *64*(5), 2312–2318.

Chapa, F., Künnecke, B., Calvo, R., Escobar del Rey, F., Morreale de Escobar, G., & Cerdán, S. (1995). Adult-onset hypothyroidism and the cerebral metabolism of (1,2-13C2) acetate as detected by 13C nuclear magnetic resonance. *Endocrinology*, *136*(1), 296–305. https://doi.org/10.1210/endo.136.1.7828544.

Chen, G., Xiao, J., Liu, B., & Wang, F. (2015). The predictive value of thyroid hormone levels on the neurological outcomes of patients with acute ischemic stroke. *Chinese Journal of Contemporary Neurology & Neurosurgery*, *15*(2), 133–136. https://doi.org/10.3969/cjcnn.v15i2.1139.

Chocron, E. S., Sayre, N. L., Holstein, D., Saelim, N., Ibdah, J. A., Dong, L. Q., et al. (2012). The trifunctional protein mediates thyroid hormone receptor-dependent stimulation of mitochondria metabolism. *Molecular Endocrinology*, *26*(7), 1117–1128. https://doi.org/10.1210/me.2011-1348.

Constantinou, C., Chrysanthopoulos, P. K., Margarity, M., & Klapa, M. I. (2011). GC-MS metabolomic analysis reveals significant alterations in cerebellar metabolic physiology in a mouse model of adult onset hypothyroidism. *Journal of Proteome Research*, *10*(2), 869–879. https://doi.org/10.1021/pr100699m.

Courtin, F., Zrouri, H., Lamirand, A., Li, W. W., Mercier, G., Schumacher, M., et al. (2005). Thyroid hormone deiodinases in the central and peripheral nervous system. *Thyroid*, *15*(8), 931–942. https://doi.org/10.1089/thy.2005.15.931.

Crantz, F. R., Silva, J. E., & Larsen, P. R. (1982). An analysis of the sources and quantity of 3,5,3′-triiodothyronine specifically bound to nuclear receptors in rat cerebral cortex and cerebellum. *Endocrinology, 110*(2), 367–375. https://doi.org/10.1210/endo-110-2-367.
Daum, G. (1985). Lipids of mitochondria. *Biochimica et Biophysica Acta, 822*(1), 1–42. https://doi.org/10.1016/0304-4157(85)90002-4.
Dhopeshwarkar, G. A., & Mead, J. F. (1970). Fatty acid uptake by the brain: III. Incorporation of [I-14C]oleic acid into the adult rat brain. *Biochimica et Biophysica Acta, 210*(2), 250–256. https://doi.org/10.1016/0005-2760(70)90169-4.
Dhopeshwarkar, G. A., Subramanian, C., & Mead, J. F. (1971). Fatty acid uptake by the brain V. Incorporation of [1-14c]linolenic acid into adult rat brain. *Biochimica et Biophysica Acta, 239*(2), 162–167. https://doi.org/10.1016/0005-2760(71)90162-7.
Diano, S., Leonard, J. L., Meli, R., Esposito, E., & Schiavo, L. (2003). Hypothalamic type II iodothyronine deiodinase: A light and electron microscopic study. *Brain Research, 976*(1), 130–134. https://doi.org/10.1016/S0006-8993(03)02692-1.
Donzelli, R., Colligiani, D., Kusmic, C., Sabatini, M., Lorenzini, L., Accorroni, A., et al. (2016). Effect of hypothyroidism and hyperthyroidism on tissue thyroid hormone concentrations in rat. *European Thyroid Journal, 5*(1), 27–34. https://doi.org/10.1159/000443523.
van Doom, J., Roelfsema, F., & van der Heide, D. (1982). Contribution from local conversion of thyroxine to 3,5,3'-triiodothyronine to intracellular 3,5,3'-triiodothyronine in several organs in hypothyroid rats at isotope equilibrium. *Acta Endocrinologica, 101*(3), 386–396. https://doi.org/10.1530/acta.0.1010386.
Eaton, S., Middleton, B., & Bartlett, K. (1998). Control of mitochondrial beta-oxidation: Sensitivity of the trifunctional protein to [NAD+]/[NADH] and [acetyl-CoA]/[CoA]. *Biochimica et Biophysica Acta, 1429*(1), 230–238.
Ebert, D., Haller, R. G., & Walton, M. E. (2003). Energy contribution of octanoate to intact rat brain metabolism measured by 13C nuclear magnetic resonance spectroscopy. *Journal of Neuroscience, 23*(13), 5928–5935.
Economidou, F., Douka, E., Tzanela, M., Nanas, S., & Kotanidou, A. (2011). Thyroid function during critical illness. *Hormones (Athens, Greece), 10*(2), 117–124.
Edmond, J., Robbins, R. A., Bergstrom, J. D., Cole, R. A., & de Vellis, J. (1987). Capacity for substrate utilization in oxidative metabolism by neurons, astrocytes, and oligodendrocytes from developing brain in primary culture. *Journal of Neuroscience Research, 18*(4), 551–561. https://doi.org/10.1002/jnr.490180407.
Esfandiari, A., Courtin, F., Lennon, A. M., Gavaret, J. M., & Pierre, M. (1992). Induction of type III deiodinase activity in astroglial cells by thyroid hormones. *Endocrinology, 131*(4), 1682–1688. https://doi.org/10.1210/endo.131.4.1396314.
Fazekas, J. F., Graves, F. B., & Alman, R. W. (1951). The influence of the thyroid on cerebral metabolism. *Endocrinology, 48*(2), 169–174. https://doi.org/10.1210/endo-48-2-169.
Fould, B., Garlatti, V., Neumann, E., Fenel, D., Gaboriaud, C., & Arlaud, G. J. (2010). Structural and functional characterization of the recombinant human mitochondrial trifunctional protein. *Biochemistry, 49*(39), 8608–8617. https://doi.org/10.1021/bi100742w.
García-Villafranca, J., Guillén, A., & Castro, J. (2003). Involvement of nitric oxide/cyclic GMP signaling pathway in the regulation of fatty acid metabolism in rat hepatocytes. *Biochemical Pharmacology, 65*(5), 807–812.
Geiger, A., Magnes, J., & Geiger, R. S. (1952). Survival of the perfused cat's brain in the absence of glucose. *Nature, 170*(4331), 754–755.
Genovese, T., Impellizzeri, D., Ahmad, A., Cornelius, C., Campolo, M., Cuzzocrea, S., et al. (2013). Post-ischaemic thyroid hormone treatment in a rat model of acute stroke. *Brain Research, 1513*, 92–102. https://doi.org/10.1016/j.brainres.2013.03.001.
Gereben, B., Zavacki, A. M., Ribich, S., Kim, B. W., Huang, S. A., Simonides, W. S., et al. (2008). Cellular and molecular basis of deiodinase-regulated thyroid hormone signaling. *Endocrine Reviews, 29*(7), 898–938. https://doi.org/10.1210/er.2008-0019.

Gjedde, S., Gormsen, L. C., Rungby, J., Nielsen, S., Jørgensen, J. O. L., Pedersen, S. B., et al. (2010). Decreased lipid intermediate levels and lipid oxidation rates despite normal lipolysis in patients with hypothyroidism. *Thyroid*, *20*(8), 843–849. https://doi.org/10.1089/thy.2009.0212.

Gonçalves Trentin, A., De Aguiar, C. B. N. M., Castilho Garcez, R., & Alvarez-Silva, M. (2003). Thyroid hormone modulates the extracellular matrix organization and expression in cerebellar astrocyte: Effects on astrocyte adhesion. *Glia*, *42*(4), 359–369. https://doi.org/10.1002/glia.10228.

Gondou, A., Toyoda, N., Nishikawa, M., Tabata, S., Yonemoto, T., Ogawa, Y., et al. (1998). Induction of type 2 deiodinase activity by cyclic guanosine 3',5'-monophosphate in cultured rat glial cells. *Thyroid: Official Journal of the American Thyroid Association*, *8*(7), 615–622. https://doi.org/10.1089/thy.1998.8.615.

Gouill, E. L., Jimenez, M., Binnert, C., Jayet, P.-Y., Thalmann, S., Nicod, P., et al. (2007). Endothelial nitric oxide synthase (eNOS) knockout mice have defective mitochondrial β-oxidation. *Diabetes*, *56*(11), 2690–2696. https://doi.org/10.2337/db06-1228.

Grenell, R. G., & Davies, P. W. (1950). Effect of nembutal on the in vivo rate of oxygen consumption of the cerebral cortex. *Transactions of the American Neurological Association*, *51*, 105–108.

Halim, N. D., Mcfate, T., Mohyeldin, A., Okagaki, P., Korotchkina, L. G., Patel, M. S., et al. (2010). Phosphorylation status of pyruvate dehydrogenase distinguishes metabolic phenotypes of cultured rat brain astrocytes and neurons. *Glia*, *58*(10), 1168–1176. https://doi.org/10.1002/glia.20996.

Haorah, J., Rump, T. J., & Xiong, H. (2013). Reduction of brain mitochondrial β-oxidation impairs complex I and V in chronic alcohol intake: The underlying mechanism for neurodegeneration. *PloS One*, *8*(8), e70833. https://doi.org/10.1371/journal.pone.0070833.

Hiroi, Y., Kim, H.-H., Ying, H., Furuya, F., Huang, Z., Simoncini, T., et al. (2006). Rapid nongenomic actions of thyroid hormone. *Proceedings of the National Academy of Sciences of the United States of America*, *103*(38), 14104–14109. https://doi.org/10.1073/pnas.0601600103.

Hoch, F. L. (1988). Lipids and thyroid hormones. *Progress in Lipid Research*, *27*(3), 199–270. https://doi.org/10.1016/0163-7827(88)90013-6.

Hostetler, K. Y. (1991). Effect of thyroxine on the activity of mitochondrial cardiolipin synthase in rat liver. *Biochimica et Biophysica Acta*, *1086*(1), 139–140. https://doi.org/10.1016/0005-2760(91)90165-E.

Itoh, Y., Esaki, T., Kaneshige, M., Suzuki, H., Cook, M., Sokoloff, L., et al. (2001). Brain glucose utilization in mice with a targeted mutation in the thyroid hormone α or β receptor gene. *Proceedings of the National Academy of Sciences of the United States of America*, *98*(17), 9913–9918. https://doi.org/10.1073/pnas.171319498.

Jackson-Hayes, L., Song, S., Lavrentyev, E. N., Jansen, M. S., Hillgartner, F. B., Tian, L., et al. (2003). A thyroid hormone response unit formed between the promoter and first intron of the carnitine palmitoyltransferase-Ialpha gene mediates the liver-specific induction by thyroid hormone. *The Journal of Biological Chemistry*, *278*(10), 7964–7972. https://doi.org/10.1074/jbc.M211062200.

Kamijo, T., Aoyama, T., Komiyama, A., & Hashimoto, T. (1994). Structural analysis of cDNAs for subunits of human mitochondrial fatty acid beta-oxidation trifunctional protein. *Biochemical and Biophysical Research Communications*, *199*(2), 818–825. https://doi.org/10.1006/bbrc.1994.1302.

Karahanian, E., Rivera-Meza, M., Quintanilla, M. E., Muñoz, D., Fernández, K., & Israel, Y. (2015). PPARα agonists reduce alcohol drinking: Do they act in the brain or in the liver? *Alcohol and Alcoholism*, *50*(6), 717–718. https://doi.org/10.1093/alcalc/agv060.

Kleindienst, A., Brabant, G., Bock, C., Maser-Gluth, C., & Buchfelder, M. (2009). Neuroendocrine function following traumatic brain injury and subsequent intensive care treatment: A prospective longitudinal evaluation. *Journal of Neurotrauma*, *26*(9), 1435–1446. https://doi.org/10.1089/neu.2008.0601.

Krausz, Y., Freedman, N., Lester, H., Newman, J. P., Barkai, G., Bocher, M., et al. (2004). Regional cerebral blood flow in patients with mild hypothyroidism. *Journal of Nuclear Medicine*, *45*(10), 1712–1715.

Krueger, J. J., Ning, X.-H., Argo, B. M., Hyyti, O., & Portman, M. A. (2001). Triidothyronine and epinephrine rapidly modify myocardial substrate selection: A 13C isotopomer analysis. *American Journal of Physiology—Endocrinology and Metabolism*, *281*(5), E983–E990.

Kudo, M., & Nagayama, T. (1988). Light and electron microscopic study of lipid accumulation along margins of experimental cerebral infarcts in rats. *Stroke*, *19*(12), 1544–1549. https://doi.org/10.1161/01.STR.19.12.1544.

Kuge, Y., Yajima, K., Kawashima, H., Yamazaki, H., Hashimoto, N., & Miyake, Y. (1995). Brain uptake and metabolism of [1-11C]octanoate in rats: Pharmacokinetic basis for its application as a radiopharmaceutical for studying brain fatty acid metabolism. *Annals of Nuclear Medicine*, *9*(3), 137–142. https://doi.org/10.1007/BF03165040.

Lehninger, A. L., & Ray, B. L. (1957). Oxidation-reduction state of rat liver mitochondria and the action of thyroxine. *Biochimica et Biophysica Acta*, *26*(3), 643–644. https://doi.org/10.1016/0006-3002(57)901154.

Leonard, J. L., Siegrist-Kaiser, C. A., & Zuckerman, C. J. (1990). Regulation of type II iodothyronine 5'-deiodinase by thyroid hormone. Inhibition of actin polymerization blocks enzyme inactivation in cAMP-stimulated glial cells. *The Journal of Biological Chemistry*, *265*(2), 940–946.

Lesch, P., Schmidt, E., & Schmidt, F. W. (1973). Effects of chronic alcohol abuse on the fatty acid composition of major lipids in the human brain. Hepatocerebral degeneration II. *Zeitschrift Fur Klinische Chemie und Klinische Biochemie*, *11*(4), 159–166.

Liang, D. S. (1991). Stroke and thyroid hormones. *Chinese Journal of Neurology and Psychiatry*, *24*(6), 352–354. 384.

Liu, J., Wang, D., Xiong, Y., Yuan, R., Tao, W., & Liu, M. (2016). Low free triiodothyronine levels are related to symptomatic intracranial hemorrhage and poor functional outcomes after intravenous thrombolysis in acute ischemic stroke patients. *Neurological Research*, *38*(5), 429–433. https://doi.org/10.1080/01616412.2016.1178480.

Liu, X., Wu, L., Deng, G., Li, N., Chu, X., Guo, F., et al. (2008). Characterization of mitochondrial trifunctional protein and its inactivation study for medicine development. *Biochimica et Biophysica Acta*, *1784*(11), 1742–1749. https://doi.org/10.1016/j.bbapap.2008.06.018.

Lombardi, A., De, L., Silvestri, E., Busiello, R. A., Lanni, A., Goglia, F., et al. (2009). 3,5-Diiodo-L-thyronine rapidly enhances mitochondrial fatty acid oxidation rate and thermogenesis in rat skeletal muscle: AMP-activated protein kinase involvement. *American Journal of Physiology. Endocrinology and Metabolism*, *296*(3), E497–E502. https://doi.org/10.1152/ajpendo.90642.2008.

Madden, T. (2003). *The BLAST Sequence Analysis Tool*. USA: National Center for Biotechnology Information. Retrieved from https://www.ncbi.nlm.nih.gov/books/NBK21097/.

Malekpour, B., Mehrafshan, A., Saki, F., Malekmohammadi, Z., & Saki, N. (2012). Effect of posttraumatic serum thyroid hormone levels on severity and mortality of patients with severe traumatic brain injury. *Acta Medica Iranica*, *50*(2), 113–116.

Manzano, J., Bernal, J., & Morte, B. (2007). Influence of thyroid hormones on maturation of rat cerebellar astrocytes. *International Journal of Developmental Neuroscience*, *25*(3), 171–179. https://doi.org/10.1016/j.ijdevneu.2007.01.003.

Martinez, R., & Gomes, F. C. A. (2002). Neuritogenesis induced by thyroid hormone-treated astrocytes is mediated by epidermal growth factor/mitogen-activated protein kinase-phosphatidylinositol 3-kinase pathways and involves modulation of extracellular matrix proteins. *Journal of Biological Chemistry*, 277(51), 49311–49318. https://doi.org/10.1074/jbc.M209284200.

Martius, C., & Hess, B. (1951). The mode of action of thyroxin. *Archives of Biochemistry and Biophysics*, *33*(3), 486–487. https://doi.org/10.1016/0003-9861(51)90137-3.

Mavinakere, M. S., Powers, J. M., Subramanian, K. S., Roggero, V. R., & Allison, L. A. (2012). Multiple novel signals mediate thyroid hormone receptor nuclear import and export. *Journal of Biological Chemistry*, *287*(37), 31280–31297. https://doi.org/10.1074/jbc.M112.397745.

McGarry, J. D., Leatherman, G. F., & Foster, D. W. (1978). Carnitine palmitoyltransferase I. The site of inhibition of hepatic fatty acid oxidation by malonyl-CoA. *The Journal of Biological Chemistry*, *253*(12), 4128–4136.

Mdzinarishvili, A., Sutariya, V., Talasila, P. K., Geldenhuys, W. J., & Sadana, P. (2013). Engineering triiodothyronine (T3) nanoparticle for use in ischemic brain stroke. *Drug Delivery and Translational Research*, *3*(4), 309–317. https://doi.org/10.1007/s13346-012-0117-8.

Mendes-de-Aguiar, C. B. N., Alchini, R., Zucco, J. K., Costa-Silva, B., Decker, H., Alvarez-Silva, M., et al. (2010). Impaired astrocytic extracellular matrix distribution under congenital hypothyroidism affects neuronal development in vitro. *Journal of Neuroscience Research*, *88*(15), 3350–3360. https://doi.org/10.1002/jnr.22481.

Mileykovskaya, E., & Dowhan, W. (2014). Cardiolipin-dependent formation of mitochondrial respiratory supercomplexes. *Chemistry and Physics of Lipids*, *179*, 42–48. https://doi.org/10.1016/j.chemphyslip.2013.10.012.

Mitchell, R. W., On, N. H., Del Bigio, M. R., Miller, D. W., & Hatch, G. M. (2011). Fatty acid transport protein expression in human brain and potential role in fatty acid transport across human brain microvessel endothelial cells. *Journal of Neurochemistry*, *117*(4), 735–746. https://doi.org/10.1111/j.1471-4159.2011.07245.x.

Müller, M. J., & Seitz, H. J. (1984). Thyroid hormone action on intermediary metabolism—Part II: Lipid metabolism in hypo- and hyperthyroidism. *Klinische Wochenschrift*, *62*(2), 49–55. https://doi.org/10.1007/BF01769663.

Mutter, T., Dolinsky, V. W., Ma, B. J., Taylor, W. A., & Hatch, G. M. (2000). Thyroxine regulation of monolysocardiolipin acyltransferase activity in rat heart. *Biochemical Journal*, *346*, 403–406. https://doi.org/10.1042/0264-6021:3460403.

Mynatt, R. L., Park, E. A., Thorngate, F. E., Das, H. K., & Cook, G. A. (1994). Changes in carnitine palmitoyltransferase-I mRNA abundance produced by hyperthyroidism and hypothyroidism parallel changes in activity. *Biochemical and Biophysical Research Communications*, *201*(2), 932–937. https://doi.org/10.1006/bbrc.1994.1791.

Nouws, J., Te Brinke, H., Nijtmans, L. G., & Houten, S. M. (2014). ACAD9, a complex I assembly factor with a moonlighting function in fatty acid oxidation deficiencies. *Human Molecular Genetics*, *23*(5), 1311–1319. https://doi.org/10.1093/hmg/ddt521.

O'Keefe, L. M., Conway, S. E., Czap, A., Malchoff, C. D., Benashski, S., Fortunato, G., et al. (2015). Thyroid hormones and functional outcomes after ischemic stroke. *Thyroid Research*, *8*, 9. https://doi.org/10.1186/s13044-015-0021-7.

Olivecrona, Z., Dahlqvist, P., & Koskinen, L.-O. D. (2013). Acute neuro-endocrine profile and prediction of outcome after severe brain injury. *Scandinavian Journal of Trauma, Resuscitation and Emergency Medicine*, *21*, 33. https://doi.org/10.1186/1757-7241-21-33.

Orii, K. E., Orii, K. O., Souri, M., Oriii, T., Kondo, N., Hashimoto, T., et al. (1999). Genes for the human mitochondrial trifunctional protein α- and β-subunits are divergently transcribed from a common promoter region. *Journal of Biological Chemistry*, *274*(12), 8077–8084. http://dx.doi.org/10.1074/jbc.274.12.8077.

Osterweil, D., Syndulko, K., Cohen, S. N., Pettier-Jennings, P. D., Hershman, J. M., Cummings, J. L., et al. (1992). Cognitive function in non-demented older adults with hypothyroidism. *Journal of the American Geriatrics Society*, *40*(4), 325–335. https://doi.org/10.1111/j.1532-5415.1992.tb02130.x.

Park, K. S., Jeong, G.-C., Cho, S.-G., Jang, K. S., Kim, J., Ha, J.-M., et al. (2014). The effect of thyroid hormone withdrawal on brain glucose metabolism in differentiated thyroid cancer patients. *Journal of Nuclear Medicine*, *55*(Suppl. 1), 1881.

Raichle, M. E., & Gusnard, D. A. (2002). Appraising the brain's energy budget. *Proceedings of the National Academy of Sciences*, *99*(16), 10237–10239. https://doi.org/10.1073/pnas.172399499.

Raju, K., Doulias, P.-T., Evans, P., Krizman, E. N., Jackson, J. G., Horyn, O., et al. (2015). Regulation of brain glutamate metabolism by nitric oxide and S-nitrosylation. *Science Signaling*, *8*(384), ra68. https://doi.org/10.1126/scisignal.aaa4312.

Rami, A., & Krieglstein, J. (1992). Thyroxine attenuates hippocampal neuronal damage caused by ischemia in the rat. *Life Sciences*, *50*(9), 645–650. https://doi.org/10.1016/0024-3205(92)90251-J.

Saelim, N., Holstein, D., Chocron, E. S., Camacho, P., & Lechleiter, J. D. (2007). Inhibition of apoptotic potency by ligand stimulated thyroid hormone receptors located in mitochondria. *Apoptosis*, *12*(10), 1781–1794. https://doi.org/10.1007/s10495-007-0109-1.

Saelim, N., John, L. M., Wu, J., Park, J. S., Bai, Y. D., Camacho, P., et al. (2004). Nontranscriptional modulation of intracellular Ca2+ signaling by ligand stimulated thyroid hormone receptor. *Journal of Cell Biology*, *167*(5), 915–924. https://doi.org/10.1083/jcb.200409011.

Saunier, B., Pierre, M., Jacquemin, C., & Courtin, F. (1993). Evidence for cAMP-independent thyrotropin effects on astroglial cells. *European Journal of Biochemistry*, *218*(3), 1091–1094.

Sayre, N. L., & Lechleiter, J. D. (2012). Fatty acid metabolism and thyroid hormones. *Current Trends in Endocrinology*, *6*, 65–76.

Sayre, N. L., Sifuentes, M., Holstein, D., Cheng, S., Zhu, X., & Lechleiter, J. D. (2016). Stimulation of astrocyte fatty acid oxidation by thyroid hormone is protective against ischemic stroke-induced damage. *Journal of Cerebral Blood Flow & Metabolism*, *37*, 514–527. https://doi.org/10.1177/0271678X16629153.

Schild, L., Dombrowski, F., Lendeckel, U., Schulz, C., Gardemann, A., & Keilhoff, G. (2008). Impairment of endothelial nitric oxide synthase causes abnormal fat and glycogen deposition in liver. *Biochimica et Biophysica Acta*, *1782*(3), 180–187. https://doi.org/10.1016/j.bbadis.2007.12.007.

Schulz, J. G., Laranjeira, A., Van Huffel, L., Gärtner, A., Vilain, S., Bastianen, J., et al. (2015). Glial β-oxidation regulates drosophila energy metabolism. *Scientific Reports*, *5*, 7805. https://doi.org/10.1038/srep07805.

Serfozo, Z., Kiss, P. B., Kukor, Z., Lontay, B., Palatka, K., Varga, V., et al. (2008). Thyroid hormones affect the level and activity of nitric oxide synthase in rat cerebral cortex during postnatal development. *Neurochemical Research*, *33*(3), 569–578. https://doi.org/10.1007/s11064-007-9480-0.

Shamaeva, K. K., Savin, I. A., Tenedieva, V. D., Goriachev, A. S., Oshorov, A. V., Sychev, A. A., et al. (2012). Analysis of thyroid homeostasis disorders in patients with severe traumatic brain injury. *Zhurnal Voprosy Neirokhirurgii Imeni N. N. Burdenko*, *76*(4), 26–30. discussion 30–31.

Sierra, A. Y., Gratacós, E., Carrasco, P., Clotet, J., Ureña, J., Serra, D., et al. (2008). CPT1c is localized in endoplasmic reticulum of neurons and has carnitine palmitoyltransferase activity. *Journal of Biological Chemistry*, *283*(11), 6878–6885. https://doi.org/10.1074/jbc.M707965200.

Smith, C. D., & Ain, K. B. (1995). Brain metabolism in hypothyroidism studied with 31P magnetic-resonance spectroscopy. *The Lancet*, *345*(8950), 619–620. https://doi.org/10.1016/S0140-6736(95)90522-7.

Song, S., Attia, R. R., Connaughton, S., Niesen, M. I., Ness, G. C., Elam, M. B., et al. (2010). Peroxisome proliferator activated receptor alpha (PPARalpha) and PPAR gamma coactivator (PGC-1alpha) induce carnitine palmitoyltransferase IA (CPT-1A) via independent gene elements. *Molecular and Cellular Endocrinology*, *325*(1–2), 54–63. https://doi.org/10.1016/j.mce.2010.05.019.

Spiekerkoetter, U., Khuchua, Z., Yue, Z., Bennett, M. J., & Strauss, A. W. (2004). General mitochondrial trifunctional protein (TFP) deficiency as a result of either α- or β-subunit mutations exhibits similar phenotypes because mutations in either subunit alter TFP complex expression and subunit turnover. *Pediatric Research*, *55*(2), 190–196. https://doi.org/10.1203/01.PDR.0000103931.80055.06.
Stakkestad, J. A., & Lund, H. (1984). The effect of nutritional and thyroid state on the distribution of fatty acids between oxidation and esterification in isolated rat hepatocytes. *Biochimica et Biophysica Acta*, *793*(1), 1–9. https://doi.org/10.1016/0005-2760(84)90046-8.
Stathatos, N., Levetan, C., Burman, K. D., & Wartofsky, L. (2001). The controversy of the treatment of critically ill patients with thyroid hormone. *Best Practice & Research Clinical Endocrinology & Metabolism*, *15*(4), 465–477. https://doi.org/10.1053/beem.2001.0164.
Sterling, K. (1991). Thyroid hormone action: Identification of the mitochondrial thyroid hormone receptor as adenine nucleotide translocase. *Thyroid*, *1*(2), 167–171.
Sterling, K., Brenner, M. A., & Sakurada, T. (1980). Rapid effect of triiodothyronine on the mitochondrial pathway in rat liver in vivo. *Science*, *210*(4467), 340–342. https://doi.org/10.1126/science.7423197.
Sterling, K., Campbell, G. A., & Brenner, M. A. (1983). Purification of the mitochondrial triiodothyronine (T3) receptor from rat liver. *Transactions of the Association of American Physicians*, *96*, 324–335.
Sterling, K., & Milch, P. O. (1975). Thyroid hormone binding by a component of mitochondrial membrane. *Proceedings of the National Academy of Sciences of the United States of America*, *72*(8), 3225–3229.
Stobart, J. L. L., Lu, L., Anderson, H. D. I., Mori, H., & Anderson, C. M. (2013). Astrocyte-induced cortical vasodilation is mediated by D-serine and endothelial nitric oxide synthase. *Proceedings of the National Academy of Sciences of the United States of America*, *110*(8), 3149–3154. https://doi.org/10.1073/pnas.1215929110.
Suda, S., Muraga, K., Kanamaru, T., Okubo, S., Abe, A., Aoki, J., et al. (2016). Low free triiodothyronine predicts poor functional outcome after acute ischemic stroke. *Journal of the Neurological Sciences*, *368*, 89–93. https://doi.org/10.1016/j.jns.2016.06.063.
Sumegi, B., & Srere, P. A. (1984). Complex I binds several mitochondrial NAD-coupled dehydrogenases. *The Journal of Biological Chemistry*, *259*(24), 15040–15045.
Suzuki, A., Iwasaki, M., Kato, M., & Wagai, N. (1997). Sequential operation of ceramide synthesis and ICE cascade in CPT-11-initiated apoptotic death signaling. *Experimental Cell Research*, *233*(1), 41–47. https://doi.org/10.1006/excr.1997.3498.
Taïb, B., Bouyakdan, K., Hryhorczuk, C., Rodaros, D., Fulton, S., & Alquier, T. (2013). Glucose regulates hypothalamic long-chain fatty acid metabolism via AMP-activated kinase (AMPK) in neurons and astrocytes. *The Journal of Biological Chemistry*, *288*(52), 37216–37229. https://doi.org/10.1074/jbc.M113.506238.
Tapley, D. F., Cooper, C., & Lehninger, A. L. (1955). The action of thyroxine on mitochondria and oxidative phosphorylation. *Biochimica et Biophysica Acta*, *18*(4), 597–598.
Taylor, W. A., & Hatch, G. M. (2009). Identification of the human mitochondrial linoleoyl-coenzyme A monolysocardiolipin acyltransferase (MLCL AT-1). *Journal of Biological Chemistry*, *284*(44), 30360–30371. https://doi.org/10.1074/jbc.M109.048322.
Taylor, W. A., Mejia, E. M., Mitchell, R. W., Choy, P. C., Sparagna, G. C., & Hatch, G. M. (2012). Human trifunctional protein alpha links cardiolipin remodeling to beta-oxidation. *PloS One*, 7(11). https://doi.org/10.1371/journal.pone.0048628.
Trentin, A. G. (2006). Thyroid hormone and astrocyte morphogenesis. *Journal of Endocrinology*, *189*(2), 189–197. https://doi.org/10.1677/joe.1.06680.
Videla, L. A. (2014). T3-induced liver AMP-activated protein kinase signaling: Redox dependency and upregulation of downstream targets. *World Journal of Gastroenterology*, *20*(46), 17416. https://doi.org/10.3748/wjg.v20.i46.17416.

Wallis, K., Dudazy, S., van Hogerlinden, M., Nordström, K., Mittag, J., & Vennström, B. (2010). The thyroid hormone receptor α1 protein is expressed in embryonic postmitotic neurons and persists in most adult neurons. *Molecular Endocrinology*, *24*(10), 1904–1916. https://doi.org/10.1210/me.2010-0175.

Wang, Y., Mohsen, A.-W., Mihalik, S. J., Goetzman, E. S., & Vockley, J. (2010). Evidence for physical association of mitochondrial fatty acid oxidation and oxidative phosphorylation complexes. *Journal of Biological Chemistry*, *285*(39), 29834–29841. https://doi.org/10.1074/jbc.M110.139493.

Warden, A., Truitt, J., Merriman, M., Ponomareva, O., Jameson, K., Ferguson, L. B., et al. (2016). Localization of PPAR isotypes in the adult mouse and human brain. *Scientific Reports*, *6*, 27618. https://doi.org/10.1038/srep27618.

Weinberger, M. J., Rinaldo, P., Strauss, A. W., & Bennett, M. J. (1995). Intact alpha-subunit is required for membrane-binding of human mitochondrial trifunctional beta-oxidation protein, but is not necessary for conferring 3-ketoacyl-CoA thiolase activity to the beta-subunit. *Biochemical and Biophysical Research Communications*, *209*(1), 47–52. https://doi.org/10.1006/bbrc.1995.1468.

Wiencken, A. E., & Casagrande, V. A. (1999). Endothelial nitric oxide synthetase (eNOS) in astrocytes: Another source of nitric oxide in neocortex. *Glia*, *26*(4), 280–290.

Woolf, P. D., Lee, L. A., Hamill, R. W., & McDonald, J. V. (1988). Thyroid test abnormalities in traumatic brain injury: Correlation with neurologic impairment and sympathetic nervous system activation. *The American Journal of Medicine*, *84*(2), 201–208.

Wrutniak, C., Cassar-Malek, I., Marchal, S., Rascle, A., Heusser, S., Keller, J.-M., et al. (1995). A 43-kDa protein related to c-Erb a 1 is located in the mitochondrial matrix of rat liver. *Journal of Biological Chemistry*, *270*(27), 16347–16354. https://doi.org/10.1074/jbc.270.27.16347.

Xu, X., Li, W., & Hu, X. (2016). Alteration of thyroid-related hormones within normal ranges and early functional outcomes in patients with acute ischemic stroke. *International Journal of Endocrinology*, *2016*, 1–5. https://doi.org/10.1155/2016/3470490.

Yamauchi, M., Kambe, F., Cao, X., Lu, X., Kozaki, Y., Oiso, Y., et al. (2008). Thyroid hormone activates adenosine 5'-monophosphate-activated protein kinase via intracellular calcium mobilization and activation of calcium/calmodulin-dependent protein kinase kinase-beta. *Molecular Endocrinology*, *22*(4), 893–903. https://doi.org/10.1210/me.2007-0249.

Yu, A. C., Drejer, J., Hertz, L., & Schousboe, A. (1983). Pyruvate carboxylase activity in primary cultures of astrocytes and neurons. *Journal of Neurochemistry*, *41*(5), 1484–1487.

Zhang, Y., & Meyer, M. A. (2010). Clinical analysis on alteration of thyroid hormones in the serum of patients with acute ischemic stroke. *Stroke Research and Treatment*, *2010*, e290678. https://doi.org/10.4061/2010/290678.

Zhang, M., Mileykovskaya, E., & Dowhan, W. (2002). Gluing the respiratory chain together cardiolipin is required for supercomplex formation in the inner mitochondrial membrane. *Journal of Biological Chemistry*, *277*(46), 43553–43556. https://doi.org/10.1074/jbc.C200551200.

CHAPTER NINE

The NF-κB Family of Transcription Factors and Its Role in Thyroid Physiology

Carla Reale*, Tiziana Zotti†, Ivan Scudiero*, Pasquale Vito*,†,‡,1, Romania Stilo‡

*Biogem Consortium, Ariano Irpino, Italy
†Genus Biotechnology, Università del Sannio, Benevento, Italy
‡Università del Sannio, Benevento, Italy
1Corresponding author: e-mail address: vito@unisannio.it

Contents

1. Introduction 196
2. The NF-κB Family 196
   2.1 Dimers Repertoire 198
   2.2 Variety of κB Sites 199
   2.3 Posttranslational Modifications of NF-κB 200
3. Canonical Activation of NF-κB 201
4. NEMO/IKKγ-Dependent IKK Complex Activation 202
5. NF-κB in Thyroid Physiology 204
6. Conclusion 206
References 207

Abstract

The nuclear factor (NF)-κB signaling pathway controls a variety of important biological functions, including immune and inflammatory responses, differentiation, cell growth, tumorigenesis, and apoptosis. Two distinct pathways of NF-κB activation are known. The classical, canonical pathway is found virtually in all mammalian cells and NF-κB activation is mediated by the IKK complex, consisting of the IKK1/IKKα and IKK2/IKKβ catalytic kinase subunits and the NF-κB essential modulator (NEMO)/IKKγ protein. The NF-κB-driven transcriptional responses to many different stimuli have been widely characterized in the pathophysiology of the mammalian immune system, mainly because this transcription factor regulates the expression of cytokines, growth factors, and effector enzymes in response to ligation of cellular receptors involved in immunity and inflammation. However, an impressive literature produced in the last two decades shows that NF-κB signaling plays an important role also outside of the immune system, performing different roles and functions depending on the type of tissue and organ. In thyroid, NF-κB signaling is crucial for thyrocytes survival and expression of critical thyroid

Vitamins and Hormones, Volume 106
ISSN 0083-6729
http://dx.doi.org/10.1016/bs.vh.2017.05.003

markers, including Nis, Ttf1, Pax8, Tpo, and thyroglobulin, making this transcription factor essential for maintenance of normal thyroid function.

1. INTRODUCTION

Nuclear factor (NF)-κB is an inducible transcription factor with pleiotropic functions, actively and critically involved in many biological processes, including cell proliferation and survival, differentiation, embryonic development, immune response, inflammation, and tumoral transformation and progression (DiDonato, Mercurio, & Karin, 2012; Ghosh, 2006).

The NF-κB-driven transcriptional responses to many different stimuli have been widely characterized in the pathophysiology of the mammalian immune system, mainly because this transcription factor regulates the expression of cytokines, growth factors, and effector enzymes in response to ligation of cellular receptors involved in immunity and inflammation, including T-cell receptor and B-cell receptor, CD40, B-cell-activating factor receptor (BAFF-R), interleukin 1-receptor, tumor necrosis factor receptors, and pattern recognition receptors. Nevertheless, NF-κB also controls the expression of critical genes outside of the immune system and, hence, can influence multiple aspects of normal physiology and disease (Curtois & Gilmore, 2006). After 30 years since the publication of its discovery (Sen & Baltimore, 1986) and plenty of sophisticated studies, many signaling mechanisms have been dissected and clearly described, but the full understanding of how the transcriptional potential, activity, and selectivity of NF-κB are regulated is still an intriguing object of intense investigation.

2. THE NF-κB FAMILY

The NF-κB family of transcription factors in *Mammalia* consists of five members, p50 and p52 (Class I members), p65/RelA, c-Rel, and RelB (Class II members), encoded by *NFKB1*, *NFKB2*, *RELA*, *REL*, and *RELB*, respectively (Nabel & Verma, 1993). These monomers share an N-terminal Rel homology domain (RHD) responsible for both homo- and heterodimerization and DNA binding due to their ability to recognize a common sequence on DNA called the κB motif. The RHD (IPR011539) is found in many eukaryotic transcription factors and is structurally composed of two distinct subdomains: an N-terminal domain with two beta-barrels for DNA "gripping" and a C-terminal domain with an immunoglobulin-like fold that serves for dimer formation (Ghosh, Wang, Huang, & Fusco, 2012; Zheng, Yin, & Wu, 2011). NF-κB active complexes are formed from homo- and heterodimers of the five members and bind to κB sites within the promoters/enhancers of the target genes, thereby regulating their transcription also through recruitment of coactivators and corepressors. The transcription activation domain (TAD) necessary for the positive regulation of gene expression is present only in the Class II members p65, c-Rel, and RelB. Class I members p50 and p52 are generated following proteolytic cleavage of the longer precursors p105 and p100, respectively. The C-terminus of p100 and p105 displays multiple copies of the ankyrin repeat motif (ANK; IPR002110), a domain found in a large variety of eukaryotic proteins and involved in protein–protein interaction (Al-Khodor, Price, Kalia, & Abu Kwaik, 2010; Bork, 1993; Ghosh et al., 2012) (Fig. 1).

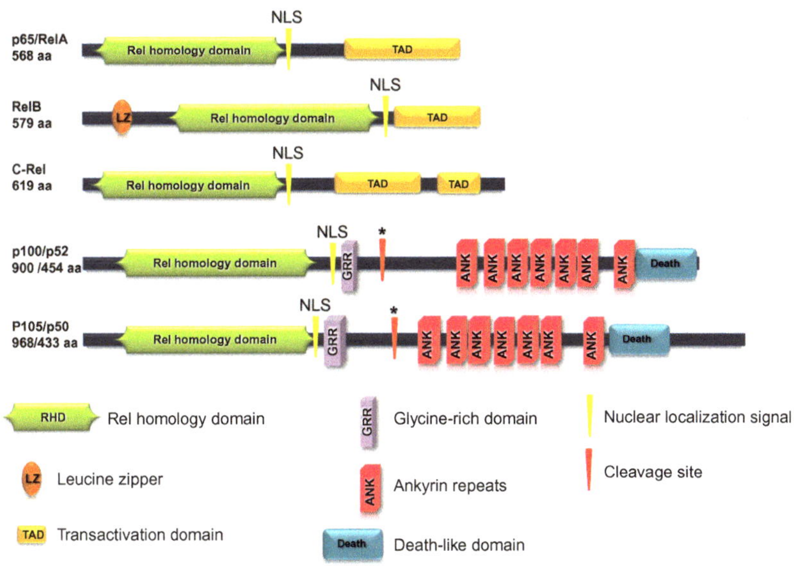

Fig. 1 The NF-κB family of transcription factors. Schematic representation of domain organization of NF-κB subunits. In Mammalia, NF-κB family is formed by five members characterized by an N-terminal Rel homology domain (RHD) that is responsible for DNA binding and dimerization. Class II members, RelA/p65, c-Rel, and RelB, contain a transactivation domain (TAD) at the C-terminus that is required for activation of gene expression. Class I members p100 and p105 display several inhibitory ankyrin repeats that are removed after their proteolytical processing triggered by activation signals.

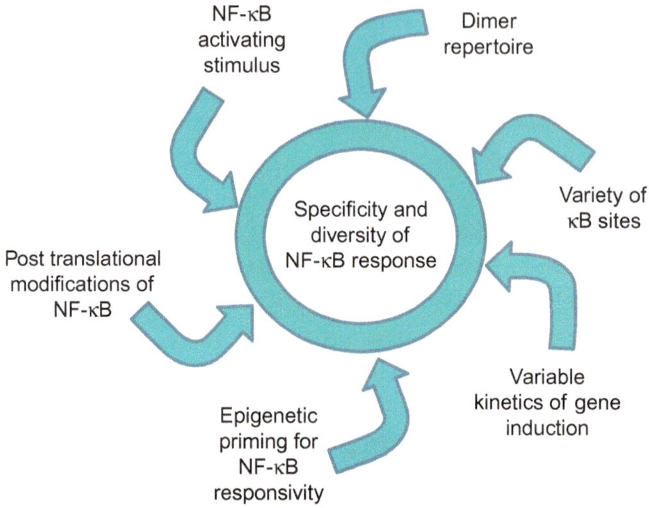

Fig. 2 NF-κB response. The wide diversity of cellular outcomes resulting from NF-κB program activation is achieved through many factors.

The mechanisms of subunit association and DNA recognition by the NF-κB transcription factors have been broadly described by biochemical, kinetic, and structural studies. These works have contributed to unveil how variety and specificity of NF-κB-dependent biological responses are achieved in the cell following a specific stimulus. Essentially, the huge diversity of cellular outcomes resulting from NF-κB program activation depends on a number of factors, such as dimers repertoire, the presence of different κB sites on promoters and enhancers, the existence of a cell-specific set of genes epigenetically primed for responsiveness to NF-κB, variable kinetics of gene induction, and posttranslational modifications of NF-κB dimers (Fig. 2).

2.1 Dimers Repertoire

NF-κB family members function as dimers since some monomeric subunits are considered highly unstable (Huang, Vu, & Ghosh, 2005). The NF-κB complex generation system determines the relative abundance of specific dimers according to both dimerization affinities among distinct subunits and expression pattern of single members in different cell types. NF-κB monomers have different dimerization affinities ranging from nM to μM and can form 15 diverse (homo- and hetero-) dimers. At least 12 of them have been shown to bind DNA and potentially regulate transcription (Hoffmann, Natoli, & Ghosh, 2006; Huang et al., 2005). The common term "NF-κB" conventionally refers to the p50/p65 heterodimer that is the most abundant dimer present in the vast majority of cell types, whereas p65/p65, p65/c-Rel, p65/p52, c-Rel/c-Rel, p52/c-Rel, p50/c-Rel, p50/p50, RelB/p50, and RelB/p52 have been described with different relative abundance, and some of them have been found only in few subsets of cell types (Oeckinghaus & Ghosh, 2009). Notably, p50 and p52 homodimers cannot activate transcription due to the absence of a TAD in their C-terminal halves. Therefore, the p50–p50 and p52–p52 homodimers seem to be able to act as negative regulators of transcription by competing with the active dimers for binding to κB sites (Grundström, Anderson, Scheipers, & Sunstedt, 2004; Plaksin, Baeuerle, & Eisenbach, 1993). In addition, due to the combinatorial nature of NF-κB dimerization, the generation of specific dimers is also affected by chemical competition among monomers for binding to a specific subunit (Mitchell, Vargas, & Hoffmann, 2016). Another extrinsic factor influencing the relative abundance of distinct NF-κB dimers is the association with proteins that act as molecular chaperones such as I-κBβ that stabilizes dimers, and protect them from degradation, thereby representing an additional way to regulate NF-κB functions (Tsui et al., 2015).

2.2 Variety of κB Sites

NF-κB can activate expression of many hundred genes. Expanded analysis of their *cis*-regulatory elements (promoter/enhancer) has led to the definition of a pseudosymmetric κB consensus as 5′-GGGRNWYYCC-3′ (R = purines, N = any nucleotide, W = adenine or thymine, and Y = pyrimidine). Structural studies of X-ray crystals formed by distinct NF-κB homo- and heterodimers bound to different κB sites have suggested that the RHD of each monomer binds one half-site at the DNA major groove through base-specific connections, wherein the flanking (G)GG/(C)CC sequences are contacted by a set of conserved residues. The inner nucleotides participate in more variable interactions with the NF-κB subunits (Wang et al., 2012). A number of studies carried out on mutant mice and knockout cells have demonstrated that each target gene requires a specific combination of NF-κB subunits for complete activation, and a unique succession of events affecting *cis*-acting elements. In addition, the κB sequence may also affect the conformation of the bound dimers, thus influencing their ability to recruit molecular partners (Natoli & De Santa, 2006; Oeckinghaus & Ghosh, 2009).

Specific NF-κB inducers can generate activation profiles of different duration and intensity, suggesting the existence of a direct link between temporal control of NF-κB activity and selective gene induction. Many studies focused on the kinetics of NF-κB-induced gene expression have revealed that NF-κB dimers initially bind only to a subset of genes. Other genes are bound at later time points or upon a longer stimulation, suggesting that their consensus sequences are not readily accessible and may require further events of promoter/enhancer remodeling (Bhatt & Ghosh, 2014; Saccani, Pantano, & Natoli, 2001).

2.3 Posttranslational Modifications of NF-κB

The NF-κB subunits undergo a wide range of regulatory posttranslational modifications, including phosphorylation, acetylation, nitrosylation, and ubiquitination. Some of these modifications are crucial for cross talking among different signaling pathways and for signal modulation (Huang, Yang, Lamb, & Chen, 2010). p65/RelA is phosphorylated by many kinase proteins, with different effects on its transcriptional activity. Many studies have described p65/RelA-activating phosphorylation at Ser276 by the catalytic subunit of protein kinase A and mitogen- and stress-activated protein kinase-1 and -2 (MSK1 and MSK2); at Ser311 by the atypical PKCζ; at Ser468 by I-κB kinase β and ε (IKKβ and ε) and glycogen synthase kinase-3β (GSK3β); at Ser529 by casein kinase 2 (CK2); at Ser 535 by the calcium/calmodulin-dependent protein kinase type IV (CaMKIV); and at Ser536 by I-κB kinase β, α, ε, TANK-binding kinase-1 (TBK1), and p90 ribosomal protein S6 kinase (RSK1/p90S6K) (Christian, Smith, & Carmody, 2016; Ghosh, 2006; Perkins, 2007). These modifications can all be considered as stimulatory phosphorylations that enhance the transcriptional activity of p65/RelA and/or its ability to interact with transcriptional coactivators, such as p300 and CBP. By contrast, ATM- and Rad3-related kinase can induce CHK1-mediated inhibitory phosphorylation of p65/RelA at Thr505 in response to ARF tumor suppressor induction or following DNA damage (Campbell, Witty, Rocha, & Perkins, 2006; Msaki et al., 2011; Perkins, 2007). Also acetylation is important for orchestrating biological outcomes following NF-κB activation. p65/RelA can be acetylated at the lysine residues in positions 122, 123, 218, 221, 310, and 314, whereas p50 is acetylated on lysines 431, 440, and 441.

For both monomers, acetylations are stimulus-coupled and have different impacts (both negative and positive) on NF-κB-mediated transcription (Ghosh, 2006). The reversible acetylation of p65/RelA plays also a role in regulating the assembly of NF-κB with its specific inhibitor I-κBα (Chen & Green, 2003). In addition to phosphorylation and acetylation, NF-κB activation can be regulated through the covalent addition of different linked polyubiquitin chains by several E3 ubiquitin ligases. Although very little is known about polyubiquitin chains with activating functions on NF-κB dimers, many studies have reported the existence of a ubiquitin-dependent degradation pathway, which can modulate inappropriate or excessive activity during NF-κB-mediated inflammatory disease and in cancer (Xu, You, Shi, & Hou, 2015). In particular, several E3 ubiquitin ligases, including PDLIM2, ORF73, SOCS2-COMMD1, HSV-1 ICP0, cIAPs, HERC3, Peli1, ING4, PPARγ, and TRAF7, have been shown to target NF-κB subunits for nuclear ubiquitination and degradation through both proteasomal- and lysosomal-dependent pathways (Collins, Mitxitorena, & Carmody, 2016; Hou, Moreau, & Chadee, 2012; Maine, Mao, Komarck, & Burstein, 2007; Tanaka, Grusby, & Kaisho, 2007; Zotti et al., 2011). Conversely, only two deubiquitinases (DUB), USP48/USP31 and USP7/HAUSP, are known so far to remove polyubiquitin chains from p65/RelA and other NF-κB monomers. However, the mechanism by which E3 ubiquitin ligases and DUBs modulate nuclear NF-κB signals is still unclear (Collins et al., 2016).

Finally, protein S-nitrosylation is now recognized to be integral in coordinating the immune response by regulating numerous proteins in a multitude of signaling pathways. The effect of S-nitrosylation is primarily inhibitory, resulting in the deactivation of proinflammatory pathways, including the NF-κB pathway (Marshall & Stamler, 2001). In the nucleus, both subunits of the p50–p65 heterodimer are targeted for S-nitrosylation at a conserved cysteine in the Rel DNA-binding domain with modification of either monomer disrupting DNA binding and inhibiting κB-dependent transcription (Kelleher, Matsumoto, Stamler, & Marshall, 2007).

3. CANONICAL ACTIVATION OF NF-κB

The activation of NF-κB complexes is achieved by two major signal transduction pathways: the canonical, NEMO/IKKγ-dependent pathway and the noncanonical, NEMO/IKKγ-independent, pathway. The canonical activation pathway is mediated by the IKK complex, formed by the scaffold/adaptor protein NEMO/IKKγ and two catalytic subunits, the I-κB kinases IKKα/IKK1 and IKKβ/IKK2. The canonical signaling machinery is virtually present in all cell types and is triggered following cell exposure to a number of ligands, including proinflammatory molecules, such as LPS, TNFα, and IL1.

In the canonical pathway, the inducing stimuli promote activation of the IKK complex that leads to phosphorylation, ubiquitination, and degradation of the cytosolic NF-κB inhibitory proteins I-κBs. In fact, I-κBs proteins sequester NF-κB dimers in the cytosol in an inactive state, until a stimulatory signal initiates the pathway that induces their proteosomal degradation (Beg, Finco, Nantermet, & Baldwin, 1993; Ghosh & Baltimore, 1990). This event results in release of the NF-κB complexes that expose the NLS (nuclear localization signal) and are free to enter the nucleus, where they promote transcription of target genes.

IKKα/IKK1 and IKKβ/IKK2 show a similar domain organization sharing 50% of amino acidic composition. They both include an amino-terminal kinase domain, a leucine zipper (LZ) for kinase homo- and heterodimerization, and a NEMO/IKKγ-binding domain at the C-terminus. NEMO/IKKγ has no enzymatic activity but is absolutely required for the canonical activation pathway (Israël, 2010). Human NEMO/IKKγ is a 419-aa-long dimeric protein whose structure has been investigated by both X-ray crystallography and NMR spectroscopy. This protein contains a ubiquitin-binding domain (UBD) formed by two coiled-coil domains (CC1 and CC2) separated by a linker, a LZ, and a carboxy-terminal zinc finger domain. The amino-terminal part of the protein is required for binding to IKKα/IKK1 and IKKβ/IKK2 (Rushe et al., 2008). As suggested by structural data, NEMO/IKKγ folds in a series of parallel intermolecular coiled coils and its dimeric structure requires interacting partners such as IKK kinases and polyubiquitins to achieve protein stability (Israël, 2010; Rushe et al., 2008) (Fig. 3).

4. NEMO/IKKγ-DEPENDENT IKK COMPLEX ACTIVATION

Many models have been hypothesized to explain activation of the IKK complex. Targeted disruption of each of the IKK genes, as well as transgenic and conditional knockout animals, has shed some light on our understanding regarding the role of each component of the complex (Li & Verma, 2002; Pasparakis, Luedde, & Schmidt-Supprian, 2006). Due to the

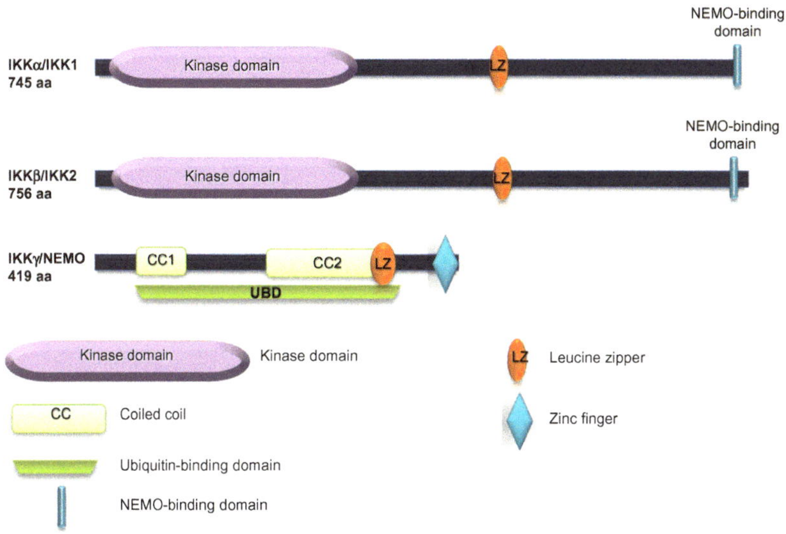

Fig. 3 The IKK complex subunits. Schematic representation of domain organization of the IKK complex subunits. This trimeric complex is the core component of the canonical, NEMO/IKKγ-dependent, activation of the NF-κB signaling pathway and is formed by two catalytic subunits, the kinases IKKα/IKK1 and IKKβ/IKK2, and a regulatory subunit, the scaffold/adaptor protein NEMO/IKKγ. IKKα/IKK1 and IKKβ/IKK2 mediate I-κB phosphorylation through their kinase domains. NEMO/IKKγ has no enzymatic activity but its coiled-coil regions (CC1 and CC2), by forming a large ubiquitin-binding domain (UBD), are absolutely required for the canonical activation pathway.

similarity between the IKKβ/IKK2 knockout and p65/RelA knockout phenotypes—embryonic lethality due to extensive liver damage at around days 13–15—a central role for IKKβ/IKK2 in activation of p65/RelA dimers via I-κBα phosphorylation was proposed (Beg, Sha, Bronson, Ghosh, & Baltimore, 1995; Li, Van Antwerp, Mercurio, Lee, & Verma, 1999; Tanaka et al., 1999). Mice deficient in IKKα/IKK1 survive embryonic development but die perinatally due to multiple morphological defects, including aborted epidermal and skeletal development (Hu et al., 1999). While initially it appeared that IKKα/IKK1 was dispensable for classical NF-κB activation, subsequent reports indicated a requirement for IKKα in multiple NF-κB signaling pathways (Sun, 2011, 2012). NEMO/IKKγ is undoubtedly necessary for signaling in all canonical NF-κB pathways, and NEMO/IKKγ-deficient mice also die at embryonic stage due to massive hepatocyte apoptosis (Rudolph et al., 2000; Schmidt-Supprian et al., 2000).

Moreover, NEMO/IKKγ-deficient cells exhibit a more severe and broader loss of NF-κB activation respect to IKKβ knockout cells, demonstrating that in the absence of IKKβ some NF-κB functions are intact (Pasparakis et al., 2006).

The IKK kinases are activated by phosphorylation of serine residues in the activation T-loop, and three activation mechanisms, all NEMO/IKKγ-dependent, have been proposed (Mitchell et al., 2016):

1. Receptor-induced posttranslational modifications trigger conformational changes in NEMO/IKKγ that in turn induces multimerization of IKK subunits and activation via trans-autophosphorylation;
2. NEMO/IKKγ modifications can also mediate the association between IKK kinases and upstream kinases, leading to mutual activation in a positive feedback circuit;
3. Linear ubiquitin chains attached by LUBAC enzyme on NEMO/IKKγ facilitate kinase transactivation and formation of transient activating signaling complexes.
4. Also phosphorylation of NEMO/IKKγ at multiple serine residues has been suggested to regulate NF-κB activity in response to various stimuli.

5. NF-κB IN THYROID PHYSIOLOGY

Given the crucial role that NF-κB plays in tumor progression, it is not surprising that much of the existing literature on NF-κB and thyroid focuses on the role of this transcription factor in thyroid cancer. An early observation reports that the neoplastic potential of several human thyroid carcinoma cell lines requires functional p65/RelA (Visconti et al., 1997), whereas a number of later observations show that NF-κB inhibitors, alone or in combination with radio- or chemotherapy, result in enhanced sensitivity of thyroid cancer cells to drug-induced apoptosis, to the loss of the ability of these cells to form colonies in soft agar, and to induce tumor growth in nude mice (Li, Abdel-Mageed, Mondal, & Kandil, 2013; Pacifico et al., 2004). On the other hand, much less known is the role that NF-κB plays in normal thyroid physiology. There is evidence indicating a role for NF-κB in the transcriptional regulation of the Na^+/I^- symporter Nis, a membrane protein responsible for the active transport of iodide in the thyroid, gastric mucosa, intestinal epithelium, and lactating mammary glands (Dohán et al., 2003). Indeed, it has demonstrated the existence of a conserved binding site for NF-κB (5′-GGACAGTCCC-3′) within the enhancer region present in the Nis promoter that is responsible for the LPS-enhanced Nis gene expression in thyroid cells (Nicola, Nazar, Mascanfroni, Pellizas, & Masini-Repiso, 2010). In fact, in thyroid cells NF-κB controls the expression of Nis together with the thyroid-specific transcription factor paired box 8 (Pax8), and a physical interaction between p65 subunit and Pax8 has been observed. Interestingly, p65-mediated transactivation of the Nis promoter is further regulated by a posttranslational modification of p65, as nitric oxide donors promote S-nitrosylation of p65 on Cys-38, thereby reducing the transcriptional activity of p65, with consequent reduction of Nis gene expression (Nicola et al., 2015).

In addition to Nis, NF-κB also controls the expression of thyroid peroxidase (Tpo), a thyroid-specific enzyme that catalyzes covalent incorporation of oxidized iodine into tyrosines in thyroglobulin and coupling of iodinated tyrosines to form thyroid hormones (Nazar, Nicola, Vélez, Pellizas, & Masini-Repiso, 2012). Bioinformatic analysis of the Tpo promoter uncovered a putative conserved κB-binding site (5′-AGAAACTCCC-3′). In addition, phosphorylation of p65 on S-536 constitutes an essential step to induce p65-dependent transcriptional expression of Tpo.

Important information on the role of the canonical pathway of NF-κB in thyroid development, differentiation, and function has come from the analysis of a mouse strain bearing a thyroid-specific genetic inactivation of NEMO/IKKγ ($NEMO^{TS-KO}$ mice) (Reale et al., 2016). In these mice, development and differentiation of the thyroid gland appears normal. However, starting soon after birth, $NEMO^{TS-KO}$ mice develop a progressive loss of thyroid cellularity and thyroid markers expression. In fact, in $NEMO^{TS-KO}$ mice the thyroid follicular architecture appears highly variable in diameter with irregular outlines, and significant thyroid cell loss accompanied with massive apoptosis is observed. Significantly, the expression level of several thyroid-specific molecules, such as the transcription factors Pax8 and Ttf1, thyroglobulin, and the Na^+/I^- symporter Nis steadily declines over time. When compared to control mice, $NEMO^{TS-KO}$ mice present significantly lower levels of serum-free T4 hormone, and increased thyroid-stimulating hormone levels, which is indicative of hypothyroidism. Overall, these data indicate that in thyroid canonical NF-κB signaling is required for at least two essential aspects related to thyrocytes survival and maintenance of thyroid markers expression (Fig. 4).

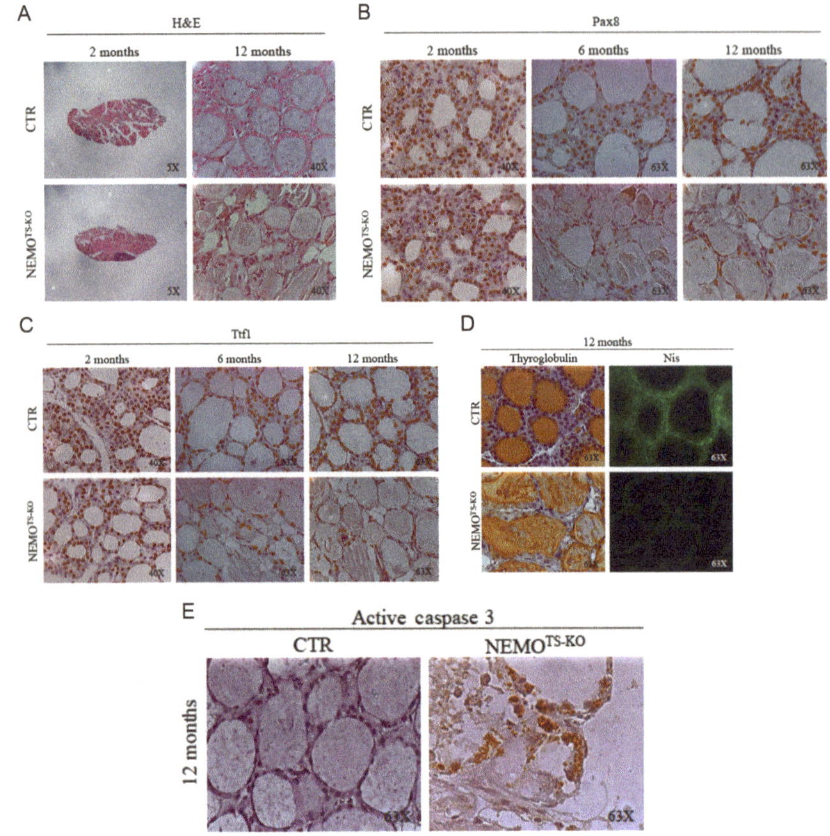

Fig. 4 IκB proteins.

6. CONCLUSION

The NF-κB signaling pathway controls the expression of a large number of genes that have mainly functions in the regulation of immune and inflammatory responses, including cytokines, chemokines, adhesion molecules, and other immunoregulatory proteins. In addition, canonical NF-κB controls the expression of proteins with antiapoptotic, proproliferative, and antioxidant activities.

Experiments based on tissue- and/or organ-specific deletion of key components of the NF-κB signaling pathway have given diverse results regarding the physiological role of this transcription factor outside of the immune system. In fact, functional ablation of NF-κB in liver parenchymal cells causes spontaneous development of steatohepatitis and hepatocellular carcinoma (Luedde et al., 2007). Instead, intestinal epithelial-cell-specific deletion of NEMO/IKKγ results in severe chronic intestinal inflammation due to apoptosis of colonic epithelial cells, impaired expression of antimicrobial peptides, and translocation of bacteria into the mucosa (Nenci et al., 2007). Central nervous system-specific ablation of NEMO/IKKγ results in no apparent abnormalities, but rather ameliorates inflammatory and autoimmune pathologies (van Loo et al., 2006). NEMO/IKKγ inactivation in the heart did not affect embryonic cardiac development but led to spontaneous and progressive impairment of cardiac function, progression to dilated cardiomyopathy, and heart failure (Kratsios et al., 2010). Thus, it appears that canonical NF-κB signaling performs different roles and functions depending on the type of tissue and organ. In this context, consistent data indicate that NF-κB-dependent gene expression is essential for maintaining normal thyroid gland structure and function.

REFERENCES

Al-Khodor, S., Price, C., Kalia, A., & Abu Kwaik, Y. (2010). Functional diversity of ankyrin repeats in microbial proteins. *Trends in Microbiology*, *18*, 132–139.

Beg, A., Finco, T., Nantermet, P., & Baldwin, A. (1993). Tumor necrosis factor and interleukin-1 lead to phosphorylation and loss of IKB-a. *Molecular and Cellular Biology*, *13*, 3301–3310.

Beg, A., Sha, W., Bronson, R., Ghosh, S., & Baltimore, D. (1995). Embryonic lethality and liver degeneration in mice lacking the RelA component of NF-kappa B. *Nature*, *376*, 167–170.

Bhatt, D., & Ghosh, S. (2014). Regulation of the NF-κB-mediated transcription of inflammatory genes. *Frontiers in Immunology*, *5*, 71.

Bork, P. (1993). Hundreds of ankyrin-like repeats in functionally diverse proteins: Mobile modules that cross phyla horizontally? *Proteins*, *17*, 363–374.

Campbell, K., Witty, J., Rocha, S., & Perkins, N. (2006). Cisplatin mimics ARF tumor suppressor regulation of RelA (p65) nuclear factor-kappaB transactivation. *Cancer Research*, *66*, 929–935.

Chen, L., & Green, W. (2003). Regulation of distinct biological activities of the NF-κB transcription factor complex by acetylation. *Journal of Molecular Medicine*, *81*, 549–557.

Christian, F., Smith, E., & Carmody, R. (2016). The regulation of NF-κB subunits by phosphorylation. *Cell*, *5*, E12.

Collins, P., Mitxitorena, I., & Carmody, R. (2016). The ubiquitination of NF-kappaB subunits in the control of transcription. *Cell*, *5*, E23.

Curtois, G., & Gilmore, T. (2006). Mutations in the NF-kappaB signaling pathway: Implications for human disease. *Oncogene*, *25*, 6831–6843.

DiDonato, J., Mercurio, F., & Karin, M. (2012). NF-κB and the link between inflammation and cancer. *Immunological Reviews*, *246*, 379–400.

Dohán, O., De la Vieja, A., Paroder, V., Riedel, C., Artani, M., Reed, M., et al. (2003). The sodium/iodide symporter (NIS): Characterization, regulation, and medical significance. *Endocrine Reviews*, *24*, 48–77.

Ghosh, S. (2006). *Handbook of transcription factor NF-kappaB*. Boca Raton, FL, USA: CRC Press.

Ghosh, S., & Baltimore, D. (1990). Activation in vitro of NF-κB by phosphorylation of its inhibitor IKB. *Nature*, *344*, 678–682.

Ghosh, G., Wang, V., Huang, D., & Fusco, A. (2012). NF-κB regulation: Lessons from structures. *Immunological Reviews*, *246*, 36–58.

Grundström, S., Anderson, P., Scheipers, P., & Sunstedt, A. (2004). Bcl−3 and NF-kappaB p50-p50 homodimers act as transcriptional repressors in tolerant CD4+ T cells. *The Journal of Biological Chemistry*, *279*, 8460–8468.

Hoffmann, A., Natoli, G., & Ghosh, G. (2006). Transcriptional regulation via the NF-kappaB signaling module. *Oncogene*, *25*, 6706–6716.

Hou, Y., Moreau, F., & Chadee, K. (2012). PPARγ is an E3 ligase that induces the degradation of NFκB/p65. *Nature Communications*, *3*, 1300.

Hu, Y., Baud, V., Delhase, M., Zhang, P., Deerinck, T., Ellisman, M., et al. (1999). Abnormal morphogenesis but intact IKK activation in mice lacking the IKKα subunit of the IκB kinase. *Science*, *284*, 316–320.

Huang, D., Vu, D., & Ghosh, G. (2005). NF-κB RelB forms an intertwined homodimer. *Structure*, *13*, 1365–1373.

Huang, B., Yang, X., Lamb, A., & Chen, L. (2010). Posttranslational modifications of NF-κB: Another layer of regulation for NF-κB signaling pathway. *Cellular Signaling*, *22*, 1282–1290.

Israël, A. (2010). The IKK complex, a central regulator of NF-kappaB activation. *Cold Spring Harbor Perspective in Biology*, *2*, a000158.

Kelleher, Z. T., Matsumoto, A., Stamler, J. S., & Marshall, H. E. (2007). NOS2 regulation of NF-κB by S-nitrosylation of p65. *The Journal of Biological Chemistry*, *282*, 30667–30672.

Kratsios, P., Huth, M., Temmerman, L., Salimova, E., Al Banchaabouchi, M., Sgoifo, A., et al. (2010). Antioxidant amelioration of dilated cardiomyopathy caused by conditional deletion of NEMO/IKKgamma in cardiomyocytes. *Circulation Research*, *106*, 133–144.

Li, X., Abdel-Mageed, A. B., Mondal, D., & Kandil, E. (2013). The nuclear factor kappa-B signaling pathway as a therapeutic target against thyroid cancers. *Thyroid*, *23*, 209–218.

Li, Q., Van Antwerp, D., Mercurio, F., Lee, K., & Verma, I. (1999). Severe liver degeneration in mice lacking the IkappaB kinase 2 gene. *Science*, *284*, 321–325.

Li, Q., & Verma, I. (2002). NF-κB regulation in the immune system. *Nature Reviews. Immunology*, *2*, 725–734.

Luedde, T., Beraza, N., Kotsikoris, V., van Loo, G., Nenci, A., De Vos, R., et al. (2007). Deletion of NEMO/IKKgamma in liver parenchymal cells causes steatohepatitis and hepatocellular carcinoma. *Cancer Cell*, *11*, 119–132.

Maine, G., Mao, X., Komarck, C., & Burstein, E. (2007). COMMD1 promotes the ubiquitination of NF-kappaB subunits through a cullin-containing ubiquitin ligase. *EMBO Journal*, *26*, 436–447.

Marshall, H. E., & Stamler, J. S. (2001). Inhibition of NF-κB by S-nitrosylation. *Biochemistry*, *40*, 1688–1693.

Mitchell, S., Vargas, J., & Hoffmann, A. (2016). Signaling via the NFκB system. *WIREs Systems Biology and Medicine*, *8*, 227–241.

Msaki, A., Sanchez, A., Koh, L., Barré, B., Rocha, S., Perkins, N., et al. (2011). The role of RelA (p65) threonine 505 phosphorylation in the regulation of cell growth, survival, and migration. *Molecular Biology of the Cell*, *22*(17), 3032–3040.

Nabel, G., & Verma, I. (1993). Proposed NF-kappa B/I kappa B family nomenclature. *Genes & Development*, *7*, 2063.

Natoli, G., & De Santa, F. (2006). Shaping alternative NF-kappaB-dependent gene expression programs: New clues to specificity. *Cell Death and Differentiation, 13*, 693–696.
Nazar, M., Nicola, J., Vélez, M., Pellizas, C., & Masini-Repiso, A. (2012). Thyroid peroxidase gene expression is induced by lipopolysaccharide involving nuclear factor (NF)-κB p65 subunit phosphorylation. *Endocrinology, 153*, 6114–6125.
Nenci, A., Becker, C., Wullaert, A., Gareus, R., van Loo, G., Danese, S., et al. (2007). Epithelial NEMO links innate immunity to chronic intestinal inflammation. *Nature, 446*, 557–561.
Nicola, J., Nazar, M., Mascanfroni, I., Pellizas, C., & Masini-Repiso, A. (2010). NF-kappaB p65 subunit mediates lipopolysaccharide-induced Na(+)/I(-) symporter gene expression by involving functional interaction with the paired domain transcription factor Pax8. *Molecular Endocinology, 24*, 1846–1862.
Nicola, J. P., Peyret, V., Nazar, M., Romero, J. M., Lucero, A. M., Montesinos, M., et al. (2015). S-Nitrosylation of NF-κB p65 inhibits TSH-induced Na(+)/I(-) symporter expression. *Endocrinology, 156*, 4741–4754.
Oeckinghaus, A., & Ghosh, S. (2009). The NF-κB family of transcription factors and its regulation. *Cold Spring Harbor Perspectives in Biology, 1*(4), a000034.
Pacifico, F., Mauro, C., Barone, C., Crescenzi, E., Mellone, S., Monaco, M., et al. (2004). Oncogenic and anti-apoptotic activity of NF-kappa B in human thyroid carcinomas. *The Journal of Biological Chemistry, 279*, 54610–54619.
Pasparakis, M., Luedde, T., & Schmidt-Supprian, M. (2006). Dissection of the NF-kappaB signalling cascade in transgenic and knockout mice. *Cell Death and Differentiation, 13*, 861–872.
Perkins, N. (2007). Integrating cell-signalling pathways with NF-kappaB and IKK function. *Nature Reviews Molecular Cell Biology, 8*, 49–62.
Plaksin, D., Baeuerle, P., & Eisenbach, L. (1993). KBF1 (p50 NF-kappa B homodimer) acts as a repressor of H-2Kb gene expression in metastatic tumor cells. *The Journal of Experimental Medicine*, 77, 165116–165162.
Reale, C., Iervolino, A., Scudiero, I., Ferravante, A., D'Andrea, L. E., Mazzone, P., et al. (2016). NF κB essential modulator (NEMO) is critical for thyroid function. *The Journal of Biological Chemistry, 291*, 5765–5773.
Rudolph, D., Yeh, W., Wakeham, A., Rudolph, B., Nallainathan, D., Potter, J., et al. (2000). Severe liver degeneration and lack of NF-kappaB activation in NEMO/IKKgamma-deficient mice. *Genes & Development, 14*, 854–862.
Rushe, M., Silvian, L., Bixler, S., Chen, L., Cheung, A., Bowes, S., et al. (2008). Structure of a NEMO/IKK-associating domain reveals architecture of the interaction site. *Structure, 16*, 798–808.
Saccani, S., Pantano, S., & Natoli, G. (2001). Two waves of nuclear factor kappaB recruitment to target promoters. *Journal of Experimental Medicine, 193*, 1351–1359.
Schmidt-Supprian, M., Bloch, W., Courtois, G., Addicks, K., Israel, A., Rajewsky, K., et al. (2000). NEMO/IKK gamma-deficient mice model incontinentia. *Molecular Cell, 5*, 981–992.
Sen, R., & Baltimore, D. (1986). Multiple nuclear factors interact with the immunoglobulin enhancer sequences. *Cell, 46*, 705–716.
Sun, S. (2011). Non-canonical NF-κB signaling pathway. *Cell Research, 21*, 71–85.
Sun, S. (2012). The non-canonical NF-κB pathway. *Immunological Reviews, 246*, 125–140.
Tanaka, M., Fuentes, M., Yamguchi, K., Durnin, M., Dalrymple, S., Hardy, K., et al. (1999). Embryonic lethality, liver degeneration, and impaired NF-kappa B activation in IKK-beta-deficient mice. *Immunity, 10*, 421–429.
Tanaka, T., Grusby, M., & Kaisho, T. (2007). PDLIM2-mediated termination of transcription factor NF-kappa B activation by intranuclear sequestration and degradation of the p65 subunit. *Nature Immunology, 8*, 584–591.

Tsui, R., Kearns, J., Lynch, C., Vu, D., Ngo, K., Basak, S., et al. (2015). IκBβ enhances the generation of the low-affinity NFκB/RelA homodimer. *Nature Communications*, *6*, 7068.
van Loo, G., De Lorenzi, R., Schmidt, H., Huth, M., Mildner, A., Schmidt-Supprian, M., et al. (2006). Inhibition of transcription factor NF-kappaB in the central nervous system ameliorates autoimmune encephalomyelitis in mice. *Nature Immunology*, 7, 954–961.
Visconti, R., Cerutti, J., Battista, S., Fedele, M., Trapasso, F., Zeki, K., et al. (1997). Expression of the neoplastic phenotype by human thyroid carcinoma cell lines requires NFkappaB p65 protein expression. *Oncogene*, *15*, 1987–1994.
Wang, V., Huang, W., Asagiri, M., Spann, N., Hoffmann, A., Glass, C., et al. (2012). The transcriptional specificity of NF-κB dimers is coded within the κB DNA response elements. *Cell Reports*, *2*, 824–839.
Xu, H., You, M., Shi, H., & Hou, Y. (2015). Ubiquitin-mediated NFkappaB degradation pathway. *Cellular and molecular immunology*, *12*, 653–655.
Zheng, C., Yin, Q., & Wu, H. (2011). Structural studies of NF-κB signaling. *Cell Research*, *21*, 183–195.
Zotti, T., Uva, A., Ferravante, A., Vessichelli, M., Scudiero, I., Ceccarelli, M., et al. (2011). TRAF7 protein promotes Lys-29-linked polyubiquitination of IkappaB kinase (IKKgamma)/NF-kappaB essential modulator (NEMO) and p65/RelA protein and represses NF-kappaB activation. *The Journal of Biological Chemistry*, *286*, 22924–22933.

CHAPTER TEN

Thyroid Hormone Regulation of Adult Neurogenesis

Sashaina E. Fanibunda*,[1], Lynette A. Desouza*,[1], Richa Kapoor*, Rama A. Vaidya†, Vidita A. Vaidya*,[2]

*Tata Institute of Fundamental Research, Mumbai, India
†Medical Research Centre, Kasturba Health Society, Mumbai, India
[2]Corresponding author: e-mail address: vvaidya@tifr.res.in

Contents

1. Thyroid Hormone and the Adult Brain 213
2. Making New Neurons in the Adult Brain: A Role for Thyroid Hormone 215
3. Thyroid Hormone and Adult Hippocampal Neurogenesis 216
   3.1 From Hippocampal Progenitors to Mature Neurons 216
   3.2 Thyroid Hormone Perturbations and Adult Hippocampal Neurogenesis: Insights From *in vivo* Studies 218
   3.3 Influence of Thyroid Hormone on Hippocampal Progenitors *In Vitro* 221
   3.4 Contribution of TRs to Adult Hippocampal Neurogenesis 223
4. Thyroid Hormone Influence on Adult SVZ Neurogenesis 227
   4.1 Progression of SVZ Progenitor Development 227
   4.2 Thyroid Hormone, TRs, and SVZ Neurogenesis 228
5. Underlying Molecular Mediators for the Regulation of Neurogenesis by Thyroid Hormone 232
6. Functional Implications of the Neurogenic Effects of Thyroid Hormone 235
   6.1 Thyroid Hormone Regulation of Behavior in Animal Models 235
   6.2 Clinical Relevance of the Neurogenic Actions of Thyroid Hormone 236
7. Conclusion 239
Acknowledgments 240
References 240

Abstract

Thyroid hormone is classically known to play a crucial role in neurodevelopment. The potent effects that thyroid hormone exerts on the adult mammalian brain have been uncovered relatively recently, including an important role in the modulation of progenitor development in adult neurogenic niches. This chapter extensively reviews the current understanding of the influence of thyroid hormone on distinct stages of adult progenitor development in the subgranular zone (SGZ) of the hippocampus and

[1] These authors contributed equally to the chapter.

Vitamins and Hormones, Volume 106
ISSN 0083-6729
http://dx.doi.org/10.1016/bs.vh.2017.04.006

subventricular zone (SVZ) that lines the lateral ventricles. We discuss the role of specific thyroid hormone receptor isoforms, in particular TRα1, which modulates cell cycle exit in neural stem cells, progenitor survival, and cell fate choice, with both a discrete and overlapping nature of regulation noted in SGZ and SVZ progenitors. The balance between liganded and unliganded TRα1 can evoke differing consequences for adult progenitor development, and the relevance of this to conditions such as adult-onset hypothyroidism, wherein unliganded thyroid hormone receptors (TRs) dominate, is also a focus of discussion. Although a detailed molecular understanding of the specific thyroid hormone target genes that contribute to the neurogenic actions of thyroid hormone is currently lacking, we highlight the current state of knowledge and discuss avenues for future investigation. The goal of this chapter is to provide a comprehensive and detailed analysis of the effects of thyroid hormone on adult neurogenesis, to discuss putative molecular mechanisms that mediate these effects, and the behavioral, functional, and clinical implications of the neurogenic actions of thyroid hormone.

Thyroid hormone plays a seminal role in shaping the development of the mammalian nervous system (Bernal, 2007; Preau, Fini, Morvan-Dubois, & Demeneix, 2015; Rovet, 2014). While the role of thyroid hormone has been best appreciated and studied during neurodevelopment, a growing body of knowledge also highlights its continued role in the regulation of plasticity within the adult mammalian brain (Bauer, Goetz, Glenn, & Whybrow, 2008; Calza, Aloe, & Giardino, 1997; Koromilas et al., 2010; Sarkar, 2002). The critical influence of thyroid hormone during neurodevelopment was the primary focus of several studies due to the clinical observations that highlighted the severe neurological consequences of perturbations of euthyroid status during gestational and early postnatal development (de Escobar, Obregon, & del Rey, 2007; Dugbartey, 1998; Williams, 2008; Zoeller & Crofton, 2005). Relatively few studies focussed on a continued role for thyroid hormone in the regulation of adult brain structure, plasticity, and function, only further contributing to the common idea that the effects of thyroid hormone on the mammalian brain are largely restricted to "critical periods" of neurodevelopment (Axelstad et al., 2008; Bernal, 2002; Gilbert & Sui, 2006; Mastorakos, Karoutsou, Mizamtsidi, & Creatsas, 2007; Zoeller & Rovet, 2004). This perception has now been revised, due to several studies that highlight the important role of thyroid hormone within the adult brain (Henley & Koehnle, 1997; Koromilas et al., 2010; Schroeder & Privalsky, 2014), in particular, a key influence of thyroid hormone in the regulation of progenitor development within the neurogenic niches of the adult brain (Kapoor, Fanibunda, Desouza, Guha, & Vaidya, 2015; Remaud, Gothie, Morvan-Dubois, & Demeneix, 2014). The focus of this chapter is to review in detail the literature focussed on the influence of thyroid hormone on adult mammalian neurogenesis, and to discuss the functional implications of the neurogenic effects of thyroid hormone.

1. THYROID HORMONE AND THE ADULT BRAIN

The synthesis of thyroid hormone is under the control of the hypothalamic–pituitary–thyroid axis. The hypothalamus secretes thyrotropin-releasing hormone (TRH), which in turn stimulates the release of thyroid-stimulating hormone (TSH) from the pituitary. TSH results in the synthesis and secretion of thyroid hormone from the follicular cells of the thyroid gland (Chiamolera & Wondisford, 2009; Costa-e-Sousa & Hollenberg, 2012; Joseph-Bravo, Jaimes-Hoy, Uribe, & Charli, 2015). The secreted thyroid hormone is largely in the thyroxine (3,3′,5,5′-tetraiodothyronine, T4) precursor form and in smaller quantities in its active form (3,3′,5-triiodothyronine, T3) and is transported in the plasma bound to passive carriers, including albumin, thyroxine-binding globulin, and transthyretin (TTR). At the blood–brain barrier, the active entry of thyroid hormone into the brain is facilitated via TTR in the choroid plexus and the organic anion transporter proteins (Bernal, Guadano-Ferraz, & Morte, 2015; Jansen, Friesema, Milici, & Visser, 2005; Richardson, Wijayagunaratne, D'Souza, Darras, & Van Herck, 2015; Wirth, Schweizer, & Kohrle, 2014). On entry into the brain, the local availability of thyroid hormone is further modulated by the activity of three types of iodothyronine deiodinases—I, II, and III (D1, D2, and D3). In particular, coordination of the relative expression and activity of D2, which converts the prohormone T4 to active T3, and D3, which is responsible for inactivation of T3 and T4, can dynamically modulate the levels of thyroid hormone available within local brain regions (Bianco, 2011; Bianco & Kim, 2006).

Thyroid hormone mediates its actions through thyroid hormone receptors (TRs), which act as transcription factors to regulate gene expression of target genes (Cheng, Leonard, & Davis, 2010; Harvey & Williams, 2002). Alternative splicing of two TR genes—alpha and beta results in multiple different TR isoforms that exhibit considerable similarity in their DNA-binding domains, but differ in their transactivation and ligand-binding domains. The TRα1, TRα2, TRβ1, and TRβ2 isoforms are expressed in the mammalian brain (Bradley, Towle, & Young, 1992; Forrest & Vennstrom, 2000; Mellstrom, Naranjo, Santos, Gonzalez, & Bernal, 1991), with TRα1 accounting for 70% of the total TR expression (Schwartz, Strait, Ling, & Oppenheimer, 1992; Wallis et al., 2010). The TRα2 isoform due to an absence of thyroid hormone-binding ability is suggested to exert a dominant-negative role via thyroid hormone response elements (TREs) within target genes (Guissouma et al., 2014; Koenig et al., 1989). Of the TRβ isoforms, TRβ1 is widely expressed across the brain (Forrest & Vennstrom, 2000; Murata, 1998; Zhang & Lazar, 2000), while TRβ2 expression is limited to the hypothalamus and pituitary (Abel, Ahima, Boers, Elmquist, & Wondisford, 2001; Cook, Kakucska, Lechan, & Koenig, 1992; Lechan, Qi, Jackson, & Mahdavi, 1994), with expression levels increasing across postnatal development. The expression of thyroid hormone target genes can be regulated by both ligand-bound TRs as well as unliganded aporeceptors, often in a very distinct fashion (Bernal & Morte, 2013; Chassande, 2003; Zhang & Lazar, 2000). In general, the presence of thyroid hormone results in the recruitment of coactivators that mediate the acetylation of histones, and subsequent binding of RNA Polymerase II (Bernal et al., 2015; Cheng et al., 2010; Harvey & Williams, 2002; Liu & Brent, 2010). In contrast, the absence of thyroid hormone resulting in a TR aporeceptor form often causes the repression of thyroid hormone target genes due to the recruitment of corepressors that contribute to histone deacetylation (Astapova & Hollenberg, 2013; Flamant, Gauthier, & Samarut, 2007; Hu & Lazar, 2000; Jepsen & Rosenfeld, 2002; Privalsky, 2004; Shi, 2009). In addition, to the genomic effects of thyroid hormone mediated via the TRs, thyroid hormone can also exert nongenomic actions mediated by membrane-bound receptors or through modulation of signaling pathways (Cao, Kambe, Moeller, Refetoff, & Seo, 2005; Davis, Leonard, & Davis, 2008; Davis et al., 2010; Furuya, Lu, Guigon, & Cheng, 2009; Lanni, Moreno, & Goglia, 2016; Silva, Giannocco, Santos, & Nunes, 2006). The effects of thyroid hormone on the adult mammalian brain are subject to regulation at diverse levels. These span from changes in local availability of thyroid hormone through the regulation of transport and the enzymatic activity of different deiodinases, the differential expression of TR isoforms, the interaction of TRs with coactivator or corepressor complexes and chromatin remodeling machinery, the balance of ligand-dependent and independent changes in gene expression, as well as the nongenomic actions of thyroid hormone.

2. MAKING NEW NEURONS IN THE ADULT BRAIN: A ROLE FOR THYROID HORMONE

The two predominant sites within the adult mammalian brain which house the progenitors that give rise to new neurons include the subgranular zone (SGZ) in the dentate gyrus (DG) subfield of the hippocampus, and the subventricular zone (SVZ) that lines the lateral ventricles (Fig. 1)

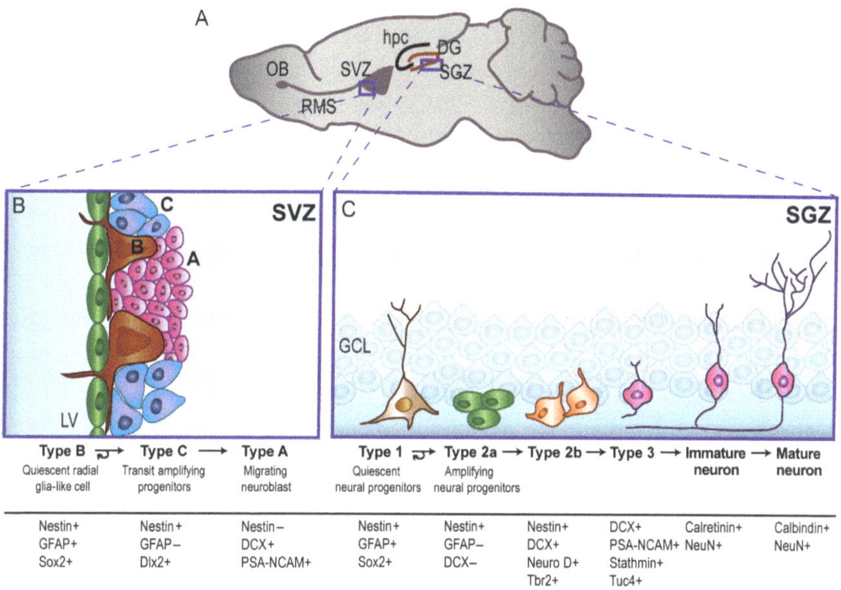

Fig. 1 Adult neurogenesis within the major neurogenic niches of the mammalian brain. (A) Shown is a schematic depicting a sagittal section through an (A) adult rodent brain highlighting the major neurogenic niches of (B) the subventricular zone (SVZ) lining the lateral ventricles (LV) from which progenitors traverse along the rostral migratory stream (RMS) to the olfactory bulb (OB) and (C) the subgranular zone (SGZ) in the dentate gyrus (DG) subfield of the hippocampus (Hpc). (B) The illustration highlights the distinct stages of progression for SVZ progenitors and the markers that characterize specific stages of SVZ progenitor development, starting with the Type B quiescent radial glia-like stem cell, the Type C transit amplifying progenitors progressing to the Type A migratory neuroblasts that proceed along the RMS to the OB. (C) The schematic highlights the stages of SGZ progenitor and markers associated with specific stages in the hippocampal neurogenic niche. Shown are the Type 1 quiescent neural progenitors which self-renew and also generate Type 2a amplifying neural progenitors which give rise to Type 2b cells that further mature through the Type 3 neuroblast stage to then become immature and finally mature neurons within the granule cell layer (GCL) of the DG hippocampal subfield. Shown is the expression of the stage-specific markers associated with progenitor development in both the SVZ and SGZ.

(Ming & Song, 2011). These progenitors retain the capacity to divide and give rise to daughter cells that undergo structural and functional maturation, eventually integrating into mature neuronal networks (Sailor, Schinder, & Lledo, 2016). Progenitor development and maturation are subject to regulation by diverse intrinsic and extrinsic factors at all the stages that the progenitor traverses en route to forming a mature neuron. Among the extrinsic factors, adult neurogenesis is highly sensitive to hormones, neurotransmitters, and growth factors present in the neurogenic niche (Mahmoud, Wainwright, & Galea, 2016; Perez-Domper, Gradari, & Trejo, 2013; Vaidya, Vadodaria, & Jha, 2007). Thyroid hormone has been shown to exert a strong influence on the development of progenitors both within the SGZ as well as the SVZ (Kapoor et al., 2015; Remaud et al., 2014). These studies have examined direct effects of thyroid hormone on neuronal progenitors *in vitro* (Desouza et al., 2005; Kapoor, Desouza, Nanavaty, Kernie, & Vaidya, 2012), as well as the consequences of perturbing the levels of circulating thyroid hormone *in vivo* using pharmacological and surgical approaches (Ambrogini et al., 2005; Desouza et al., 2005; Lemkine et al., 2005; Montero-Pedrazuela et al., 2006). More recently, studies have also capitalized on the use of TR isoform-specific mutant mouse lines to examine effects on adult neurogenesis (Kapoor, Ghosh, Nordstrom, Vennstrom, & Vaidya, 2011; Kapoor et al., 2010; Lemkine et al., 2005; Lopez-Juarez et al., 2012). We have a much better understanding of the progenitor stage-specific effects of thyroid hormone, although the molecular mechanisms that underlie these neurogenic effects and the target genes, both at the level of the progenitors and through influences within the neurogenic niche, remain poorly elucidated. In addition, the physiological and behavioral consequences of the neurogenic effects of thyroid hormone remain to be determined. We will describe at length the effects of thyroid hormone on both hippocampal and SVZ progenitors, highlighting the insights currently available based on the existent scientific literature and also point out lacunae in our current understanding.

3. THYROID HORMONE AND ADULT HIPPOCAMPAL NEUROGENESIS

3.1 From Hippocampal Progenitors to Mature Neurons

The entire process of generation of mature granule cell neurons within the hippocampal DG subfield from progenitors that lie within the SGZ and their functional integration into the existing neurocircuitry takes around

4–6 weeks (Goncalves, Schafer, & Gage, 2016; Ming & Song, 2011). Around 9000 progenitors have been estimated to be generated everyday within the rodent SGZ; however, only around half of these actually go on to survive and contribute to new neuron addition to the granule cell layer (GCL). The surviving progenitors are largely committed to take on the identity of excitatory granule cell neurons, with a minor fraction observed to acquire a glial fate. The milestones that a hippocampal progenitor progresses through on its journey from a neural stem cell to a mature granule cell neuron are characterized by the presence of specific markers (Fig. 1) (Kempermann, Jessberger, Steiner, & Kronenberg, 2004; Kempermann, Song, & Gage, 2015; Kuhn, Eisch, Spalding, & Peterson, 2016). The stem cell/quiescent neural progenitor (QNP) or Type 1 cell is a radial glial-like cell that expresses the intermediate filament protein nestin, the glial fibrillary acidic protein (GFAP), and the transcription factor Sox2. QNPs exhibit distinct characteristics, including the fact that they undergo asymmetric, slow division replenishing their own pool and simultaneously generating daughter cells characterized by the ability to rapidly amplify their numbers. These highly mitotic, transit amplifying neural progenitors (ANPs) also called Type 2a cells continue to express nestin, but lose GFAP and Sox2 expression. The ANPs divide rapidly to generate Type 2b neuroblasts with limited proliferative capacity, which retain nestin immunopositivity. These Type 2b hippocampal progenitors acquire expression of the microtubule-associated protein doublecortin (DCX), as well as the basic helix–loop–helix transcription factor NeuroD, that contributes to neuronal cell fate determination. Type 2b cells then go on to migrate into the GCL and undergo maturation to form DCX-positive, nestin-negative postmitotic Type 3 cells that also express polysialylated neural cell adhesion molecule (PSA-NCAM) and markers such as Stathmin and TUC-4 (Fig. 1) (Kempermann et al., 2004; Kuhn et al., 2016; Ming & Song, 2011; Zhao, Deng, & Gage, 2008). As these immature neurons undergo maturation, they extend their dendritic arbors, receive synaptic inputs, send out axonal projections to the CA3 hippocampal subfield, and undergo a switch from the transient expression of the calcium-binding protein calretinin to the expression of calbindin, which is associated with mature granule cell neurons. This entire process of maturation from a hippocampal progenitor to an integrated mature neuron takes about 4–6 weeks and is highly sensitive to perturbations of the environment both external and within the neurogenic niche (Kempermann et al., 2015; Ming & Song, 2011). During this process, immature neurons that exhibit distinct electrophysiological characteristics of high excitability and a lower threshold for firing can disproportionately influence hippocampal network function (Song, Christian, Ming, & Song, 2012), though the total numbers of new neurons remain a relatively small fraction of all granule cell neurons. The process of adult hippocampal neurogenesis has been linked to hippocampal dependent functions such as pattern separation, spatial learning and memory, and the modulation of mood and anxiety (Besnard & Sahay, 2016; Hage & Azar, 2012; Lieberwirth, Pan, Liu, Zhang, & Wang, 2016; Miller & Hen, 2014; Sahay et al., 2011).

3.2 Thyroid Hormone Perturbations and Adult Hippocampal Neurogenesis: Insights From *in vivo* Studies

The influence of thyroid hormone on adult hippocampal neurogenesis was first demonstrated in rodent models (Fig. 2). Hypothyroid status was induced using goitrogen treatment or thyroidectomy, and hyperthyroidism through the administration of exogenous thyroid hormone (Ambrogini et al., 2005; Desouza et al., 2005; Montero-Pedrazuela et al., 2006). Adult-onset hypothyroidism in rats significantly decreased the postmitotic survival and neuronal differentiation of hippocampal progenitors, without affecting their proliferation (Ambrogini et al., 2005; Desouza et al., 2005). This reduction in progenitor survival was likely mediated through increased apoptotic cell death. Both the decline in progenitor survival and neuronal differentiation were normalized in hypothyroid animals by restoration of euthyroid status through thyroid hormone replacement therapy (Desouza et al., 2005). Another report in rats indicated delayed neuronal morphological maturation and the prolonged expression of the immature neuronal marker TUC-4, following goitrogen-mediated hypothyroidism (Ambrogini et al., 2005). While both these studies using goitrogens in adult rats did not alter hippocampal progenitor proliferation, findings from a study in thyroidectomized rats demonstrated a decrease in progenitor proliferation in the hippocampus (Montero-Pedrazuela et al., 2006). There was also a decrease in DCX-positive immature neurons and decreased dendritic complexity noted in this study. Interestingly, these hypothyroid effects correlated with a depressive phenotype in the forced swim test without altering cognitive performance in the novel object recognition test.

While all the studies detailed earlier were in agreement about the hypothyroidism induced decrease in hippocampal neurogenesis, the specific progenitor stages that are affected by alterations in thyroid hormone remained

A

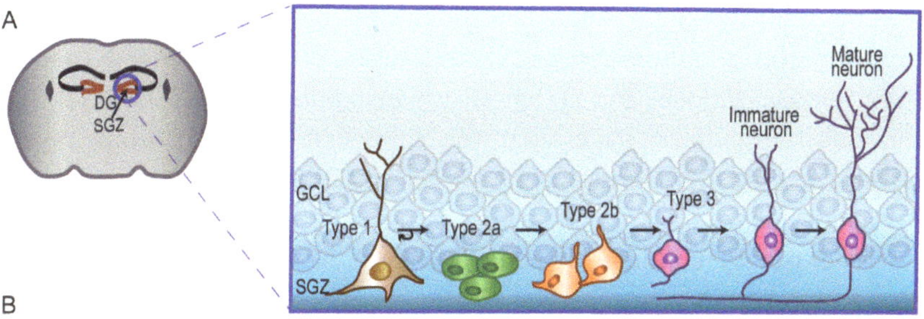

B

Condition	Type 1	Type 2a	Type 2b	Type 3	Immature Neuron	References
Adult-onset hyperthyroidism	n.e.	n.e.	↑	↑	↑	Kapoor et al. (2012)
Adult-onset hypothyroidism	n.e.	n.e.	↓	↓	↓	Desouza et al. (2005); Ambrogini et al. (2005); Montero-Pedrazuela et al. (2006); Sanchez-Huerta et al. (2010)
$TR\alpha1^{-/-}$	n.e.	n.e.	↑	↑	↑	Kapoor et al. (2010)
$TR\alpha2^{-/-}$	n.e.	n.e.	↓	↓	↓	Kapoor et al. (2010)
$TR\alpha1^{+/m}$	n.e.	n.e.	↓	↓	↓	Kapoor et al. (2010)
$TR\beta^{-/-}$	↑	↑	↑	n.e.	n.e.	Kapoor et al. (2011)

Fig. 2 Thyroid hormone regulation of adult hippocampal neurogenesis. (A) Shown is a schematic of a coronal section through the adult rodent brain at the level of the hippocampus highlighting the neurogenic niche of the subgranular zone (SGZ) within the dentate gyrus (DG) subfield. The schematic illustrates the developmental progression of adult SGZ progenitors from Type 1 quiescent neural progenitors to mature neurons integrated into the hippocampal network. (B) Shown is a table describing the effects of thyroid hormone on distinct stages of SGZ progenitor development, including studies from adult-onset perturbations of thyroid hormone levels and TR isoform-specific mutant mouse models. *n.e.* refers to no reported effect.

unclear. Experiments by Ambrogini et al. (2005) and Desouza et al. (2005) suggested that hippocampal progenitor survival and their differentiation into neurons are affected by perturbations of thyroid hormone status. In contrast, studies by Montero-Pedrazuela et al. (2006) suggested that the proliferative stage of hippocampal progenitors is also sensitive to thyroid hormone perturbations. These discrepancies are possibly due to the differences in treatment paradigms as well as methods used to perturb thyroid hormone status. In addition, the use of bromodeoxyuridine (BrdU) to assess progenitor turnover is complicated by the fact that the BrdU administration paradigm and the time of sacrifice can result in the labeling of multiple stages of progenitor development, making it difficult to delineate effects on cell turnover, cell cycle duration, and short-term survival (Taupin, 2007). The advent of transgenic mouse models and a better understanding of the markers that distinguish individual stages that the hippocampal progenitors traverse during their maturation process have greatly helped the study of effects on specific stages of hippocampal progenitor development. For example, the use of transgenic Nestin-green fluorescent protein (GFP) reporter mice (Yu, Dandekar, Monteggia, Parada, & Kernie, 2005) along with triple immunofluorescence labeling has made it possible to distinguish between the first three stages (Type 1, 2a, and 2b) of hippocampal progenitor development. Using these Nestin-GFP mice, studies have demonstrated that while adult-onset hypothyroidism did not affect the total proliferative pool (Type 1 and 2a) of hippocampal progenitors, there was a significant decrease in the number of Nestin-GFP and DCX double-labeled cells (Type 2b neuroblasts). In addition, the postmitotic survival of these hippocampal progenitors (Type 3, namely those that were DCX- and NeuroD-positive) was significantly decreased in adult-onset hypothyroid Nestin-GFP mice (Kapoor et al., 2012). Recent evidence also corroborates the view that it is predominantly postmitotic hippocampal progenitors that are sensitive to a decline in thyroid hormone levels, with no change in proliferation (Sanchez-Huerta, Garcia-Martinez, Vergara, Segovia, & Pacheco-Rosado, 2016) with results suggestive of an important role for thyroid hormone in maintenance and differentiation of the progenitor pool once they exit the cell cycle (Fig. 2). A recent report, using a model for moderate adult-onset thyroid hormone insufficiency, noted no effects on numbers of DCX-positive immature neurons, suggesting that the degree of thyroid hormone insufficiency may also be a critical factor in determining the extent of regulation of adult hippocampal neurogenesis (Gilbert, Goodman, Gomez, Johnstone, & Ramos, 2016).

In contrast to the clear similarities in observations from studies of the effects of adult-onset hypothyroidism in rat and mouse models, adult-onset hyperthyroidism evokes distinct effects on hippocampal neurogenesis in these rodent models. While, in rat models, increased circulating thyroid hormone levels had no effect on hippocampal progenitor proliferation, survival, or differentiation (Desouza et al., 2005), in adult-onset hyperthyroid mice a significant increase was noted in postmitotic survival and differentiation (Kapoor et al., 2012) (Fig. 2). Adult mice administered thyroid hormone exhibited significant increases in DCX- and NeuroD-positive hippocampal progenitors (Kapoor et al., 2012). Further, studies using

Nestin-GFP mice demonstrated that the maturation of hippocampal progenitors into neurons was accelerated with both a total increase noted in the number of DCX-positive (Type 3) immature neurons and a speeding up of the acquisition of markers of differentiation by hippocampal progenitors observed using a BrdU pulse-chase experiment (Kapoor et al., 2012). The differences noted in the effects of exogenous thyroid hormone treatment on adult hippocampal progenitors in the mouse vs rat brain raise certain possibilities. Cells within the hippocampal neurogenic niche or progenitor TRs in the murine brain may not be saturated under euthyroid conditions, thus retaining the possibility for additional effects (enhanced survival, accelerated maturation) upon exogenous thyroid hormone administration. In contrast, the studies thus far suggest that in case of the rat hippocampal neurogenic niche or progenitors, it is possible that TRs are already completely saturated under the baseline state of euthyroidism, leaving no scope for further effects on progenitor survival and differentiation on addition of thyroid hormone.

3.3 Influence of Thyroid Hormone on Hippocampal Progenitors *In Vitro*

While *in vivo* studies clearly indicate an important influence of thyroid hormone on adult hippocampal neurogenesis (Kapoor et al., 2015; Remaud et al., 2014), thus far these studies do not allow a clear distinction of whether the neurogenic effects of thyroid hormone are mediated directly on progenitors or via an influence on the neurogenic niche. In this regard, studies that have examined the influence of thyroid hormone on hippocampal progenitors in culture provide specific insights and also highlight avenues for future investigation.

Thyroid hormone treatment of dispersed hippocampal progenitor cultures indicates effects on enhanced progenitor proliferation and survival, as well as a shift toward increased glial differentiation (Desouza et al., 2005). The decline in progenitor cell death *in vitro* is consistent with the improved survival of DCX-positive progenitors noted upon exogenous thyroid administration *in vivo*. In contrast, both the enhanced proliferation and glial differentiation are only observed *in vitro*, suggesting that removal of hippocampal progenitors from their neurogenic niche may also alter the nature of regulation in response to thyroid hormone (Desouza et al., 2005). In an alternate *in vitro* model for neural stem cells, namely the hippocampal neurosphere assay, thyroid hormone treatment induced no change in the number of neurospheres, but a shift was noted from larger to smaller neurospheres, suggesting the possibility of enhanced differentiation. Indeed, this was confirmed by the greater expression of the neuronal marker, Tuj-1, in neurospheres treated with thyroid hormone (Kapoor et al., 2012). While dispersed hippocampal progenitors and neurosphere cultures both appear to express the different TR isoforms (Desouza et al., 2005; Kapoor et al., 2012), their relative abundance is yet unknown and the presence of thyroid hormone transporters and the deiodinases has not been examined.

Although there is a certain degree of overlap between the effects of thyroid hormone observed *in vitro* and *in vivo*, in particular the enhanced progenitor survival, the range of effects *in vivo* suggests an important influence mediated via the neurogenic niche. This opens up an important area for investigation to delineate direct effects mediated on hippocampal progenitors and those that involve cellular components of the neurogenic niche. Within the neurogenic niche, thyroid hormone may exert an influence on neurons, astrocytes, as well as microglia, which in turn may then modulate hippocampal progenitor development and maturation. In this regard, astrocytes are of particular interest as they are known to provide important cues for progenitor development (Ashton et al., 2012; Magnusson & Frisen, 2016; Seri, Garcia-Verdugo, McEwen, & Alvarez-Buylla, 2001). While the effects of thyroid hormone on astrocytes are well studied (Dezonne, Lima, Trentin, & Gomes, 2015; Martinez & Gomes, 2002; Mohacsik, Zeold, Bianco, & Gereben, 2011; Morte & Bernal, 2014) and thyroid hormone levels are known to be locally regulated by D2 present in astrocytes (Mohacsik et al., 2011; Morte & Bernal, 2014), the contribution of astrocytes in mediating the neurogenic effects of thyroid hormone remains unclear. Thyroid hormone could exert effects on hippocampal progenitors by modulating paracrine release of growth factors such as basic fibroblast growth factor and neurotrophin-3 from astrocytes (Igelhorst, Niederkinkhaus, Karus, Lange, & Dietzel, 2015; Niederkinkhaus, Marx, Hoffmann, & Dietzel, 2009; Tseng et al., 2006). In addition, thyroid hormone also influences the regulation of diverse signaling pathways within the hippocampus. For example, thyroid hormone has been reported to regulate the expression of the developmental signaling morphogen, sonic hedgehog, and its receptors, which may in turn be relevant to the neurogenic actions of thyroid hormone (Desouza et al., 2011). Further, thyroid hormone is known to exert nongenomic effects on critical signaling pathways such as the MAP kinase cascade (Bitiktas et al., 2016), in diverse cell types including in microglia influencing their migration and phagocytosis (Mori et al., 2015). Although the contributions of these effects of thyroid hormone on neurons and glia in the hippocampus in mediating the neurogenic effects of thyroid hormone remain poorly understood, these studies motivate future research to determine the effects of thyroid hormone that arise through direct regulation of hippocampal progenitors and those mediated by components of the neurogenic niche.

3.4 Contribution of TRs to Adult Hippocampal Neurogenesis

Thus far, studies have identified the postmitotic Type 2b and 3 hippocampal progenitor stages as the critical stages of progenitor development that are particularly responsive to thyroid hormone-mediated regulation. These predominantly postmitotic stages in the developmental progression for hippocampal progenitors in some sense can be viewed as a gate at which specific signals and cues may serve to determine whether these cells survive and differentiate, thus resulting in their integration into the hippocampal network or end up as slated for death (Fig. 3) (Kapoor et al., 2012). In this regard, thyroid hormone has been speculated to be such a putative cue that may determine the cell cycle exit, eventual survival, and differentiation of adult hippocampal progenitors. This notion is inspired by several studies with other progenitor types, namely, oligodendrocytic progenitors, embryonic stem cells, and myogenic progenitors wherein a role for thyroid hormone has been identified as a key timing switch to initiate cell cycle exit and promote differentiation (Baas et al., 1997; Billon, Jolicoeur, Tokumoto, Vennstrom, & Raff, 2002; Chen et al., 2012; Daury et al., 2001; Fernandez, Pirondi, Manservigi, Giardino, & Calza, 2004; Pascual & Aranda, 2013). Studies with whole-body TR isoform-specific knockout and transgenic mice (Kapoor et al., 2012, 2011, 2010) have provided insights into the role of thyroid hormone in hippocampal progenitor development; at the same time the findings of these studies have also clearly highlighted the need and importance for future experiments using conditional TR mutant mice to gain a deeper understanding of the role of TRs in adult hippocampal neurogenesis.

The contribution of TRα1, the dominant TR isoform within the mammalian brain (Schwartz et al., 1992; Wallis et al., 2010), in the regulation of adult hippocampal neurogenesis has been analyzed through the use of specific mutant mouse lines namely, the TR$\alpha1^{-/-}$ mice which lack TRα1 expression, TR$\alpha2^{-/-}$ mice which demonstrate a compensatory

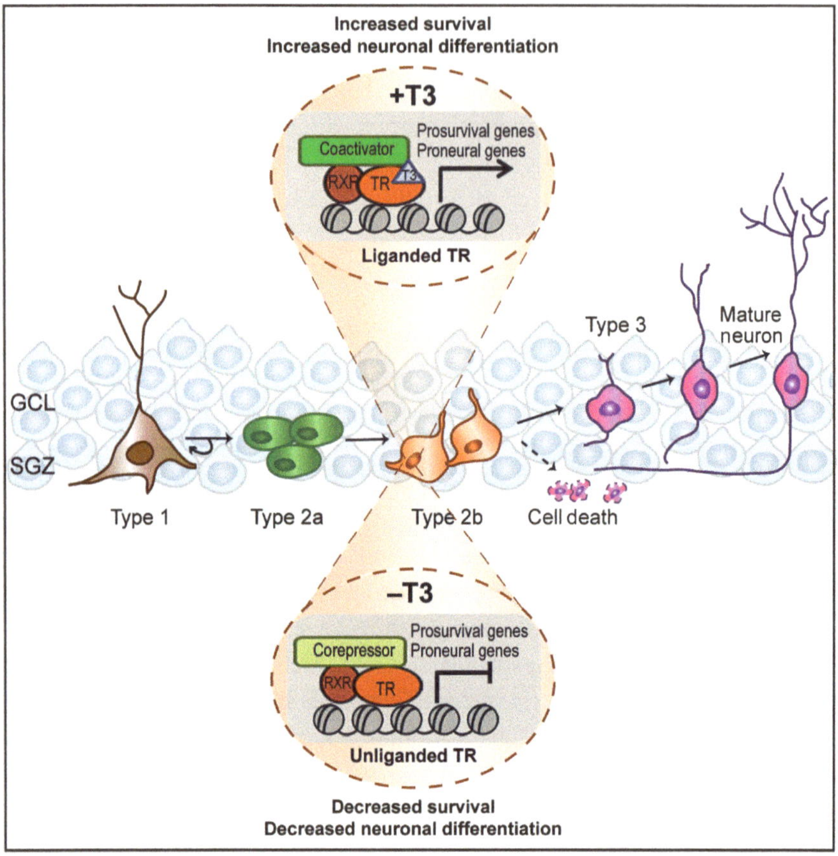

Fig. 3 Putative mechanism of thyroid hormone action on adult hippocampal progenitors. Shown is a schematic of the putative underlying mechanism for thyroid hormone action during hippocampal progenitor development from a Type 1 quiescent neural progenitor in the subgranular zone (SGZ) to a full-fledged mature neuron integrated into the granule cell layer (GCL). Studies support a working model for thyroid hormone action, in particular in Type 2b hippocampal progenitors, where in the presence of thyroid hormone (T3), liganded thyroid hormone receptors (TRs) heterodimerize with retinoid-X-receptor (RXR) and recruit coactivators to target genes, enhancing the transcription of genes associated with cell survival and neuronal differentiation. Thyroid hormone promotes cell survival and progression toward a neuronal fate in the hippocampal neurogenic niche. In contrast, in the absence of thyroid hormone, unliganded TRs recruit corepressors to TRE-containing promoters of cell survival and proneural genes, resulting in repression of gene expression and a decline in progenitor survival and neuronal differentiation.

overexpression of TRα1, the TRα1$^{+/m}$ heterozygous mice harboring a point mutation (TRα1R384C) in TRα1, resulting in a 10-fold lower affinity of the receptor for the ligand, and the TRα1-GFP knockin mouse line (Kapoor et al., 2010; Salto et al., 2001; Tinnikov et al., 2002; Wallis et al.,

2010; Wikstrom et al., 1998). TRα1$^{-/-}$-null mice exhibited an enhanced postmitotic survival of hippocampal progenitors (Kapoor et al., 2010), a phenotype that overlaps with changes noted in adult-onset hyperthyroid mice (Kapoor et al., 2012). In contrast, TRα2$^{-/-}$ mice (Salto et al., 2001) that overexpress the TRα1 receptor several fold exhibited a decrease in the survival of progenitors with a decline also noted in the NeuroD-positive pool of progenitors (Kapoor et al., 2010). These observations have contributed to the hypothesis that an overexpression of TRα1 may result in a shift in balance toward an aporeceptor state, paralleling the state that arises in adult-onset hypothyroidism (Ambrogini et al., 2005; Desouza et al., 2005; Montero-Pedrazuela et al., 2006), of a shift toward unliganded TRs; and that TRα1 aporeceptor bias may then predispose postmitotic hippocampal progenitors toward cell death (Fig. 3). Further evidence in support of this hypothesis stems from studies using the dominant-negative TRα1$^{+/m}$ mouse line (Tinnikov et al., 2002) that harbor a mutant TRα1 (point mutation—TRα1R384C) with a 10-fold lower binding affinity for thyroid hormone, thus biasing toward a TRα1 aporeceptor form. TRα1$^{+/m}$ mice also exhibited a significant decline in hippocampal progenitor survival and neuronal differentiation (Kapoor et al., 2010). The decreased progenitor survival and neuronal differentiation noted in both these mutant mouse lines that cause a shift toward enhanced TRα1 aporeceptor activity, namely the TRα2$^{-/-}$ and TRα1$^{+/m}$ mice, thus recapitulating a hypothyroid-like state (Desouza et al., 2005; Kapoor et al., 2012), could be completely rescued to wild-type levels of adult neurogenesis simply through a rescue of circulating thyroid hormone levels via T3 administration (Kapoor et al., 2010). Taken together, these studies indicate that an unliganded aporeceptor TRα1 may contribute to the deficits in hippocampal neurogenesis observed following adult-onset hypothyroidism and raise the possibility that this may contribute to the cognitive and mood-related disturbances that arise due to such disruptions of euthyroid status.

While *in vitro* studies suggest that TRα1 is expressed by proliferating progenitors, based on qPCR and antibody staining in both dispersed hippocampal progenitor cultures and neurospheres (Desouza et al., 2005; Kapoor et al., 2012), *in vivo* results suggest a differential distribution of TR isoforms across progenitor development (Desouza et al., 2005). Given the paucity of high-quality antibodies that allow the ease of distinction across different TR isoforms, thus far expression studies have not revealed the relative expression of specific TRs across individual stages of hippocampal progenitor development. In this regard, the TRα1–GFP knockin mouse model (Wallis et al., 2010) provides an important tool to determine expression of this major TR isoform within adult hippocampal progenitors and in the hippocampal neurogenic niche. Strikingly, studies with this reporter mouse line indicate a lack of GFP expression in proliferating (BrdU positive) hippocampal progenitors, with expression noted in postmitotic, progenitors committed to a neuronal fate (NeuroD-positive) (Kapoor et al., 2012). This then raises the intriguing possibility that TRα1 expression may only switch on once hippocampal progenitors exit the cell cycle. The current working model suggests that TRα1 expression switches on in postmitotic hippocampal progenitors and that TRα1 aporeceptor states would bias progenitors toward cell death, whereas liganded TRα1 activity would enhance cell survival and neuronal cell fate acquisition (Fig. 3). What remains unclear though is whether thyroid hormone through specific TR isoforms serves in anyway the role of an "intrinsic timer" determining the time point for cell cycle exit for hippocampal progenitors, as has been suggested for other progenitor cell types (Billon et al., 2002; Fernandez et al., 2004; Gao, Apperly, & Raff, 1998).

Although the data thus far have largely focussed on TRα1 receptors and their role in adult hippocampal progenitor development, TRβ isoforms are also expressed by hippocampal progenitors both *in vivo* and *in vitro* and are reported to regulate progenitor turnover (Desouza et al., 2005; Kapoor et al., 2012, 2011). $TR\beta^{-/-}$ mice show increased proliferation of hippocampal progenitors as well as enhanced numbers of NeuroD-positive progenitor cells; however, this does not result in a consequent increase in DCX-positive immature neurons. These results are suggestive of a negative role of TRβ (either the apo- or holoreceptor form) in regulating hippocampal progenitor cell division. An increase in circulating levels of thyroid hormone in $TR\beta^{-/-}$ mice (Forrest et al., 1996) may also contribute to neurogenic alterations observed in $TR\beta^{-/-}$ mice. Also, the ability of TRβ isoforms to offset the mitogenic action of growth factors such as epidermal growth factors and IGF-1 in tumor cell lines (Martinez-Iglesias et al., 2009) opens the speculative possibility that the loss of TRβ may modify growth factor action on progenitors.

All of the above studies delineating the role of specific TR isoforms in hippocampal neurogenesis have employed mutants that are whole-body knockouts. These come with the inherent drawback of loss of function in embryonic and developmental stages as well, which may confound the effects observed on neurogenesis in adulthood. Further, since these animals are whole-body knockouts, they may exhibit alterations in T3, T4, and TSH levels, which could further confound the interpretation of the effects of TRs on adult neurogenesis (Flamant & Gauthier, 2013; O'Shea & Williams,

2002). To overcome these drawbacks and gain a mechanistic understanding of TR contributions to different stages of neurogenesis, future studies require the development of conditional TR-specific knockout mice, which would allow spatiotemporal control over TR expression at specific stages of hippocampal progenitor development. Double mutants that disturb the stoichiometry of TR isoforms would also add deeper mechanistic insights to the interaction and function of the diverse TRs. Further, data from mutant mouse lines that disrupt deiodinase expression, as well as thyroid hormone transporter function, would provide a more complete picture of the effects of thyroid hormone on adult neurogenesis. To further elucidate the effects of thyroid hormone on different cell types of the neurogenic niche, studies in TR mutant mice are required that provide temporal and spatial control of TR expression in different subpopulations of the hippocampal niche namely astrocytes, mature granule cells, and microglia. These studies are needed to test the working idea that thyroid hormone may serve as a gating mechanism to promote cell cycle exit in dividing hippocampal progenitors, and the balance between TR aporeceptors and holoreceptors may then dynamically influence the numbers of progenitors that survive, undergo differentiation, and integrate into the hippocampal network.

4. THYROID HORMONE INFLUENCE ON ADULT SVZ NEUROGENESIS

4.1 Progression of SVZ Progenitor Development

Along the walls of the lateral ventricles lies the SVZ, which contains the largest number of dividing progenitors within the adult mammalian brain. The proliferative zone of the SVZ contains three types of progenitor cells (Fig. 1). The radial glia-like cells are the quiescent, multipotent neural stem cells or the type B cells. These divide asymmetrically to self-renew and give rise to the transit amplifying cells or type C cells, which in turn generate the neuroblasts or type A cells. The neuroblasts form a chain and migrate along the rostral migratory stream (RMS) to the olfactory bulb (OB), where they differentiate into different subtypes of interneurons—granule cells and periglomerular neurons, eventually integrating into existing OB neurocircuitry (Lim & Alvarez-Buylla, 2016; Ming & Song, 2011; Sakamoto, Kageyama, & Imayoshi, 2014). The proliferating type B cells express GFAP, Nestin, and Sox2. The type C cells express the homeobox transcription factor Dlx2 and the type A migrating neuroblasts are DCX and PSA-NCAM positive (Fig. 1) (Braun & Jessberger, 2014; Lim & Alvarez-Buylla, 2016). About

30,000 progenitors are produced in the SVZ everyday (Cameron & McKay, 2001; Ming & Song, 2005), and while many of these die, a reasonable fraction are slated to migrate to the OB and give rise to OB interneurons (Ming & Song, 2011; Zhao et al., 2008). Survival of SVZ progenitors is regulated at the neuroblast stage (Type A cells) and also at the time of immature neuron integration into OB circuitry (Lim & Alvarez-Buylla, 2016). SVZ neurogenesis is precisely regulated by both cell autonomous and noncell autonomous cues that adapt to alterations in the immediate local SVZ environment, as well as changes in the environment of the animal (Kuhn, Cooper-Kuhn, Eriksson, & Nilsson, 2005). The process of OB neurogenesis may contribute to the structural plasticity that is necessary for adaptations in olfactory discrimination, as newborn OB neurons have been reported to respond to novel odorant cues (Alvarez-Buylla & Garcia-Verdugo, 2002; Zhao et al., 2008). Among the major differences between SVZ and SGZ neurogenesis is that neuroblasts must migrate long distances tangentially and then radially to reach their final destination in the OB and develop into OB interneurons, while in the SGZ neuroblasts traverse a short trajectory within the DG subfield. It has only more recently been appreciated that adult neural stem cells both within the SVZ and SGZ not only exhibit clear differences between these neurogenic niches (Chaker, Codega, & Doetsch, 2016; Curtis, Low, & Faull, 2012), but also within any single niche neural stem cells are not homogeneous but rather exhibit heterogeneity at the level of proliferation potential, transcriptional expression, and diverse differentiation potential (Bonaguidi et al., 2016; Gebara et al., 2016; Giachino & Taylor, 2014; Goncalves et al., 2016; Jhaveri et al., 2015; Jhaveri, Taylor, & Bartlett, 2012; Llorens-Bobadilla et al., 2015).

4.2 Thyroid Hormone, TRs, and SVZ Neurogenesis

Distinct aspects of SVZ progenitor development have been reported to be sensitive to thyroid hormone fluctuations and regulated by specific TR isoforms (Fig. 4) (Lemkine et al., 2005; Lopez-Juarez et al., 2012). Adult-onset hypothyroidism results in a decline in SVZ progenitor proliferation with a failure noted in entry to cell cycle, resulting in an overall impairment of SVZ neurogenesis (Lemkine et al., 2005). Hypothyroid status is associated with a higher fraction of SVZ progenitors that continue to remain in interphase, resulting in an overall decline in proliferation. This reduction in SVZ neurogenesis and the decline in numbers of cycling progenitors are restored by exogenous thyroid hormone treatment. This has led to the view that T3

A

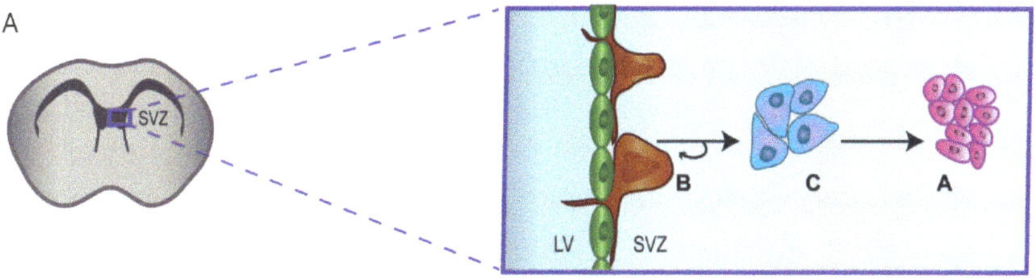

B

Condition	B	C	A	References
Adult-onset hypothyroidism	Persistence in interphase Accumulation of quiescent progenitors	Decrease in Dlx2-positive cell numbers	Decrease in migratory neuroblasts	Lemkine et al. (2005); Lopez-Juarez et al. (2012)
TRα$^{0/0}$	Increase in Sox2+ cells Reduced cell cycle entry	Consequent decline in neuroblast numbers		Lemkine et al. (2005); Lopez-Juarez et al. (2012)
TRα1 siRNA	Increase in Sox2+ cells	Consequent decline in neuroblast numbers		Lopez-Juarez et al. (2012)
TRα1 overexpression	Decrease in Sox2 and Cyclin D1 expression, leading to progression to neuronal commitment		Increase in numbers of DCX+ cells	Lopez-Juarez et al. (2012)

Fig. 4 Thyroid hormone regulation of adult SVZ neurogenesis. (A) Shown is a schematic of a coronal section through the adult rodent brain highlighting the neurogenic niche of the subventricular zone (SVZ) lining the lateral ventricles (LV). The schematic illustrates the developmental progression of adult SVZ progenitors from Type B quiescent neural progenitors to Type A migratory neuroblasts that travel along the rostral migratory stream to the olfactory bulb. (B) Shown is a table describing the effects of thyroid hormone on distinct stages of SVZ progenitor development, including studies from adult-onset perturbations of thyroid hormone levels and from TR isoform-specific mutant mouse models.

may be important for exit from a quiescent stem cell-like state, and indeed, hypothyroid animals show a reduction in Dlx2 positive, rapidly amplifying Type C cells. In addition, adult-onset hypothyroidism is also associated with decreased apoptotic cell death in the SVZ and an overall decline in migratory neuroblasts in the RMS (Lemkine et al., 2005; Lopez-Juarez et al., 2012). SVZ progenitor survival is also regulated by TTR, which is a choroid plexus-secreted protein that functions to regulate thyroid hormone active transport into the brain. TTR loss-of-function mice display lowered thyroid hormone levels in the brain, with a decrease observed in apoptotic cell loss and reduced cell death in the SVZ (Richardson, Lemkine, Alfama, Hassani, & Demeneix, 2007), phenocopying specific deficits in adult-onset hypothyroidism. In the influence on progenitor cell death, a very different phenotype is noted upon adult-onset hypothyroidism in the SVZ and SGZ with enhanced cell survival in the SVZ and increased cell death in the SGZ, indicating that thyroid hormone exerts both distinct and overlapping effects in the regulation of progenitors in these two neurogenic niches. Taken together, these studies indicate that thyroid hormone influences diverse aspects of SVZ neurogenesis from regulating entry of quiescent stem cells into the cell cycle, cell death within proliferating progenitors in the niche and acquisition of a neuronal cell fate (Fig. 4).

Thus far, reports indicate that the sole TR isoform expressed in the SVZ is TRα1 (Lemkine et al., 2005; Lopez-Juarez et al., 2012), with reports indicating that TRα1 is not present in the quiescent neural stem cells, but rather appears in the Dlx2-positive, rapidly amplifying type C cells and is present at high levels in the DCX-positive migratory neuroblast (Lopez-Juarez et al., 2012). The absence of TRα1 in type B progenitor pool, and its expression in the type C and type A cells, suggests the possibility that TRα1 may drive progenitors to differentiate and commit to a neuronal fate. As a corollary to this, TRα1 overexpression was found to increase the number of neuroblast cells (Fig. 4) (Lopez-Juarez et al., 2012). In contrast, siRNA knockdown of TRα1 resulted in an increase in the type B progenitor pool, thought to arise due to a loss of repressive effects of thyroid hormone/TRα1 on the expression of *Sox2*, which causes progenitor cells to retain their stem cell identity and prevents differentiation (Lopez-Juarez et al., 2012). This is also phenocopied in the $TR\alpha^{0/0}$ mice, wherein the progenitors do not progress toward neuronal commitment, leading to an accumulation of quiescent, nonproliferating progenitors (Lemkine et al., 2005; Lopez-Juarez et al., 2012). The prevailing view is that the liganded TRα1 receptor enhances SVZ progenitor maturation toward a neuronal lineage through the repression of stem cell identity genes like *Sox2* and also the repression of cell cycle genes such as *cyclinD1* and *c-myc* (Lemkine et al., 2005; Lopez-Juarez et al., 2012), causing proliferating progenitors to exit cell cycle and progress toward a neuroblast identity (Fig. 5). This is similar to the pattern of TRα1 expression in the SGZ progenitors within the hippocampal neurogenic niche, suggesting a certain common theme between both of these adult progenitors wherein TRα1 may play the role of a gatekeeper during progenitor development promoting acquisition of a neuronal fate.

Studies on SVZ progenitor cells *in vitro* have demonstrated that when neural stem cells are cultured from hyperthyroid animals, they show increased oligodendrocyte differentiation (Fernandez et al., 2004). This suggests that much like SGZ progenitors, SVZ progenitors when removed from the neurogenic niche seem to respond to thyroid hormone quite differently. Both SGZ and SVZ progenitors show a greater propensity for glial fates when treated with thyroid hormone *in vitro* (Desouza et al., 2005;

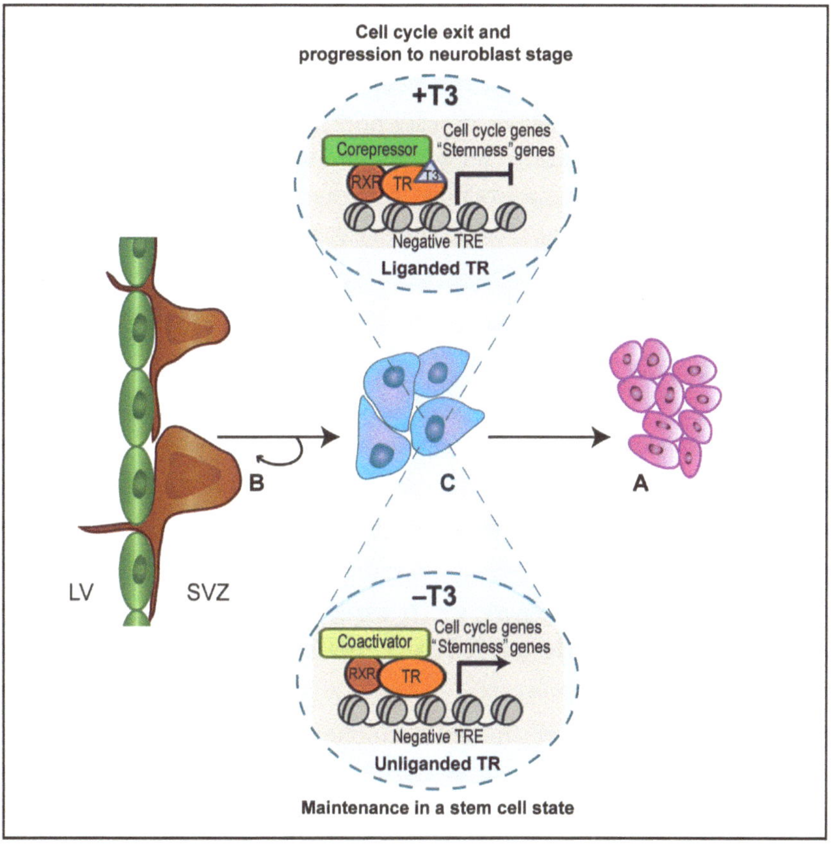

Fig. 5 Putative mechanism of action of thyroid hormone on SVZ progenitors. Shown is a schematic of the putative underlying mechanism for thyroid hormone action during SVZ progenitor development from a Type B quiescent neural progenitor to a Type A migratory neuroblast destined to integrate into the olfactory bulb neuronal network. Studies support a working model for thyroid hormone action, likely in the transition from Type B to Type C cells, where in the presence of thyroid hormone (T3), liganded thyroid hormone receptors (TRs) heterodimerize with retinoid-X-receptor (RXR) and recruit corepressors to negative TRE-containing thyroid hormone target genes, repressing the transcription of genes associated with "stemness" and cell cycle. Thyroid hormone promotes cell cycle exit and progression toward a neuronal fate in SVZ progenitors. In contrast, in the absence of thyroid hormone, unliganded TRs recruit coactivators to negative TRE-containing promoters of stem cell and cell cycle genes, resulting in activation of gene expression such as Sox2, and a maintenance of stem cell state, with a decline noted in progression toward differentiation.

Fernandez et al., 2004), whereas *in vivo* studies (Kapoor et al., 2012, 2010; Lemkine et al., 2005; Lopez-Juarez et al., 2012) indicate enhanced neuronal differentiation of both SVZ and SGZ progenitors in response to thyroid hormone. This opens up the possibility of distinct effects on progenitors while present within their neurogenic niches and when in isolated conditions, thus motivating experiments to address the effects of thyroid hormone on diverse aspects of the neurogenic niche. Further, there are no reports investigating the presence of deiodinases or thyroid hormone transporters in the SVZ. Given the key role that astrocytes within the SVZ niche have been reported to play, it is important to study how local thyroid hormone levels may be dynamically influenced via D2/D3 activity balance within diverse cell types of the SVZ neurogenic niche. Studies on TR isoforms thus far have focussed predominantly on the TRα1 isoform, and while other isoforms have not been reported in the SVZ so far, a systematic analysis of their expression merits investigation. Additionally, mutant mouse lines that afford spatial and temporal control of TRα1 expression, within specific stages of SVZ progenitor development, would help to elucidate the effects of thyroid hormone at each stage. This would also enable the delineation of thyroid hormone target genes at individual stages of progenitor development.

5. UNDERLYING MOLECULAR MEDIATORS FOR THE REGULATION OF NEUROGENESIS BY THYROID HORMONE

At the nuclear level, thyroid hormone mediates its effects via TRs, which serve as transcription factors at TREs driving the regulation of thyroid hormone target genes. TRs toggle between an unliganded aporeceptor state and a ligand-bound state, regulating gene transcription differentially in both forms (Bernal et al., 2015; Brent, 2012; Chassande, 2003; Cheng et al., 2010; Harvey & Williams, 2002). The TRs bound to TREs at thyroid hormone target genes exist in complexes with coactivators or corepressors, switching on or off gene transcription (Bernal & Morte, 2013; Bianco, 2011; Bianco & Kim, 2006; Lee & Privalsky, 2005; Schroeder & Privalsky, 2014). Depending on the nature of the TRE, they can either activate or repress transcription. Positive TREs are more common, where TR aporeceptors repress transcription by complexing with corepressors that possess histone deacetylase activity (Chassande, 2003; Zhang & Lazar, 2000). The repression by TR aporeceptors is lifted by thyroid hormone binding to the aporeceptor in conjunction with recruitment of coactivators that exhibit histone acetylase activity and mediate recruitment of RNA polymerase II, thereby initiating transcription (Astapova & Hollenberg, 2013; Cheng et al., 2010; Harvey & Williams, 2002; Liu & Brent, 2010). On the other hand, negative TREs are those wherein aporeceptor activity leads to switching on of target genes and ligand binding serves to repress gene transcription (Wu & Koenig, 2000). The coactivators (Goodman & Smolik, 2000; Ito & Roeder, 2001) or corepressors (Astapova & Hollenberg, 2013; Hu & Lazar, 2000; Privalsky, 2004) that communicate with transcriptional machinery are part of multi-subunit complexes with enzyme activities. Over and above histone acetylation/deacetylation activities, these complexes can also have enzymes such as methylases, kinases, or phosphatases which act in concert with TR isoforms, to fine tune gene expression (Bernal & Morte, 2013; Cheng et al., 2010). Thyroid hormone also exerts nongenomic actions through modulation of membrane receptors and signaling pathways (Bitiktas et al., 2016; Davis et al., 2010; Lanni et al., 2016).

While several studies have reported thyroid hormone target genes (Chatonnet, Flamant, & Morte, 2015) during neurodevelopment and in neuronal cells, including *Reelin* (Alvarez-Dolado et al., 1999), *Stat3* (Chen et al., 2012), *Tag1* (Alvarez-Dolado et al., 2001), *Sox2*, *Jun*, *Notch1*, *Notch4*, *Klf7*, *Ngf*, *Pax9*, *Slit2* (Chatonnet, Guyot, Benoit, & Flamant, 2013), and *Dab* (Alvarez-Dolado et al., 1999), thus far only a few bona fide target genes have been identified in adult progenitors. At present, there is a paucity of understanding of the molecular mechanisms that mediate the neurogenic actions of thyroid hormone within the adult mammalian brain. In the SVZ, the current model proposes that TRα1 aporeceptor activity at negative TREs activates the expression of genes that regulate stem cell identity, as well as genes regulating cell cycle entry to maintain cells in a proliferating undifferentiated state (Fig. 5) (Lemkine et al., 2005; Lopez-Juarez et al., 2012). Thyroid hormone binding to its aporeceptor would cause repression of these genes that likely contain negative TREs, leading to cell cycle exit and differentiation to neuroblasts (Fig. 5). In this context, the stem cell-associated gene *Sox2* and the cell cycle genes (*cyclin D1* and *c-myc*) have been identified as thyroid hormone target genes that are repressed in response to thyroid hormone in neural stem cells of the SVZ (Lemkine et al., 2005; Lopez-Juarez et al., 2012). Thyroid hormone repression of cell cycle regulatory genes and genes associated with "stemness" through TRα1 raises the possibility that thyroid hormone may exert a permissive role on SVZ neural stem cells enhancing cell cycle exit and acquisition of a differentiated phenotype. What is unclear at present is whether thyroid hormone also contributes to the activation of specific proneural genes in SVZ stem cells, thus also actively promoting the acquisition of a neuronal cell fate. It is likely that *Sox2*, *c-myc*, and *cyclin D1* represent only a subset of the thyroid hormone responsive genes. A more detailed understanding of the transcriptional targets for thyroid hormone in SVZ progenitors across their developmental progression is required to determine whether thyroid hormone plays a more passive permissive role in this niche, or whether it activately targets gene programs to drive neuronal cell fate determination.

In the hippocampal SGZ, the current model proposes that TRα1 expression may switch on as neural stem cells exit cell cycle. TRα1 aporeceptor activity in postmitotic hippocampal progenitors would then result in repression of prosurvival genes and eventual cell death in the absence of thyroid hormone (Fig. 3). In contrast, TRα1 holoreceptor would activate gene expression of both prosurvival and proneural genes, enhancing progenitor survival and differentiation into neurons (Fig. 3). While thus far no bona fide thyroid hormone target genes have been demonstrated in hippocampal progenitors, both *in vivo* and *in vitro* studies have highlighted putative gene targets. In response to thyroid hormone administration *in vivo*, the transcripts of multiple genes associated with neuronal progenitor development such as *Tis21*, *Tlx*, *Dlx2*, *Math-1*, and *Ngn1* were robustly upregulated (Kapoor et al., 2012). Of these, activation of *Tis 21*, which is expressed in Type 2 and 3 progenitors, is known to hasten neuronal differentiation of progenitors (Attardo et al., 2010; Farioli-Vecchioli et al., 2009), akin to that seen following thyroid hormone treatment. Interestingly, treatment of hippocampal derived neurospheres by thyroid hormone enhanced expression of proneural genes *Emx2* and *Klf9*, indicating a distinct pattern of regulation from *in vivo* studies (Kapoor et al., 2012). It is important to note that at present whether any of these genes serve as genuine thyroid hormone target genes with recruitment of specific TRs to TRE sequences in their promoters remains to be further ascertained.

Chromatin immunoprecipitation sequencing (ChIP-seq) data using TR isoform-specific antibodies is essential to elucidate the specific transcriptional targets of thyroid hormone across the genome at different stages of SGZ and SVZ progenitor development. Currently, these studies have been hindered by the nonavailability of TR isoform-specific ChIP grade antibodies. Nonetheless, studies have taken advantage of ChAP experiments with GST-tagged TR isoforms or pan-RXR ChIPs (Chatonnet et al., 2013) or made use of available GFP-labeled TR isoform-specific knockin mice (Dudazy-Gralla et al., 2013), to investigate TR-specific target genes in neural cells. These nature of experiments open up the possibility of gaining important understanding of the sites where TRs bind within the genome of SVZ and SGZ progenitors, and the manner of regulation mediated by thyroid hormone at target genes through both negative and positive

TREs. Further, the nongenomic actions of thyroid hormone within neurogenic niches and in adult neuronal progenitors also warrant further investigation. In toto, the molecular effects that arise downstream of thyroid hormone in the SGZ and SVZ progenitor pool remain poorly delineated and need further investigation to determine the mechanistic underpinnings of the effects of thyroid hormone on regulation of neural stem cell quiescence, cell cycle entry and exit, progenitor survival, cell fate acquisition, progenitor maturation, and integration into neuronal networks.

6. FUNCTIONAL IMPLICATIONS OF THE NEUROGENIC EFFECTS OF THYROID HORMONE

6.1 Thyroid Hormone Regulation of Behavior in Animal Models

The behavioral implications of thyroid hormone-mediated regulation of adult hippocampal and OB neurogenesis remain poorly understood. Studies in rodent models implicate adult hippocampal neurogenesis in the regulation of hippocampal dependent learning and memory (Deng, Aimone, & Gage, 2010; Lieberwirth et al., 2016; Stuchlik, 2014; Yau, Li, & So, 2015) as well as in the regulation of anxiety and mood behavior (Cameron & Glover, 2015; Hage & Azar, 2012; Miller & Hen, 2014). In this regard, adult-onset hypothyroidism evokes significant impairments in learning tasks (Fundaro, 1989) and also enhances behavioral despair on tasks such as the forced swim test considered to provide a measure for depressive-like behavior (Kulikov, Torresani, & Jeanningros, 1997; Montero-Pedrazuela et al., 2006). These correlative observations suggest that the decline in hippocampal neurogenesis noted in adult-onset hypothyroid animals may contribute to the cognitive and mood-related behavioral changes observed in these animals. Interestingly, TR mutant mice, such as the dominant-negative TR$\alpha 1^{+/m}$ mice that have a significant increase in TRα1 aporeceptor activity, show enhanced anxiety behavior and also have performance deficits in recognition memory tasks (Pilhatsch et al., 2010; Venero et al., 2005). This behavioral phenotype in TR$\alpha 1^{+/m}$ mice is associated with a significant reduction in ongoing adult hippocampal neurogenesis (Kapoor et al., 2010). These studies highlight that in an unliganded state TR isoforms, in particular TRα1, evoke cellular changes of a neurogenic decline in the hippocampus accompanied by impaired cognition and perturbed mood behavior. Though at present these links remain correlative, given the building evidence linking hippocampal neurogenesis to the modulation of learning, memory, and mood (Cameron & Glover, 2015; Hage & Azar, 2012; Lieberwirth et al., 2016; Miller & Hen, 2014; Yau et al., 2015), it is tempting to speculate that there may exist causal links that warrant further investigation. SVZ neurogenesis has been linked to olfactory perception, discrimination learning and memory (Hoyk, Szilagyi, & Halasz, 1996; Mackay-Sim & Beard, 1987; Malvaut & Saghatelyan, 2016; Paternostro & Meisami, 1991, 1996; Sakamoto et al., 2014). Rodent models of adult-onset hypothyroidism exhibit a reduction in their olfactory perception (Beard & Mackay-Sim, 1987; Paternostro & Meisami, 1993, 1996); conversely hyperthyroid animals exhibit increased olfactory responsiveness (Brunjes, Schwark, & Greenough, 1982; Johanson, 1980). Although at present it is a leap to conclude that neurogenic changes brought about in the OB due to perturbations of SVZ neurogenesis contribute to the effects of thyroid hormone on olfaction, this notion requires experimental investigation to identify any possible link. The understanding of the contributions of the neurogenic effects to thyroid hormone influenced behavioral consequences remains very limited and largely correlative.

6.2 Clinical Relevance of the Neurogenic Actions of Thyroid Hormone

Clinical thyropathies span diverse conditions including overt hypo- and hyperthyroidism and their subclinical counterparts, Hashimoto's thyroiditis, thyrotoxicosis associated with Graves' disease, thyroid goiters, benign thyroid nodules, and thyroid cancers (Cooper & Biondi, 2012; Franklyn & Boelaert, 2012; Jones, May, & Geraci, 2010). Furthermore, they also include "nonthyroidal illness syndrome" and resistance to thyroid hormone—a syndrome with reduced thyroid hormone responsiveness (Cooper & Biondi, 2012; Olateju & Vanderpump, 2006). Hypothyroidism is one of the most prevalent endocrine disorders with a global incidence of 5%–9% in the general population and is particularly prevalent in menopausal women (Canaris, Manowitz, Mayor, & Ridgway, 2000; Stuenkel, 2015; Unnikrishnan & Menon, 2011). In addition to the pathophysiological conditions described earlier, it is also possible that more subtle changes to thyroid hormone-mediated actions could arise through disruption of normal thyroid hormone signaling via perturbations of local deiodinase activity within specific brain regions, perturbation of TR isoforms, and their relative stoichiometry giving rise to subclinical situations wherein thyroid hormone function is hampered.

Adult-onset thyroid disorders, both hypothyroidism and hyperthyroidism, have long been recognized to be associated with cognitive impairments, anxiety, and mood instability (Fig. 6) (Bathla, Singh, & Relan, 2016; Bauer et al., 2008; Berent, Zboralski, Orzechowska, & Galecki, 2014; Demartini et al., 2014; Dugbartey, 1998; Ittermann, Volzke, Baumeister, Appel, & Grabe, 2015; Jackson, 1998). Though overt hypothyroidism can adversely affect diverse domains of cognition (Capet et al., 2000; Smith, Evans, Costall, & Smythe, 2002), among the most sensitive is memory retrieval, in particular for verbal memory (Gobel et al., 2016; Miller et al., 2007). Further, a cross-sectional and interventional study in subclinical hypothyroid patients with elevated TSH but normal free T4 indicated mild cognitive impairments in hippocampal dependent memory tasks, which were reversed upon hormone replacement. In contrast, the cognitive impairments in overtly hypothyroid patients did not show a complete reversal, highlighting

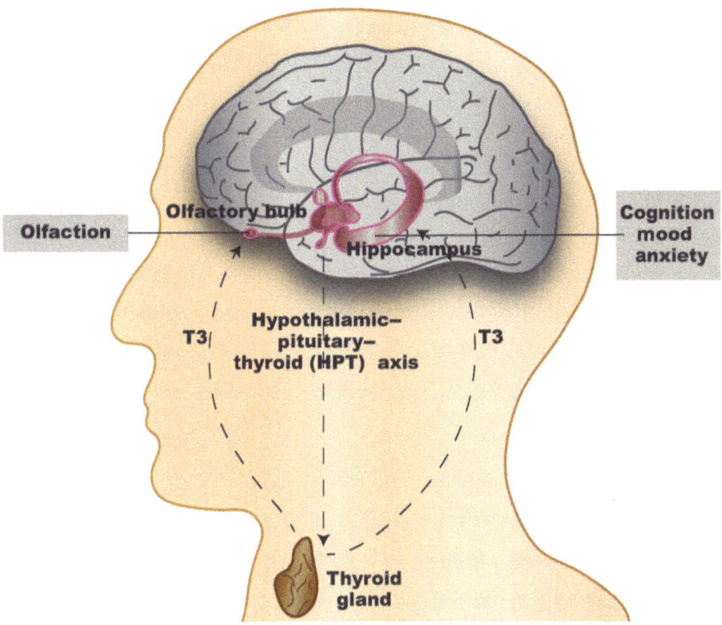

Fig. 6 Behavioral implications of thyroid hormone action on adult neurogenesis. Shown is a sagittal schematic view of the human brain, depicting the neurogenic niches of the hippocampus and olfactory bulb, and also the hypothalamic–pituitary–thyroid axis that regulates thyroid hormone secretion from the thyroid gland. Clinical evidence links thyroid hormone dysfunction to perturbations in cognition, mood, and anxiety, as well as in the sensory modality of olfaction. Studies in preclinical models indicate that thyroid hormone regulates both hippocampal and olfactory bulb neurogenesis, highlighting the importance of studies focused on examining the behavioral significance of the neurogenic actions of thyroid hormone.

the importance of early detection and treatment for full recovery of cognitive symptoms (Correia et al., 2009). Hippocampal volumetric analysis based on MRI studies indicated volumetric loss associated with adult-onset hypothyroidism in human patients (Cooke, Mullally, Correia, O'Mara, & Gibney, 2014). An interesting imaging study of mildly hypothyroid patients indicated a negative correlation of hippocampal volume with serum levels of TSH (Daghighi et al., 2016), suggesting that potential volumetric changes may arise even prior to overt hypothyroidism onset. It would be of interest to address whether such changes in hippocampal volume can be restored following hormone replacement therapy. Although animal studies highlight the neurogenic actions of thyroid hormone, clinical data either on human neural stem cells or through postmortem analysis are lacking. Such studies would address whether the effects of thyroid hormone on neural stem cells observed in animal models crossover to studies with humans.

Thyroid hormone dysfunction is also linked to affective disorders, with enhanced susceptibility to depressive symptomatology in hypothyroid patients and increased anxiety noted in hyperthyroid subjects (Dayan & Panicker, 2013; Hage & Azar, 2012; Kathol & Delahunt, 1986). Further, thyroid hormone is used as an effective adjunct treatment to pharmacological antidepressants and has been suggested to accelerate and augment their therapeutic efficacy (Bauer & Whybrow, 2001; Cooper-Kazaz et al., 2007; Haggerty & Prange, 1995; Henley & Koehnle, 1997; Uhl et al., 2014). Given clinical evidence supports a hippocampal volumetric decline in patients of major depression (Bremner et al., 2000; Videbech & Ravnkilde, 2004), imaging studies that address the importance of thyroid hormone augmentation strategies in being able to reverse such structural correlates are of interest. Preclinical studies highlight that pharmacological antidepressants (Malberg, Eisch, Nestler, & Duman, 2000) and thyroid hormone (Kapoor et al., 2015) both enhance adult hippocampal neurogenesis, albeit influencing distinct stages of hippocampal progenitor development. While antidepressants enhance turnover and morphological maturation of hippocampal progenitors (Madsen, Yeh, Valentine, & Duman, 2005; Perera et al., 2007; Santarelli et al., 2003), thyroid hormone enhances postmitotic survival and neuronal differentiation (Kapoor et al., 2012, 2010). This then suggests that in combination, these treatments may work to significantly increase diverse aspects of hippocampal neurogenesis and indeed this is borne out in preclinical studies (Eitan et al., 2010). However, there is a paucity of information from clinical studies, imaging observations, or using human neural stem cells that directly examines the importance of the neurogenic effects of thyroid hormone in aiding specific actions of antidepressant therapies.

While clinical evidence has provided more support for altered hippocampal mediated behaviors in patients with thyroid hormone dysfunction, nevertheless, thyroid hormone dysfunction has also been linked to the regulation of smell (Fig. 6). Hypothyroid patients are reported to exhibit dysosmia, which can be restored by hormone replacement (Deniz et al., 2016; Gunbey et al., 2015). While the olfactory impairments are thought to arise due to effects of thyroid hormone on olfactory receptor neurons in the olfactory epithelium (Paternostro & Meisami, 1993, 1996), given the preclinical studies linking thyroid hormone to regulation of OB neurogenesis, it seems premature to preclude a role for the neurogenic effects of thyroid hormone. In general, the extent of ongoing neurogenesis is thought to be highly restricted in the human brain with evidence, thus far supporting progenitor turnover in the human hippocampus (Eriksson et al., 1998; Spalding et al., 2013), but limited for the SVZ (Ernst & Frisen, 2015; van Strien, van den Berge, & Hol, 2011). This factor of course is important to bear in mind when extrapolating the findings of the neurogenic effects of thyroid hormone from animal models to humans. However, this only serves to highlight the lacunae and the importance of studies that address the effects of thyroid hormone on neural stem cells of human origin.

7. CONCLUSION

Thyroid hormone exerts a key instructive influence on adult hippocampal and SVZ progenitors, regulating their exit from cell cycle, survival, and commitment to a neuronal fate. These actions are mediated via TR isoforms, which in their unliganded vs liganded state can evoke differing consequences for the developmental trajectory of adult progenitors. It is critical to study the functional and behavioral consequences of these neurogenic effects of thyroid hormone, expanding preclinical studies to the use of human neural stem cells to address the relevance of findings in animal models. Thyroid hormone dysfunction is among the most prevalent of endocrine disorders, linked to both cognitive and psychiatric symptoms, highlighting the importance of studies that aid in the development of a deep mechanistic understanding of the actions of thyroid hormone.

ACKNOWLEDGMENTS

The authors gratefully acknowledge support from the Tata Institute of Fundamental Research and the Department of Science and Technology, Government of India.

Conflict of interest disclosure. The authors have no conflicts of interest.

REFERENCES

Abel, E. D., Ahima, R. S., Boers, M. E., Elmquist, J. K., & Wondisford, F. E. (2001). Critical role for thyroid hormone receptor beta2 in the regulation of paraventricular thyrotropin-releasing hormone neurons. *The Journal of Clinical Investigation, 107*(8), 1017–1023. http://dx.doi.org/10.1172/jci10858.

Alvarez-Buylla, A., & Garcia-Verdugo, J. M. (2002). Neurogenesis in adult subventricular zone. *The Journal of Neuroscience, 22*(3), 629–634.

Alvarez-Dolado, M., Figueroa, A., Kozlov, S., Sonderegger, P., Furley, A. J., & Munoz, A. (2001). Thyroid hormone regulates TAG-1 expression in the developing rat brain. *The European Journal of Neuroscience, 14*(8), 1209–1218.

Alvarez-Dolado, M., Ruiz, M., Del Rio, J. A., Alcantara, S., Burgaya, F., Sheldon, M., et al. (1999). Thyroid hormone regulates reelin and dab1 expression during brain development. *The Journal of Neuroscience, 19*(16), 6979–6993.

Ambrogini, P., Cuppini, R., Ferri, P., Mancini, C., Ciaroni, S., Voci, A., et al. (2005). Thyroid hormones affect neurogenesis in the dentate gyrus of adult rat. *Neuroendocrinology, 81*(4), 244–253. http://dx.doi.org/10.1159/000087648.

Ashton, R. S., Conway, A., Pangarkar, C., Bergen, J., Lim, K. I., Shah, P., et al. (2012). Astrocytes regulate adult hippocampal neurogenesis through ephrin-B signaling. *Nature Neuroscience, 15*(10), 1399–1406. http://dx.doi.org/10.1038/nn.3212.

Astapova, I., & Hollenberg, A. N. (2013). The in vivo role of nuclear receptor corepressors in thyroid hormone action. *Biochimica et Biophysica Acta, 1830*(7), 3876–3881. http://dx.doi.org/10.1016/j.bbagen.2012.07.001.

Attardo, A., Fabel, K., Krebs, J., Haubensak, W., Huttner, W. B., & Kempermann, G. (2010). Tis21 expression marks not only populations of neurogenic precursor cells but also new postmitotic neurons in adult hippocampal neurogenesis. *Cerebral Cortex, 20*(2), 304–314. http://dx.doi.org/10.1093/cercor/bhp100.

Axelstad, M., Hansen, P. R., Boberg, J., Bonnichsen, M., Nellemann, C., Lund, S. P., et al. (2008). Developmental neurotoxicity of propylthiouracil (PTU) in rats: Relationship between transient hypothyroxinemia during development and long-lasting behavioural and functional changes. *Toxicology and Applied Pharmacology, 232*(1), 1–13. http://dx.doi.org/10.1016/j.taap.2008.05.020.

Baas, D., Bourbeau, D., Sarlieve, L. L., Ittel, M. E., Dussault, J. H., & Puymirat, J. (1997). Oligodendrocyte maturation and progenitor cell proliferation are independently regulated by thyroid hormone. *Glia, 19*(4), 324–332.

Bathla, M., Singh, M., & Relan, P. (2016). Prevalence of anxiety and depressive symptoms among patients with hypothyroidism. *Indian Journal of Endocrinology and Metabolism, 20*(4), 468–474. http://dx.doi.org/10.4103/2230-8210.183476.

Bauer, M., Goetz, T., Glenn, T., & Whybrow, P. C. (2008). The thyroid-brain interaction in thyroid disorders and mood disorders. *Journal of Neuroendocrinology, 20*(10), 1101–1114. http://dx.doi.org/10.1111/j.1365-2826.2008.01774.x.

Bauer, M., & Whybrow, P. C. (2001). Thyroid hormone, neural tissue and mood modulation. *The World Journal of Biological Psychiatry, 2*(2), 59–69.

Beard, M. D., & Mackay-Sim, A. (1987). Loss of sense of smell in adult, hypothyroid mice. *Brain Research, 433*(2), 181–189.

Berent, D., Zboralski, K., Orzechowska, A., & Galecki, P. (2014). Thyroid hormones association with depression severity and clinical outcome in patients with major depressive disorder. *Molecular Biology Reports*, *41*(4), 2419–2425. http://dx.doi.org/10.1007/s11033-014-3097-6.

Bernal, J. (2002). Action of thyroid hormone in brain. *Journal of Endocrinological Investigation*, *25*(3), 268–288. http://dx.doi.org/10.1007/bf03344003.

Bernal, J. (2007). Thyroid hormone receptors in brain development and function. *Nature Clinical Practice. Endocrinology & Metabolism*, *3*(3), 249–259. http://dx.doi.org/10.1038/ncpendmet0424.

Bernal, J., Guadano-Ferraz, A., & Morte, B. (2015). Thyroid hormone transporters—Functions and clinical implications. *Nature Reviews. Endocrinology*, *11*(7), 406–417. http://dx.doi.org/10.1038/nrendo.2015.66.

Bernal, J., & Morte, B. (2013). Thyroid hormone receptor activity in the absence of ligand: Physiological and developmental implications. *Biochimica et Biophysica Acta*, *1830*(7), 3893–3899. http://dx.doi.org/10.1016/j.bbagen.2012.04.014.

Besnard, A., & Sahay, A. (2016). Adult hippocampal neurogenesis, fear generalization, and stress. *Neuropsychopharmacology*, *41*(1), 24–44. http://dx.doi.org/10.1038/npp.2015.167.

Bianco, A. C. (2011). Minireview: Cracking the metabolic code for thyroid hormone signaling. *Endocrinology*, *152*(9), 3306–3311. http://dx.doi.org/10.1210/en.2011-1104.

Bianco, A. C., & Kim, B. W. (2006). Deiodinases: Implications of the local control of thyroid hormone action. *The Journal of Clinical Investigation*, *116*(10), 2571–2579. http://dx.doi.org/10.1172/JCI29812.

Billon, N., Jolicoeur, C., Tokumoto, Y., Vennstrom, B., & Raff, M. (2002). Normal timing of oligodendrocyte development depends on thyroid hormone receptor alpha 1 (TRalpha1). *The EMBO Journal*, *21*(23), 6452–6460.

Bitiktas, S., Kandemir, B., Tan, B., Kavraal, S., Liman, N., Dursun, N., et al. (2016). Adult-onset hyperthyroidism impairs spatial learning: Possible involvement of mitogen-activated protein kinase signaling pathways. *Neuroreport*, *27*(11), 802–808. http://dx.doi.org/10.1097/wnr. 0000000000000612.

Bonaguidi, M. A., Stadel, R. P., Berg, D. A., Sun, J., Ming, G. L., & Song, H. (2016). Diversity of neural precursors in the adult mammalian brain. *Cold Spring Harbor Perspectives in Biology*, *8*(4), a018838. http://dx.doi.org/10.1101/cshperspect.a018838.

Bradley, D. J., Towle, H. C., & Young, W. S., 3rd. (1992). Spatial and temporal expression of alpha- and beta-thyroid hormone receptor mRNAs, including the beta 2-subtype, in the developing mammalian nervous system. *The Journal of Neuroscience*, *12*(6), 2288–2302.

Braun, S. M., & Jessberger, S. (2014). Adult neurogenesis: Mechanisms and functional significance. *Development*, *141*(10), 1983–1986. http://dx.doi.org/10.1242/dev.104596.

Bremner, J. D., Narayan, M., Anderson, E. R., Staib, L. H., Miller, H. L., & Charney, D. S. (2000). Hippocampal volume reduction in major depression. *The American Journal of Psychiatry*, *157*(1), 115–118. http://dx.doi.org/10.1176/ajp.157.1.115.

Brent, G. A. (2012). Mechanisms of thyroid hormone action. *The Journal of Clinical Investigation*, *122*(9), 3035–3043. http://dx.doi.org/10.1172/JCI60047.

Brunjes, P. C., Schwark, H. D., & Greenough, W. T. (1982). Olfactory granule cell development in normal and hyperthyroid rats. *Brain Research*, *281*(2), 149–159.

Calza, L., Aloe, L., & Giardino, L. (1997). Thyroid hormone-induced plasticity in the adult rat brain. *Brain Research Bulletin*, *44*(4), 549–557.

Cameron, H. A., & Glover, L. R. (2015). Adult neurogenesis: Beyond learning and memory. *Annual Review of Psychology*, *66*, 53–81. http://dx.doi.org/10.1146/annurev-psych-010814-015006.

Cameron, H. A., & McKay, R. D. (2001). Adult neurogenesis produces a large pool of new granule cells in the dentate gyrus. *The Journal of Comparative Neurology*, *435*(4), 406–417.

Canaris, G. J., Manowitz, N. R., Mayor, G., & Ridgway, E. C. (2000). The Colorado thyroid disease prevalence study. *Archives of Internal Medicine*, *160*(4), 526–534.

Cao, X., Kambe, F., Moeller, L. C., Refetoff, S., & Seo, H. (2005). Thyroid hormone induces rapid activation of Akt/protein kinase B-mammalian target of rapamycin-p70S6K cascade through phosphatidylinositol 3-kinase in human fibroblasts. *Molecular Endocrinology*, *19*(1), 102–112. http://dx.doi.org/10.1210/me.2004-0093.

Capet, C., Jego, A., Denis, P., Noel, D., Clerc, I., Cornier, A. C., et al. (2000). Is cognitive change related to hypothyroidism reversible with replacement therapy? *La Revue de Médecine Interne*, *21*(8), 672–678.

Chaker, Z., Codega, P., & Doetsch, F. (2016). A mosaic world: Puzzles revealed by adult neural stem cell heterogeneity. *Wiley Interdisciplinary Reviews: Developmental Biology*, *5*(6), 640–658. http://dx.doi.org/10.1002/wdev.248.

Chassande, O. (2003). Do unliganded thyroid hormone receptors have physiological functions? *Journal of Molecular Endocrinology*, *31*(1), 9–20.

Chatonnet, F., Flamant, F., & Morte, B. (2015). A temporary compendium of thyroid hormone target genes in brain. *Biochimica et Biophysica Acta*, *1849*(2), 122–129. http://dx.doi.org/10.1016/j.bbagrm.2014.05.023.

Chatonnet, F., Guyot, R., Benoit, G., & Flamant, F. (2013). Genome-wide analysis of thyroid hormone receptors shared and specific functions in neural cells. *Proceedings of the National Academy of Sciences of the United States of America*, *110*(8), E766–E775. http://dx.doi.org/10.1073/pnas.1210626110.

Chen, C., Zhou, Z., Zhong, M., Zhang, Y., Li, M., Zhang, L., et al. (2012). Thyroid hormone promotes neuronal differentiation of embryonic neural stem cells by inhibiting STAT3 signaling through TRalpha1. *Stem Cells and Development*, *21*(14), 2667–2681. http://dx.doi.org/10.1089/scd.2012.0023.

Cheng, S. Y., Leonard, J. L., & Davis, P. J. (2010). Molecular aspects of thyroid hormone actions. *Endocrine Reviews*, *31*(2), 139–170. http://dx.doi.org/10.1210/er.2009-0007.

Chiamolera, M. I., & Wondisford, F. E. (2009). Minireview: Thyrotropin-releasing hormone and the thyroid hormone feedback mechanism. *Endocrinology*, *150*(3), 1091–1096. http://dx.doi.org/10.1210/en.2008-1795.

Cook, C. B., Kakucska, I., Lechan, R. M., & Koenig, R. J. (1992). Expression of thyroid hormone receptor beta 2 in rat hypothalamus. *Endocrinology*, *130*(2), 1077–1079. http://dx.doi.org/10.1210/endo.130.2.1733708.

Cooke, G. E., Mullally, S., Correia, N., O'Mara, S. M., & Gibney, J. (2014). Hippocampal volume is decreased in adults with hypothyroidism. *Thyroid*, *24*(3), 433–440. http://dx.doi.org/10.1089/thy.2013.0058.

Cooper, D. S., & Biondi, B. (2012). Subclinical thyroid disease. *The Lancet*, *379*(9821), 1142–1154. http://dx.doi.org/10.1016/s0140-6736(11)60276-6.

Cooper-Kazaz, R., Apter, J. T., Cohen, R., Karagichev, L., Muhammed-Moussa, S., Grupper, D., et al. (2007). Combined treatment with sertraline and liothyronine in major depression: A randomized, double-blind, placebo-controlled trial. *Archives of General Psychiatry*, *64*(6), 679–688. http://dx.doi.org/10.1001/archpsyc.64.6.679.

Correia, N., Mullally, S., Cooke, G., Tun, T. K., Phelan, N., Feeney, J., et al. (2009). Evidence for a specific defect in hippocampal memory in overt and subclinical hypothyroidism. *The Journal of Clinical Endocrinology and Metabolism*, *94*(10), 3789–3797. http://dx.doi.org/10.1210/jc.2008-2702.

Costa-e-Sousa, R. H., & Hollenberg, A. N. (2012). Minireview: The neural regulation of the hypothalamic-pituitary-thyroid axis. *Endocrinology*, *153*(9), 4128–4135. http://dx.doi.org/10.1210/en.2012-1467.

Curtis, M. A., Low, V. F., & Faull, R. L. (2012). Neurogenesis and progenitor cells in the adult human brain: A comparison between hippocampal and subventricular progenitor proliferation. *Developmental Neurobiology*, *72*(7), 990–1005. http://dx.doi.org/10.1002/dneu.22028.

Daghighi, M. H., Poureisa, M., Ahmadi, P., Reshadatjoo, M., Golestani, S., Naghavi-Behzad, M., et al. (2016). Serum thyroid-stimulating hormone level and relation with size of hippocampus in patients with mild cognitive disorders. *Nigerian Medical Journal, 57*(6), 353–356. http://dx.doi.org/10.4103/0300-1652.193862.

Daury, L., Busson, M., Casas, F., Cassar-Malek, I., Wrutniak-Cabello, C., & Cabello, G. (2001). The triiodothyronine nuclear receptor c-ErbAalpha1 inhibits avian MyoD transcriptional activity in myoblasts. *FEBS Letters, 508*(2), 236–240.

Davis, P. J., Leonard, J. L., & Davis, F. B. (2008). Mechanisms of nongenomic actions of thyroid hormone. *Frontiers in Neuroendocrinology, 29*(2), 211–218. http://dx.doi.org/10.1016/j.yfrne.2007.09.003.

Davis, P. J., Zhou, M., Davis, F. B., Lansing, L., Mousa, S. A., & Lin, H. Y. (2010). Mini-review: Cell surface receptor for thyroid hormone and nongenomic regulation of ion fluxes in excitable cells. *Physiology & Behavior, 99*(2), 237–239. http://dx.doi.org/10.1016/j.physbeh.2009.02.015.

Dayan, C. M., & Panicker, V. (2013). Hypothyroidism and depression. *European Thyroid Journal, 2*(3), 168–179. http://dx.doi.org/10.1159/000353777.

de Escobar, G. M., Obregon, M. J., & del Rey, F. E. (2007). Iodine deficiency and brain development in the first half of pregnancy. *Public Health Nutrition, 10*(12a), 1554–1570. http://dx.doi.org/10.1017/s1368980007360928.

Demartini, B., Ranieri, R., Masu, A., Selle, V., Scarone, S., & Gambini, O. (2014). Depressive symptoms and major depressive disorder in patients affected by subclinical hypothyroidism: A cross-sectional study. *The Journal of Nervous and Mental Disease, 202*(8), 603–607. http://dx.doi.org/10.1097/nmd.0000000000000168.

Deng, W., Aimone, J. B., & Gage, F. H. (2010). New neurons and new memories: How does adult hippocampal neurogenesis affect learning and memory? *Nature Reviews. Neuroscience, 11*(5), 339–350. http://dx.doi.org/10.1038/nrn2822.

Deniz, F., Ay, S. A., Salihoglu, M., Kurt, O., Baskoy, K., Altundag, A., et al. (2016). Thyroid hormone replacement therapy improves olfaction and taste sensitivity in primary hypothyroid patients: A prospective randomised clinical trial. *Experimental and Clinical Endocrinology & Diabetes, 124*(9), 562–567. http://dx.doi.org/10.1055/s-0042-108446.

Desouza, L. A., Ladiwala, U., Daniel, S. M., Agashe, S., Vaidya, R. A., & Vaidya, V. A. (2005). Thyroid hormone regulates hippocampal neurogenesis in the adult rat brain. *Molecular and Cellular Neurosciences, 29*(3), 414–426. http://dx.doi.org/10.1016/j.mcn.2005.03.010.

Desouza, L. A., Sathanoori, M., Kapoor, R., Rajadhyaksha, N., Gonzalez, L. E., Kottmann, A. H., et al. (2011). Thyroid hormone regulates the expression of the sonic hedgehog signaling pathway in the embryonic and adult mammalian brain. *Endocrinology, 152*(5), 1989–2000. http://dx.doi.org/10.1210/en.2010-1396.

Dezonne, R. S., Lima, F. R., Trentin, A. G., & Gomes, F. C. (2015). Thyroid hormone and astroglia: Endocrine control of the neural environment. *Journal of Neuroendocrinology, 27*(6), 435–445. http://dx.doi.org/10.1111/jne.12283.

Dudazy-Gralla, S., Nordstrom, K., Hofmann, P. J., Meseh, D. A., Schomburg, L., Vennstrom, B., et al. (2013). Identification of thyroid hormone response elements in vivo using mice expressing a tagged thyroid hormone receptor alpha1. *Bioscience Reports, 33*(2), e00027. http://dx.doi.org/10.1042/bsr20120124.

Dugbartey, A. T. (1998). Neurocognitive aspects of hypothyroidism. *Archives of Internal Medicine, 158*(13), 1413–1418.

Eitan, R., Landshut, G., Lifschytz, T., Einstein, O., Ben-Hur, T., & Lerer, B. (2010). The thyroid hormone, triiodothyronine, enhances fluoxetine-induced neurogenesis in rats: Possible role in antidepressant-augmenting properties. *The International Journal of Neuropsychopharmacology, 13*(5), 553–561. http://dx.doi.org/10.1017/s1461145709990769.

Eriksson, P. S., Perfilieva, E., Bjork-Eriksson, T., Alborn, A. M., Nordborg, C., Peterson, D. A., et al. (1998). Neurogenesis in the adult human hippocampus. *Nature Medicine, 4*(11), 1313–1317. http://dx.doi.org/10.1038/3305.

Ernst, A., & Frisen, J. (2015). Adult neurogenesis in humans—Common and unique traits in mammals. *PLoS Biology, 13*(1), e1002045. http://dx.doi.org/10.1371/journal.pbio.1002045.

Farioli-Vecchioli, S., Saraulli, D., Costanzi, M., Leonardi, L., Cina, I., Micheli, L., et al. (2009). Impaired terminal differentiation of hippocampal granule neurons and defective contextual memory in PC3/Tis21 knockout mice. *PLoS One, 4*(12), e8339. http://dx.doi.org/10.1371/journal.pone.0008339.

Fernandez, M., Pirondi, S., Manservigi, M., Giardino, L., & Calza, L. (2004). Thyroid hormone participates in the regulation of neural stem cells and oligodendrocyte precursor cells in the central nervous system of adult rat. *The European Journal of Neuroscience, 20*(8), 2059–2070. http://dx.doi.org/10.1111/j.1460-9568.2004.03664.x.

Flamant, F., & Gauthier, K. (2013). Thyroid hormone receptors: The challenge of elucidating isotype-specific functions and cell-specific response. *Biochimica et Biophysica Acta, 1830*(7), 3900–3907. http://dx.doi.org/10.1016/j.bbagen.2012.06.003.

Flamant, F., Gauthier, K., & Samarut, J. (2007). Thyroid hormones signaling is getting more complex: STORMs are coming. *Molecular Endocrinology, 21*(2), 321–333. http://dx.doi.org/10.1210/me.2006-0035.

Forrest, D., Hanebuth, E., Smeyne, R. J., Everds, N., Stewart, C. L., Wehner, J. M., et al. (1996). Recessive resistance to thyroid hormone in mice lacking thyroid hormone receptor beta: Evidence for tissue-specific modulation of receptor function. *The EMBO Journal, 15*(12), 3006–3015.

Forrest, D., & Vennstrom, B. (2000). Functions of thyroid hormone receptors in mice. *Thyroid, 10*(1), 41–52.

Franklyn, J. A., & Boelaert, K. (2012). Thyrotoxicosis. *Lancet, 379*(9821), 1155–1166. http://dx.doi.org/10.1016/s0140-6736(11)60782-4.

Fundaro, A. (1989). Behavioral modifications in relation to hypothyroidism and hyperthyroidism in adult rats. *Progress in Neuro-Psychopharmacology & Biological Psychiatry, 13*(6), 927–940.

Furuya, F., Lu, C., Guigon, C. J., & Cheng, S. Y. (2009). Nongenomic activation of phosphatidylinositol 3-kinase signaling by thyroid hormone receptors. *Steroids, 74*(7), 628–634. http://dx.doi.org/10.1016/j.steroids.2008.10.009.

Gao, F. B., Apperly, J., & Raff, M. (1998). Cell-intrinsic timers and thyroid hormone regulate the probability of cell-cycle withdrawal and differentiation of oligodendrocyte precursor cells. *Developmental Biology, 197*(1), 54–66. http://dx.doi.org/10.1006/dbio.1998.8877.

Gebara, E., Bonaguidi, M. A., Beckervordersandforth, R., Sultan, S., Udry, F., Gijs, P. J., et al. (2016). Heterogeneity of radial glia-like cells in the adult hippocampus. *Stem Cells, 34*(4), 997–1010. http://dx.doi.org/10.1002/stem.2266.

Giachino, C., & Taylor, V. (2014). Notching up neural stem cell homogeneity in homeostasis and disease. *Frontiers in Neuroscience, 8*, 32. http://dx.doi.org/10.3389/fnins.2014.00032.

Gilbert, M. E., Goodman, J. H., Gomez, J., Johnstone, A. F., & Ramos, R. L. (2016). Adult hippocampal neurogenesis is impaired by transient and moderate developmental thyroid hormone disruption. *Neurotoxicology, 59*, 9–21. http://dx.doi.org/10.1016/j.neuro.2016.12.009.

Gilbert, M. E., & Sui, L. (2006). Dose-dependent reductions in spatial learning and synaptic function in the dentate gyrus of adult rats following developmental thyroid hormone insufficiency. *Brain Research, 1069*(1), 10–22. http://dx.doi.org/10.1016/j.brainres.2005.10.049.

Gobel, A., Heldmann, M., Gottlich, M., Dirk, A. L., Brabant, G., & Munte, T. F. (2016). Effect of mild thyrotoxicosis on performance and brain activations in a working memory task. *PLoS One*, *11*(8). e0161552. http://dx.doi.org/10.1371/journal.pone.0161552

Goncalves, J. T., Schafer, S. T., & Gage, F. H. (2016). Adult neurogenesis in the hippocampus: From stem cells to behavior. *Cell*, *167*(4), 897–914. http://dx.doi.org/10.1016/j.cell.2016.10.021.

Goodman, R. H., & Smolik, S. (2000). CBP/p300 in cell growth, transformation, and development. *Genes & Development*, *14*(13), 1553–1577.

Guissouma, H., Ghaddab-Zroud, R., Seugnet, I., Decherf, S., Demeneix, B., & Clerget-Froidevaux, M. S. (2014). TR alpha 2 exerts dominant negative effects on hypothalamic Trh transcription in vivo. *PLoS One*, *9*(4), e95064. http://dx.doi.org/10.1371/journal.pone.0095064.

Gunbey, E., Karli, R., Gokosmanoglu, F., Duzgun, B., Ayhan, E., Atmaca, H., et al. (2015). Evaluation of olfactory function in adults with primary hypothyroidism. *International Forum of Allergy & Rhinology*, *5*(10), 919–922. http://dx.doi.org/10.1002/alr.21565.

Hage, M. P., & Azar, S. T. (2012). The link between thyroid function and depression. *Journal of Thyroid Research*, *2012*, 590648. http://dx.doi.org/10.1155/2012/590648.

Haggerty, J. J., Jr., & Prange, A. J., Jr. (1995). Borderline hypothyroidism and depression. *Annual Review of Medicine*, *46*, 37–46. http://dx.doi.org/10.1146/annurev.med.46.1.37.

Harvey, C. B., & Williams, G. R. (2002). Mechanism of thyroid hormone action. *Thyroid*, *12*(6), 441–446. http://dx.doi.org/10.1089/105072502760143791.

Henley, W. N., & Koehnle, T. J. (1997). Thyroid hormones and the treatment of depression: An examination of basic hormonal actions in the mature mammalian brain. *Synapse*, *27*(1), 36–44. http://dx.doi.org/10.1002/(sici)1098-2396(199709)27:1<36::aid-syn4->3.0.co;2-e.

Hoyk, Z., Szilagyi, T., & Halasz, N. (1996). Modulation by thyroid hormones of the development of external plexiform layer in the rat olfactory bulb. *Neurobiology (Budapest, Hungary)*, *4*(1–2), 45–57.

Hu, X., & Lazar, M. A. (2000). Transcriptional repression by nuclear hormone receptors. *Trends in Endocrinology and Metabolism*, *11*(1), 6–10.

Igelhorst, B. A., Niederkinkhaus, V., Karus, C., Lange, M. D., & Dietzel, I. D. (2015). Regulation of neuronal excitability by release of proteins from glial cells. *Philosophical Transactions of the Royal Society of London. Series B, Biological Sciences*, *370*(1672), 20140194. http://dx.doi.org/10.1098/rstb.2014.0194.

Ito, M., & Roeder, R. G. (2001). The TRAP/SMCC/mediator complex and thyroid hormone receptor function. *Trends in Endocrinology and Metabolism*, *12*(3), 127–134.

Ittermann, T., Volzke, H., Baumeister, S. E., Appel, K., & Grabe, H. J. (2015). Diagnosed thyroid disorders are associated with depression and anxiety. *Social Psychiatry and Psychiatric Epidemiology*, *50*(9), 1417–1425. http://dx.doi.org/10.1007/s00127-015-1043-0.

Jackson, I. M. (1998). The thyroid axis and depression. *Thyroid*, *8*(10), 951–956.

Jansen, J., Friesema, E. C., Milici, C., & Visser, T. J. (2005). Thyroid hormone transporters in health and disease. *Thyroid*, *15*(8), 757–768. http://dx.doi.org/10.1089/thy.2005.15.757.

Jepsen, K., & Rosenfeld, M. G. (2002). Biological roles and mechanistic actions of co-repressor complexes. *Journal of Cell Science*, *115*(Pt. 4), 689–698.

Jhaveri, D. J., O'Keeffe, I., Robinson, G. J., Zhao, Q. Y., Zhang, Z. H., Nink, V., et al. (2015). Purification of neural precursor cells reveals the presence of distinct, stimulus-specific subpopulations of quiescent precursors in the adult mouse hippocampus. *The Journal of Neuroscience*, *35*(21), 8132–8144. http://dx.doi.org/10.1523/jneurosci.0504-15.2015.

Jhaveri, D. J., Taylor, C. J., & Bartlett, P. F. (2012). Activation of different neural precursor populations in the adult hippocampus: Does this lead to new neurons with discrete functions? *Developmental Neurobiology*, *72*(7), 1044–1058. http://dx.doi.org/10.1002/dneu.22027.

Johanson, I. B. (1980). Development of olfactory and thermal responsiveness in hypothyroid and hyperthyroid rat pups. *Developmental Psychobiology*, *13*(3), 343–351. http://dx.doi.org/10.1002/dev.420130309.

Jones, D. D., May, K. E., & Geraci, S. A. (2010). Subclinical thyroid disease. *The American Journal of Medicine*, *123*(6), 502–504. http://dx.doi.org/10.1016/j.amjmed.2009.12.023.

Joseph-Bravo, P., Jaimes-Hoy, L., Uribe, R. M., & Charli, J. L. (2015). 60 YEARS OF NEUROENDOCRINOLOGY: TRH, the first hypophysiotropic releasing hormone isolated: Control of the pituitary-thyroid axis. *The Journal of Endocrinology*, *226*(2), T85–T100. http://dx.doi.org/10.1530/joe-15-0124.

Kapoor, R., Desouza, L. A., Nanavaty, I. N., Kernie, S. G., & Vaidya, V. A. (2012). Thyroid hormone accelerates the differentiation of adult hippocampal progenitors. *Journal of Neuroendocrinology*, *24*(9), 1259–1271. http://dx.doi.org/10.1111/j.1365-2826.2012.02329.x.

Kapoor, R., Fanibunda, S. E., Desouza, L. A., Guha, S. K., & Vaidya, V. A. (2015). Perspectives on thyroid hormone action in adult neurogenesis. *Journal of Neurochemistry*, *133*(5), 599–616. http://dx.doi.org/10.1111/jnc.13093.

Kapoor, R., Ghosh, H., Nordstrom, K., Vennstrom, B., & Vaidya, V. A. (2011). Loss of thyroid hormone receptor beta is associated with increased progenitor proliferation and NeuroD positive cell number in the adult hippocampus. *Neuroscience Letters*, *487*(2), 199–203. http://dx.doi.org/10.1016/j.neulet.2010.10.022.

Kapoor, R., van Hogerlinden, M., Wallis, K., Ghosh, H., Nordstrom, K., Vennstrom, B., et al. (2010). Unliganded thyroid hormone receptor alpha1 impairs adult hippocampal neurogenesis. *The FASEB Journal*, *24*(12), 4793–4805. http://dx.doi.org/10.1096/fj.10-161802.

Kathol, R. G., & Delahunt, J. W. (1986). The relationship of anxiety and depression to symptoms of hyperthyroidism using operational criteria. *General Hospital Psychiatry*, *8*(1), 23–28.

Kempermann, G., Jessberger, S., Steiner, B., & Kronenberg, G. (2004). Milestones of neuronal development in the adult hippocampus. *Trends in Neurosciences*, *27*(8), 447–452. http://dx.doi.org/10.1016/j.tins.2004.05.013.

Kempermann, G., Song, H., & Gage, F. H. (2015). Neurogenesis in the adult hippocampus. *Cold Spring Harbor Perspectives in Biology*, 7(9), a018812. http://dx.doi.org/10.1101/cshperspect.a018812.

Koenig, R. J., Lazar, M. A., Hodin, R. A., Brent, G. A., Larsen, P. R., Chin, W. W., et al. (1989). Inhibition of thyroid hormone action by a non-hormone binding c-erbA protein generated by alternative mRNA splicing. *Nature*, *337*(6208), 659–661. http://dx.doi.org/10.1038/337659a0.

Koromilas, C., Liapi, C., Schulpis, K. H., Kalafatakis, K., Zarros, A., & Tsakiris, S. (2010). Structural and functional alterations in the hippocampus due to hypothyroidism. *Metabolic Brain Disease*, *25*(3), 339–354. http://dx.doi.org/10.1007/s11011-010-9208-8.

Kuhn, H. G., Cooper-Kuhn, C., Eriksson, P., & Nilsson, M. (2005). Signals regulating neurogenesis in the adult olfactory bulb. *Chemical Senses*, *30*(Suppl. 1), i109–i110. http://dx.doi.org/10.1093/chemse/bjh138.

Kuhn, H. G., Eisch, A. J., Spalding, K., & Peterson, D. A. (2016). Detection and phenotypic characterization of adult neurogenesis. *Cold Spring Harbor Perspectives in Biology*, *8*(3), a025981. http://dx.doi.org/10.1101/cshperspect.a025981.

Kulikov, A., Torresani, J., & Jeanningros, R. (1997). Experimental hypothyroidism increases immobility in rats in the forced swim paradigm. *Neuroscience Letters*, *234*(2–3), 111–114.

Lanni, A., Moreno, M., & Goglia, F. (2016). Mitochondrial actions of thyroid hormone. *Comprehensive Physiology*, *6*(4), 1591–1607. http://dx.doi.org/10.1002/cphy.c150019.

Lechan, R. M., Qi, Y., Jackson, I. M., & Mahdavi, V. (1994). Identification of thyroid hormone receptor isoforms in thyrotropin-releasing hormone neurons of the hypothalamic paraventricular nucleus. *Endocrinology*, *135*(1), 92–100. http://dx.doi.org/10.1210/endo.135.1.7516871.

Lee, S., & Privalsky, M. L. (2005). Heterodimers of retinoic acid receptors and thyroid hormone receptors display unique combinatorial regulatory properties. *Molecular Endocrinology*, *19*(4), 863–878. http://dx.doi.org/10.1210/me.2004-0210.

Lemkine, G. F., Raj, A., Alfama, G., Turque, N., Hassani, Z., Alegria-Prevot, O., et al. (2005). Adult neural stem cell cycling in vivo requires thyroid hormone and its alpha receptor. *The FASEB Journal*, *19*(7), 863–865. http://dx.doi.org/10.1096/fj.04-2916fje.

Lieberwirth, C., Pan, Y., Liu, Y., Zhang, Z., & Wang, Z. (2016). Hippocampal adult neurogenesis: Its regulation and potential role in spatial learning and memory. *Brain Research*, *1644*, 127–140. http://dx.doi.org/10.1016/j.brainres.2016.05.015.

Lim, D. A., & Alvarez-Buylla, A. (2016). The adult ventricular-subventricular zone (V-SVZ) and olfactory bulb (OB) neurogenesis. *Cold Spring Harbor Perspectives in Biology*, *8*(5). http://dx.doi.org/10.1101/cshperspect.a018820

Liu, Y. Y., & Brent, G. A. (2010). Thyroid hormone crosstalk with nuclear receptor signaling in metabolic regulation. *Trends in Endocrinology and Metabolism*, *21*(3), 166–173. http://dx.doi.org/10.1016/j.tem.2009.11.004.

Llorens-Bobadilla, E., Zhao, S., Baser, A., Saiz-Castro, G., Zwadlo, K., & Martin-Villalba, A. (2015). Single-cell transcriptomics reveals a population of dormant neural stem cells that become activated upon brain injury. *Cell Stem Cell*, *17*(3), 329–340. http://dx.doi.org/10.1016/j.stem.2015.07.002.

Lopez-Juarez, A., Remaud, S., Hassani, Z., Jolivet, P., Pierre Simons, J., Sontag, T., et al. (2012). Thyroid hormone signaling acts as a neurogenic switch by repressing Sox2 in the adult neural stem cell niche. *Cell Stem Cell*, *10*(5), 531–543. http://dx.doi.org/10.1016/j.stem.2012.04.008.

Mackay-Sim, A., & Beard, M. D. (1987). Hypothyroidism disrupts neural development in the olfactory epithelium of adult mice. *Brain Research*, *433*(2), 190–198.

Madsen, T. M., Yeh, D. D., Valentine, G. W., & Duman, R. S. (2005). Electroconvulsive seizure treatment increases cell proliferation in rat frontal cortex. *Neuropsychopharmacology*, *30*(1), 27–34. http://dx.doi.org/10.1038/sj.npp.1300565.

Magnusson, J. P., & Frisen, J. (2016). Stars from the darkest night: Unlocking the neurogenic potential of astrocytes in different brain regions. *Development*, *143*(7), 1075–1086. http://dx.doi.org/10.1242/dev.133975.

Mahmoud, R., Wainwright, S. R., & Galea, L. A. (2016). Sex hormones and adult hippocampal neurogenesis: Regulation, implications, and potential mechanisms. *Frontiers in Neuroendocrinology*, *41*, 129–152. http://dx.doi.org/10.1016/j.yfrne.2016.03.002.

Malberg, J. E., Eisch, A. J., Nestler, E. J., & Duman, R. S. (2000). Chronic antidepressant treatment increases neurogenesis in adult rat hippocampus. *The Journal of Neuroscience*, *20*(24), 9104–9110.

Malvaut, S., & Saghatelyan, A. (2016). The role of adult-born neurons in the constantly changing olfactory bulb network. *Neural Plasticity*, *2016*. http://dx.doi.org/10.1155/2016/1614329.

Martinez, R., & Gomes, F. C. (2002). Neuritogenesis induced by thyroid hormone-treated astrocytes is mediated by epidermal growth factor/mitogen-activated protein kinase-phosphatidylinositol 3-kinase pathways and involves modulation of extracellular matrix proteins. *The Journal of Biological Chemistry*, *277*(51), 49311–49318. http://dx.doi.org/10.1074/jbc.M209284200.

Martinez-Iglesias, O., Garcia-Silva, S., Tenbaum, S. P., Regadera, J., Larcher, F., Paramio, J. M., et al. (2009). Thyroid hormone receptor beta1 acts as a potent suppressor of tumor invasiveness and metastasis. *Cancer Research*, *69*(2), 501–509. http://dx.doi.org/10.1158/0008-5472.can-08-2198.

Mastorakos, G., Karoutsou, E. I., Mizamtsidi, M., & Creatsas, G. (2007). The menace of endocrine disruptors on thyroid hormone physiology and their impact on intrauterine development. *Endocrine*, *31*(3), 219–237.

Mellstrom, B., Naranjo, J. R., Santos, A., Gonzalez, A. M., & Bernal, J. (1991). Independent expression of the alpha and beta c-erbA genes in developing rat brain. *Molecular Endocrinology*, *5*(9), 1339–1350. http://dx.doi.org/10.1210/mend-5-9-1339.

Miller, B. R., & Hen, R. (2014). The current state of the neurogenic theory of depression and anxiety. *Current Opinion in Neurobiology*, *30*, 51–58. http://dx.doi.org/10.1016/j.conb.2014.08.012.

Miller, K. J., Parsons, T. D., Whybrow, P. C., Van Herle, K., Rasgon, N., Van Herle, A., et al. (2007). Verbal memory retrieval deficits associated with untreated hypothyroidism. *The Journal of Neuropsychiatry and Clinical Neurosciences*, *19*(2), 132–136. http://dx.doi.org/10.1176/jnp.2007.19.2.132.

Ming, G. L., & Song, H. (2005). Adult neurogenesis in the mammalian central nervous system. *Annual Review of Neuroscience*, *28*, 223–250. http://dx.doi.org/10.1146/annurev.neuro.28.051804.101459.

Ming, G. L., & Song, H. (2011). Adult neurogenesis in the mammalian brain: Significant answers and significant questions. *Neuron*, *70*(4), 687–702. http://dx.doi.org/10.1016/j.neuron.2011.05.001.

Mohacsik, P., Zeold, A., Bianco, A. C., & Gereben, B. (2011). Thyroid hormone and the neuroglia: Both source and target. *Journal of Thyroid Research*, *2011*, 215718. http://dx.doi.org/10.4061/2011/215718.

Montero-Pedrazuela, A., Venero, C., Lavado-Autric, R., Fernandez-Lamo, I., Garcia-Verdugo, J. M., Bernal, J., et al. (2006). Modulation of adult hippocampal neurogenesis by thyroid hormones: Implications in depressive-like behavior. *Molecular Psychiatry*, *11*(4), 361–371. http://dx.doi.org/10.1038/sj.mp.4001802.

Mori, Y., Tomonaga, D., Kalashnikova, A., Furuya, F., Akimoto, N., Ifuku, M., et al. (2015). Effects of 3,3′,5-triiodothyronine on microglial functions. *Glia*, *63*(5), 906–920. http://dx.doi.org/10.1002/glia.22792.

Morte, B., & Bernal, J. (2014). Thyroid hormone action: Astrocyte-neuron communication. *Frontiers in Endocrinology*, *5*, 82. http://dx.doi.org/10.3389/fendo.2014.00082.

Murata, Y. (1998). Multiple isoforms of thyroid hormone receptor: An analysis of their relative contribution in mediating thyroid hormone action. *Nagoya Journal of Medical Science*, *61*(3–4), 103–115.

Niederkinkhaus, V., Marx, R., Hoffmann, G., & Dietzel, I. D. (2009). Thyroid hormone (T3)-induced up-regulation of voltage-activated sodium current in cultured postnatal hippocampal neurons requires secretion of soluble factors from glial cells. *Molecular Endocrinology*, *23*(9), 1494–1504. http://dx.doi.org/10.1210/me.2009-0132.

Olateju, T. O., & Vanderpump, M. P. (2006). Thyroid hormone resistance. *Annals of Clinical Biochemistry*, *43*(Pt. 6), 431–440. http://dx.doi.org/10.1258/000456306778904678.

O'Shea, P. J., & Williams, G. R. (2002). Insight into the physiological actions of thyroid hormone receptors from genetically modified mice. *The Journal of Endocrinology*, *175*(3), 553–570.

Pascual, A., & Aranda, A. (2013). Thyroid hormone receptors, cell growth and differentiation. *Biochimica et Biophysica Acta*, *1830*(7), 3908–3916. http://dx.doi.org/10.1016/j.bbagen.2012.03.012.

Paternostro, M. A., & Meisami, E. (1991). Lack of thyroid hormones but not their excess affects the maturation of olfactory receptor neurons: A quantitative morphologic study in the postnatal rat. *International Journal of Developmental Neuroscience*, *9*(5), 439–452.

Paternostro, M. A., & Meisami, E. (1993). Developmental plasticity of the rat olfactory receptor sheet as shown by complete recovery of surface area and cell number from extensive early hypothyroid growth retardation. *Brain Research. Developmental Brain Research*, *76*(2), 151–161.

Paternostro, M. A., & Meisami, E. (1996). Marked restoration of density and total number of mature (knob-bearing) olfactory receptor neurons in rats recovering from early hypothyroid-induced growth retardation. *Brain Research. Developmental Brain Research*, *96*(1–2), 173–183.

Perera, T. D., Coplan, J. D., Lisanby, S. H., Lipira, C. M., Arif, M., Carpio, C., et al. (2007). Antidepressant-induced neurogenesis in the hippocampus of adult nonhuman primates. *The Journal of Neuroscience*, *27*(18), 4894–4901. http://dx.doi.org/10.1523/jneurosci.0237-07.2007.

Perez-Domper, P., Gradari, S., & Trejo, J. L. (2013). The growth factors cascade and the dendrito-/synapto-genesis versus cell survival in adult hippocampal neurogenesis: The chicken or the egg. *Ageing Research Reviews*, *12*(3), 777–785. http://dx.doi.org/10.1016/j.arr.2013.06.001.

Pilhatsch, M., Winter, C., Nordstrom, K., Vennstrom, B., Bauer, M., & Juckel, G. (2010). Increased depressive behaviour in mice harboring the mutant thyroid hormone receptor alpha 1. *Behavioural Brain Research*, *214*(2), 187–192. http://dx.doi.org/10.1016/j.bbr.2010.05.016.

Preau, L., Fini, J. B., Morvan-Dubois, G., & Demeneix, B. (2015). Thyroid hormone signaling during early neurogenesis and its significance as a vulnerable window for endocrine disruption. *Biochimica et Biophysica Acta*, *1849*(2), 112–121. http://dx.doi.org/10.1016/j.bbagrm.2014.06.015.

Privalsky, M. L. (2004). The role of corepressors in transcriptional regulation by nuclear hormone receptors. *Annual Review of Physiology*, *66*, 315–360. http://dx.doi.org/10.1146/annurev.physiol.66.032802.155556.

Remaud, S., Gothie, J. D., Morvan-Dubois, G., & Demeneix, B. A. (2014). Thyroid hormone signaling and adult neurogenesis in mammals. *Frontiers in Endocrinology*, *5*, 62. http://dx.doi.org/10.3389/fendo.2014.00062.

Richardson, S. J., Lemkine, G. F., Alfama, G., Hassani, Z., & Demeneix, B. A. (2007). Cell division and apoptosis in the adult neural stem cell niche are differentially affected in transthyretin null mice. *Neuroscience Letters*, *421*(3), 234–238. http://dx.doi.org/10.1016/j.neulet.2007.05.040.

Richardson, S. J., Wijayagunaratne, R. C., D'Souza, D. G., Darras, V. M., & Van Herck, S. L. (2015). Transport of thyroid hormones via the choroid plexus into the brain: The roles of transthyretin and thyroid hormone transmembrane transporters. *Frontiers in Neuroscience*, *9*, 66. http://dx.doi.org/10.3389/fnins.2015.00066.

Rovet, J. F. (2014). The role of thyroid hormones for brain development and cognitive function. *Endocrine Development*, *26*, 26–43. http://dx.doi.org/10.1159/000363153.

Sahay, A., Scobie, K. N., Hill, A. S., O'Carroll, C. M., Kheirbek, M. A., Burghardt, N. S., et al. (2011). Increasing adult hippocampal neurogenesis is sufficient to improve pattern separation. *Nature*, *472*(7344), 466–470. http://dx.doi.org/10.1038/nature09817.

Sailor, K. A., Schinder, A. F., & Lledo, P. M. (2016). Adult neurogenesis beyond the niche: Its potential for driving brain plasticity. *Current Opinion in Neurobiology*, *42*, 111–117. http://dx.doi.org/10.1016/j.conb.2016.12.001.

Sakamoto, M., Kageyama, R., & Imayoshi, I. (2014). The functional significance of newly born neurons integrated into olfactory bulb circuits. *Frontiers in Neuroscience*, *8*, 121. http://dx.doi.org/10.3389/fnins.2014.00121.

Salto, C., Kindblom, J. M., Johansson, C., Wang, Z., Gullberg, H., Nordstrom, K., et al. (2001). Ablation of TRalpha2 and a concomitant overexpression of alpha1 yields a mixed hypo- and hyperthyroid phenotype in mice. *Molecular Endocrinology*, *15*(12), 2115–2128. http://dx.doi.org/10.1210/mend.15.12.0750.

Sanchez-Huerta, K., Garcia-Martinez, Y., Vergara, P., Segovia, J., & Pacheco-Rosado, J. (2016). Thyroid hormones are essential to preserve non-proliferative cells of adult neurogenesis of the dentate gyrus. *Molecular and Cellular Neurosciences*, *76*, 1–10. http://dx.doi.org/10.1016/j.mcn.2016.08.001.

Santarelli, L., Saxe, M., Gross, C., Surget, A., Battaglia, F., Dulawa, S., et al. (2003). Requirement of hippocampal neurogenesis for the behavioral effects of antidepressants. *Science*, *301*(5634), 805–809. http://dx.doi.org/10.1126/science.1083328.

Sarkar, P. K. (2002). In quest of thyroid hormone function in mature mammalian brain. *Indian Journal of Experimental Biology*, *40*(8), 865–873.

Schroeder, A. C., & Privalsky, M. L. (2014). Thyroid hormones, t3 and t4, in the brain. *Frontiers in Endocrinology*, *5*, 40. http://dx.doi.org/10.3389/fendo.2014.00040.

Schwartz, H. L., Strait, K. A., Ling, N. C., & Oppenheimer, J. H. (1992). Quantitation of rat tissue thyroid hormone binding receptor isoforms by immunoprecipitation of nuclear triiodothyronine binding capacity. *The Journal of Biological Chemistry*, *267*(17), 11794–11799.

Seri, B., Garcia-Verdugo, J. M., McEwen, B. S., & Alvarez-Buylla, A. (2001). Astrocytes give rise to new neurons in the adult mammalian hippocampus. *The Journal of Neuroscience*, *21*(18), 7153–7160.

Shi, Y. B. (2009). Dual functions of thyroid hormone receptors in vertebrate development: The roles of histone-modifying cofactor complexes. *Thyroid*, *19*(9), 987–999. http://dx.doi.org/10.1089/thy.2009.0041.

Silva, F. G., Giannocco, G., Santos, M. F., & Nunes, M. T. (2006). Thyroid hormone induction of actin polymerization in somatotrophs of hypothyroid rats: Potential repercussions in growth hormone synthesis and secretion. *Endocrinology*, *147*(12), 5777–5785. http://dx.doi.org/10.1210/en.2006-0110.

Smith, J. W., Evans, A. T., Costall, B., & Smythe, J. W. (2002). Thyroid hormones, brain function and cognition: A brief review. *Neuroscience and Biobehavioral Reviews*, *26*(1), 45–60.

Song, J., Christian, K. M., Ming, G. L., & Song, H. (2012). Modification of hippocampal circuitry by adult neurogenesis. *Developmental Neurobiology*, *72*(7), 1032–1043. http://dx.doi.org/10.1002/dneu.22014.

Spalding, K. L., Bergmann, O., Alkass, K., Bernard, S., Salehpour, M., Huttner, H. B., et al. (2013). Dynamics of hippocampal neurogenesis in adult humans. *Cell*, *153*(6), 1219–1227. http://dx.doi.org/10.1016/j.cell.2013.05.002.

Stuchlik, A. (2014). Dynamic learning and memory, synaptic plasticity and neurogenesis: An update. *Frontiers in Behavioral Neuroscience*, *8*, 106. http://dx.doi.org/10.3389/fnbeh.2014.00106.

Stuenkel, C. A. (2015). Subclinical thyroid disorders. *Menopause*, *22*(2), 231–233. http://dx.doi.org/10.1097/gme.0000000000000407.

Taupin, P. (2007). BrdU immunohistochemistry for studying adult neurogenesis: Paradigms, pitfalls, limitations, and validation. *Brain Research Reviews*, *53*(1), 198–214. http://dx.doi.org/10.1016/j.brainresrev.2006.08.002.

Tinnikov, A., Nordstrom, K., Thoren, P., Kindblom, J. M., Malin, S., Rozell, B., et al. (2002). Retardation of post-natal development caused by a negatively acting thyroid hormone receptor alpha1. *The EMBO Journal*, *21*(19), 5079–5087.

Tseng, H. C., Ruegg, S. J., Maronski, M., Messam, C. A., Grinspan, J. B., & Dichter, M. A. (2006). Injuring neurons induces neuronal differentiation in a population of hippocampal precursor cells in culture. *Neurobiology of Disease*, *22*(1), 88–97. http://dx.doi.org/10.1016/j.nbd.2005.10.007.

Uhl, I., Bez, J. A., Stamm, T., Pilhatsch, M., Assion, H. J., Norra, C., et al. (2014). Influence of levothyroxine in augmentation therapy for bipolar depression on central serotonergic function. *Pharmacopsychiatry*, *47*(4–5), 180–183. http://dx.doi.org/10.1055/s-0034-1383654.

Unnikrishnan, A. G., & Menon, U. V. (2011). Thyroid disorders in India: An epidemiological perspective. *Indian Journal of Endocrinology and Metabolism*, *15*(Suppl. 2), S78–S81. http://dx.doi.org/10.4103/2230-8210.83329.

Vaidya, V. A., Vadodaria, K. C., & Jha, S. (2007). Neurotransmitter regulation of adult neurogenesis: Putative therapeutic targets. *CNS & Neurological Disorders Drug Targets*, *6*(5), 358–374.

van Strien, M. E., van den Berge, S. A., & Hol, E. M. (2011). Migrating neuroblasts in the adult human brain: A stream reduced to a trickle. *Cell Research*, *21*(11), 1523–1525. http://dx.doi.org/10.1038/cr.2011.101.

Venero, C., Guadano-Ferraz, A., Herrero, A. I., Nordstrom, K., Manzano, J., de Escobar, G. M., et al. (2005). Anxiety, memory impairment, and locomotor dysfunction caused by a mutant thyroid hormone receptor alpha1 can be ameliorated by T3 treatment. *Genes & Development*, *19*(18), 2152–2163. http://dx.doi.org/10.1101/gad.346105.

Videbech, P., & Ravnkilde, B. (2004). Hippocampal volume and depression: A meta-analysis of MRI studies. *The American Journal of Psychiatry*, *161*(11), 1957–1966. http://dx.doi.org/10.1176/appi.ajp.161.11.1957.

Wallis, K., Dudazy, S., van Hogerlinden, M., Nordstrom, K., Mittag, J., & Vennstrom, B. (2010). The thyroid hormone receptor alpha1 protein is expressed in embryonic postmitotic neurons and persists in most adult neurons. *Molecular Endocrinology*, *24*(10), 1904–1916. http://dx.doi.org/10.1210/me.2010-0175.

Wikstrom, L., Johansson, C., Salto, C., Barlow, C., Campos Barros, A., Baas, F., et al. (1998). Abnormal heart rate and body temperature in mice lacking thyroid hormone receptor alpha 1. *The EMBO Journal*, *17*(2), 455–461. http://dx.doi.org/10.1093/emboj/17.2.455.

Williams, G. R. (2008). Neurodevelopmental and neurophysiological actions of thyroid hormone. *Journal of Neuroendocrinology*, *20*(6), 784–794. http://dx.doi.org/10.1111/j.1365-2826.2008.01733.x.

Wirth, E. K., Schweizer, U., & Kohrle, J. (2014). Transport of thyroid hormone in brain. *Frontiers in Endocrinology*, *5*, 98. http://dx.doi.org/10.3389/fendo.2014.00098.

Wu, Y., & Koenig, R. J. (2000). Gene regulation by thyroid hormone. *Trends in Endocrinology and Metabolism*, *11*(6), 207–211.

Yau, S. Y., Li, A., & So, K. F. (2015). Involvement of adult hippocampal neurogenesis in learning and forgetting. *Neural Plasticity*, *2015*, 717958. http://dx.doi.org/10.1155/2015/717958.

Yu, T. S., Dandekar, M., Monteggia, L. M., Parada, L. F., & Kernie, S. G. (2005). Temporally regulated expression of Cre recombinase in neural stem cells. *Genesis*, *41*(4), 147–153. http://dx.doi.org/10.1002/gene.20110.

Zhang, J., & Lazar, M. A. (2000). The mechanism of action of thyroid hormones. *Annual Review of Physiology*, *62*, 439–466. http://dx.doi.org/10.1146/annurev.physiol.62.1.439.

Zhao, C., Deng, W., & Gage, F. H. (2008). Mechanisms and functional implications of adult neurogenesis. *Cell*, *132*(4), 645–660. http://dx.doi.org/10.1016/j.cell.2008.01.033.

Zoeller, R. T., & Crofton, K. M. (2005). Mode of action: Developmental thyroid hormone insufficiency—Neurological abnormalities resulting from exposure to propylthiouracil. *Critical Reviews in Toxicology*, *35*(8–9), 771–781.

Zoeller, R. T., & Rovet, J. (2004). Timing of thyroid hormone action in the developing brain: Clinical observations and experimental findings. *Journal of Neuroendocrinology*, *16*(10), 809–818. http://dx.doi.org/10.1111/j.1365-2826.2004.01243.x.

CHAPTER ELEVEN

Thyroid Hormone and the White Matter of the Central Nervous System: From Development to Repair

Laura Calzà*,†,‡,1, Vito A. Baldassarro*,†,‡, Mercedes Fernandez†, Alessandro Giuliani§, Luca Lorenzini‡, Luciana Giardino*,‡,§

*CIRI-SDV, University of Bologna, Bologna, Italy
†FaBit, University of Bologna, Bologna, Italy
‡IRET Foundation, Ozzano Emilia, Italy
§DIMEVET, University of Bologna, Bologna, Italy
[1]Corresponding author: e-mail address: laura.calza@unibo.it

Contents

1. Introduction 254
2. TH Metabolism in the CNS 255
   2.1 Intracellular Content and TH Genomic Effects 260
   2.2 Nongenomic Effects of THs 261
3. Tissue Hypothyroidism in the CNS 261
4. TH, Developmental Myelination and Neonatal Hypoxic/Ischemic Insults 265
   4.1 OPC and Developmental Myelination 266
   4.2 Implications for Myelination After Perinatal Inflammatory Insults 266
5. TH and Myelin Turnover/Repair in Adulthood 267
   5.1 OPC and Myelin Repair 268
   5.2 Implications for Myelination Repair After Inflammatory Insults 269
6. Therapeutic Implications 271
Acknowledgments 272
References 272
Further Reading 280

Abstract

The role of thyroid hormone (TH) on brain development, and particularly in myelination, is well known since many decades, as testified by the severe structural and functional consequences of congenital hypothyroidism. This role during development, the consideration that the early TH supplementation restores myelination capability, and the fact the cell responsible for developmental myelination and remyelination is the same, i.e., the oligodendrocyte precursor cell (OPC), claimed the attempt to improve myelin repair in the adulthood via TH supplementation.

Vitamins and Hormones, Volume 106
ISSN 0083-6729
http://dx.doi.org/10.1016/bs.vh.2017.04.003

In this chapter, the impact of TH on development, homeostasis, and repair of the myelin in the CNS will be reviewed, focusing on the regulation of the TH tissue signaling during physiological and pathological conditions affecting myelination and/or myelin repair during early postnatal age and during the adulthood. The impact of the tissue inflammation on molecular mediators of the TH cellular action and metabolism will be discussed, with regard to the consequences on the biology of the OPC.

1. INTRODUCTION

The importance of an appropriate thyroid hormone (TH) signaling during brain development is dramatically illustrated in humans by the severe mental retardation and growth impairment observed in conditions leading to a substantial decrease of fetal TH availability, such as congenital hypothyroidism (Anderson, Schoonover, & Jones, 2003; Bernal, 2002), severe maternal hypothyroidism (Min et al., 2016), gestational exposure to thyroid-disrupting chemicals (Fernández et al., 2010; Zoeller, 2007; Zoeller & Crofton, 2005), and the Allan–Herndon–Dudley syndrome due to mutations of the TH membrane transporter MCT8 (La Piana, Vanasse, Brais, & Bernard, 2015). Hypo/dysmyelination is a common and major feature in these conditions, as described in early clinical studies of infants with congenital hypothyroidism (Rosman, 1972) and in a number of further studies describing both abnormalities in the white matter structure and myelin composition (Jagannathan, Tandon, Raghunathan, & Kochupillai, 1998). In fact TH deeply influences brain maturation, and more specifically myelination. During the second and third trimesters, and more extensively postnatally, TH promotes gliogenesis and myelination, regulating different steps of this complex and lengthy process (Bernal, 2015; Calzà, Fernandez, & Giardino, 2015; Porterfield & Hendrich, 1993).

Myelination actually represents a seminal example of the role of THs in brain development and its spatiotemporal regulation. Myelination is a highly orchestrated process that involves different cell types and molecular signals leading, in the central nervous system (CNS), to the ensheathment of axons by the plasma membranes of oligodendrocytes (OLs) in a spiral fashion to form an alternation of internodal segments (wrapped axon), and nodes of Ranvier (unwrapped axon). This structural organization of the axon facilitates the fast saltatory propagation of action potentials. Prolonged neonatal hypothyroidism causes delayed and poor deposition of myelin and a decrease in the number of myelinated axons (Bernal, 2015), leading to motor, sensory, and cognitive deficits if not appropriately treated. These dramatic effects are related in the first place to the key role of TH in the biology of the oligodendrocyte precursor cell (OPC) (Lee & Petratos, 2016). OPCs are unipotent progenitor cells originally recognized through their positivity to the membrane-associated chondroitin sulfate proteoglycan NG2 and to receptor for platelet-derived growth factor (PDGFαR). OPCs can divide symmetrically to produce two daughter OPCs, or divide asymmetrically in order to self-renew while producing one daughter OL. OPCs are produced during development by a yet to be defined cell source in specific CNS regions. The neural precursor cells are then specified toward the OL lineage under the influence of multiple soluble and intrinsic factors, undergoing progressive maturation in four stages: OPCs, preoligodendrocytes, immature OL, and mature OL (Barateiro & Fernandes, 2014).

OPC is also responsible for myelin repair during adulthood as remyelination is the only true regenerative capability of the CNS (Crawford, Chambers, & Franklin, 2013). OPCs are present in the mature CNS, where they represent a major proliferating cell population, constituting around 5% of the total cell population (Levine, Reynolds, & Fawcett, 2001). In response to myelin loss or increased demand, "adult" OPCs have the capacity to differentiate into mature myelinating OL (Crawford, Tripathi, Richardson, & Franklin, 2016), to guarantee an appropriate white matter turnover. This significant capability, demonstrated in physiological conditions, led to the so-called recapitulation hypothesis for myelin repair which suggested that remyelination after myelin damage follows several myelination steps and molecular mechanisms occurring during developmental myelination (Franklin & Hinks, 1999).

In this chapter, the impact of TH on myelin development, homeostasis, and repair will be reviewed, focusing on the regulation of TH tissue signaling under physiological and pathological conditions affecting myelination and/or myelin repair during early postnatal age and during adulthood.

2. TH METABOLISM IN THE CNS

Tetraiodothyronine, also called L-thyroxine, or T4, and L-triiodothyronine, or T3, are blood circulating THs in which more than 99% of both T4 and T3 is bound to serum proteins. The thyroid gland mainly secretes T4, which is considered the inactive form of TH or a prohormone, while T3 is mainly produced by the target tissues and has a much greater biological activity (about 10×) than T4. In humans, the thyroid secretes about 80 μg of T4 and only 5 μg of T3 per day. An additional 25 μg per day of T3 is produced by peripheral tissues from T4, mainly in the liver and kidneys. In addition to T3, an equal amount of "reverse T3" which has no biological activity may also be formed.

Discrepancies between plasma and tissue concentration of TH have been described in rare diseases, such as MCT8 mutations, also known as Allan–Herndon–Dudley syndrome (Fu, Refetoff, & Dumitrescu, 2013), but also in common clinical conditions such as cardiovascular diseases (heart failure, acute myocardial infarction, and acute coronary syndrome; Wang et al., 2017). Moreover, it has been known for many decades that patients in intensive care and during long-term critical illness typically present decreased concentrations of plasma T3, T4, and a normal range or slightly decreased concentration of thyroid-stimulating hormone, a condition known as non-thyroidal illness syndrome (NTIS) (Fliers, Bianco, Langouche, & Boelen, 2015). Inflammatory cytokines, and in particular IL6 and TNFα, released during local or systemic inflammation are the main molecular candidates in driving TH tissue alterations (reviewed by Calzà et al., 2015; de Vries, Fliers, & Boelen, 2015). Thus, the tissue concentration of THs and their molecular mediators possibly more effectively reflects the functional state of the thyroid system (TS) at target than circulating hormones (Little, 2016). A summary of total T3 and T4 concentrations in the plasma and several tissues in different animal species and human is reported in Table 1.

In the brain, it has been estimated that only 20% of the T3 concentration derives from circulation, since the ratio of T3 to T4 is much higher than in peripheral tissues (Donzelli et al., 2016). Moreover, T3 concentration equilibrates more slowly in the CNS than in peripheral tissues (Schroeder & Privalsky, 2014). This is linked to the extremely complex metabolism of THs in the CNS, and particularly to the interplay between TH membrane transporters and deiodinase activity. The THs actually cross the BBB partially via gaps among the cells that compose the blood–brain barrier, but mainly using transporters, thus through a saturable process (Bernal, Guadaño-Ferraz, & Morte, 2015). Membrane transport systems for TH include organic anion transporters and amino acid transporters, and several TH cell-membrane transporters (THCMTs) (Bernal et al., 2015). The monocarboxylate transporter (MCT) 8 is strongly expressed in the choroid plexus, capillary endothelial cells, and selected neuronal populations (Heuer & Visser, 2009). T4 exhibits greater affinity than T3 for the transporters and thus predominantly enters the CNS (Heuer, 2007).

Table 1 Total T3 and T4 Plasma and Tissue Concentration in Different Animal Species

Specie	Sex	Substrate	Total T4 (ng/mL)	Total T3 (ng/mL)	Method	References
Rat (Long-Evans)	Female	Serum	510 ± 6.1	4.7 ± 0.6	RIA	Christian and Trenton (2003)
Rat (Sprague-Dawley)	Male	Serum	38 ± 5	0.825 ± 0.087	RIA	Chang et al. (2008)
	Female	Serum	31.8 ± 2.1	0.66 ± 0.027		
Rat (Wistar)	Female	Plasma	35	0.909	RIA	Ayala-Moreno, Racotta, Anguiano, Aceves, and Quevedo (2013)
Rat (Dark Agouti)	Female	Plasma	55.2 ± 3.64	0.92 ± 0.09	Fluorescent ELISA-based multiparametric assay	Fernández, Baldassarro, Sivilia, Giardino, and Calzà (2016)
Rat (Wistar) pmol g^{-1}	Male + female	Heart	27.9 ± 4.4	5.18 ± 0.33	HPLC coupled to tandem MS	Donzelli et al. (2016)
		Liver	124.3 ± 8.2	12.9 ± 0.81		
		Kidney	90.9 ± 4.19	19.7 ± 1.26		
		WAT visceral	2.26 ± 0.16	0.21 ± 0.02		
		WAT subcutaneous	6.45 ± 0.95	0.39 ± 0.03		
		Brain	3.13 ± 0.35	2.23 ± 0.17		

Continued

Table 1 Total T3 and T4 Plasma and Tissue Concentration in Different Animal Species—cont'd

Specie	Sex	Substrate	Total T4 (ng/mL)	Total T3 (ng/mL)	Method	References
Rat (Sprague-Dawley)	Female	Heart	5.02 ± 0.4	1.47 ± 0.04	RIA, $ng\,g^{-1}$	Escobar-Morreale, Obregón, Escobar del Rey, and Morreale de Escobar (1995)
		Liver	18.35 ± 0.81	2.81 ± 0.23		
		Kidney	24.79 ± 1.26	4.78 ± 0.5		
		BAT	5.31 ± 0.43	1.79 ± 0.22		
		Lung	18.35 ± 0.81	2.81 ± 0.23		
		Skeletal muscle	0.89 ± 0.07	0.44 ± 0.01		
		Spleen	2.79 ± 0.23	0.76 ± 0.02		
		Cortex	1.51 ± 0.17	1.92 ± 0.11		
		Cerebellum	7.43 ± 0.7	1.63 ± 0.04		
		Plasma	37 ± 3	0.91 ± 0.05		
Rat (Sprague-Dawley)	Male	Cerebral cortex	10–15.0	10–12.0	RIA, $ng\,g^{-1}$	Broedel et al. (2003)
		Hippocampus	20	10		
		Hypothalamo	30	18		
		Midbrain	18	18		
		Cerebellum	20	17		
		Medulla	40	12		
		Serum	39–42	0.5–0.61		

Marmoset	Male + female	Plasma	167.5 ± 3.5	1.72 ± 0.27	RIA, $ng\,g^{-1}$	D'intino et al. (2011)
Human	Male + female	Serum	4.5–12.5 μg/dL	80–200 ng/dL	Several techniques	Lee (2013)
			58–161 nmol/L (SI)	0.8–2.0 ng/mL (SI)		

Original values have been converted into $ng\,mL^{-1}$ for plasma/serum or $ng\,g^{-1}$ for tissues, in the case they were expressed in other different units. Data included in the table without the statistical error are derived from graphs reported in the papers.

2.1 Intracellular Content and TH Genomic Effects

In the brain, T4 is taken up by astrocytes, probably by OATP1C1 transporter. Intracellularly, the enzyme type-2 iodothyronine deiodinase (D2) generates T3 by catalyzing the removal of ring iodine atoms in the cytosol, from which T3 easily reaches the nucleus. In other cell types, such as neurons, D3 catalyzes the removal of the outer- or inner-ring iodine atoms to produce T2 or the inactive form rT3, respectively (reviewed by Calzà et al., 2015). T3-target cells also include neurons, oligodendrocytes, and many other cell types in the nervous tissue.

The most classic cellular actions of TH are genomically mediated by T3 binding to nuclear receptors (TRs). TRs belong to the superfamily of nuclear receptors (NRs), which are ligand-activated transcription factors (Yen, 2001). They are expressed widely in the brain, and their concentration during development differs from that in adulthood. TRs are encoded by two genes, TR-alpha (TR-α) and TR-beta (TR-β) (Flamant et al., 2006). The transcripts of these genes undergo alternative splicing, generating the protein isoforms. In mammals, TR-α encodes three C-terminal variants: TR-α1, a functional receptor that binds both DNA and T3; TR-α2 and TR-α3 fail to bind T3 and act as antagonists *in vitro* (Harvey & Williams, 2002). In mouse, a promoter within intron 7 of TR-α gives rise to two truncated variants, TRΔ-α1 and TRΔ-α2, which act as potent dominant antagonists *in vitro* too. The TR-β gene encodes two N-terminal variants, TR-β1 and TR-β2 functional receptors. In rat, two further transcripts have also been cloned, TR-β3 and TRΔ-β3, although their physiological role is not fully understood (Harvey, Bassett, Maruvada, Yen, & Williams, 2007). TR-α1 and TR-β1 are predominant targets of tissue T3.

NRs form homo- and heterodimers. TRs may act either as homodimers or as heterodimers with retinoid X receptor (RXR). T3 preferentially binds TR–RXR heterodimers (Fattori et al., 2015). Like most NRs, TRs can bind their response elements both in the absence (aporeceptors) and in the presence (holoreceptor) of ligand (Bernal & Morte, 2013). Unliganded TRs have been shown to have critical regulatory roles in specific tissues during development in different tissues (Hasimoto et al., 2001). TR action is also regulated by small molecule ligands which induce a conformational change in the receptor suitable for the binding of "transcription coregulators" (coactivators or corepressors) that facilitate or inhibit the transcription of the associated target gene, thus controlling the cellular mechanisms of signal transduction. This mechanism probably confers cell and tissue specificity through differential expression of coactivators and corepressor.

Moreover, TRs also orchestrate modifications to local chromatin structure by recruiting histone-modifying enzymes (Cheng, Leonard, & Davis, 2010) and influencing DNA methylation (Kyono et al., 2016).

2.2 Nongenomic Effects of THs

The best known nongenomic effect of TH is probably the so-called calorigenic effect due to an increased metabolic rate. Although many aspects of this effect are still obscure, it is probably produced by T3 at mitochondrial level, where it induces a rapid stimulation of the respiration (Cioffi, Senese, Lanni, & Goglia, 2013; Davis, Goglia, & Leonard, 2016). This effect is possibly produced via TH binding sites to the inner membrane of the mitochondria (Sterling & Milch, 1975), or indirectly by binding to a mitochondrial TR, thus affecting mitochondrial transcription directly. TRα1, for example, has been demonstrated to be expressed in mitochondria in HeLa cells, and sequences similar to that of nuclear hormone-responsive elements located at various sites of the mitochondrial genome have been identified (Psarra, Solakidi, & Sekeris, 2006). Finally, TH also mediates mitochondrial biogenesis and changes in mitochondrial mass through the binding to T3-responsive nuclear genes (Cioffi et al., 2013).

TH also acts at the cellular level through a cell surface receptor that has been identified as the integrin $\alpha_V\beta_3$, a structural protein of the plasma membrane of virtually all cells, which interacts as a heterodimer with many proteins of the extracellular matrix (ECM) (Davis, Davis, & Cody, 2005; Davis, Davis, Mousa, Luidens, & Lin, 2011). This protein may function bidirectionally, converging ECM signals toward the cell, and cellular signals toward the ECM. Both T3 and T4 bind $\alpha_V\beta_3$, since T4 affinity is higher than that of T3. Downstream effects are multiples, tissue and cell specific (Davis et al., 2011).

3. TISSUE HYPOTHYROIDISM IN THE CNS

Apart from the cellular consequences of clinical hypo- and hyperthyroidism, the cellular effects of TH are thus relatively independent of TH blood concentration and are strictly dependent on the availability and activity of membrane transporters, deiodinases, and the expression of intracellular receptors at target. Therefore, systemic and local conditions that alter the TH cellular signaling regardless of the thyroid function possibly also modify TH cellular availability and downstream genomic and nongenomic effects. In particular, the regulation of deiodinase tissue activity has been extensively investigated, showing that both D expression and activity are regulated by numerous nonthyroid, physiological, and pathological factors, including fasting, illness, inflammation, and injury (Bianco & Kim, 2006). For example, the synthesis of the D2 protein exhibits a circadian rhythmicity, is inhibited by stress, and posttranscriptionally inactivated by ubiquitination (Bianco, 2011). The Dio2 gene is a highly sensitive cAMP-responsive gene that is also positively regulated by the nuclear factor κB and forkhead box (Fox)O3. Several chemicals regulate Dio3 gene expression, such as retinoic acid, 12-*O*-tetradecanoylphorbol-13-acetate, basic fibroblast growth factor, TGFβ, the hedgehog-GLI family zinc finger 2 pathway, and *conditions* (reviewed in Dentice & Salvatore, 2011).

Tissue inflammation and tissue hypoxia are powerful regulators of TH cellular signaling. Ectopic D3 expression is observed in peripheral tissues (myocardium, liver, and skeletal muscle) (Simonides et al., 2008) and in the brain (Fernández et al., 2016; Freitas et al., 2010) in inflammation and lesion models. Moreover, D3 is transcriptionally induced by hypoxia-inducible transcriptional factors (HIFs) (Simonides et al., 2008), and the intracellular trafficking of D3 is dependent on O_2 availability. Under conditions of normal O_2 availability, D3 is effectively transported through the Golgi system and sorted to the plasma membrane, while under hypoxia D3 is imported to the nucleus directly from the endoplasmic reticulum through a mechanism that depends on the cochaperone heat-shock protein 40 kDa. Also the expression of TRs is severely affected by tissue inflammation, as demonstrated in several tissues including the CNS (Castro, Quisenberry, Calvo, Obregon, & Lado-Abeal, 2013; Fernández et al., 2016), possibly through epigenetic mechanisms (Mitchell et al., 2013). Moreover, HIF also downregulates the expression of T3-target genes (Freitas et al., 2010).

Tissue inflammation and tissue hypoxia are also pathology landmarks in diseases characterized by white matter injury (WMI) both during development (as in neonatal hypoxic/ischemic injury) and in the adult (as in multiple sclerosis, MS). We recently reported that a complex dysregulation of the TH tissue signaling is present in experimental allergic encephalomyelitis (EAE), an experimental model of inflammatory/demyelinating disease extensively used as a model for MS. A downregulation of all TRs, including TRα, which is responsible for oligodendroglial lineage induction from neural stem cell (NSC) (Marziali, Garcia, & Pasquini, 2015), as well as TRβ, which is responsible for OPC maturation into oligodendrocytes (Baxi et al., 2014), was found. In parallel, a 30-fold upregulation was found in the T3-inactivating enzyme D3 mRNA, and the ectopic expression of D3 protein was observed also in astrocytes, NG2-IR OPCs, and CNPase-IR OL (Fernández et al., 2016). D3 protein strongly upregulates also *in vitro* in astrocytes and oligodendrocytes exposed to proinflammatory cytokines, as illustrated in Fig. 1 (see the figure legend for brief experimental details). Conversely, no significant changes were found in the expression level of D2 gene encoding for the T3 synthesis enzyme. Notably, alterations in TR and D expression correlate with the rise of inflammatory cytokines released by peripheral cells and resident microglia (Borjini, Fernández, Giardino, & Calzà, 2016; Fernández et al., 2016).

An increase in D3 activity and mRNA expression level has been described in other neurological conditions characterized by inflammation and demyelination, i.e., ischemia-induced hypoxic brain damage (Jo et al., 2012) and nerve lesion (Li et al., 2001). After experimental cerebral ischemia, TH plasma levels are reduced for 2 weeks and TR receptors differentially regulate around the ischemic infarct (Lourbopoulos et al., 2014). TRα1 is expressed by macrophages and TRβ by reactive astrocytes within the infarction; neurons around the infarct exhibit increased nuclear localization of TRα1, and a novel localization of TRβ in the paranodal/juxtaparanodal area of Ranvier nodes has been described.

The overall scenario emerging from *in vivo* data collected in animal models of inflammatory diseases like that from human patients, namely, D3 overexpression and TR downregulation, supports the hypothesis that inflammation leads to a substantial tissue hypothyroidism. Inflammatory cytokines and HIF are regarded as potential candidates for mediating at least the D regulation in many experimental models (Gereben et al., 2008; Simonides et al., 2008). Thus, tissue hypothyroidism during inflammation could impact on cellular processes, requiring an appropriate TH drive, including OPC differentiation into myelinating OL (Calzà, Fernandez, & Giardino, 2010; Fernández et al., 2016).

These results and considerations form the basis of preclinical studies attempting to correct myelination and remyelination failure in experimental conditions characterized by intense tissue inflammation and TS signaling dysregulation, starting from the fact that the block of TH-dependent OPC differentiation has been described in both perinatal (Buser et al., 2012) and adult (Kuhlmann et al., 2008) animal and human samples. In perinatal WMI, such as in MS, an increased number of premyelinating OLs is in fact observed in the acute phase, which is not followed by an increased number of myelinating OLs and appropriate myelin formation/repair, thus suggesting the idea that to overcome OPC differentiation block might represent a reliable target for novel therapies.

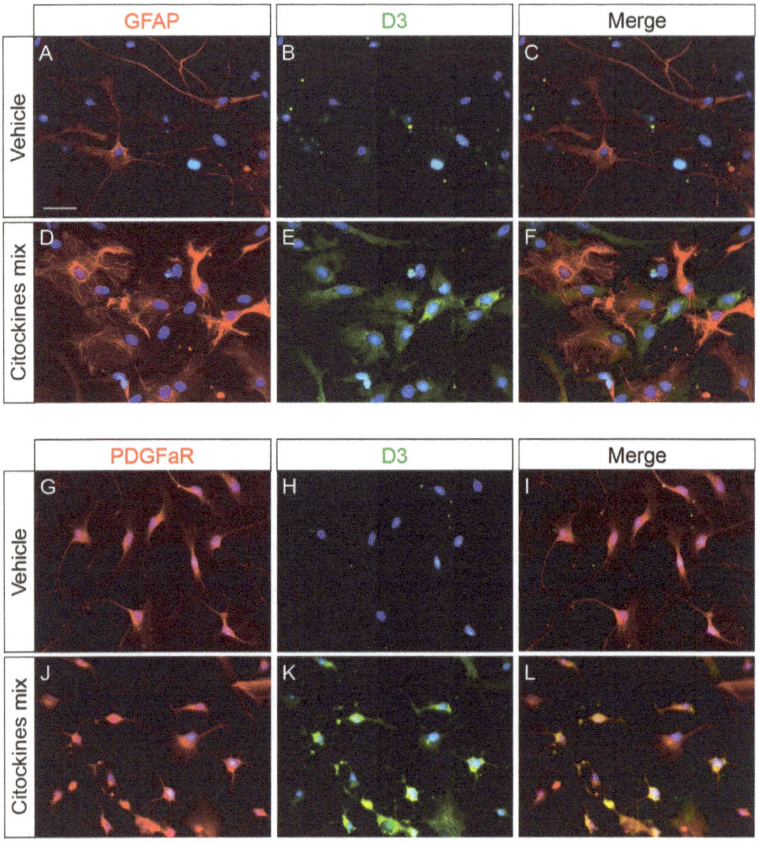

Fig. 1 Figure shows the expression of D3 in astrocytes (GFAP-positive cells) and OPC (PDGFaR-positive cells) in control conditions (vehicle) and after cytokines exposure, at 12 DIV. Exposure to cytokines induces a strong increase in D3 expression in both astrocytes and OPC. The culture system derives from neural precursor cells isolated from the subventricular zone of adult rats, cultured in the presence of T3 to isolate an OPC-enriched population (Fernández et al., 2016). Briefly, primary neurospheres were maintained adding bFGF and EGF (20 ng/mL each) in the culture medium. After splitting, oligospheres were obtained exposing cells to PDGF-AA and bFGF (20 ng/mL). In this proliferation phase that last 7 days, cells were exposed to a cytokine mix (TGF-β1, TNF-α, IL-1β, IL-6, IL-17, and IFN-γ; 20 ng/mL each) or to vehicle (0.04% of the cytokines solvent: 10% glycerol/100 nM glycine/25 nM Tris, pH 7.3). After splitting and plating on poly-D,L-ornithine/laminin, cells were exposed to T3 (50 nM), CNTF (ciliary neurotrophic factor, 10 ng/mL), and *N*-acetyl-L-cysteine, in order to induce OL differentiation. After 12 DIV of differentiation specific markers were identified by immunocytochemistry. Bar: 20 μm.

4. TH, DEVELOPMENTAL MYELINATION AND NEONATAL HYPOXIC/ISCHEMIC INSULTS

TS plays a major role in almost all the steps leading to myelination, as indicated by human clinical observations, *in vivo* studies in hypo-, hyperthyroid and transgenic animals and by *in vitro* studies on OPCs (Baas, Legrand, Samarut, & Flamant, 2002; Calzà et al., 2010, 2015; Obregon, Calvo, Del Rey, & de Escobar, 2007; O'Shea & Williams, 2002).

THs are present in the embryos already 3 days after implantation (Bernal, 2015), and the TR mRNA and proteins appear very early in the brain, their concentration progressively increasing until birth (Bernal, 2015; Howdeshell, 2002). The fetal thyroid starts to produce TH at about half the gestational age, when the iodine uptake in follicular cells starts and reaches functional maturation at the end of the gestational period, when the autoregulation of iodine begins (Howdeshell, 2002). Thus, transplacental T4 and T3 transfer from maternal to fetal circulation guarantees the appropriate TH signaling for development and maturation (reviewed in Moog et al., 2017). Maternal thyroid dysfunctions that might severely impact the brain development of the fetus include hypothyroxinemia, overt hypothyroidism, subclinical hypothyroidism, and thyroid autoimmunity. Moreover, gestational states, including stress and depression, may also alter maternal thyroid function, thus impacting on TH availability for fetus development. More disputed, but raising increasing interest also as regards public health, is the possible role of thyroid-disrupting chemicals, which have been shown to interact with virtually all elements of TH homeostasis and tissue metabolism (Gore et al., 2015; Gutleb, Cambier, & Serchi, 2016). Thyroid-disrupting chemicals belong to the large family of endocrine disrupting chemicals, and an extensive literature supports a major role of endocrine disruption in several human diseases, including neurodevelopmental disorders (Gore et al., 2015). Numerous environmental pollutants produced by industrial waste or derived from the massive use of pesticides in agriculture, including polychlorinated biphenyl, polybrominated diphenyl ethers, certain phthalates, perchlorates, and dioxins, lead to marginal reductions of circulating TH, thus impacting on brain processing requiring an accurate TH homeostasis in the maternal-fetal unit (Gilbert, Rovet, Chen, & Koibuchi, 2012).

During brain development, 500–1000 genes are directly or indirectly regulated by TH (Chatonnet, Flamant, & Morte, 2015) and some of them are implicated in oligodendrogenesis and myelinogenesis. T3 strongly induces the expression of KLF9, which is necessary and sufficient to promote OPC differentiation *in vitro* into myelinating OL (Denver & Williamson, 2009). Moreover, TH regulates the expression of practically all main myelin protein genes, including proteolipid protein (PLP), myelin basic protein (MBP), and myelin-associated glycoprotein (MAG). Notably, the sensitivity of these genes to TH follows a temporal and regional pattern mirroring myelination waves (Rodriguez-Pena, Ibarrola, Iñiguez, Muñoz, & Bernal, 1993).

4.1 OPC and Developmental Myelination

THs have multiple effects on OPC and developmental myelination via genomic, epigenetic, and nongenomic mechanisms (Lee & Petratos, 2016). THs induce more OLs to form from multipotent NSCs (Fernández, Pirondi, Manservigi, Giardino, & Calzà, 2004; Kang et al., 2007; Rogister, Ben-Hur, & Dubois-Dalcq, 1999; Whittemore, Morassutti, Walters, Liu, & Magnuson, 1999). T3 is one of the best characterized differentiating agents of OPCs *in vitro*, binding the nuclear T3 receptor TRα1, which also synchronizes the complete arrest of OPC proliferation (Picou, Fauquier, Chatonnet, & Flamant, 2012). T3 acts on the mitogen-dependent molecular machinery that regulates OPC proliferation and differentiation (Durand & Raff, 2000; Raff, 2006). When OPCs proliferate, they become sensitive to T3 after an appropriate time possibly corresponding to a definite number of cell divisions, probably because of a cell cycle-dependent expression of TRs (Maruvada, Dmitrieva, East-Palmer, & Yen, 2004). At this stage, TH functions as an instructive agent, triggering OPC (O-2A cell) cell cycle exit in close cooperation with platelet-derived growth factor (PDGF; Billon, Jolicoeur, Ying, Smith, & Raff, 2002; Durand, Gao, & Raff, 1997; Durand & Raff, 2000). Furthermore, T3 favors OPC terminal differentiation also through the transcription regulation of premyelinating genes (Baxi et al., 2014; Casaccia-Bonnefil & Liu, 2003; Dugas, Ibrahim, & Barres, 2012). Studies in knock-out mice have indicated that the absence of all TRs results in the persistence of OPC proliferation and the absence of differentiation in the adult optic nerves, leading to a defect in myelination (Baas et al., 2002).

4.2 Implications for Myelination After Perinatal Inflammatory Insults

As already discussed, myelination is a lengthy process that starts prenatally and undergoes a powerful maternal-to-fetal switch in TH availability as soon as the fetal thyroid function is mature, i.e., at birth. Conditions altering the physiological development of the fetus also impact on developmental myelination, and might be responsible for the WMI causing long-term motor and cognitive disabilities, a still unsolved and challenging problem, e.g., in very low birth weight infants (Elitt & Rosenberg, 2014).

It is well known that more than 50% of premature infants develop transient hypothyroxinemia, a condition characterized by low concentrations of serum T3 and T4 but normal concentrations of thyroid-stimulating hormone (Elitt & Rosenberg, 2014; Reuss, Leviton, Paneth, & Susser, 1997; van Tilborg et al., 2015), and a reduction in the MCT8 expression has been reported in the brain from human fetuses with intrauterine growth restriction (Chan et al., 2014). In these conditions, a combined insult of infection/inflammation plus hypoxia–ischemia appears to impact on premyelinating OPC leading to injury and death, but also to the failure of proper OL maturation (Billiards et al., 2008; Buser et al., 2012).

This hormone profile described in many premature infants resembles the NTIS described in severe and chronic illness in adult, thus transferring also to pediatric medicine the discussion on the clinical management of NTIS (Schang, Gressens, & Fleiss, 2014). Indeed conflicting results have been obtained by TH treatment in experimental perinatal inflammation-induced WMI, whether positive (perinatal hypoxia–ischemia, Hung, Huang, Huang, Tu, & Chang, 2013; intraventricular hemorrhage, Vose et al., 2013), or with no effects (WMI induced by interleukin-1β, Schang, Van Steenwinckel, et al., 2014). The substantial differences in these models make it quite difficult to compare these data sets, also considering the delicate balance between maternal-derived TH and the perinatal development of fetal thyroid function (Moog et al., 2017). Moreover, results from clinical trials using T4 in preterm infants with hypothyroidism, considering the neurological maturation as primary end-point, point to the need for further studies, to better stratify patients as a function of gestation age, inflammatory status, specific tissue injury (e.g., intraventricular hemorrhage vs focal or diffuse WMI) (Schang, Gressens, et al., 2014).

5. TH AND MYELIN TURNOVER/REPAIR IN ADULTHOOD

In mammals, myelination is protracted postnatally and continues throughout life, being characterized by a high degree of plasticity (Boulanger & Messier, 2014; Miller et al., 2012; Purger, Gibson, & Monje, 2016). For example, intense training seems to produce increased myelination, thus favoring an improvement in conduction velocity

(Bengtsson et al., 2005), and a myelin remodeling, thus involving a reset of intermodal distances, might also occur to optimize the related neural function (Young et al., 2013). More generally, a number of physiological conditions including daylight, gender and pregnancy, voluntary physical exercise, environmental and social enrichment, motor learning, and cognitive training affect "myelin plasticity," which severely declines with aging (Wang & Young, 2013).

5.1 OPC and Myelin Repair

Adult OPCs actively proliferate and migrate in response to injury, in particular during inflammatory/demyelination conditions, and are also usually able to differentiate into mature myelinating OL to fully restore the white matter structure and function (Crawford et al., 2013). This capability is guaranteed by OPCs generated during development that populate the white and gray matter of the adult CNS. Moreover, new OPCs can be formed by the neurogenic areas (mainly the subventricular zone) that persist in the adult brain (Menn et al., 2006), which may directly contribute to myelin repair (Islam, Tatsumi, Okuda, Shiosaka, & Wanaka, 2009). These cells maintain the significant capability to differentiate into myelinating OLs also thanks to the electrical signaling from axons (De Castro, Bribián, & Ortega, 2013; Purger et al., 2016).

However, for unknown reasons, this process fails in several conditions characterized by intense and chronic inflammation. Pivotal studies in MS, the most diffuse inflammatory/demyelinating disease in human, have indicated that during the acute, focal phase of the disease, remyelination may be robust and efficient and early lesions of the white matter (fresh plaques) may be completely and correctly repaired and their function restored (Rodgers, Robinson, & Miller, 2013). However, this process progressively fails in the chronic phase of the disease, in spite of the fact that a significant number of OPCs, also newly generated from NSCs, are found in early lesions in MS (Boyd, Zhang, & Williams, 2013; Lassmann, van Horssen, & Mahad, 2012). A blockage of OPCs is currently regarded as a preeminent cause of remyelination failure in MS (Kuhlmann et al., 2008; Miller & Mi, 2007).

Remyelination failure is also a pathological landmark in other neurological conditions characterized by intense inflammation and is a major factor in chronic sensory, motor, and cognitive disabilities. There is general consensus that acute demyelination after spinal cord injury is caused by an oligodendrocyte loss (Myers, Bankston, Burke, Ohri, & Whittemore, 2016) and impaired OPC differentiation (Franklin & Ffrench-Constant, 2008; Mekhail, Almazan, & Tabrizian, 2012) both in experimental animals and humans. Also in stroke models local OPCs in the white matter proliferate and a portion differentiates into astrocytes, with the remaining cells mainly blocked in a progenitor phenotype (Sozmen et al., 2016). Thus, remyelination is considered a reliable therapeutic target in neurological conditions, in which the myelin protection and repair represents a neuroprotective strategy (Verden & Macklin, 2016).

5.2 Implications for Myelination Repair After Inflammatory Insults

On the basis of the original *in vivo* and *in vitro* work from our lab (Calzà et al., 2005, 2010, 2015), numerous results obtained by different laboratories indicate that short protocols of TH supplementation favor myelin protection/repair in different demyelination models characterized by intense tissue inflammation. Moreover, numerous *in vitro* and *in vivo* studies have indicated that several inflammatory cytokines, including IFN-γ, IL-17, IL-1β, IL-6, TGF-β1, and TNF-α, impair OPC differentiation, affecting cell cycle exit and the expression of myelin-related genes (Chew, King, Kennedy, & Gallo, 2005; Falahati et al., 2013; Kang et al., 2013; Su et al., 2011; Tanner, Cherry, & Mayer-Proschel, 2011).

Our work in EAE demonstrated that short *in vivo* TH administration during the acute phase of EAE, when proliferating, undifferentiated cells increase in the subventricular zone and spinal cord, promotes OPC maturation.

T4 *in vivo* administration reduces proliferation and the number of undifferentiated precursors and upregulates expression of markers for OL progenitors (polysialylated-neural cell adhesion molecule, O4, A2B5) and mature OL (MBP) in the subventricular zone and spinal cord of EAE animals (Calzà, Fernandez, Giuliani, Aloe, & Giardino, 2002). *In vitro* we also showed that T3 in the presence of mitogens reduced the proliferation rate of neural precursor cells derived from adult brain, shifting the cell fate toward oligodendroglial lineage as indicated by increased expression of Olig-1. Moreover, *in vitro* T3 favors OL maturation, as indicated by the expression of antigens associated with different OL maturation stages (Fernández, Pirondi, et al., 2004), and T3 *in vitro* exposure completely restores the OL generation that is lost in neural precursors derived from EAE brains (Fernandez, Paradisi, Del Vecchio, Giardino, & Calzà, 2009). We then demonstrated that TH treatment not only favors OPC maturation, but also results in the protection of the myelin sheath and axons and ultimately improves functional and clinical outcome. This was demonstrated both in rats and in the nonhuman primate *Callithrix jacchus*. In rat (both Wistar and Dark-Agouti), we showed that pulsed TH treatment increases the MBP content, restores myelin sheath thickness (Fernandez, Giuliani, et al., 2004), normalizes neurofilament immunoreactivity in the fasciculus gracilis and tau hyperphosphorylation in the lumbar spinal cord, finally favoring a better outcome of clinical and neurophysiological parameters, as indicated by somatosensory evoked potentials (Dell'Acqua et al., 2012).

These results were then confirmed in the nonhuman primate *C. jacchus*, where we reported that short protocols of T3 treatment shifts the demyelination/remyelination balance toward remyelination (D'Intino et al., 2011).

These results have been replicated in the cuprizone demyelination model in mouse (Harsan et al., 2008) and rat (Franco, Silvestroff, Soto, & Pasquini, 2008; Silvestroff, Bartucci, Pasquini, & Franco, 2012), where it has been suggested that THs could regulate the emergence of remyelinating oligodendrocytes from the pool of proliferating cells via Shh. TH has also been coadministered with volporic acid in EAE, confirming amelioration of the clinical course and pathology in DA rats and suggesting the involvement of epigenetic mechanism (Castelo-Branco et al., 2014). TH also improves remyelination in demyelination due to CNS hypoxia (Hung et al., 2013) and intraventricular hemorrhage (Vose et al., 2013).

We also found that in EAE animals a severe alteration of deiodinase and thyroid hormone receptor mRNA expression level is found in the spinal cord (see also "Tissue hypothyroidism in the CNS"; D'Intino et al., 2011; Fernández et al., 2016). In a recent paper, we also suggested a direct link between cytokines, D3 expression, and OPC differentiation failure (Fernández et al., 2016). By *in vitro* experiments, we demonstrated that under cytokine exposure a dramatic rise in the number of D3-positive cells was observed in OPC derived from neural precursors. We then used the deiodinase blocker iopanocic acid (IOP) in order to decrease D3 activity and then increase T3 intracellular availability. This treatment was able to restore the capability of the OPC culture to express differentiation markers (CNPase and MBP), also restoring OL mature morphology. It may thus be concluded that IOP inhibits D3, thereby restoring an appropriate T3 content in OPCs that overcome the cytokine-induced OPC differentiation block.

This effect could explain the positive results obtained by *in vivo* TH supplementation in experimental models of inflammatory demyelination. Indeed, these studies indicated that TH supplementation restores the expression on a number of TH-dependent genes which are critical for OPC biology (Zhang, Ma, Qin, & Yao, 2015). Among others: Olig1/2 (D'Intino et al., 2011; Harsan et al., 2008), which are key determinants for OPC-maturation (Meijer et al., 2012; Zhu et al., 2012); sonic hedgehog, the activation of which is necessary to promote remyelination by fingolimod in EAE (Zhang, Zhang, et al., 2015); Klf9, the activation of which is necessary and sufficient to promote OL differentiation *in vitro* (Denver & Williamson, 2009); and remyelination *in vivo* (Dugas et al., 2012).

6. THERAPEUTIC IMPLICATIONS

The results summarized in this review highlight three major points:

1. The remyelination failure in neurological conditions characterized by intense inflammation is likely due not only to a loss of OL, but also to a OPC differentiation block;
2. The intense inflammation also causes a complex dysregulation of TH signaling in the CNS, characterized by D3 upregulation and TR downregulation, possibly leading to tissue hypothyroidism;
3. Since T3 is a critical player for OPC differentiation, the decreased availability of the active T3 in the tissue, and particularly in OPC during inflammation, might contribute to OPC differentiation failure.

Thus, the possibility of including TH as adjuvant therapy in diseases and injuries during the acute inflammatory phase might well be considered. The obvious criticism concerning the translation impact of these results is the risk of hyperthyroidism, which must be carefully considered. We raise three considerations in this discussion:

1. In the experiments supporting a role of exogenous TH administration to improve remyelination, we used pulsed administrations of T4 or T3 for short periods in the acute phase of the disease. The conventional treatments in rat were T4, 0.2mg s.c. at 11, 13, and 15 days post-immunization; T3, 10μg/kg bw, s.c., twice a day starting at 10 days postimmunization, then every 3 days; in marmoset: 10μg/kg, s.c., at

days 12, 13, 14, 25, 26, 27 postimmunization. Similar protocols were used by the other authors. None of these treatments induced stable hyperthyroidism.

2. Clinical experiences in nonthyroid diseases in which TH administration is used could provide an important forum also for neurological diseases and injuries. In particular, NTIS in chronic, noncritical patients offers an important and consolidated opportunity for this discussion, although it has not yet led to any guidelines. Moreover, supraphysiological doses of T4 (100 μg per day during week 1, 200 μg per day during week 2, and 300 μg per day during weeks 3–6) added to standard treatment for bipolar depression show promising results (Bauer et al., 2016) and the current hypothesis is that it corrects cellular hypothyroidism (Kelly, 2016).
3. Novel solutions for targeted drug delivery in the CNS based on biomaterials are emerging. For example, T3 administration via engineered T3 nanoparticles crossing the blood–brain barrier resulted in significant protection in experimental brain stroke (Mdzinarishvili, Sutariya, Talasila, Geldenhuys, & Sadana, 2013); polymeric nanofibers topically deliver T4 (Azarbayjani et al., 2010) and T3 (Wang, Liu, Hao, Zhang, & Zhang, 2013), such as chitosan conduits combined with T3 (Mohammadi et al., 2013) improved peripheral nerve regeneration. Finally, the T4 analog TETRAC acting as antagonist for the integrin receptor conjugated to nanoparticles is emerging as a promising therapy for cancer (Mousa et al., 2012; Yalcin et al., 2010)

ACKNOWLEDGMENTS

This work has been supported by Regione Emilia Romagna, POR-FESR program, "Step-by-Step" project (Pg/2015/737703), and Ministero dell'Università e Ricerca Scientifica (MIUR), cluster tecnologici nazionali program, IRMI project (CTN01_00177_888744).

REFERENCES

Anderson, G. W., Schoonover, C. M., & Jones, S. A. (2003). Control of thyroid hormone action in the developing rat brain. *Thyroid*, *13*, 1039–1056.

Ayala-Moreno, R., Racotta, R., Anguiano, B., Aceves, C., & Quevedo, L. (2013). Perinatal undernutrition programmes thyroid function in the adult rat offspring. *British Journal of Nutrition*, *110*, 2207–2215.

Azarbayjani, A. F., Venugopal, J. R., Ramakrishna, S., Lim, P. F., Chan, Y. W., & Chan, S. Y. (2010). Smart polymeric nanofibers for topical delivery of levothyroxine. *Journal of Pharmacy & Pharmaceutical Sciences*, *13*, 400–410.

Baas, D., Legrand, C., Samarut, J., & Flamant, F. (2002). Persistence of oligodendrocyte precursor cells and altered myelination in optic nerve associated to retina degeneration in mice devoid of all thyroid hormone receptors. *Proceedings of the National Academy of Sciences of the United States of America*, *99*, 2907–2911.

Barateiro, A., & Fernandes, A. (2014). Temporal oligodendrocyte lineage progression: In vitro models of proliferation, differentiation and myelination. *Biochimica et Biophysica Acta*, *1843*, 1917–1929.

Bauer, M., Berman, S., Stamm, T., Plotkin, M., Adli, M., Pilhatsch, M., et al. (2016). Levothyroxine effects on depressive symptoms and limbic glucose metabolism in bipolar disorder: A randomized, placebo-controlled positron emission tomography study. *Molecular Psychiatry*, *21*, 229–236.

Baxi, E. G., Schott, J. T., Fairchild, A. N., Kirby, L. A., Karani, R., Uapinyoying, P., et al. (2014). A selective thyroid hormone β receptor agonist enhances human and rodent oligodendrocyte differentiation. *Glia*, *62*, 1513–1529.

Bengtsson, S. L., Nagy, Z., Skare, S., Forsman, L., Forssberg, H., & Ullén, F. (2005). Extensive piano practicing has regionally specific effects on white matter development. *Nature Neuroscience*, *8*, 1148–1150.

Bernal, J. (2002). Action of thyroid hormone in brain. *Journal of Endocrinological Investigation*, *25*, 268–288.

Bernal, J. (2015). Thyroid hormones in brain development and function. In L. J. De Groot, G. Chrousos, K. Dungan, K. R. Feingold, A. Grossman, J. M. Hershman, C. Koch, M. Korbonits, R. McLachlan, M. New, J. Purnell, R. Rebar, F. Singer, & A. Vink (Eds.), *Source endotext [Internet]*. South Dartmouth (MA): MDText.com, Inc.

Bernal, J., Guadaño-Ferraz, A., & Morte, B. (2015). Thyroid hormone transporters—Functions and clinical implications. *Nature Reviews Endocrinology*, *11*, 406–417.

Bernal, J., & Morte, B. (2013). Thyroid hormone receptor activity in the absence of ligand: Physiological and developmental implications. *Biochimica et Biophysica Acta*, *1830*, 3893–3899.

Bianco, A. C. (2011). Cracking the metabolic code for thyroid hormone signaling. *Endocrinology*, *152*, 3306–3311.

Bianco, A. C., & Kim, B. W. (2006). Deiodinases: Implications of the local control of thyroid hormone action. *Journal of Clinical Investigation*, *116*, 2571–2579.

Billiards, S. S., Haynes, R. L., Folkerth, R. D., Borenstein, N. S., Trachtenberg, F. L., Rowitch, D. H., et al. (2008). Myelin abnormalities without oligodendrocyte loss in periventricular leukomalacia. *Brain Pathology*, *18*, 153–163.

Billon, N., Jolicoeur, C., Ying, Q. L., Smith, A., & Raff, M. (2002). Normal timing of oligodendrocyte development from genetically engineered, lineage-selectable mouse ES cells. *Journal of Cell Science*, *115*, 3657–3665.

Borjini, N., Fernández, M., Giardino, L., & Calzà, L. (2016). Cytokine and chemokine alterations in tissue, CSF, and plasma in early presymptomatic phase of experimental allergic encephalomyelitis (EAE), in a rat model of multiple sclerosis. *Journal of Neuroinflammation*, *13*, 191.

Boulanger, J. J., & Messier, C. (2014). From precursors to myelinating oligodendrocytes: Contribution of intrinsic and extrinsic factors to white matter plasticity in the adult brain. *Neuroscience*, *269*, 343–366.

Boyd, A., Zhang, H., & Williams, A. (2013). Insufficient OPC migration into demyelinated lesions is a cause of poor remyelination in MS and mouse models. *Acta Neuropathologica*, *125*, 841–859.

Broedel, O., Eravci, M., Fuxius, S., Smolarz, T., Jeitner, A., Grau, H., et al. (2003). Effects of hyper- and hypothyroidism on thyroid hormone concentrations in regions of the rat brain. *The American Journal of Physiology-Endocrinology and Metabolism*, *285*, E470–E480.

Buser, J. R., Maire, J., Riddle, A., Gong, X., Nguyen, T., Nelson, K., et al. (2012). Arrested preoligodendrocyte maturation contributes to myelination failure in premature infants. *Annals of Neurology*, *71*, 93–109.

Calzà, L., Fernandez, M., & Giardino, L. (2010). Cellular approaches for stimulating CNS remyelination: Thyroid hormone to promote myelin repair via endogenous stem and precursor cells. *Journal of Molecular Endocrinology*, *44*, 13–23.

Calzà, L., Fernandez, M., & Giardino, L. (2015). Role of the thyroid system in myelination and neural connectivity. *Comprehensive Physiology*, *5*, 1405–1421.
Calzà, L., Fernandez, M., Giuliani, A., Aloe, L., & Giardino, L. (2002). Thyroid hormone activates oligodendrocyte precursors and increases a myelin-forming protein and NGF content in the spinal cord during experimental allergic encephalomyelitis. *Proceedings of the National Academy of Sciences of the United States of America*, *99*, 3258–3263.
Calzà, L., Fernandez, M., Giuliani, A., D'Intino, G., Pirondi, S., Sivilia, S., et al. (2005). Thyroid hormone and remyelination in adult central nervous system: A lesson from an inflammatory-demyelinating disease. *Brain Research Reviews*, *48*, 339–346.
Casaccia-Bonnefil, P., & Liu, A. (2003). Relationship between cell cycle molecules and onset of oligodendrocyte differentiation. *Journal of Neuroscience Research*, *72*, 1–11.
Castelo-Branco, G., Stridh, P., Guerreiro-Cacais, A. O., Adzemovic, M. Z., Falcão, A. M., Marta, M., et al. (2014). Acute treatment with valproic acid and l-thyroxine ameliorates clinical signs of experimental autoimmune encephalomyelitis and prevents brain pathology in DA rats. *Neurobiology of Disease*, *71*, 220–233.
Castro, I., Quisenberry, L., Calvo, R. M., Obregon, M. J., & Lado-Abeal, J. (2013). Septic shock non-thyroidal illness syndrome causes hypothyroidism and conditions for reduced sensitivity to thyroid hormone. *Journal of Molecular Endocrinology*, *50*, 255–266.
Chan, S. Y., Hancox, L. A., Martín-Santos, A., Loubière, L. S., Walter, M. N., González, A. M., et al. (2014). MCT8 expression in human fetal cerebral cortex is reduced in severe intrauterine growth restriction. *Journal of Endocrinology*, *220*, 85–95.
Chang, S. C., Thibodeaux, J. R., Eastvold, M. L., Ehresman, D. J., Bjork, J. A., Froehlich, J. W., et al. (2008). Thyroid hormone status and pituitary function in adult rats given oral doses of perfluorooctanesulfonate (PFOS). *Toxicology*, *243*, 330–339.
Chatonnet, F., Flamant, F., & Morte, B. (2015). A temporary compendium of thyroid hormone target genes in brain. *Biochimica et Biophysica Acta*, *1849*, 122–129.
Cheng, S. Y., Leonard, J. L., & Davis, P. J. (2010). Molecular aspects of thyroid hormone actions. *Endocrine Reviews*, *31*, 139–170.
Chew, L. J., King, W. C., Kennedy, A., & Gallo, V. (2005). Interferon-gamma inhibits cell cycle exit in differentiating oligodendrocyte progenitor cells. *Glia*, *52*, 127–143.
Christian, M. S., & Trenton, N. A. (2003). Evaluation of thyroid function in neonatal and adult rats: The neglected endocrine mode of action. *Pure and Applied Chemistry*, *75*, 2055–2068.
Cioffi, F., Senese, R., Lanni, A., & Goglia, F. (2013). Thyroid hormones and mitochondria: With a brief look at derivatives and analogues. *Molecular and Cellular Endocrinology*, *379*, 51–61.
Crawford, A. H., Chambers, C., & Franklin, R. J. (2013). Remyelination: The true regeneration of the central nervous system. *Journal of Comparative Pathology*, *149*, 242–254.
Crawford, A. H., Tripathi, R. B., Richardson, W. D., & Franklin, R. J. (2016). Developmental origin of oligodendrocyte lineage cells determines response to demyelination and susceptibility to age-associated functional decline. *Cell Reports*, 761–773. http://dx.doi.org/10.1016/j.celrep.2016.03.069.
Davis, P. J., Davis, F. B., & Cody, V. (2005). Membrane receptors mediating thyroid hormone action. *Trends in Endocrinology and Metabolism*, *16*, 429–435.
Davis, P. J., Davis, F. B., Mousa, S. A., Luidens, M. K., & Lin, H. Y. (2011). Membrane receptor for thyroid hormone: Physiologic and pharmacologic implications. *Annual Review of Pharmacology and Toxicology*, *51*, 99–115.
Davis, P. J., Goglia, F., & Leonard, J. L. (2016). Nongenomic actions of thyroid hormone. *Nature Reviews Endocrinology*, *12*, 111–121.
De Castro, F., Bribián, A., & Ortega, M. C. (2013). Regulation of oligodendrocyte precursor migration during development, in adulthood and in pathology. *Cellular and Molecular Life Sciences*, *70*, 4355–4368.

de Vries, E. M., Fliers, E., & Boelen, A. (2015). The molecular basis of the non-thyroidal illness syndrome. *Journal of Endocrinology, 225*, R67–R81.

Dell'Acqua, M. L., Lorenzini, L., D'Intino, G., Sivilia, S., Pasqualetti, P., Panetta, V., et al. (2012). Functional and molecular evidence of myelin- and neuroprotection by thyroid hormone administration in experimental allergic encephalomyelitis. *Neuropathology and Applied Neurobiology, 38*, 454–470.

Dentice, M., & Salvatore, D. (2011). Deiodinases: The balance of thyroid hormone: Local impact of thyroid hormone inactivation. *Journal of Endocrinology, 209*, 273–282.

Denver, R. J., & Williamson, K. E. (2009). Identification of a thyroid hormone response element in the mouse Kruppel-like factor 9 gene to explain its postnatal expression in the brain. *Endocrinology, 150*, 3935–3943.

D'Intino, G., Lorenzini, L., Fernández, M., Taglioni, A., Perretta, G., Del Vecchio, G., et al. (2011). T3 administration ameliorates the demyelination/remyelination ratio in a non-human primate model of multiple sclerosis by correcting tissue hypothyroidism. *Journal of Neuroendocrinology, 23*, 778–790.

Donzelli, R., Colligiani, D., Kusmic, C., Sabatini, M., Lorenzini, L., Accorroni, A., et al. (2016). Effect of hypothyroidism and hyperthyroidism on tissue thyroid hormone concentrations in rat. *European Thyroid Journal, 5*, 27–34.

Dugas, J. C., Ibrahim, A., & Barres, B. A. (2012). The T3-induced gene KLF9 regulates oligodendrocyte differentiation and myelin regeneration. *Molecular and Cellular Neuroscience, 50*, 45–57.

Durand, B., Gao, F. B., & Raff, M. (1997). Accumulation of the cyclin-dependent kinase inhibitor p27/Kip1 and the timing of oligodendrocyte differentiation. *EMBO Journal, 16*, 306–317.

Durand, B., & Raff, M. (2000). A cell-intrinsic timer that operates during oligodendrocyte development. *BioEssays, 22*, 64–71.

Elitt, C. M., & Rosenberg, P. A. (2014). The challenge of understanding cerebral white matter injury in the premature infant. *Neuroscience, 276*, 216–238.

Escobar-Morreale, H. F., Obregón, M. J., Escobar del Rey, F., & Morreale de Escobar, G. (1995). Replacement therapy for hypothyroidism with thyroxine alone does not ensure euthyroidism in all tissues, as studied in thyroidectomized rats. *Journal of Clinical Investigation, 96*, 2828–2838.

Falahati, S., Breu, M., Waickman, A. T., Phillips, A. W., Arauz, E. J., Snyder, S., et al. (2013). Ischemia-induced neuroinflammation is associated with disrupted development of oligodendrocyte progenitors in a model of periventricular leukomalacia. *Developmental Neuroscience, 35*, 182–196.

Fattori, J., Campos, J. L., Doratioto, T. R., Assis, L. M., Vitorino, M. T., Polikarpov, I., et al. (2015). RXR agonist modulates TR: Corepressor dissociation upon 9-cis retinoic acid treatment. *Molecular Endocrinology, 29*, 258–273.

Fernández, M., Baldassarro, V. A., Sivilia, S., Giardino, L., & Calzà, L. (2016). Inflammation severely alters thyroid hormone signaling in the central nervous system during experimental allergic encephalomyelitis in rat: Direct impact on OPCs differentiation failure. *Glia, 64*, 1573–1589.

Fernandez, M., Giuliani, A., Pirondi, S., D'Intino, G., Giardino, L., Aloe, L., et al. (2004b). Thyroid hormone administration enhances remyelination in chronic demyelinating inflammatory disease. *Proceedings of the National Academy of Sciences of the United States of America, 101*, 16363–16368.

Fernandez, M., Paradisi, M., Del Vecchio, G., Giardino, L., & Calzà, L. (2009). Thyroid hormone induces glial lineage of primary neurospheres derived from non-pathological and pathological rat brain: Implications for remyelination-enhancing therapies. *International Journal of Developmental Neuroscience, 27*, 769–778.

Fernández, M., Paradisi, M., D'Intino, G., Del Vecchio, G., Sivilia, S., Giardino, L., et al. (2010). A single prenatal exposure to the endocrine disruptor 2,3,7,8-

tetrachlorodibenzo-p-dioxin alters developmental myelination and remyelination potential in the rat brain. *Journal of Neurochemistry*, *115*, 897–909.
Fernández, M., Pirondi, S., Manservigi, M., Giardino, L., & Calzà, L. (2004a). Thyroid hormone participates in the regulation of neural stem cells and oligodendrocyte precursor cells in the central nervous system of adult rat. *European Journal of Neuroscience*, *20*, 2059–2070.
Flamant, F., Baxter, J. D., Forrest, D., Refetoff, S., Samuels, H., Scanlan, T. S., et al. (2006). International union of pharmacology. LIX. The pharmacology and classification of the nuclear receptor superfamily: Thyroid hormone receptors. *Pharmacological Reviews*, *58*, 705–711.
Fliers, E., Bianco, A. C., Langouche, L., & Boelen, A. (2015). Thyroid function in critically ill patients. *The Lancet Diabetes & Endocrinology*, *3*, 816–825.
Franco, P. G., Silvestroff, L., Soto, E. F., & Pasquini, J. M. (2008). Thyroid hormones promote differentiation of oligodendrocyte progenitor cells and improve remyelination after cuprizone-induced demyelination. *Experimental Neurology*, *212*, 458–467.
Franklin, R. J., & Ffrench-Constant, C. (2008). Remyelination in the CNS: From biology to therapy. *Nature Reviews Neuroscience*, *9*, 839–855.
Franklin, R. J., & Hinks, G. L. (1999). Understanding CNS remyelination: Clues from developmental and regeneration biology. *Journal of Neuroscience Research*, *58*, 207–213.
Freitas, B. C., Gereben, B., Castillo, M., Kalló, I., Zeöld, A., Egri, P., et al. (2010). Paracrine signaling by glial cell-derived triiodothyronine activates neuronal gene expression in the rodent brain and human cells. *The Journal of Clinical Investigation*, *120*, 2206–2217.
Fu, J., Refetoff, S., & Dumitrescu, A. M. (2013). Inherited defects of thyroid hormone-cell-membrane transport: Review of recent findings. *Current Opinion in Endocrinology, Diabetes and Obesity*, *20*, 434–440.
Gereben, B., Zavacki, A. M., Ribich, S., Kim, B. W., Huang, S. A., Simonides, W. S., et al. (2008). Cellular and molecular basis of deiodinase-regulated thyroid hormone signaling. *Endocrine Reviews*, *29*, 898–938.
Gilbert, M. E., Rovet, J., Chen, Z., & Koibuchi, N. (2012). Developmental thyroid hormone disruption: Prevalence, environmental contaminants and neurodevelopmental consequences. *Neurotoxicology*, *33*, 842–852.
Gore, A. C., Chappell, V. A., Fenton, S. E., Flaws, J. A., Nadal, A., Prins, G. S., et al. (2015). Executive summary to EDC-2: The Endocrine Society's second scientific statement on endocrine-disrupting chemicals. *Endocrine Reviews*, *36*, 593–602.
Gutleb, A. C., Cambier, S., & Serchi, T. (2016). Impact of endocrine disruptors on the thyroid hormone system. *Hormone Research in Pædiatrics*, *86*, 271–278.
Harsan, L. A., Steibel, J., Zaremba, A., Agin, A., Sapin, R., Poulet, P., et al. (2008). Recovery from chronic demyelination by thyroid hormone therapy: Myelinogenesis induction and assessment by diffusion tensor magnetic resonance imaging. *Journal of Neuroscience*, *28*, 14189–14201.
Harvey, C. B., Bassett, J. H., Maruvada, P., Yen, P. M., & Williams, G. R. (2007). The rat thyroid hormone receptor (TR) Deltabeta3 displays cell-, TR isoform-, and thyroid hormone response element-specific actions. *Endocrinology*, *148*, 1764–1773.
Harvey, C. B., & Williams, G. R. (2002). Mechanism of thyroid hormone action. *Thyroid*, *12*, 441–446.
Hasimoto, K., Curty, F. H., Borges, P. P., Lee, C. E., Abel, E. D., Elmquist, J. K., et al. (2001). An unliganded thyroid hormone receptor causes severe neurological dysfunction. *Proceedings of the National Academy of Sciences of the United States of America*, *98*, 3998–4003.
Heuer, H. (2007). The importance of thyroid hormone transporters for brain development and function. *Best Practice & Research Clinical Endocrinology & Metabolism*, *21*, 265–276.
Heuer, H., & Visser, T. J. (2009). Minireview: Pathophysiological importance of thyroid hormone transporters. *Endocrinology*, *150*, 1078–1083.

Howdeshell, K. L. (2002). A model of the development of the brain as a construct of the thyroid system. *Environmental Health Perspectives, 110*(Suppl. 3), 337–348.

Hung, P. L., Huang, C. C., Huang, H. M., Tu, D. G., & Chang, Y. C. (2013). Thyroxin treatment protects against white matter injury in the immature brain via brain-derived neurotrophic factor. *Stroke, 44*, 2275–2283.

Islam, M. S., Tatsumi, K., Okuda, H., Shiosaka, S., & Wanaka, A. (2009). Olig2-expressing progenitor cells preferentially differentiate into oligodendrocytes in cuprizone-induced demyelinated lesions. *Neurochemistry International, 54*, 192–198.

Jagannathan, N. R., Tandon, N., Raghunathan, P., & Kochupillai, N. (1998). Reversal of abnormalities of myelination by thyroxine therapy in congenital hypothyroidism: Localized in vivo proton magnetic resonance spectroscopy (MRS) study. *Developmental Brain Research, 109*, 179–186.

Jo, S., Kallò, I., Bardoczi, Z., Arrojo e Drigo, R., Zeold, A., Liposits, Z., et al. (2012). Neuronal hypoxia induces Hsp40-mediated nuclear import of type 3 deiodinase as an adaptive mechanism to reduce cellular metabolism. *Journal of Neuroscience, 32*, 8491–8500.

Kang, S. M., Cho, M. S., Seo, H., Yoon, C. j., Oh, S. K., Choi, Y. M., et al. (2007). Efficient induction of oligodendrocytes from human embryonic stem cells. *Stem Cells, 25*, 419–424.

Kang, Z., Wang, C., Zepp, J., Wu, L., Sun, K., Zhao, J., et al. (2013). Act1 mediates IL-17-induced EAE pathogenesis selectively in NG2+ glial cells. *Nature Neuroscience, 16*, 1401–1408.

Kelly, T. (2016). A hypothesis on the mechanism of action of high-dose thyroid in refractory mood disorders. *Medical Hypotheses, 97*, 16–21.

Kuhlmann, T., Miron, V., Cui, Q., Wegner, C., Antel, J., & Brück, W. (2008). Differentiation block of oligodendroglial progenitor cells as a cause for remyelination failure in chronic multiple sclerosis. *Brain, 131*, 1749–1758.

Kyono, Y., Subramani, A., Ramadoss, P., Hollenberg, A. N., Bonett, R. M., & Denver, R. J. (2016). Liganded thyroid hormone receptors transactivate the DNA methyltransferase 3a gene in mouse neuronal cells. *Endocrinology, 157*, 3647–3657.

La Piana, R., Vanasse, M., Brais, B., & Bernard, G. (2015). Myelination delay and Allan-Herndon-Dudley syndrome caused by a novel mutation in the SLC16A2 Gene. *Journal of Child Neurology, 30*, 1371–1374.

Lassmann, H., van Horssen, J., & Mahad, D. (2012). Progressive multiple sclerosis: Pathology and pathogenesis. *Nature Reviews Neurology, 8*, 647–656.

Lee, M. (Ed.), (2013). *Basic skills in interpreting laboratory data.* (5th ed.): Bethesda, MD: American Society of Health-System Pharmacists, ISBN: 1585283436.

Lee, J. Y., & Petratos, S. (2016). Thyroid hormone signaling in oligodendrocytes: From extracellular transport to intracellular signal. *Molecular Neurobiology, 53*, 6568–6583.

Levine, J. M., Reynolds, R., & Fawcett, J. W. (2001). The oligodendrocyte precursor cell in health and disease. *Trends in Neurosciences, 24*, 39–47.

Li, W. W., Le Goascogne, C., Ramaugé, M., Schumacher, M., Pierre, M., & Courtin, F. (2001). Induction of type 3 iodothyronine deiodinase by nerve injury in the rat peripheral nervous system. *Endocrinology, 142*, 5190–5197.

Little, A. G. (2016). A review of the peripheral levels of regulation by thyroid hormone. *Journal of Comparative Physiology B, 186*, 677–688.

Lourbopoulos, A., Mourouzis, I., Karapanayiotides, T., Nousiopoulou, E., Chatzigeorgiou, S., Mavridis, T., et al. (2014). Changes in thyroid hormone receptors after permanent cerebral ischemia in male rats. *Journal of Molecular Neuroscience, 54*, 78–91.

Maruvada, P., Dmitrieva, N. I., East-Palmer, J., & Yen, P. M. (2004). Cell cycle-dependent expression of thyroid hormone receptor-beta is a mechanism for variable hormone sensitivity. *Molecular Biology of the Cell, 15*, 1895–1903.

Marziali, L. N., Garcia, C. I., & Pasquini, J. M. (2015). Transferrin and thyroid hormone converge in the control of myelinogenesis. *Experimental Neurology*, *265*, 129–141.

Mdzinarishvili, A., Sutariya, V., Talasila, P. K., Geldenhuys, W. J., & Sadana, P. (2013). Engineering triiodothyronine (T3) nanoparticle for use in ischemic brain stroke. *Drug Delivery and Translational Research*, *3*, 309–317.

Meijer, D. H., Kane, M. F., Mehta, S., Liu, H., Harrington, E., Taylor, C. M., et al. (2012). Separated at birth? The functional and molecular divergence of OLIG1 and OLIG2. *Nature Reviews Neuroscience*, *13*, 819–831.

Mekhail, M., Almazan, G., & Tabrizian, M. (2012). Oligodendrocyte-protection and remyelination post-spinal cord injuries: A review. *Progress in Neurobiology*, *96*, 322–339.

Menn, B., Garcia-Verdugo, J. M., Yaschine, C., Gonzalez-Perez, O., Rowitch, D., & Alvarez-Buylla, A. (2006). Origin of oligodendrocytes in the subventricular zone of the adult brain. *The Journal of Neuroscience*, *26*, 7907–7918.

Miller, D. J., Duka, T., Stimpson, C. D., Schapiro, S. J., Baze, W. B., McArthur, M. J., et al. (2012). Prolonged myelination in human neocortical evolution. *Proceedings of the National Academy of Sciences of the United States of America*, *109*, 16480–16485.

Miller, R. H., & Mi, S. (2007). Dissecting demyelination. *Nature Neuroscience*, *10*, 1351–1354.

Min, H., Dong, J., Wang, Y., Wang, Y., Teng, W., Xi, Q., et al. (2016). Maternal hypothyroxinemia-induced neurodevelopmental impairments in the progeny. *Molecular Neurobiology*, *53*, 1613–1624.

Mitchell, C. M., Sykes, S. D., Pan, X., Pringle, K. G., Lumbers, E. R., Hirst, J. J., et al. (2013). Inflammatory and steroid receptor gene methylation in the human amnion and decidua. *Journal of Molecular Endocrinology*, *50*, 267–277.

Mohammadi, R., Amini, K., Yousefi, A., Abdollahi-Pirbazari, M., Belbasi, A., & Abedi, F. (2013). Functional effects of local administration of thyroid hormone combined with chitosan conduit after sciatic nerve transection in rats. *Journal of Oral and Maxillofacial Surgery*, *71*, 1763–1776.

Moog, N. K., Entringer, S., Heim, C., Wadhwa, P. D., Kathmann, N., & Buss, C. (2017). Influence of maternal thyroid hormones during gestation on fetal brain development. *Neuroscience*, *342*, 68–100.

Mousa, S. A., Yalcin, M., Bharali, D. J., Meng, R., Tang, H. Y., Lin, H. Y., et al. (2012). Tetraiodothyroacetic acid and its nanoformulation inhibit thyroid hormone stimulation of non-small cell lung cancer cells in vitro and its growth in xenografts. *Lung Cancer*, *76*, 39–45.

Myers, S. A., Bankston, A. N., Burke, D. A., Ohri, S. S., & Whittemore, S. R. (2016). Does the preclinical evidence for functional remyelination following myelinating cell engraftment into the injured spinal cord support progression to clinical trials? *Experimental Neurology*, *283*, 560–572.

Obregon, M. J., Calvo, R. M., Del Rey, F. E., & de Escobar, G. M. (2007). Ontogenesis of thyroid function and interactions with maternal function. *Endocrine Development*, *10*, 86–98.

O'Shea, P. J., & Williams, G. R. (2002). Insight into the physiological actions of thyroid hormone receptors from genetically modified mice. *Journal of Endocrinology*, *175*, 553–570.

Picou, F., Fauquier, T., Chatonnet, F., & Flamant, F. (2012). A bimodal influence of thyroid hormone on cerebellum oligodendrocyte differentiation. *Molecular Endocrinology*, *26*, 608–618.

Porterfield, S. P., & Hendrich, C. E. (1993). The role of thyroid hormones in prenatal and neonatal neurological development—Current perspectives. *Endocrine Reviews*, *14*, 94–106.

Psarra, A. M., Solakidi, S., & Sekeris, C. E. (2006). The mitochondrion as a primary site of action of steroid and thyroid hormones: Presence and action of steroid and thyroid hormone receptors in mitochondria of animal cells. *Molecular and Cellular Endocrinology*, *246*, 21–33.

Purger, D., Gibson, E. M., & Monje, M. (2016). Myelin plasticity in the central nervous system. *Neuropharmacology*, *110*, 563–573.

Raff, M. (2006). The mystery of intracellular developmental programmes and timers. *Biochemical Society Transactions*, *34*, 663–670.

Reuss, M. L., Leviton, A., Paneth, N., & Susser, M. (1997). Thyroxine values from newborn screening of 919 infants born before 29 weeks' gestation. *American Journal of Public Health*, *87*, 1693–1697.

Rodgers, J. M., Robinson, A. P., & Miller, S. D. (2013). Strategies for protecting oligodendrocytes and enhancing remyelination in multiple sclerosis. *Discovery Medicine*, *16*, 53–63.

Rodriguez-Pena, A., Ibarrola, N., Iñiguez, M. A., Muñoz, A., & Bernal, J. (1993). Neonatal hypothyroidism affects the timely expression of myelin-associated glycoprotein in the rat brain. *Journal of Clinical Investigation*, *91*, 812–818.

Rogister, B., Ben-Hur, T., & Dubois-Dalcq, M. (1999). From neural stem cells to myelinating oligodendrocytes. *Molecular and Cellular Neuroscience*, *14*, 287–300.

Rosman, N. P. (1972). The neuropathology of congenital hypothyroidism. *Advances in Experimental Medicine and Biology*, *30*, 337–366.

Schang, A. L., Gressens, P., & Fleiss, B. (2014a). Revisiting thyroid hormone treatment to prevent brain damage of prematurity. *Journal of Neuroscience Research*, *92*, 1609–1610.

Schang, A. L., Van Steenwinckel, J., Chevenne, D., Alkmark, M., Hagberg, H., Gressens, P., et al. (2014b). Failure of thyroid hormone treatment to prevent inflammation-induced white matter injury in the immature brain. *Brain, Behavior, and Immunity*, *37*, 95–102.

Schroeder, A. C., & Privalsky, M. L. (2014). Thyroid hormones, t3 and t4, in the brain. *Frontiers in Endocrinology*, *5*, 40.

Silvestroff, L., Bartucci, S., Pasquini, J., & Franco, P. (2012). Cuprizone-induced demyelination in the rat cerebral cortex and thyroid hormone effects on cortical remyelination. *Experimental Neurology*, *235*, 357–367.

Simonides, W. S., Mulcahey, M. A., Redout, E. M., Muller, A., Zuidwijk, M. J., Visser, T. J., et al. (2008). Hypoxia-inducible factor induces local thyroid hormone inactivation during hypoxic-ischemic disease in rats. *The Journal of Clinical Investigation*, *118*, 975–983.

Sozmen, E. G., Rosenzweig, S., Llorente, I. L., DiTullio, D. J., Machnicki, M., Vinters, H. V., et al. (2016). Nogo receptor blockade overcomes remyelination failure after white matter stroke and stimulates functional recovery in aged mice. *Proceedings of the National Academy of Sciences of the United States of America*, *113*, E8453–E8462. Dec 12. pii: 201615322.

Sterling, K., & Milch, P. O. (1975). Thyroid hormone binding by a component of mitochondrial membrane. *Proceedings of the National Academy of Sciences of the United States of America*, *72*, 3225–3229.

Su, Z., Yuan, Y., Chen, J., Zhu, Y., Qiu, Y., Zhu, F., et al. (2011). Reactive astrocytes inhibit the survival and differentiation of oligodendrocyte precursor cells by secreted TNF-α. *Journal of Neurotrauma*, *28*, 1089–10100.

Tanner, D. C., Cherry, J. D., & Mayer-Proschel, M. (2011). Oligodendrocyte progenitors reversibly exit the cell cycle and give rise to astrocytes in response to interferon-γ. *Journal of Neuroscience*, *31*, 6235–6246.

van Tilborg, E., Heijnen, C. J., Benders, M. J., van Bel, F., Fleiss, B., Gressens, P., et al. (2015). Impaired oligodendrocyte maturation in preterm infants: Potential therapeutic targets. *Progress in Neurobiology, 136*, 28–49.

Verden, D., & Macklin, W. B. (2016). Neuroprotection by central nervous system remyelination: Molecular, cellular, and functional considerations. *Journal of Neuroscience Research, 94*, 1411–1420.

Vose, L. R., Vinukonda, G., Jo, S., Miry, O., Diamond, D., Korumilli, R., et al. (2013). Treatment with thyroxine restores myelination and clinical recovery after intraventricular hemorrhage. *Journal of Neuroscience, 33*, 17232–17246.

Wang, M., Liu, C., Hao, Y., Zhang, Y., & Zhang, X. (2013). Poly (dextrogyr-levogyr) lactide acid-triiodothyronine scaffold for peripheral nerve regeneration. *Neuroendocrinology Letters, 34*, 62–70.

Wang, B., Liu, S., Li, L., Yao, Q., Song, R., Shao, X., et al. (2017). Non-thyroidal illness syndrome in patients with cardiovascular diseases: A systematic review and meta-analysis. *International Journal of Cardiology, 226*, 1–10.

Wang, S., & Young, K. M. (2013). White matter plasticity in adulthood. *Neuroscience, 276*, 148–160.

Whittemore, S. R., Morassutti, D. J., Walters, W. M., Liu, R. H., & Magnuson, D. S. (1999). Mitogen and substrate differentially affect the lineage restriction of adult rat subventricular zone neural precursor cell populations. *Experimental Cell Research, 252*, 75–95.

Yalcin, M., Dyskin, E., Lansing, L., Bharali, D. J., Mousa, S. S., Bridoux, A., et al. (2010). Tetraiodothyroacetic acid (tetrac) and nanoparticulate tetrac arrest growth of medullary carcinoma of the thyroid. *The Journal of Clinical Endocrinology & Metabolism, 95*, 1972–1980.

Yen, P. M. (2001). Physiological and molecular basis of thyroid hormone action. *Physiological Reviews, 81*, 1097–1142.

Young, K. M., Psachoulia, K., Tripathi, R. B., Dunn, S. J., Cossell, L., Attwell, D., et al. (2013). Oligodendrocyte dynamics in the healthy adult CNS: Evidence for myelin remodeling. *Neuron*, 77, 873–885.

Zhang, M., Ma, Z., Qin, H., & Yao, Z. (2015a). Thyroid hormone potentially benefits multiple sclerosis via facilitating remyelination. *Molecular Neurobiology, 53*, 4406–4416.

Zhang, J., Zhang, Z. G., Li, Y., Ding, X., Shang, X., Lu, M., et al. (2015b). Fingolimod treatment promotes proliferation and differentiation of oligodendrocyte progenitor cells in mice with experimental autoimmune encephalomyelitis. *Neurobiology of Disease, 76*, 57–66.

Zhu, X., Zuo, H., Maher, B. J., Serwanski, D. R., LoTurco, J. J., Lu, Q. R., et al. (2012). Olig2- dependent developmental fate switch of NG2 cells. *Development, 139*, 2299–2307.

Zoeller, R. T. (2007). Environmental chemicals impacting the thyroid: Targets and consequences. *Thyroid*, 17, 811–817.

Zoeller, R. T., & Crofton, K. M. (2005). Mode of action: Developmental thyroid hormone insufficiency—Neurological abnormalities resulting from exposure to propylthiouracil. *Critical Reviews in Toxicology, 35*, 771–781.

FURTHER READING

Barca-Mayo, O., & Lu, Q. R. (2012). Fine-tuning oligodendrocyte development by microRNAs. *Frontiers in Neuroscience, 6*, 13.

Dong, H., You, S. E., Williams, A., Wade, M. G., Yauk, C. L., & Zoeller, R. T. (2015). Transient maternal hypothyroxinemia potentiates the transcriptional response to exogenous thyroid hormone in the fetal cerebral cortex before the onset of fetal thyroid function: A messenger and MicroRNA profiling study. *Cerebral Cortex*, *25*, 1735–1745.

Dugas, J. C., Cuellar, T. L., Scholze, A., Ason, B., Ibrahim, A., Emery, B., et al. (2010). Dicer1 and miR-219 are required for normal oligodendrocyte differentiation and myelination. *Neuron*, *65*, 597–611.

Letzen, B. S., Liu, C., Thakor, N. V., Gearhart, J. D., All, A. H., & Kerr, C. L. (2010). MicroRNA expression profiling of oligodendrocyte differentiation from human embryonic stem cells. *PLoS One*, *5*, e10480.

CHAPTER TWELVE

Thyroid Hormone and Astrocyte Differentiation

Moitreyi Das, Mausam Ghosh, Kusumika Gharami, Sumantra Das[1]
Indian Institute of Chemical Biology, Kolkata, India
[1]Corresponding author: e-mail addresses: sdas@iicb.res.in; sumantra00@gmail.com

Contents

1. Introduction 283
2. Morphology of Astrocytes During Brain Development 285
3. Morphology of Astrocytes in Diseases 287
4. Regulators of Astrocyte Morphology 288
5. THs on Astrocyte Function 291
6. Effect of THs on Astrocyte Morphogenesis 293
7. Mechanism of TH Action on Astrocyte Morphogenesis 297
8. Conclusion 305
References 306

Abstract

Thyroid hormones (THs) have important contributions to the development of the mammalian brain, targeting its actions on both neurons and glial cells. Astrocytes, which constitute about half of the glial cells, characteristically undergo dramatic changes in their morphology during development and such changes become necessary for the proper development of the brain. Interestingly, a large number of studies have suggested that THs play a profound role in such morphological maturation of the astrocytes. This review discusses the present knowledge on the mechanisms by which THs elicit progressive differentiation and maturation of the astrocytes. As a prelude, information on astrocyte morphology during development and its regulations, the role of THs in the various functions of astrocyte shall be dealt with for a thorough understanding of the subject of this review.

1. INTRODUCTION

Thyroid hormones (THs) are essential prerequisites for the proper development of the mammalian brain. Impairment of neuronal connections, cognitive deficits, and mental disorders have been linked to a deficiency of

Vitamins and Hormones, Volume 106
ISSN 0083-6729
http://dx.doi.org/10.1016/bs.vh.2017.05.004

THs during childhood. For a long time, the major focus of research on the effects of THs in the brain centered on the neurons, and it was observed that THs played key roles in all the important stages of neurogenesis like proliferation, migration, synaptogenesis, myelinogenesis, etc. Studies on the effect of the hormones on astrocytes were initiated much later as these cells were traditionally been considered as playing a supportive role for neurons.

Astrocytes, however, are the most abundant cells in the mammalian brain. While both neuronal and glial cells are developed from common neuroepithelial (NE) cells, neuronal cells appear first after which the two types of glial cells, astrocytes and oligodendrocytes, are differentiated from these multipotential and self-renewable progenitor cells (Bayer & Altman, 1991; Davis & Temple, 1994). A plethora of information on newer roles of astrocytes has made scientists reconsider these cells to be as important as the neuron in maintaining major functions of the brain in nerve tissue growth and morphogenesis, behavior, survival, and ability to protect against toxic insults. The important functions of astrocytes include providing structural support to neuron; providing functional neuron–glia–vascular microenvironment; providing metabolic support to neuron by acting as the glycogen depot of the brain; maintaining blood–brain barrier; regulating neurogenesis, neuronal development (growth factors secreted by astrocytes), and maturation (radial glia provide axonal guidance); secreting and absorbing gliotransmitters important for brain signaling; controlling extracellular homeostasis and K^+ ionic balance in the brain which is important in water homeostasis in the brain; regulating formation, maturation, maintenance, and stability of synapses and synaptic transmission; and supporting immune defense in the brain, to name a few (Allaman, Bélanger, & Magistretti, 2011; Araque et al., 2014; De Pitta, Brunel, & Volterra, 2016; Freeman, 2010; Janzer & Raff, 1987; Perea, Navarrete, & Araque, 2009; Santello, Cali, & Bezzi, 2012; Wiese, Karus, & Faissner, 2012). Astrocyte–neuron cross talk is an inevitable occurrence in all CNS functions and any aberrations may lead to neurological disorders (Blanco-Suárez, Caldwell, & Allen, 2017; Fields, Woo, & Basser, 2015; Zorec, Parpura, Vardjan, & Verkhratsky, 2017). Astrocytes regulate the formation of dendritic spines in neurons, and evidence indicate that by releasing D-serine, astrocytes support the dendritic maturation and survival of neurons that cross their territories (Sultan et al., 2015). Astrocytes have been implicated in higher functions like sleep homeostasis (Halassa et al., 2009), memory consolidation (Gibbs, Hutchinson, & Hertz, 2008), and in the regulation of breathing (Gourine et al., 2010), and recent evidence have demonstrated that astrocytes can also release functional mitochondria that enter neurons (Hayakawa et al., 2016).

This chapter shall review the effects of THs on the multifarious functions of astrocytes, in general, and the various findings of the role of the hormones on astrocyte morphology, in particular.

2. MORPHOLOGY OF ASTROCYTES DURING BRAIN DEVELOPMENT

Historically, it was in 1872 that Camillo Golgi first described the morphology of glial cells by using the silver–chromate technique observing some cells displaying endfeet on their processes, attached to the blood vessels. In 1893, Michael von Lenhossek coined the term "astrocyte," based on the morphology displayed by these cells. These astroglial cells undergo changes in morphology during development (Fedoroff, 1986), and it has long been known that these various shapes appear to be highly correlated with specific neuronal behavior and positioning. Elongated forms, such as radial glia in embryonic brain or Bergmann glia in the postnatal cerebellum, guide migration of postmitotic neurons (Rakic, 1971; Sidman & Rakic, 1973). The multipolar stellate forms found in the white and gray matter of the mature nervous system harbor neuronal perikarya, axons, and synaptic arrangements (Palay & Chan-Palay, 1974). Astrocyte shape conversion plays a role in guiding migration and axonal growth of neurons during embryonic development (Mason, Edmondson, & Hatten, 1988).

In the developing cerebral cortex and the hippocampus of the mouse, glial fibrillary acidic protein (GFAP)-positive radial glia have been shown to appear in association with migrating neurons in the later stages of cortical plate formation (Woodhams, Bascó, Hajós, Csillág, & Balázs, 1981). Ontogenic studies of the postnatal development of astrocytes in the CA1 region of rat hippocampus showed that major proliferation of astrocytes occurred between postnatal days 8 and 16. Astrocytes became larger during the first three postnatal weeks, after which there appeared to be a reduction of the size and branching pattern until adulthood when the cells became polarized (Nixdorf-Bergweiler, Albrecht, & Heinemann, 1994). The impregnation techniques using silver or Golgi stain and immunohistochemical techniques using astrocyte markers used for studying the morphology of astrocytes *in vivo* during early years have now been replaced by more improved imaging techniques supported by superresolution microscopes and improved stains, which has enabled better viewing of the cells in the brain. Based on morphology, astrocytes, throughout the brain, can be classified into nine separate groups (Emsley & Macklis, 2006). The protoplasmic astrocytes in the gray matter differ distinctively from the fibrous astrocytes of the white matter. In rodents, cell bodies of fibrous astrocytes are small and their processes are aligned with the myelinated fibers, giving an elongated morphology while the protoplasmic astrocytes are typically larger and have more sheath-like and branched processes (Lundgaard, Osório, Kress, Sanggaard, & Nedergaard, 2014). Studies using confocal imaging combined with immunolabeling and dye-filling techniques have demonstrated complex morphological features of protoplasmic astrocytes having bushy or spongiform morphology with both primary and secondary processes (Bushong, Martone, & Ellisman, 2004; Ogata & Kosaka, 2002). During early postnatal development, the astrocytes extend stringy, filopodial processes and later on the fine, spongiform processes appear. Human astrocytes are much more complex and diverse, and their morphology is drastically different from rodent and primate astrocytes. In the human neocortex, several anatomically defined subclasses of astrocytes have been identified using confocal laser scanning microscopy-based immunocytochemistry. In addition to fibrous and protoplasmic astrocytes, a population of astrocytes were observed to reside in layers 5–6 and extend long fibers characterized by regularly spaced varicosities. Another specialized type of astrocyte, the interlaminar astrocyte, abundantly populates the superficial cortical layers and extends long processes without varicosities to cortical layers 3 and 4 (Oberheim et al., 2009). Moreover, the protoplasmic astrocytes in the human neocortex were much larger in diameter and extend manyfold more primary processes than those in rodents (Oberheim et al., 2009).

The high degree of plasticity of astrocytes *in vivo* to carry out varied functions in the brain throughout life has been mimicked using *in vitro* cultures. One of the earliest *in vitro* studies highlighting the complex shapes of astrocytes was based on a progressive culture of cerebellar astroglia (Mason et al., 1988). Two principal astroglial forms were observed in these studies, viz., stellate astrocytes and bipolar cells having fewer, longer processes reminiscent of Bergmann glia *in vivo*. With time, while the processes of the stellate cells only slightly increased in their length, they were strikingly increased in the bipolar cells. The stellate astrocytes participate in the formation of a neuronal network by providing neurotrophic support (Hatten et al., 1988). Astroglia cultured from adult human cerebral temporal lobe exhibited flat, bipolar or multipolar shape, and variable size with the presence of long (600 μm) processes with few branches or only terminal branches in some cells (Davies, Niesman, Boop, & Phelan, 2000). Using astrocytes from the neocortex of newborn rat pups, a number of laboratories including ours have demonstrated the morphological changes associated with the maturation of the astrocytic cells in primary culture (Gavaret, Toru-Delbauffe, Baghdassarian-Chalaye, Pomerance, & Pierre, 1991; Lim, Mitsunobu, & Li, 1973; Moonen, Cam, Sensenbrenner, & Mandel, 1975; Paul, Das, Poddar, & Sarkar, 1996). These studies suggest that astrocytes, in culture, undergo progressive differentiation, over a period of 10–15 days, initially from immature radial glia into polygonal cells with flat epithelioid morphology with large cell body and no processes and finally into mature single long process-bearing stellate form.

How are such changes of astrocyte morphology accomplished? Investigations along these lines have unequivocally identified the functional role of the cell cytoskeletons involving the microtubule and stress fibers during shape conversion of astrocytes derived from murine (Goldman & Abramson, 1990; Koyama & Baba, 1996) and human (Liu et al., 1994). Ultrastructural studies suggest that the microtubules are progressively reduced in number during astrocyte differentiation and were gradually replaced by intermediate filaments like GFAP (Trimmer, Reier, Oh, & Eng, 1982). While the levels of both the IF proteins, vimentin and GFAP, increased dramatically during dBcAMP-induced differentiation of astrocytes, the actin content exhibited a decrease (Goldman & Chiu, 1984). GFAP phosphorylation/dephosphorylation induces changes in the shape of astrocytes (Gottfried et al., 1999).

3. MORPHOLOGY OF ASTROCYTES IN DISEASES

Astrocytes respond to insults through a change in their morphology, often referred as astrogliosis, and the cells are termed as reactive astrocytes. A pathological hallmark of many diseases is the formation of the reactive astrocytes. Such conversion of astrocyte morphology has also been demonstrated when brains are damaged by hypoxia, ischemia, seizures (Eng & Ghirnikar, 1994; Mathewson & Berry, 1985; Nathaniel & Nathaniel, 1981; Norton, Aquino, Hozumi, Chiu, & Brosnan, 1992), Alzheimer's disease, or Down syndrome (Jorgensen, Brooksbank, & Balazs, 1990). These so-called reactive astrocytes have more prominent processes with more branching than normal astrocytes (Shao & McCarthy, 1994). Morphologically reactive astrocytes look closely similar to normal stellate astrocytes, except that these appear to be hypertrophied, having increased expression

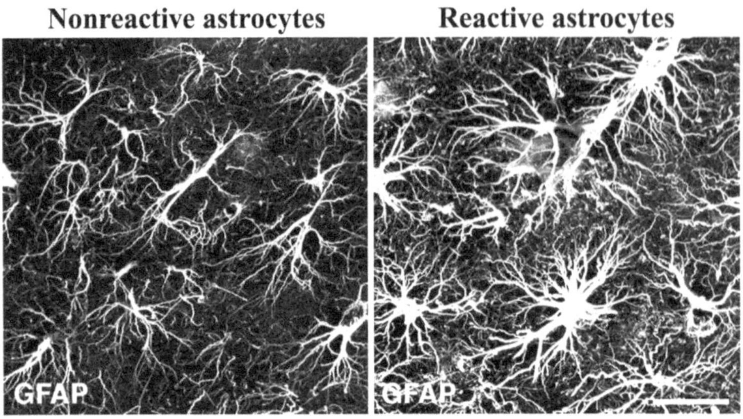

Fig. 1 *Right panel*: Reactive gliosis after entorhinal cortex lesion in the hippocampus. *Left panel*: Nonreactive astrocytes on the contralateral side. Antibodies against GFAP used to visualize bundles of intermediate filaments predominantly found in the soma and the main cellular processes of astrocytes. *Courtesy: Wilhelmsson, U., Bushong, E.A., Price, D.L., Smarr, B.L., Phung, V., Terada, M., et al., 2006. Redefining the concept of reactive astrocytes as cells that remain within their unique domains upon reaction to injury.* Proceedings of the National Academy of Sciences of the United States of America, 103, *17513–17518, Copyright (2006) National Academy of Sciences, USA.*

of GFAP as well as the number and length of GFAP-positive processes, compared to the stellate cells (Fig. 1).

Much effort has been directed in recent years to understand the functions and mechanisms of reactive astrogliosis to encounter a pathological condition. Accumulated evidence indicates that astrogliosis may result in either loss of function or gain of abnormal function which may adversely affect the CNS leading to diseases normal CNS (Sofroniew & Vinters, 2010).

4. REGULATORS OF ASTROCYTE MORPHOLOGY

Considerable efforts have been directed toward delineating the mechanism for shape conversion of brain astrocytes cultured from neonatal brains (Baba, 1998; Shain, Bausback, Fiero, Madelian, & Turner, 1992; Shao & McCarthy, 1994). Since stellate astrocytes resemble differentiated astrocytes *in vivo*, the stellation response in culture has been widely studied as an *in vitro* model of astrocyte differentiation.

The extracellular environment has crucial contributions toward the maintenance of astrocyte morphology (Gottfried, Cechin, Gonzalez,

Vaccaro, & Rodnight, 2003). The glycosaminoglycan, hyaluronan, a major component of extracellular matrix (ECM) proteins in the brain, actively regulates astrocyte morphology through downstream signaling via the small GTPase of the Rho family, Rac1 (Konopka et al., 2016). Enzymatic cleavage of the protein or knockdown of its receptor, CD44, in the astrocytes results in the stellate conformation of the cells, while overexpression of CD44 causes an increase in cell body size and changes the shape of astrocytes into flattened cells. The extensive restructuring of the cytoskeleton, involving both microtubules and microfilaments, and the influence of Rac1 in the stellation process have been reported by others (John et al., 2004; Racchetti, D'Alessandro, & Meldolesi, 2012). By some elegant studies, it has been demonstrated that inhibition of endogenous Rac1 and RhoA GTPases can be triggered by the activation of c-Ha-Ras GTPase when astrocytes are exposed to basic fibroblast growth factor (bFGF) (Kalman, Gomperts, Hardy, Kitamura, & Bishop, 1999). This results in actin reorganization in the cell soma and actin polymerization in the growing processes, thereby facilitating growth of the astrocytic processes.

Of the various growth factors, while epidermal growth factor (EGF) or nerve growth factor regulates proliferation of astrocytes (Cragnolini, Huang, Gokina, & Friedman, 2009; Mayer, Rössler, Endo, Charnay, & Thiel, 2009), bFGF induces process formation in the cells and has been extensively studied. In addition to inhibition of Rac1, described earlier, subsequent studies revealed opposing roles of ERK and p38 MAP kinases in bFGF-induced process formation (Heffron & Mandell, 2005). Bone morphogenetic proteins also influence the morphology of astrocytes by downregulating EGF receptors (EGFRs) (Scholze, Foo, Mulinyawe, & Barres, 2014).

Increasing the intracellular cAMP levels in cultured astrocytes by treatments with forskolin, dibutyryl cAMP (dBcAMP) can convert flat polygonal-shaped cells into process-bearing stellate cells (Abe & Saito, 1997; Moonen et al., 1975). Considerable regional heterogeneity exists in the astrocytes with regard to its response to morphological changes in response to cAMP stimulation (Won & Oh, 2000). While high intracellular c-AMP concentrations in astrocyte cultures, derived from most of the brain regions, caused 95% of the cells to undergo stellation, only 56% of the hypothalamic astrocytes were transformed into stellate morphology. At lower intracellular cAMP concentrations, cell stellation most effectively occurred in cerebellar astrocytes. Stellation induced by dBcAMP is associated with a drastic decrease in the phosphorylation of myosin light chains as well as with a decrease in an actin-depolymerizing factor, resulting in a partial disassembly of the actin cytoskeleton (Baorto, Mellado, & Shelanski, 1992; Padmanabhan & Shelanski, 1998). β-Adrenergic receptor (β-AR) stimulation by the agonist, isoproterenol (ISP), also induces intracellular cAMP levels and facilitate stellation (Abe & Saito, 1997; Gharami & Das, 2000; Hatton, Luckman, & Bicknell, 1991) which could be suppressed by ATP (Abe & Saito, 1999). ATP, by itself, induced stellation by breaking down into adenosine, which stimulated P1 purinoceptors (Abe & Saito, 1999). Using primary cultures of mouse astrocytes, it has been demonstrated that at low concentrations, ISP acts on β_2 receptors to activate protein kinase A (PKA) activity which is required for Gs/Gi switching (Du et al., 2010). This is followed by Ca^{2+} release from intracellular stores and $G_{i\alpha}$- and metalloproteinase-dependent transactivation of the EGFR, resulting in β-arrestin 1/2 recruitment and MAPK/ERK kinase-dependent ERK1/2 phosphorylation. However, the inability of the specific PKA blocker, Rp-cAMP, to prevent stellation induced by the intracellular c-AMP inducing agents, forskolin or pituitary adenylate cyclase-activating polypeptide, questioned the role of PKA (Perez, Bouschet, Fernandez, Bockaert, & Journot, 2005). It appeared from the investigations carried out in our laboratory that ISP induced an immediate activation of PKA for a short duration which was sufficient to drive the process of differentiation of the astrocytes to completion (Gharami & Das, 2004). Furthermore, ISP treatment caused a biphasic response of cellular p-ERK levels marked by an initial decline followed by a sustained induction at a later stage and the latter appeared to be critical for the differentiation of the astrocytes. Moreover, exposure to ISP altered the expression of actin and its intracellular organization in a way conducive to the morphological transformation of the cells (Gharami & Das, 2000). Compounds acting on common signal transduction pathways have been used to delineate the signaling mechanisms involved in astrocyte differentiation. Both the membrane-permeable cyclic AMP analog dBcAMP or the protein kinase C (PKC) activator phorbol 12-myristate 13-acetate (PMA) caused transformation of astrocytes having flattened, polygonal morphology into process-bearing stellate cells (Abe & Saito, 2000). However, U0126 and PD98059, specific inhibitors of the MAPK-activating enzyme MEK, selectively prevented the PMA-induced astrocyte stellation, suggesting the involvement of MEK/ERK cascade in both induction and maintenance of astrocyte stellation mediated by PKC (Abe & Saito, 2000). The small GTP-binding protein Rho is involved in the regulation of cell adhesion and morphology in various types of cells, and studies show that Rho kinases are in an activated state in the absence of stimuli and contribute to the maintenance of polygonal morphology of cultured astrocytes (Abe & Misawa, 2003). Pituicyte, a modified astrocyte in the neurohypophysis, undergo changes in morphology from flat to stellate shape when treated with adenosine, isoproterenol, and dBcAMP accompanied with rearrangement of the stress fibers and microtubules (Miyata, Furuya, Nakai, Bun, & Kiyohara, 1999). Pretreatment with the microtubule-disrupting agent, colchicines, and the tyrosine phosphatase inhibitor, orthovanadate, prevented the dBcAMP-induced stellation. Sphingosine, which inhibits PKC; W-7, which inhibits calmodulin-dependent protein kinase; ML-9, which inhibits myosin light chain kinase; and cytochalasin B, which disrupts microfilament, all of these induced stellation of the pituicyte astrocytes.

5. THS ON ASTROCYTE FUNCTION

Astrocytes are the principal transporters of thyroxine from the blood. Unlike deiodinase 3, which degrades T_4 and T_3 to rT_3 and T_2, respectively, and is selectively expressed in neurons (Tu et al., 1999), the enzyme deiodinase 2, which is required to generate the active T_3 from T_4, is expressed mainly in astrocytes throughout the brain (Guadaño-Ferraz, Obregón, St-Germain, & Bernal, 1997) and is considered as one of the marker genes of astrocytes (Cahoy et al., 2008). Thus, astrocytes function as the major supplier of the biologically active form of the hormone to the neural tissues.

Like neurons, astrocytes too possess TH receptors (TRs) as suggested by *in vitro* studies (Ortiz-Caro, Yusta, & Montiel, 1986; Puymirat, 1992). TR has also been shown to be present in glial nuclei, purified from postnatal rat brain (Hubank, Sinha, Gullo, & Ekins, 1990). At the initial period, some controversy was generated because immunofluorescence localization of the TRs using specific antisera to TR isoforms, TR alpha 1, TR beta 1, and TR beta 2, showed lack of presence of TR isoforms in astroglial cells (Carlson, Strait, Schwartz, & Oppenheimer, 1994), but this has not been corroborated by other studies (Carlson, Strait, Schwartz, & Oppenheimer, 1996; Lebel, Lherault, Dussault, & Puymirat, 1993; Leonard, Farwell, Yen, Chin, & Stula, 1994; Luo, Puymirat, & Dussault, 1989).

Early studies to identify a putative role of THs on astrocytes came up with a number of observations. THs were found to influence the transport of glucose across plasma membranes of astrocytes (Roeder, Williams, &

Tildon, 1985). THs are reported to contribute to astrocyte development by transcriptionally regulating the astroglial protein, GFAP, as observed in cultures prepared from neocortex and cerebellum (Andres-Barquin, Fages, Le Prince, Rolland, & Tardy, 1994). Our laboratory has demonstrated that THs regulate the level of β-AR in astrocytes of the developing brain (Das & Paul, 1994). Hypothyroidism resulted in a decline in the β-AR in astrocytes prepared from rat brain or in primary astrocyte cultures.

THs play a key role in astrocyte–neuron interactions. Neural progenitors cells when cultured onto T_3-treated astrocyte monolayers increased neuronal population followed by enhanced neuronal maturation, arborization, and neurite outgrowth, suggesting a key mediatory role of astrocytes in the action of THs on neuronal development (Dezonne et al., 2013). Astrocytes secrete a large number of proteins, and the release of some of these is augmented by T_3 as observed in the culture medium (Gavaret et al., 1991; Trentin, Alvarez-Silva, & Moura Neto, 2001). Conditioned medium obtained from T_3-treated astrocytes significantly increased cerebellar neuronal population in culture, and further studies suggested that T_3 induced astrocytes to secrete growth factors like tumor necrosis factor-beta and EGF that can interfere with neuronal proliferation via a paracrine pathway (Gomes, Maia, de Menezes, & Neto, 1999). Increased release of mitogenic factors like fibroblast growth factor-2 (FGF2) by T_3, induced production of the ECM proteins, laminin, and fibronectin by the astrocytes, which could attribute to the increase in the number of cerebellar neurons and longer neurites (Trentin, De Aguiar, Garcez, & Alvarez-Silva, 2003). Similarly, bFGF and neurotrophin-3, which are released from astrocytes by exposure to THs, influence each other to alter hippocampal neuronal excitability by enhancing Na^+ current density (Igelhorst, Niederkinkhaus, Karus, Lange, & Dietzel, 2015). Observations that THs induce EGF secretion to promote neuronal migration through Bergmann glia elongation may be an explanation for the migration deficits of granule cells of the cerebellum during hypothyroidism (Martinez, Eller, Viana, & Gomes, 2011).

T_3 significantly enhanced glutamate uptake by cerebellar astrocytes' transcriptional activation of the glutamate transporters, GLAST and GLT-1. Such an increase in astrocytic glutamate improves the astrocytic microenvironment, thereby protecting both astrocytes and neurons against glutamate toxicity (Mendes-de-Aguiar et al., 2008). Studies from this laboratory have demonstrated that THs protect the brain from oxidative stress by activating glutamate cysteine ligase in astrocytes, resulting in a rapid increase in the level of glutathione (Dasgupta, Das, & Sarkar, 2007). In another study, we have shown that alteration of glutathione homeostasis by THs protects astrocytes from morphine by regulating nitric oxide and pERK1/2 pathways in the cells (Deb & Das, 2011).

THs also regulate astrocytic activity in pathophysiological conditions. Studies demonstrate that fatty acid oxidation and ATP production are stimulated by T_3 in astrocytes, resulting in neuroprotection after cerebral ischemic stroke (Sayre et al., 2017). Apolipoprotein E (Apo E) protein is involved in lipid metabolism and is associated with several neurodegenerative disorders including Alzheimer's disease. It has been shown that THs upregulate gene expression of Apo E in astrocytes by facilitating the binding of the TRβ/RXRα complex to the regulatory element ME.2 of the Apo E gene (Roman et al., 2015).

6. EFFECT OF THS ON ASTROCYTE MORPHOGENESIS

The profound effect of THs on morphological changes in astrocytes stemmed from a number of studies. *In vivo*, thyroidal regulation of astrocyte number and maturation of Bergmann glia in the cerebellum have been reported (Clos & Legrand, 1973; Clos, Legrand, & Legrand, 1980). *In vitro* studies with cerebellar glial-enriched cultures demonstrated that T_3 treatment increases the proportion of Bergmann-like cells to velate-like astrocytes (Messer, Snodgrass, & Maskin, 1985). In rat neocortex also, exposure to T_3 accelerated the morphological and biochemical differentiation of astrocytes (Aizenman & de Vellis, 1987; Gavaret et al., 1991; Gould, Frankfurt, Westlind-Danielsson, & McEwen, 1990). Studies on Zucker lean and Zucker diabetic fatty rats suggest that while diabetes had no effect on the morphology of astrocytes of the hippocampal CA1 region and dentate gyrus, hypothyroidism during the onset of diabetes prominently reduced the processes of astrocytes in these regions (Nam et al., 2013).

We undertook detailed studies and observed the progression of morphological changes in the astrocytes cultured under normal and TH-deficient, hypothyroid conditions (Paul et al., 1996). Two distinct stages of morphological transformation were observed in the normal astrocyte cultures. First, the radial glia at day 5 transformed into a flat epithelioid polygonal form at day 7 and finally to the mature process-bearing stellate form by day 10 and continued to remain in this form thereafter. However, under TH-deficient conditions (hypothyroid), the initial transformation to the flat polygonal form was delayed, appearing at day 10 of culture, while the final transformation to the mature process-bearing form was completely arrested (Fig. 2). Interestingly, unlike such changes in morphology induced by T_3 in primary astrocyte cultures from newborn rat cerebral hemispheres, T_3 induced proliferation in cerebellar astrocyte cultures (Lima, Goncalves, Gomes, de Freitas, & Moura Neto, 1998; Lima, Trentin, Rosenthal, Chagas, & Moura Neto, 1997; Trentin, Gomes, Lima, & Moura Neto, 1998; Trentin & Moura Neto, 1995; Trentin, Rosenthal, & Moura Neto, 1995). Seeding density and THs appear to be pivotal factors in the development of rat cerebral astrocyte cultures of embryonic origin (Adachi et al., 2002).

Ultrastructural studies have demonstrated that neonatal TH treatment resulted in a premature disappearance of vimentin-immunoreactive radial glia with a concomitant appearance of GFAP-immunoreactive astrocytes with a mature morphological characteristic in the basal forebrain and hippocampus, regions where the hormone also exerts influence on the neurons (Gould et al., 1990). While in the euthyroid animals, the differentiated astrocytes appeared in the superior medullary velum by postnatal day 2 and reached the white matter and internal granular layer by postnatal day 4, in the hypothyroid rats; however, a lack of differentiated astrocytes in the internal granular layer was observed at postnatal day 4 and most of the differentiated astrocytes remained confined to the white matter. Overall, TH deficiency induced a delay and a partial arrest of astrocyte differentiation (Manzano, Bernal, & Morte, 2007).

Alteration in astrocyte morphology is invariably associated with changes in actin dynamics and reorganization of actin filaments as described earlier. Using FITC-phalloidin staining, we observed that astrocytes cultured under hypothyroid displayed a flat polygonal structure with prominent stress fibers or actin cables running across the cytoplasm from the proximal to the distal end of the cell body (Fig. 3A, right panel) which remained unaltered at day 20 except that the number and thickness of the fibers appeared to increase (Paul et al., 1996). On the other hand, astrocytes cultured under normal conditions displayed a flat epithelioid morphology and possessed stress fibers or actin cables only until day 7 (Fig. 3A, left panel). At day 10, when these cells underwent transformation into the mature process-bearing form, visible reorganization of the cytoskeletal actin filaments took place and the normal astrocytes at day 20 displayed filaments arising mostly from the perinuclear region and extending into the cell processes (Fig. 3A, left panel), which became progressively longer with maturation. Use of an improved DNase I assay to quantitate the various forms of intracellular actin (Paul, Das, & Sarkar, 1992) demonstrated alterations in the level of cytoskeletal

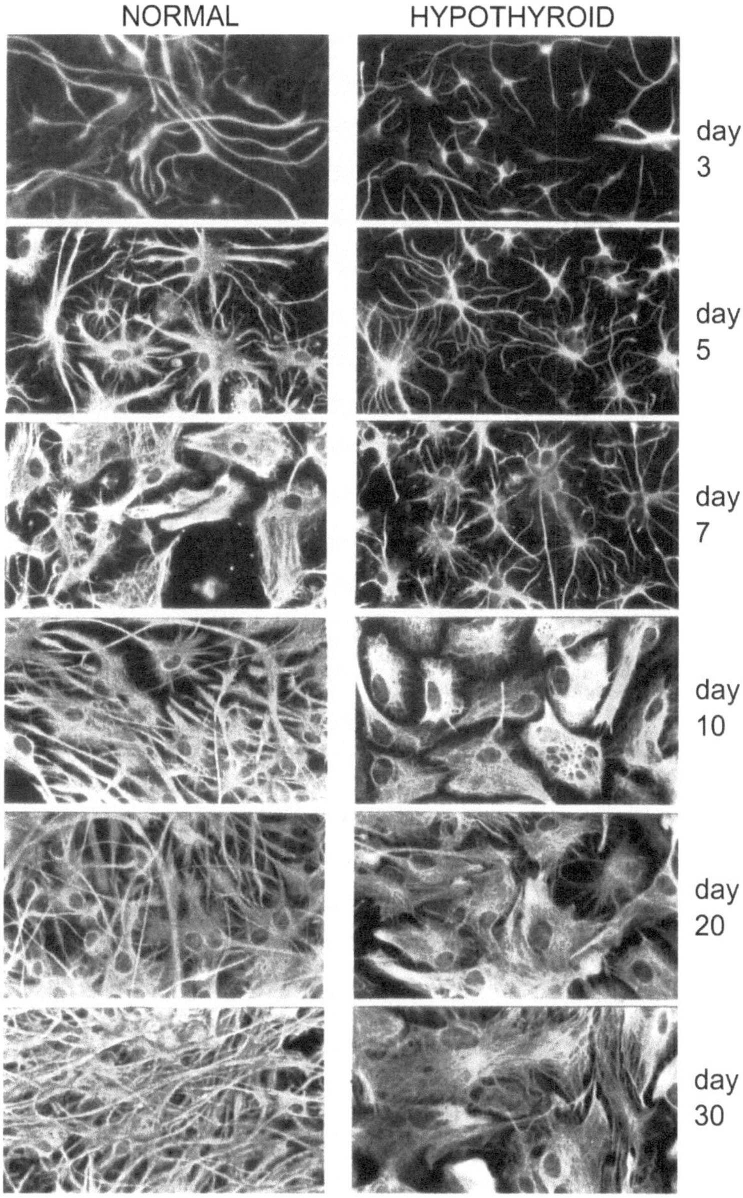

Fig. 2 Morphology of normal and hypothyroid astrocytes from cerebra of neonatal normal and hypothyroid rats, respectively, and cultured in normal and TH-deficient serum-containing medium, respectively. Cells, cultured for 3, 5, 7, 10, 20, and 30 days, were viewed after immunofluorescent staining using anti-GFAP. *Courtesy: Paul, S., Das, S., Poddar, R., Sarkar, P.K., 1996. Role of thyroid hormone in the morphological differentiation and maturation of astrocytes: Temporal correlation with synthesis and organization of actin.* European Journal of Neuroscience, 8, *2361–2370, Copyright (1996) John Wiley & Sons, Inc.*

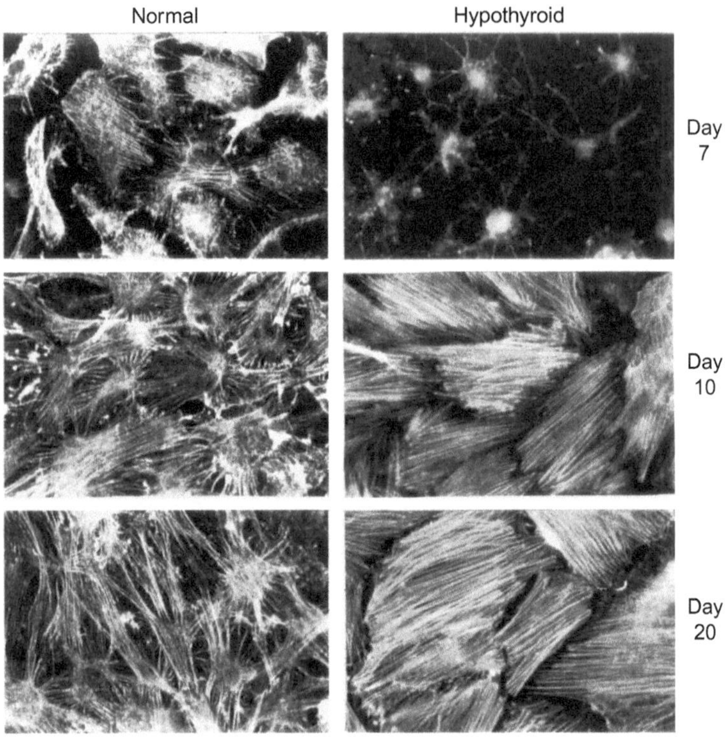

Fig. 3 Organization of the cytoskeletal actin filaments of astrocytes cultured under normal (*left panel*) and TH-deficient conditions (*right panel*) at days 7, 10, and 20. The cytoskeletal actin filaments were visualized by FTTC-phalloidin. Magnification, ×400. *Courtesy: Paul, S., Das, S., Poddar, R., Sarkar, P.K., 1996. Role of thyroid hormone in the morphological differentiation and maturation of astrocytes: Temporal correlation with synthesis and organization of actin.* European Journal of Neuroscience, 8, *2361–2370, Copyright (1996) John Wiley & Sons, Inc.*

and noncytoskeletal filamentous (F) actin associated with the maturation of astroglial cells when exposed to THs (Paul et al., 1996).

In the above study, the morphogenesis of the hypothyroid astrocytes was investigated using TH-deficient serum by repeatedly absorbing T_3 and T_4 of the normal serum in the ion-exchange resin, Dowex 1X8, as per published protocol (Samuels, Stanley, & Casanova, 1979). Although this protocol negates absorption of other mitogenic factors by the resin, it could not be ruled out. We measured concentration of T_3 and T_4 in the serum before and after treatment with the resin and added the difference in the resin-treated serum to bring it up to euthyroid level. Astrocytes cultured in a medium containing the euthyroid serum showed progressive transformation, similar to that observed in normal serum (Fig. 2) thereby confirming that the astrocytic differentiation observed in culture was under the control of THs. Furthermore, exposure of physiological concentrations of THs to 10-day-old hypothyroid astrocytes resulted in complete transformation of the polygonal cells to process-bearing stellate cells within 48 h (Gharami & Das, 2000).

7. MECHANISM OF TH ACTION ON ASTROCYTE MORPHOGENESIS

TR isoforms play an important role in astroglial cell maturation. Cerebellar slices of $TR_{\alpha 1}$-deficient mice exhibited reduced GFAP and nestin immunoreactivity in the mature astrocytes and their Bergmann processes in contrast to wild-type mice, suggesting a possible influence of the receptor isoform (Morte, Manzano, Scanlan, Vennström, & Bernal, 2004). $TR_{\beta 1}$, on the other hand, was detrimental on astroglial cell differentiation.

Implication of growth factors secreted by astrocytes in response to T_3 treatment for the maturation of astrocytes has been suggested (Lima et al., 1997; Trentin et al., 1998, 1995). Comparison of the intermediate filament expression of rat cerebellar astrocyte in normal and hypothyroid animals indicated that unlike in the normal development of astroglial cells which sequentially express nestin, vimentin, and GFAP, the vimentin–GFAP transition was delayed during hypothyroidism (Manzano et al., 2007). A number of laboratories have reported that the increased GFAP expression during TH treatment induces a change in morphology of the cultured astrocytes, although these effects differ according to the age and region of tissue origin (Aizenman & de Vellis, 1987; Gavaret et al., 1991; Lima et al., 1998). One study demonstrated that the transformation to process-bearing morphology in C6 cells and cortical astrocytes upon TH exposure was accompanied with reorganization of actin and GFAP cytoskeleton due to phosphorylation of GFAP possibly mediated through the RhoA signaling pathway (Zamoner et al., 2007). With regard to vimentin, our observations indicate that THs induced the expression of its phosphorylated forms in astrocytes which may be essential for the intracellular organization of the cytoskeleton in a way to facilitate the morphological maturation of astrocytes (Paul, Gharami, Das, & Sarkar, 1999). The ECM protein, fibronectin, which had a punctate distribution on the cell surface became diffused during T_3-induced proliferation of cerebellar astrocytes (Trentin & Moura Neto, 1995). On the other hand, cortical cultures when grown in TH-deficient medium showed decreased expression of GFAP and fibronectin (Dezonne, Stipursky, & Gomes, 2009).

A slew of evidence of the interaction of THs with β-AR system (Bahouth, 1991; Das & Paul, 1994; Gross, Brodde, & Schumann, 1980; Rozanov & Dratman, 1996; Tejani-Butt & Yang, 1994; Viticchi, Grinta, & Piantanelli, 1990) together with the ability of ISP to initiate morphological transformation of astrocytes prompted us to explore possible involvement of the β-AR system in TH-induced differentiation of the astrocytes. It was observed that β-AR antagonist, propranolol, comprehensively blocked the transformation of the flat polygonal cells to process-bearing stellate cells when induced by THs (Gharami & Das, 2000). The unique properties of the β-AR system in promoting TH-induced differentiation were evident from the fact that the cholinergic antagonist, atropine, and the α1-AR antagonist, prazosin, could not block the hormone-induced differentiation, whereas the α-AR antagonist, phenoxybenzamine, could partially antagonize the same (Gharami & Das, 2000). Furthermore, TH-induced maturation appears to be selectively regulated through the β_2-AR. Such obligatory role of the β-AR system in TH-induced differentiation of astrocytes led to exploring further downstream events like induction of PKA activity. PKA activity in the cells was monitored from 30 min to 48 h after addition of THs to 10-day-old hypothyroid cultures, to cover the entire time span required for THs to cause transformation of the polygonal astrocytes to stellate morphology. Interestingly, exposure to THs caused an initial increase in PKA activity with a peak at 2 h after which it came back to basal value (Ghosh, Gharami, Paul, & Das, 2005). However, this initial surge of PKA appeared to be crucial for the transformation of all the cells by 48 h, since removal of the hormones after 2 h from the media failed to block the transformation process. Simultaneously, TH treatment resulted in a biphasic response on the cellular phospho-MAP kinase (p-MAPK or p-ERK) level: an initial decline in the p-ERK level followed by an induction at 18–24 h, both of which could be blocked by a PKA inhibitor (Fig. 4). Here again, the delayed but sustained activation of p-ERK levels by THs at this later stage appeared to be critical and coincided with initiation of differentiation of the astrocytes. Magnitude and duration of ERK activation in response to a specific extracellular stimulus are critical factors governing distinct biological outcomes in a normal cell (Marshall, 1995). Unlike EGF, NGF causes differentiation of PC12 cells primarily because NGF treatment results in prolonged activation of ERK which is not observed in the case of EGF (Kao, Jaiswal, Kolch, & Landreth, 2001). Recent studies have identified two distinct signaling pathways for the c-AMP-induced ERK activations, one responsible for transient activation of ERK through Ras signaling and the

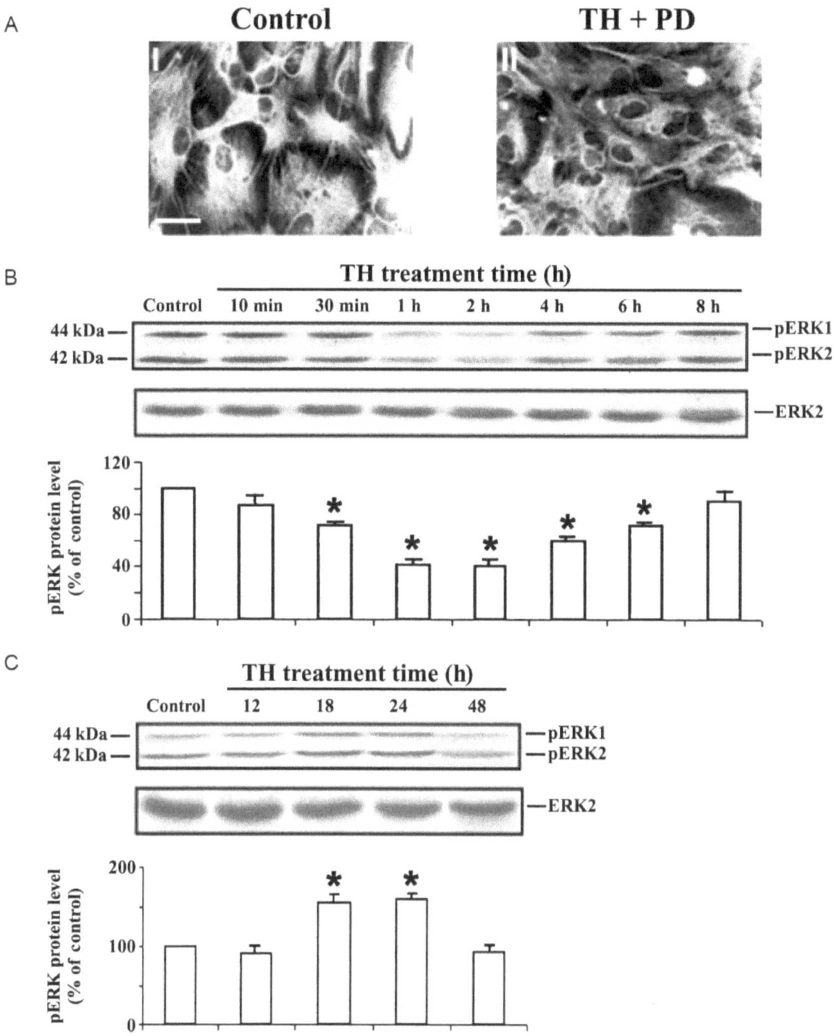

Fig. 4 Involvement of ERK in TH-induced transformation of astrocytes. (A) Inhibition of the MEK–MAPK pathway by PD098059 (PD) completely blocked TH-induced stellation in 10-day-old astrocyte cultures in TH-deficient medium. ERK expression for (B) short- and (C) long-term exposure to THs using Western blotting. $^{*}P < 0.05$ vs untreated control. Scale bar, 50 μm. *Modified from Ghosh, M., Gharami, K., Paul, S., Das, S., 2005. Thyroid hormone-induced morphological differentiation and maturation of astrocytes involves activation of protein kinase A and ERK signalling pathway.* European Journal of Neuroscience, 22, *1609–1617, Copyright (2005) John Wiley & Sons, Inc.*

other resulting in sustained activation of ERK involving Rap1 signaling (Takahashi, Li, Dillon, & Stork, 2017). The sustained induction of p-ERK activity by THs, resulting in differentiation of the astrocytes, was preceded by a PKI-sensitive decline in p-ERK levels at the initial stages, which, in turn, could trigger cessation of proliferation of the cells due to the possible involvement of the Ras–Raf-1 pathway. Another important outcome of the study was the identification of phospho-CREB as a downstream effector of TH action since the induction of p-ERK by the hormone was accompanied by a parallel increase in phospho-CREB activity which, however, persisted even at the end of the transformation of the astroglial cells (Ghosh et al., 2005).

A detailed study of the contributions of the subtypes of β-AR in the TH-induced transformation process in astrocytes was also carried out (Ghosh & Das, 2007). Specific [^{125}I]Pindolol ([^{125}I]PIN) binding to β_2-AR selectively increased with days of culture in both hypothyroid and TH-treated astrocytes, although the increase in binding at all ages was much higher in the TH-treated astrocytes when compared to TH-deficient controls possibly due to an increase in the affinity of the receptors (K_d) and not in the maximum binding sites (B_{max}) in the TH-treated cells (Fig. 5).

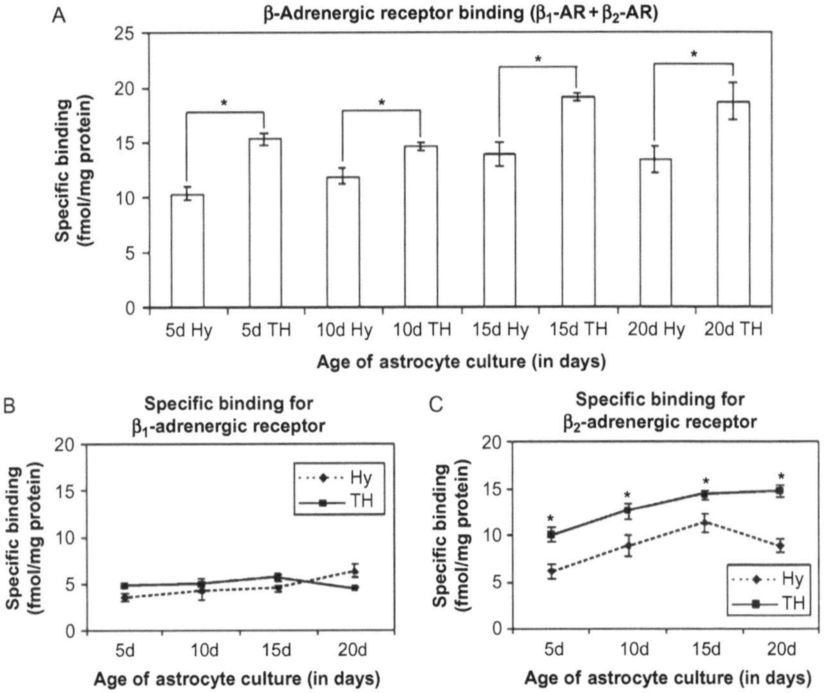

Fig. 5 Selective increase in [^{125}I]-PIN binding to β_2-AR in membranes of primary astrocyte cultured at various days in TH-deficient medium without (Hy) or with TH (0.31 nM T_3 and 22 nM T_4). *$P < 0.001$ from corresponding age-matched hypothyroid controls. *Courtesy: Ghosh, M., Das, S., 2007. Increased β2-adrenergic receptor activity by thyroid hormone possibly leads to differentiation and maturation of astrocytes in culture.* Cellular and Molecular Neurobiology, 27, *1007–1021, Copyright (2007) Springer.*

TH treatment failed to cause any significant change in the β_2-AR mRNA levels when compared corresponding hypothyroid controls.

$[^{125}I]$PIN binding was also determined after addition of THs at various time intervals in 10-day-old hypothyroid astrocytes to ascertain the fate of the β-AR subtypes during the process of transformation from polygonal to process-bearing morphology. Compared to hypothyroid controls, there was selective decline in the specific $[^{125}I]$PIN binding to β_2-AR at the initial phase (2–12h) which came back to the basal level by 24h (Das, Ghosh, & Das, 2016). Such decreased availability of the β_2-AR on the membrane could be attributed to endocytosis of the receptors since Western blot analysis following immunoprecipitation demonstrated that at 6h of TH supplementation, β_2-AR was significantly decreased in the membrane (Fig. 6A i) but increased in the cytosol (Fig. 6A ii) compared to untreated controls.

Prior treatment with the β_2-AR antagonists ICI-118,551 to block the receptors (Fig. 6B i and ii) or with the endocytosis inhibitor dynasore to block the endocytosis of β_2-AR (Fig. 6D i and ii), attenuated the TH-induced alteration of β_2-AR levels in the membrane as well as in the

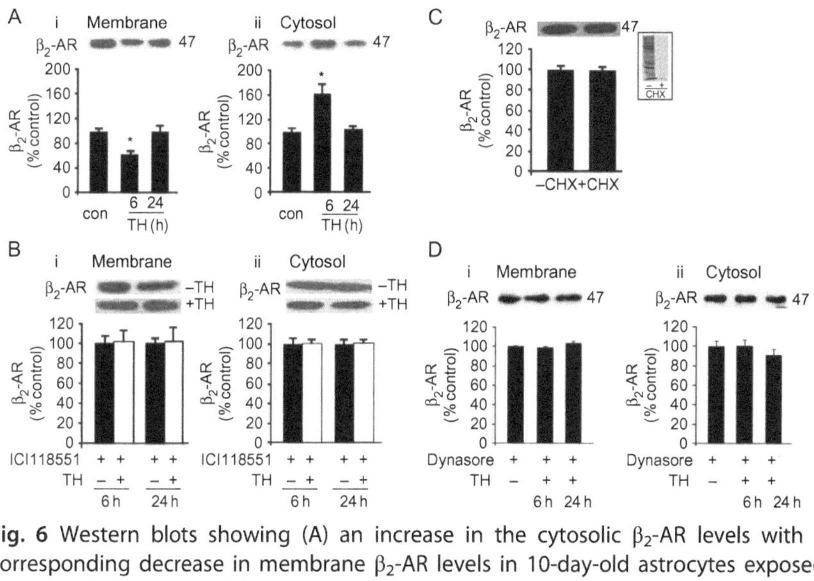

Fig. 6 Western blots showing (A) an increase in the cytosolic β_2-AR levels with a corresponding decrease in membrane β_2-AR levels in 10-day-old astrocytes exposed to TH at 6h which was blocked by (B) pretreatment with the β_2-AR antagonist, ICI-118,551, and (D) the endocytosis inhibitor, dynasore. (C) Prior addition of cycloheximide to inhibit any new protein synthesis failed to antagonize the increase in the cytosolic β_2-AR levels by TH as observed in (A). $^*P < 0.01$ from hypothyroid controls. *Courtesy: Das, M., Ghosh, M., Das, S., 2016. Thyroid hormone-induced differentiation of astrocytes is associated with transcriptional upregulation of β-arrestin-1 and β-adrenergic receptor-mediated endosomal signaling.* Molecular Neurobiology, 53, *5178–5190, Copyright (2016) Springer.*

cytosolic fraction at both 6 and 24 h. The observed increase in β_2-AR in the cytosol by THs at 6 h was unaffected in the presence of the protein synthesis inhibitor, cycloheximide (Fig. 6C), further corroborating that THs promoted increased endocytosis of the receptors from the membrane and not increasing synthesis of new receptors.

Endocytosis of the GPCR receptors like β_2-AR is known to be regulated by β-arrestins and β-AR kinases, and our subsequent studies identified a key role of β-arrestin-1 in the endocytosis of β_2-AR in the astrocytes when exposed to THs (Das et al., 2016). β-Arrestin 1/2 protein levels increased gradually at different time intervals (2, 4, 6, 12, 24, and 48 h) after the addition of THs (Fig. 7A). More specifically, β-arrestin-1 levels increased

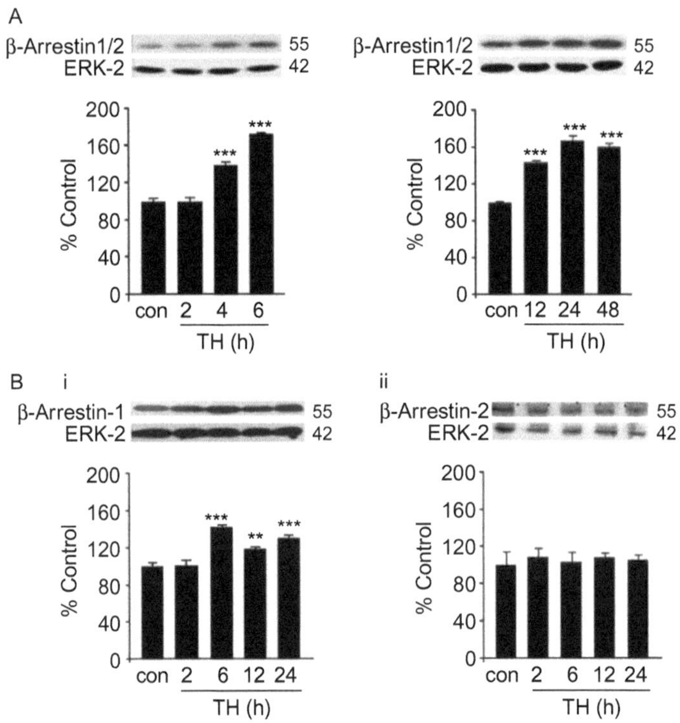

Fig. 7 Effect of THs on the β-arrestin levels in 10-day-old astrocytes, cultured under a TH-deficient condition for the indicated time points (2–24 h) using Western blot analysis. THs selectively induced β-arrestin-1 levels (bi) which contributed to the overall increase in total β-arrestin level in the cells (A) with no effect on β-arrestin-2 (bii). $^{**}P < 0.01$ and $^{***}P < 0.001$ indicate significant differences from hypothyroid control. *Courtesy: Das, M., Ghosh, M., Das, S., 2016. Thyroid hormone-induced differentiation of astrocytes is associated with transcriptional upregulation of β-arrestin-1 and β-adrenergic receptor-mediated endosomal signaling.* Molecular Neurobiology, 53, *5178–5190, Copyright (2016) Springer.*

(Fig. 7B i), but no change was found in β-arrestin-2 levels (Fig. 7B ii). A significant decline in p-β-arrestin, the inactive form of β-arrestin, was observed at 2 h of exposure to THs, which continued up to 12 h, and the ratio of the total:p-β-arrestin, a measure of overall activity, exhibited a peak at 6 h, coinciding with the period of endocytosis induced by THs in the astrocytes.

Overexpression on β-arrestin-1 in the 10-day-old hypothyroid astrocytes induced differentiation in the absence of THs by 24 h whereby the flat polygonal cells underwent transformation into stellate morphology (Das et al., 2016). The β-arrestin-2-overexpressing astrocytes continued to exhibit polygonal morphology on exposure to THs for 24 h but readily differentiated at 48 h, suggesting a crucial role of β-arrestin-1 in the differentiation of astrocytes by THs. The importance of β-arrestin in driving cell differentiation has been highlighted in recent studies in osteoblast (Boutin, Eliseeva, Gershengorn, & Neumann, 2014); adipocytes (Santos-Zas et al., 2013); human mesenchymal stem cell (Radio, Doctor, Jockers, & Witt-Enderby, 2006); PC12 cells (Rakhit, Pyne, & Pyne, 2001); etc. In our study, the effect of overexpression of β-arrestin-1 causing differentiation of the cells, even in the absence of THs, supported the theory that TH-induced astrocyte differentiation occurred through the exclusive upregulation of β-arrestin-1 isoform. β-Arrestin-2 did not appear to have any influence since it was neither regulated by THs nor its overexpression in the cells had any effect on differentiation.

Knockdown studies were also carried out with the isoforms of β-arrestin (Das et al., 2016). Interestingly, unlike β-arrestin-1 siRNA-transfected astrocytes, β-arrestin-2 siRNA-transfected cells exhibited marked inhibition of β_2-AR endocytosis when treated with THs for 6 h. This suggests an important contribution of the β-arrestin 2, as well, in the TH-induced differentiation of the cells. The exact role of both arrestins is not clear, and a positive cooperation between the two isoforms cannot be ruled out for executing the differentiation of the cells.

Overall, an immediate endocytosis of β_2-AR receptors during TH supplementation appears to be essential for the delayed but sustained activation of p-ERK at 18–24 h of TH treatment, so critical for astrocyte differentiation. Of the various endocytic inhibitors, dynasore, an inhibitor of dynamin-mediated endocytosis of β_2-AR receptors, as well as bafilomycin A1 and brefeldin A, both being inhibitors of exocytosis of β_2-AR receptors, inhibited the differentiation of the astrocytes on exposure to THs at 48 h (Fig. 8B i, C i, and D i, respectively). Interestingly, cells pretreated with dynasore, bafilomycin A1, and brefeldin A also significantly inhibited

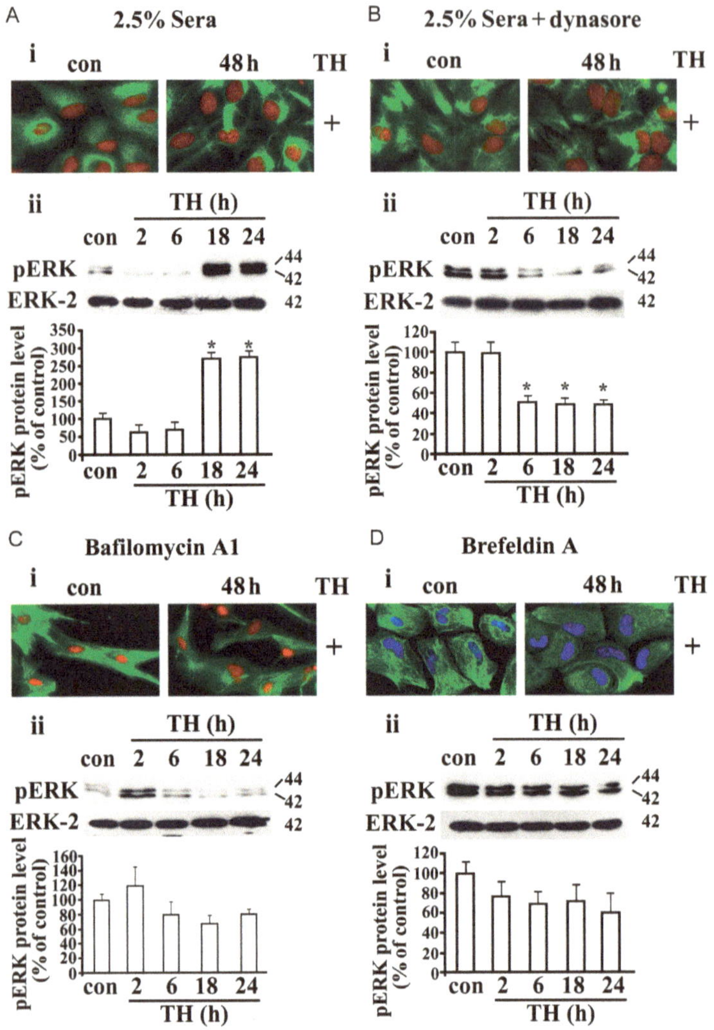

Fig. 8 Effect of endocytosis pathway inhibitors on TH-induced astrocyte differentiation and ERK activation. Ten-day-old hypothyroid astrocytes were pretreated with (A) 2.5% TH-deficient sera together with (B) dynasore, (C) bafilomycin A1, and (D) brefeldin A before addition of TH for the indicated time (h). *Upper panels*: morphology of cells stained with GFAP-FITC. Cell nucleus was either stained with propidium iodide (*red*) (A i, B i, C i) or with Hoechst 33342 (*blue*) (D i). *Lower panel*: Western blots of total cell lysate against p-ERK. *$P < 0.05$, significantly different from hypothyroid control. *Courtesy: Das, M., Ghosh, M., Das, S., 2016. Thyroid hormone-induced differentiation of astrocytes is associated with transcriptional upregulation of β-arrestin-1 and β-adrenergic receptor-mediated endosomal signaling.* Molecular Neurobiology, 53, *5178–5190, Copyright (2016) Springer.*

TH-induced activation of p-ERK in the late phase (Fig. 8B ii, C ii, and D ii, respectively) as observed in the control cells exposed to THs at 18 h (Fig. 8A ii). Taken together, results indicated that a delayed activation of ERK during TH-induced astrocyte differentiation is purely endosomal and essential for cellular differentiation.

8. CONCLUSION

A model (Fig. 9) is proposed describing the molecular mechanisms of TH-mediated differentiation of astrocytes from flat polygonal morphology to the stellate one. TH causes transcriptional upregulation of β-arrestin-1, thereby increasing β-arrestin-1 mRNA as well as its protein levels. Increased β-arrestin-1 facilitates increased endocytosis of β_2-AR, which in turn induces the activation of ERK inside the endosome. This is the second wave of signaling where a critical level of ERK signaling promotes the cellular morphogenesis. Overexpression of β-arrestin-1 also did the same and it can be said that TH actually mimics the β-arrestin-1 to drive the process of differentiation in the astrocytes.

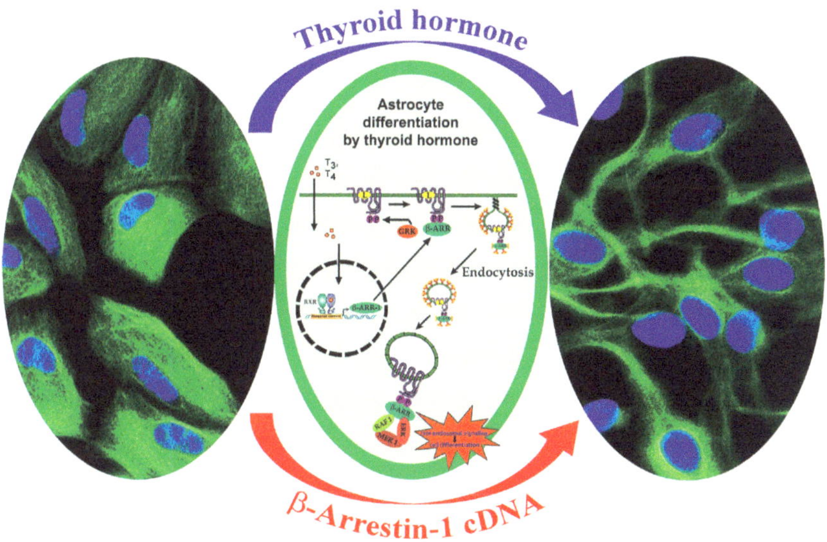

Fig. 9 Model of TH-mediated signaling leading to differentiation of astrocytes.

REFERENCES

Abe, K., & Misawa, M. (2003). A strocyte stellation induced by Rho kinase inhibitors in culture. *Developmental Brain Research*, *143*, 99–104.

Abe, K., & Saito, H. (1997). Developmental changes in cyclic AMP-stimulated stellation of cultured rat cortical astrocytes. *Japanese Journal of Pharmacology*, *75*, 433–438.

Abe, K., & Saito, H. (1999). Effect of ATP on astrocyte stellation is switched from suppressive to stimulatory during development. *Brain Research*, *850*, 150–157.

Abe, K., & Saito, H. (2000). The p44/42 mitogen-activated protein kinase cascade is involved in the induction and maintenance of astrocyte stellation mediated by protein kinase C. *Neuroscience Research*, *36*, 251–257.

Adachi, T., Takanaga, H., Sakurai, Y., Ishido, M., Kunimoto, M., & Asou, H. (2002). Influence of cell density and thyroid hormone on glial cell development in primary cultures of embryonic rat cerebral hemisphere. *Journal of Neuroscience Research*, *69*, 61–71.

Aizenman, Y., & de Vellis, J. (1987). Synergistic action of thyroid hormone, insulin, and hydrocortisone on astrocyte differentiation. *Brain Research*, *414*, 301–308.

Allaman, I., Bélanger, M., & Magistretti, P. J. (2011). Astrocyte-neuron metabolic relationships: For better and for worse. *Trends in Neurosciences*, *34*, 76–87.

Andres-Barquin, P. J., Fages, C., Le Prince, G., Rolland, B., & Tardy, M. (1994). Thyroid hormones influence the astroglial plasticity: Changes in the expression of glial fibrillary acidic protein (GFAP) and of its encoding message. *Neurochemical Research*, *19*, 65–69.

Araque, A., Carmignoto, G., Haydon, P. G., Oliet, S. H., Robitaille, R., & Volterra, A. (2014). Gliotransmitters travel in time and space. *Neuron*, *81*, 728–739.

Baba, A. (1998). Role of endotheline B receptor signals in reactive astrocytes. *Life Sciences*, *62*, 1711–1715.

Bahouth, S. W. (1991). Thyroid hormones transcriptionally regulate the β1-adrenergic receptor gene in cultured ventricular myocytes. *Journal of Biological Chemistry*, *266*, 15863–15869.

Baorto, D. M., Mellado, W., & Shelanski, M. L. (1992). Astrocyte process growth induction by actin breakdown. *Journal of Cell Biology*, *117*, 357–367.

Bayer, S. A., & Altman, J. (1991). Development of the endopiriform nucleus and the claustrum in the rat brain. *Neuroscience*, *45*, 391–412.

Blanco-Suárez, E., Caldwell, A. L., & Allen, N. J. (2017). Role of astrocyte-synapse interactions in CNS disorders. *Journal of Physiology*, *595*, 1903–1916. http://dx.doi.org/10.1113/JP270988.

Boutin, A., Eliseeva, E., Gershengorn, M. C., & Neumann, S. (2014). β-Arrestin-1 mediates thyrotropin-enhanced osteoblast differentiation. *FASEB Journal*, *28*, 3446–3455.

Bushong, E. A., Martone, M. E., & Ellisman, M. H. (2004). Maturation of astrocyte morphology and the establishment of astrocyte domains during postnatal hippocampal development. *International Journal of Developmental Neuroscience*, *22*, 73–86.

Cahoy, J. D., Emery, B., Kaushal, A., Foo, L. C., Zamanian, J. L., Christopherson, K. S., et al. (2008). A transcriptome database for astrocytes, neurons, and oligodendrocytes: A new resource for understanding brain development and function. *Journal of Neuroscience*, *28*, 264–278.

Carlson, D. J., Strait, K. A., Schwartz, H. L., & Oppenheimer, J. H. (1994). Immunofluorescent localization of thyroid hormone receptor isoforms in glial cells of rat brain. *Endocrinology*, *135*, 1831–1836.

Carlson, D. J., Strait, K. A., Schwartz, H. L., & Oppenheimer, J. H. (1996). Thyroid hormone receptor isoform content in cultured type 1 and type 2 astrocytes. *Endocrinology*, *137*, 911–917.

Clos, J., & Legrand, J. (1973). Effects of thyroid deficiency on the different cell populations of the cerebellum in the young rat. *Brain Research*, *63*, 450–455.

Clos, J., Legrand, C., & Legrand, J. (1980). Effects of thyroid state on the formation and early morphological development of Bergmann glia in the developing cerebellum. *Developmental Neuroscience*, *3*, 199–208.

Cragnolini, A. B., Huang, Y., Gokina, P., & Friedman, W. J. (2009). Nerve growth factor attenuates proliferation of astrocytes via the p75 neurotrophin receptor. *Glia*, *57*, 1386–1392.

Das, M., Ghosh, M., & Das, S. (2016). Thyroid hormone-induced differentiation of astrocytes is associated with transcriptional upregulation of β-arrestin-1 and β-adrenergic receptor-mediated endosomal signaling. *Molecular Neurobiology*, *53*, 5178–5190.

Das, S., & Paul, S. (1994). Decrease in beta-adrenergic receptors of cerebral astrocytes in hypothyroid rat brain. *Life Sciences*, *54*, 621–629.

Dasgupta, A., Das, S., & Sarkar, P. K. (2007). Thyroid hormone promotes glutathione synthesis in astrocytes by up regulation of glutamate cysteine ligase through differential stimulation of its catalytic and modulator subunit mRNAs. *Free Radical Biology and Medicine*, *42*, 617–626.

Davies, D. L., Niesman, I. R., Boop, F. A., & Phelan, K. D. (2000). Heterogeneity of astroglia cultured from adult human temporal lobe. *International Journal of Developmental Neuroscience*, *18*, 151–160.

Davis, A. A., & Temple, S. (1994). A self-renewing multipotential stem cell in embryonic rat cerebral cortex. *Nature*, *372*, 263–266.

Deb, I., & Das, S. (2011). Thyroid hormones protect astrocytes from morphine-induced apoptosis by regulating nitric oxide and pERK 1/2 pathways. *Neurochemistry International*, *58*, 861–871.

De Pitta, M., Brunel, N., & Volterra, A. (2016). Astrocytes: Orchestrating synaptic plasticity? *Neuroscience*, *323*, 43–61.

Dezonne, R. S., Stipursky, J., Araujo, A. P., Nones, J., Pavão, M. S., Porcionatto, M., et al. (2013). Thyroid hormone treated astrocytes induce maturation of cerebral cortical neurons through modulation of proteoglycan levels. *Frontiers in Cellular Neuroscience*, 7, 125. http://dx.doi.org/10.3389/fncel.2013.00125.

Dezonne, R. S., Stipursky, J., & Gomes, F. C. (2009). Effect of thyroid hormone depletion on cultured murine cerebral cortex astrocytes. *Neuroscience Letters*, *467*, 58–62.

Du, T., Li, B., Li, H., Li, M., Hertz, L., & Peng, L. (2010). Signaling pathways of isoproterenol-induced ERK1/2 phosphorylation in primary cultures of astrocytes are concentration-dependent. *Journal of Neurochemistry*, *115*, 1007–1023.

Emsley, J. G., & Macklis, J. D. (2006). Astroglial heterogeneity closely reflects the neuronal-defined anatomy of the adult murine CNS. *Neuron Glia Biology*, *2*, 175–186.

Eng, L. F., & Ghirnikar, R. S. (1994). GFAP and astrogliosis. *Brain Patholology*, *4*, 229–237.

Fedoroff, S. (1986). Prenatal ontogenesis of astrocytes. In S. Fedoroff & A. Vernadakis (Eds.), *Astrocytes: Development, morphology, and regional specialization of astrocytes: Vol. 1* (pp. 35–74). Orlando: Academic.

Fields, R. D., Woo, D. H., & Basser, P. J. (2015). Glial regulation of the neuronal connectome through local and long-distant communication. *Neuron*, *86*, 374–386.

Freeman, M. R. (2010). Specification and morphogenesis of astrocytes. *Science*, *330*, 774–778.

Gavaret, J. M., Toru-Delbauffe, D., Baghdassarian-Chalaye, D., Pomerance, M., & Pierre, M. (1991). Thyroid hormone action: Induction of morphological changes and protein secretion in astroglial cell cultures. *Developmental Brain Research*, *58*, 43–49.

Gharami, K., & Das, S. (2000). Thyroid hormone-induced morphological differentiation and maturation of astrocytes are mediated through the beta-adrenergic receptor. *Journal of Neurochemistry*, *75*, 1962–1969.

Gharami, K., & Das, S. (2004). Delayed but sustained induction of mitogen-activated protein kinase activity is associated with beta-adrenergic receptor-mediated morphological differentiation of astrocytes. *Journal of Neurochemistry*, *88*, 12–22.

Ghosh, M., & Das, S. (2007). Increased β2-adrenergic receptor activity by thyroid hormone possibly leads to differentiation and maturation of astrocytes in culture. *Cellular and Molecular Neurobiology*, *27*, 1007–1021.

Ghosh, M., Gharami, K., Paul, S., & Das, S. (2005). Thyroid hormone-induced morphological differentiation and maturation of astrocytes involves activation of protein kinase A and ERK signalling pathway. *European Journal of Neuroscience*, *22*, 1609–1617.
Gibbs, M. E., Hutchinson, D., & Hertz, L. (2008). Astrocytic involvement in learning and memory consolidation. *Neuroscience and Biobehavioral Reviews*, *32*, 927–944.
Goldman, J. E., & Abramson, B. (1990). Cyclic AMP-induced shape changes of astrocytes are accompanied by rapid depolymerization of actin. *Brain Research*, *528*, 189–196.
Goldman, J. E., & Chiu, F. C. (1984). Dibutyryl cyclic AMP causes intermediate filament accumulation and actin reorganization in astrocytes. *Brain Research*, *306*, 85–95.
Gomes, F. C., Maia, C. G., de Menezes, J. R., & Neto, V. M. (1999). Cerebellar astrocytes treated by thyroid hormone modulate neuronal proliferation. *Glia*, *25*, 247–255.
Gottfried, C., Cechin, S. R., Gonzalez, M. A., Vaccaro, T. S., & Rodnight, R. (2003). The influence of the extracellular matrix on the morphology and intracellular pH of cultured astrocytes exposed to media lacking bicarbonate. *Neuroscience*, *121*, 553–562.
Gottfried, C., Valentim, L., Salbego, C., Karl, J., Wofchuk, S. T., & Rodnight, R. (1999). Regulation of protein phosphorylation in astrocyte cultures by external calcium ions: Specific effects on the phosphorylation of glial fibrillary acidic protein (GFAP), vimentin and heat shock protein 27 (HSP27). *Brain Research*, *833*, 142–149.
Gould, E., Frankfurt, M., Westlind-Danielsson, A., & McEwen, B. S. (1990). Developing forebrain astrocytes are sensitive to thyroid hormone. *Glia*, *3*, 283–292.
Gourine, A. V., Kasymov, V., Marina, N., Tang, F., Figueiredo, M. F., Lane, S., et al. (2010). Astrocytes control breathing through pH dependent release of ATP. *Science*, *329*, 571–575.
Gross, G., Brodde, O. E., & Schumann, D. G. (1980). Effects of thyroid hormone deficiency on pre- and postsynaptic noradrenergic mechanisms in the rat cerebral cortex. *Archives of International Pharmacodynamics and Therapeutics*, *244*, 219–230.
Guadaño-Ferraz, A., Obregón, M. J., St-Germain, D., & Bernal, J. (1997). The type 2 iodothyronine deiodinase is expressed primarily in glial cells in the neonatal rat brain. *Proceedings of the National Academy of Sciences of the United States of America*, *94*, 10391–10396.
Halassa, M. M., Florian, C., Fellin, T., Munoz, J. R., Lee, S. Y., Abel, T., et al. (2009). Astrocytic modulation of sleep homeostasis and cognitive consequences of sleep loss. *Neuron*, *61*, 213–219.
Hatten, M. E., Lynch, M., Rydel, R. E., Sanchez, J., Joseph-Silverstein, J., Moscatelli, D., et al. (1988). In vitro neurite extension by granule neurons is dependent upon astroglial-derived fibroblast growth factor. *Developmental Biology*, *125*, 280–289.
Hatton, G. I., Luckman, S. M., & Bicknell, R. J. (1991). Adrenalin activation of beta 2-adrenoceptors stimulates morphological changes in astrocytes (pituicytes) cultured from adult rat neurohypophyses. *Brain Research Bulletin*, *26*, 765–769.
Hayakawa, K., Esposito, E., Wang, X., Terasaki, Y., Liu, Y., Xing, C., et al. (2016). Transfer of mitochondria from astrocytes to neurons after stroke. *Nature*, *535*, 551–555.
Heffron, D. S., & Mandell, J. W. (2005). Opposing roles of ERK and p38 MAP kinases in FGF2-induced astroglial process extension. *Molecular and Cellular Neuroscience*, *28*, 779–790.
Hubank, M., Sinha, A. K., Gullo, D., & Ekins, R. P. (1990). Nuclear tri-iodothyronine (T3) binding in neonatal rat brain suggests a direct glial requirement for T3 during development. *Journal of Endocrinology*, *126*, 409–415.
Igelhorst, B. A., Niederkinkhaus, V., Karus, C., Lange, M. D., & Dietzel, I. D. (2015). Regulation of neuronal excitability by release of proteins from glial cells. *Philosophical Transactions of the Royal Society of London. Series B, Biological Sciences*, *370*(1672), 1–10. http://dx.doi.org/10.1098/rstb.2014.0194.
Janzer, R. C., & Raff, M. C. (1987). Astrocytes induce blood–brain barrier properties in endothelial cells. *Nature*, *325*, 253–257.

John, G. R., Chen, L., Riviccio, M. A., Melendez-Vasquez, C. V., Hartley, A., & Brosnan, C. F. (2004). Interleukin-1b induces reactive astroglial phenotype via deactivation of the Rho GTPase-Rock axis. *Journal of Neuroscience, 24*, 2837–2845.

Jorgensen, O. S., Brooksbank, B. W. L., & Balazs, R. (1990). Neuronal plasticity and astrocytic reaction in Down syndrome and Alzheimer disease. *Journal of Neurological Sciences, 98*, 63–79.

Kalman, D., Gomperts, S. N., Hardy, S., Kitamura, M., & Bishop, J. M. (1999). Ras family GTPases control growth of astrocyte processes. *Molecular Biology of the Cell, 10*, 1665–1683.

Kao, S., Jaiswal, R. K., Kolch, W., & Landreth, G. E. (2001). Identification of the mechanisms regulating the differential activation of the MAPK cascade by epidermal growth factor and nerve growth factor in PC12 cells. *Journal of Biological Chemistry, 276*, 18169–18177.

Konopka, A., Zeug, A., Skupien, A., Kaza, B., Mueller, F., Chwedorowicz, A., et al. (2016). Cleavage of hyaluronan and CD44 adhesion molecule regulate astrocyte morphology via Rac1 signalling. *PLoS One, 11*, e0155053.

Koyama, Y., & Baba, A. (1996). Endothelin-induced cytoskeletal actin re-organization in cultured astrocytes: Inhibition by C3 ADP-ribosyltransferase. *Glia, 16*, 342–350.

Lebel, J. M., Lherault, S., Dussault, J. H., & Puymirat, J. (1993). Thyroid hormone up-regulates thyroid hormone receptor beta gene expression in rat cerebral hemisphere astrocyte cultures. *Glia, 9*, 105–112.

Leonard, J. L., Farwell, A. P., Yen, P. M., Chin, W. W., & Stula, M. (1994). Differential expression of thyroid hormone receptor isoforms in neurons and astroglial cells. *Endocrinology, 135*, 548–555.

Lim, R., Mitsunobu, K., & Li, W. K. P. (1973). Maturation-stimulating effect of brain extract and dibutyryl cyclic AMP on dissociated enibryonic brain cells in culture. *Experimental Cell Research, 79*, 243–246.

Lima, F. R. S., Goncalves, N., Gomes, F. C. A., de Freitas, M. S., & Moura Neto, V. (1998). Thyroid hormone action on astroglial cells from distinct brain regions during development. *International Journal of Developmental Neuroscience: The Official Journal of the International Society for Developmental Neuroscience, 16*, 19–27.

Lima, F. R. S., Trentin, A. G., Rosenthal, D., Chagas, C., & Moura Neto, V. (1997). Thyroid hormone induces protein secretion and morphological changes in astroglial cells with an increase in expression of glial fibrillary acidic protein. *Journal of Endocrinology, 154*, 167–175.

Liu, W., Shafit-Zagardo, B., Aquino, D. A., Zhao, M.-L., Dickson, D. W., Brosnan, C. F., et al. (1994). Cytoskeletal alterations in human fetal astrocytes induced by interleukin-1 beta. *Journal of Neurochemistry, 63*, 1625–1634.

Lundgaard, I., Osório, M. J., Kress, B. T., Sanggaard, S., & Nedergaard, M. (2014). White matter astrocytes in health and disease. *Neuroscience, 276*, 161–173.

Luo, M., Puymirat, J., & Dussault, J. H. (1989). Immunocytochemical localization of nuclear 3,5,3-triiodothyronine (L-T3) receptors in astrocyte cultures. *Developmental Brain Research, 46*, 131–136.

Manzano, J., Bernal, J., & Morte, B. (2007). Influence of thyroid hormones on maturation of rat cerebellar astrocytes. *International Journal of Developmental Neuroscience: The Official Journal of the International Society for Developmental Neuroscience, 25*, 171–179.

Marshall, C. J. (1995). Specificity of receptor tyrosine kinase signaling: Transient versus sustained extracellular signal-regulated kinase activation. *Cell, 80*, 179–185.

Martinez, R., Eller, C., Viana, N. B., & Gomes, F. C. (2011). Thyroid hormone induces cerebellar neuronal migration and Bergmann glia differentiation through epidermal growth factor/mitogen-activated protein kinase pathway. *European Journal of Neuroscience, 33*, 26–35.

Mason, C. A., Edmondson, J. C., & Hatten, M. E. (1988). The extending astroglial process: Development of glial cell shape, the growing tip, and interactions with neurons. *Journal of Neuroscience*, *8*, 3124–3134.

Mathewson, A. J., & Berry, M. (1985). Observation on the astrocyte response to a cerebral stab wound in adult rats. *Brain Research*, *327*, 61–69.

Mayer, S. I., Rössler, O. G., Endo, T., Charnay, P., & Thiel, G. (2009). Epidermal-growth-factor-induced proliferation of astrocytes requires Egr transcription factors. *Journal of Cell Science*, *122*, 3340–3350.

Mendes-de-Aguiar, C. B., Alchini, R., Decker, H., Alvarez-Silva, M., Tasca, C. I., & Trentin, A. G. (2008). Thyroid hormone increases astrocytic glutamate uptake and protects astrocytes and neurons against glutamate toxicity. *Journal of Neuroscience Research*, *86*, 3117–3125.

Messer, A., Snodgrass, G. L., & Maskin, P. (1985). Timecourse of effects of triiodothyronine on mouse cerebellar cells cultured by two different methods. *International Journal of Developmental Neuroscience*, *3*, 291–299.

Miyata, S., Furuya, K., Nakai, S., Bun, H., & Kiyohara, T. (1999). Morphological plasticity and rearrangement of cytoskeletons in pituicytes cultured from adult rat neurohypophysis. *Neuroscience Research*, *33*, 299–306.

Moonen, G., Cam, Y., Sensenbrenner, M., & Mandel, P. (1975). Variability of the effects of serum-free medium, dibutyryl-cyclic AMP or theophylline on the morphology of cultured new-born rat astroblasts. *Cell and Tissue Research*, *163*, 365–372.

Morte, B., Manzano, J., Scanlan, T. S., Vennström, B., & Bernal, J. (2004). Aberrant maturation of astrocytes in thyroid hormone receptor alpha 1 knockout mice reveals an interplay between thyroid hormone receptor isoforms. *Endocrinology*, *145*, 1386–1391.

Nam, S. M., Kim, Y. N., Yoo, D. Y., Yi, S. S., Choi, J. H., Hwang, I. K., et al. (2013). Hypothyroidism affects astrocyte and microglial morphology in type 2 diabetes. *Neural Regeneration Research*, *8*, 2458–2467.

Nathaniel, E. J. H., & Nathaniel, D. R. (1981). The reactive astrocyte. *Advances in Cellular Neurobiology*, *2*, 249–301.

Nixdorf-Bergweiler, B. E., Albrecht, D., & Heinemann, U. (1994). Developmental changes in the number, size, and orientation of GFAP-positive cells in the CA1 region of rat hippocampus. *Glia*, *12*, 180–195.

Norton, W. T., Aquino, D. A., Hozumi, I., Chiu, F. C., & Brosnan, C. F. (1992). Quantitative aspects of reactive gliosis: A review. *Neurochemical Research*, *17*, 877–885.

Oberheim, N. A., Takano, T., Han, X., He, W., Lin, J. H., Wang, F., et al. (2009). Uniquely hominid features of adult human astrocytes. *Journal of Neuroscience*, *29*, 3276–3287.

Ogata, K., & Kosaka, T. (2002). Structural and quantitative analysis of astrocytes in the mouse hippocampus. *Neuroscience*, *113*, 221–233.

Ortiz-Caro, J., Yusta, B., & Montiel, F. (1986). Identification and characterization of L-triiodothyronine receptors in cells of glial and neuronal origin. *Endocrinology*, *119*, 2163–2167.

Padmanabhan, J., & Shelanski, M. L. (1998). Process formation in astrocytes: Modulation of cytoskeletal proteins. *Neurochemical Research*, *23*, 377–384.

Palay, S. L., & Chan-Palay, V. (1974). *Cerebellar cortex: Cytology and organization*. Berlin: Springer-Verlag [Chapter 10].

Paul, S., Das, S., Poddar, R., & Sarkar, P. K. (1996). Role of thyroid hormone in the morphological differentiation and maturation of astrocytes: Temporal correlation with synthesis and organization of actin. *European Journal of Neuroscience*, *8*, 2361–2370.

Paul, S., Das, S., & Sarkar, P. K. (1992). Effect of hypothyroidism on different forms of actin in rat cerebral neuronal cultures studied by an improved DNase I inhibition assay. *Journal of Neurochemistry*, *59*, 701–707.

Paul, S., Gharami, K., Das, S., & Sarkar, P. K. (1999). Thyroid hormone-induced maturation of astrocytes is associated with the expression of new variants of vimentin and their phosphorylation. *Journal of Neurochemistry*, *73*, 1964–1972.
Perea, G., Navarrete, M., & Araque, A. (2009). Tripartite synapses: Astrocytes process and control synaptic information. *Trends in Neurosciences*, *32*, 421–431.
Perez, V., Bouschet, T., Fernandez, C., Bockaert, J., & Journot, L. (2005). Dynamic reorganization of the astrocyte actin cytoskeleton elicited by cAMP and PACAP: A role for phosphatidylinositol 3-kinase inhibition. *European Journal of Neuroscience*, *21*, 26–32.
Puymirat, J. (1992). Thyroid receptors in the rat brain. *Progress in Neurobiology*, *39*, 281–294.
Racchetti, G., D'Alessandro, R., & Meldolesi, J. (2012). Astrocyte stellation, a process dependent on Rac1 is sustained by the regulated exocytosis of enlargeosomes. *Glia*, *60*, 465–475.
Radio, N., Doctor, J., Jockers, R., & Witt-Enderby, P. (2006). MT2 melatonin receptors couple to Gi protein, beta arrestin, and MEK/ERK (1/2) in an agonist-dependent manner in human adult mesenchymal stem cells differentiated into osteoblasts. *FASEB Journal*, *20*, A250–A251.
Rakhit, S., Pyne, S., & Pyne, N. J. (2001). Nerve growth factor stimulation of p42/p44 mitogen-activated protein kinase in PC12 cells: Role of Gi/o, G protein-coupled receptor kinase 2, β-arrestin 1 and endocytic processing. *Molecular Pharmacology*, *60*, 63–70.
Rakic, P. (1971). Neuron-glia relationships during granule cell migration in developing cerebellar cortex. A Golgi and electron microscopic study in Macacus rhesus. *The Journal of Comparative Neurology*, *141*, 282–312.
Roeder, L. M., Williams, I. B., & Tildon, J. T. (1985). Glucose transport in astrocytes: Regulation by thyroid hormone. *Journal of Neurochemistry*, *45*, 1653–1657.
Roman, C., Fuior, E. V., Trusca, V. G., Kardassis, D., Simionescu, M., & Gafencu, A. V. (2015). Thyroid hormones upregulate apolipoprotein E gene expression in astrocytes. *Biochemical and Biophysical Research Communications*, *468*, 190–195.
Rozanov, C. B., & Dratman, M. B. (1996). Immunohistochemical mapping of brain triiodothyronine reveals prominent localization in central noradrenergic systems. *Neuroscience*, *74*, 897–915.
Samuels, N. H., Stanley, F., & Casanova, Z. (1979). Depletion of L-3,5,3′-triiodothyronine and L-thyroxine in euthyroid calf serum for use in cell culture studies of the action of thyroid hormone. *Endocrinology*, *105*, 80–85.
Santello, M., Cali, C., & Bezzi, P. (2012). Gliotransmission and the tripartite synapse. *Advances in Experimental Medicine and Biology*, *970*, 307–331.
Santos-Zas, I., Lodeiro, M., Gurriarán-Rodríguez, U., Bouzo-Lorenzo, M., Mosteiro, C. S., Casanueva, F. F., et al. (2013). β-Arrestin signal complex plays a critical role in adipose differentiation. *International Journal of Biochemistry and Cell Biology*, *45*, 1281–1292.
Sayre, N. L., Sifuentes, M., Holstein, D., Cheng, S. Y., Zhu, X., & Lechleiter, J. D. (2017). Stimulation of astrocyte fatty acid oxidation by thyroid hormone is protective against ischemic stroke-induced damage. *Journal of Cerebral Blood Flow and Metabolism: Official Journal of the International Society of Cerebral Blood Flow and Metabolism*, *37*, 514–527.
Scholze, A. R., Foo, L. C., Mulinyawe, S., & Barres, B. A. (2014). BMP signaling in astrocytes downregulates EGFR to modulate survival and maturation. *PLoS One*, *9*, e110668.
Shain, W., Bausback, D., Fiero, A., Madelian, V., & Turner, J. N. (1992). Regulation of receptor-mediated shape change in astroglial cells. *Glia*, *5*, 223–238.
Shao, Y., & McCarthy, K. (1994). Plasticity of astrocytes. *Glia*, *11*, 147–155.
Sidman, R. L., & Rakic, P. (1973). Neuronal migration with special reference to the developing human brain: A review. *Brain Research*, *62*, 1–35.
Sofroniew, M. V., & Vinters, H. V. (2010). Astrocytes: Biology and pathology. *Acta Neuropathologica*, *119*, 7–35.

Sultan, S., Li, L., Moss, J., Petrelli, F., Cassé, F., Gebara, E., et al. (2015). Synaptic integration of adult-born hippocampal neurons is locally controlled by astrocytes. *Neuron*, *88*, 957–972.

Takahashi, M., Li, Y., Dillon, T. J., & Stork, P. J. (2017). Phosphorylation of Rap1 by cAMP-dependent protein kinase A (PKA) creates a binding site for KSR to sustain ERK activation by cAMP. *The Journal of Biological Chemistry*, *292*, 1449–1461.

Tejani-Butt, S. M., & Yang, J. (1994). A time course of altered thyroid states on the noradrenergic system in rat brain by quantitative autoradiography. *Neuroendocrinology*, *59*, 235–244.

Trentin, A. G., Alvarez-Silva, M., & Moura Neto, V. (2001). Thyroid hormone induces cerebellar astrocytes and C6 glioma cells to secrete mitogenic growth factors. *The American Journal of Physiology—Endocrinology and Metabolism*, *281*, E1088–E1094.

Trentin, A. G., De Aguiar, C. B., Garcez, R. C., & Alvarez-Silva, M. (2003). Thyroid hormone modulates the extracellular matrix organization and expression in cerebellar astrocyte: Effects on astrocyte adhesion. *Glia*, *42*, 359–369.

Trentin, A. G., Gomes, F. C. A., Lima, F. R. S., & Moura Neto, V. (1998). Thyroid hormone induces secretion of factors and progressive morphological changes on primary and subcultured astrocytes. *In Vitro Cellular and Developmental Biology*, *34*, 280–282.

Trentin, A. G., & Moura Neto, V. (1995). Triiodothyronine affects cerebellar astrocyte proliferation, GFAP, and fibronectin organization. *Neuroreport*, *6*, 293–296.

Trentin, A. G., Rosenthal, D., & Moura Neto, V. (1995). Thyroid hormone and conditioned medium effects on astroglial cells from hypothyroid and normal rat brain: Factor secretion, cell, differentiation, and proliferation. *Journal of Neuroscience Research*, *41*, 409–417.

Trimmer, P. A., Reier, P. J., Oh, T. H., & Eng, L. F. (1982). An ultrastructural and immunocytochemical study of astrocyte differentiation in vitro. *Journal of Neuroimmunology*, *2*, 235–260.

Tu, H. M., Legradi, G., Bartha, T., Salvatore, D., Lechan, R. M., & Larsen, P. R. (1999). Regional expression of the type 3 iodothyronine deiodinase messenger ribonucleic acid in the rat central nervous system and its regulation by thyroid hormone. *Endocrinology*, *140*, 784–790.

Viticchi, C., Grinta, R., & Piantanelli, L. (1990). Influence of age on the thyroid hormone-induced up-regulation of β-adrenoceptors in mouse brain cortex. *Gerontology*, *36*, 286–292.

Wiese, S., Karus, M., & Faissner, A. (2012). Astrocytes as a source for extracellular matrix molecules and cytokines. *Frontiers in Pharmacology*, *3*, 1–13, Article 120.

Won, C.-K., & Oh, Y. S. (2000). cAMP-induced stellation in primary astrocyte cultures with regional heterogeneity. *Brain Research*, *887*, 250–258.

Woodhams, P. L., Bascó, E., Hajós, F., Csillág, A., & Balázs, R. (1981). Radial glia in the developing mouse cerebral cortex and hippocampus. *Anatomy and Embryology (Berlin)*, *163*, 331–343.

Zamoner, A., Funchal, C., Jacques-Silva, M. C., Gottfried, C., Barreto Silva, F. R., & Pessoa-Pureur, R. (2007). Thyroid hormones reorganize the cytoskeleton of glial cells through Gfap phosphorylation and Rhoa-dependent mechanisms. *Cellular and Molecular Neurobiology*, *27*, 845–865.

Zorec, R., Parpura, V., Vardjan, N., & Verkhratsky, A. (2017). Astrocytic face of Alzheimer's disease. *Behavioural Brain Research*, *322*, 250–257. http://dx.doi.org/10.1016/j.bbr.2016.05.021.

CHAPTER THIRTEEN

Thyroid Hormone in the CNS: Contribution of Neuron–Glia Interaction

Mami Noda[1]
Laboratory of Pathophysiology, Graduate School of Pharmaceutical Sciences, Kyushu University, Fukuoka, Japan
[1]Corresponding author: e-mail address: noda@phar.kyushu-u.ac.jp

Contents

1. Introduction 314
2. Receptors and Function of TH in the CNS 314
3. Abnormal Level of TH in the CNS 315
4. Transportation of THs to the Brain and Metabolism in Astrocytes 317
5. Functional Role of TH in Microglia 317
   5.1 Effects of TH on Microglia *In Vitro* 317
   5.2 Effects of TH on Microglia *In Vivo* 318
   5.3 Signaling Cascade of TH in Microglia 320
6. Astrocytes and Oligodendrocytes Differentiation by T3 322
7. Pathological Effects of TH on Glial Cells 323
8. Sex- and Age-Dependent Effects of Hyperthyroidism 323
9. Possible Influence of TH-Induced Neuron–Glia Interaction on Psychiatric Symptoms 324
10. Conclusion and Future Directions 325
Acknowledgment 326
References 326

Abstract

The endocrine system and the central nervous system (CNS) are intimately linked. Among hormones closely related to the nervous system, thyroid hormones (THs) are critical for the regulation of development and differentiation of neurons and neuroglia and hence for development and function of the CNS. T3 (3,3′,5-triiodothyronine), an active form of TH, is important not only for neuronal development but also for differentiation of astrocytes and oligodendrocytes, and for microglial development. In adult brain, T3 affects glial morphology with sex- and age-dependent manner and therefore may affect their function, leading to influence on neuron–glia interaction. T3 is an important signaling factor that affects microglial functions such as migration and phagocytosis via complex mechanisms. Therefore, dysfunction of THs may impair glial function as well as neuronal function and thus disturb the brain, which may cause

Vitamins and Hormones, Volume 106
ISSN 0083-6729
http://dx.doi.org/10.1016/bs.vh.2017.05.005

mental disorders. Investigations on molecular and cellular basis of hyperthyroidism and hypothyroidism will help us to understand changes in neuron–glia interaction and therefore consequent psychiatric symptoms.

1. INTRODUCTION

Thyroid hormones (THs) are important for the development and function not only in periphery but also in the central nervous system (CNS) (Alkemade, 2010; Stenzel & Huttner, 2013; Zoeller & Rovet, 2004). In the CNS, THs are not only important for neuronal development (Gould & Butcher, 1989; Rami, Patel, & Rabie, 1986) but it also supports development of microglia (Lima et al., 2001), astrocytes (Gould, Frankfurt, Westlind-Danielsson, & McEwen, 1990; Manzano, Bernal, & Morte, 2007) including radial glial cells (Martinez-Galan, Escobar del Rey, Morreale de Escobar, Santacana, & Ruiz-Marcos, 2004), and oligodendrocytes (Jones, Jolson, Cuta, Mariash, & Anderson, 2003; Walravens & Chase, 1969). Due to recent advances in glial cell biology, the role of THs in glial cells is becoming evident. Though the research on THs–glia interaction may be still behind compared to that on THs–neuron interaction, the effect of THs on glial cells and its impact on brain function is gaining an attention, because it may give us a hint why abnormal THs level gives not only physical but psychological symptoms. In addition, it may help us to innovate a new strategy to prevent cognitive impairment in elderly people due to subclinical hypo- and hyperthyroidism, as high prevalence syndrome. In this chapter, the effects of THs on glial cells and how glial cells respond to abnormal level of THs are summarized.

2. RECEPTORS AND FUNCTION OF TH IN THE CNS

In mammals, there are several TR isoforms: TRα1, TRα2, TRβ1, TRβ2a, and TRβ3. Among these subtypes, TRα1, TRβ1, TRβ2, and TRβ3 are reported to act as functional TH binding receptors. TRα2 acts as antagonist of other TRs by competing for DNA binding, which does not have T3-binding activity. TRα1 is predominantly and widely expressed in the developing brain (Koenig et al., 1989; Macchia et al., 2001).

As for the functions of THs, in addition to genomic effects of THs, nongenomic signaling of THs through a plasma membrane-localized receptor has been described (Kalyanaraman et al., 2014; Mori et al., 2015).

Potential mechanism for integrating regulation of development and metabolism by THs and receptor tyrosine kinases through association of TRβ with PI3K was also suggested (Martin et al., 2014). These nongenomic effects of T3 seem to be important for glial function. In addition, it was suggested that the heterogeneity of TR expression throughout brain regions and between different cell types might lead to diverse effects on neuronal morphogenesis. The complexity may also result from not only the direct action of the hormone on neurons but also from indirect actions triggered by astrocytes (Dezonne, Lima, Trentin, & Gomes, 2015) or other glial cell type including microglia.

3. ABNORMAL LEVEL OF TH IN THE CNS

Hyperthyroidism and hypothyroidism result from overactivation or suppression of thyroid gland, leading to either excessive or insufficient production of THs. The prevalence of subclinical hyperthyroidism ranges from 1% to 15%, and of subclinical hypothyroidism from 3% to 16% in individuals aged 60 years and older was reported. In addition, there are differences in age, gender, and dietary iodine intake in the populations studied (Biondi & Cooper, 2008).

Under abnormal level of THs, development is disturbed not only in periphery but also in the CNS, causing severe and irreversible changes to the overall architecture and function of human brain and leading to various neurological dysfunctions (Di Liegro, 2008; Duntas & Maillis, 2013; Henrichs et al., 2010). In adulthood, abnormal level of THs often causes psychological and behavioral abnormality, suggesting abnormality or dysfunction of the CNS.

In many respects the hypothyroid brain appears morphologically normal. However, clinical observations reported that hypothyroidism may be associated with both neurological and behavioral abnormalities with mental retardation, ataxia, and spasticity (Thompson & Potter, 2000). As an influence of hypothyroidism during development, decrease in cognitive performance by decreasing hippocampal synaptic plasticity is reported (Salazar, Cisternas, Codocedo, & Inestrosa, 2017). Adult-onset hypothyroidism is one of the most common causes of cognitive impairment (Dugbartey, 1998; Mallett et al., 1995). Chronic hypothyroidism can lead to an insidious onset of psychiatric symptoms in patients, such as dementia, delirium, psychosis, hallucinations, and coma (Mavroson, Patel, & Akker, 2017). Psychiatric symptoms of hypothyroidism can include psychosis, mood instability, mania, hypersomnia, apathy, anergia, impaired memory mimicking dementia (Goh, Chiu, & Shen, 2014; Osterweil et al., 1992), psychomotor slowing, and attentional problems (Awad, 2000). The incidence of hypothyroidism increases with age.

When it comes to the metabolism and balance of TH levels in hypothyroidism, it is not simple in the CNS, unlike in periphery. In an animal model of hypothyroidism using propylthiouracil, a TH inhibitor, D2 in heart and testis is rather upregulated (Wagner et al., 2003), suggesting D2 in astrocyte as a major deiodinase in the brain is upregulated as well. In fact, both mRNA concentration and the activity of D2 are increased in hypothyroidism (van Doorn, Roelfsema, & van der Heide, 1982; van Doorn, van der Heide, & Roelfsema, 1983) and decreased in hyperthyroidism (Burmeister, Pachucki, & St Germain, 1997; Croteau, Davey, Galton, & St Germain, 1996; Leonard, Kaplan, Visser, Silva, & Larsen, 1981; van Doorn, van der Heide, & Roelfsema, 1984). Thus, D2 is believed to serve a protective role to preserve the concentration of intracerebral T3 during states of thyroid hormone deficiency. This may explain, in part, why both hypo- and hyperthyroidism cause similar neurological symptoms.

Hyperthyroidism may induce emotional lability, impatience and irritability, distractible overactivity, exaggerated sensitivity to noise, and problems with sleep and the appetite (Awad, 2000), or depression and anxiety (Demet et al., 2002). Importantly, even at subclinical level, hyperthyroidism, as well as hypothyroidism, in the elderly is suggested to increase the risk of cognitive decline, dementia, and Alzheimer's disease (AD) (Kalmijn et al., 2000; van Osch, Hogervorst, Combrinck, & Smith, 2004; Wijsman et al., 2013).

It has been reported that both hyper- and hypothyroidism can affect the immune system (De Vito et al., 2012; Klecha, Arcos, Frick, Genaro, & Cremaschi, 2008). Hyperthyroidism decreases the proinflammatory activities of monocytes and macrophages. On the other hand, during hypothyroidism enhancement of phagocytosis and increased levels of reactive oxygen species may occur so that the expression of proinflammatory molecules such as macrophage inflammatory protein-1α and interleukin-1β increases (De Vito et al., 2011). In contrast, hypothyroidism is reported to produce opposite effects on parameters of the immune function, such as decrease in immune response, antibody production, cell migration, and lymphocyte proliferation markers (Klecha et al., 2006, 2000), antioxidant enzymes, and their activity (De Vito et al., 2011). It seems that the relationship between THs and immune cells is complex, depending on the cell type.

Studies in the periphery suggest the similar change of immune system in the CNS. The role of microglia, an immune cell population in the CNS, in the relationship between thyroid dysfunctions and neuropsychological disorders is still under investigation.

4. TRANSPORTATION OF THS TO THE BRAIN AND METABOLISM IN ASTROCYTES

How do THs work in the CNS? Circulating T4 (L-thyroxine) crosses blood–brain barrier via specific transporters such as organic anion-transporting polypeptides (OATPs) containing OATP14/SLCO1C1 (OATP1c1) (Sugiyama, Kusuhara, Taniguchi, et al., 2003; Tohyama, Kusuhara, & Sugiyama, 2004) and OATP1a2 (Hagenbuch, 2007; Lee et al., 2005), L-type amino acid transporters (LAT), mainly LAT1 (Taylor & Ritchie, 2007), and monocarboxylate transporters 8 (MCT8) (SLC16A2) (Roberts et al., 2008). In the brain parenchyma, T4 is taken up by astrocytes via LAT2, where it is deiodinated by type 2-deiodinase (D2) to produce T3. Subsequently T3 is released via LAT (Blondeau, Beslin, Chantoux, & Francon, 1993; Francon, Chantoux, & Blondeau, 1989), presumably LAT2, and taken by other cell types via distinct transporters; for example, neurons express MCT8, while microglial cells express OATP4a1, LAT2, and MCT10 (Braun et al., 2011). Then T3 exerts various effects supporting CNS development, such as axonal myelination and dendrites growth (Bernal & Nunez, 1995; Legrand, 1982; Porterfield & Hendrich, 1993), astrocytes and oligodendrocytes differentiation (Baxi et al., 2014; Jones et al., 2003; Manzano et al., 2007; Schoonover et al., 2004), and microglial development (Lima et al., 2001) (Fig. 1).

5. FUNCTIONAL ROLE OF TH IN MICROGLIA

The expressions of TRα1 and TRβ1 have been identified in primary cultured rat microglia (Lima et al., 2001). Nongenomic effects of T3 on microglial functions were also reported, both *in vitro* and *in vivo* (Mori et al., 2015).

5.1 Effects of TH on Microglia *In Vitro*

T3 acts as an inducer of microglial migration and phagocytosis. Primary cultured rodent microglial cells, the purity of the cell population from mixed glial culture being usually more than 90%, are often used to study the

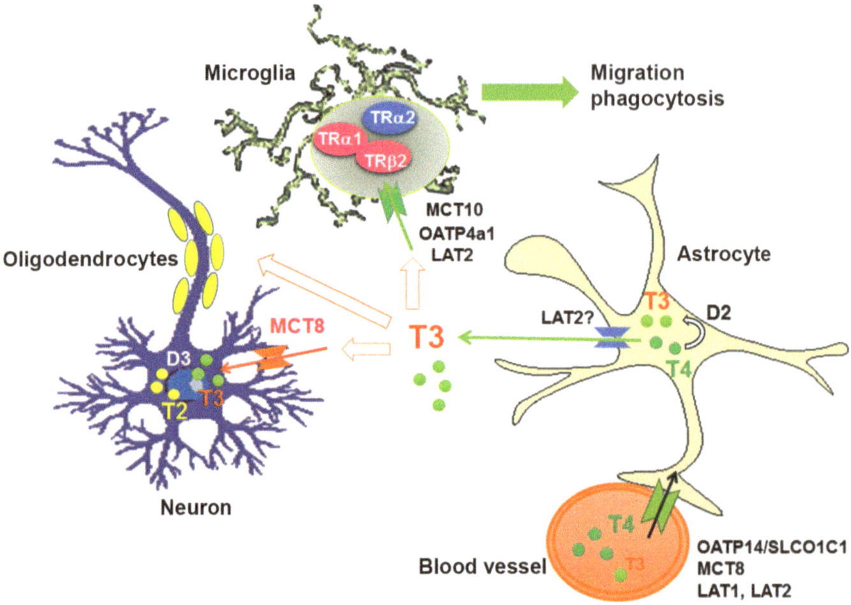

Fig. 1 Transport of thyroid hormones (THs) to the brain and their metabolism (*D*, iodothyronine deiodinases; *LAT*, L-type amino acid transporter; *MCT*, monocarboxylate transporter; *OATP*, organic anion-transporting polypeptide; *T2*, diiodothyronines; *T3*, triiodothyronine; *T4*, thyroxine; *TR*, thyroid hormone receptor). *From Noda, M. (2015). Possible role of glial cells in the relationship between thyroid dysfunction and mental disorders.* Frontiers in Cellular Neuroscience, 9, *194.*

function of microglia (Fig. 2A). T3 increases microglial chemotaxis dose dependently (Fig. 2B and C). Phagocytosis is able to analyze using fluorescent beads. By staining microglial cells with different color of fluorescence, three-dimensional (3D)-reconstructed images from *Z*-stack images observed by confocal laser microscope can tell whether beads are phagocytosed or just sitting on the cells (Fig. 3A and B). T3 increases the number of phagocytosed beads in microglia, though the effect is not concentration dependent (Fig. 3C and D). T3 also induces membrane ruffling which is closely related to microglial migration (Mori et al., 2015).

5.2 Effects of TH on Microglia *In Vivo*

Microglial cells are highly motile *in vivo* as well. Especially microglial processes are constantly motile, extending and retracting as if they are surveying the circumstance (Davalos et al., 2005; Nimmerjahn, Kirchhoff, & Helmchen, 2005). The first chemoattractant for microglia even under physiological conditions was adenosine triphosphate (ATP). More endogenous

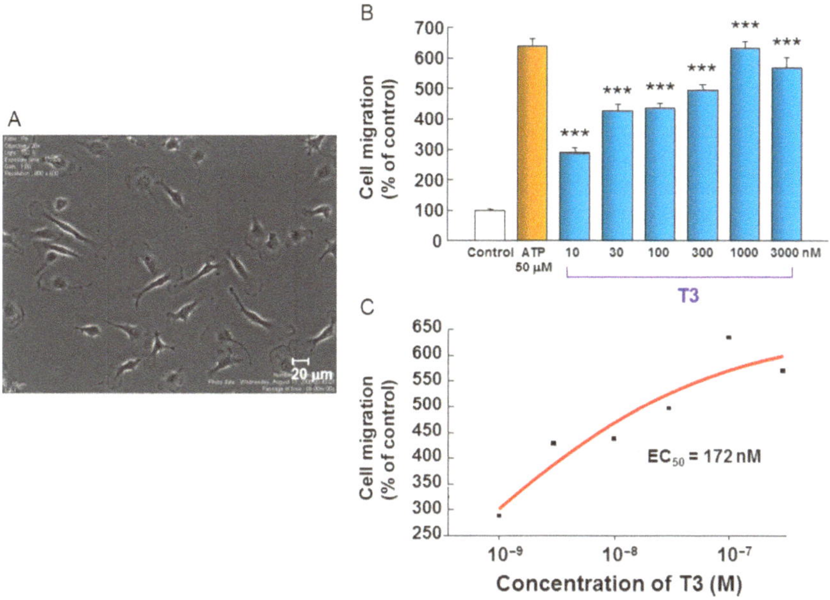

Fig. 2 L-Triiodothyronine (T3) acts as an inducer of microglial migration *in vitro*. (A) Differential interference contrast microscope (DIC) image of cultured mouse microglia in the presence of 10 μM T3. (B) Relative increase in microglial chemotaxis induced by T3 (10, 30, 100, 300 nM; 1 and 3 μM). ATP (50 μM) is used as a positive control. Microglial chemotactic migration was tested with 48-well microchemotaxis Boyden chamber. $^{***}P < 0.001$ vs control (Scheffe's test). (C) Concentration-dependent effect of T3 on microglial chemotaxis with half maximum effective concentration (EC_{50}) of 172 nM. *Panel (B): From Mori, Y., Tomonaga, D., Kalashnikova, A., Furuya, F., Akimoto, N., Ifuku, M., et al. (2015). Effects of 3, 3′, 5-triiodothyronine on microglial functions.* Glia, *63, 906–920.*

microglial chemoattractants have been shown, such as bradykinin (Ifuku et al., 2007; Shi et al., 2013) and galanin (Ifuku et al., 2011), many of them belonging to neuropeptides (Fleisher-Berkovich et al., 2010; Noda et al., 2011).

Microglial migration *in vivo* is often shown in stab wound model (Fig. 4A). In the brain lesion area, microglial cells, presumably together with invading macrophages, are attracted by various substances including ATP. Injection of T3 further increases the number of microglia around lesion (Fig. 4B and C). These results suggest that T3 may play an important role in the CNS by influencing microglial migration.

T3 increases microglial phagocytosis *in vivo* as well. Using fluorescent-labeled amyloid β peptide ($A\beta_{1-42}$), microglial phagocytosis was significantly stimulated in T4-injected mouse brain (Noda, Mori, & Yoshioka, 2016).

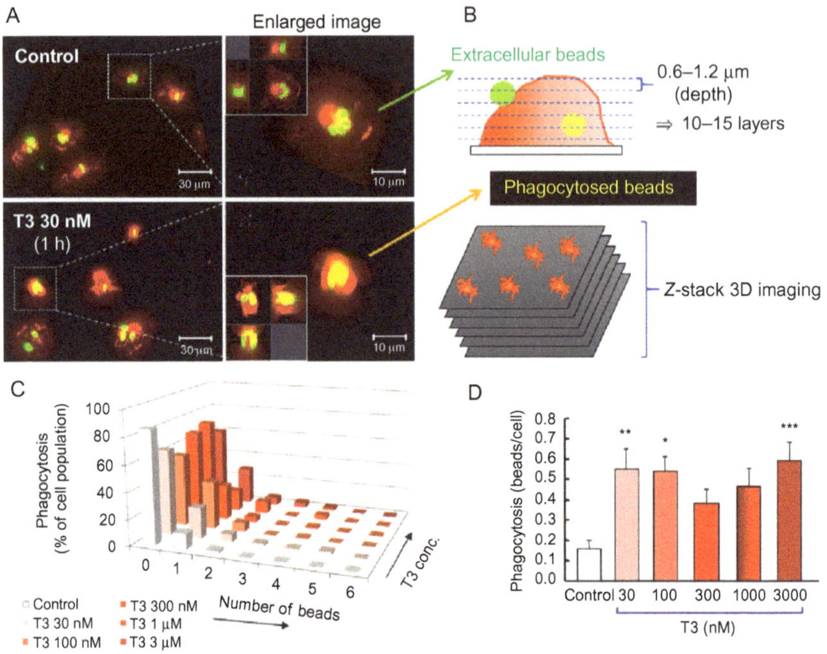

Fig. 3 T3 increases microglial phagocytosis *in vitro*. (A) 3D images of phagocytic microglia of control and T3 (30 nM)-treated groups with fluorescent zymosan beads. (B) Schematic diagram showing how extracellular beads and phagocytosed beads are distinguished with *Z*-stack 3D imaging using confocal laser microscope. (C) The relationship of relative cell population with number of phagocytosed beads with T3 (0, 10, 30, 100, 300 nM; 1 and 3 μM). (D) Mean ± S.E.M. of the number of phagocytosed beads with T3 (0, 10, 30, 100, 300 nM; 1 and 3 μM) shown in (C). $^{*}P < 0.05$, $^{**}P < 0.01$, $^{***}P < 0.005$ vs control (Bonferroni/Dunnett test). *Partially from Mori, Y., Tomonaga, D., Kalashnikova, A., Furuya, F., Akimoto, N., Ifuku, M., et al. (2015). Effects of 3, 3′, 5-triiodothyronine on microglial functions.* Glia, 63, *906–920.*

It may suggest the possibility of increased Aβ clearance in hyperthyroidism, though there is rather opposite report that hyperthyroidism, even at subclinical level, in the elderly is suggested to remarkably increase the risk of cognitive decline, dementia, and AD (Kalmijn et al., 2000; van Osch et al., 2004; Wijsman et al., 2013).

5.3 Signaling Cascade of TH in Microglia

T3 might interact with a plasma membrane-associated TH receptor α variant (Kalyanaraman et al., 2014), and with cytoplasmic TH receptor β (Martin et al., 2014), while T4 interacts with integrin $\alpha_v\beta_3$ (Cheng,

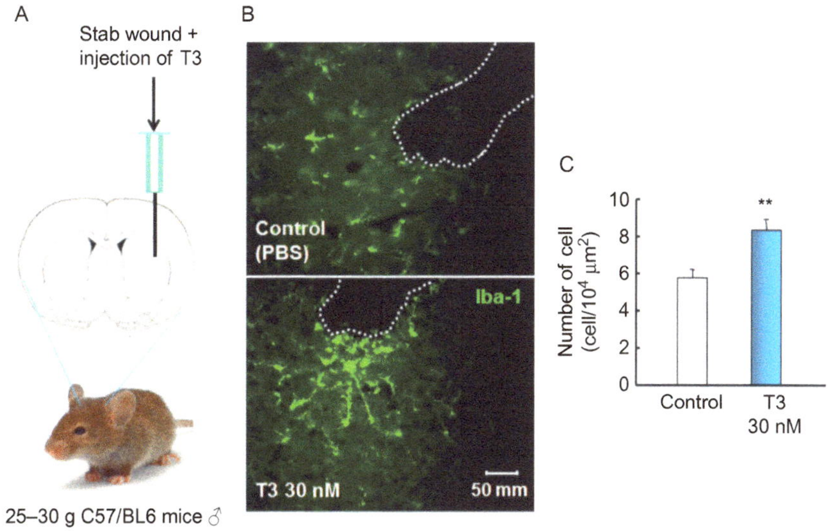

Fig. 4 THs increase microglial migration *in vivo*. (A) Experimental diagram. Stab wound is made by inserting a 2-μL Hamilton syringe with a needle into the right cortex through a small hole drilled through the skull. Either T3 (30 nM) or PBS is injected. After 18 h, each mouse is deeply anesthetized, the brain is removed, postfixed, and is followed by immunohistochemical procedure. (B) Confocal images of microglia in the sections containing lesion. Microglial cells are stained using rabbit antimouse Iba-1 IgG. (C) The number of microglia around lesion is increased in T3-injected brain compared to that in the control (PBS-injected) brain. $^{**}P < 0.01$ vs control (Scheffe's test). *Partly from Mori, Y., Tomonaga, D., Kalashnikova, A., Furuya, F., Akimoto, N., Ifuku, M., et al. (2015). Effects of 3, 3′, 5-triiodothyronine on microglial functions.* Glia, 63, *906–920.*

Leonard, & Davis, 2010; Davis, Leonard, & Davis, 2008; Sugiyama, Kusuhara, Lee, et al., 2003; Zhao, Gao, Wang, & Goldman, 2000), though both nongenomic and genomic effects can overlap in the nucleus (Zhao et al., 2000). From pharmacological investigation *in vitro*, microglial migration is mediated through T3 uptake by TH transporters and binding to the TRs. The TH signaling in microglia involved several signaling pathways including PI3K, and MAPK/ERK (Mori et al., 2015), similar to the one for ATP-induced microglial migration (Honda et al., 2001). T3-induced nitric oxide signaling (Kalyanaraman et al., 2014) is also present in microglia. In addition, Na^+/K^+-ATPase, Na^+/Ca^{2+} exchanger, operating in the reverse mode and consequently inducing Ca^{2+} influx, are involved in T3-induced microglial migration (Mori et al., 2015) (Fig. 5).

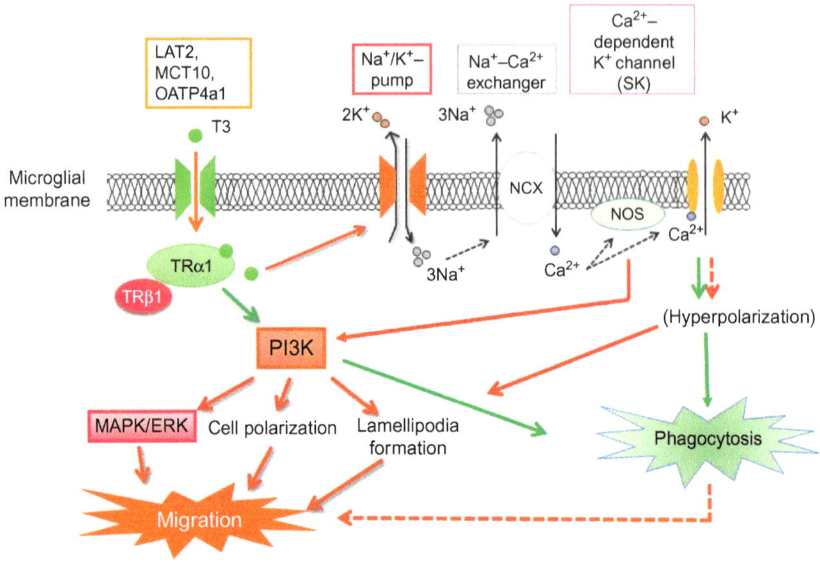

Fig. 5 T3-induced signal transduction in microglia. *PI3K*, phosphoinositide 3-kinase; *MAPK/ERK*, mitogen-activated protein kinase/extracellular signal-regulated kinase; *NCX*, Na^+/Ca^{2+} exchanger; *NOS*, nitric oxide synthase. *Modified from Mori, Y., Tomonaga, D., Kalashnikova, A., Furuya, F., Akimoto, N., Ifuku, M., et al. (2015). Effects of 3, 3′, 5-triiodothyronine on microglial functions.* Glia, 63, *906–920.*

6. ASTROCYTES AND OLIGODENDROCYTES DIFFERENTIATION BY T3

There are not many reports on the effects of T3 on other glial cells, mainly astrocytes and oligodendrocytes, in the adult brain. In developing CNS, the relationship between THs and astrocytes is summarized in a review (Dezonne et al., 2015). Briefly, T3 exerts numerous effects regulating axonal myelination and dendritic growth (Bernal & Nunez, 1995; Legrand, 1982; Porterfield & Hendrich, 1993; Vose et al., 2013; Walravens & Chase, 1969). Astrocyte and oligodendrocyte differentiation are also influenced by THs (Baxi et al., 2014; Dezonne, Stipursky, & Gomes, 2009; Jones et al., 2003; Manzano et al., 2007; Martinez-Galan et al., 2004, 1997; Schoonover et al., 2004). Differentiation of human-cultured CD34+ stem cells into oligodendrocyte precursors under THs action was also reported (Venkatesh et al., 2014). Expression alterations of genes using hypothyroidism model rats showed that immature astrocytes immunoreactivity for vimentin and glial fibrillary acidic protein (GFAP) was increased, while oligodendrocyte lineage transcription factor 2 was decreased in the corpus callosum (Shiraki et al., 2014). Effects and molecular mechanisms of T3 action on astrocytes and oligodendrocytes in matured or aged brain remain to be investigated (Noda, 2015).

7. PATHOLOGICAL EFFECTS OF TH ON GLIAL CELLS

Since impairment of TH leads to various neurological dysfunctions, THs may have protective effects nongenomically or genomically on neurons and glial cells, for example, in the setting of acute brain ischemia (Henrichs et al., 2010). Abnormal level of THs affects glial morphology and function in mouse model of thyroidism. For example, hyperthyroidism activates glial cells, at least morphologically, in mouse model (Noda et al., 2016). Hyperthyroid model is relatively easily made, giving T4 either orally or by intraperitonially for a couple of weeks (Fig. 6A).

8. SEX- AND AGE-DEPENDENT EFFECTS OF HYPERTHYROIDISM

In the immunohistochemical analyses in the cerebral cortex and hippocampus, fluorescent intensity of cell markers for both microglia (Iba-1) and astrocytes (GFAP) is stronger in hyperthyroidism only in male mice (Fig. 6B). On the contrary, fluorescent intensity for both Iba-1 and GFAP is significantly decreased in female mice (Fig. 6B). Interestingly the effects of high level of THs are not only sex dependent as mentioned earlier but also age dependent. Though glial activation was observed in young male mice, the Iba-1 and GFAP immunoreactivities in cerebral cortex are decreased in old (>1-year-old) male hyperthyroid mice: both Iba-1 and GFAP in cortex and only GFAP in hippocampus (Noda et al., 2016). In old female mice, no apparent change (GFAP) or rather decrease (Iba-1) in immunofluorescence is observed.

Though the molecular or genetic background of sex- and age-dependent effects of THs needs to be investigated, it will be explained why the prevalence of hyperthyroidism or hypothyroidism is much higher in female. The age-dependent effect of THs may also explain the increased risk of dementia in elderly people with thyroidism.

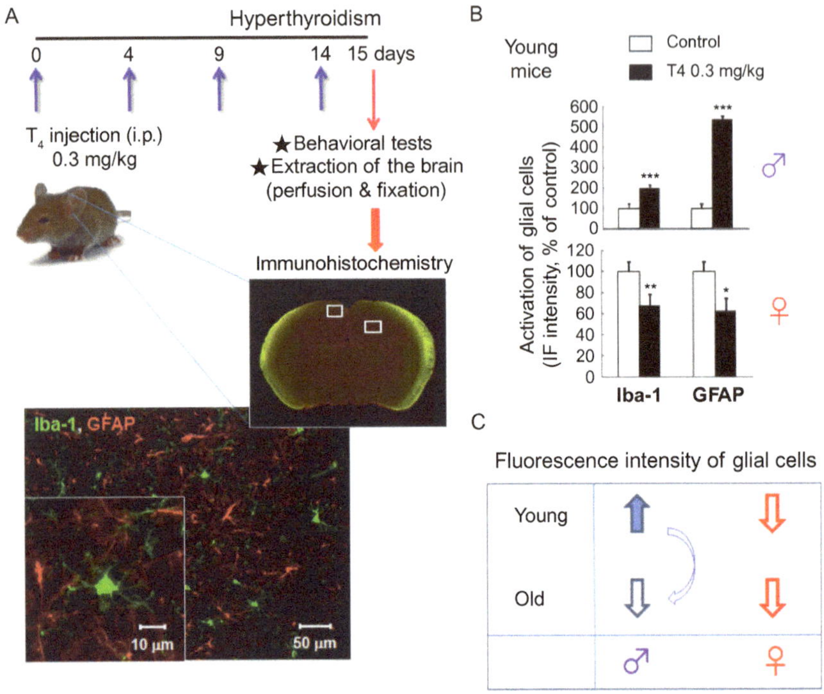

Fig. 6 Sex-dependent glial activation in hyperthyroid model of young mice. (A) Experimental diagram. Hyperthyroidism is induced by intraperitoneally injecting T4 (0.3 mg/kg) four times during 2 weeks. Immunohistochemical images show microglia (Iba-1, *green*) and astrocyte (GFAP, *red*). (B) The relative fluorescence intensity of Iba-1 and GFAP immunoreactivity in cerebral cortex of young male (♂: *upper panels*) and female (♀: *lower panels*) with or without treatment of T4. *$P < 0.05$, **$P < 0.01$, ***$P < 0.005$ vs control (Scheffe's test). (C) Table showing increase or decrease of fluorescent intensity of glial cells (both microglia and astrocytes) in male and female mouse brain when they are young (8 week old) and old (over 1 year old). *Partially from Noda, M., Mori, Y., Yoshioka, Y. (2016). Sex- and age-dependent effects of thyroid hormone on glial morphology and function.* Opera Medica et Physiologica. *2(2), 164–171.*

9. POSSIBLE INFLUENCE OF TH-INDUCED NEURON–GLIA INTERACTION ON PSYCHIATRIC SYMPTOMS

Therapeutic use of THs in psychiatric disorders, mainly in depression, came to the light, thus contributing to better understanding the action of THs in the brain (Weissel, 1999). In an animal model, it is shown that TH regulates anxiety in male mice (Buras, Battle, Landers, Nguyen, & Vasudevan, 2014). Perhaps the major role of thyroxine therapy on depression might be due to hypothalamus–pituitary–thyroid axis activity and serotonin function in depressive episodes (Gomes, Lima, Trentin, & Moura Neto, 2001). From a transcranial magnetic stimulation study, it was shown that hypothyroidism patients and patients with resistance to TH, with elevated THs level and unsuppressed thyroid-stimulating hormone, showed that the inhibitory cortical circuits were affected compared with euthyroid controls, but in opposite ways, suggesting THs may influence cortical excitability and cortical inhibitory circuits (Terranova et al., 2016). Glial cells may also contribute to these changes in neural excitability and therefore psychiatric symptoms. For example, in activated microglia and astrocytes, immunologically induced fatigue was reported (Ifuku, Hossain, Noda, & Katafuchi, 2014). Microglial oxidative reactions in schizophrenia (Kato et al., 2011; Monji et al., 2013) and alteration of astrocytes or oligodendrocyte function in bipolar disorder (Dong & Zhen, 2015) have also been reported. On the other hand, decreased glial density in association with glial hypotrophy in bipolar disorder or major depression (Bowley, Drevets, Ongur, & Price, 2002; Rajkowska, Halaris, & Selemon, 2001) is reported. Considering these reports, it is likely that indirect actions of THs through glial cells are important for neuronal activity.

Interestingly there was no significant difference between the quality of life of patients with hypothyroidism and that of the normal people; however, the comparison of mental health level of hypothyroidism patients and normal people showed a significant difference at all aspects (Rakhshan, Ghanbari, Rahimi, & Mostafavi, 2017). Influence of THs on neuron–glia interaction in higher brain function may give us a hint to understand these observations.

10. CONCLUSION AND FUTURE DIRECTIONS

T3 is important not only for neuronal development but also for differentiation of astrocytes and oligodendrocytes, and for microglial development. In adult brain, T3 affects glial morphology and therefore may affect function, leading to influence on neuron–glia interaction. T3 is an important signaling factor that affects microglial functions via complex mechanisms. Therefore, dysfunction of THs may impair glial function as well as neuronal function and thus disturb the brain, which may cause mental disorders. Therefore, contribution of glial cells to neuropsychological disorders, especially in the TH-sensitive regions of the brain (Fonseca et al., 2013), needs to be considered. Investigations on molecular and cellular basis of hyperthyroidism and hypothyroidism will help us to understand changes in neuron–glia interaction and therefore consequent psychiatric symptoms.

ACKNOWLEDGMENT

We thank Research Support Center, Graduate School of Medical Sciences, Kyushu University for their experimental help.

Conflicting interests: The authors declare no conflicting interests.

REFERENCES

Alkemade, A. (2010). Central and peripheral effects of thyroid hormone signalling in the control of energy metabolism. *Journal of Neuroendocrinology*, *22*(1), 56–63.

Awad, A. G. (2000). The thyroid and the mind and emotions/thyroid dysfunction and mental disorders. *Thyroid Foundation in Canada, reprinted from Thyrobulletin*, 7(3). http://www.thyroid.ca/e10f.php.

Baxi, E. G., Schott, J. T., Fairchild, A. N., Kirby, L. A., Karani, R., Uapinyoying, P., et al. (2014). A selective thyroid hormone beta receptor agonist enhances human and rodent oligodendrocyte differentiation. *Glia*, *62*(9), 1513–1529.

Bernal, J., & Nunez, J. (1995). Thyroid hormones and brain development. *European Journal of Endocrinology*, *133*(4), 390–398.

Biondi, B., & Cooper, D. S. (2008). The clinical significance of subclinical thyroid dysfunction. *Endocrine Reviews*, *29*(1), 76–131.

Blondeau, J. P., Beslin, A., Chantoux, F., & Francon, J. (1993). Triiodothyronine is a high-affinity inhibitor of amino acid transport system L1 in cultured astrocytes. *Journal of Neurochemistry*, *60*(4), 1407–1413.

Bowley, M. P., Drevets, W. C., Ongur, D., & Price, J. L. (2002). Low glial numbers in the amygdala in major depressive disorder. *Biological Psychiatry*, *52*(5), 404–412.

Braun, D., Kinne, A., Brauer, A. U., Sapin, R., Klein, M. O., Kohrle, J., et al. (2011). Developmental and cell type-specific expression of thyroid hormone transporters in the mouse brain and in primary brain cells. *Glia*, *59*(3), 463–471.

Buras, A., Battle, L., Landers, E., Nguyen, T., & Vasudevan, N. (2014). Thyroid hormones regulate anxiety in the male mouse. *Hormones and Behavior*, *65*, 88–96.

Burmeister, L. A., Pachucki, J., & St Germain, D. L. (1997). Thyroid hormones inhibit type 2 iodothyronine deiodinase in the rat cerebral cortex by both pre- and posttranslational mechanisms. *Endocrinology*, *138*(12), 5231–5237.

Cheng, S. Y., Leonard, J. L., & Davis, P. J. (2010). Molecular aspects of thyroid hormone actions. *Endocrine Reviews*, *31*(2), 139–170.

Croteau, W., Davey, J. C., Galton, V. A., & St Germain, D. L. (1996). Cloning of the mammalian type II iodothyronine deiodinase. A selenoprotein differentially expressed and regulated in human and rat brain and other tissues. *The Journal of Clinical Investigation*, *98*(2), 405–417.

Davalos, D., Grutzendler, J., Yang, G., Kim, J. V., Zuo, Y., Jung, S., et al. (2005). ATP mediates rapid microglial response to local brain injury in vivo. *Nature Neuroscience*, *8*(6), 752–758.

Davis, P. J., Leonard, J. L., & Davis, F. B. (2008). Mechanisms of nongenomic actions of thyroid hormone. *Frontiers in Neuroendocrinology*, *29*(2), 211–218.

De Vito, P., Balducci, V., Leone, S., Percario, Z., Mangino, G., Davis, P. J., et al. (2012). Nongenomic effects of thyroid hormones on the immune system cells: New targets, old players. *Steroids*, 77(10), 988–995.

De Vito, P., Incerpi, S., Pedersen, J. Z., Luly, P., Davis, F. B., & Davis, P. J. (2011). Thyroid hormones as modulators of immune activities at the cellular level. *Thyroid*, *21*(8), 879–890.

Demet, M. M., Ozmen, B., Deveci, A., Boyvada, S., Adiguzel, H., & Aydemir, O. (2002). Depression and anxiety in hyperthyroidism. *Archives of Medical Research*, *33*(6), 552–556.

Dezonne, R. S., Lima, F. R., Trentin, A. G., & Gomes, F. C. (2015). Thyroid hormone and astroglia: Endocrine control of the neural environment. *Journal of Neuroendocrinology*, *27*(6), 435–445.

Dezonne, R. S., Stipursky, J., & Gomes, F. C. (2009). Effect of thyroid hormone depletion on cultured murine cerebral cortex astrocytes. *Neuroscience Letters*, *467*(2), 58–62.

Di Liegro, I. (2008). Thyroid hormones and the central nervous system of mammals (review). *Molecular Medicine Reports*, *1*(3), 279–295.

Dong, X. H., & Zhen, X. C. (2015). Glial pathology in bipolar disorder: Potential therapeutic implications. *CNS Neuroscience & Therapeutics*, *21*(5), 393–397.

Dugbartey, A. T. (1998). Neurocognitive aspects of hypothyroidism. *Archives of Internal Medicine*, *158*(13), 1413–1418.

Duntas, L. H., & Maillis, A. (2013). Hypothyroidism and depression: Salient aspects of pathogenesis and management. *Minerva Endocrinologica*, *38*(4), 365–377.

Fleisher-Berkovich, S., Filipovich-Rimon, T., Ben-Shmuel, S., Hulsmann, C., Kummer, M. P., & Heneka, M. T. (2010). Distinct modulation of microglial amyloid beta phagocytosis and migration by neuropeptides (i). *Journal of Neuroinflammation*, *7*, 61.

Fonseca, T. L., Correa-Medina, M., Campos, M. P. O., Wittmann, G., Werneck-de-Castro, J. P., Drigo, R. A. E., et al. (2013). Coordination of hypothalamic and pituitary T3 production regulates TSH expression. *Journal of Clinical Investigation*, *123*(4), 1492–1500.

Francon, J., Chantoux, F., & Blondeau, J. P. (1989). Carrier-mediated transport of thyroid hormones into rat glial cells in primary culture. *Journal of Neurochemistry*, *53*(5), 1456–1463.

Goh, K. K., Chiu, Y. H., & Shen, W. W. (2014). Hashimoto's encephalopathy mimicking presenile dementia. *General Hospital Psychiatry*, *36*(3). 360.e369–311.

Gomes, F. C., Lima, F. R., Trentin, A. G., & Moura Neto, V. (2001). Thyroid hormone role in nervous system morphogenesis. *Progress in Brain Research*, *132*, 41–50.

Gould, E., & Butcher, L. L. (1989). Developing cholinergic basal forebrain neurons are sensitive to thyroid hormone. *The Journal of Neuroscience*, *9*(9), 3347–3358.

Gould, E., Frankfurt, M., Westlind-Danielsson, A., & McEwen, B. S. (1990). Developing forebrain astrocytes are sensitive to thyroid hormone. *Glia*, *3*(4), 283–292.

Hagenbuch, B. (2007). Cellular entry of thyroid hormones by organic anion transporting polypeptides. *Best Practice & Research. Clinical Endocrinology & Metabolism*, *21*(2), 209–221.

Henrichs, J., Bongers-Schokking, J. J., Schenk, J. J., Ghassabian, A., Schmidt, H. G., Visser, T. J., et al. (2010). Maternal thyroid function during early pregnancy and cognitive functioning in early childhood: The generation R study. *The Journal of Clinical Endocrinology and Metabolism*, *95*(9), 4227–4234.

Honda, S., Sasaki, Y., Ohsawa, K., Imai, Y., Nakamura, Y., Inoue, K., et al. (2001). Extracellular ATP or ADP induce chemotaxis of cultured microglia through Gi/o-coupled P2Y receptors. *The Journal of Neuroscience*, *21*(6), 1975–1982.

Ifuku, M., Farber, K., Okuno, Y., Yamakawa, Y., Miyamoto, T., Nolte, C., et al. (2007). Bradykinin-induced microglial migration mediated by B1-bradykinin receptors depends on Ca2+ influx via reverse-mode activity of the Na+/Ca2+ exchanger. *The Journal of Neuroscience*, *27*(48), 13065–13073.

Ifuku, M., Hossain, S. M., Noda, M., & Katafuchi, T. (2014). Induction of interleukin-1beta by activated microglia is a prerequisite for immunologically induced fatigue. *The European Journal of Neuroscience*, *40*(8), 3253–3263.

Ifuku, M., Okuno, Y., Yamakawa, Y., Izumi, K., Seifert, S., Kettenmann, H., et al. (2011). Functional importance of inositol-1,4,5-triphosphate-induced intracellular Ca2+ mobilization in galanin-induced microglial migration. *Journal of Neurochemistry*, *117*(1), 61–70.

Jones, S. A., Jolson, D. M., Cuta, K. K., Mariash, C. N., & Anderson, G. W. (2003). Triiodothyronine is a survival factor for developing oligodendrocytes. *Molecular and Cellular Endocrinology*, *199*(1–2), 49–60.

Kalmijn, S., Mehta, K. M., Pols, H. A., Hofman, A., Drexhage, H. A., & Breteler, M. M. (2000). Subclinical hyperthyroidism and the risk of dementia. The Rotterdam study. *Clinical Endocrinology*, *53*(6), 733–737.

Kalyanaraman, H., Schwappacher, R., Joshua, J., Zhuang, S., Scott, B. T., Klos, M., et al. (2014). Nongenomic thyroid hormone signaling occurs through a plasma membrane-localized receptor. *Science Signaling*, 7(326), ra48.

Kato, T. A., Monji, A., Yasukawa, K., Mizoguchi, Y., Horikawa, H., Seki, Y., et al. (2011). Aripiprazole inhibits superoxide generation from phorbol-myristate-acetate (PMA)-stimulated microglia in vitro: Implication for antioxidative psychotropic actions via microglia. *Schizophrenia Research*, *129*(2–3), 172–182.

Klecha, A. J., Arcos, M. L. B., Frick, L., Genaro, A. M., & Cremaschi, G. (2008). Immune-endocrine interactions in autoimmune thyroid diseases. *Neuroimmunomodulation*, *15*(1), 68–75.

Klecha, A. J., Genaro, A. M., Gorelik, G., Barreiro Arcos, M. L., Silberman, D. M., Schuman, M., et al. (2006). Integrative study of hypothalamus-pituitary-thyroid-immune system interaction: Thyroid hormone-mediated modulation of lymphocyte activity through the protein kinase C signaling pathway. *The Journal of Endocrinology*, *189*(1), 45–55.

Klecha, A. J., Genaro, A. M., Lysionek, A. E., Caro, R. A., Coluccia, A. G., & Cremaschi, G. A. (2000). Experimental evidence pointing to the bidirectional interaction between the immune system and the thyroid axis. *International Journal of Immunopharmacology*, *22*(7), 491–500.

Koenig, R. J., Lazar, M. A., Hodin, R. A., Brent, G. A., Larsen, P. R., Chin, W. W., et al. (1989). Inhibition of thyroid hormone action by a non-hormone binding c-erbA protein generated by alternative mRNA splicing. *Nature*, *337*(6208), 659–661.

Lee, W., Glaeser, H., Smith, L. H., Roberts, R. L., Moeckel, G. W., Gervasini, G., et al. (2005). Polymorphisms in human organic anion-transporting polypeptide 1A2 (OATP1A2): Implications for altered drug disposition and central nervous system drug entry. *The Journal of Biological Chemistry*, *280*(10), 9610–9617.

Legrand, J. (1982). Thyroid hormones and maturation of the nervous system. *Journal of Physiology, Paris*, *78*(7), 603–652.

Leonard, J. L., Kaplan, M. M., Visser, T. J., Silva, J. E., & Larsen, P. R. (1981). Cerebral cortex responds rapidly to thyroid hormones. *Science*, *214*(4520), 571–573.

Lima, F. R., Gervais, A., Colin, C., Izembart, M., Neto, V. M., & Mallat, M. (2001). Regulation of microglial development: A novel role for thyroid hormone. *The Journal of Neuroscience*, *21*(6), 2028–2038.

Macchia, P. E., Takeuchi, Y., Kawai, T., Cua, K., Gauthier, K., Chassande, O., et al. (2001). Increased sensitivity to thyroid hormone in mice with complete deficiency of thyroid hormone receptor alpha. *Proceedings of the National Academy of Sciences of the United States of America*, *98*(1), 349–354.

Mallett, P., Andrew, M., Hunter, C., Smith, J., Richards, C., Othman, S., et al. (1995). Cognitive function, thyroid status and postpartum depression. *Acta Psychiatrica Scandinavica*, *91*(4), 243–246.

Manzano, J., Bernal, J., & Morte, B. (2007). Influence of thyroid hormones on maturation of rat cerebellar astrocytes. *International Journal of Developmental Neuroscience*, *25*(3), 171–179.

Martin, N. P., Marron Fernandez de Velasco, E., Mizuno, F., Scappini, E. L., Gloss, B., Erxleben, C., et al. (2014). A rapid cytoplasmic mechanism for PI3 kinase regulation by the nuclear thyroid hormone receptor, TRbeta, and genetic evidence for its role in the maturation of mouse hippocampal synapses in vivo. *Endocrinology*, *155*(9), 3713–3724.

Martinez-Galan, J. R., Escobar del Rey, F., Morreale de Escobar, G., Santacana, M., & Ruiz-Marcos, A. (2004). Hypothyroidism alters the development of radial glial cells in the term fetal and postnatal neocortex of the rat. *Brain Research. Developmental Brain Research*, *153*(1), 109–114.

Martinez-Galan, J. R., Pedraza, P., Santacana, M., Escobar del Ray, F., Morreale de Escobar, G., & Ruiz-Marcos, A. (1997). Early effects of iodine deficiency on radial glial cells of the hippocampus of the rat fetus. A model of neurological cretinism. *The Journal of Clinical Investigation*, *99*(11), 2701–2709.

Mavroson, M. M., Patel, N., & Akker, E. (2017). Myxedema psychosis in a patient with undiagnosed Hashimoto thyroiditis. *The Journal of the American Osteopathic Association*, *117*(1), 50–54.

Monji, A., Kato, T. A., Mizoguchi, Y., Horikawa, H., Seki, Y., Kasai, M., et al. (2013). Neuroinflammation in schizophrenia especially focused on the role of microglia. *Progress in Neuro-Psychopharmacology & Biological Psychiatry*, *42*, 115–121.

Mori, Y., Tomonaga, D., Kalashnikova, A., Furuya, F., Akimoto, N., Ifuku, M., et al. (2015). Effects of 3, 3′, 5-triiodothyronine on microglial functions. *Glia*, *63*(5), 906–920.

Nimmerjahn, A., Kirchhoff, F., & Helmchen, F. (2005). Resting microglial cells are highly dynamic surveillants of brain parenchyma in vivo. *Science*, *308*(5726), 1314–1318.

Noda, M. (2015). Possible role of glial cells in the relationship between thyroid dysfunction and mental disorders. *Frontiers in Cellular Neuroscience*, *9*, 194.

Noda, M., Ifuku, M., Okuno, Y., Beppu, K., Mori, Y., & Naoe, S. (2011). Neuropeptides as attractants of immune cells in the brain and their distinct signaling. *Advances in Neuroimmune Biology*, *1*, 53–62.

Noda, M., Mori, Y., & Yoshioka, Y. (2016). Sex- and age-dependent effects of thyroid hormone on glial morphology and function. *Opera Medica et Physiologica*, *2*(2), 164–171.

Osterweil, D., Syndulko, K., Cohen, S. N., Pettler-Jennings, P. D., Hershman, J. M., Cummings, J. L., et al. (1992). Cognitive function in non-demented older adults with hypothyroidism. *Journal of the American Geriatrics Society*, *40*(4), 325–335.

Porterfield, S. P., & Hendrich, C. E. (1993). The role of thyroid hormones in prenatal and neonatal neurological development—Current perspectives. *Endocrine Reviews*, *14*(1), 94–106.

Rajkowska, G., Halaris, A., & Selemon, L. D. (2001). Reductions in neuronal and glial density characterize the dorsolateral prefrontal cortex in bipolar disorder. *Biological Psychiatry*, *49*(9), 741–752.

Rakhshan, M. P., Ghanbari, A. B., Rahimi, A. M., & Mostafavi, I. B. (2017). A comparison between the quality of life and mental health of patients with hypothyroidism and normal people referred to Motahari clinic of Shiraz University of Medical Sciences. *International Journal of Community Based Nursing and Midwifery*, *5*(1), 30–37.

Rami, A., Patel, A. J., & Rabie, A. (1986). Thyroid hormone and development of the rat hippocampus: Morphological alterations in granule and pyramidal cells. *Neuroscience*, *19*(4), 1217–1226.

Roberts, L. M., Woodford, K., Zhou, M., Black, D. S., Haggerty, J. E., Tate, E. H., et al. (2008). Expression of the thyroid hormone transporters monocarboxylate transporter-8 (SLC16A2) and organic ion transporter-14 (SLCO1C1) at the blood-brain barrier. *Endocrinology*, *149*(12), 6251–6261.

Salazar, P., Cisternas, P., Codocedo, J. F., & Inestrosa, N. C. (2017). Induction of hypothyroidism during early postnatal stages triggers a decrease in cognitive performance by decreasing hippocampal synaptic plasticity. *Biochimica et Biophysica Acta*, *1863*(4), 870–883.

Schoonover, C. M., Seibel, M. M., Jolson, D. M., Stack, M. J., Rahman, R. J., Jones, S. A., et al. (2004). Thyroid hormone regulates oligodendrocyte accumulation in developing rat brain white matter tracts. *Endocrinology*, *145*(11), 5013–5020.

Shi, Y., Yuan, H., Kim, D., Chanana, V., Baba, A., Matsuda, T., et al. (2013). Stimulation of Na(+)/H(+) exchanger isoform 1 promotes microglial migration. *PLoS One*, *8*(8), e74201.

Shiraki, A., Saito, F., Akane, H., Takeyoshi, M., Imatanaka, N., Itahashi, M., et al. (2014). Expression alterations of genes on both neuronal and glial development in rats after developmental exposure to 6-propyl-2-thiouracil. *Toxicology Letters*, *228*(3), 225–234.

Stenzel, D., & Huttner, W. B. (2013). Role of maternal thyroid hormones in the developing neocortex and during human evolution. *Frontiers in Neuroanatomy*, 7, 19.
Sugiyama, D., Kusuhara, H., Lee, Y. J., & Sugiyama, Y. (2003). Involvement of multidrug resistance associated protein 1 (Mrp1) in the efflux transport of 17beta estradiol-D-17beta-glucuronide (E217betaG) across the blood-brain barrier. *Pharmaceutical Research*, *20*(9), 1394–1400.
Sugiyama, D., Kusuhara, H., Taniguchi, H., Ishikawa, S., Nozaki, Y., Aburatani, H., et al. (2003). Functional characterization of rat brain-specific organic anion transporter (Oatp14) at the blood-brain barrier: High affinity transporter for thyroxine. *The Journal of Biological Chemistry*, *278*(44), 43489–43495.
Taylor, P. M., & Ritchie, J. W. (2007). Tissue uptake of thyroid hormone by amino acid transporters. *Best Practice & Research. Clinical Endocrinology & Metabolism*, *21*(2), 237–251.
Terranova, C., Rizzo, V., Calamuneri, A., Bartolone, L., Morgante, F., Bruschetta, D., et al. (2016). Cortical excitability in patients with resistance to thyroid hormone compared to patients with hypothyroidism and euthyroid controls: A transcranial magnetic stimulation study. *Archives Italiennes de Biologie*, *154*(2–3), 68–77.
Thompson, C. C., & Potter, G. B. (2000). Thyroid hormone action in neural development. *Cerebral Cortex*, *10*(10), 939–945.
Tohyama, K., Kusuhara, H., & Sugiyama, Y. (2004). Involvement of multispecific organic anion transporter, Oatp14 (Slc21a14), in the transport of thyroxine across the blood-brain barrier. *Endocrinology*, *145*(9), 4384–4391.
van Doorn, J., Roelfsema, F., & van der Heide, D. (1982). Contribution fron local conversion of thyroxine to 3,5,3′-triiodothyronine to intracellular 3,5,3′-triiodothyronine in several organs in hypothyroid rats at isotope equilibrium. *Acta Endocrinologica*, *101*(3), 386–396.
van Doorn, J., van der Heide, D., & Roelfsema, F. (1983). Sources and quantity of 3,5,3′-triiodothyronine in several tissues of the rat. *The Journal of Clinical Investigation*, *72*(5), 1778–1792.
van Doorn, J., van der Heide, D., & Roelfsema, F. (1984). The contribution of local thyroxine monodeiodination to intracellular 3,5, 3′-triiodothyronine in several tissues of hyperthyroid rats at isotopic equilibrium. *Endocrinology*, *115*(1), 174–182.
van Osch, L. A., Hogervorst, E., Combrinck, M., & Smith, A. D. (2004). Low thyroid-stimulating hormone as an independent risk factor for Alzheimer disease. *Neurology*, *62*(11), 1967–1971.
Venkatesh, K., Srikanth, L., Vengamma, B., Chandrasekhar, C., Prasad, B. C., & Sarma, P. V. (2014). In vitro transdifferentiation of human cultured CD34+ stem cells into oligodendrocyte precursors using thyroid hormones. *Neuroscience Letters*, *588*, 36–41.
Vose, L. R., Vinukonda, G., Jo, S., Miry, O., Diamond, D., Korumilli, R., et al. (2013). Treatment with thyroxine restores myelination and clinical recovery after intraventricular hemorrhage. *The Journal of Neuroscience*, *33*(44), 17232–17246.
Wagner, M. S., Morimoto, R., Dora, J. M., Benneman, A., Pavan, R., & Maia, A. L. (2003). Hypothyroidism induces type 2 iodothyronine deiodinase expression in mouse heart and testis. *Journal of Molecular Endocrinology*, *31*(3), 541–550.
Walravens, P., & Chase, H. P. (1969). Influence of thyroid on formation of myelin lipids. *Journal of Neurochemistry*, *16*(10), 1477–1484.
Weissel, M. (1999). Administration of thyroid hormones in therapy of psychiatric illnesses. *Acta Medica Austriaca*, *26*(4), 129–131.
Wijsman, L. W., de Craen, A. J., Trompet, S., Gussekloo, J., Stott, D. J., Rodondi, N., et al. (2013). Subclinical thyroid dysfunction and cognitive decline in old age. *PLoS One*, *8*(3), e59199.

Zhao, R., Gao, F., Wang, P. J., & Goldman, I. D. (2000). Role of the amino acid 45 residue in reduced folate carrier function and ion-dependent transport as characterized by site-directed mutagenesis. *Molecular Pharmacology*, *57*(2), 317–323.

Zoeller, R. T., & Rovet, J. (2004). Timing of thyroid hormone action in the developing brain: Clinical observations and experimental findings. *Journal of Neuroendocrinology*, *16*(10), 809–818.

CHAPTER FOURTEEN

Thyroid Hormone Signaling in Retinal Development, Survival, and Disease

Fan Yang, Hongwei Ma, Xi-Qin Ding[1]
University of Oklahoma Health Sciences Center, Oklahoma City, OK, United States
[1]Corresponding author: e-mail address: xi-qin-ding@ouhsc.edu

Contents

1. Introduction 334
2. Retinal Structure and Function 335
3. TH and Retinal Development 337
   3.1 TH Regulation of Neuroretina/Cone Photoreceptor Differentiation 338
   3.2 TRβ2-Mediated Regulation of Cone Opsin Expression 338
   3.3 TH Regulation of Cone Opsin Expression 341
   3.4 TH Regulation of Neuroretinogenesis 342
4. TH and Photoreceptor Survival 342
   4.1 Stimulating TH Signaling Promotes Death of Developing Photoreceptors 342
   4.2 Suppressing TH Signaling Preserves Cones in Mouse Models of Retinal Degeneration 343
5. TH and Retinal Diseases 344
   5.1 Impaired Cone Function Resulting From a Compound Heterozygosity Mutation in TRβ2 344
   5.2 Thyroid Eye Disease 344
   5.3 TH and Age-Related Macular Degeneration 344
6. Conclusion and Future Perspectives 345
Acknowledgments 346
References 346

Abstract

Thyroid hormone (TH) is essential in numerous physiological functions and developmental processes. It acts through TH receptors (TRs) to regulate gene expression. The retina is the light-sensitive tissue lining the back of the eye and functions as the first step of the visual process. Rod and cone photoreceptors are specialized sensory neurons in the retina that initiate phototransduction. Rods are responsible for dim light vision, whereas cones are responsible for daytime vision, color vision, and visual acuity. TH signaling regulates retinal development and maintenance. The requirement of TH signaling is typically manifested as its regulation in the cone maturation and expression

Vitamins and Hormones, Volume 106
ISSN 0083-6729
http://dx.doi.org/10.1016/bs.vh.2017.05.001

of the light-sensing pigment protein (cone opsin). There are two components of this regulation. First, TRβ2, a TH-activated transcription factor, is expressed in immature cones and regulates cone differentiation and cone opsin expression; activation of TRβ2 suppresses the expression of short-wave-sensitive opsin 1, induces the expression of medium-wave-sensitive opsin 1, and promotes dorsal–ventral opsin patterning. Second, hypothyroid mouse models display abnormalities in cone opsin expression, supporting the necessity of TH itself in retinal development. TH has been linked to photoreceptor survival. Excessive TH signaling leads to death of developing photoreceptors in healthy and diseased retina, whereas suppressing TH signaling preserves cones in mouse models of retinal degeneration. Some eye diseases, including age-related macular degeneration, have been associated with elevated circulation TH levels. Future work should aim to better understand how TH regulates retinal development, functionality, and survival, to examine the role of TH signaling in the pathogenesis of retinal degeneration, and to explore the potential of TH signaling manipulation for photoreceptor protection. Hopefully, these knowledge bases will lead to the identification of novel strategies for retinal disease prevention and management.

1. INTRODUCTION

Thyroid hormone (TH) regulates diverse developmental processes, including neuronal development. Perturbation of TH signaling causes a variety of diseases. The thyroid gland predominantly produces the prohormone thyroxine (T4; ~95%), along with a small amount of the bioactive hormone T3 (~5%). In the peripheral tissues, T4 and T3 are transported to cells, where T4 is converted to T3 by the type 2 iodothyronine deiodinases (DIO2). T3 is then transferred to the nucleus and binds to TH receptor (TR), initiating downstream gene expression responses, and intracellular T4 and T3 are degraded by the type 3 iodothyronine deiodinases (DIO3) to reverse T3 (rT3) and 3,5-diiodo-l-thyronine (T2), respectively (Bianco & Kim, 2006; Gereben et al., 2008) (Fig. 1). TRs belong to the nuclear hormone receptor superfamily and function as ligand-dependent transcription factors (Brent, 2012; Cheng, Leonard, & Davis, 2010; Flamant et al., 2006; Forrest, Reh, & Rusch, 2002; Yen, 2001). There are three forms of TRs that are able to bind T3. TRα1, encoded by *THRA*, and TRβ1 and TRβ2, encoded by *THRB*, have identical DNA-binding and T3-binding domains, but divergent amino termini (Flamant et al., 2006; Forrest et al., 2002). Although the T4 and T3 levels in the circulation are fairly stable and they constantly enter cells in all tissues, intracellular TH homeostasis is a process that is highly locally regulated

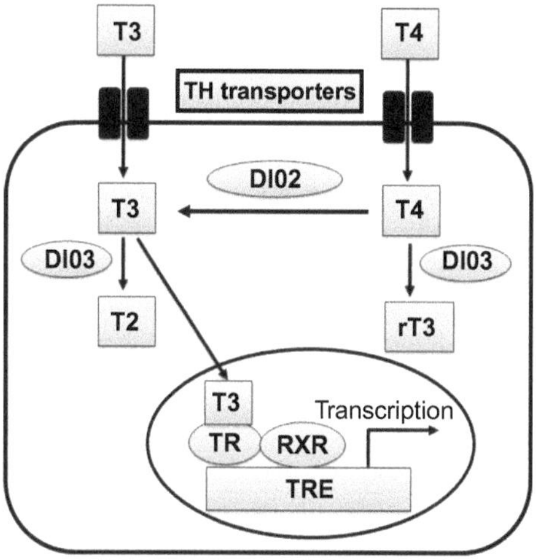

Fig. 1 Cellular TH signaling. Circulating T4 and T3 are transported into cells, where T4 is converted to T3 by DIO2. T3 is then transferred to the nucleus and binds to TRs to initiate downstream gene expression alterations. T4 and T3 are degraded to T2 and reverse T3 (rT3) by DIO3.

by DIOs and TR (Cheng et al., 2010; Dentice & Salvatore, 2011). Rod and cone photoreceptors are specialized sensory neurons that initiate phototransduction. Rods are the most abundant cell type in the retina and are responsible for dark-adapted vision. Cones, constituting only 3%–5% of the total photoreceptor population, function in bright light, and mediate color vision and visual acuity. TH regulates retinal development and maturation (Forrest et al., 2002; Harpavat & Cepko, 2003; Swaroop, Kim, & Forrest, 2010). Significant progress has been made toward understanding TH regulation in the retina. Here, we review the current knowledge on the TH regulation of retinal development/photoreceptor maturation, cone opsin expression, and photoreceptor survival, and the association of TH signaling with retinal disease.

2. RETINAL STRUCTURE AND FUNCTION

The vertebrate retina is a light-sensitive tissue that comprises several neuronal cell types (rod and cone photoreceptors, bipolar, horizontal, amacrine, and ganglion cells) and nonneuronal *Müller* glial cells (Grossniklaus, Geisert, & Nickerson, 2015). Phototransduction occurs in the retina. Photoreceptors capture and convert light to biological signals, which are then processed by interneurons (bipolar, horizontal, amacrine cells) and transmitted to the brain via the axons of ganglion cells. *Müller* glial cells play a supportive role for other neurons (Fig. 2). The two classic photoreceptor cells are rods and cones, located at the outermost layer of the retina. They constitute over 70% of the retina cells in humans and mice. Rods and cones have a similar basic structure. Rods outnumber cones by 20:1 in humans, with 120 million rods and 6 million cones, and by 30:1 in mice. Each rod and cone has an outer segment that contains a large number of pancake-like disks, an inner segment where the mitochondrion are located, a nucleus, and an axon terminal that conveys signals to the next neuron (Fig. 3). Vertebrate photoreceptors can respond to light because they contain a visual pigment embedded in their outer segment. The visual pigment consists of a protein called opsin, the light-sensing protein, and a polyene chromophore known as retinal, derived from vitamin A. When the light hits the pigment in the photoreceptors, the pigment changes its conformation. This conformational change leads to the alteration of the cell membrane's electrical polarization and ultimately

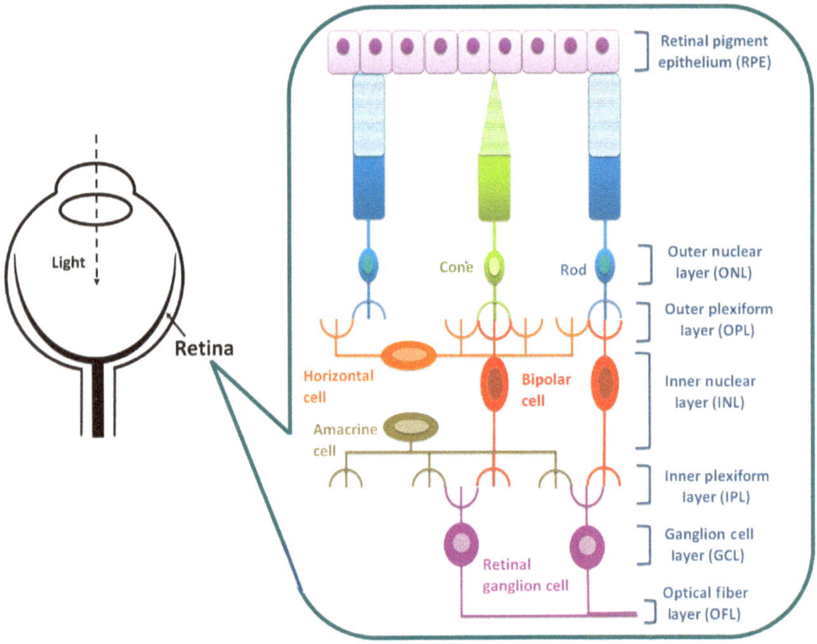

Fig. 2 Schematic diagram of the retinal layers. The retina is located at the back of the eye and made up of seven different cell types. Photoreceptors (rods and cones), located the farthest from the eye lens, convert the light into electrical signals that are processed by interneurons (bipolar, horizontal, and amacrine cells) and transmitted to the brain by ganglion cells. *Müller* glial cells play a supportive role for other neurons (not shown).

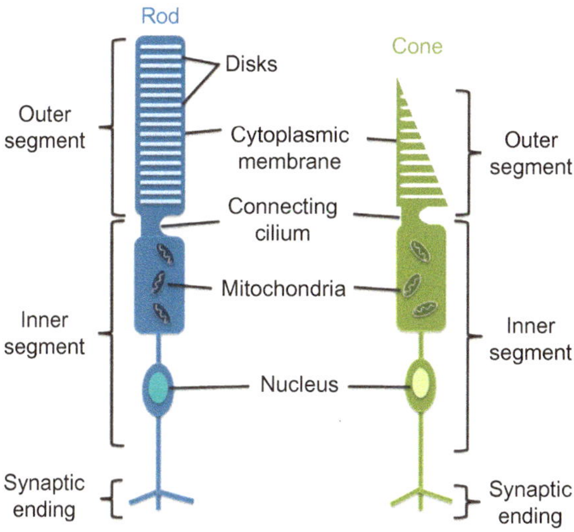

Fig. 3 The structure of rod and cone photoreceptors.

causes the transmission of a neural signal to the brain via the optic nerve. There are major functional differences between rods and cones. Rods containing only one opsin, rhodopsin, function in dim light and are responsible for night vision. Cones function in bright light and are responsible for color vision and visual acuity. Most mammals have two types of cones that mediate dichromatic color vision: S-cones and M-cones. S-cones containing mainly short-wave-sensitive opsin 1 (S-opsin) (also known as blue opsin) respond to short-wavelength light, whereas M-cones consisting mainly of medium-wave-sensitive opsin 1 (M-opsin) (also known as green opsin) are sensitive to medium-wavelength light. Humans have an additional opsin, L-opsin (also known as red opsin) which is sensitive to long-wavelength light. Therefore, humans are trichromatic.

3. TH AND RETINAL DEVELOPMENT

TH is essential in the development of many organs/tissues, including the retina. Several studies have significantly enhanced our understanding of TH regulation in retinal development, particularly in the regulation of cone photoreceptor differentiation/maturation and cone opsin expression.

3.1 TH Regulation of Neuroretina/Cone Photoreceptor Differentiation

Perhaps the earliest understanding of TH regulation in retinal development came from the work on the regulation of cone differentiation. In the mid-1990s, a study using rat embryonic retinal cells demonstrated that the presence of TH was sufficient to direct differentiation of the retinal progenitor cells into cone photoreceptors (Kelley, Turner, & Reh, 1995a). The effect of TH was more specific for cone photoreceptors, since the addition of T3 did not induce amacrine cells or rod photoreceptors. Similar results were obtained when human fetal retinal cells were used (Kelley, Turner, & Reh, 1995b). Thus, these early studies provided the groundwork for researchers to examine the role of TH in cone photoreceptor differentiation. A further understanding of TH regulation came from a study by the Cepko group. Using a chick retina, Trimarchi et al. showed that the TH components, including DIO2, DIO3, and TRs, were expressed in dynamic patterns with three sequential waves in cycling progenitor cells and photoreceptors cells during development. The first wave occurred as progenitor cells began to make photoreceptors, the second wave followed as some cell types took on a more mature location and differentiation state, and the third wave was initiated as *Müller* glia began to generate (Trimarchi, Harpavat, Billings, & Cepko, 2008). These findings support the notion that TH signaling affects the fate of progenitor cells and differentiation of the photoreceptors and supporting cells. A follow-up study by Fischer et al. further demonstrated that TH stimulates photoreceptor maturation in the chick retina. The maturation of photoreceptors in the chick retina was accelerated by treatment with T4 or T3 (Fischer, Bongini, Bastaki, & Sherwood, 2011).

3.2 TRβ2-Mediated Regulation of Cone Opsin Expression

3.2.1 The Expression Patterns of TH Receptors in the Developing Retina

T3 acts via its binding to TRs in the nucleus and triggering the downstream gene expression alterations. Using in situ hybridization, Sjoberg et al. detected the mRNA expression of TRα1 and two TRβ variants in the chick retina, and demonstrated that they were differentially expressed in the developing chick retina (Sjoberg, Vennstrom, & Forrest, 1992). *Thrα1* was expressed over all retinal layers at intermediate and later developmental stages. *Thrβ1* appeared at a low level in the outer and inner nuclear layers during development. *Thrβ2* was localized only in the outer nuclear layer

(ONL), the site of photoreceptor cells, and its expression level was high during retinal development and decreased as the retina matured. These early findings suggested that TRα1 and TRβ1 may have a general function in many retinal cell types, whereas TRβ2 may mediate specific events in photoreceptor development. Later, TRβ2 expression was also detected at both mRNA and protein levels in the mouse retina. Using Northern-blotting, the Forrest group showed that the expression pattern of TRβ2 mRNA was similar to that described in the chick retina (Ng et al., 2001). It was further shown that the TRβ2 protein level in the retina peaked in the late embryonic stage and declined in the postnatal period, which is consistent with the mRNA expression observed in the chick and mouse retinas (Ng, Ma, Curran, & Forrest, 2009). The immunolabeling analysis showed that the TRβ2-positive cells were initially dispersed, but gradually moved and aligned at the edge of the outer neuroblastic layer in the late prenatal stage. Postnatally, these positive cells migrated inward before postnatal day 10 (P10), then outward to the edge of the ONL. Together, the following evidence supports the idea that TRβ2 is specifically expressed in cones in the retina: (i) the relative small number of TRβ2-positive cells matches the small cone population in the mouse retina (Carter-Dawson & LaVail, 1979a), (ii) the accumulation of TRβ2-positive cells in the embryonic stage correlates with the period of cone generation (Applebury et al., 2007; Carter-Dawson & LaVail, 1979b), (iii) the migration pattern of TRβ2-positive cells is similar to that of the maturing cones (Rich, Zhan, & Blanks, 1997), and (iv) TRβ2 colocalized in the cells that express a cone marker transgene (Ng et al., 2009).

3.2.2 The TRβ2 Regulation of Cone Opsin Expression

The critical role of TH signaling in cone opsin expression was first evidenced by the functional activity of TRβ2. The Forrest group deleted a part of the *Thrβ2*-specific exon in mice to obtain the *Thrb2*$^{-/-}$ model (Ng et al., 2001). In wild-type mice, S-cones appear near birth and are much more abundant in the ventral (inferior) half of the retina. In contrast, M-cones appear about a week after birth, predominantly in the dorsal (superior) half of the retina (Szel et al., 1992; Szel, Rohlich, Mieziewska, Aguirre, & van Veen, 1993). *Thrb2*$^{-/-}$ mice had normal numbers of cones and showed normal rod function. However, M-opsin was not detectable in any retinal regions in these mice, whereas S-opsin appeared in all cones in both dorsal and ventral regions, that is, the pattern of cone opsin distribution was disturbed. Moreover, electroretinogram, which measures the function of photoreceptors and interneurons, showed normal cone responses to short-wave light, indicating that S-cones were functional, and reduced sensitivity to longer-wave light, which was consistent with the significant loss of M-cones. Thus, TRβ2 is required for M-opsin expression and for maintenance of the dorsal–ventral cone opsin expression patterning. The controlling role of TRβ2 was also supported by research with a transgenic mouse line, which expresses mutant TRβ1 and TRβ2 that cannot bind T3 (Pessoa et al., 2008; Roberts, Srinivas, Forrest, Morreale de Escobar, & Reh, 2006). The transgenic mice had reduced expression of M-opsin and enhanced expression of S-opsin/loss of the dorsal–ventral patterning, a phenotype identical to *Thrb2*$^{-/-}$ mice. With a mouse model that lacks the rod differentiation-regulating gene *Nrl* and expresses a *Thrb2* gene (*Nrl*$^{b2/b2}$, generated by targeted replacement of *Nrl* gene with a *Thrb2* gene), Ng et al. demonstrated that TRβ2 regulates the plasticity of the precursor cells and controls the M- vs S-cone identity (Ng, Lu, Swaroop, Sharlin, & Forrest, 2011). Consistent with these findings, Iwagawa et al. showed that activation of TRβ2 stimulates the promoter activities of the human and mouse M-cone opsin (Iwagawa, Tanaka, Iida, Itoh, & Watanabe, 2013).

TRs often form heterodimers with the retinoid X receptors (RXRs) to regulate gene expression. RXRγ is highly expressed in the retina and localized in the developing cones (Kelley et al., 1995a, 1995b). The phenotype in *Rxrγ*$^{-/-}$ mice was similar to that of *Thrb2*$^{-/-}$ mice, with S-opsin expressed in all retinal regions and diminished expression of M-opsin (Roberts, Hendrickson, McGuire, & Reh, 2005; Srinivas, Ng, Liu, Jia, & Forrest, 2006). The findings from *Rxrγ*$^{-/-}$ mice suggest that RXRγ, along with TRβ2, is required to suppress S-opsin in dorsal cones, but is not necessary for M-opsin expression. The suppression of S-opsin expression by RXR was also observed in the retinas of *staggerer* mice that carry a null mutation of RORα (retinoic acid receptor-related orphan receptor alpha) (Fujieda, Bremner, Mears, & Sasaki, 2009). It appears that both RXRγ and RORα are involved in the regulation of cone opsin expression during retinal development.

3.2.3 The Regulation of TRβ2 Expression

The identification of the TRβ2 regulation in programming cone development and cone opsin expression promoted the investigations on the regulation of TRβ2 expression. Analysis of the *Thrb* gene identified an intron control region that directs the peak expression of the receptor gene in immature cones (Jones, Ng, Liu, & Forrest, 2007). NeuroD1, a proneural bHLH (basic helix–loop–helix) transcription factor that binds the intron control region, was shown to be necessary for sustained expression of TRβ2 in immature cones (Liu et al., 2008). The opsin expression phenotype in NeuroD1-deficient mice was identical to that in TRβ2-deficient mice: S-opsin was present in all retinal regions, and no M-opsin was expressed. Moreover, transfection of NeuroD1 into the retinal explants restored the expression of TRβ2 (Liu et al., 2008).

3.3 TH Regulation of Cone Opsin Expression

The discovery that TRβ2 is essential for the retinal development and cone opsin expression raised questions about the role of TH itself. Consequently, many efforts have been made to investigate the requirement for TH in retinal development and photoreceptor differentiation. The findings from a number of hypothyroid mouse models have established the regulatory role of TH in cone opsin expression. In TSH (thyroid-stimulating hormone) receptor-deficient ($Tshr^{-/-}$) mice that have severe early-onset hypothyroidism with low levels of T4 and T3, M-opsin expression was dramatically delayed at the premature stage and did not reach normal levels at adult ages, whereas S-opsin expression showed a slightly increased distribution pattern over the dorsal region of the retina. Moreover, addition of T3 to the $Tshr^{-/-}$ mice recovered M-opsin expression (Lu et al., 2009). Using another hypothyroid mouse line, $Pax8^{-/-}$ mice, which do not develop thyroid gland follicles and thus are unable to produce T4 and T3, Glaschke et al. showed that the wild-type pattern of cone opsin expression was abolished, with M-opsin downregulated and S-opsin upregulated throughout the entire retina (Glaschke, Glosmann, & Peichl, 2010). The phenotype observed in these hypothyroid mice was similar, but somewhat milder than that in TRβ2-deficient mice. Moreover, wild-type mice with chemically induced hypothyroidism showed similar abnormalities of cone opsin expression (Lu et al., 2009). Thus, the phenotype in $Thrb2^{-/-}$ and hypothyroid model mice established the controlling role of TH signaling in cone differentiation and cone opsin expression. TH has also been shown to regulate/maintain opsin expression patterning in mature cones in addition to its regulation during development. Adult wild-type mice receiving 12 weeks of antithyroid drug treatment displayed altered cone patterns with increased S-opsin expression and decreased M-opsin expression (Glaschke et al., 2011), resembling the phenotypes observed in $Thrb2^{-/-}$ and hypothyroid model mice. Furthermore, $Pax8^{-/-}$ mice that were treated with T4 for the first 12 postnatal weeks developed and maintained the normal cone opsin patterning. However, if the T4 treatment was terminated at postnatal week 4, when the mature pattern of cones is already established, the opsin distribution in Pax8$^{-/-}$ mice returned to the mutant pattern within 8 weeks (Glaschke et al., 2011).

3.4 TH Regulation of Neuroretinogenesis

It should be noted that *Thrb2*$^{-/-}$ and hypothyroid mice did not show apparent changes in eye size or retinal morphology, including retinal thickness and cone numbers (Glaschke et al., 2010, 2011; Ma et al., 2014; Ng et al., 2001), suggesting that the major role of TH in mice is to regulate cone maturation and opsin expression, but not neuroretinogenesis. However, there were some reports in which TH has been shown to regulate neuroretinogenesis. Most of these studies were conducted with rats. Compared with control rats, the TH-depleted rats showed a significant reduction in the retinal thickness and layering, retinal volume, cell number, and nuclear volumes in all cell layers (Gamborino et al., 2001; Pinazo-Duran et al., 2005; Sevilla-Romero, Munoz, & Pinazo-Duran, 2002), and supplementation with TH rescued the phenotype (Pinazo-Duran et al., 2005). These data suggest that the reduced TH signaling delays the retinal developmental processes in the rats, and TH signaling is critical for rat neuroretinogenesis. It seems that the rat retinal progenitor cells are more dependent on TH than are mouse retinal progenitor cells, and the mechanism(s) underlying this species difference remains to be determined.

4. TH AND PHOTORECEPTOR SURVIVAL

4.1 Stimulating TH Signaling Promotes Death of Developing Photoreceptors

In addition to the regulation of photoreceptor maturation, TH has been shown to affect photoreceptor survival, and excessive TH signaling harms developing photoreceptors. A study with *Dio3*$^{-/-}$ mice provides evidence that exemplifies these characteristics (Ng et al., 2010). DIO3 is a TH-inactivating enzyme that degrades T4 and T3 (see Fig. 1). *Dio3*$^{-/-}$ mice showed severe cone loss (~80%). Analysis of cell death markers revealed that *Dio3*$^{-/-}$ mice displayed cone selective death in the developmental stage. Deletion of TRβ2 in *Dio3*$^{-/-}$ mice restored the cone number. Moreover, the wild-type mice (C57BL/6) that received a T3 treatment between P0 and P3 showed complete loss of cones, with a moderate loss of rods (Ng et al., 2010). These findings support the view that developing photoreceptors, particularly cones, are sensitive to excess TH. Similar findings were obtained with mouse models of retinal degeneration. Treatment with T3 significantly accelerated cone and rod death in *Cngb3*$^{-/-}$ mice, a model of achromatopsia with defects of the cone cyclic nucleotide-gated channel B subunit, and *Gucy2e*$^{-/-}$ mice, a model of congenital amaurosis (LCA) with a defect in retinal guanylate cyclase 1 (Ma et al., 2014). After T3 treatment, cone density and the ONL (outer nuclear layer) thickness in these mice were reduced by about 30%, compared with controls. The idea that TH affects developing cone survival was also supported by studies with juvenile rainbow trout. The ultraviolet-sensitive (UVS) cone is one distinct cone type in the salmonid fishes (trout and salmon) and degenerates naturally during development. The plasma T4 level was found to be higher before this natural process, and treatment with T4 promoted loss of UVS cones in the juvenile rainbow trout (Raine, Coffin, & Hawryshyn, 2010). Furthermore, chemically blocking endogenous T4 production delayed the UVS cone loss in these fish. Thus, systemic TH is essential to induce UVS cone loss in the rainbow trout and initiates retinal remodeling during development.

4.2 Suppressing TH Signaling Preserves Cones in Mouse Models of Retinal Degeneration

In line with the findings that excessive TH leads to cone death, suppressing TH signaling was shown to preserve cones in mouse models of retinal degeneration. Treatment with antithyroid drug significantly improved cone survival in retinol isomerase RPE65-deficient (*Rpe65*$^{-/-}$) mice, a model of LCA, and cone photoreceptor function loss type 1 (*Cpfl1*) mice with defects in *Pde6c* (encoding the cone cGMP-specific 3′,5′-cyclic phosphodiesterase subunit α′), a model of achromatopsia (Ma et al., 2014). The protective effects were observed in both developing and adult mice. The protective effect of TH signaling suppression was also demonstrated by targeting the cellular DIOs locally in the retina. Subretinal delivery of AAV5-IRBP/GNAT2-*DIO3*, which directs expression of human DIO3 specifically in cones, increased cone density by 30%–40% in *Rpe65*$^{-/-}$ mice and *Cpfl1* mice, compared with their respective untreated controls. Intravitreal and topical delivery of a DIO2 inhibitor (iopanoic acid) significantly improved cone survival in *Rpe65*$^{-/-}$ mice (Yang et al., 2016). A similar finding was observed in zebrafish. Treatment with iopanoic acid was shown to accelerate reinnervation of the optic tectum following optic nerve crush in the fish (Bhumika, Lemmens, Vancamp, Moons, & Darras, 2015). The finding that suppressing TH signaling protects cones is promising, as it may represent a novel strategy for retinal degeneration management.

5. TH AND RETINAL DISEASES

5.1 Impaired Cone Function Resulting From a Compound Heterozygosity Mutation in TRβ2

Given the importance of TRβ2 to the determination of cone opsin expression and patterning, a mutation in TRβ2 may cause cone photoreceptor disorders. A case of impaired cone function has been reported in a patient with a compound heterozygosity in the *THRB2* gene, showing two mutations within the ligand-binding domain of *THRB2*, confirmed by DNA sequencing (Weiss, Kelly, Bisset, & Deeb, 2012). The electroretinogram examination showed a normal dark-adapted (scotopic) response, indicating that the rod function was not affected, and a significantly reduced light-adapted (photopic) response. Further spectral testing showed that the patient had a small response to the red flash and a larger response to the blue flash, indicating that the patient had reduced L- and M-, but increased S-, cone functions. The phenotype in this patient was similar to that in *Thrb2*$^{-/-}$ mice. This is the first reported case of cone disorder associated with mutations in *THRB2* gene. Indeed, there are about 150 mutations that have been identified in *THRB* genes, and we know little about the retinal phenotype in patients with these mutations. It will be worthwhile to profile the retinal function and morphology in these patients.

5.2 Thyroid Eye Disease

Thyroid eye disease (TED) in patients with Graves' disease, which is an autoimmune hyperthyroid disease associated with an overactive thyroid gland, is a chronic inflammatory disorder in the eye tissues resulting from hyperthyroidism. Excessive TH is produced by the stimulation of the binding of autoantibodies to the thyrotropin receptor on thyroid follicular endothelial cells. In patients with TED, the eye muscles and fatty tissues behind the eye become inflamed, which causes the eyes to be pushed forward and the eyes and eyelids to become swollen and red (Bahn, 2010). The priority for treatment is to restore normal thyroid condition. Eye condition should be evaluated and treated simultaneously.

5.3 TH and Age-Related Macular Degeneration

In what is probably one of the most interesting findings with regard to TH and retinal disease, TH has recently been associated with age-related macular degeneration (AMD). AMD is the most prevalent form of irreversible blindness among the aged population worldwide. Dry AMD, also known as geographic atrophy, is a form of slowly progressing geographic atrophy of the macula and comprises a majority of AMD cases (~90%; Ambati & Fowler, 2012; Friedman et al., 2004). There is no treatment to halt or reverse dry AMD. Pathologically, dry AMD is characterized by a progressive macular degeneration of the retinal pigment epithelium (RPE) and photoreceptors. It is generally recognized that multiple factors, including aging, oxidative stress, chronic inflammation, and genetic defects, are involved in RPE and photoreceptor dystrophy/AMD lesions (Ambati & Fowler, 2012; Cruz-Guilloty & Perez, 2011). Prospective population-based cohort studies revealed that higher free serum TH values are associated with increased risk of AMD (Chaker et al., 2015; Gopinath, Liew, Kifley, & Mitchell, 2016). The study ($n = 5573$) by Chaker et al. showed that higher free serum TH values were associated with increased risk of AMD (Chaker et al., 2015). Another study ($n = 906$) by Gopinath et al. assessed the prospective associations with serum thyroid-stimulating hormone and free thyroxine (T4) levels, and showed that overt hyperthyroidism was associated with an increased risk of incident AMD, and thyroxine use in older adults was positively associated with incidence of AMD (Gopinath et al., 2016). Based on the essential role of TH signaling in cellular metabolism/oxidative reactions, it is likely that TH signaling plays a role in oxidative damage-associated AMD lesions. Elevated TH signaling may lead to increased basal metabolic rate, oxidative metabolism, and generation of reactive oxygen species. RPE and photoreceptors may become more sensitive to TH during the aging process. These clinical findings provide a strong rationale for pursuing an experimental study to determine whether TH signaling plays a role in the pathogenesis of AMD.

6. CONCLUSION AND FUTURE PERSPECTIVES

TH is an important regulator of neurodevelopment. In the retina, TH is critical for retinal development, photoreceptor differentiation, and cone opsin expression, and regulates photoreceptor survival in healthy and diseased conditions. TRβ2 is the primary receptor form that is involved in the regulation of cones. The finding that suppressing TH signaling locally in the retina protects cones is promising, and the accessibility of eyes to topically delivered therapeutics (eye drops) provides potential for retinal disease management. Thus, the development of TH component inhibitors that can be delivered locally to the retina represents a valuable direction for translational research. This investigation should be combined with efforts to determine how TH signaling regulates photoreceptor survival. Clinical observations have linked TH signaling with AMD. However, these clinical correlative findings must be verified with experimental evidence. Compared with our knowledge of the TRβ2 regulation of retinal development, we know little about the role of the other TH components, including DIOs and TRα. Future research should also include efforts to understand the regulatory role of these TH components in retinal development and maintenance.

ACKNOWLEDGMENTS

This work was supported by Grants from the National Eye Institute (P30EY021725, T32EY023202, R01EY019490, and R21EY024583), the Foundation Fighting Blindness, and the Knights Templar Eye Foundation.

REFERENCES

Ambati, J., & Fowler, B. J. (2012). Mechanisms of age-related macular degeneration. *Neuron*, *75*, 26–39.

Applebury, M. L., Farhangfar, F., Glosmann, M., Hashimoto, K., Kage, K., Robbins, J. T., et al. (2007). Transient expression of thyroid hormone nuclear receptor TRbeta2 sets S opsin patterning during cone photoreceptor genesis. *Developmental Dynamics*, *236*, 1203–1212.

Bahn, R. S. (2010). Graves' ophthalmopathy. *The New England Journal of Medicine*, *362*, 726–738.

Bhumika, S., Lemmens, K., Vancamp, P., Moons, L., & Darras, V. M. (2015). Decreased thyroid hormone signaling accelerates the reinnervation of the optic tectum following optic nerve crush in adult zebrafish. *Molecular and Cellular Neurosciences*, *68*, 92–102.

Bianco, A. C., & Kim, B. W. (2006). Deiodinases: Implications of the local control of thyroid hormone action. *The Journal of Clinical Investigation*, *116*, 2571–2579.

Brent, G. A. (2012). Mechanisms of thyroid hormone action. *The Journal of Clinical Investigation*, *122*, 3035–3043.

Carter-Dawson, L. D., & LaVail, M. M. (1979a). Rods and cones in the mouse retina. I. Structural analysis using light and electron microscopy. *The Journal of Comparative Neurology*, *188*, 245–262.

Carter-Dawson, L. D., & LaVail, M. M. (1979b). Rods and cones in the mouse retina. II. Autoradiographic analysis of cell generation using tritiated thymidine. *The Journal of Comparative Neurology*, *188*, 263–272.

Chaker, L., Buitendijk, G. H., Dehghan, A., Medici, M., Hofman, A., Vingerling, J. R., et al. (2015). Thyroid function and age-related macular degeneration: A prospective population-based cohort study—The Rotterdam study. *BMC Medicine*, *13*, 94.

Cheng, S. Y., Leonard, J. L., & Davis, P. J. (2010). Molecular aspects of thyroid hormone actions. *Endocrine Reviews*, *31*, 139–170.

Cruz-Guilloty, F., & Perez, V. L. (2011). Molecular medicine: Defence against oxidative damage. *Nature*, *478*, 42–43.

Dentice, M., & Salvatore, D. (2011). Deiodinases: The balance of thyroid hormone: Local impact of thyroid hormone inactivation. *The Journal of Endocrinology*, *209*, 273–282.

Fischer, A. J., Bongini, R., Bastaki, N., & Sherwood, P. (2011). The maturation of photoreceptors in the avian retina is stimulated by thyroid hormone. *Neuroscience*, *178*, 250–260.

Flamant, F., Baxter, J. D., Forrest, D., Refetoff, S., Samuels, H., Scanlan, T. S., et al. (2006). International Union of Pharmacology. LIX. The pharmacology and classification of the nuclear receptor superfamily: Thyroid hormone receptors. *Pharmacological Reviews*, *58*, 705–711.

Forrest, D., Reh, T. A., & Rusch, A. (2002). Neurodevelopmental control by thyroid hormone receptors. *Current Opinion in Neurobiology*, *12*, 49–56.

Friedman, D. S., O'Colmain, B. J., Munoz, B., Tomany, S. C., McCarty, C., de Jong, P. T., et al. (2004). Prevalence of age-related macular degeneration in the United States. *Archives of Ophthalmology*, *122*, 564–572.

Fujieda, H., Bremner, R., Mears, A. J., & Sasaki, H. (2009). Retinoic acid receptor-related orphan receptor alpha regulates a subset of cone genes during mouse retinal development. *Journal of Neurochemistry*, *108*, 91–101.

Gamborino, M. J., Sevilla-Romero, E., Munoz, A., Hernandez-Yago, J., Renau-Piqueras, J., & Pinazo-Duran, M. D. (2001). Role of thyroid hormone in craniofacial and eye development using a rat model. *Ophthalmic Research*, *33*, 283–291.

Gereben, B., Zavacki, A. M., Ribich, S., Kim, B. W., Huang, S. A., Simonides, W. S., et al. (2008). Cellular and molecular basis of deiodinase-regulated thyroid hormone signaling. *Endocrine Reviews*, *29*, 898–938.

Glaschke, A., Glosmann, M., & Peichl, L. (2010). Developmental changes of cone opsin expression but not retinal morphology in the hypothyroid Pax8 knockout mouse. *Investigative Ophthalmology & Visual Science*, *51*, 1719–1727.

Glaschke, A., Weiland, J., Del Turco, D., Steiner, M., Peichl, L., & Glosmann, M. (2011). Thyroid hormone controls cone opsin expression in the retina of adult rodents. *The Journal of Neuroscience*, *31*, 4844–4851.

Gopinath, B., Liew, G., Kifley, A., & Mitchell, P. (2016). Thyroid dysfunction and ten-year incidence of age-related macular degeneration. *Investigative Ophthalmology & Visual Science*, *57*, 5273–5277.

Grossniklaus, H. E., Geisert, E. E., & Nickerson, J. M. (2015). Introduction to the retina. *Progress in Molecular Biology and Translational Science*, *134*, 383–396.

Harpavat, S., & Cepko, C. L. (2003). Thyroid hormone and retinal development: An emerging field. *Thyroid*, *13*, 1013–1019.

Iwagawa, T., Tanaka, Y., Iida, A., Itoh, T., & Watanabe, S. (2013). Enhancer/promoter activities of the long/middle wavelength-sensitive opsins of vertebrates mediated by thyroid hormone receptor beta2 and COUP-TFII. *PloS One*, *8*, e72065.

Jones, I., Ng, L., Liu, H., & Forrest, D. (2007). An intron control region differentially regulates expression of thyroid hormone receptor beta2 in the cochlea, pituitary, and cone photoreceptors. *Molecular Endocrinology*, *21*, 1108–1119.

Kelley, M. W., Turner, J. K., & Reh, T. A. (1995a). Ligands of steroid/thyroid receptors induce cone photoreceptors in vertebrate retina. *Development*, *121*, 3777–3785.

Kelley, M. W., Turner, J. K., & Reh, T. A. (1995b). Regulation of proliferation and photoreceptor differentiation in fetal human retinal cell cultures. *Investigative Ophthalmology & Visual Science*, *36*, 1280–1289.

Liu, H., Etter, P., Hayes, S., Jones, I., Nelson, B., Hartman, B., et al. (2008). NeuroD1 regulates expression of thyroid hormone receptor 2 and cone opsins in the developing mouse retina. *The Journal of Neuroscience*, *28*, 749–756.

Lu, A., Ng, L., Ma, M., Kefas, B., Davies, T. F., Hernandez, A., et al. (2009). Retarded developmental expression and patterning of retinal cone opsins in hypothyroid mice. *Endocrinology*, *150*, 1536–1544.

Ma, H., Thapa, A., Morris, L., Redmond, T. M., Baehr, W., & Ding, X. Q. (2014). Suppressing thyroid hormone signaling preserves cone photoreceptors in mouse models of retinal degeneration. *Proceedings of the National Academy of Sciences of the United States of America, 111*, 3602–3607.

Ng, L., Hurley, L. B., Dierks, B., Srinivas, M., Salto, C., Vennstrom, B., et al. (2001). A thyroid hormone receptor that is required for the development of green cone photoreceptors. *Nature Genetics, 27*, 94–98.

Ng, L., Lu, A., Swaroop, A., Sharlin, D. S., & Forrest, D. (2011). Two transcription factors can direct three photoreceptor outcomes from rod precursor cells in mouse retinal development. *The Journal of Neuroscience, 31*, 11118–11125.

Ng, L., Lyubarsky, A., Nikonov, S. S., Ma, M., Srinivas, M., Kefas, B., et al. (2010). Type 3 deiodinase, a thyroid-hormone-inactivating enzyme, controls survival and maturation of cone photoreceptors. *The Journal of Neuroscience, 30*, 3347–3357.

Ng, L., Ma, M., Curran, T., & Forrest, D. (2009). Developmental expression of thyroid hormone receptor beta2 protein in cone photoreceptors in the mouse. *Neuroreport, 20*, 627–631.

Pessoa, C. N., Santiago, L. A., Santiago, D. A., Machado, D. S., Rocha, F. A., Ventura, D. F., et al. (2008). Thyroid hormone action is required for normal cone opsin expression during mouse retinal development. *Investigative Ophthalmology & Visual Science, 49*, 2039–2045.

Pinazo-Duran, M. D., Iborra, F. J., Pons, S., Sevilla-Romero, E., Gallego-Pinazo, R., & Munoz, A. (2005). Postnatal thyroid hormone supplementation rescues developmental abnormalities induced by congenital-neonatal hypothyroidism in the rat retina. *Ophthalmic Research, 37*, 225–234.

Raine, J. C., Coffin, A. B., & Hawryshyn, C. W. (2010). Systemic thyroid hormone is necessary and sufficient to induce ultraviolet-sensitive cone loss in the juvenile rainbow trout retina. *The Journal of Experimental Biology, 213*, 493–501.

Rich, K. A., Zhan, Y., & Blanks, J. C. (1997). Migration and synaptogenesis of cone photoreceptors in the developing mouse retina. *The Journal of Comparative Neurology, 388*, 47–63.

Roberts, M. R., Hendrickson, A., McGuire, C. R., & Reh, T. A. (2005). Retinoid X receptor (gamma) is necessary to establish the S-opsin gradient in cone photoreceptors of the developing mouse retina. *Investigative Ophthalmology & Visual Science, 46*, 2897–2904.

Roberts, M. R., Srinivas, M., Forrest, D., Morreale de Escobar, G., & Reh, T. A. (2006). Making the gradient: Thyroid hormone regulates cone opsin expression in the developing mouse retina. *Proceedings of the National Academy of Sciences of the United States of America, 103*, 6218–6223.

Sevilla-Romero, E., Munoz, A., & Pinazo-Duran, M. D. (2002). Low thyroid hormone levels impair the perinatal development of the rat retina. *Ophthalmic Research, 34*, 181–191.

Sjoberg, M., Vennstrom, B., & Forrest, D. (1992). Thyroid-hormone receptors in Chick retinal development—Differential expression of messenger-RNAs for alpha and N-terminal variant beta-receptors. *Development, 114*, 39–47.

Srinivas, M., Ng, L., Liu, H., Jia, L., & Forrest, D. (2006). Activation of the blue opsin gene in cone photoreceptor development by retinoid-related orphan receptor beta. *Molecular Endocrinology, 20*, 1728–1741.

Swaroop, A., Kim, D., & Forrest, D. (2010). Transcriptional regulation of photoreceptor development and homeostasis in the mammalian retina. *Nature Reviews. Neuroscience, 11*, 563–576.

Szel, A., Rohlich, P., Caffe, A. R., Juliusson, B., Aguirre, G., & Van Veen, T. (1992). Unique topographic separation of two spectral classes of cones in the mouse retina. *The Journal of Comparative Neurology*, *325*, 327–342.

Szel, A., Rohlich, P., Mieziewska, K., Aguirre, G., & van Veen, T. (1993). Spatial and temporal differences between the expression of short- and middle-wave sensitive cone pigments in the mouse retina: A developmental study. *The Journal of Comparative Neurology*, *331*, 564–577.

Trimarchi, J. M., Harpavat, S., Billings, N. A., & Cepko, C. L. (2008). Thyroid hormone components are expressed in three sequential waves during development of the chick retina. *BMC Developmental Biology*, *8*, 101.

Weiss, A. H., Kelly, J. P., Bisset, D., & Deeb, S. S. (2012). Reduced L- and M- and increased S-cone functions in an infant with thyroid hormone resistance due to mutations in the THRbeta2 gene. *Ophthalmic Genetics*, *33*, 187–195.

Yang, F., Ma, H., Belcher, J., Butler, M. R., Redmond, T. M., Boye, S. L., et al. (2016). Targeting iodothyronine deiodinases locally in the retina is a therapeutic strategy for retinal degeneration. *The FASEB Journal*, *30*, 4313–4325.

Yen, P. M. (2001). Physiological and molecular basis of thyroid hormone action. *Physiological Reviews*, *81*, 1097–1142.

CHAPTER FIFTEEN

Thyroid Hormone Signaling in the Development of the Endochondral Skeleton

Richard C. Lindsey*,†, Patrick Aghajanian*, Subburaman Mohan*,†,1

*Musculoskeletal Disease Center, Loma Linda VA Healthcare System, Loma Linda, CA, United States
†Loma Linda University School of Medicine, Loma Linda, CA, United States
[1]Corresponding author: e-mail address: subburaman.mohan@va.gov

Contents

1. Introduction 352
2. TH Physiology 353
   2.1 Regulation and Production 353
   2.2 TH Transport and Metabolism 355
   2.3 TH Receptors and Actions 356
3. TH in the Developing Skeleton 357
   3.1 Skeletal Development 357
   3.2 TH in Postnatal Growth 358
   3.3 TH in Secondary Ossification 362
4. TH in the Adult Skeleton 364
5. Skeletal Effects of TR Signaling 365
   5.1 TRα Signaling in Postnatal Growth 366
   5.2 TRβ Signaling in Postnatal Growth 367
   5.3 Nongenomic TH Signaling 368
6. Summary and Conclusions 368
Acknowledgments 370
References 370

Abstract

Thyroid hormone (TH) is an established regulator of skeletal growth and maintenance both in clinical studies and in laboratory models. The clinical consequences of altered thyroid status on the skeleton during development and in adulthood are well known, and genetic mouse models in which elements of the TH signaling axis have been manipulated illuminate the mechanisms which underlie TH regulation of the skeleton. TH is involved in the regulation of the balance between proliferation and differentiation in several skeletal cell types including chondrocytes, osteoblasts, and osteoclasts. The effects of TH are mediated primarily via the thyroid hormone receptors

Vitamins and Hormones, Volume 106
ISSN 0083-6729
http://dx.doi.org/10.1016/bs.vh.2017.06.003

(TRs) α and β, ligand-inducible nuclear receptors which act as transcription factors to regulate target gene expression. Both TRα and TRβ signaling are important for different stages of skeletal development. The molecular mechanisms of TH action in bone are complex and include interaction with a number of growth factor signaling pathways. This review provides an overview of the regulation and mechanisms of TH action in bone, focusing particularly on the role of TH in endochondral bone formation during postnatal growth.

1. INTRODUCTION

Thyroid hormones (TH) exert specific effects on growth, development, and metabolism in a variety of tissues in humans as well as experimental animal models. The presence of TH is required during development of the skeleton, linear bone growth, maintenance of bone mass, and efficient healing of fractures (Bassett & Williams, 2016). Congenital hypothyroidism, the most common congenital endocrine disorder, has an increasing incidence which has recently risen as high as 1 in 1400 (Albert et al., 2012; Bassett & Williams, 2016; Corbetta et al., 2009; Deladoëy, Ruel, Giguère, & Van Vliet, 2011; Mengreli et al., 2010; Mitchell, Hsu, Sahai, & Massachusetts Pediatric Endocrine Work Group, 2011; Wassner & Brown, 2015). Congenital and acquired juvenile hypothyroidism are known causes of short stature due to arrested bone growth, development, and mineralization; left untreated, these patients can develop a skeletal dysplasia which includes coarse facial features, truncal shortening, scoliosis, patent fontanelles, vertebral deformities, and absent femoral ossification centers (Hüffmeier, Tietze, & Rauch, 2007). However, therapeutic TH replacement in these cases can prevent dysplasia and result in corrected growth and final height (Delvecchio et al., 2015; Rivkees, Bode, & Crawford, 1988). Furthermore, juvenile thyrotoxicosis, most commonly caused by Graves' disease, leads to accelerated bone formation and early closure of the growth plates, thus resulting in short stature (Bassett & Williams, 2016; Segni, Leonardi, Mazzoncini, Pucarelli, & Pasquino, 1999). Evidence for the importance of TH for skeletal development is strong, but the molecular mechanisms underlying TH action in bone are not fully understood. This review discusses the regulation and mechanisms of TH action in bone, focusing especially on recently uncovered developments in our understanding of the cellular and molecular mechanisms of TH regulation of endochondral bone formation during postnatal growth.

2. TH PHYSIOLOGY

2.1 Regulation and Production

The hypothalamus–pituitary–thyroid (HPT) axis is responsible for maintaining systemic serum TH levels via a classical negative feedback loop (Fig. 1). The paraventricular nucleus of the hypothalamus produces thyrotropin-releasing hormone (TRH). Through the hypothalamic–pituitary portal circulation, TRH travels to the anterior pituitary gland and triggers thyrotrophs to release thyroid-stimulating hormone (TSH). The TSH receptor (TSHR) on thyroid follicular cells responds to TSH, stimulating cellular proliferation and the synthesis and secretion of 3,5,3′,5′-L-tetraiodothyronine (thyroxine, T4), the primary hormone secreted by the thyroid gland, and 3,5,3′-L-triiodothyronine (T3), the more active form of the hormone which is primarily produced by peripheral conversion of T4 to T3 (Fig. 2; Bassett & Williams, 2016; Kopp, 2001). Once secreted into the circulation, the vast majority of T3 and T4 are bound to transport proteins such as thyroid hormone-binding globulin, transthyretin, and albumin; only ~0.2% of serum T3 and ~0.02% of

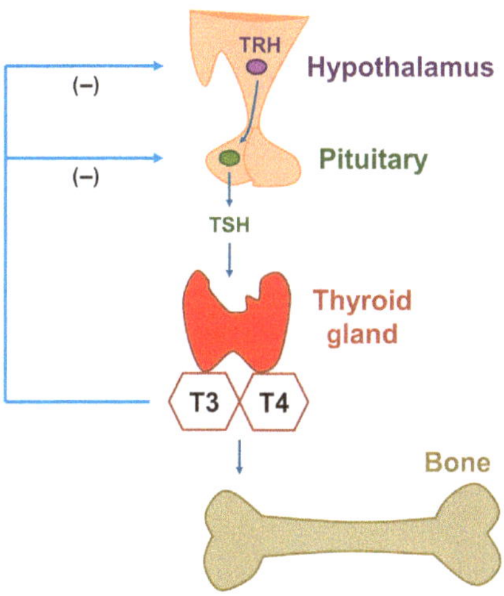

Fig. 1 Negative feedback establishes the HPT axis set point. Thyrotropin-releasing hormone (TRH) is secreted by the hypothalamus. Via the hypothalamic–pituitary portal circulation, TRH signals anterior pituitary thyrotrophs to release thyroid-stimulating hormone (TSH). TSH stimulates the thyroid gland to produce and release T3 and T4, which travel through the circulation to exert effects on peripheral tissues such as bone. T3 exerts negative feedback via TRβ2 to the hypothalamus and pituitary, inhibiting release of TRH and TSH and stabilizing serum TH levels.

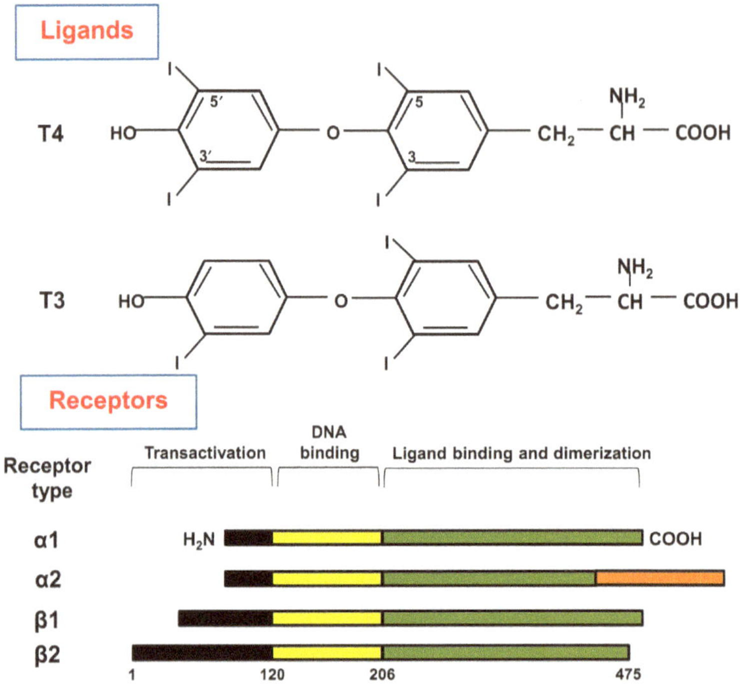

Fig. 2 T4 (3,5,3′,5′-L-tetraiodothyronine, thyroxine) is the major form of TH found in circulation. In peripheral tissues, T4 is converted to T3 (3,5,3′-L-triiodothyronine), the TH with higher affinity for TH receptors (TRs), by deiodinases. Two genes, *THRA* and *THRB*, encode the TRs TRα and TRβ, each of which produces multiple splice variants. The differential expression of the various forms of TR mediates the tissue-specific effects of TH.

serum T4 are free and unbound in the circulation, available to act on tissues. Serum T3 feeds back upon the hypothalamus and pituitary gland, signaling via TH receptor β (TRβ) to inhibit TRH and TSH synthesis and release in order to maintain serum TH levels within a physiological range (Abel, Ahima, Boers, Elmquist, & Wondisford, 2001; Forrest et al., 1996; Macchia et al., 2001; Nikrodhanond et al., 2006). Thus, the inverse relationship between TSH and TH levels which defines the HPT axis set point is maintained by a negative feedback loop (Andersen, Bruun, Pedersen, & Laurberg, 2003; Bassett & Williams, 2008).

Although T3 has a much higher affinity for the thyroid hormone receptors (TRs) than T4 and is therefore classically considered the biologically active form of TH, the fact that T4 exists in much higher circulating concentrations has prompted investigation into whether T4 itself may have biological actions. Indeed, recent studies of T4, T3, and their metabolites have revealed that, at circulating concentrations and in the absence of deiodinases, T3 and T4 were equally able to induce osteoblast differentiation (Cheng, Xing, Pourteymoor, & Mohan, 2016). Further exploration of possible direct T4 effects on specific cell types such as osteoblasts could provide therapeutic targets that can affect the skeleton without off-target T3 effects on other tissues.

2.2 TH Transport and Metabolism

Conversion of T4 to T3 occurs in peripheral tissues via the action of deiodinases (Bianco & Kim, 2006; Lin, Fukuda, & Cheng, 1990). Type 1 deiodinase (D1) can activate or inactivate TH, catalyzing the removal of 5- or 5′-iodine atoms to form three possible products: T4 can be converted to either active T3 or inactive 3,3′,5′-triiodothyronine (reverse T3, rT3), and both T3 and rT3 can be converted to inactive 3,3′-diiodothyronine (T2). The majority of circulating T3 is produced by conversion of T4 to T3 by D1. Type 2 deiodinase (D2) primarily converts T4 to active T3, providing another source of circulating T3 and regulating intracellular T3 concentrations. Type 3 deiodinase (D3), however, is primarily responsible for inactivation of THs, irreversibly converting T3 and T4 to inactive T2 and rT3, respectively (Gereben et al., 2008; Maia, Kim, Huang, Harney, & Larsen, 2005; St Germain, Galton, & Hernandez, 2009). D1 is not expressed in the skeleton or cartilage (Gouveia et al., 2005; LeBron, Pekary, Mirell, Hahn, & Hershman, 1989; Waung, Bassett, & Williams, 2012; Williams et al., 2008), but D2 is expressed in osteoblasts (Bassett et al., 2010; Williams et al., 2008), and D3 is expressed in cells of all skeletal lineages (Capelo et al., 2008; Williams et al., 2008). The specific roles of deiodinases in cartilage and bone development *in vivo* remain to be determined.

While THs are relatively lipophilic and thus can passively diffuse across cell membranes, there are also several specific membrane transporter proteins which actively regulate TH transport into and out of cells. These membrane transporters include the monocarboxylate transporters 8 and 10 (MCT8 and MCT10), sodium-dependent organic anion cotransporting polypeptide 1 (OATP1), the sodium taurocholate cotransporting polypeptide (NTCP), and the L-type amino acid transporters 1 and 2 (LAT1 and LAT2) (Friesema, Jansen, Milici, & Visser, 2005; Heuer & Visser, 2009; Visser, 2013; Visser, Friesema, & Visser, 2011). Studies have revealed that

MCT8, LAT1, and LAT2 are expressed in mouse skeletal tissues (Capelo et al., 2008), suggesting that the levels of active T3 available for intracellular signaling via the nuclear TRs in bone depend upon both the relative activities of activating D2 and inactivating D3 and the expression and efficiency of TH membrane transport proteins.

2.3 TH Receptors and Actions

The primary mechanism by which THs exert their actions is via the nuclear TRs, ligand-inducible transcription factors which regulate expression of target genes (Fig. 2). Two types of TR have been identified: TRα and TRβ. The mammalian *THRA* gene produces three splice variants of TRα in addition to two truncated TRα proteins (TRΔα1 and TRΔα2). Of these TRα splice variants, only TRα1 binds both T3 and DNA to regulate target gene expression. TRα2 and TRα3 do not bind T3; they act as antagonists of TH signaling *in vivo* (Fraichard et al., 1997; Harvey & Williams, 2002; Koenig et al., 1989; Plateroti et al., 2001; Williams, 2000). TRΔα1 and TRΔα2 have been shown to play a role in intestinal development (Chassande et al., 1997; Plateroti et al., 2001), but their role in other tissues including bone remains to be elucidated. The *THRB* gene produces two splice variants, TRβ1 and TRβ2, both of which bind T3 and act as functional receptors. Transcripts of TRβ3 and TRΔβ3 have been detected, but any possible physiological action is unknown (Harvey, Bassett, Maruvada, Yen, & Williams, 2007; Williams, 2000).

TRα1 and TRβ1 are widely expressed across all tissue types; however, their relative abundance and functions depend upon stage of development and tissue type, with the result that different tissues are more or less responsive to a specific TR (Forrest, Sjöberg, & Vennström, 1990). TRα1, for example, is highly expressed in the heart, the brain, and bone, while TRβ1 exhibits high expression in the liver and the pituitary gland (Cheng, 2005). TRβ2, however, is restricted in expression to a few tissues; in the hypothalamus and pituitary gland, TRβ2 mediates the negative feedback inhibition of TRH and TSH to regulate the HPT axis, and in the retina and cochlea, TRβ2 is responsible for regulating development of sensory organs (Abel et al., 2001, 1999; Jones, Ng, Liu, & Forrest, 2007; Ng, Hurley, et al., 2001).

As with other steroid hormones, recent studies have discovered that T3 can exert nongenomic actions in addition to its canonical genomic actions.

These nongenomic effects of T3 act via mechanisms other than direct regulation of nuclear gene transcription; in fact, the nongenomic effects can be measured before changes in nuclear gene transcription would have had time to occur (Lösel & Wehling, 2003; Martin et al., 2014). Nongenomic TH actions can occur via a number of mechanisms. TH can bind directly to the αVβ3 integrin, triggering a downstream signaling cascade. Alternatively, the TRs themselves can mediate nongenomic actions by binding to proteins in other signaling pathways.

3. TH IN THE DEVELOPING SKELETON

3.1 Skeletal Development

In endochondral bone formation, the process by which long bones develop, a cartilage model is replaced by bone tissue. Chondrocytes differentiate from condensations of mesenchymal precursor cells and secrete matrix proteins including type II collagen and proteoglycans to produce a cartilage template. In the traditional model of endochondral bone formation, chondrocytes in the mid-shaft hypertrophy, and blood vessels infiltrate the cartilage model to bring osteoclasts and osteoblast precursors which form the primary ossification center. As osteoclasts degrade and resorb the cartilaginous matrix and osteoblasts replace that cartilage with bone, the primary ossification center expands outward toward the ends of the developing bone. In long bones, secondary ossification centers (SOCs) develop within the epiphyses; during growth, the primary and SOCs are separated by the epiphyseal growth plates, where linear growth due to growth plate chondrocyte proliferation and hypertrophy followed by new bone formation continues until adulthood. When the primary and SOCs meet, obliterating the growth plate, skeletal maturity is achieved (Mackie, Tatarczuch, & Mirams, 2011).

Regulation of endochondral bone formation and linear growth is the result of the endocrine, paracrine, and autocrine actions of many systemic hormones, cytokines, and growth factors including TH, growth hormone (GH), glucocorticoids, sex steroids, insulin-like growth factor I, parathyroid hormone-related peptide, Indian hedgehog (Ihh), bone morphogenetic proteins (BMPs), fibroblast growth factors (FGFs), and vascular endothelial growth factors (Kronenberg, 2003; Lui, Nilsson, & Baron, 2014). Although linear growth ends during puberty with the closure of the growth plates, the processes of mineralization and accumulation of bone mass continue until early adulthood when peak bone mass is reached (Bonjour & Chevalley, 2014;

Bonjour, Theintz, Law, Slosman, & Rizzoli, 1994; Karsenty & Wagner, 2002). Indeed, TH is critical for the development of peak bone mass during childhood and the maintenance of bone mass in adults. However, thyroid status is a balancing act; hyperthyroidism in adults causes an increase in bone turnover and fracture incidence (Vestergaard & Mosekilde, 2002; Vestergaard, Rejnmark, & Mosekilde, 2005).

3.2 TH in Postnatal Growth

Several different genetic mouse models with mutations affecting TSH and TH levels have been studied to investigate the contribution of TH signaling to skeletal development (Table 1). For example, mice lacking the *Pax8* gene, a thyroid-specific transcription factor necessary for development of thyroid follicular cells (Friedrichsen et al., 2003; Mansouri, Chowdhury, & Gruss, 1998), fail to develop a thyroid gland. $Pax8^{-/-}$ mice have undetectably low TH levels and produce extremely high amounts of TSH in an attempt to compensate. In the absence of TH, these mice exhibit impaired linear growth, delayed endochondral bone formation, defective chondrocyte differentiation, reduced cortical bone, impaired trabecular bone remodeling, and decreased bone mineralization. Additionally, mice with a nonfunctioning TSHR ($Tshr^{hyt/hyt}$ mice) have high levels of TSH, extremely low levels of TH, and a similarly impaired skeletal phenotype (Beamer & Cresswell, 1982; Beamer, Eicher, Maltais, & Southard, 1981; Xing et al., 2012). Furthermore, dual oxidase-2 (*Duox2*) mutant mice with hypothyroidism exhibited reduced bone mass at the end of the prepubertal growth period (Xing et al., 2012). The skeletal phenotypes of these mice appear to be due primarily to the effects of TH deficiency, not to the presence of excess TSH (Bassett et al., 2008). TSHR knockout ($Tshr^{-/-}$) mice, however, do display a bone phenotype with severe growth retardation, impaired linear growth, reduced bone mineral density, and increased bone resorption (Marians et al., 2002). The fact that this low bone mass could not be rescued by TH supplementation later in life led to the suggestion that TSH may be an independent negative regulator of bone remodeling (Abe et al., 2003). Clinical congenital hypothyroidism in humans due to a loss-of-function TSHR mutation results in prolonged jaundice, constipation, poor feeding, umbilical hernia, macroglossia, a wide-open posterior fontanelle, and edematous and dry skin (Papadimitriou, Papadimitriou, Papadopoulou, Nicolaidou, & Fretzayas, 2007; Persani et al., 2010).

Table 1 Mutant Mouse Models Showing Phenotypes of Disrupted TH Signaling Pathways in the Developing Skeleton

Mouse Model	Genetic Alteration	TH Axis Phenotype	Developing Skeletal Phenotype	References
$Pax8^{-/-}$	Impaired thyroid follicular cell development due to loss of transcription factor Pax8	Lack of thyroid, ~2000-fold elevation of TSH, undetectable T3/T4, functional TSHR	Compromised linear growth, late endochondral bone formation, decreased cortical bone and bone mineralization	Bassett et al. (2008), Flamant et al. (2002), Friedrichsen et al. (2003), and Mansouri et al. (1998)
$Tshr^{hyt/hyt}$	Loss of function in TSHR	~2000-Fold elevation of TSH, low T3/T4, no TSHR signaling	Compromised linear growth, late endochondral bone formation, decreased cortical bone and mineralization	Bassett et al. (2008), Beamer and Cresswell (1982), Gu et al. (1995), and Stein et al. (1994)
$Tshr^{-/-}$	Deleted Tshr	Hypoplastic thyroid, undetectable T3/T4, large elevation in TSH	Severely impaired linear and overall growth, decreased mineral density, increased bone resorption, altered bone formation, death if not treated with TH after weaning	Abe et al. (2003), Baliram et al. (2012), Marians et al. (2002), and Sun et al. (2013)
$Dio2^{-/-}$	Deleted deiodinase 2	Slight increase in T4 and TSH, normal T3	Normal growth, reduced bone formation, increased mineralization, brittle bones with age	Bassett et al. (2010), Liao et al. (2011), and Schneider et al. (2001)
$Dio3^{-/-}$	Deleted deiodinase 3	Perinatal thyrotoxicosis, hypothyroidism in adulthood	Reduced body length	Hernandez et al. (2006, 2007)
$TR\alpha^{-/-}$	Deleted TRα1/2, TRΔα1/2 preserved	Severe reduction in T3/T4, elevated TSH (normal GH)	Severe growth retardation, late endochondral bone formation, impaired chondrocyte differentiation, reduced mineralization	Chassande et al. (1997), Fraichard et al. (1997), and Gauthier et al. (1999)

Continued

Table 1 Mutant Mouse Models Showing Phenotypes of Disrupted TH Signaling Pathways in the Developing Skeleton—cont'd

Mouse Model	Genetic Alteration	TH Axis Phenotype	Developing Skeletal Phenotype	References
$TR\alpha1^{-/-}$	Deleted $TR\alpha1/\Delta\alpha1$, $TR\alpha2/\Delta\alpha2$ preserved	Mild decreased T4 in males only	No change in growth	Wikström et al. (1998)
$TR\alpha2^{-/-}$	Deleted $TR\alpha2/\Delta\alpha2$, increased $TR\alpha1/\Delta\alpha1$	Mild decrease in T3/T4 (low IGF-I)	No change in growth, low BMD/cortical bone by adulthood	Saltó et al. (2001)
$TR\alpha^{0/0}$	Deleted $TR\alpha1/2$ and $TR\Delta\alpha1/2$	Slight decrease in T4, normal T3 and TSH (normal GH)	Transient growth retardation, late intramembranous and endochondral bone formation, compromised chondrocyte differentiation and mineralization	Bassett, O'Shea, et al. (2007) and Gauthier et al. (2001)
$TR\alpha^{PV/+}$	Heterozygous dominant-negative $TR\alpha$	Mild thyroid failure (normal GH, reduced IGF-I)	Transient growth retardation, late endochondral bone formation, reduced chondrocyte differentiation and mineralization	Bassett et al. (2014), Kaneshige et al. (2001), and O'Shea et al. (2005)
$TR\beta^{-/-}$	Deleted $TR\beta1/2$	RTH, goiter, increased T3/T4, TSH	Persistent short stature, accelerated intramembranous and endochondral bone formation, increased mineralization	Bassett, Nordström, et al. (2007), Bassett, O'Shea, et al. (2007), Forrest et al. (1996), Gauthier et al. (2001), and Ng, Hurley, et al. (2001)
$TR\beta2^{-/-}$	Deleted $TR\beta2$	Mild RTH, slight increase in T3/T4, TSH	No change in growth	Abel et al. (1999) and Ng, Rüsch, et al. (2001)

Table 1 Mutant Mouse Models Showing Phenotypes of Disrupted TH Signaling Pathways in the Developing Skeleton—cont'd

Mouse Model	Genetic Alteration	TH Axis Phenotype	Developing Skeletal Phenotype	References
$TR\beta^{PV/PV}$	Homozygous dominant-negative TRβ	Severe RTH, goiter, ~400-fold increase in TSH, ~15-fold increase T4, ~9-fold increase in T3 (reduced GH, increased IGF-I)	Accelerated prenatal growth, persistent postnatal growth retardation, accelerated intramembranous and endochondral bone formation, increased mineralization	Bassett, Nordström, et al. (2007), Kaneshige et al. (2000), and O'Shea et al. (2005, 2003)
$TR\alpha^{-/-}$ $TR\beta^{-/-}$	Deleted TRα1/2 and TRβ1/2	RTH, small goiter, increased T3/T4, TSH	Disorganized growth plate, late endochondral bone formation, compromised chondrocyte differentiation and mineralization, death within 5 weeks	Gauthier et al. (1999)
$TR\alpha1^{-/-}$ $TR\beta^{-/-}$	Deleted TRα1/Δα1 and TRβ1/2, TRα2/Δα2 preserved	RTH, large goiter, increased T3/T4, TSH (reduced GH/IGF-I)	Persistent growth retardation, late endochondral bone formation, reduced mineralization	Göthe et al. (1999) and Kindblom et al. (2001, 2005)
$TR\alpha2^{-/-}$ $TR\beta^{-/-}$	Deleted TRα2/Δα2 and TRβ1/2, increased TRα1/Δα1	Mild decrease in T3/T4, normal TSH	Transient growth delay	Ng, Hurley, et al. (2001) and Ng, Rüsch, et al. (2001)
$TR\alpha^{0/0}$ $TR\beta^{-/-}$	Deleted TRα1/2, TRΔα1/2, and TRβ1/2	RTH, goiter, >10-fold increase in T3/T4, >200-fold increase in TSH (low GH/IGF-I)	Disorganized growth plate, delayed growth, late endochondral bone formation, compromised chondrocyte differentiation and mineralization	Bassett, Swinhoe, Chassande, Samarut, and Williams (2006), Flamant et al. (2002), and Gauthier et al. (2001)

Studies investigating the effect of knocking out deiodinase activity on the developing skeleton have largely found no effect (Berry et al., 1993; Liao et al., 2011; Schneider et al., 2006; Schoenmakers, Pigmans, Poland, & Visser, 1993), although D2-null mice show reduced bone formation, increased mineralization, and brittle bones with age (Bassett et al., 2010; Liao et al., 2011; Schneider et al., 2001), and D3-null mice exhibit thyrotoxicosis when they are very young and have reduced body length (Hernandez, Martinez, Fiering, Galton, & St Germain, 2006; Hernandez et al., 2007). Furthermore, the only reports of an effect on the developing skeleton due to knocking out TH transporters involve combining *Mct8* knockout with *Oatp1c1*, D1, or D2 knockout (Liao et al., 2011; Mayerl et al., 2014; Morte et al., 2010). Clinical cases involving mutant *Mct8* genes exhibit normal linear growth (Dumitrescu, Liao, Best, Brockmann, & Refetoff, 2004; Friesema et al., 2004); however, patients with a mutation in the pathway necessary to produce functional deiodinases have presented with short stature and transient growth retardation (Dumitrescu et al., 2005; Refetoff & Dumitrescu, 2007).

3.3 TH in Secondary Ossification

In mice, there is a window from postnatal days 5–14 when serum T3 levels rapidly increase. This window is followed by a shift from longitudinal bone growth to GH-independent bone mass acquisition and increase in BMD (Xing et al., 2012). This shift from linear bone growth to accelerated bone mass acquisition coincides with the initiation of secondary ossification in the epiphyses in mice. In humans, the identification of a proximal humeral epiphyseal ossification center occurs around a gestational age of 38 weeks which coincides with attainment of peak levels of TH (36–42 weeks). Thus, the time of appearance of secondary ossification in several species including humans, rats, and mice coincides with the time when peak levels of TH are attained (Xing, Cheng, Wergedal, & Mohan, 2014).

The epiphyseal region in long bones contains many regions of interest, including the chondrocyte growth plate, the SOC, and the articular cartilage. All of these regions become specified during different periods, although the entire region is populated by immature chondrocytes during embryonic and very early postnatal development. Chondrocytes in the growth plate then form a hierarchical organization involving slowly dividing stem-like cells, rapidly proliferating cells, and finally prehypertrophic and hypertrophic cells. A similar region surrounds the SOC, albeit in a circular, not linear, formation. Unlike primary ossification, which occurs during embryonic development, secondary ossification occurs during early postnatal development; however, both primary and secondary ossification take place via endochondral bone formation, and the method by which endochondral bone formation initiates during both stages of development has recently been brought into question. Of note, given recent studies of chondrogenic to osteogenic transdifferentiation in regions of primary ossification (Ono, Ono, Nagasawa, & Kronenberg, 2014; Yang, Tsang, Tang, Chan, & Cheah, 2014; Zhou et al., 2014), it may be likely that early secondary ossification involves a similar mechanism, given that ossification begins prior to vascular invasion.

Many studies have suggested that TH plays a key role in the development of these regions and is required for their ossification. Furthermore, our studies have suggested that the increase in TH levels during the prepubertal growth period is essential for initiation and progression of SOC formation (Xing et al., 2016, 2014). TH is a known factor for promoting the terminal differentiation of growth plate chondrocytes (Ishikawa, Genge, Wuthier, & Wu, 1998; Robson, Siebler, Stevens, Shalet, & Williams, 2000) and osteoblasts. Indeed, hypothyroid mice exhibit delayed or incomplete ossification of the SOC as well as a disrupted growth plate such that it inhibits differentiation (Fig. 3; Bassett et al., 2008). Conversely, hyperthyroidism can lead to adverse phenotypes such as premature epiphyseal fusion caused by increased early ossification prior to growth plate proliferation (Segni & Gorman, 2001).

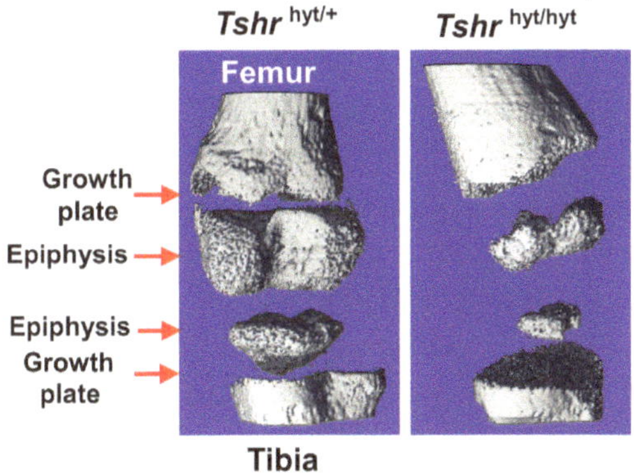

Fig. 3 TH deficiency in *Tshr*$^{hyt/hyt}$ mice impairs formation of the epiphysis. Micro-computed tomography images of the distal femurs and proximal tibiae of 14-day-old euthyroid *Tshr*$^{hyt/+}$ and hypothyroid *Tshr*$^{hyt/hyt}$ mice show epiphyseal defects in the femur and tibia of *Tshr*$^{hyt/hyt}$ mice, indicated by *arrowheads*.

While TRα is ubiquitously expressed in chondrocytes and osteoblasts, TRβ is expressed only in prehypertrophic cells post-TH treatment, which necessitates mediation through TRα (Stevens et al., 2000). Moreover, recent findings suggest that TRβ, not TRα, plays a key role in thyroid signaling to the SOC and that Ihh plays an important role as a downstream effector of this process (Xing et al., 2016, 2014). Unlike sonic hedgehog (Shh), Ihh is typically involved in promoting differentiation. In a recent study of zebrafish jawbone regeneration, Ihh regulated the differentiation of chondrocytes to osteoblasts (Paul et al., 2016). Notably, TH also increases the expression of several osteogenic factors such as osteocalcin and alkaline phosphatase in chondrocytes and in euthyroid mice in part via promotion of osterix expression in hypertrophic chondrocytes (Xing et al., 2014).

4. TH IN THE ADULT SKELETON

Although this review primarily focuses on studies involving the developing skeleton, we will include a brief overview of the effects of TH in the adult skeleton for the purpose of contrast. In the developing skeleton, the effects of TH are broadly anabolic; juvenile hypothyroidism leads to short stature due to a lack of skeletal growth, and juvenile thyrotoxicosis leads to short stature due to accelerated skeletal growth and premature closure of the growth plates. By contrast, TH affects adult bone maintenance by increasing bone turnover leading to net bone resorption. Adult hyperthyroidism is associated with increased markers of both bone formation and resorption (Garnero, Vassy, Bertholin, Riou, & Delmas, 1994; Guo, Weetman, & Eastell, 1997; Harvey et al., 1991; Toivonen, Tähtelä, Laitinen, Risteli, & Välimäki, 1998), and initiation of therapy for thyrotoxicosis in premenopausal women led to a 4% increase in BMD within the first year of treatment (Udayakumar, Chandrasekaran, Rasheed, Suresh, & Sivaprakash, 2006). Moreover, thyrotoxicosis leads to increased risk of hip fracture (Vestergaard & Mosekilde, 2003). Furthermore, adult hypothyroidism is associated with decreased osteoblast and osteoclast activity (Eriksen, Mosekilde, & Melsen, 1985, 1986; Mosekilde & Melsen, 1978), and, while hypothyroidism is not directly related to fracture risk, overly aggressive TH supplementation in hypothyroidism may lead to an increased risk of fracture (Ko et al., 2014). For a more comprehensive discussion of the TH effects on the adult skeleton, see the recent review by Bassett and Williams (2016).

5. SKELETAL EFFECTS OF TR SIGNALING

Before ligation by T3, TRs are located in the nucleus where they can form homodimers or heterodimers with retinoid X receptors (RXR). These heterodimers can bind to specific genomic TH-response element (TRE) sequences which exist in the promoter regions of T3 target genes (Fig. 4). In the absence of T3, the TR–RXR complex associates with corepressor proteins (e.g., nuclear receptor corepressor and the silencing mediator for retinoid and TR) which can recruit histone deacetylases and inhibit gene transcription (Astapova et al., 2008, 2011; Chassande, 2003; Hashimoto et al., 2001; Venero et al., 2005; Wallis et al., 2008). When T3 enters the cell, either directly or via conversion from T4 by D1/D2, the corepressors are released, and coactivator proteins (e.g., CBP/p300, pCAF, and steroid receptor coactivator 1) are recruited to activate transcription of T3 target genes. Indeed, the fact that TRs both inhibit and activate transcription of target genes depending on whether they are T3 bound is

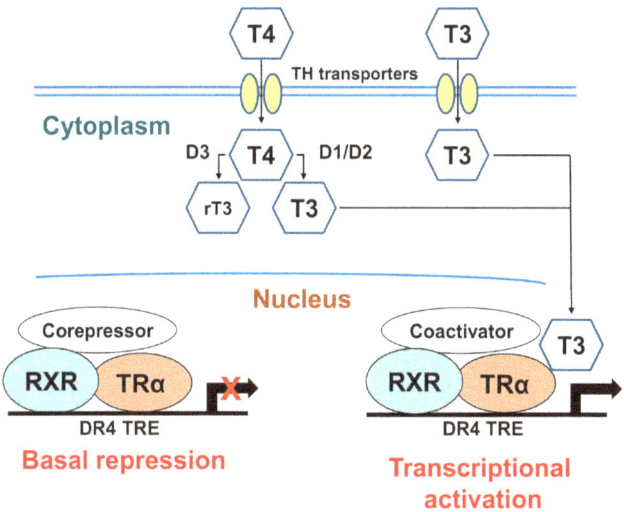

Fig. 4 T3 and T4 enter the cell via a combination of direct diffusion and membrane transport proteins. The levels of active T3 in the cell are regulated by deiodinase conversion of T4 to T3 or inactivation to rT3 and T2. Before ligation with T3, TRα in a heterodimer with RXR binds to specific TH-response element (TRE) sequences, recruiting corepressors to inhibit transcription of target genes. When T3 binds TRα, the corepressors are replaced by coactivators, allowing target gene transcription.

responsible for the large effect size observed in T3-mediated transcriptional modulation (Bassett, Harvey, & Williams, 2003; Brent, 2012; Cheng, Leonard, & Davis, 2010; Harvey & Williams, 2002; Kim et al., 2005; Vella et al., 2014).

5.1 TRα Signaling in Postnatal Growth

Genetic mouse models have been used to study the role of TRα in skeletal development (Table 1). As per the discussion in Section 2.3, the *THRA* gene encodes several TRα splice variants and truncated isoforms. Mice lacking all TRα isoforms (TR$\alpha^{0/0}$ mice) are systemically euthyroid; however, these mice exhibit transient developmental growth retardation and delayed endochondral bone formation including impaired chondrocyte differentiation and reduced mineral deposition (Bassett et al., 2008; Gauthier et al., 2001). Moreover, mice with dominant-negative TRα1 mutations do experience some amount of systemic hypothyroidism and display more severe skeletal development phenotypes than TR$\alpha^{0/0}$ mice. For example, TR$\alpha 1^{PV/+}$ mice have severe and unremitting postnatal growth retardation with delayed endochondral bone formation and impaired bone mineralization (Bassett & Williams, 2009; Kaneshige et al., 2001). The fact that the presence of a dominant-negative TRα causes a more severe skeletal phenotype than the simple absence of the TRα demonstrates the powerful antagonistic effect of the unbound TRα1 receptor which, when unable to bind T3, recruits repressor proteins and represses expression of T3 target genes such as fibroblast growth factor receptors 1 and 3 (Barnard et al., 2005; Bassett, Nordström, et al., 2007; Chassande, 2003). In clinical cases of a heterozygous dominant-negative TRα receptor, patients present with a type of resistance to thyroid hormone (RTH) which is accompanied by short stature and delayed bone development, among other symptoms (Bochukova et al., 2012; van Mullem et al., 2012).

Mechanistically, TRα1 is required for longitudinal growth as it promotes expression of Shh, an important regulatory factor which is known to favor chondrocyte proliferation and survival, which lead to longitudinal growth, over differentiation, and hypertrophy. Shh increases expression of master chondrocyte regulator Sox9, and overexpression of Shh in chondrocytes led to impaired endochondral bone formation (Desouza et al., 2011; Gil-Ibáñez, Bernal, & Morte, 2014; Kim & Mohan, 2013; Tavella et al., 2004).

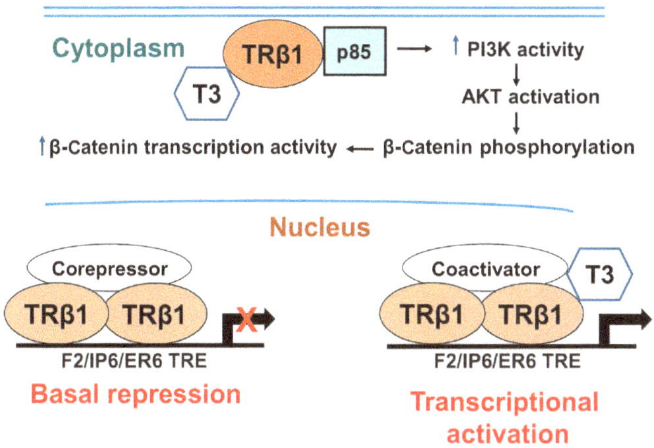

Fig. 5 TRβ as a homodimer exerts genomic actions via a similar mechanism to TRα, binding to different TRE sequences to regulate transcription of different target genes. In addition, TRs can also exert nongenomic actions. For example, TRβ can interact directly with the p85 subunit of phosphoinositide 3-kinase (PI3K). In the presence of T3, TRβ releases PI3K to activate downstream signaling cascades involving AKT and β-catenin.

5.2 TRβ Signaling in Postnatal Growth

As with TRα, mutant mouse models (Table 1) have been studied to investigate the effect of TRβ signaling in skeletal development (Fig. 5). Since TRβ is responsible for sending negative feedback to the hypothalamus to regulate the HPT axis set point, TRβ-null ($TR\beta^{-/-}$) mice lack that negative feedback and exhibit RTH. These mice display the effects of skeletal hyperthyroidism including persistent short stature due to premature closure of the growth plates, advanced endochondral and intramembranous bone formation, accelerated chondrocyte differentiation, and increased mineral deposition (Abel et al., 1999; Bassett et al., 2008; Forrest et al., 1996). Furthermore, mice with a dominant-negative TRβ mutation ($TR\beta^{PV/PV}$ mice) have a more severe RTH and skeletal hyperthyroid phenotype characterized by accelerated prenatal growth and advanced endochondral and intramembranous bone formation (Kaneshige et al., 2000; O'Shea et al., 2005, 2003). Since these mice have high levels of T3 which results in increased TRα signaling in bone, skeletal expression of T3 target genes FGFR1 and 3 is increased (Barnard et al., 2005; Bassett, Nordström, et al., 2007; Bassett et al., 2008). Clinical reports of patients with RTH due to dominant-negative TRβ mutations include developmental and growth delay

(Ferrara et al., 2012; Refetoff, DeWind, & DeGroot, 1967; Refetoff & Dumitrescu, 2007; Weiss, 2008).

In contrast to TRα regulation of longitudinal growth via upregulation of Shh and the resulting increase in chondrocyte proliferation and survival, TRβ promotes chondrocyte differentiation by stimulating Ihh signaling. Increased Ihh induces a number of growth and transcription factor signaling pathways including osterix, Runx2, β-catenin, IGF-I, and RANKL, leading to chondrocyte differentiation, hypertrophy, and even transdifferentiation into osteoblasts to carry out endochondral bone formation (Amano, Densmore, Nishimura, & Lanske, 2014; Paul et al., 2016; Xing et al., 2016, 2014).

5.3 Nongenomic TH Signaling

Nongenomic TH actions have been observed in the region of the plasma membrane, within the cytoplasm, and at the mitochondria. For example, T4 has been shown to act via the MAPK pathway to stimulate proliferation and angiogenesis, an effect mediated by the αVβ3 integrin. Furthermore, palmitoylated TRα activates the nitric oxide/protein kinase G2/Src pathway to increase MAPK and PI3K/AKT signaling which mediate rapid T3 actions in osteoblasts, and T3 acts via TRβ to activate the PI3K/AKT pathway (Fig. 5; Bassett et al., 2003; Bergh et al., 2005; Cao, Kambe, Moeller, Refetoff, & Seo, 2005; Cheng et al., 2010; Davis, Goglia, & Leonard, 2016; Davis et al., 2006; Furuya, Hanover, & Cheng, 2006; Gauthier & Flamant, 2014; Kalyanaraman et al., 2014; Martin et al., 2014; Moeller & Broecker-Preuss, 2011; Moeller, Dumitrescu, & Refetoff, 2005; Storey et al., 2006). AKT-mediated phosphorylation of β-catenin leads to nuclear accumulation of β-catenin to promote its transcriptional activity (Fang et al., 2007). Thus, nongenomic signaling by TH adds to the complexity needed for the multitude of TH effects in various cell types.

6. SUMMARY AND CONCLUSIONS

In summary, TH is a critical regulator of skeletal development and postnatal growth (Fig. 6). During the very early stages of development (the first few postnatal days in mice), low levels of T3 stimulate constitutive TRα in bone to upregulate expression of Shh and Sox9. Under these conditions, chondrocytes remain immature and are stimulated to proliferate, leading to longitudinal growth. However, when serum T3 levels rise during postnatal days 5–14, TRβ expression is induced. T3 then binds to TRβ,

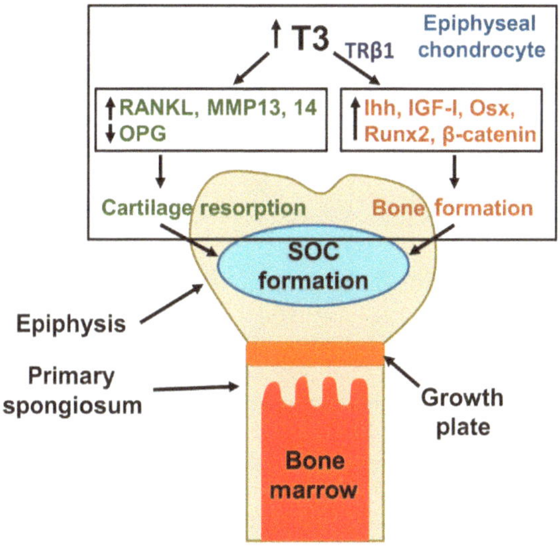

Fig. 6 Model of TH regulation of the secondary ossification center in the epiphysis. The increase in serum levels of THs during the early postnatal growth period leads to an increase in TRβ1 expression in epiphyseal chondrocytes. Activation of TH signaling in epiphyseal chondrocytes via TRβ1 leads to an increase in Ihh and other growth factor pathways (IGF-I, Wnt, BMPs, and RANKL) resulting in chondrocyte to osteoblast transdifferentiation and bone formation.

shifting the balance from predominant Shh expression in chondrocytes to Ihh expression (Aghajanian, Xing, Cheng, Pourteymoor, & Mohan, 2016). Ihh induces Osx, Runx2, β-catenin, IGF-I, and RANKL, causing chondrocytes to differentiate, hypertrophy, and transdifferentiate into endochondral bone-forming osteoblasts. While much progress has been made in recent years in our understanding of the roles played by TH in skeletal development, much still remains to be discovered. In order to determine the specific roles of each TR during endochondral bone formation, conditional knockout of the various TRs in specific cell types during specific stages of development will be required. Furthermore, the implications of chondrocyte to osteoblast transdifferentiation for the traditional model of endochondral bone formation are still in the early stages of being elucidated, and whether TH regulation of postnatal development in other tissues involves transdifferentiation is an open question. Additionally, the role of nongenomic TH signaling in bone development is still largely unexplored. Continuing to develop our understanding of these mechanisms of TH regulation of bone will open the door to development of novel therapies for musculoskeletal conditions which plague so many worldwide.

ACKNOWLEDGMENTS

This study was supported by funding from the National Institutes of Arthritis and Musculoskeletal Skin Diseases RO1 Grant (AR048139) and the Department of Veterans Affairs. S.M. is a recipient of a Senior Research Career Scientist Award from the Department of Veterans Affairs.

REFERENCES

Abe, E., Marians, R. C., Yu, W., Wu, X. B., Ando, T., Li, Y., et al. (2003). TSH is a negative regulator of skeletal remodeling. *Cell, 115*(2), 151–162.

Abel, E. D., Ahima, R. S., Boers, M. E., Elmquist, J. K., & Wondisford, F. E. (2001). Critical role for thyroid hormone receptor beta2 in the regulation of paraventricular thyrotropin-releasing hormone neurons. *The Journal of Clinical Investigation, 107*(8), 1017–1023. http://dx.doi.org/10.1172/JCI10858.

Abel, E. D., Boers, M. E., Pazos-Moura, C., Moura, E., Kaulbach, H., Zakaria, M., et al. (1999). Divergent roles for thyroid hormone receptor beta isoforms in the endocrine axis and auditory system. *The Journal of Clinical Investigation, 104*(3), 291–300. http://dx.doi.org/10.1172/JCI6397.

Aghajanian, P., Xing, W., Cheng, S., Pourteymoor, S., & Mohan, S. (2016). Opposing actions of SHH and IHH control transition of proliferating immature chondrocytes into mature hypertrophic chondrocytes during secondary center ossification. Presented at the American Society for Bone and Mineral Research Annual Meeting, Atlanta, GA.

Albert, B. B., Cutfield, W. S., Webster, D., Carll, J., Derraik, J. G. B., Jefferies, C., et al. (2012). Etiology of increasing incidence of congenital hypothyroidism in New Zealand from 1993–2010. *The Journal of Clinical Endocrinology and Metabolism, 97*(9), 3155–3160. http://dx.doi.org/10.1210/jc.2012-1562.

Amano, K., Densmore, M., Nishimura, R., & Lanske, B. (2014). Indian hedgehog signaling regulates transcription and expression of collagen type X via Runx2/Smads interactions. *The Journal of Biological Chemistry, 289*(36), 24898–24910. http://dx.doi.org/10.1074/jbc.M114.570507.

Andersen, S., Bruun, N. H., Pedersen, K. M., & Laurberg, P. (2003). Biologic variation is important for interpretation of thyroid function tests. *Thyroid: Official Journal of the American Thyroid Association, 13*(11), 1069–1078. http://dx.doi.org/10.1089/105072503770867237.

Astapova, I., Lee, L. J., Morales, C., Tauber, S., Bilban, M., & Hollenberg, A. N. (2008). The nuclear corepressor, NCoR, regulates thyroid hormone action in vivo. *Proceedings of the National Academy of Sciences of the United States of America, 105*(49), 19544–19549. http://dx.doi.org/10.1073/pnas.0804604105.

Astapova, I., Vella, K. R., Ramadoss, P., Holtz, K. A., Rodwin, B. A., Liao, X.-H., et al. (2011). The nuclear receptor corepressor (NCoR) controls thyroid hormone sensitivity and the set point of the hypothalamic-pituitary-thyroid axis. *Molecular Endocrinology (Baltimore, Md.), 25*(2), 212–224. http://dx.doi.org/10.1210/me.2010-0462.

Baliram, R., Sun, L., Cao, J., Li, J., Latif, R., Huber, A. K., et al. (2012). Hyperthyroid-associated osteoporosis is exacerbated by the loss of TSH signaling. *The Journal of Clinical Investigation, 122*(10), 3737–3741. http://dx.doi.org/10.1172/JCI63948.

Barnard, J. C., Williams, A. J., Rabier, B., Chassande, O., Samarut, J., Cheng, S.-Y., et al. (2005). Thyroid hormones regulate fibroblast growth factor receptor signaling during chondrogenesis. *Endocrinology, 146*(12), 5568–5580. http://dx.doi.org/10.1210/en.2005-0762.

Bassett, J. H. D., Boyde, A., Howell, P. G. T., Bassett, R. H., Galliford, T. M., Archanco, M., et al. (2010). Optimal bone strength and mineralization requires the type 2 iodothyronine deiodinase in osteoblasts. *Proceedings of the National Academy of Sciences of the United States of America, 107*(16), 7604–7609. http://dx.doi.org/10.1073/pnas.0911346107.
Bassett, J. H. D., Boyde, A., Zikmund, T., Evans, H., Croucher, P. I., Zhu, X., et al. (2014). Thyroid hormone receptor α mutation causes a severe and thyroxine-resistant skeletal dysplasia in female mice. *Endocrinology, 155*(9), 3699–3712. http://dx.doi.org/10.1210/en.2013-2156.
Bassett, J. H. D., Harvey, C. B., & Williams, G. R. (2003). Mechanisms of thyroid hormone receptor-specific nuclear and extra nuclear actions. *Molecular and Cellular Endocrinology, 213*(1), 1–11. http://dx.doi.org/10.1016/j.mce.2003.10.033.
Bassett, J. H. D., Nordström, K., Boyde, A., Howell, P. G. T., Kelly, S., Vennström, B., et al. (2007). Thyroid status during skeletal development determines adult bone structure and mineralization. *Molecular Endocrinology (Baltimore, Md), 21*(8), 1893–1904. http://dx.doi.org/10.1210/me.2007-0157.
Bassett, J. H. D., O'Shea, P. J., Sriskantharajah, S., Rabier, B., Boyde, A., Howell, P. G. T., et al. (2007). Thyroid hormone excess rather than thyrotropin deficiency induces osteoporosis in hyperthyroidism. *Molecular Endocrinology (Baltimore, Md), 21*(5), 1095–1107. http://dx.doi.org/10.1210/me.2007-0033.
Bassett, J. H. D., Swinhoe, R., Chassande, O., Samarut, J., & Williams, G. R. (2006). Thyroid hormone regulates heparan sulfate proteoglycan expression in the growth plate. *Endocrinology, 147*(1), 295–305. http://dx.doi.org/10.1210/en.2005-0485.
Bassett, J. H. D., & Williams, G. R. (2008). Critical role of the hypothalamic-pituitary-thyroid axis in bone. *Bone, 43*(3), 418–426. http://dx.doi.org/10.1016/j.bone.2008.05.007.
Bassett, J. H. D., & Williams, G. R. (2009). The skeletal phenotypes of TRalpha and TRbeta mutant mice. *Journal of Molecular Endocrinology, 42*(4), 269–282. http://dx.doi.org/10.1677/JME-08-0142.
Bassett, J. H. D., & Williams, G. R. (2016). Role of thyroid hormones in skeletal development and bone maintenance. *Endocrine Reviews, 37*(2), 135–187. http://dx.doi.org/10.1210/er.2015-1106.
Bassett, J. H. D., Williams, A. J., Murphy, E., Boyde, A., Howell, P. G. T., Swinhoe, R., et al. (2008). A lack of thyroid hormones rather than excess thyrotropin causes abnormal skeletal development in hypothyroidism. *Molecular Endocrinology (Baltimore, Md.), 22*(2), 501–512. http://dx.doi.org/10.1210/me.2007-0221.
Beamer, W. G., & Cresswell, L. A. (1982). Defective thyroid ontogenesis in fetal hypothyroid (hyt/hyt) mice. *The Anatomical Record, 202*(3), 387–393. http://dx.doi.org/10.1002/ar.1092020311.
Beamer, W. J., Eicher, E. M., Maltais, L. J., & Southard, J. L. (1981). Inherited primary hypothyroidism in mice. *Science (New York, N.Y.), 212*(4490), 61–63.
Bergh, J. J., Lin, H.-Y., Lansing, L., Mohamed, S. N., Davis, F. B., Mousa, S., et al. (2005). Integrin alphaVbeta3 contains a cell surface receptor site for thyroid hormone that is linked to activation of mitogen-activated protein kinase and induction of angiogenesis. *Endocrinology, 146*(7), 2864–2871. http://dx.doi.org/10.1210/en.2005-0102.
Berry, M. J., Grieco, D., Taylor, B. A., Maia, A. L., Kieffer, J. D., Beamer, W., et al. (1993). Physiological and genetic analyses of inbred mouse strains with a type I iodothyronine 5′ deiodinase deficiency. *The Journal of Clinical Investigation, 92*(3), 1517–1528. http://dx.doi.org/10.1172/JCI116730.
Bianco, A. C., & Kim, B. W. (2006). Deiodinases: Implications of the local control of thyroid hormone action. *The Journal of Clinical Investigation, 116*(10), 2571–2579. http://dx.doi.org/10.1172/JCI29812.

Bochukova, E., Schoenmakers, N., Agostini, M., Schoenmakers, E., Rajanayagam, O., Keogh, J. M., et al. (2012). A mutation in the thyroid hormone receptor alpha gene. *The New England Journal of Medicine*, *366*(3), 243–249. http://dx.doi.org/10.1056/NEJMoa1110296.

Bonjour, J.-P., & Chevalley, T. (2014). Pubertal timing, bone acquisition, and risk of fracture throughout life. *Endocrine Reviews*, *35*(5), 820–847. http://dx.doi.org/10.1210/er.2014-1007.

Bonjour, J.-P., Theintz, G., Law, F., Slosman, D., & Rizzoli, R. (1994). Peak bone mass. *Osteoporosis International: A Journal Established as Result of Cooperation Between the European Foundation for Osteoporosis and the National Osteoporosis Foundation of the USA*, *4*(Suppl. 1), 7–13.

Brent, G. A. (2012). Mechanisms of thyroid hormone action. *The Journal of Clinical Investigation*, *122*(9), 3035–3043. http://dx.doi.org/10.1172/JCI60047.

Cao, X., Kambe, F., Moeller, L. C., Refetoff, S., & Seo, H. (2005). Thyroid hormone induces rapid activation of Akt/protein kinase B-mammalian target of rapamycin-p70S6K cascade through phosphatidylinositol 3-kinase in human fibroblasts. *Molecular Endocrinology (Baltimore, Md.)*, *19*(1), 102–112. http://dx.doi.org/10.1210/me.2004-0093.

Capelo, L. P., Beber, E. H., Huang, S. A., Zorn, T. M. T., Bianco, A. C., & Gouveia, C. H. A. (2008). Deiodinase-mediated thyroid hormone inactivation minimizes thyroid hormone signaling in the early development of fetal skeleton. *Bone*, *43*(5), 921–930. http://dx.doi.org/10.1016/j.bone.2008.06.020.

Chassande, O. (2003). Do unliganded thyroid hormone receptors have physiological functions? *Journal of Molecular Endocrinology*, *31*(1), 9–20.

Chassande, O., Fraichard, A., Gauthier, K., Flamant, F., Legrand, C., Savatier, P., et al. (1997). Identification of transcripts initiated from an internal promoter in the c-erbA alpha locus that encode inhibitors of retinoic acid receptor-alpha and triiodothyronine receptor activities. *Molecular Endocrinology (Baltimore, Md.)*, *11*(9), 1278–1290. http://dx.doi.org/10.1210/mend.11.9.9972.

Cheng, S.-Y. (2005). Isoform-dependent actions of thyroid hormone nuclear receptors: Lessons from knockin mutant mice. *Steroids*, *70*(5–7), 450–454. http://dx.doi.org/10.1016/j.steroids.2005.02.003.

Cheng, S.-Y., Leonard, J. L., & Davis, P. J. (2010). Molecular aspects of thyroid hormone actions. *Endocrine Reviews*, *31*(2), 139–170. http://dx.doi.org/10.1210/er.2009-0007.

Cheng, S., Xing, W., Pourteymoor, S., & Mohan, S. (2016). Effects of thyroxine (T4), 3,5,3′-triiodo-L-thyronine (T3) and their metabolites on osteoblast differentiation. *Calcified Tissue International*, *99*(4), 435–442. http://dx.doi.org/10.1007/s00223-016-0159-x.

Corbetta, C., Weber, G., Cortinovis, F., Calebiro, D., Passoni, A., Vigone, M. C., et al. (2009). A 7-year experience with low blood TSH cutoff levels for neonatal screening reveals an unsuspected frequency of congenital hypothyroidism (CH). *Clinical Endocrinology*, *71*(5), 739–745. http://dx.doi.org/10.1111/j.1365-2265.2009.03568.x.

Davis, P. J., Goglia, F., & Leonard, J. L. (2016). Nongenomic actions of thyroid hormone. *Nature Reviews. Endocrinology*, *12*(2), 111–121. http://dx.doi.org/10.1038/nrendo.2015.205.

Davis, F. B., Tang, H.-Y., Shih, A., Keating, T., Lansing, L., Hercbergs, A., et al. (2006). Acting via a cell surface receptor, thyroid hormone is a growth factor for glioma cells. *Cancer Research*, *66*(14), 7270–7275. http://dx.doi.org/10.1158/0008-5472.CAN-05-4365.

Deladoëy, J., Ruel, J., Giguère, Y., & Van Vliet, G. (2011). Is the incidence of congenital hypothyroidism really increasing? A 20-year retrospective population-based study in Québec. *The Journal of Clinical Endocrinology and Metabolism*, *96*(8), 2422–2429. http://dx.doi.org/10.1210/jc.2011-1073.

Delvecchio, M., Vigone, M. C., Wasniewska, M., Weber, G., Lapolla, R., Popolo, P. P., et al. (2015). Final height in Italian patients with congenital hypothyroidism detected by neonatal screening: A 20-year observational study. *Italian Journal of Pediatrics, 41*, 82. http://dx.doi.org/10.1186/s13052-015-0190-y.

Desouza, L. A., Sathanoori, M., Kapoor, R., Rajadhyaksha, N., Gonzalez, L. E., Kottmann, A. H., et al. (2011). Thyroid hormone regulates the expression of the sonic hedgehog signaling pathway in the embryonic and adult mammalian brain. *Endocrinology, 152*(5), 1989–2000. http://dx.doi.org/10.1210/en.2010-1396.

Dumitrescu, A. M., Liao, X.-H., Abdullah, M. S. Y., Lado-Abeal, J., Majed, F. A., Moeller, L. C., et al. (2005). Mutations in SECISBP2 result in abnormal thyroid hormone metabolism. *Nature Genetics, 37*(11), 1247–1252. http://dx.doi.org/10.1038/ng1654.

Dumitrescu, A. M., Liao, X.-H., Best, T. B., Brockmann, K., & Refetoff, S. (2004). A novel syndrome combining thyroid and neurological abnormalities is associated with mutations in a monocarboxylate transporter gene. *American Journal of Human Genetics, 74*(1), 168–175. http://dx.doi.org/10.1086/380999.

Eriksen, E. F., Mosekilde, L., & Melsen, F. (1985). Trabecular bone remodeling and bone balance in hyperthyroidism. *Bone, 6*(6), 421–428.

Eriksen, E. F., Mosekilde, L., & Melsen, F. (1986). Kinetics of trabecular bone resorption and formation in hypothyroidism: Evidence for a positive balance per remodeling cycle. *Bone, 7*(2), 101–108.

Fang, D., Hawke, D., Zheng, Y., Xia, Y., Meisenhelder, J., Nika, H., et al. (2007). Phosphorylation of beta-catenin by AKT promotes beta-catenin transcriptional activity. *The Journal of Biological Chemistry, 282*(15), 11221–11229. http://dx.doi.org/10.1074/jbc.M611871200.

Ferrara, A. M., Onigata, K., Ercan, O., Woodhead, H., Weiss, R. E., & Refetoff, S. (2012). Homozygous thyroid hormone receptor β-gene mutations in resistance to thyroid hormone: Three new cases and review of the literature. *The Journal of Clinical Endocrinology and Metabolism, 97*(4), 1328–1336. http://dx.doi.org/10.1210/jc.2011-2642.

Flamant, F., Poguet, A.-L., Plateroti, M., Chassande, O., Gauthier, K., Streichenberger, N., et al. (2002). Congenital hypothyroid Pax8(-/-) mutant mice can be rescued by inactivating the TRalpha gene. *Molecular Endocrinology (Baltimore, Md), 16*(1), 24–32. http://dx.doi.org/10.1210/mend.16.1.0766.

Forrest, D., Hanebuth, E., Smeyne, R. J., Everds, N., Stewart, C. L., Wehner, J. M., et al. (1996). Recessive resistance to thyroid hormone in mice lacking thyroid hormone receptor beta: Evidence for tissue-specific modulation of receptor function. *The EMBO Journal, 15*(12), 3006–3015.

Forrest, D., Sjöberg, M., & Vennström, B. (1990). Contrasting developmental and tissue-specific expression of alpha and beta thyroid hormone receptor genes. *The EMBO Journal, 9*(5), 1519–1528.

Fraichard, A., Chassande, O., Plateroti, M., Roux, J. P., Trouillas, J., Dehay, C., et al. (1997). The T3R alpha gene encoding a thyroid hormone receptor is essential for post-natal development and thyroid hormone production. *The EMBO Journal, 16*(14), 4412–4420. http://dx.doi.org/10.1093/emboj/16.14.4412.

Friedrichsen, S., Christ, S., Heuer, H., Schäfer, M. K. H., Mansouri, A., Bauer, K., et al. (2003). Regulation of iodothyronine deiodinases in the Pax8-/- mouse model of congenital hypothyroidism. *Endocrinology, 144*(3), 777–784. http://dx.doi.org/10.1210/en.2002-220715.

Friesema, E. C. H., Grueters, A., Biebermann, H., Krude, H., von Moers, A., Reeser, M., et al. (2004). Association between mutations in a thyroid hormone transporter and severe X-linked psychomotor retardation. *Lancet (London, England), 364*(9443), 1435–1437. http://dx.doi.org/10.1016/S0140-6736(04)17226-7.

Friesema, E. C. H., Jansen, J., Milici, C., & Visser, T. J. (2005). Thyroid hormone transporters. *Vitamins and Hormones*, *70*, 137–167. http://dx.doi.org/10.1016/S0083-6729(05)70005-4.

Furuya, F., Hanover, J. A., & Cheng, S. (2006). Activation of phosphatidylinositol 3-kinase signaling by a mutant thyroid hormone beta receptor. *Proceedings of the National Academy of Sciences of the United States of America*, *103*(6), 1780–1785. http://dx.doi.org/10.1073/pnas.0510849103.

Garnero, P., Vassy, V., Bertholin, A., Riou, J. P., & Delmas, P. D. (1994). Markers of bone turnover in hyperthyroidism and the effects of treatment. *The Journal of Clinical Endocrinology and Metabolism*, *78*(4), 955–959. http://dx.doi.org/10.1210/jcem.78.4.8157727.

Gauthier, K., Chassande, O., Plateroti, M., Roux, J. P., Legrand, C., Pain, B., et al. (1999). Different functions for the thyroid hormone receptors TRalpha and TRbeta in the control of thyroid hormone production and post-natal development. *The EMBO Journal*, *18*(3), 623–631. http://dx.doi.org/10.1093/emboj/18.3.623.

Gauthier, K., & Flamant, F. (2014). Nongenomic, TRβ-dependent, thyroid hormone response gets genetic support. *Endocrinology*, *155*(9), 3206–3209. http://dx.doi.org/10.1210/en.2014-1597.

Gauthier, K., Plateroti, M., Harvey, C. B., Williams, G. R., Weiss, R. E., Refetoff, S., et al. (2001). Genetic analysis reveals different functions for the products of the thyroid hormone receptor alpha locus. *Molecular and Cellular Biology*, *21*(14), 4748–4760. http://dx.doi.org/10.1128/MCB.21.14.4748-4760.2001.

Gereben, B., Zavacki, A. M., Ribich, S., Kim, B. W., Huang, S. A., Simonides, W. S., et al. (2008). Cellular and molecular basis of deiodinase-regulated thyroid hormone signaling. *Endocrine Reviews*, *29*(7), 898–938. http://dx.doi.org/10.1210/er.2008-0019.

Gil-Ibáñez, P., Bernal, J., & Morte, B. (2014). Thyroid hormone regulation of gene expression in primary cerebrocortical cells: Role of thyroid hormone receptor subtypes and interactions with retinoic acid and glucocorticoids. *PLoS One*, *9*(3), e91692. http://dx.doi.org/10.1371/journal.pone.0091692.

Göthe, S., Wang, Z., Ng, L., Kindblom, J. M., Barros, A. C., Ohlsson, C., et al. (1999). Mice devoid of all known thyroid hormone receptors are viable but exhibit disorders of the pituitary-thyroid axis, growth, and bone maturation. *Genes & Development*, *13*(10), 1329–1341.

Gouveia, C. H. A., Christoffolete, M. A., Zaitune, C. R., Dora, J. M., Harney, J. W., Maia, A. L., et al. (2005). Type 2 iodothyronine selenodeiodinase is expressed throughout the mouse skeleton and in the MC3T3-E1 mouse osteoblastic cell line during differentiation. *Endocrinology*, *146*(1), 195–200. http://dx.doi.org/10.1210/en.2004-1043.

Gu, W. X., Du, G. G., Kopp, P., Rentoumis, A., Albanese, C., Kohn, L. D., et al. (1995). The thyrotropin (TSH) receptor transmembrane domain mutation (Pro556-Leu) in the hypothyroid hyt/hyt mouse results in plasma membrane targeting but defective TSH binding. *Endocrinology*, *136*(7), 3146–3153. http://dx.doi.org/10.1210/endo.136.7.7789342.

Guo, C. Y., Weetman, A. P., & Eastell, R. (1997). Longitudinal changes of bone mineral density and bone turnover in postmenopausal women on thyroxine. *Clinical Endocrinology*, *46*(3), 301–307.

Harvey, C. B., Bassett, J. H. D., Maruvada, P., Yen, P. M., & Williams, G. R. (2007). The rat thyroid hormone receptor (TR) deltabeta3 displays cell-, TR isoform-, and thyroid hormone response element-specific actions. *Endocrinology*, *148*(4), 1764–1773. http://dx.doi.org/10.1210/en.2006-1248.

Harvey, R. D., McHardy, K. C., Reid, I. W., Paterson, F., Bewsher, P. D., Duncan, A., et al. (1991). Measurement of bone collagen degradation in hyperthyroidism and during thyroxine replacement therapy using pyridinium cross-links as specific urinary markers. *The Journal of Clinical Endocrinology and Metabolism*, *72*(6), 1189–1194. http://dx.doi.org/10.1210/jcem-72-6-1189.

Harvey, C. B., & Williams, G. R. (2002). Mechanism of thyroid hormone action. *Thyroid: Official Journal of the American Thyroid Association, 12*(6), 441–446. http://dx.doi.org/10.1089/105072502760143791.

Hashimoto, K., Curty, F. H., Borges, P. P., Lee, C. E., Abel, E. D., Elmquist, J. K., et al. (2001). An unliganded thyroid hormone receptor causes severe neurological dysfunction. *Proceedings of the National Academy of Sciences of the United States of America, 98*(7), 3998–4003. http://dx.doi.org/10.1073/pnas.051454698.

Hernandez, A., Martinez, M. E., Fiering, S., Galton, V. A., & St Germain, D. (2006). Type 3 deiodinase is critical for the maturation and function of the thyroid axis. *The Journal of Clinical Investigation, 116*(2), 476–484. http://dx.doi.org/10.1172/JCI26240.

Hernandez, A., Martinez, M. E., Liao, X.-H., Van Sande, J., Refetoff, S., Galton, V. A., et al. (2007). Type 3 deiodinase deficiency results in functional abnormalities at multiple levels of the thyroid axis. *Endocrinology, 148*(12), 5680–5687. http://dx.doi.org/10.1210/en.2007-0652.

Heuer, H., & Visser, T. J. (2009). Minireview: Pathophysiological importance of thyroid hormone transporters. *Endocrinology, 150*(3), 1078–1083. http://dx.doi.org/10.1210/en.2008-1518.

Hüffmeier, U., Tietze, H.-U., & Rauch, A. (2007). Severe skeletal dysplasia caused by undiagnosed hypothyroidism. *European Journal of Medical Genetics, 50*(3), 209–215. http://dx.doi.org/10.1016/j.ejmg.2007.02.002.

Ishikawa, Y., Genge, B. R., Wuthier, R. E., & Wu, L. N. (1998). Thyroid hormone inhibits growth and stimulates terminal differentiation of epiphyseal growth plate chondrocytes. *Journal of Bone and Mineral Research: The Official Journal of the American Society for Bone and Mineral Research, 13*(9), 1398–1411. http://dx.doi.org/10.1359/jbmr.1998.13.9.1398.

Jones, I., Ng, L., Liu, H., & Forrest, D. (2007). An intron control region differentially regulates expression of thyroid hormone receptor beta2 in the cochlea, pituitary, and cone photoreceptors. *Molecular Endocrinology (Baltimore, Md.), 21*(5), 1108–1119. http://dx.doi.org/10.1210/me.2007-0037.

Kalyanaraman, H., Schwappacher, R., Joshua, J., Zhuang, S., Scott, B. T., Klos, M., et al. (2014). Nongenomic thyroid hormone signaling occurs through a plasma membrane-localized receptor. *Science Signaling*, 7(326), ra48. http://dx.doi.org/10.1126/scisignal.2004911.

Kaneshige, M., Kaneshige, K., Zhu, X., Dace, A., Garrett, L., Carter, T. A., et al. (2000). Mice with a targeted mutation in the thyroid hormone beta receptor gene exhibit impaired growth and resistance to thyroid hormone. *Proceedings of the National Academy of Sciences of the United States of America, 97*(24), 13209–13214. http://dx.doi.org/10.1073/pnas.230285997.

Kaneshige, M., Suzuki, H., Kaneshige, K., Cheng, J., Wimbrow, H., Barlow, C., et al. (2001). A targeted dominant negative mutation of the thyroid hormone alpha 1 receptor causes increased mortality, infertility, and dwarfism in mice. *Proceedings of the National Academy of Sciences of the United States of America, 98*(26), 15095–15100. http://dx.doi.org/10.1073/pnas.261565798.

Karsenty, G., & Wagner, E. F. (2002). Reaching a genetic and molecular understanding of skeletal development. *Developmental Cell, 2*(4), 389–406.

Kim, S.-W., Ho, S.-C., Hong, S.-J., Kim, K. M., So, E. C., Christoffolete, M., et al. (2005). A novel mechanism of thyroid hormone-dependent negative regulation by thyroid hormone receptor, nuclear receptor corepressor (NCoR), and GAGA-binding factor on the rat cD44 promoter. *The Journal of Biological Chemistry, 280*(15), 14545–14555. http://dx.doi.org/10.1074/jbc.M411517200.

Kim, H.-Y., & Mohan, S. (2013). Role and mechanisms of actions of thyroid hormone on the skeletal development. *Bone Research*, *1*(2), 146–161. http://dx.doi.org/10.4248/BR201302004.

Kindblom, J. M., Gevers, E. F., Skrtic, S. M., Lindberg, M. K., Göthe, S., Törnell, J., et al. (2005). Increased adipogenesis in bone marrow but decreased bone mineral density in mice devoid of thyroid hormone receptors. *Bone*, *36*(4), 607–616. http://dx.doi.org/10.1016/j.bone.2005.01.017.

Kindblom, J. M., Göthe, S., Forrest, D., Törnell, J., Törnell, J., Vennström, B., et al. (2001). GH substitution reverses the growth phenotype but not the defective ossification in thyroid hormone receptor alpha 1-/-beta-/- mice. *The Journal of Endocrinology*, *171*(1), 15–22.

Ko, Y.-J., Kim, J. Y., Lee, J., Song, H.-J., Kim, J.-Y., Choi, N.-K., et al. (2014). Levothyroxine dose and fracture risk according to the osteoporosis status in elderly women. *Journal of Preventive Medicine and Public Health = Yebang Uihakhoe Chi*, *47*(1), 36–46. http://dx.doi.org/10.3961/jpmph.2014.47.1.36.

Koenig, R. J., Lazar, M. A., Hodin, R. A., Brent, G. A., Larsen, P. R., Chin, W. W., et al. (1989). Inhibition of thyroid hormone action by a non-hormone binding c-erbA protein generated by alternative mRNA splicing. *Nature*, *337*(6208), 659–661. http://dx.doi.org/10.1038/337659a0.

Kopp, P. (2001). The TSH receptor and its role in thyroid disease. *Cellular and Molecular Life Sciences: CMLS*, *58*(9), 1301–1322.

Kronenberg, H. M. (2003). Developmental regulation of the growth plate. *Nature*, *423*(6937), 332–336. http://dx.doi.org/10.1038/nature01657.

LeBron, B. A., Pekary, A. E., Mirell, C., Hahn, T. J., & Hershman, J. M. (1989). Thyroid hormone 5′-deiodinase activity, nuclear binding, and effects on mitogenesis in UMR-106 osteoblastic osteosarcoma cells. *Journal of Bone and Mineral Research: The Official Journal of the American Society for Bone and Mineral Research*, *4*(2), 173–178. http://dx.doi.org/10.1002/jbmr.5650040207.

Liao, X.-H., Di Cosmo, C., Dumitrescu, A. M., Hernandez, A., Van Sande, J., St Germain, D. L., et al. (2011). Distinct roles of deiodinases on the phenotype of Mct8 defect: A comparison of eight different mouse genotypes. *Endocrinology*, *152*(3), 1180–1191. http://dx.doi.org/10.1210/en.2010-0900.

Lin, K. H., Fukuda, T., & Cheng, S. Y. (1990). Hormone and DNA binding activity of a purified human thyroid hormone nuclear receptor expressed in Escherichia coli. *The Journal of Biological Chemistry*, *265*(9), 5161–5165.

Lösel, R., & Wehling, M. (2003). Nongenomic actions of steroid hormones. *Nature Reviews Molecular Cell Biology*, *4*(1), 46–56. http://dx.doi.org/10.1038/nrm1009.

Lui, J. C., Nilsson, O., & Baron, J. (2014). Recent research on the growth plate: Recent insights into the regulation of the growth plate. *Journal of Molecular Endocrinology*, *53*(1), T1–T9. http://dx.doi.org/10.1530/JME-14-0022.

Macchia, P. E., Takeuchi, Y., Kawai, T., Cua, K., Gauthier, K., Chassande, O., et al. (2001). Increased sensitivity to thyroid hormone in mice with complete deficiency of thyroid hormone receptor alpha. *Proceedings of the National Academy of Sciences of the United States of America*, *98*(1), 349–354. http://dx.doi.org/10.1073/pnas.011306998.

Mackie, E. J., Tatarczuch, L., & Mirams, M. (2011). The skeleton: A multi-functional complex organ: The growth plate chondrocyte and endochondral ossification. *The Journal of Endocrinology*, *211*(2), 109–121. http://dx.doi.org/10.1530/JOE-11-0048.

Maia, A. L., Kim, B. W., Huang, S. A., Harney, J. W., & Larsen, P. R. (2005). Type 2 iodothyronine deiodinase is the major source of plasma T3 in euthyroid humans. *The Journal of Clinical Investigation*, *115*(9), 2524–2533. http://dx.doi.org/10.1172/JCI25083.

Mansouri, A., Chowdhury, K., & Gruss, P. (1998). Follicular cells of the thyroid gland require Pax8 gene function. *Nature Genetics*, *19*(1), 87–90. http://dx.doi.org/10.1038/ng0598-87.

Marians, R. C., Ng, L., Blair, H. C., Unger, P., Graves, P. N., & Davies, T. F. (2002). Defining thyrotropin-dependent and -independent steps of thyroid hormone synthesis by using thyrotropin receptor-null mice. *Proceedings of the National Academy of Sciences of the United States of America*, *99*(24), 15776–15781. http://dx.doi.org/10.1073/pnas.242322099.

Martin, N. P., Marron Fernandez de Velasco, E., Mizuno, F., Scappini, E. L., Gloss, B., Erxleben, C., et al. (2014). A rapid cytoplasmic mechanism for PI3 kinase regulation by the nuclear thyroid hormone receptor, TRβ, and genetic evidence for its role in the maturation of mouse hippocampal synapses in vivo. *Endocrinology*, *155*(9), 3713–3724. http://dx.doi.org/10.1210/en.2013-2058.

Mayerl, S., Müller, J., Bauer, R., Richert, S., Kassmann, C. M., Darras, V. M., et al. (2014). Transporters MCT8 and OATP1C1 maintain murine brain thyroid hormone homeostasis. *The Journal of Clinical Investigation*, *124*(5), 1987–1999. http://dx.doi.org/10.1172/JCI70324.

Mengreli, C., Kanaka-Gantenbein, C., Girginoudis, P., Magiakou, M.-A., Christakopoulou, I., Giannoulia-Karantana, A., et al. (2010). Screening for congenital hypothyroidism: The significance of threshold limit in false-negative results. *The Journal of Clinical Endocrinology and Metabolism*, *95*(9), 4283–4290. http://dx.doi.org/10.1210/jc.2010-0057.

Mitchell, M. L., Hsu, H.-W., Sahai, I., & Massachusetts Pediatric Endocrine Work Group (2011). The increased incidence of congenital hypothyroidism: Fact or fancy? *Clinical Endocrinology*, *75*(6), 806–810. http://dx.doi.org/10.1111/j.1365-2265.2011.04128.x.

Moeller, L. C., & Broecker-Preuss, M. (2011). Transcriptional regulation by nonclassical action of thyroid hormone. *Thyroid Research*, *4*(Suppl. 1), S6. http://dx.doi.org/10.1186/1756-6614-4-S1-S6.

Moeller, L. C., Dumitrescu, A. M., & Refetoff, S. (2005). Cytosolic action of thyroid hormone leads to induction of hypoxia-inducible factor-1alpha and glycolytic genes. *Molecular Endocrinology (Baltimore, Md.)*, *19*(12), 2955–2963. http://dx.doi.org/10.1210/me.2004-0542.

Morte, B., Ceballos, A., Diez, D., Grijota-Martínez, C., Dumitrescu, A. M., Di Cosmo, C., et al. (2010). Thyroid hormone-regulated mouse cerebral cortex genes are differentially dependent on the source of the hormone: A study in monocarboxylate transporter-8- and deiodinase-2-deficient mice. *Endocrinology*, *151*(5), 2381–2387. http://dx.doi.org/10.1210/en.2009-0944.

Mosekilde, L., & Melsen, F. (1978). Morphometric and dynamic studies of bone changes in hypothyroidism. *Acta Pathologica Et Microbiologica Scandinavica. Section A, Pathology*, *86*(1), 56–62.

Ng, L., Hurley, J. B., Dierks, B., Srinivas, M., Saltó, C., Vennström, B., et al. (2001). A thyroid hormone receptor that is required for the development of green cone photoreceptors. *Nature Genetics*, *27*(1), 94–98. http://dx.doi.org/10.1038/83829.

Ng, L., Rüsch, A., Amma, L. L., Nordström, K., Erway, L. C., Vennström, B., et al. (2001). Suppression of the deafness and thyroid dysfunction in Thrb-null mice by an independent mutation in the Thra thyroid hormone receptor alpha gene. *Human Molecular Genetics*, *10*(23), 2701–2708.

Nikrodhanond, A. A., Ortiga-Carvalho, T. M., Shibusawa, N., Hashimoto, K., Liao, X. H., Refetoff, S., et al. (2006). Dominant role of thyrotropin-releasing hormone in the hypothalamic-pituitary-thyroid axis. *The Journal of Biological Chemistry*, *281*(8), 5000–5007. http://dx.doi.org/10.1074/jbc.M511530200.

O'Shea, P. J., Bassett, J. H. D., Sriskantharajah, S., Ying, H., Cheng, S., & Williams, G. R. (2005). Contrasting skeletal phenotypes in mice with an identical mutation targeted to thyroid hormone receptor alpha1 or beta. *Molecular Endocrinology (Baltimore, Md), 19*(12), 3045–3059. http://dx.doi.org/10.1210/me.2005-0224.

O'Shea, P. J., Harvey, C. B., Suzuki, H., Kaneshige, M., Kaneshige, K., Cheng, S.-Y., et al. (2003). A thyrotoxic skeletal phenotype of advanced bone formation in mice with resistance to thyroid hormone. *Molecular Endocrinology (Baltimore, Md.), 17*(7), 1410–1424. http://dx.doi.org/10.1210/me.2002-0296.

Ono, N., Ono, W., Nagasawa, T., & Kronenberg, H. M. (2014). A subset of chondrogenic cells provides early mesenchymal progenitors in growing bones. *Nature Cell Biology, 16*(12), 1157–1167. http://dx.doi.org/10.1038/ncb3067.

Papadimitriou, A., Papadimitriou, D. T., Papadopoulou, A., Nicolaidou, P., & Fretzayas, A. (2007). Low TSH levels are not associated with osteoporosis in childhood. *European Journal of Endocrinology, 157*(2), 221–223. http://dx.doi.org/10.1530/EJE-07-0247.

Paul, S., Schindler, S., Giovannone, D., de Millo Terrazzani, A., Mariani, F. V., & Crump, J. G. (2016). Ihha induces hybrid cartilage-bone cells during zebrafish jawbone regeneration. *Development (Cambridge, England), 143*(12), 2066–2076. http://dx.doi.org/10.1242/dev.131292.

Persani, L., Calebiro, D., Cordella, D., Weber, G., Gelmini, G., Libri, D., et al. (2010). Genetics and phenomics of hypothyroidism due to TSH resistance. *Molecular and Cellular Endocrinology, 322*(1–2), 72–82. http://dx.doi.org/10.1016/j.mce.2010.01.008.

Plateroti, M., Gauthier, K., Domon-Dell, C., Freund, J. N., Samarut, J., & Chassande, O. (2001). Functional interference between thyroid hormone receptor alpha (TRalpha) and natural truncated TRDeltaalpha isoforms in the control of intestine development. *Molecular and Cellular Biology, 21*(14), 4761–4772. http://dx.doi.org/10.1128/MCB.21.14.4761-4772.2001.

Refetoff, S., DeWind, L. T., & DeGroot, L. J. (1967). Familial syndrome combining deaf-mutism, stuppled epiphyses, goiter and abnormally high PBI: Possible target organ refractoriness to thyroid hormone. *The Journal of Clinical Endocrinology and Metabolism, 27*(2), 279–294. http://dx.doi.org/10.1210/jcem-27-2-279.

Refetoff, S., & Dumitrescu, A. M. (2007). Syndromes of reduced sensitivity to thyroid hormone: Genetic defects in hormone receptors, cell transporters and deiodination. *Best Practice & Research. Clinical Endocrinology & Metabolism, 21*(2), 277–305. http://dx.doi.org/10.1016/j.beem.2007.03.005.

Rivkees, S. A., Bode, H. H., & Crawford, J. D. (1988). Long-term growth in juvenile acquired hypothyroidism: The failure to achieve normal adult stature. *The New England Journal of Medicine, 318*(10), 599–602. http://dx.doi.org/10.1056/NEJM198803103181003.

Robson, H., Siebler, T., Stevens, D. A., Shalet, S. M., & Williams, G. R. (2000). Thyroid hormone acts directly on growth plate chondrocytes to promote hypertrophic differentiation and inhibit clonal expansion and cell proliferation. *Endocrinology, 141*(10), 3887–3897. http://dx.doi.org/10.1210/endo.141.10.7733.

Saltó, C., Kindblom, J. M., Johansson, C., Wang, Z., Gullberg, H., Nordström, K., et al. (2001). Ablation of TRalpha2 and a concomitant overexpression of alpha1 yields a mixed hypo- and hyperthyroid phenotype in mice. *Molecular Endocrinology (Baltimore, Md.), 15*(12), 2115–2128. http://dx.doi.org/10.1210/mend.15.12.0750.

Schneider, M. J., Fiering, S. N., Pallud, S. E., Parlow, A. F., St Germain, D. L., & Galton, V. A. (2001). Targeted disruption of the type 2 selenodeiodinase gene (DIO2) results in a phenotype of pituitary resistance to T4. *Molecular Endocrinology (Baltimore, Md.), 15*(12), 2137–2148. http://dx.doi.org/10.1210/mend.15.12.0740.

Schneider, M. J., Fiering, S. N., Thai, B., Wu, S., St Germain, E., Parlow, A. F., et al. (2006). Targeted disruption of the type 1 selenodeiodinase gene (Dio1) results in marked changes in thyroid hormone economy in mice. *Endocrinology, 147*(1), 580–589. http://dx.doi.org/10.1210/en.2005-0739.

Schoenmakers, C. H., Pigmans, I. G., Poland, A., & Visser, T. J. (1993). Impairment of the selenoenzyme type I iodothyronine deiodinase in C3H/He mice. *Endocrinology, 132*(1), 357–361. http://dx.doi.org/10.1210/endo.132.1.8419134.

Segni, M., & Gorman, C. A. (2001). The aftermath of childhood hyperthyroidism. *Journal of Pediatric Endocrinology & Metabolism: JPEM, 14*(Suppl. 5), 1277–1282. discussion 1297–1298.

Segni, M., Leonardi, E., Mazzoncini, B., Pucarelli, I., & Pasquino, A. M. (1999). Special features of Graves' disease in early childhood. *Thyroid: Official Journal of the American Thyroid Association, 9*(9), 871–877. http://dx.doi.org/10.1089/thy.1999.9.871.

Stein, S. A., Oates, E. L., Hall, C. R., Grumbles, R. M., Fernandez, L. M., Taylor, N. A., et al. (1994). Identification of a point mutation in the thyrotropin receptor of the hyt/hyt hypothyroid mouse. *Molecular Endocrinology (Baltimore, Md.), 8*(2), 129–138. http://dx.doi.org/10.1210/mend.8.2.8170469.

Stevens, D. A., Hasserjian, R. P., Robson, H., Siebler, T., Shalet, S. M., & Williams, G. R. (2000). Thyroid hormones regulate hypertrophic chondrocyte differentiation and expression of parathyroid hormone-related peptide and its receptor during endochondral bone formation. *Journal of Bone and Mineral Research: The Official Journal of the American Society for Bone and Mineral Research, 15*(12), 2431–2442. http://dx.doi.org/10.1359/jbmr.2000.15.12.2431.

St Germain, D. L., Galton, V. A., & Hernandez, A. (2009). Minireview: Defining the roles of the iodothyronine deiodinases: Current concepts and challenges. *Endocrinology, 150*(3), 1097–1107. http://dx.doi.org/10.1210/en.2008-1588.

Storey, N. M., Gentile, S., Ullah, H., Russo, A., Muessel, M., Erxleben, C., et al. (2006). Rapid signaling at the plasma membrane by a nuclear receptor for thyroid hormone. *Proceedings of the National Academy of Sciences of the United States of America, 103*(13), 5197–5201. http://dx.doi.org/10.1073/pnas.0600089103.

Sun, L., Zhu, L.-L., Lu, P., Yuen, T., Li, J., Ma, R., et al. (2013). Genetic confirmation for a central role for TNFα in the direct action of thyroid stimulating hormone on the skeleton. *Proceedings of the National Academy of Sciences of the United States of America, 110*(24), 9891–9896. http://dx.doi.org/10.1073/pnas.1308336110.

Tavella, S., Biticchi, R., Schito, A., Minina, E., Di Martino, D., Pagano, A., et al. (2004). Targeted expression of SHH affects chondrocyte differentiation, growth plate organization, and Sox9 expression. *Journal of Bone and Mineral Research: The Official Journal of the American Society for Bone and Mineral Research, 19*(10), 1678–1688. http://dx.doi.org/10.1359/JBMR.040706.

Toivonen, J., Tähtelä, R., Laitinen, K., Risteli, J., & Välimäki, M. J. (1998). Markers of bone turnover in patients with differentiated thyroid cancer with and following withdrawal of thyroxine suppressive therapy. *European Journal of Endocrinology, 138*(6), 667–673.

Udayakumar, N., Chandrasekaran, M., Rasheed, M. H., Suresh, R. V., & Sivaprakash, S. (2006). Evaluation of bone mineral density in thyrotoxicosis. *Singapore Medical Journal, 47*(11), 947–950.

van Mullem, A., van Heerebeek, R., Chrysis, D., Visser, E., Medici, M., Andrikoula, M., et al. (2012). Clinical phenotype and mutant TRα1. *The New England Journal of Medicine, 366*(15), 1451–1453. http://dx.doi.org/10.1056/NEJMc1113940.

Vella, K. R., Ramadoss, P., Costa E Sousa, R. H., Astapova, I., Ye, F. D., Holtz, K. A., et al. (2014). Thyroid hormone signaling in vivo requires a balance between coactivators and corepressors. *Molecular and Cellular Biology, 34*(9), 1564–1575. http://dx.doi.org/10.1128/MCB.00129-14.

Venero, C., Guadaño-Ferraz, A., Herrero, A. I., Nordström, K., Manzano, J., de Escobar, G. M., et al. (2005). Anxiety, memory impairment, and locomotor dysfunction caused by a mutant thyroid hormone receptor alpha1 can be ameliorated by T3 treatment. *Genes & Development, 19*(18), 2152–2163. http://dx.doi.org/10.1101/gad.346105.

Vestergaard, P., & Mosekilde, L. (2002). Fractures in patients with hyperthyroidism and hypothyroidism: A nationwide follow-up study in 16,249 patients. *Thyroid: Official Journal of the American Thyroid Association, 12*(5), 411–419. http://dx.doi.org/10.1089/105072502760043503.

Vestergaard, P., & Mosekilde, L. (2003). Hyperthyroidism, bone mineral, and fracture risk—A meta-analysis. *Thyroid: Official Journal of the American Thyroid Association, 13*(6), 585–593. http://dx.doi.org/10.1089/105072503322238854.

Vestergaard, P., Rejnmark, L., & Mosekilde, L. (2005). Influence of hyper- and hypothyroidism, and the effects of treatment with antithyroid drugs and levothyroxine on fracture risk. *Calcified Tissue International, 77*(3), 139–144. http://dx.doi.org/10.1007/s00223-005-0068-x.

Visser, T. J. (2013). Thyroid hormone transporters and resistance. *Endocrine Development, 24*, 1–10. http://dx.doi.org/10.1159/000343695.

Visser, W. E., Friesema, E. C. H., & Visser, T. J. (2011). Minireview: Thyroid hormone transporters: The knowns and the unknowns. *Molecular Endocrinology (Baltimore, Md.), 25*(1), 1–14. http://dx.doi.org/10.1210/me.2010-0095.

Wallis, K., Sjögren, M., van Hogerlinden, M., Silberberg, G., Fisahn, A., Nordström, K., et al. (2008). Locomotor deficiencies and aberrant development of subtype-specific GABAergic interneurons caused by an unliganded thyroid hormone receptor alpha1. *The Journal of Neuroscience: The Official Journal of the Society for Neuroscience, 28*(8), 1904–1915. http://dx.doi.org/10.1523/JNEUROSCI.5163-07.2008.

Wassner, A. J., & Brown, R. S. (2015). Congenital hypothyroidism: Recent advances. *Current Opinion in Endocrinology, Diabetes, and Obesity, 22*(5), 407–412. http://dx.doi.org/10.1097/MED.0000000000000181.

Waung, J. A., Bassett, J. H. D., & Williams, G. R. (2012). Thyroid hormone metabolism in skeletal development and adult bone maintenance. *Trends in Endocrinology and Metabolism: TEM, 23*(4), 155–162. http://dx.doi.org/10.1016/j.tem.2011.11.002.

Weiss, R. E. (2008). "They have ears but do not hear" (Psalms 135:17): Non-thyroid hormone receptor beta (non-TRbeta) resistance to thyroid hormone. *Thyroid: Official Journal of the American Thyroid Association, 18*(1), 3–5. http://dx.doi.org/10.1089/thy.2007.0373.

Wikström, L., Johansson, C., Saltó, C., Barlow, C., Campos Barros, A., Baas, F., et al. (1998). Abnormal heart rate and body temperature in mice lacking thyroid hormone receptor alpha 1. *The EMBO Journal, 17*(2), 455–461. http://dx.doi.org/10.1093/emboj/17.2.455.

Williams, G. R. (2000). Cloning and characterization of two novel thyroid hormone receptor beta isoforms. *Molecular and Cellular Biology, 20*(22), 8329–8342.

Williams, A. J., Robson, H., Kester, M. H. A., van Leeuwen, J. P. T. M., Shalet, S. M., Visser, T. J., et al. (2008). Iodothyronine deiodinase enzyme activities in bone. *Bone, 43*(1), 126–134. http://dx.doi.org/10.1016/j.bone.2008.03.019.

Xing, W., Aghajanian, P., Goodluck, H., Kesavan, C., Cheng, S., Pourteymoor, S., et al. (2016). Thyroid hormone receptor-β1 signaling is critically involved in regulating secondary ossification via promoting transcription of the Ihh gene in the epiphysis. *American Journal of Physiology. Endocrinology and Metabolism, 310*(10), E846–E854. http://dx.doi.org/10.1152/ajpendo.00541.2015.

Xing, W., Cheng, S., Wergedal, J., & Mohan, S. (2014). Epiphyseal chondrocyte secondary ossification centers require thyroid hormone activation of Indian hedgehog and osterix signaling. *Journal of Bone and Mineral Research: The Official Journal of the American Society for Bone and Mineral Research, 29*(10), 2262–2275. http://dx.doi.org/10.1002/jbmr.2256.

Xing, W., Govoni, K. E., Donahue, L. R., Kesavan, C., Wergedal, J., Long, C., et al. (2012). Genetic evidence that thyroid hormone is indispensable for prepubertal insulin-like growth factor-I expression and bone acquisition in mice. *Journal of Bone and Mineral Research: The Official Journal of the American Society for Bone and Mineral Research*, *27*(5), 1067–1079. http://dx.doi.org/10.1002/jbmr.1551.

Yang, L., Tsang, K. Y., Tang, H. C., Chan, D., & Cheah, K. S. E. (2014). Hypertrophic chondrocytes can become osteoblasts and osteocytes in endochondral bone formation. *Proceedings of the National Academy of Sciences of the United States of America*, *111*(33), 12097–12102. http://dx.doi.org/10.1073/pnas.1302703111.

Zhou, X., von der Mark, K., Henry, S., Norton, W., Adams, H., & de Crombrugghe, B. (2014). Chondrocytes transdifferentiate into osteoblasts in endochondral bone during development, postnatal growth and fracture healing in mice. *PLoS Genetics*, *10*(12), e1004820. http://dx.doi.org/10.1371/journal.pgen.1004820.

CHAPTER SIXTEEN

Thyroid Hormone and Skeletal Development

Cecilia H.A. Gouveia*,†,1, Manuela Miranda-Rodrigues*, Gisele M. Martins*,†,‡, Bianca Neofiti-Papi*,†

*Institute of Biomedical Sciences, University of São Paulo, São Paulo, SP, Brazil
†Experimental Pathophysiology Program, School of Medicine, University of São Paulo, São Paulo, SP, Brazil
‡Federal University of Espírito Santo, Vitória, ES, Brazil
[1]Corresponding author: e-mail address: cgouveia@usp.br

Contents

1. Introduction 384
2. Overview of Skeletal Development 386
   2.1 Osteogenesis 387
   2.2 The Growth Plate and the Longitudinal Bone Growth 388
3. Overview of TH Physiology 389
4. Expression of TH Physiology Components in the Skeleton 393
   4.1 Expression of TR 393
   4.2 Expression of TSHR 395
   4.3 Expression and Activity of Iodothyronine Deiodinases 395
   4.4 Expression of TH Transporters 397
5. The Effects of TH 398
   5.1 Effects in Osteoblast Lineage Cells 398
   5.2 Effects in Osteoclasts 399
   5.3 Effects in Chondrocytes 402
   5.4 Effect on Ossification 405
   5.5 Effects in the Growth Plate and Longitudinal Bone Growth 408
6. Mechanisms of Action of TH to Promote Skeletal Development 410
   6.1 The Role of GH/IGF-1 Axis 410
   6.2 The Role of the Ihh/PTHrP Axis 415
   6.3 The Role of the FGF/FGFR Signaling Pathway 416
   6.4 The Role of the Wnt/β-Catenin Signaling Pathway 420
   6.5 The Role of the PPARδ Signaling Pathway 426
   6.6 The Role of the BMP Signaling Pathway 427
   6.7 The Role of the ECM 428
7. The Skeletal Phenotype and Development in TH-Physiology-Linked-Mutant Mice 429
   7.1 TR Mutant Mice 429
   7.2 TSHR Mutant Mice 436
   7.3 Deiodinase Mutant Mice 440

Vitamins and Hormones, Volume 106
ISSN 0083-6729
http://dx.doi.org/10.1016/bs.vh.2017.06.002

7.4 TH Transporters Mutant Mice 441
8. The Skeletal Phenotype and Development in Patients With RTH 444
8.1 RTHα Patients 445
8.2 RTHβ Patients 448
9. Concluding Remarks 449
10. Final Conclusions and Future Directions 452
Acknowledgments 452
References 453

Abstract

Thyroid hormone (TH) is essential for skeletal development from the late fetal life to the onset of puberty. During this large window of actions, TH has key roles in endochondral and intramembranous ossifications and in the longitudinal bone growth. There is evidence that TH acts directly in skeletal cells but also indirectly, specially via the growth hormone/insulin-like growth factor-1 axis, to control the linear skeletal growth and maturation. The presence of receptors, plasma membrane transporters, and activating and inactivating enzymes of TH in skeletal cells suggests that direct actions of TH in these cells are crucial for skeletal development, which has been confirmed by several *in vitro* and *in vivo* studies, including mouse genetic studies, and clinical studies in patients with resistance to thyroid hormone due to dominant-negative mutations in TH receptors. This review examines progress made on understanding the mechanisms by which TH regulates the skeletal development.

1. INTRODUCTION

Thyroid hormone (TH) exerts several effects on development, growth, and metabolism of a wide variety of tissues, organs, and systems, including the skeleton. TH actions on skeletal development are among the most expressive effects of this hormone (Asling, Simpson, Li, & Evans, 1954; Ray et al., 1954; Shao, Wang, & Ballock, 2006). Deficiency of TH in children is often associated with severe ossification delay (delayed bone age), short stature, and slipped capital femoral epiphysis (Wells, King, Roe, & Kaufman, 1993). In animal models, TH deficiency results in delayed intramembranous and endochondral ossification (EO) coupled with important alterations in the growth plate (Fig. 1), such as reduction in thickness, disorganized columns of chondrocytes, and a significant impairment of the differentiation of proliferating chondrocytes (PCs) into hypertrophic chondrocytes (HCs; Freitas et al., 2005; Lewinson, Harel, Shenzer, Silbermann, & Hochberg, 1989; Robson, Siebler, Stevens, Shalet, & Williams, 2000). On the other hand, thyrotoxicosis may result in accelerated

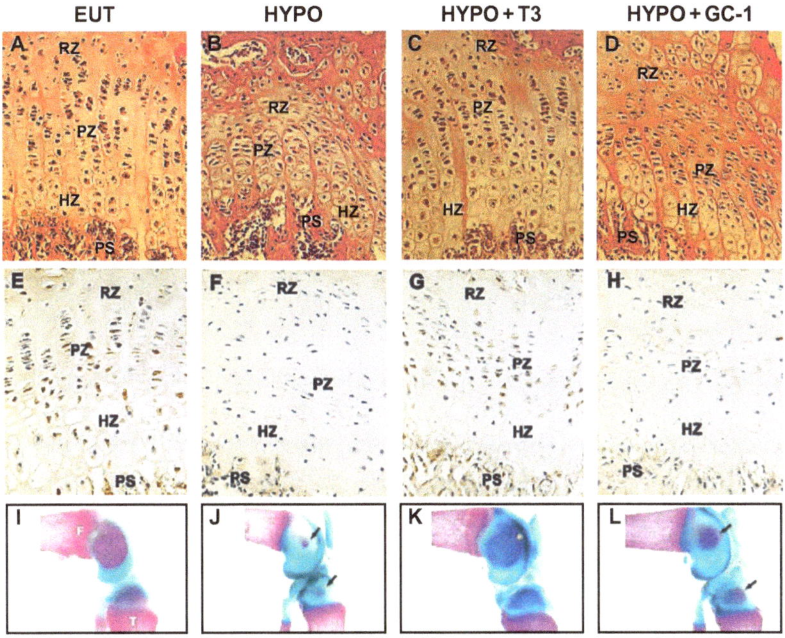

Fig. 1 Growth plates and endochondral ossification in euthyroid and hypothyroid rats. (A–H) Sections of the distal femur from euthyroid (Eut), hypothyroid (Hypo), T3-treated Hypo, and GC-1-treated Hypo rats, showing the growth plate zones: reserve zone (RZ); proliferative zone (PZ); hypertrophic zone (HZ); primary spongiosa (PS). (A–D) H&E staining shows a thinner and disorganized growth plate, with a severe decrease in number and size of HC in Hypo (B), which were fully restored by T3 (C), whereas GC-1 only induced HC differentiation (D). (E–H) Expression of IGF-I by immunohistochemistry in Eut (E) and T3-treated (G), but not in Hypo (F) and GC-1-treated (H) growth plates. Hypo was induced by treating 21-day-old rats with methimazole and sodium perchlorate (MM + SP) for 5 weeks. Treatment with T3 (1.5 μg/100 g BW) or equimolar doses of GC-1 started 1 week after initiation of Hypo induction and lasted 4 weeks. Magnification, ×200. (I–L) Distal and proximal epiphyses of femur (F) and tibia (T), respectively, stained with alcian blue (cartilage) and alizarin red (bone). *Arrows* indicate grossly reduced ossification centers (OC) in Hypo (J) and a partial reversion by GC-1 (L), whereas T3 fully restored the size of the OC. For EO analysis, Hypo was induced from P1 to P14, by treating their dams with MM + P. Neonatal rats were treated with T3 or GC-1 from P7 to P14 (Freitas et al., 2005).

skeletal maturation, premature closure of the growth plates, and a subsequent decrease in body weight gain and linear body growth (Ross, 1994).

TH also regulates bone turnover, stimulating both osteoblastic and osteoclastic activities (Allain, Thomas, McGregor, & Salisbury, 1995; Mosekilde & Melsen, 1978a, 1978b). Thyrotoxicosis in adults leads to increased bone turnover, favoring osteoclast-mediated bone resorption, negative balance of calcium, and bone loss (Mosekilde, Melsen, Bagger, Myhre-Jensen, & Schwartz Sorensen, 1977). In contrast, hypothyroidism results in decreased bone metabolism, with no change or a slight increase in bone mass (Eriksen, Mosekilde, & Melsen, 1986). Animal studies show that early-onset hypothyroidism significantly limits bone mass acquisition, due to an important decrease in bone formation (Bassett, O'Shea, et al., 2007; Freitas et al., 2005), indicating that TH exerts crucial anabolic actions in the bone tissue during skeletal development.

TH, along with growth hormone (GH), insulin-like growth factors (IGFs), and glucocorticoids, represents the main systemic factors that influence skeletal growth and maturation before puberty (Underwood & Van Wyk, 1992; Weiss & Refetoff, 1996). During adolescence, sex steroids and the GH/IGF-1 axis drive the pubertal growth spurt. The ending of this rapid growth in height is a consequence of the ossification of the growth plates, mainly due to the actions of the sex steroids (Murray & Clayton, 2013). During prenatal development, lack of TH barely affects skeletal development (Capelo et al., 2008), whereas TH excess may cause important detrimental alterations, such as craniosynostosis—the premature fusion of one or more calvarial sutures prior to the cessation of brain expansion (Howie et al., 2016; Rasmussen et al., 2007). TH becomes essential for skeletal development only postnatally (Capelo et al., 2008), when it is considered the main prerequisite for normal skeletal maturation (Underwood & Van Wyk, 1992).

TH regulates skeletal development via direct and indirect mechanisms. The presence of all the factors necessary for local actions of TH in the skeleton suggests that direct mechanisms are relevant, not only during development but also in the adulthood. Advances in the understanding of TH physiology and bone development and biology (Brent, 2012) have been the basis for revealing the mechanisms of action of TH in the skeleton. This review aims to summarize the *in vivo* and *in vitro* findings related to the actions and effects of TH on skeletal growth and maturation.

2. OVERVIEW OF SKELETAL DEVELOPMENT

Skeletal development refers to several skeletal-related processes that occur from the early days of pregnancy until the skeleton has reached full development in late puberty. These skeletal-related processes include (i) skeletogenesis, which requires recruitment, migration, and condensation of precursor cells to the future skeletal sites, followed by their differentiation to form the skeletal tissues; (ii) skeletal patterning, a process that determines bone number and shape; (iii) skeletal growth in length and width, which depends on proliferation, survival, and differentiation of chondrogenic and/or osteogenic cells; (iv) and maturation of the skeleton, determined mainly by ossification of a primordial skeleton through two processes, endochondral and intramembranous ossification (IO) (Hall & Miyake, 2000; Horton, 2003). Skeletal development is extremely complex, involves multiple local signaling pathways and diverse systemic factors, which actions are highly coordinated. There is evidence that TH is not required for the initial steps of skeletal development (Capelo et al., 2008), but has key roles on osteogenesis and skeletal growth, from the late fetal development until puberty. Therefore, this review will focus on the effects of TH on osteogenesis (ossification) and longitudinal bone growth.

2.1 Osteogenesis

Osteogenesis, also called ossification, occurs through two distinct processes: endochondral and IO. In both processes, a preexisting mesenchymal tissue is transformed into bone tissue. Groups of mesenchymal cells in these mesenchymal tissues condense and initiate a characteristic genetic program that leads to ossification. In IO, mesenchymal cells condense at the same time as angiogenesis occurs, forming a mesenchymal membrane. The cells of these mesenchymal condensations differentiate directly into osteoblasts (the bone-forming cells), which will synthesize the extracellular matrix (ECM) and control its mineralization. Consequently, the mesenchymal membrane is replaced by bone tissue (Rice & Rice, 2008).

In EO, there is an intermediate stage, where mesenchymal cells condense and differentiate into chondrocytes, which form a cartilaginous template that is later replaced by bone. Therefore, the formation of endochondral bone begins with the condensation of the mesenchyme cells, which differentiate into a population of centrally located PCs expressing collagen type II (Col II); and into a population of more peripherally located osteoprogenitor cells, which express collagen type I (Col I) (Kosher, Kulyk, & Gay, 1986; Stevens et al., 2003). At this stage, chondrocytes begin to synthesize a specialized ECM, containing mainly Col II and aggrecan. At the central region of the cartilaginous template, chondrocytes stop proliferating, undergo hypertrophy, and begin to synthesize collagen type X (Col X), in addition to a series of proteins and factors, composing a hypertrophic region (Schmid & Linsenmayer, 1985). These HCs then signal the adjacent perichondral cells to differentiate into osteoblasts, inducing the formation of the periosteum and a bone collar around the cartilaginous template. At the same time, HCs mediate the mineralization of the surrounding ECM, release the vascular endothelial growth factor (VEGF), which induces angiogenesis, and undergo apoptosis leaving gaps in the cartilaginous matrix. Blood capillaries invade these gaps, resulting in the vascularization of the avascular cartilaginous mold. With this vascularization, endothelial cells, osteoclast precursors, and osteoprogenitor cells invade the hypertrophic region, secrete Col I in the ECM, and mediate the formation of bone tissue within the cartilaginous mold, composing the primary spongiosa (consisting of osteocartilaginous spicules), which constitutes a primary center of ossification (POC). The HCs also secrete the receptor activator of nuclear factor kappa B ligand (RANKL), a factor that induces osteoclastogenesis. Therefore, HCs contribute to the recruitment of osteoclasts (the bone resorbing cells) to the cartilaginous template (Williams, 2013). These cells, in turn, digest the ECM synthesized by HCs. As the ossification process proceeds, this primary ossification expands toward the bone ends. In the cartilaginous mold of the long bones, the POC occurs in the diaphysis and expands toward both epiphyses (Ornitz & Marie, 2002). In the cartilaginous epiphyses, round chondrocytes continue to proliferate, and those near the POC flatten and form columns of PCs. These cells stop proliferating and differentiate into HCs, which are located even closer to the POC. Subsequently, secondary ossification centers are formed in the epiphyses. Rounded chondrocytes in the center of the cartilaginous epiphyses stop dividing, undergo hypertrophy, and stimulate angiogenesis, which results in a secondary ossification center. Cartilage is maintained at the ends of the bones, constituting the articular cartilage (AC), and between the primary and secondary ossification centers, composing the epiphyseal growth plate or growth cartilage, which contains the machinery for the longitudinal bone growth.

2.2 The Growth Plate and the Longitudinal Bone Growth

The mammalian growth plate is composed of four well-defined main zones that follow an elegant developmental program: the reserve, proliferative, prehypertrophic, and hypertrophic zones (Mackie, Ahmed, Tatarczuch, Chen, & Mirams, 2008). The reserve zone or the resting zone (RZ), the one closest to the epiphysis, consists of slow-growing undifferentiated chondroprogenitor cells, the reserve or resting chondrocytes (RC). These cells are rounded, relatively small, and are dispersed in an ECM rich in Col II and proteoglycans. These cells differentiate into disc-shaped (flattened) PCs, organized in columns parallel to the long axis of the long bones, which characterizes the proliferative zone (PZ). PCs secrete mainly Col II, which is the main component of the ECM of the PZ, and differentiate into prehypertrophic chondrocytes (PHCs), chondrocytes immediately before hypertrophy, composing the prehypertrophic zone (PHZ). It is worth mentioning that the PZ plays a crucial role in EO and bone growth, since it is the active cell replication site of the growth plate. When PC divides, the two daughter cells align to the longitudinal axis of the bone (Mackie et al., 2008; Mackie, Tatarczuch, & Mirams, 2011). This spatial organization directs bone growth and is thus responsible for the elongated shape of various endochondral bones. HCs, which compose the hypertrophic zone (HZ), also assume a columnar organization. These cells, as well as in the primary and secondary ossification centers, secrete Col X and VEGF, among other proteins and factors, and control the EO of the growth plate (Ballock & O'Keefe, 2003). Thus, the longitudinal bone growth occurs in the growth plate, by chondrocyte proliferation and hypertrophy, cartilage deposition in the epiphysis–diaphysis direction, and by EO of this cartilage, in the diaphysis–epiphysis direction (Ballock & O'Keefe, 2003; Mackie et al., 2008, 2011). In adolescence, EO overcomes the cartilage deposition in the growth plate, leading to its total ossification; therefore, epiphysis and diaphysis are fused together (the epiphyseal fusion), which ceases the longitudinal bone growth, around 18 years for women and 21 years for men (Shim, 2015).

3. OVERVIEW OF TH PHYSIOLOGY

The main products secreted by the thyroid gland are the iodothyronines, compounds that result from binding of two molecules of iodinated tyrosine. In humans, about 80% of thyroid production is 3,5,3′,5′-tetraiodothyronine (thyroxine or T4), <20% is 3,5,3′-triiodothyronine (T3), and <1% is represented by other biologically inactive iodothyronines, including 3,3′,5′-T3 (reverse T3 or rT3) and 3,3′-diiodothyronine (T2) (Abdalla & Bianco, 2014). T4 basically functions as a prohormone and is converted to T3, the active form of TH, by the action of cellular enzymes, known as iodothyronine deiodinases. This conversion occurs in the thyroid gland and, especially, in T3 target tissues (Engler & Burger, 1984). Three types of iodothyronine deiodinases were identified in mammalian tissues: types I, II and III (D1, D2, and D3, respectively). D1 and D2 convert T4 to T3, activating the iodothyronines. D1 also converts T4 to rT3, and rT3 and T3 to T2. Considering that both rT3 and T2 are inactive compounds, D1 is capable of both activating and inactivating the iodothyronines. D3 catalyzes the degradation of T4 to rT3, and T3 to T2 (Bianco, Salvatore, Gereben, Berry, & Larsen, 2002; Fekkes, Hennemann, & Visser, 1982). We and others have shown that iodothyronine deiodinases are expressed in the skeleton during pre- and postnatal skeletal development, suggesting a role of these enzymes during these processes (Capelo et al., 2008; Gouveia et al., 2005; Williams et al., 2008), which will be discussed in Section 4.3 and 7.3.

TH receptors (TRs) belong to the nuclear receptor superfamily, which includes receptors for thyroid and steroid hormones, retinoids, and vitamin D (VD), as well as different "orphan" receptors of unknown ligands (Aranda & Pascual, 2001). Unliganded TRs are located in the cellular nucleus, bound to TH response elements (TREs), normally present in the promoter of TH target genes, through a DNA-binding domain. Unliganded TRs interact with TREs as homodimers (TR/TR), or as heterodimers with retinoid X-receptors (TR/RXR and TR/RAR), and associate with corepressors, a complex of proteins that inhibit gene transcription. TR binds T3 through its ligand-binding domain (LBD), which promotes a TR molecular rearrangement resulting in dissociation of corepressors and binding of coactivators to the RAR/TR/TRE, RXR/TR/TRE, or TR/TR/TRE complex, which activates gene transcription (Apriletti et al., 1998; Brent, 2012). Therefore, unliganded TRs repress basal gene transcription, whereas T3–TR interactions revert this gene suppression and induce gene transcription.

The TR affinity for T3 is approximately 10-fold greater than T4, which makes T3 the most potent TH, and T4 a prohormone. It is noteworthy that T4, at high concentrations, can bind TR and induce gene transcription; however, T4 does not enter the nucleus at high enough concentrations to effectively bind TR. Therefore, T4–TR interactions have minimal physiological relevance (Brent, 2012). There are four classical TR isoforms encoded by two genes (Freedman, 1992), THRA and THRB, located on human chromosomes 17 and 3, respectively. THRB encodes TRβ1 and TRβ2, while THRA encodes TRα1 and TRα2. This latter isoform does not bind T3 and functions, at least *in vitro*, as an antagonist of TRα1 and TRβ1 (Katz & Lazar, 1993). TRα1, TRα2, and TRβ1 are expressed in the main skeletal cells, the osteoblasts, osteoclasts, and chondrocytes, respectively (Abu, Bord, Horner, Chatterjee, & Compston, 1997; Williams, Bland, & Sheppard, 1994); however, TRα1 is the predominant isoform in the skeleton (Gauthier et al., 2001; O'Shea et al., 2005, 2003; Stevens et al., 2000). The functional participation of the different TR isoforms will be discussed in Section 4.1 and 7.1.

The T3–TR interaction depends on the cellular availability of T3, which not only depends on the local TH activation–inactivation by iodothyronine deiodinases but also depends on the cellular influx–efflux of TH mediated by cell membrane transporters (Refetoff, Weiss, & Usala, 1993). Several plasma membrane transporters have been shown to mediate the cellular influx and or efflux of iodothyronines, including the sodium-independent organic anion cotransporting polypeptide 1 (OATP1), the sodium taurocholate cotransporting polypeptide (NTCP), the L-type amino acid transporter 1 (LAT1) and 2 (LAT2), and the monocarboxylate transporter 8 and 10 (MCT8 and MCT10) (Friesema, Jansen, Milici, & Visser, 2005). Among the cell membrane transporters, MCT8 was demonstrated to be a very active and specific TH transporter highly expressed in the brain, liver, heart, placenta, and other TH-responsive tissues (Friesema et al., 2005; Halestrap, 2012). The relevance of MCT8 as a TH transporter has been demonstrated by carriers of an X-linked syndrome called Allan–Herndon–Dudley syndrome, caused by a mutation on the MCT8 gene, which present with severe mental retardation, only mild growth failure and abnormal circulating TH levels (Dumitrescu, Liao, Best, Brockmann, & Refetoff, 2004; Heuer et al., 2005; Maranduba et al., 2006). The generation of MCT8 knockout (KO) mice by Dumitrescu, Liao, Weiss, Millen, and Refetoff (2006) demonstrated that the absence of active MCT8 caused morphological and functional alterations related to TH restriction in the central nervous system (CNS), but not in the liver, where a thyrotoxic phenotype was observed. The presence of TH transporters was identified in skeletal cells and will be discussed in Sections 4.4 and 7.4.

Serum levels of T4 and T3 are maintained within a narrow range by a negative feedback loop (Fig. 2) involving the hypothalamus–pituitary–thyroid (HPT) axis (Costa-e-Sousa & Hollenberg, 2012). Thyrotropin-releasing hormone (TRH), secreted by the paraventricular nucleus of the hypothalamus, stimulates the pituitary thyrotrophs to produce and release thyrotropin or thyroid-stimulating hormone (TSH). Afterward, TSH binds its receptor (TSHR) in the thyroid follicular cells, to stimulate all steps of the

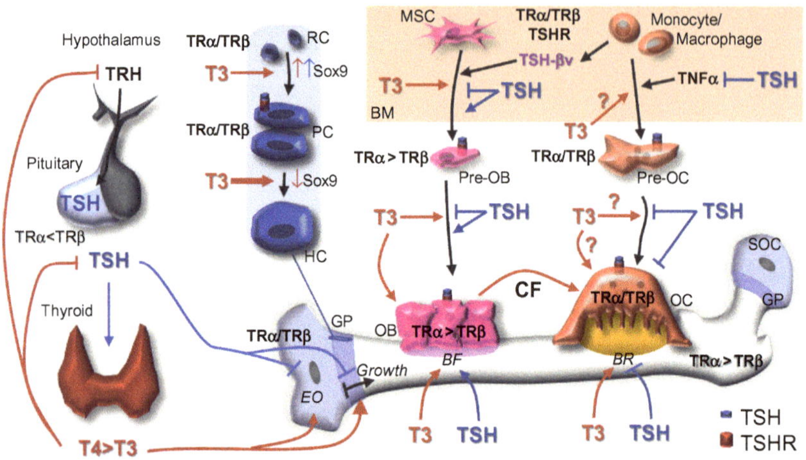

Fig. 2 Schematic representation of T3 and TSH actions on the hypothalamus–pituitary–thyroid (HPT) axis and on skeletal cells. TSH induces TH synthesis and secretion by thyroid gland, which inhibits TRH and TSH synthesis and secretion by hypothalamus and pituitary, respectively. TH induces endochondral ossification (EO) and linear bone growth (growth), whereas TSH impairs these processes. TH acts directly in reserve and proliferating chondrocytes (RC and PC) of the growth plate (GP), which express TRs, to induce differentiation. TRs are not expressed in hypertrophic chondrocytes (HC). TSH actions in the GP are unknown. TRs and TSHR are expressed in bone marrow (BM), pre- and mature osteoblasts (OB), and osteoclasts (OC). TRα predominates in bones and in osteoblastic cells. T3 induces osteoblast differentiation and activity, increasing bone formation (BF). TSH also induces BF, but its effects on differentiation are contradictory. T3 induces bone resorption (BR), but it is unclear if T3 directly induces osteoclastogenesis and osteoclastic activity. T3 induces these processes indirectly, acting in OBs, probably modulating the expression of OB–OC coupling factors (CF). TSH directly limits osteoclastogenesis and osteoclastic activity, in a mechanism that involves tumor necrosis factor α (TNFα). A novel functional form of TSH-β (TSH-βv) is produced by macrophages, which induce osteoblastogenesis. *Red* and *blue lines* indicate T3 and TSH actions, respectively. Secondary ossification center (SOC); mesenchymal stem cell (MSC).

biosynthesis and secretion of T4 and T3. TH, in turn, exerts negative feedback control over the hypothalamus and pituitary, where T3 interacts with TRs to inhibit the synthesis and secretion of TRH and TSH, respectively (Ortiga-Carvalho, Chiamolera, Pazos-Moura, & Wondisford, 2016). Consequently, physiological levels of the HPT axis hormones are achieved. Reduction of circulating T4 and T3 levels results in increased TRH and TSH production, whereas the opposite occurs when circulating T4 and T3 are increase. Therefore, there is an inverse relationship between serum levels of TSH and TH (Costa-e-Sousa & Hollenberg, 2012; Ortiga-Carvalho et al., 2016).

4. EXPRESSION OF TH PHYSIOLOGY COMPONENTS IN THE SKELETON

4.1 Expression of TR

The presence of TRα1, TRα2, and TRβ1 was identified in rodent and human osteoblasts (Abu et al., 1997; Abu, Horner, Teti, Chatterjee, & Compston, 2000; Milne et al., 1999; Rizzoli, Poser, & Burgi, 1986). All these isoforms were detected, by immunolocalization, in a few osteocytes in remodeling sites of human osteophytes and heterotopic bone (Abu et al., 1997). TRs were also identified in the rib and growth plate chondrocytes of rodents (Ballock, Mita, Zhou, Chen, & Mink, 1999; Rabier, Williams, Mallein-Gerin, Williams, & Chassande, 2006; Stevens et al., 2000). In mice, growth plate chondrocytes of both RZ and PZ have been shown to express TRα1, TRα2, and TRβ1, whereas TRs were not detected in HCs (Stevens et al., 2000). In sites of EO in human osteophytes, heterotopic bone or osteoclastoma tissue, protein and mRNA expression of TRα1, TRα2, and TRβ1 were identified in undifferentiated chondrocytes, PCs and HCs (Abu et al., 1997, 2000). The responsiveness of osteoblasts and chondrocytes to T3, when isolated in culture (which will be discussed in Sections 5.1 and 5.3), suggests direct actions of TH on bone formation and cartilage biology. The functionality of TRs in osteocytes has not been investigated yet. The presence of all known TR isoforms has been demonstrated in human osteoclasts (Abu et al., 1997, 2000; Allain, Yen, Flanagan, & McGregor, 1996), but the functionality and relevance of this expression are still unclear (discussed in Section 5.2). Interestingly, TRβ2 mRNA was detected in osteoblasts, osteoclasts, and chondrocytes of humans (Abu et al., 2000), but not in the growth plate of rats (Robson et al., 2000), and there are no data regarding a role of TRβ2 in mediating TH actions in the skeleton. An important finding was the identification of TRα and TRβ isoforms in the bone marrow (BM) of mice, especially in stromal cells (Gruber et al., 1999). The BM is an extremely important component of bone physiology since it contains hematopoietic and stromal cells (Fig. 2), which include precursor cells of osteoclasts and osteoblasts, respectively. The presence of TRs in these cells suggests direct actions of TH on osteoclastic and osteoblastic differentiation.

It is noteworthy that a 12-fold greater gene expression of TRα1 over TRβ1 was detected in the bone (pools of femurs and tibias) of mice (O'Shea et al., 2003). Accordingly, TRα1 was 6–14 times and 15–100 times more expressed than TRβ1 in rat (ROS17/2.8) and mouse (MC3T3-E1) osteoblastic cell lines, respectively, depending on the cell differentiation stage and on the presence or absence of T3 treatment (Beber et al., 2009). These findings call attention to a predominant role of TRα1 in mediating the effects of T3 in the bone tissue, especially in osteoblasts. Interestingly, the expression of TRβ1 was shown to increase as ROS17/2.8 and MC3T3-E1 cells differentiate an effect barely observed for TRα (Beber et al., 2009). In addition, T3 treatment was shown to enhance TRβ1 expression, especially in rat osteoblast-like cells (Beber et al., 2009). TH induction of TRβ1 expression was also observed in rib chondrocytes (Rabier et al., 2006), primary cultures of epiphyseal chondrocytes, proximal tibia, and distal femur epiphyses (Xing et al., 2016) of mice, and in tadpoles (*Xenopus laevis*) (Opitz, Lutz, Nguyen, Scanlan, & Kloas, 2006). These findings are consistent with the identification of TREs in the promoter of the THRB gene (Sakurai, Miyamoto, & DeGroot, 1992) and suggest that T3 can modulate its skeletal responsiveness via TRβ1. Interestingly, treatment with T3 but not GC-1, a TRβ1-specific agonist, increased TRβ1 expression in epiphyseal chondrocytes, suggesting that TRα1 signaling is involved in the T3-induced increase in TRβ1 expression (Xing et al., 2016). In contrast, T3 has been shown to decrease TRα1 mRNA expression in rib chondrocytes (Rabier et al., 2006) or to have no effects (Xing et al., 2016). Another interesting observation was a 12-fold higher expression of TRα1 in mouse osteoblastic cells over rat osteoblastic cells, suggesting that the relevance of TRα1 in bone tissue may be species dependent.

The coexpression of TR with RXR, RAR, and VD receptor (VDR) in skeletal cells (Williams et al., 1994), along with the demonstration that there is a complex interaction between retinoic acid and 1.25(OH)2VD3 (VD3) to regulate the specificity of T3 actions in bone cells, suggests that there is a fine balance between the physiological levels of the respective nuclear receptors and ligands (Williams et al., 1994). It is possible that this balance is one of the main determinants of the specificity of the bone cell responses to different hormonal conditions.

In conclusion, TRs are widely expressed in the skeleton. TR isoforms have been described in the main components of bones: bone marrow, cartilage, and bone tissue (Fig. 2). TRα1, TRα2, TRβ1, and TRβ2 were detected in osteoblasts, osteoclasts, and chondrocytes, but TRα1 is the predominant functional isoform in osteoblasts. In contrast to osteoblasts and chondrocytes, the functionality of TRs in osteoclasts has not been proved yet. The presence of TRs in osteocytes is less clear and needs further investigation. TRs were also detected the BM, suggesting direct actions of TH on osteoblastic and osteoclastic differentiation.

4.2 Expression of TSHR

It has been shown that cells and organs other than thyroid gland express thyrotropin receptor (TSHR) (Davies, Marians, & Latif, 2002), including skeletal cells (Abe et al., 2003; Inoue, Tawata, Yokomori, Endo, & Onaya, 1998). TSHR was initially detected in osteoblast-like rat osteosarcoma cells (UMR-106), human osteoblast-like osteosarcoma cells (MG63), and mouse primary osteoblastic (POB) cells. In addition, it was shown that recombinant human TSH (rhTSH) increases cyclic adenosine monophosphate (cAMP) production in a concentration-dependent manner in these cells, suggesting that osteoblasts possess functional TSHR (Inoue et al., 1998). TSHR was also identified in osteoblast and osteoclast precursors, and TSH was shown to inhibit (Abe et al., 2003) or induce (Baliram et al., 2011) osteoblast differentiation, and to inhibit osteoclastogenesis from stem cells (Abe et al., 2003). The same was observed in RAW-C3, an osteoclast precursor cell line (Abe et al., 2003). In addition, TSH was shown to induce osteoblastic activity and inhibit osteoclastic resorption (Fig. 2; Abe et al., 2003). Recently, TSHR was also identified in chondrocytes of the AC, growth plate, and other cartilage tissues of mice. Moreover, *in vitro* studies showed that rhTSH dose dependently increases cAMP production in isolated chondrocytes derived from femoral head cartilage, suggesting that TSHR is also functional in chondrocytes (Endo & Kobayashi, 2013).

In conclusion, functional TSHR has been described in osteoblasts, osteoclasts, and chondrocytes (Fig. 2). The role of TSH–TSHR interaction on skeletal morphophysiology and development will be discussed in Section 7.2.

4.3 Expression and Activity of Iodothyronine Deiodinases

A series of studies have shown that D1 is not expressed in bone cells and that its activity is not present in pre- or postnatal skeleton of mice (Capelo et al., 2008; Gouveia et al., 2005; Williams et al., 2008). D2 mRNA expression was detected in growth plate chondrocytes of mice (Williams et al., 2008) and in AC of rats and humans (Bos et al., 2012; Nagase et al., 2013). D2 mRNA was also detected in osteoblasts of rodents (Gouveia et al., 2005; Miura et al., 2002b; Williams et al., 2008) and humans (Morimura et al., 2005), and in osteoclasts of mice (Williams et al.,

2008). D2 activity was identified in the bone tissue and bone marrow of various skeletal sites of adult mice (Gouveia et al., 2005) and in the femurs of E18.5 mouse fetuses, where hypothyroidism increased its activity in 70% (Capelo et al., 2008). However, D2 activity was only detected in osteoblasts (Gouveia et al., 2005; Miura et al., 2002b; Williams et al., 2008), but not in chondrocytes and osteoclasts (Williams et al., 2008) of mice, and was not investigated in osteocytes. In MC3T3-E1 cells, D2 activity was shown to exhibit the biochemical and cellular properties characteristic of other tissues (Gouveia et al., 2005), where D2 is involved in intracellular regulation of thyroid status. D3 mRNA was detected in growth plate chondrocytes, osteoblasts, and osteoclasts of mice (Williams et al., 2008) and in AC of rats (Nagase et al., 2013); and D3 protein expression was detected in AC chondrocytes of humans (Nagase et al., 2013). D3 activity was detected in the femurs of E18.5 mouse fetuses, where hypothyroidism decreased its activity in 52% (Capelo et al., 2008), and was also detected in the growth plate chondrocytes (Williams et al., 2008), osteoblasts (Gouveia et al., 2005; Williams et al., 2008), and osteoclasts (Williams et al., 2008) of mice.

Interestingly, skeletal gene expression of D2 has been shown to be low during the onset of mice fetal life, increasing during prenatal development, and reaching a peak around postnatal day 14 (P14). In contrast, D3 mRNA is easily detected around prenatal day 14.5 (E14.5), but its expression markedly decreases during prenatal and early postnatal development, becoming undetectable after P14 (Capelo et al., 2008). It has been observed, therefore, that there is a coordination of the regulation of D2 and D3 gene expression in the skeleton during prenatal development, which seems to be important to keep thyroid signaling at low levels during the early stages of skeletal development (Capelo et al., 2008). Interestingly, studies in the growth plate of chicken show that hedgehog signaling reciprocally increases D3 gene expression and inactivates D2 via induction of WSB-1, the E3 ligase Ub adaptor that ubiquitinates D2 (Dentice et al., 2005, 2007). It is worth mentioning that hypothyroidism practically does not affect the skeleton during the early stages of prenatal development (Capelo et al., 2008), which reinforces the hypothesis of a low TH signaling. Altogether, these findings suggest that D3 is important for prenatal skeletal development, keeping low levels of TH; and that D2 plays a more important role at the end of prenatal development, during postnatal development and in mature skeleton, to maintain the local skeletal euthyroidism. This hypothesis is corroborated by the phenotype of D2-deficient mice (discussed in Section 7.3), which present normal serum levels of T3 and normal bone development, but present a reduction in bone resistance to fractures, a reduction in bone formation, and a generalized increase in bone mineralization (Bassett et al., 2010).

In conclusion, D1 is not expressed in the skeleton, whereas D2 and D3 mRNAs are expressed in chondrocytes, osteoblasts, and osteoclasts. D2 activity was detected only in osteoblasts, whereas D3 activity was detected in growth plate chondrocytes, osteoblasts, and osteoclasts. It seems that D2 and D3 expression and activity in the skeleton are reciprocally regulate to fine tuning the appropriate skeletal levels of TH that ensure normal bone development and physiology. D3 is mainly expressed during prenatal development, to avoid premature TH exposure, which is deleterious to skeletal development. On the other hand, D2 is mainly expressed postnatally and lifelong, to guarantee an euthyroid skeleton.

4.4 Expression of TH Transporters

The gene expression of TH membrane transporters has been demonstrated in skeletal cells. Among these transporters, MCT8, which is a very active and specific transporter of iodothyronines, had its expression identified in osteoblasts, osteoclasts, and chondrocytes and in pre- and postnatal bone femurs (Capelo, Beber, Fonseca, & Gouveia, 2009; Williams et al., 2008). In addition, MCT8 mRNA expression was found to decrease during skeletal development and to be negatively regulated by T3 in the femurs and in osteoblastic cells (Capelo et al., 2009). Skeletal expression of other transporters involved in the influx–efflux of iodothyronines through the plasmatic membrane, such as LAT1 and LAT2, NTCP, and OATP1 was also investigated. Among these transporters, only LAT1 and LAT2 were identified in bone tissue and osteoblastic cells; however, their expression was practically unaffected by TH status or by cell differentiation or bone maturation stage (Capelo et al., 2009). MCT10, MCT8, and LAT1 mRNA expressions were detected in mouse chondrogenic ATDC5 cells and in murine growth plates (Abe et al., 2012). MCT10 was mainly detected in RZ, whereas MCT18 and LAT1 were most expressed in HZ. MCT10 was the most abundant TH transporter expressed, as compared with MCT8 and LAT1 in ATDC5 cells. In these cells, MCT8 mRNA levels increased and MCT10 mRNA levels diminished with chondrocyte differentiation. MCT10 played a significant role in transporting TH into ATC5 cells, whereas LAT1 played a minimal role. Small interfering RNA-mediated KO of MCT10 mRNA decreased T3 uptake up to 44% and abolished the known T3 effects, such as suppression of cell proliferation and enhancement of chondrocyte differentiation in ATDC5 cells (Abe et al., 2012).

In conclusion, the mRNA expression of MCT8 was detected in the femur, osteoblasts, osteoclasts, and chondrocytes of mice. LAT1 and LAT2 were also detected in the femur and osteoblasts of mice. MCT8, LAT1, and MCT10 were detected in mouse growth plate and ATDC5 cells, where MCT10 was the most abundant transporter expressed and was shown to play a significant role in TH transport. Possible roles of these transporters in skeletal development will be discussed in Section 7.4.

5. THE EFFECTS OF TH

5.1 Effects in Osteoblast Lineage Cells

It is well known that TH stimulates osteoblastic activity (Allain et al., 1995; Mosekilde & Melsen, 1978a, 1978b), but the mechanisms are still unclear. T3 stimulates the mRNA and protein expression of bone matrix formation-associated factors, such as alkaline phosphatase (ALP), osteocalcin (OCN), and collagen in isolated osteoblasts (Egrise, Martin, Neve, Verhas, & Schoutens, 1990; Kasono et al., 1988; Rizzoli et al., 1986; Sato, Han, Fujii, Tsushima, & Shizume, 1987). IGF-1 mRNA levels are also increased by T3 in osteoblastic cells of mice (Varga, Rumpler, & Klaushofer, 1994) and rats (Milne, Kang, Quail, & Baran, 1998). Although these genes are responsive to TH in osteoblasts, it is not clear if they are directly regulated by T3. In ROS 17/2.8 cells, T3 has been shown to stimulate the expression of OCN at the transcriptional and posttranscriptional levels (Gouveia, Schultz, Bianco, & Brent, 2001); however, in transfection experiments, T3 was unable to activate the rat OCN promoter (Gouveia et al., 2001). On the other hand, a TRE was identified in the promoter of the mouse OCN gene (Varga, Spitzer, Rumpler, & Klaushofer, 2003), suggesting that it is directly regulated by T3 in mouse osteoblasts. Two other T3-responsive genes in osteoblasts are the matrix metalloproteinases (MMPs) collagenase 3 (MMP-13) and gelatinase B (MMP-9) (Pereira, Jorgetti, & Canalis, 1999). MMPs include a family of proteolytic enzymes involved in bone remodeling and morphogenesis (Filanti et al., 2000). MMP-13 appears to have the function of digesting the nonmineralized bone matrix (osteoid) to initiate the osteoclast-resorptive action (Kusano et al., 1998). It is believed that this process is important, since osteoclasts are unable to attach to nonmineralized bone surfaces. Although MMP-13 and MMP-9 have been shown to be T3-regulated at the transcriptional level, TREs have not yet been identified in their genes (Pereira et al., 1999). TH has also been shown to have effects on osteoblastic proliferation and differentiation. Some studies have shown that T3 induces proliferation (Ernst & Froesch, 1987; Kassem, Mosekilde, & Eriksen, 1993), others have shown an inhibitory effect (Beber et al., 2009; Kasono et al., 1988), and others have not found an effect (LeBron, Pekary, Mirell, Hahn, & Hershman, 1989; Sato et al., 1987; Williams et al., 1994). Varga et al. (1997) have shown that T3 stimulates osteoblastic differentiation by depressing genes specifically related to cell proliferation (histone H4 and c-Fos/c-Jun) and inducing the expression of phenotypic genes, such as OCN and ALP. In contrast, Ohishi et al. (1994) have shown that T3 suppresses the osteoblast differentiation, but increases the functional activity of mature osteoblasts. These divergent findings are probably explained by the different osteoblastic cell lines studied and the different culture conditions, such as number of passages, cell confluence, cell culture media, and dose of TH. Given the fact that TH induces the expression of genes related to the maturity of osteoblasts gradually in a time-dependent fashion (Beber et al., 2009), a positive action of TH on osteoblast differentiation is practically a certainty.

In conclusion, the effects of TH on osteoblast proliferation are inconsistent, whereas stimulatory actions of TH on osteoblastic activity are clear. The expression of TRs in osteoblasts and the responsiveness of isolated osteoblasts to T3 suggest that direct actions of TH in osteoblasts are important to positively regulate bone formation and remodeling.

5.2 Effects in Osteoclasts

It is well known that thyrotoxicosis leads to increased bone resorption and bone loss (Reddy, Harinarayan, Sachan, Suresh, & Rajagopal, 2012), and experimental data clearly show that TH acts directly in the bone to induce osteoclastic activity. TH stimulates bone resorption in organ cultures of neonatal mouse calvarias (Klaushofer et al., 1989) and fetal rat limb bones (Hoffmann et al., 1986; Mundy, Shapiro, Bandelin, Canalis, & Raisz, 1976). At this point, however, it is not clear if TR isoforms expressed in osteoclast precursors or osteoclastic cells (Allain et al., 1996; Kanatani et al., 2004) are functional and, therefore, whether TH acts directly in these cells. Allain et al. found that there was no effect of T3 on bone resorption by osteoclasts isolated from neonatal rat long bones incubated alone or cocultured with UMR-106 osteoblast-like cells (Allain, Chambers, Flanagan, & McGregor, 1992). However, in cultures with mixed bone cells, there was a significant relationship between the concentration of T3 and bone resorption, suggesting that TH acts indirectly on osteoclasts by a mechanism not involving osteoblasts, but requiring another cell type present in bone (Allain et al., 1992). Later, Britto, Fenton, Holloway, and Nicholson (1994) also showed that T3 produced no stimulation of bone resorption in isolated osteoclasts. However, T3 increased resorption in cocultures of osteoclasts and osteoblasts (UMR-106-01 or calvarial cells). Furthermore, stimulation of bone resorption also occurred when the osteoblastic cells were pretreated with T3 and the subsequent osteoblast–osteoclast cocultures were conducted for 24 h in the absence of T3. Thus, the direct exposure of osteoclasts to T3 was not required for the stimulatory effect. Overall, these findings indicate that TH can act on osteoblasts or other bone cells to indirectly stimulate osteoclastic bone resorption (Fig. 2). In contrast, findings of Kanatani et al. (2004) showed that T3 stimulates osteoclast differentiation and activity, in the absence of osteoblasts, acting directly on osteoclast precursor cells—hematopoietic blast cells derived from spleen cells of mice, which were also shown to express TRα1 mRNA. In another study, T3 was shown to increase the expression of integrin β3 and β5 in isolated cultures of osteoclasts (Holly et al., 1993). Considering that these proteins are important for the attachment of osteoclasts to bone surfaces, it is possible that such direct action may facilitate bone resorption.

Miura et al. (2002a) found that T3 alone did not induce osteoclast differentiation in coculture of BM cells with POB cells; however, T3 enhanced the VD3-induced osteoclast formation. In addition, they showed that T3 induced the mRNA expression of RANKL in POB cells, which is a key molecule for osteoclastogenesis and osteoclastic activation. The binding of RANKL to its receptor, RANK, expressed in osteoclast precursor cells and mature osteoclasts, activates osteoclastogenesis and increases osteoclastic activity and survival, respectively (Burgess et al., 1999; Hsu et al., 1999). Accordingly, 30 days of treatment with supraphysiological doses of T3 increased tartrate-resistant acid phosphatase (TRAP) mRNA expression, a marker of osteoclastic activity, in addition to enhance the mRNA expressions of RANK and RANKL (2.5- and 2.6-fold) and to decrease the mRNA expression of osteoprotegerin (OPG) by 50% in the femur of wild-type (WT) mice *in vivo* (Cruz Grecco Teixeira et al., 2016). OPG, produced by the osteoblasts, is the natural inhibitor of osteoclastic activity, since it binds RANKL and thereby impairs RANKL/RANK association (Burgess et al., 1999; Hsu et al., 1999). When T3 treatment was increased to 90 days, only a further decrease in OPG mRNA expression was detected (93%) in the femurs, drastically increasing the RANKL/OPG ratio (12 times), which was followed by an additional increase in TRAP mRNA expression (Cruz Grecco Teixeira et al., 2016). These findings suggest that the RANKL–RANK–OPG system may be a mechanism by which TH controls osteoclastic differentiation and activity *in vivo*.

On the other hand, *in vitro* studies by Katany et al. showed that T3 increased OPG mRNA levels and had no effect on RANKL mRNA expression in unfractionated bone cells (containing both osteoblasts and osteoclasts) derived from femurs and tibias of mice (Kanatani et al., 2004). In hematopoietic blast cells derived from spleen cells, T3 did not affect RANK mRNA levels nor NF-κB or mitogen-activated protein kinase (MAPK) activation, which occurs as a consequence of RANK-mediated signals (Li et al., 2002; Takayanagi et al., 2002), but T3 increased c-Fos and Fra-1 mRNA levels and induced c-Fos activation by immunoblot analysis (Kanatani et al., 2004). It has been reported that Fra-1, which transcription is induced in a c-Fos-dependent manner in osteoclast precursor cells, has a key role in osteoclast differentiation (Matsuo et al., 2000; Owens, Matsuo, Nicholson, Wagner, & Chambers, 1999). Furthermore, antisense oligodeoxynucleotides (as-ODN) complementary to c-Fos mRNA significantly inhibited T3-induced osteoclast-like cell formation from osteoclast precursors in the absence of stromal cells, whereas sense ODN did not affect T3-induced osteoclast formation (Kanatani et al., 2004). Altogether, these *in vitro* findings suggest that the RANKL–RANK–OPG system is not the major regulator of the T3-induced bone resorption and that T3 directly stimulates osteoclast differentiation at least in part by upregulating c-Fos protein in osteoclast precursor cells.

Prostaglandins (PGs) have been considered as a mediator of TH actions on bone resorption. Klaushofer et al. (1989) reported that the TH-induced increase in bone-resorbing activity in organ cultures of neonatal mouse calvaria is followed by a gradual increase in medium concentrations of PGs. The inhibition of PG synthesis by indomethacin did reduce the TH-induced bone resorption, but only by 50%, indicating that TH action in osteoclasts is also mediated by PG-independent mechanisms. In other studies, indomethacin had no effect on the TH-induced bone resorption in cultures of fetal rat limb bones (Hoffmann et al., 1986; Mundy et al., 1976). In mouse marrow cultures, T3 alone was not able to induce osteoclastogenesis but clearly increased the VD3-induced osteoclast differentiation (Schiller et al., 1998). The indomethacin inhibition of PGs significantly inhibited the VD3-induced osteoclastogenesis and the synergistic T3 effects (Schiller et al., 1998). IL-6, a cytokine highly induced by PGs in bone cells (Holt, Davie, Braidman, & Marshall, 1994), was induced by VD3, but not by T3 in the mouse marrow cultures. In addition, T3 had no additional effect on the VD3 induction of IL-6 synthesis. On the other hand, antibodies for IL-6, IL-6 receptor, and IL-11 blocked the stimulatory effect of VD3 on osteoclastogenesis and prevented the T3-induced augmentation of the VD3-induced osteoclastogenesis (Schiller et al., 1998). It seems, therefore, that TH facilitates the IL-6 dependent action of VD3 on osteoclast differentiation and, in turn, on bone resorption.

In conclusion, TH acts directly in the skeleton to induce bone resorption; however, it is not clear if TH acts directly in osteoclastic cells (including osteoclast precursor cells and mature osteoclasts). *In vitro* studies suggest that TH acts mainly indirectly, via osteoblast-mediated mechanisms, to induce osteoclastogenesis and osteoclastic activity (Fig. 2). However, the presence of TR isoforms in osteoclasts and a certain responsiveness of these cells to T3 in isolated cultures suggest that a direct, but less significant, mechanism of T3 action in osteoclasts may occur, which remains to be further investigated. There is evidence that there is a participation of the RANKL–RANK–OPG system, PGs, and IL-6 in mediating TH actions on bone resorption; however, the relevance and mechanism of action of each one of these factors in mediating TH actions in osteoclasts need to be clarified.

5.3 Effects in Chondrocytes

The presence of TRs in chondrocytes (discussed in Section 4.1) and the pronounced effects of T4 and T3 in isolated chondrocytes *in vitro* are strong evidence that direct actions of TH in these cells are critical for cartilage development and physiology. A series of studies have shown that addition of TH to the medium of chondrocyte three-dimensional (3D) pellet cultures leads to the organization of chondrocytes into a columnar pattern, which is a characteristic of the growth plate architecture, and induces chondrocyte hypertrophy (Ballock & Reddi, 1994; Ballock et al., 2000; Wang, Shao, & Ballock, 2007, 2009, 2010, 2011), demonstrating a direct and critical role of TH in chondrocyte terminal differentiation and in growth plate morphogenesis. Similar effects were observed in suspension primary cultures of chondrocytes derived from rat tibias, where cellular proliferation was inhibited by T3, which also induced Col X mRNA expression, ALP activity, and deposition of an alcian blue-positive matrix, suggesting that TH reciprocally inhibits proliferation and induces HC differentiation (Robson et al., 2000). Further experiments using the 3D pellet cultures of rat epiphyseal chondrocytes suggested that TH regulates terminal differentiation in part by arresting cell cycle progression through induction of cyclin-dependent kinase inhibitors (Ballock et al., 2000). In contrast, later studies showed a positive effect of TH on rat chondrocyte proliferation and on cyclin D1 expression, a cell proliferation factor (Shao et al., 2006; Wang et al., 2010; Wang, Shao, et al., 2011). Accordingly, *in vivo* studies (Fig. 1) have shown decreased number of PCs and HCs, thinner PZ and HZ in the growth plate of hypothyroid rats, and a reversion of this condition after TH treatment (Freitas et al., 2005; Lewinson et al., 1989). Notably, chondrocyte proliferation was shown to be significantly decreased in the growth plate of mice that expresses a TRα mutant (TRα^{L400R}) specifically in chondrocytes (Desjardin et al., 2014). TRα^{L400R} exerts dominant-negative activity on functional TRα1 and TRβ, resulting in a skeletal phenotype of hypothyroidism. Additionally, in cultured human fetal epiphyseal chondrocytes, T3 had no effect on cell proliferation but again induced chondrocyte differentiation, shown by increased ALP activity (Carrascosa, Ferrandez, Audi, & Ballabriga, 1992). Therefore, effects of TH on chondrocyte proliferation have some inconsistency and may reflect a local interaction with the IGF-1/IGF-1R system, which will be discussed in Section 6.1. Additionally, TH may have a dual effect on proliferation.

On the other hand, data regarding the TH induction of chondrocyte columnar architecture and chondrocyte terminal differentiation are consistent *in vitro* (Ballock & Reddi, 1994; Ballock et al., 2000; Carrascosa et al., 1992; Okubo & Reddi, 2003; Robson et al., 2000) and are supported by *in vivo* studies in rat models of hypothyroidism (Freitas et al., 2005; Lewinson et al., 1989; Shao et al., 2006), and by studies in mouse models of "functional skeletal hypothyroidism," due to KO of TRα (Bassett, O'Shea, et al., 2007; Fraichard et al., 1997; Gauthier et al., 2001; Gothe et al., 1999), or due to dominant target mutations in the THRA locus (Bassett, Nordstrom, et al., 2007; Desjardin et al., 2014; O'Shea et al., 2005), suggesting a predominant role of TRα in the skeletal development of mice. Interestingly, T3 inhibited proliferation and induced HC differentiation in primary cultures of chondrocytes isolated from the ribs of newborn mice with lack of all TRα isoforms (TR$\alpha^{0/0}$). In contrast, these T3 effects were abrogated in TRβ-deficient chondrocytes (TR$\beta^{-/-}$), indicating that T3-stimulated differentiation of primary rib chondrocytes *in vitro* requires TRβ (Rabier et al., 2006). The importance of TRβ1 in modulating the T3 induction of chondrocyte terminal differentiation is supported by *in vivo* studies using the relative TRβ-selective agonist GC-1 (Freitas et al., 2005). GC-1 binds TRβ with the same affinity as T3 but binds TRα1 with an affinity that is 10 times lower compared with T3 (Chiellini et al., 1998). Accordingly, GC-1 presents selective actions *in vitro* (Chiellini et al., 1998) and *in vivo* (Freitas et al., 2003; Morte, Manzano, Scanlan, Vennstrom, & Bernal, 2002; Ribeiro et al., 2001; Trost et al., 2000). In young hypothyroid rats, GC-1, such as T3, restored HC differentiation (Fig. 1) and normal expression of Col X in the growth plate. However, unlike T3, GC-1 was unable to restore serum levels of IGF-1 and protein expression of IGF-1 in the growth plate, suggesting that the GC-1 actions were independent of the GH/IGF-1 axis and TRβ-mediated (Freitas et al., 2005), reinforcing a role of TRβ on HC differentiation of rats.

In pellet cultures of chondrocytes derived from mouse ribs, T4 was again shown to induce the organization of chondrocytes into a columnar pattern, HC differentiation, and ECM synthesis. On the other hand, T4 significantly downregulated the mRNA expression of SRY-related high-mobility group-box gene 9 (Sox9) (Okubo & Reddi, 2003), which is a pivotal transcription factor in chondrogenesis, expressed in all chondroprogenitor cells and all differentiated chondrocytes, except HCs (Ng et al., 1997; Zhao, Eberspaecher, Lefebvre, & De Crombrugghe, 1997). Sox9 promotes chondrogenic differentiation of mesenchymal stem cells (MSCs) and, therefore, induces expression of genes encoding chondrocyte-specific ECM proteins, such as Col II and aggrecan (Lefebvre, Huang, Harley, Goodfellow, & de Crombrugghe, 1997); on the other hand, this transcription factor inhibits chondrocyte terminal differentiation, preventing transition of PCs into HCs (Li & Dong, 2016). Addition of GH to the pellet cultures of rib chondrocytes has not modified the action of T4 on HC differentiation and Sox9 mRNA expression inhibition, suggesting that TH directly induces chondrocyte terminal differentiation, independently of GH, and involving Sox9 signaling pathway (Okubo & Reddi, 2003). In contrast, a recent study showed a positive effect of physiological doses of T3 on Sox9 mRNA expression and on chondrogenesis, but not on terminal differentiation in spheroid cultures of MSCs derived from human umbilical cord (Fernandez-Pernas et al., 2016). Besides Sox9, T3 treatment increased the mRNA expression of other genes linked to chondrogenesis, such as Col II and integrin alpha 10 beta 1, in a steroid receptor coactivator 2 (SRC2)-dependent way. The discrepancy of this study regarding chondrocyte terminal differentiation and Sox9 expression might be explained by different culture conditions, including days in culture, cell source, and dose of

TH used. On the other hand, TH probably has a dual effect on Sox9, increasing its expression in less differentiated cells, to induce chondrogenesis, and decreasing Sox9 expression later, in more differentiated chondrocytes, to promote chondrocyte terminal differentiation (Fig. 2). Nevertheless, there is evidence that TH regulation of chondrocyte differentiation involves several other signaling pathways, which will be discussed in Section 6.

In conclusion, growth plate chondrocytes are highly responsive to TH, which acts directly in these cells to promote terminal differentiation, in addition to directly induce the organization of chondrocytes into a columnar pattern in 3D pellet culture systems. It is not clear if TH decreases, induces, or promotes both effects on chondrocyte proliferation.

5.4 Effect on Ossification

5.4.1 Effect on Endochondral Ossification

During development, TH deficiency generally impairs EO (Allain & McGregor, 1993), resulting in delayed bone maturation and decreased bone age, characterized by a decrease in the number and size of ossification centers (Fig. 1), which may be substantiated by "stippled epiphyses" on X-rays (Bassett & Williams, 2016). In contrast, thyrotoxicosis accelerates EO, leading to premature ossification of the cartilaginous skeleton and closure of the growth plates, resulting in increased bone age and short stature (Allain & McGregor, 1993; Bassett & Williams, 2016). TH is one of the factors that initiate EO by promoting chondrocyte hypertrophy (discussed in the last section), leading to the development of ossification centers in the cartilaginous skeleton, and to the ossification of the growth plates. Desjardin et al. (2014), using an elegant gene-targeting approach, developed a mouse model that expresses a TRα mutant (TRα^{L400R}) specifically in chondrocytes. TRα^{L400R} exerts dominant-negative activity on functional TRα1 and TRβ. These animals present profound delay in EO and bone growth, reinforcing the importance of direct actions of TH in chondrocytes to induce bone maturation and growth. A body of *in vivo* studies with genetically modified mice (discussed in Section 7.1) has shown that TH actions on EO are mainly mediated by TRα1. Interestingly, as mentioned in Section 5.3, T3 treatment was unable to induce HC differentiation in primary cultures of mouse rib chondrocytes isolated from TRβ-deficient newborn mice, suggesting a role of TRβ1 in mediating EO. *In vivo* studies using the TRβ-selective agonist GC-1 support this hypothesis. GC-1 could clearly induce EO in several skeletal sites of young hypothyroid rats (Fig. 1), independently of the GH/IGF-1 axis (Freitas et al., 2005). However, GC-1 effects were partial, whereas equimolar doses of T3 totally rescued EO, suggesting that TRβ contributes to EO but that TRα is required for complete EO in rats (Freitas et al., 2005). The mechanism of TH action on EO is still poorly understood, but there is evidence that TH mainly regulates this process independently of the GH/IGF-1 axis. TH has been shown to rescue the ossification delay observed in hypophysectomized–thyroidectomized rats, whereas GH alone does not affect this process (Ray et al., 1954). Accordingly, GH was able to restore bone elongation, but not the ossification delay seen in TRα1- and TRβ-null mice ($TR\alpha1^{-/-}TR\beta^{-/-}$) (Kindblom et al., 2001), showing that TR-mediated actions of T3 are essential for EO, but not the GH/IGF-1 axis. Findings of Xing, Cheng, Wergedal, and Mohan (2014) indicate that TH regulates the initiation and progression of the secondary ossification centers via differentiating chondrocytes into bone matrix-producing osteoblasts by stimulating expression of osterix (Osx), a key transcription factor for osteoblast differentiation, in a process that requires previous TH induction of Indian hedgehog (Ihh) expression, which is a master regulator of osteoblast specification during EO (Long et al., 2004). Interestingly, gel mobility shift assay demonstrated that a putative TRE in the distal promoter of mouse Ihh gene interacted with TRβ1, and that GC-1 treatment increased Ihh distal promoter activity 20-fold after 48 h in chondrocytes, showing that TRβ1 mediates transcriptional upregulation of Ihh gene, which reinforces a role for TRβ1 in secondary ossification of the epiphyses (Xing et al., 2016). Several other signaling pathways may modulate TH action on EO, which will be discussed in Section 6.

5.4.2 Effect on Intramembranous Ossification

IO is also delayed in hypothyroidism and accelerated in thyrotoxicosis. In humans and rodents, early-onset hypothyroidism results in broader sutures and fontanels (Bassett & Williams, 2016; Riggs, Wilroy, & Etteldorf, 1972), which are membranes rich in fibroblast and mesenchymal cells. These membranes articulate cranial bones and allow normal cranial and brain growth (Miura et al., 2009). In contrast, hyperthyroidism is associated with craniofacial deformities (Hirano, Akita, & Fujii, 1995) and craniosynostosis (Hashmi et al., 2012; Menking et al., 1972) due to accelerated IO. This effect was shown to occur during pre- and postnatal development. There is evidence that thyrotoxicosis in pregnancy, due to TH replacement for maternal hypothyroidism, maternal Graves' disease, or the presence of thyrotoxic goiter (Radetti, Zavallone, Gentili, Beck-Peccoz, & Bona, 2002), results in a 2.5-fold increase in the risk of craniosynostosis (Rasmussen et al., 2007). Accordingly, mouse litters exposed to T4 *in utero* present craniofacial dysmorphology (brachycranic), with diminished area of the coronal and widening of the sagittal sutures, indicative of premature fusion and compensatory growth (Howie et al., 2016). The mechanism through which TH excess induces premature suture fusion is practically unknown, but it is supposed that TH acts directly in suture cells to increase osteogenesis, in a process involving the fibroblast growth factors 2 (FGF2)/FGF receptor 1 (FGFR1)/MAPK signaling pathway (Stevens et al., 2003), which will be discussed in Section 6.3. In fact, both TRα and TRβ were detected in the cranial sutures of mice, and TRβ was shown to have a greater presence in both coronal and sagittal sutures compared to TRα (Cray, Khaksarfard, Weinberg, Elsalanty, & Yu, 2013). Nevertheless, a series of studies with different TR mutant mice suggest that TRα1 is the main TR isoform to mediate TH actions on IO (discussed in Section 7.1). Studies with the murine-derived calvaria MC3T3-E1 cells support the hypothesis that TH excess stimulates osteogenesis. Exposure of these cells to TH increases ALP activity, and the mRNA expression of several markers of osteoblastogenesis, detected by gene-chip array analysis, including ALP, OCN, MMP-9, fibroblast growth factors 1 and 2 (FGF1 and FGF2), IGF-1, and high-temperature requirement serine protease A1 (HtrA1) (Cray et al., 2013). The upregulation of IGF-1 and HtrA1, a serine peptidase that regulates the availability of IGF (Zumbrunn & Trueb, 1996), is consistent with the known interaction of TH with IGF-1 in the bone tissue (discussed in Section 6.1). This upregulation was also observed *in vivo*. The intensity of immunohistochemical staining of IGF-1 was markedly increased by T3 treatment in the cranial suture margins of 10-day-old rats, which was accompanied by narrowing of the sagittal suture (Akita, Hirano, & Fujii, 1996). The sutures of mouse litters exposed to T4 *in utero* presented increased expression of markers of proliferation and osteogenesis and increased gene expression of HtrA1 and IGF-1 gene expression (Howie et al., 2016). In addition, pathways related to the HtrA1/IGF axis, specifically Wnt and Akt, also demonstrated increased activity. T4 treatment significantly upregulated several targets within the osteogenic Wnt/β-catenin signaling pathway (Howie et al., 2016), suggesting that this important osteogenic signaling pathway (Krishnan, Bryant, & Macdougald, 2006) may mediate the ossification actions of TH in the sutures of mice.

In conclusion, TH has a crucial role in EO and IO. TH effects on EO are primarily the result of direct actions of T3 in chondrocytes, mainly promoting chondrocyte terminal differentiation. The mechanisms by which TH promotes IO are less clear, but seem to involve direct actions of T3 in the cranial suture cells, promoting osteoblastogenesis and inducing osteoblastic activity. Further studies are necessary to identify and characterize the signaling cascades required for TH actions on both EO and IO.

5.5 Effects in the Growth Plate and Longitudinal Bone Growth

The growth plate is one of the most sensitive targets for TH action, where T3 has crucial roles regulating its normal morphogenesis, physiology, and function (Mackie et al., 2011; Shao et al., 2006). As discussed in Section 4.1, TRs were identified in chondrocytes of both RZ and PZ, but were not detected in HC of rodents (Stevens et al., 2000). *In vitro* studies using primary cultures of chondrocytes derived from growth plates of rat and mice (Section 5.3) indicate that TH acts directly in the growth plate. Many *in vivo* and vitro studies have shown that TH is the main regulator of the differentiation of PCs into HCs in the growth plate (Ballock & Reddi, 1994; Lewinson et al., 1989; Shao et al., 2006). Another important role of TH is the organization of the growth plate chondrocytes in columns parallel to the long axis of the bones (Ballock & Reddi, 1994; Lewinson et al., 1989; Shao et al., 2006). Therefore, T3 deficiency results in prominent alterations in the growth plate, such as disruption of the chondrocyte columnar architecture and a significant impairment of the terminal chondrocyte differentiation (Fig. 1; Shao et al., 2006), which are frequently manifested as severe growth retardation and slipped capital femoral epiphysis (Rivkees, Bode, & Crawford, 1988; Van Vliet, Larroque, Bubuteishvili, Supernant, & Leger, 2003). TH replacement reverts growth plate alteration in experimental animals (Freitas et al., 2005; Lewinson et al., 1989) and induces a rapid catch-up growth, which reflects the importance of TH to longitudinal bone growth (Rivkees et al., 1988). It is noteworthy that in humans, the predicted adult height may not be reached after TH replacement therapy, which will depend on the duration of untreated hypothyroidism (Rivkees et al., 1988).

Considering that TH regulates the synthesis and secretion of GH (Hervas, Morreale de Escobar, & Escobar Del Rey, 1975), it was initially thought that the growth plate defects and the bone elongation failure in hypothyroidism were merely consequences of the GH deficiency that follows TH deficiency. However, classical studies with hypophysectomized or hypophysectomized–thyroidectomized rats have shown that TH or GH alone is able to promote linear bone growth, although incomplete, whereas only the combination of both hormones nearly or fully rescues the longitudinal bone growth (Ray et al., 1954; Thorngren & Hansson, 1973), indicating an interaction between these hormones to promote linear bone growth, which will be discussed in Section 6.1. It has also been shown that GH without TH is unable to induce the columnar organization of the growth plate and the terminal chondrocyte differentiation *in vivo* and *in vitro* (Ballock & Reddi, 1994; Lewinson et al., 1989). On the other hand, TH promotes these effects in GH-deficient conditions (Ray et al., 1954), showing that TH acts independently of the GH/IGF-1 axis in the growth plate, as it will be further discussed in Section 6.1. Interestingly, the TRβ-selective agonist GC-1 had selective effects in the growth plate of hypothyroid young rats, where it clearly induced chondrocyte terminal differentiation (increased HC number, HZ thickness, and Col X mRNA and protein expression), but was unable to induce IGF-1 mRNA expression and the chondrocyte columnar organization (Fig. 1; Freitas et al., 2005), indicating that the positive effect of GC-1 on HC differentiation was not IGF-1 mediated. These findings suggest that TRβ1 may have a role in HC differentiation, whereas TRα1 is essential for the T3 induction of the chondrocyte columnar organization in the growth plate of rats (Freitas et al., 2005).

The terminal differentiation of growth plate chondrocytes is characterized by an increase of up to 10-fold in the cellular volume, which requires the degradation and remodeling of the pericellular ECM. TH was shown to play a crucial role in the control of proteoglycan degradation by the growth plate chondrocytes of rats, pigs, and rabbits, a process that seems to be mediated by T3 induction of the proteoglycan-degrading enzymes aggrecanase-2 and ADAMTS-5 (a disintegrin and metalloproteinase with thrombospondin motifs 5) (Makihira et al., 2003). On the other hand, T3 is essential for synthesis and normal distribution of the ECM components of the growth plate (Bassett, Swinhoe, Chassande, Samarut, & Williams, 2006), including proteoglycans (Shao et al., 2006; Stevens et al., 2000). It has been shown that hypothyroidism results in a heparan sulfate proteoglycan (HSPG)-rich matrix in the growth plate cartilage of rodents (Bassett, O'Shea, et al., 2006; Stevens et al., 2000). Considering that HS binds to a variety of protein ligands and regulates several biological processes (Forsberg & Kjellen, 2001), it was raised the hypothesis that TH may modify growth plate paracrine signaling pathways by modifying the cartilage matrix composition (Bassett,

O'Shea, et al., 2006; Stevens et al., 2000; Williams, O'Shea, & Williams, 2007), which will be discussed in Section 6.7.

In conclusion, TH promotes longitudinal bone growth acting directly in the growth plate, inducing the terminal chondrocyte differentiation, the columnar chondrocyte organization, and the synthesis and normal distribution of ECM components. Although TH can promote linear bone growth independently of the GH/IGF-1 axis, TH depends on this axis to fully induce bone growth. The mechanisms through which TH controls proliferation, differentiation, and the columnar organization of chondrocytes in the growth plate are still poorly understood, but a series of studies have shown that these mechanisms involve interaction of TH not only with the GH/IGF-1 axis but also with other major regulatory pathways of the growth plate, which will be discussed in the next section.

6. MECHANISMS OF ACTION OF TH TO PROMOTE SKELETAL DEVELOPMENT

6.1 The Role of GH/IGF-1 Axis

The GH/IGF-1 axis has anabolic effects in the skeleton and is the major regulator of longitudinal bone growth (Butler & Le Roith, 2001). GH, produced and secreted by the anterior pituitary (Fig. 3), acts in the liver where it binds its receptor (GHR) to induce the synthesis and secretion of IGF-1, which is the main regulator of linear bone growth (Yakar, Courtland, & Clemmons, 2010). GH also interacts with GHR expressed in several cells, including osteoblasts and chondrocytes, to induce IGF-1 expression locally in the skeleton (Yakar et al., 2010). IGF-1 binds its receptor (IGF-1R), expressed in osteoblasts and chondrocytes, and promotes linear bone growth primarily inducing chondrocyte proliferation (Yakar et al., 2010). There is evidence that locally produced IGF-1 has a crucial role in linear bone growth (Wu et al., 2013), whereas liver-derived IGF-1 is also important, but contributes to only 30% of adult body size (Stratikopoulos, Szabolcs, Dragatsis, Klinakis, & Efstratiadis, 2008). GH also acts directly in the skeleton to promote bone elongation, independently of IGF-1, but these effects are much milder than the IGF-1 effects (Green, Morikawa, & Nixon, 1985). According to a dual effector theory (Green et al., 1985), GH would act mainly on less differentiated chondrocytes (Fig. 3), present in the RZ of the growth plate, to stimulate chondrocyte differentiation and local IGF-1 synthesis, which, in turn, would act on more differentiated cells, inducing clonal expansion of PCs and chondrocyte

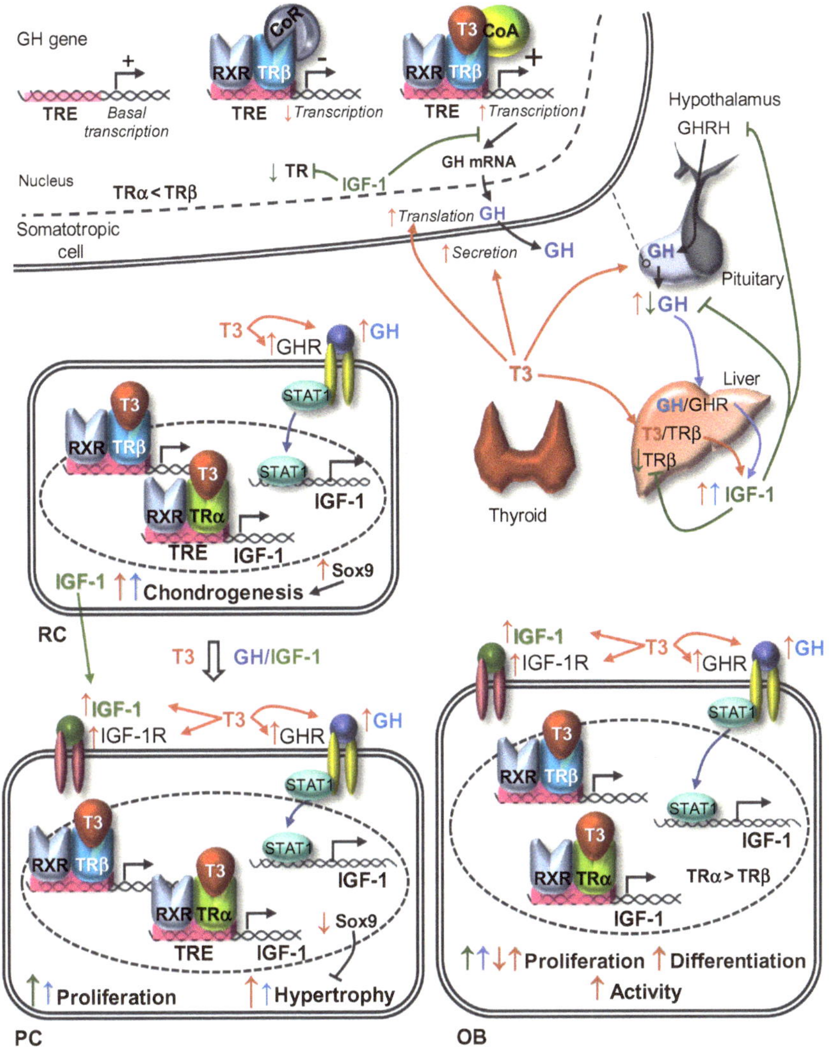

Fig. 3 Schematic representation of T3 interactions with the GH/IGF-1 axis at the systemic and skeletal levels. T3 stimulates gene transcription, translation, and secretion of GH by the anterior pituitary. T3 also stimulates synthesis and secretion of IGF-1 by the liver and induces expression of IGF-1, IGF-1R, and GHR in osteoblasts (OB) and chondrocytes in a TRα-dependent way. In turn, IGF-1 decreases the number of TRs in the somatotropic cells and in the liver. GH mainly acts in reserve chondrocytes (RC) to induce chondrogenesis, and in proliferating chondrocytes (PC) to induce IGF-1 expression. IGF-1 mainly induces proliferation in chondrocytes and OBs. T3 acts in RC and PC to induce chondrocyte differentiation. *Red*, *blue*, and *green arrows* indicate actions of T3, GH, and IGF-1, respectively.

hypertrophy, in both autocrine and paracrine manners (Ahmed & Farquharson, 2010; Ohlsson, Bengtsson, Isaksson, Andreassen, & Slootweg, 1998). However, it is now known that GHR has a broad distribution in the growth plate, suggesting roles of GH in more differentiated growth plate cells (Gevers et al., 2002). GH also interacts with GH-binding proteins (GHBPs), which are expressed in several tissue cells, including growth plate chondrocytes (Carlsson, Billig, Rymo, & Isaksson, 1990; Gevers et al., 2002). GHPBs modulate GH action sequestering free GH and impairing GHR binding. IGF-binding proteins (IGFBPs) also modulate IGF-1 activity, impairing IGF-1 degradation (Firth & Baxter, 2002), avoiding IGF-1–IGF-1R binding (Mohan et al., 1995), or storing IGF-1 in the skeletal ECM (Campbell & Andress, 1997).

One of the mechanisms by means of which TH stimulates bone growth and maturation is indirectly regulating the synthesis and secretion of GH and IGF-1 (Fig. 3; Hervas et al., 1975). The GH gene contains a positive TRE in its promoter, what makes its transcription to be directly and greatly stimulated by interactions of T3 mainly with TRβ in the anterior pituitary (Glass et al., 1987; Koenig, Brent, Warne, Larsen, & Moore, 1987), where TRβ is the predominant TR isoform (Fig. 3). Besides inducing GH gene transcription, TH also regulates GH synthesis and secretion by the pituitary (Giustina & Veldhuis, 1998; Hervas et al., 1975; Kamegai, Tamura, Ishii, Sugihara, & Wakabayashi, 2001). Furthermore, TH also directly stimulates the expression of IGF-1 in the liver (Fig. 3), osteoblasts, and chondrocytes (Harakawa et al., 1990; Huang, Golden, Tarjan, Madison, & Stern, 2000; Varga et al., 1994). A TRE was identified in intron 1 of IGF-1 gene, and *in vitro* studies showed that T3–TRα1–TRE interactions result in IGF-1 transcription in osteoblasts (Xing et al., 2012). Accordingly, GC-1, a TRβ-selective agonist, was not able to induce IGF-1 protein expression (Fig. 1) in the growth plate of euthyroid and hypothyroid rats (Freitas et al., 2005). This is supported by the observation that $TR\alpha^{0/0}$ mice and $TR\alpha1^{PV/+}$ mice (which harbor the dominant PV target mutation in TRα1) present diminished mRNA expression of GHR, IGF-1, and IGF-1R in growth plate chondrocytes (Gauthier et al., 2001; O'Shea et al., 2005). It has been clear that TH also increases GHR and IGF-1R expression in growth plate chondrocytes (Gevers et al., 2002; Ohlsson, Nilsson, Isaksson, Bentham, & Lindahl, 1992) and osteoblasts (Pepene et al., 2001), suggesting that TH increases the skeletal responsiveness to GH and IGF-1. A series of studies have shown that serum levels of IGFBPs are also regulated by TH–GH interactions and that IGFBP serum levels are decreased in hypothyroidism (Burstein, Grumbach, Kaplan, & Li, 1978; Daughaday & Reeder, 1966; Schmid et al., 2006). Besides, TH was shown to increase expression of IGFBP-4 and -5 in osteoblast cell lines (Glantschnig, Varga, & Klaushofer, 1996; Pepene et al., 2001). Therefore, TH is a key regulator of both local and systemic IGF-1 levels. In addition, experimental data have demonstrated that TH is the major regulator of IGF-1 serum levels, independently of GH, during the prepubertal growth period in mice (Xing et al., 2012). Thus, in hypothyroidism, there is a reduction in serum and tissue levels of both GH and IGF-1, in addition to a decrease in the skeletal responsiveness to these factors, which, on their own, impairs a normal skeletal development. Therefore, the well-recognized suppression of the GH/IGF-1 axis that follows hypothyroidism impairs bone growth and maturation and makes it difficult to discriminate the independent actions and effects of TH and GH/IGF-1 in the skeleton.

In this regard, classical studies have shown that T4 alone is able to promote linear bone growth similar to GH alone in hypophysectomized or hypophysectomized–thyroidectomized rats, whereas only the combination of both hormones nearly or completely rescues the longitudinal bone growth (Ray et al., 1954). In hypophysectomized rats, GH therapy normalizes liver and serum IGF-1, but not body weight and tail length (King, Smith, Meller, Dahlenburg, & Lineham, 1988). These findings suggest that TH is able to promote linear bone growth independently of GH and that TH depends on GH, and vice versa, to completely induce this process (Ray et al., 1954; Thorngren & Hansson, 1973).

Contrasting studies that have shown inhibitory effects of TH on chondrocyte proliferation *in vitro* (Ishikawa, Genge, Wuthier, & Wu, 1998; Robson et al., 2000), chondrocyte-specific expression of $TR\alpha^{L400R}$ (a mutant receptor that exerts dominant-negative activity on TRα1 and TRβ) decreases growth plate chondrocyte proliferation *in vivo* (Desjardin et al., 2014). Additionally, other *in vivo* studies have shown that hypothyroidism decreases the number of PCs and the thickness of the PZ, which are restored by TH reposition (Freitas et al., 2005; Lewinson et al., 1989). Considering that GH and IGF-1 primarily induce chondrocyte proliferation *in vivo* (Ishikawa et al., 1998; Sims et al., 2000) and *in vitro* (Lindahl, Isgaard, Carlsson, & Isaksson, 1987), the reduced number of PCs in hypothyroid animals is probably due to the suppression of the GH/IGF-1 axis that follows T3 deficiency. Accordingly, hypothyroidism impairs IGF-1 protein expression in the growth plate of rats (Freitas et al., 2005). Furthermore, $TR\alpha^{0/0}$ and $TR\alpha1^{PV/+}$ mice, which present a skeletal phenotype of hypothyroidism, display diminished mRNA expression of GHR, IGF-1, and IGF-1R in growth plate chondrocytes (Gauthier et al., 2001; O'Shea et al., 2005). Otherwise, IGF-1R was shown to be increased in the growth plate of chondrocyte-specific $TR\alpha^{L400R}$ ($TR\alpha^{L400R/C2}$) mice, but IGF2BP2, a protein known to counteract IGF-1 signaling, was also increased (Desjardin et al., 2014). Therefore, the positive effect of TH on PC number and growth plate thickness seems to involve a local interaction of TH with the GH/IGF-1 axis, suggesting a predominant role of the GH/IGF-1 axis on chondrocyte proliferation *in vivo*, over a possible negative effect of TH on proliferation. It is noteworthy that an enlarged RZ was observed in 2- to 4-week-old $TR\alpha^{0/0}$, $TR\alpha1^{+/m}$ (which harbor the R384C dominant-negative mutation in THRA), and $TR\alpha^{PV/+}$ mice, which probably occurred from impaired differentiation of RCs into PCs.

As discussed previously in this review, EO, chondrocyte terminal differentiation, and columnar morphogenesis of the growth plate seem to be predominantly regulated by TH. However, a synergistic or additive action of GH/IGF-1 axis and TH on these processes should be considered, since mice with inactivated GHR or IGF-1R present delayed ossification, severely disorganized growth plates, and decreased chondrocyte differentiation, besides decreased cell proliferation (Liu, Baker, Perkins, Robertson, & Efstratiadis, 1993; Sjogren et al., 2000; Wang, Cheng, et al., 2011). In addition, in Snell GH-deficient mice, GH has a clear effect, although less intense than T4 effect, on the size of the ossification centers (Smeets & van Buul-Offers, 1983). Besides, in hypophysectomized–thyroidectomized rats, T4 greatly increases GH effects on EO (Hoskins & Asling, 1977).

It is noteworthy that the GH/IGF-1 axis can also modulate TH actions. In primary culture of pituitary cells, IGF-1 inhibits the T3 induction of GH gene expression and GH secretion (Melmed & Yamashita, 1986), which may occur due to the inhibitory effect of IGF-1 on the number of TRs (Fig. 3). Accordingly, Pellizas, Coleoni, Costamagna, Di Fulvio, and Masini-Repiso (1998) have shown that IGF-1 modulates T3 actions in the liver of mice by a mechanism that negatively affects, at least in part, the expression of TRs (Fig. 3). Therefore, it is possible that there is a negative feedback system, in which IGF-1 limits the T3-induced synthesis of IGF-1 downregulating TR expression. Interestingly, IGFBP-6 was also shown to act as an inhibitory mediator of TH effects in osteoblast differentiation, independently of IGF-1. It was shown that IGFBP-6 interacts with $TR\alpha$ in the nucleus and inhibits TR–RXR heterodimerization, inhibiting

TRE-mediated gene expression of OCN and ALP (Qiu, Ma, Wang, Chen, & Huang, 2012).

All these studies clearly show that TH–GH/IGF-1 interactions are complex and may occur at multiple levels, both systemically and locally. For example (as discussed in Section 6.4), TH locally interacts with the Wnt signaling pathway to control growth plate chondrocyte terminal differentiation, in a mechanism that involves IGF-1 and IGF-1R. The blockage of IGF-1R impairs TH activation of chondrocyte differentiation. On the other hand, IGF-1 activates this signaling pathway and is locally regulated by TH. Thus, there is a clear TH–IGF-1 interplay in the Wnt pathway that may also occur in other TH-modulated systemic and local signaling pathways, which remains to be clarified.

In conclusion, TH regulation of bone elongation and EO involves both GH/IGF-1-independent and -dependent pathways. TH, but not the GH/IGF-1 axis, is crucial for chondrocyte terminal differentiation, columnar organization of the growth plate, and EO. On the other hand, the GH/IGF-1 axis is essential for bone elongation, primarily inducing chondrocyte proliferation. Thus, TH–GH/IGF-1 interactions are necessary for full linear bone growth. These interactions are complex and occur at multiple levels and in both directions.

6.2 The Role of the Ihh/PTHrP Axis

Two signaling proteins, Ihh and parathyroid hormone-related protein (PTHrP), have been identified as crucial to modulate the terminal differentiation of growth plate chondrocytes (Vortkamp et al., 1996). Ihh, produced by PHCs and early HCs, diffuses through the growth plate to stimulate the production of PTHrP by the articular cartilage and periarticular cells, through interaction with Smoothened (SMO), a seven-pass transmembrane domain protein (Long & Ornitz, 2013). PTHrP also acts far away from its source, interacting with PTHrP receptors (PTHrP-R), expressed in late PCs and in PHCs, to keep them proliferating and delay their terminal differentiation (Lanske et al., 1999). Chondrocytes far enough from the PTHrP source are not reached by PTHrP, which stops their proliferation and allows their differentiation into Ihh-producing PHCs. Besides acting in the overlying perichondrial cells to induced PTHrP expression, Ihh also signals to the PCs to induce proliferation (Long et al., 2001) and to regulate the formation of columnar chondrocytes (Hilton, Tu, & Long, 2007; Kobayashi et al., 2005). The result of the Ihh/PTHrP feedback system is the opposite of the well-known T3 effect of inducing chondrocyte terminal differentiation (Ballock & Reddi, 1994; Robson et al., 2000). *In vivo*, it was shown, by in situ hybridization (ISH), that hypothyroidism increased PTHrP mRNA expression, whereas thyrotoxicosis resulted in undetectable PTHrP-R mRNA in the growth plate of rats (Kindblom et al., 1999). These effects correlate with the growth plate dysgenesis and growth retardation seen in hypothyroidism and with the accelerated growth rate seen in thyrotoxicosis (Shao et al., 2006).

In conclusion, these findings suggest that TH regulates the pace of growth plate chondrocyte differentiation through the modulation of the Ihh/PTHrP-R loop. Further studies, however, are necessary to elucidate if T3 directly and/or indirectly modulate the Ihh/PTHrP-R loop, how this modulation occurs, and how relevant this modulation is to the TH effects in the growth plate.

6.3 The Role of the FGF/FGFR Signaling Pathway

The mammalian FGF family includes 18 secreted signaling proteins that bind to and activate 4 FGF receptors (FGFR1–4), which are membrane tyrosine kinase molecules (Ornitz & Itoh, 2015). FGFs and FGFRs are widely expressed during embryogenesis and skeletal development and are key factors for these processes (Ornitz & Marie, 2015). The functions of each FGF isoform are poorly known, especially because several may have redundant roles. The biological activity of canonical FGFs is enhanced by heparan sulfate (HS), which acts as a coreceptor, interacting with FGFs and FGFRs to increase FGF–FGFR affinity and enhance receptor activation (Ornitz & Itoh, 2015; Shimokawa et al., 2011). FGFs, FGFRs, and HS are expressed in a time- and site-dependent way during skeletal development. At the initiation of EO, the condensing mesenchyme expresses FGFR2, but once mesenchymal cells begin to differentiate into chondrocytes, FGFR2 expression decreases and FGFR3 expression is activated, whereas cells that will form the perichondrium and periosteum, in the periphery of the condensation, express both FGFR1 and FGFR2. As central chondrocytes begin to hypertrophy, FGFR3 expression decreases and FGFR1 increases. The forming perichondrium, bone collar, and trabecular bone express FGFR1 in mesenchymal progenitors and FGFR2 in differentiating osteoblasts (Ornitz & Itoh, 2015). *In vitro*, immature osteoblasts express mainly FGFR1, whereas mature osteoblasts express high levels of FGFR2 (Rice, Rice, & Thesleff, 2003). In long bones, FGFR3 is strongly expressed in mature osteoblasts (Delezoide et al., 1998). Growth plate displays very low levels of FGFR2 in RCs, high levels of FGFR3 in the PZ and PHZ, and high levels of FGFR1 in PCs and HCs, overlapping FGFR3 in the PZ (Ornitz & Itoh, 2015).

Genetic and functional analyses support the concept that FGF/FGFR signaling activates multiple pathways to control osteoblast proliferation, differentiation, and survival through life (Ornitz & Itoh, 2015). FGFR2 seems irrelevant to the growth plate biology (Yu et al., 2003), whereas the role of FGFR1 has not been defined yet (Fig. 4). At early stages of development, FGFR1 conditional KO (Dermo1-Cre) impairs chondrocyte hypertrophy (Hung, Yu, Lavine, & Ornitz, 2007) but, at later stages (Col2a1-Cre), results in expanded HZ (Jacob, Smith, Partanen, & Ornitz, 2006). However, these two conditional KOs also targeted osteoblasts besides osteoclasts, which may have interfered in the development defects (Jacob et al., 2006; Karolak, Yang, & Elefteriou, 2015). Activating mutations of FGFR1 and FGFR2 cause craniosynostosis syndromes that differ according to the sutures affected

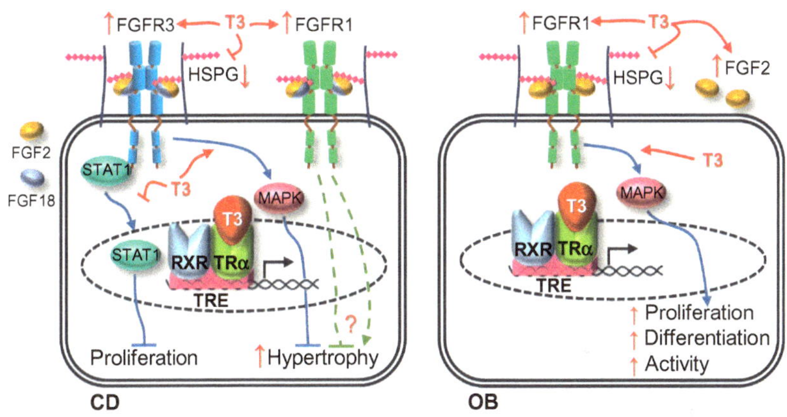

Fig. 4 Schematic representation of T3 effects on FGF/FGFR signaling pathway in chondrocytes (CD) and osteoblasts (OB). FGFR3 activation inhibits CD hypertrophy and proliferation via MAPK and STAT1 signaling pathways, respectively, whereas the effect of FGFR1 activation is uncertain. In CDs, T3 increases FGFR1 and FGFR3 mRNA expression, enhances both FGF2 and FGF18 activation of MAPK signaling pathway, and inhibits the activation of STAT1 signaling pathway. In OBs, T3 increases expression of FGFR1 and FGF2 and enhances FGF2/FGFR1 activation of the MAPK signaling pathway, which induces OB proliferation, differentiation, and activity. These T3 effects are dependent on TRα. HSPG, which acts as an FGF coreceptor, is downregulated by thyroid hormone (TH). *Blue* and *green lines* refer to FGFR3 and FGFR1 signaling pathways, respectively. *Red lines* indicate TH actions on FGFR signaling pathways.

and/or the presence of associated skeletal abnormalities (Coumoul & Deng, 2003). Activating mutations in FGFR1 may also result in severe dwarfism in humans (White et al., 2005). On the other hand, FGFR3 negatively regulates longitudinal bone growth during postnatal skeletal growth, inhibiting chondrocyte proliferation and decreasing chondrocyte hypertrophy, mainly acting on PCs, to inhibit their differentiation (Colvin, Bohne, Harding, McEwen, & Ornitz, 1996). Evidence suggests that these effects are regulated by different branches of the FGFR3 signaling pathway, although these branches may not be completely distinct. STAT1 (signal transducers and activators of transcription 1) branch inhibits chondrocyte proliferation, whereas MAPK branch suppresses terminal chondrocyte differentiation (Fig. 4) and matrix production (Murakami et al., 2004; Yasoda et al., 2004). Accordingly, FGFR3-activating mutations cause achondroplasia dwarfism syndromes (Wilkie & Morriss-Kay, 2001), whereas FGR3-deficient mice ($FGR3^{-/-}$) present overgrowth of long bones (Colvin et al., 1996). Interestingly, achondroplasia due to FGFR3 mutations may be accompanied by partial premature fusion of the coronal sutures (partial craniosynostosis), suggesting that FGFR3 has a stimulatory role in osteoblasts and in IO (Di Rocco et al., 2014; Twigg et al., 2009). Interestingly, a mouse model with chondrocyte-specific activation of FGFR3 shows premature closure of several synchondroses (cranial base, vertebrae, pelvis, and sternum) due to enhanced osteoblast differentiation and bone formation in areas surrounding the synchondroses. Evidence suggests that FGFR3, via MAPK activation, inhibits bone morphogenetic protein (BMP) inhibitors and increased BMP expression, which, in turn, activates osteoblastogenesis and osteoblastic activity in a paracrine fashion (Matsushita et al., 2009). Others have also shown that FGFR3 mutation in chondrocytes, but not in osteoblasts, disturbs bone remodeling by a paracrine mechanism, leading to anomalies of primary spongiosa, associated with increased osteoclast recruitment and defective osteoblast-mediated mineralization, resulting in decreased trabecular bone mass, indicating that chondrocyte FGFR3 signaling regulates trabecular bone development in a paracrine way (Mugniery et al., 2012).

T3 was shown to stimulate the mRNA expression of FGFR1 (6.9-fold), FGFR2 (6.1-fold), and FGFR3 (17.5-fold) in ATDC5 chondrogenic cells, in addition to decrease cell number and induce cellular differentiation (Barnard et al., 2005). In addition, FGFR3 mRNA expression was shown to be markedly reduced in the hypothyroid growth plates of TRα-null mice ($TR\alpha^{0/0}$) and increased in the thyrotoxic growth plates of $TR\beta^{pv/pv}$ mice

(Barnard et al., 2005). FGFR1 mRNA was also shown to be increased in growth plate chondrocytes of TRβ$^{pv/pv}$ mice, where its expression extended through the growth plate and was no longer restricted to the PHZ and HZs (O'Shea et al., 2003). In contrast, growth plates of TRα1$^{pv/+}$ mice, which present a hypothyroid phenotype, displayed low FGFR1 mRNA expression (O'Shea et al., 2005). In addition (Fig. 4), T3 enhanced the FGF2 and FGF18 activation of MAPK signaling pathway and inhibited the FGF2 and FGF18 activation of STAT1 signaling pathway in ATDC5 cells (Barnard et al., 2005), which were shown to prevent chondrocyte hypertrophy and chondrocyte proliferation (Murakami et al., 2004; Yasoda et al., 2004), respectively. Altogether, these studies clearly show that TH modulates FGF/FGFR signaling pathways in chondrocytes, exerting differential effects on FGFR downstream responses to FGF. However, at this point, it is difficult to conciliate this modulation with the effects of TH in the growth plate and in EO, since TH modulates FGFR3 signaling pathways in ATDC5 cells in a way that would have opposite effects than the known effects of T3 in the growth plate and in chondrocytes *in vitro* (Fig. 4). In addition, the still undefined functions of FGFR1 in chondrocytes add difficulty to the understanding of the consequences of TH–FGF/FGFR interactions to skeletal development. On the other hand, it is likely that, in the growth plate, TH modulates the FGF/FGFR3 signaling pathways in a different way than that observed in isolated ATDC5 cells (Barnard et al., 2005). One possibility is that TH activates FGF/FGFR3 signaling pathways that reduce chondrocyte proliferation and inhibits pathways that inhibit chondrocyte differentiation, facilitating its own action on chondrocyte terminal differentiation. Alternatively, it is also possible that TH activates FGF/FGFR signaling pathways to limit, balance, or terminate its own activity in the growth plate. Further studies are necessary to investigate these hypotheses.

In rat and murine osteoblast-like cells, FGFR2 and FGFR3 mRNAs were expressed at low levels and were not responsive to T3. On the other hand, T3 was shown to induce FGFR1 mRNA and protein expression and to increase the effect of FGF2 on the MAPK signaling pathway in these osteoblastic cells (Stevens et al., 2003). In addition, further *in vitro* studies proved that the FGF2-stimulated MAPK signaling and its enhancement by T3 were mediated by FGFR1 (Fig. 4). Interestingly, these effects were abrogated in primary cultures of TRα-null osteoblasts (Stevens et al., 2003). *In vivo*, FGFR1 mRNA expression was reduced in osteoblasts and osteocytes of TRα-null mice (TRα$^{0/0}$) (Stevens et al., 2003) and in osteoblasts of TRα1$^{pv/+}$ mice (O'Shea et al., 2005), both of which exhibit skeletal hypothyroidism. In contrast, increased FGFR1 mRNA expression was detected in osteoblasts of TRβ$^{pv/pv}$ mice, which present a dominant TRβ-negative mutation and skeletal thyrotoxicosis (O'Shea et al., 2003). Considering that FGF signaling induces osteoblast differentiation and activity and has a major role in regulating IO in the skull (Ornitz & Marie, 2002), these findings suggest that FGFR1 signaling mediates TH actions that regulate bone formation and IO. Thus, an interplay between TH and the FGF/FGFR signaling pathway, in a TRα-dependent mechanism, is probably involved in the thyrotoxicosis-induced craniosynostosis.

Interestingly, protein expression of HSPG (which contains HS attached to a core protein) was shown to be upregulated in hypothyroid growth plates (Bassett, Swinhoe, et al., 2006; Stevens et al., 2000), whereas mRNA expression of HSPG core proteins was shown to be downregulated by TH treatment in ATDC5 cells (Bassett, Swinhoe, et al., 2006). Considering that HS acts as an FGF coreceptor, activating FGF–FGFR functional interactions (Coumoul & Deng, 2003), it is possible that one mechanism through which TH regulates bone development is by modulating cartilage matrix composition, specially HSPG expression, which, in turn, regulates the FGF/FGFR signaling pathways.

In conclusion, these studies suggest that FGF/FGFR1 signaling pathways are mediators of the T3-stimulatory actions on bone formation and IO. A series of studies also suggest that TH interacts with FGF/FGFR in chondrocytes to control skeletal development; however, at this point, it is not clear which effects of TH on growth plate or EO are mediated by FGF–FGFR signaling pathways.

6.4 The Role of the Wnt/β-Catenin Signaling Pathway

Wnt is a family of secreted glycoproteins which signaling pathways have an important role in the development and maintenance of various organs and tissues, including bone (Cadigan & Nusse, 1997). Wnt proteins signal through various pathways; however, the Wnt/β-catenin pathway (Fig. 5), also known as canonical, seems to be the most important for bone biology (Krishnan et al., 2006). Activation of this pathway occurs through binding of Wnt proteins to the Wnt coreceptors seven-transmembrane domain-spanning frizzled receptors (Frz) and low-density lipoprotein receptor-related protein 5 and 6 (LRP5/6). This binding activates a mechanism, involving Axin, Frat-1, and disheveled (Dsh) proteins, that inhibits glycogen synthase kinase 3 (GSK3) activity, an enzyme that binds and phosphorylates

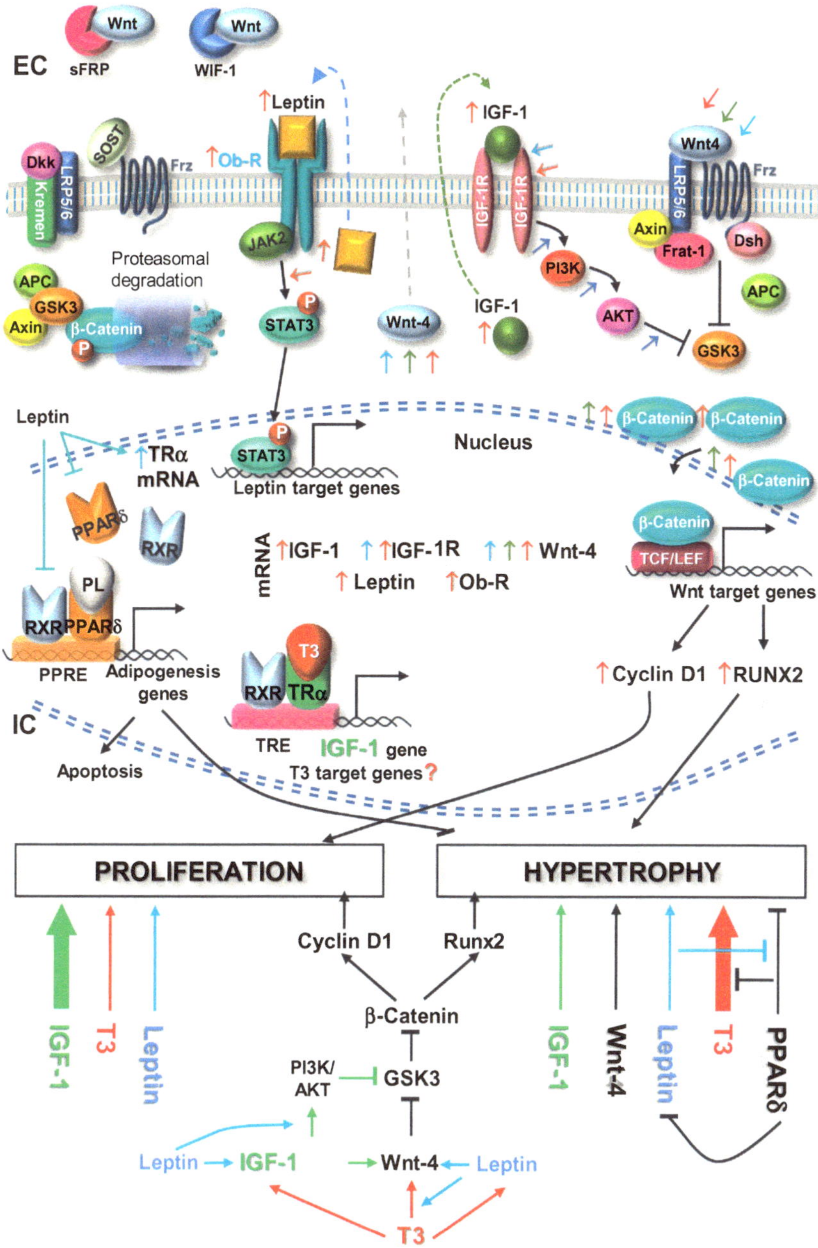

Fig. 5 Schematic representation of T3 effects on the Wnt/β-catenin signaling pathway in growth plate chondrocytes. T3 increases Wnt-4 mRNA and protein expression, the cellular accumulation of β-catenin, the TCF/LEF transcriptional activity, and the expression of the Wnt target genes Runx2 (a key factor for chondrocyte hypertrophy) and

(Continued)

β-catenin, tagging it for degradation by the ubiquitin/proteasome pathway (Hay et al., 2005). Therefore, the interaction of Wnt proteins with its coreceptors results in β-catenin stabilization and accumulation, which translocates to the nucleus, where it interacts with T-cell factor/lymphoid enhancer binding factor (TCF/LEF) transcription factors and activates the transcription of Wnt-responsive genes (Fig. 5; Levy et al., 2004). If Wnt proteins are not expressed or if the Wnt-LRP5/6-Frz interaction is inhibited, the degradation of β-catenin is facilitated through its interaction with a complex of proteins composed by GSK3, endometrial adenomatous polyposis coli (APC), and axina (Fig. 5). Wnt signaling is also regulated by antagonists, such as secreted frizzled-related protein (sFRP) family and Wnt inhibitory factor 1 (WIF-1), which bind Wnt and impair its interaction with Wnt coreceptors (Fig. 5). Wnt signaling is also inhibited by members of the sclerostin (SOST gene product) and Dickkopf (Dkk) families, which bind Frz and LRP5/6, respectively, impairing LRP5/6-Frz coreceptor activation (Krishnan et al., 2006). Binding of the Dkk/LRP complex to Kremen, a transmembrane protein, causes internalization of the Dkk/LRP complex

Fig. 5—Cont'd cyclin D (a proliferation marker); and prominently induces chondrocyte hypertrophy and modestly increases proliferation. Wnt antagonists sFRP3 or Dkk1 inhibit the T3 effects. IGF-1/IGF-1R also stimulates Wnt-4 expression and β-catenin activation and has prominent effects on chondrocyte proliferation and modest effects on terminal chondrocyte differentiation, which are partially inhibited by sFRP3 and Dkk1. Effects of IGF-1/IGF-1R are mediated by the PI3K/Akt/GSK3 pathway, which inhibition prevents the T3-induced increase in cyclin D1 mRNA expression, and in chondrocyte proliferation and differentiation, but not in chondrocyte terminal differentiation. Leptin increases Wnt-4 expression, activates the Wnt/β-catenin signaling, and promotes cell proliferation and terminal differentiation. Additionally, leptin increases TRα and IGF-1R expression, activates the IGF-1/IGF-1R-mediated PI3K/Akt signaling pathway, and enhances the T3-mediated increases in Wnt-4 mRNA expression and TCF/LEF transcriptional activity. TH increases both leptin ligand and receptor expression (Ob-R), and the leptin-mediated phosphorylation and activation of STAT3. PPAR heterodimers with RXR bind to PPREs in the targets genes. The interaction of PPAR with ciglitazone, a PPAR ligand (PL), activates apoptosis and inhibits the T3-induced chondrocyte hypertrophy, partially by RXR-binding competition. Leptin mildly downregulates PPARδ mRNA expression and attenuates the suppressive effects of PPARγ on chondrogenic differentiation and on T3-mediated chondrocyte hypertrophy. PPARγ negatively modulates leptin signaling in growth plate cells. The scheme shows that TH controls chondrocyte hypertrophy and proliferation modulating the Wnt/β-catenin signaling pathway in a mechanism that involves IGF-1R and leptin. *Red*, *green*, and *blue arrows* indicate T3, IGF-1, and leptin actions, respectively. The Wnt/β-catenin signaling pathway scheme is based on data from Krishnan et al. (2006).

for degradation, reducing the number of Wnt coreceptors for signaling (Mao et al., 2002).

Activation of the Wnt/β-catenin pathway is known to have anabolic effects on bone tissue and, therefore, to increase bone mass through a number of mechanisms including renewal of stem cells, stimulation of preosteoblast replication, induction of osteoblastogenesis, and inhibition of osteoblast and osteocyte apoptosis (Krishnan et al., 2006). Recently, members of the Wnt family have been recognized as key regulators of proliferation and differentiation of chondrocytes during limb development. The inactivation of the Wnt/β-catenin pathway in chondrocytes of mice embryos produces a phenotype of dwarfism with decreased proliferation of chondrocytes, and a delay in HC differentiation and EO (Akiyama et al., 2004). The overexpression of Wnt-8c or β-catenin induces terminal differentiation of chicken chondrocytes (Dong, do Soung, Schwarz, O'Keefe, & Drissi, 2006) and positively regulates the expression of runt-related transcription factor 2 (Runx2 or Cbfa1, core-binding factor a1) (Dong et al., 2005), a well-known master transcription factor for osteoblast differentiation that has also been shown to play essential roles during chondrocyte hypertrophy (Long & Ornitz, 2013).

Interestingly, Wang et al. (2007) showed that TH treatment of rat growth plate chondrocytes maintained in 3D pellets upregulates both Wnt-4 mRNA and protein expression, increases cellular accumulation of stabilized β-catenin and TCF/LEF transcriptional activity, and stimulates the expression of Runx2/CBFA1 and cyclin D1, which are Wnt target genes (Fig. 5). Overexpression of either Wnt-4 or a stabilized form of β-catenin promotes growth plate chondrocyte terminal differentiation. Blocking Wnt ligand/receptor interactions with the secreted Wnt antagonists Frzb/sFRP3 or Dkk1 inhibits these T3-induced increases in β-catenin accumulation and Runx2 gene expression and inhibits the maturation-promoting effects of T3 in growth plate cells. These data suggest that TH regulates terminal differentiation of growth plate chondrocytes in part modulating the canonical Wnt/β-catenin pathway (Wang et al., 2007). Using the same 3D pellet system, it was shown that TH modulates Wnt signaling pathways through the induction of carboxypeptidase Z (CPZ), an enzyme that removes carboxyl-terminal basic amino acid residues from proteins (Wang et al., 2009). CPZ contains a cysteine-rich domain (CRD) like the CRD found in the Frz family of Wnt receptors. Interestingly, Wnt-4 protein contains a C-terminal arginine residue and binds to CPZ through the CRD. Knockdown of either Wnt-4 or CPZ mRNA levels using an RNA interference technique or blocking CPZ enzymatic activity with a carboxypeptidase inhibitor reduces TH effect on both ALP activity and Col X mRNA expression. On the other hand, adenoviral overexpression of CPZ activates Wnt/β-catenin signaling and promotes terminal differentiation of growth plate cells. These findings show that TH may regulate terminal differentiation of growth plate chondrocytes in part by modulating Wnt signaling pathways through the induction of CPZ and subsequent CPZ-enhanced activation of Wnt-4 (Wang et al., 2009). It was also demonstrated that IGF-1/IGF-1R stimulates Wnt-4 expression and β-catenin activation in growth plate chondrocytes (Fig. 5; Wang et al., 2010). In addition, it was shown that IGF-1/IGF-1R has prominent effects on chondrocyte proliferation and modest effects on terminal chondrocyte differentiation, both of which were shown to be partially inhibited by the Wnt antagonists sFRP3 and Dkk1 (Wang et al., 2010). Evidence indicates that the positive effects of IGF-1/IGF-1R on chondrocyte proliferation and differentiation are mediated by the PI3K/Akt pathway, which inhibits GSK3 activity (Fig. 5; Ciarmatori, Kiepe, Haarmann, Huegel, & Tonshoff, 2007; Kiepe, Ciarmatori, Hoeflich, Wolf, & Tonshoff, 2005; Phornphutkul, Wu, Yang, Chen, & Gruppuso, 2004). Inhibition of the PI3K/Akt/GSK3 signaling pathway prevented the T3-induced increase in cyclin D1 mRNA expression (a cell proliferation factor) and the T3-induced chondrocyte differentiation, but was unable to suppress terminal differentiation in chondrocytes already committed to hypertrophy, suggesting that the PI3K/Akt/GSK3 signaling pathway mainly mediates the effect of IGF-1/IGF-1R on chondrocyte proliferation, whereas the IGF-1-induced increase in Wnt-4 (known to promote chondrocyte differentiation), which occurs through PI3K/Akt/GSK3-depedendet and independent mechanisms, may account for its effect on chondrocyte differentiation (Wang et al., 2010). Nevertheless, in this study, T3 positively regulated proliferation in addition to induce Wnt-4 expression, β-catenin activation, and terminal differentiation of growth plate chondrocytes, but these effects were partially prevented by the IGF-1R inhibitor picropodophyllin, as well as by PI3K/Akt signaling inhibitors (Wang et al., 2010). These data indicate that the interactions between TH and Wnt/β-catenin signaling pathways in regulating growth plate chondrocyte proliferation and terminal differentiation involve T3 actions on IGF-1/IGF-1R pathways, which includes both PI3K/Akt/GSK3 and Wnt-4 signaling pathways (Wang et al., 2010). Leptin, a polypeptide hormone mainly produced by adipocytes that primarily regulates food intake and energy balance, was shown to have direct effects on skeletal bone growth and maturation (Gat-Yablonski & Phillip, 2008; Houseknecht, Baile, Matteri, & Spurlock, 1998). On the other hand, chondrocytes were shown to both synthesize and secrete leptin, and express leptin long-form receptors (Ob-Rb) (Gat-Yablonski & Phillip, 2008; Kishida et al., 2005). Leptin signaling in chondrocytes occurs typically through the Janus kinases (JAK)/STAT pathway (Gat-Yablonski & Phillip, 2008). Leptin/Ob-R interaction activates JAK2, leading to tyrosine phosphorylation of STAT3, which then translocates to the nucleus and acts as a transcription factor by binding to response elements in the promoter of leptin target genes (Fig. 5; Fruhbeck, 2006). Interestingly, leptin was shown to increase IGF-1R expression in growth plate cells (Maor, Rochwerger, Segev, & Phillip, 2002; Wang, Shai, et al., 2011). In 3D pellet cultures of growth plate chondrocytes, leptin was also shown to activate the IGF-1/IGF-1R-mediated PI3K/Akt signaling pathway, to increase Wnt-4 expression, and to activate Wnt/β-catenin signaling (Wang, Shai, et al., 2011). Besides, the T3-mediated increases in Wnt-4 mRNA expression and TCF/LEF transcriptional activity were enhanced by cotreatment with leptin (Wang, Shai, et al., 2011). The *in vitro* studies using the 3D pellet system confirmed that leptin promotes cell proliferation and terminal differentiation in pellet cultures of rat growth plate chondrocytes, either alone or in the presence of TH. The effects of leptin alone on terminal chondrocyte differentiation were shown to be relatively limited when compared to the effects of TH treatment (Fig. 5), but comparable to the effects of IGF-1/IGF-1R or Wnt-4 treatment (Wang et al., 2007, 2010). Interestingly, the T3-induced increases in ALP activity, Col X expression, and chondrocyte proliferation in the 3D pellet culture system were shown to be enhanced by coaddition of leptin. Leptin was also shown to upregulate TRα expression locally in growth plate chondrocytes, resulting in increased TR transcriptional activity (Fig. 5). On the other hand, TH increases both leptin ligand and receptor expression, and the leptin-mediated phosphorylation and activation of STAT3 in the growth plate chondrocytes (Wang, Shai, et al., 2011). These findings suggest that leptin modulates TH signaling through Wnt/β-catenin signaling and that leptin synergically functions with TH in promoting growth plate chondrocyte proliferation and terminal differentiation.

In conclusion, these studies suggest that TH regulates bone elongation and maturation partially through the modulation of the canonical Wnt/β-catenin signaling pathway. This modulation occurs locally in chondrocytes and involves an interplay of T3 with the IGF-1/IGF-1R signaling pathways and a synergistic interaction between TH and leptin. Considering that T3 effects on chondrocyte hypertrophy are much more prominent than those of IGF-1/IGF-1R, leptin, and Wnt-4, other signaling pathways are certainly involved in the mechanisms of action of TH on chondrocyte terminal differentiation.

6.5 The Role of the PPARδ Signaling Pathway

Like other members of the nuclear hormone receptor superfamily, which includes TRs, peroxisome proliferator-activated receptors (PPARs) form heterodimers with RXR and bind to specific peroxisome proliferator response elements (PPREs), in the promoter regions of targets genes to regulate transcriptional activity (Willson, Brown, Sternbach, & Henke, 2000). PPARγ, a critical transcription factor for adipocyte differentiation, highly expressed in white adipose tissue and immune system (Wang et al., 2014), has also been shown to be expressed in preosteoblastic cells (Lotz et al., 1995), rat articular cartilage (Bordji et al., 2000), and growth plate chondrocytes (Shao, Wang, Hicks, Tarr, & Ballock, 2005). PPARγ regulates lineage commitment of mesenchymal cells toward osteoblasts and adipocytes. Activation of PPARγ suppresses osteoblastogenesis and induces adipocyte differentiation, resulting in decreased bone formation and accumulation of adipocytes in the skeleton (Lazarenko et al., 2007). Interestingly, the interaction of PPARγ with ciglitazone, one of its ligands, activates a pathological form of apoptosis and inhibits the TH-induced terminal differentiation of rat growth plate chondrocytes (Fig. 5; Wang, Shao, & Ballock, 2005, 2006). It is believed that this inhibition occurs in part by the competition of PPARγ and TR by RXR binding (Wang et al., 2005). Thus, PPARγ is considered a negative regulator of TH-induced gene transcription and, consequently, a negative regulator of skeletal maturation. It is possible that obesity may induce the expression of PPAR isoforms in growth plate chondrocytes, which then interfere with normal TH-mediated regulation of skeletal maturation (Wang et al., 2006). This delay in maturation at the growth plate, combined with the mechanical stress resulting from increased body weight, may combine to cause the slipped capital femoral epiphysis, an obesity-related hip disease that also has hypothyroidism as a risk factor (Heyerman & Weiner, 1984; Loder, Wittenberg, & DeSilva, 1995). Interestingly, Wang, Shao, and Ballock (2012) showed that leptin leads to a mild downregulation of PPARδ mRNA expression and a significant MAPK/ERK-dependent PPARγ phosphorylation, which decreases PPARδ transcriptional activity (Fig. 5). Therefore, leptin antagonizes PPARγ function in growth plate chondrocytes mainly through the MAPK/ERK signaling pathway. In fact, leptin was shown to attenuate the suppressive effects of PPARγ on chondrogenic differentiation and on T3-mediated chondrocyte hypertrophy. On the other hand, leptin signaling in growth plate cells is also negatively modulated by activation of PPARγ. Therefore, these two signaling pathways are mutually regulated in growth plate chondrocytes (Wang et al., 2012).

In conclusion, PPARγ is a negative regulator of TH-induced skeletal development. Leptin antagonizes PPARγ function in growth plate and vice versa. Therefore, an unbalance between these two pathways may interfere with the effects of TH on skeletal development.

6.6 The Role of the BMP Signaling Pathway

BMPs constitute a large family of molecules, within the transforming growth factor β (TGFβ) superfamily, that have important roles in the biology of bone and cartilage tissues, including an important role as regulators of late differentiation stages of EO (Murray, Brochmann Murray, Wang, & Duarte, 2016). The chondrocyte-specific overexpression of Noggin, under control of α2(XI) collagen gene promoter/enhancer sequences, resulted in transgenic mice (Col11 α2-Nog) that lacked most of their cartilaginous components. Since Noggin is a factor that binds BMP and prevents its interaction with BMP receptors, the phenotype of Col11 α2-Nog mice indicates that BMP signaling is essential for cartilage development. In contrast, the overexpression of BMP4 in cartilage chondrocytes, under control of the same cartilage-specific promoter/enhancer sequences derived from Col11 α2 (Col11 α2-BMP-4 mice), increased cartilage production and HC differentiation (Tsumaki et al., 2002), confirming the concept that BMPs induce the terminal differentiation of growth plate chondrocytes (Grimsrud et al., 1999; Leboy, Sullivan, Nooreyazdan, & Venezian, 1997; Suzuki, 1992). Lassova, Niu, Golden, Cohen, and Adams (2009) showed, in cultured chondrocytes, that T3 rapidly stimulates production of BMP4, with a concomitant decrease in Noggin, suggesting that TH potentially increases the BMP signaling. On the other hand, inhibition of BMP signaling with exogenous Noggin prevents T3 stimulation of Col X expression (Lassova et al., 2009). Further *in vitro* studies, using chondrogenic differentiating MSCs, showed that T3 had a pronounced effect on chondrocyte differentiation and on

BMP4 expression (Karl et al., 2014). In contrast, Noggin treatment inhibited the T3-induced increase in ALP activity and Col X gene expression and prevented the T3-mediated increase in number and size of HCs (Karl et al., 2014).

In conclusion, these *in vivo* and *in vitro* findings suggest that the BMP signaling pathway mediates, at least partially, the TH induction of HC differentiation.

6.7 The Role of the ECM

There is evidence that TH is essential for normal ECM deposition in the growth plate. As discussed in Sections 5.5 and 6.3, TH regulates the synthesis and degradation of HSPGs in the growth plate of rodents (Bassett, Swinhoe, et al., 2006; Stevens et al., 2000). The chondrocyte pericellular matrix degradation, a process necessary for chondrocyte hypertrophy, seems to be mediated by T3 induction of aggrecanase-2 and ADAMTS-5, two proteoglycan-degrading enzymes (Makihira et al., 2003). In addition, TH has also been shown to downregulate protein expression of HSPG in the growth plate of rodents *in vivo* (Bassett, Swinhoe, et al., 2006; Stevens et al., 2000) (as discussed in Section 5.5 and 6.3).

HSPGs are cell surface and ECM proteins that contain HS covalently attached to a protein core. HSPGs include syndecans, which are cell surface transmembrane proteins; glypicans, which cell surface glycerophosphatidylinositide-anchored proteins; and perlecan and agrin, which are diffusible protein components of the ECM (Ornitz & Itoh, 2015). Cell surface and extracellular matrix HSPGs interact with FGFs, acting as low-affinity coreceptors to enhance FGF/FGFR binding and signaling (Ornitz & Itoh, 2015; Schlessinger et al., 2000), in addition to modulate diffusion and degradation of FGFs in the growth plate (Jeong & McMahon, 2002; Schlessinger et al., 2000). HSPGs are also essential for normal Ihh growth plate diffusion (Jeong & McMahon, 2002), an essential process in the PTHrP/Ihh loop (Vortkamp et al., 1996). Interestingly, it was shown, in ATDC5 cell cultures, that T3 negatively regulates the mRNA expression of the HSPG core proteins glypican 6 (GPc6), glycosyltransferases exostosin (Ext1), and Hs6sT2 (Bassett, Swinhoe, et al., 2006). GPc6 has a critical role in development morphogenesis and has been associated with Ihh diffusion (Filmus & Selleck, 2001; Han, Belenkaya, Wang, & Lin, 2004). Ex1 is responsible for HS synthesis (Busse et al., 2007) and has been associated with chondrocyte columnar organization, chondrocyte differentiation (Sgariglia et al., 2013), and distribution of Ihh within the long-bone growth plates (Koziel, Kunath, Kelly, & Vortkamp, 2004). On the other hand, Hs6sT2 has been associated with the generation of FGF-binding sites within the growth plate matrix, which implies in the modulation of FGF sequestration, diffusion, and activity (Bassett, Swinhoe, et al., 2006; Habuchi et al., 2003, 2000).

In conclusion, these findings raise the hypothesis that one mechanism through which TH controls bone growth and maturation is modulating paracrine signaling pathways in the growth plate by modifying its cartilage matrix composition.

7. THE SKELETAL PHENOTYPE AND DEVELOPMENT IN TH-PHYSIOLOGY-LINKED-MUTANT MICE

Several transgenic animal models involving TH signaling-related genes have been powerful tools to reveal the roles of these genes in TH physiology and in TH action in several tissues, including actions on growth, maturation, and physiology.

7.1 TR Mutant Mice

Mice with single-gene inactivation of TRα1 (TRα1$^{-/-}$) do not present skeletal abnormalities (Wikstrom et al., 1998). On the other hand, TRα$^{-/-}$ mice, which do not express TRα1 and TRα2, present delayed skeletal growth and maturation (Fraichard et al., 1997). A very similar phenotype is observed in TRα$^{0/0}$ mice (Gauthier et al., 2001), which are devoid of all known isoforms produced from the THRA gene (TRα1, TRα2, TRΔα1, and TRΔα2). TRα$^{0/0}$ mice present growth retardation, delayed ossification, impaired chondrocyte differentiation, and reduced bone mineralization in juveniles, but increased trabecular bone mass in adults (Gauthier et al., 2001). These findings suggest that TRα2 has a modulatory role during bone development. Interestingly, mice with single-gene disruption of TRα2 (TRα2$^{-/-}$) overexpress TRα1 and present low levels of free T3 and T4, in addition to low serum levels of IGF-1 (Salto et al., 2001). Accordingly, these animals exhibit signs of hypothyroidism, such as retarded growth spurt (with no significant reduction in longitudinal bone growth), increased fat content, decreased cortical bone dimensions, and decreased tra becular BMD. TSH serum levels, however, showed to be oddly normal, suggesting an alteration in the HPT axis. In addition, when compared with WT mice, these mutants have a higher heart rate and body temperature, parameters usually associated with thyrotoxicosis. The results indicate that the loss of TRα2, the overexpression of TRα1, and/or the altered balance of TRα1/TRα2 lead to a mixed phenotype of hyper- and hypothyroidism. In addition, they suggest a role for TRα2 fine tuning the widespread mammalian physiology and bone development (Salto et al., 2001).

Mice with double deletion of TRα and TRβ, $TR\alpha1^{-/-}TR\beta^{-/-}$ (Gothe et al., 1999), $TR\alpha^{-/-}TR\beta^{-/-}$ (Gauthier et al., 1999), and $TR\alpha^{0/0}TR\beta^{-/-}$ (Flamant et al., 2002) basically present the same bone development abnormalities as $TR\alpha^{-/-}$ (Fraichard et al., 1997) and $TR\alpha^{0/0}$ mice (Gauthier et al., 2001), which resemble those arising in congenital hypothyroidism, but slightly more pronounced than in $TR\alpha^{0/0}$ mice (Flamant et al., 2002) and less pronounced than in hypothyroidism. $TR\alpha1^{-/-}TR\beta^{-/-}$ mice show significant decreases in bone lengths (femur, tibia, and the sixth lumbar vertebra), disorganized growth plates, and delayed ossification of the epiphysis, with decreased bone area but normal BMD (Gothe et al., 1999). Since $TR\alpha1^{-/-}TR\beta^{-/-}$ mice present marked decreases in GH mRNA and protein levels in the pituitary and decreased serum levels of IGF-1 (Gothe et al., 1999), whereas pituitary mRNA expression of GH is unaltered in $TR\alpha1^{-/-}$ or $TR\beta^{-/-}$ mice (Gothe et al., 1999), the deficiency of GH and/or IGF-1 was considered as a possible cause of the bone development delay observed in $TR\alpha1^{-/-}TR\beta^{-/-}$ mice (Gothe et al., 1999). However, intrinsic functions of TRs in bone could not be discarded, since these receptors are expressed in skeletal cells and some bone defects observed in $TR\alpha1^{-/-}TR\beta^{-/-}$ mice could not be solely explained by GH/IGF-1 deficiency, such as impaired chondrocyte differentiation, which is mainly regulated by TH (1993; Ballock & Reddi, 1994; Lewinson et al., 1989; Robson et al., 2000; Stevens et al., 2000). Unlike $TR\alpha1^{-/-}TR\beta^{-/-}$ (Gothe et al., 1999) and $TR\alpha^{-/-}TR\beta^{-/-}$ (Gauthier et al., 1999) that overproduce THs, $TR\alpha^{-/-}$ mice present progressive hypothyroidism (Fraichard et al., 1997). Considering these observations, it was initially thought that the growth retardation observed in $TR\alpha^{-/-}$ animals reflected this hypothyroidism. Later, the skeletal development of 2-week-old $TR\alpha^{-/-}$ and $TR\alpha^{-/-}TR\beta^{-/-}$ mice was compared (Gauthier et al., 1999). At this age, $TR\alpha^{-/-}$ and WT animals have similar TH plasma concentrations, avoiding any interference of hypothyroidism with the phenotype. Similar skeletal development defects were observed in $TR\alpha^{-/-}$ and $TR\alpha^{-/-}TR\beta^{-/-}$ mice (Gauthier et al., 1999), suggesting a "functional hypothyroidism," caused by the absence of TRs. The importance of TRs in mediating TH actions directly on bone is supported by the phenotype of $TR\alpha^{0/0}$ mice. As discussed before, these animals are biochemically euthyroid and show normal pituitary levels of GH, but present important defects in skeletal development (Gauthier et al., 2001). Altogether, these studies suggest direct effects of TH in the skeleton, mediated by TRs, and independent of the GH/IGF-1 axis.

In contrast to TR$\alpha^{0/0}$ mice, TH and TSH are both elevated in TR$\beta^{-/-}$ mice, due to disruption of the HPT axis (Forrest et al., 1996). Initially, evaluation of these mice did not detect skeletal alterations (Gauthier et al., 1999). Later, a more detailed analysis revealed that these mutant mice have advanced ossification with persistent short stature, increased bone mineralization, and accelerated growth plate chondrocyte differentiation (Bassett, O'Shea, et al., 2007). Interestingly, the mRNA expression of FGFR1 and FGFR3, which are T3 target genes (Stevens et al., 2003), was shown to be increased in the growth plate chondrocytes of TR$\beta^{-/-}$ juvenile mice, but decreased in TR$\alpha^{0/0}$ animals. The same was seen for GHR and IGF-1R genes, which are responsive to TH in growth plate chondrocytes (Bassett, O'Shea, et al., 2007). Altogether, these findings show that TR$\beta^{-/-}$ mice, which lack TRβ but express TRα, have skeletal thyrotoxicosis, whereas TR$\alpha^{0/0}$ mice, which lack TRα but express TRβ, have skeletal hypothyroidism, suggesting that TRα mediates TH actions on skeletal development. Remarkably, the juvenile TR$\alpha^{0/0}$ phenotype of delayed ossification and reduced bone mineralization was associated with an adult phenotype of osteosclerosis, whereas advanced ossification and increased calcified bone in juvenile TR$\beta^{-/-}$ mice preceded an osteoporotic phenotype in adults. These data indicate that TH, acting mainly via TRα, promotes anabolic effects during postnatal bone growth but has catabolic actions in the adult skeleton. Accordingly, chronic treatment of female rats with a high doses of GC-1, a TRβ-selective agonist, does not result in the typical thyrotoxicosis-induced osteopenia, but reduces TSH serum levels as much as equimolar doses of T3 (Freitas et al., 2003), which is consistent with a predominant expression of TRβ in the pituitary and a major role of TRβ in mediating TH suppression of TSH (Abel et al., 1999). An important role of TRα in the skeleton is supported by a 2.3- to 12-fold higher mRNA levels of TRα than TRβ in bone (Bookout et al., 2006; O'Shea et al., 2003). It is noteworthy, however, that thyrotoxicosis, induced by T4 treatment, caused loss of trabecular and cortical bone in TR$\alpha^{0/0}$ mice, indicating that TRβ can partially mediate TH actions on bone, or at least partly compensate for the lack of TRα (Bassett, O'Shea, et al., 2007). Remarkably, Monfoulet et al. (2011) showed that TRα deficiency does not prevent thyrotoxic effects on bone volume, bone mineralization, bone formation, or bone resorption in adult mice. In contrast, TRβ deficiency blocked all the early osteopenic effects of TH excess observed in WT mice. Interestingly, prolonged exposure of $TR\beta^{-/-}$ mice to low or high TH concentrations induced mild modifications in bone structure and remodeling parameters, suggesting that TRβ mediates the acute effects produced by transient changes of TH concentrations on bone remodeling, whereas TRα mediates long-term effects of chronic alterations of TH metabolism. These data indicate that TRβ has a relevant role in bone remodeling in adult mice and suggest that it may also have a relevant role in the developing skeleton. This is further supported by normal body growth in $TR\alpha1^{-/-}$ mice; by the observation that $TR\alpha^{0/0}TR\beta^{-/-}$ mice have a more severe skeletal phenotype than $TR\alpha^{0/0}$ mice (Bassett & Williams, 2009; Flamant et al., 2002); and by the fact that T3 was unable to induce chondrocyte terminal differentiation in cultures of cells derived from ribs of $TR\beta^{-/-}$ mice, but was able to induce this process in cells derived from $TR\alpha^{0/0}$ mice (Rabier et al., 2006) (discussed in Section 5.3). Accordingly, GC-1, a TRβ-selective agonist (Fig. 1), promoted selective effects on the growth plate and partially induced EO in rats (Freitas et al., 2005) (discussed in Sections 5.4 and 5.5). In addition, GC-1 was able to inhibit proliferation and induce differentiation in mouse and rat osteoblastic cells *in vitro* (Beber et al., 2009) (discussed in Section 5.2).

A very interesting model to evaluate the roles of TR isoforms in skeletal development is Pax8 KO mice ($Pax8^{-/-}$). Pax8 encodes a paired-box-containing protein that is necessary for the differentiation of thyroid cells (Pasca di Magliano, Di Lauro, & Zannini, 2000). The only known primary defect in $Pax8^{-/-}$ mice is the absence of thyroid follicular cells, resulting in almost complete absence of T3 and T4 production and in a severe hypothyroidism in postnatal life, when autonomous hormone production normally replaces maternal supply. (Mansouri, Chowdhury, & Gruss, 1998). A systematic comparison between $Pax8^{-/-}$ mice and $TR\alpha^{0/0}TR\beta^{-/-}$ mice, which are devoid of all TR isoforms, revealed that the former have similar but more severe defects than the latter, including a drastic delay in body growth and ossification (Flamant et al., 2002). This is interpreted as the result of a negative effect of the unliganded TR isoforms on the TH target genes in $Pax8^{-/-}$ mice, named aporeceptor activity (Hu & Lazar, 2000). Interestingly, the combination of TRβ with Pax8 gene inactivation ($Pax8^{-/-}TR\beta^{-/-}$ mice) does not modify $Pax8^{-/-}$ phenotype. In contrast, double deletion of Pax8 and TRα ($Pax8^{-/-}TR\alpha^{0/0}$ mice) results in a significant and partial reversion of the bone development defects. It is predicted that genes activated by T3 are actively repressed in $Pax8^{-/-}$ skeleton predominantly by unliganded TRα (Flamant et al., 2002). These findings demonstrate that the TRα aporeceptor activity is more detrimental to the skeleton than the lack of TRα and reveals the relevance of TRα aporeceptor (unliganded state) activity to skeletal development.

Kaneshige et al. (2000) generated mice with a targeted mutation in the THRB gene, the PV mutation, derived from a patient with severe resistance to TH (RTHβ). This mutant receptor, named PV ($TR\beta^{PV}$), is one of the most potent dominant-negative TRβ isoforms. It does not bind T3, has no transcriptional activity, and interferes with the actions of functional TR isoforms, affecting the transcriptional control of T3-target genes (Zhu, Yu, McPhie, Wong, & Cheng, 1996). Such as RTHβ patients (discussed in Section 8.2), homozygous PV mice ($TR\beta^{PV/PV}$) exhibit severe dysfunction of the HPT axis, with elevated TSH (412-fold), T4 (15-fold), and T3 (9-fold) serum levels, followed by decreased pituitary GH mRNA expression, impaired body weight gain, reduced growth spurt, and shorter femurs and tibias. These skeletal defects were initially interpreted as delayed bone development, suggesting that TRβ would be required to mediate TH actions on skeletal development (Kaneshige et al., 2000). Later, a more detailed analysis revealed that $TR\beta^{PV/PV}$ mice display accelerated growth *in utero*, associated with advanced EO and IO (O'Shea et al., 2003). The advanced ossification resulted in reduced postnatal growth, premature quiescence of the growth plates, shortened bone length, and craniosynostosis, indicating a thyrotoxic skeletal phenotype in $TR\beta^{PV/PV}$ mice, which was further confirmed by increased mRNA expression of T3-target genes (GHR, IGF-1R, and FGFR1) in the growth plate (O'Shea et al., 2003). Therefore, the skeletal thyrotoxicosis in $TR\beta^{PV/PV}$ mice supports data that point out TRα as the predominant functional TR isoform in the developing skeleton of mice.

To further investigate this hypothesis, the skeletal phenotype of $TR\alpha1^{PV}$ mice, which carry the PV mutation in the THRA gene (Kaneshige et al., 2001), was evaluated (Kaneshige et al., 2001; O'Shea et al., 2005). $TR\alpha1^{PV}$ protein is also a potent dominant-negative antagonist able to interfere with the activities of functional TRα1 and TRβ. The homozygous mutation is lethal and heterozygous $TR\alpha1^{PV/+}$ mice show only mild thyroid failure, with normal T4, and slightly elevated T3 (1.2- to 1.5-fold) and TSH (1.7- to 6-fold) serum levels, displaying the reduced T4:T3 ratio identified in RTHα patients (Bassett et al., 2014; Kaneshige et al., 2001). In contrast to $TR\beta^{PV}$, $TR\alpha1^{PV/+}$ mice display skeletal hypothyroidism, characterized by dwarfism, severe delayed EO and IO, postnatal growth retardation, normal pituitary mRNA expression of GH, but decreased growth plate mRNA expression of GHR, IGF-1R, and FGFR1, which are T3-responsive genes (Kaneshige et al., 2001; O'Shea et al., 2005). A similar phenotype of delayed skeletal development was observed in mice with conditional expression of TRα1^{L400R}, another TRα mutant with potent dominant-negative activity, due to the L400R mutation in the THRA locus (Quignodon, Vincent, Winter, Samarut, & Flamant, 2007). TRα^{AMI}xS mice, which express TRα1^{L400R} from the beginning of development, present severe growth retardation and delayed ossification, with decreased mRNA expression of GH in the pituitary. Tamoxifen induction of TRα1^{L400R} expression at late fetal stage (E17.5) significantly affected postnatal skeletal development (TRα^{AMI}xC mice) as well, although the development defects were less severe than those observed in TRα^{AMI}xS mice (Quignodon et al., 2007). These findings strengthen the conclusion that TRα1 aporeceptor activity is detrimental to skeletal development. An interestingly observation was that the thyrotropes, hepatocytes, and retina cones, which highly express THRB, were barely affected by the dominant-negative activity of TRα1^{L400R} in TRα^{AMI}xS mice. In addition, these mice survive beyond weaning, whereas Pax8$^{-/-}$ mice usually die within 3 weeks after birth. These observations suggest that liganded TRβ can balance the down-regulation exerted by TRα1^{L400R}. Such compensation cannot occur in hypothyroid Pax8$^{-/-}$ mice, where unliganded TRβ might repress gene transcription and further impair postnatal development (Quignodon et al., 2007). The fact that the TRα deletion in Pax8$^{-/-}$ mice (Pax8$^{-/-}$TRα$^{0/0}$) results in a significant but partial reversion of the skeletal development defects raises the hypothesis that unliganded TRβ may account, to a lesser extent than TRα1, for the detrimental effects of hypothyroidism in skeletal development. If this kind of TRβ participation occurs in the skeleton of RTHα mouse models remains to be investigated.

TRα1$^{R384C/+}$ (or TRα1$^{+/m}$) mice harbor a milder heterozygous THRA mutation (R384C) than PV mutation, which allows T3 binding, albeit at a lower affinity (Tinnikov et al., 2002). These animals present transiently delayed ossification and growth retardation, with decreased pituitary GH mRNA (Tinnikov et al., 2002). Interestingly, additional deletion of THRB gene in TRα1$^{+/m}$ mice (TRα1$^{+/m}$β$^{-/-}$) causes a 10-fold increase in serum T4, which probably overcomes the reduced T3-binding affinity of TRα1$^{R384C/+}$, restoring GH mRNA expression and almost completely rescuing body weight gain (Bassett, Nordstrom, et al., 2007; Tinnikov et al.,

2002). Thus, the skeletal development delay observed in TRα1$^{+/m}$ mice may be partially explained by a local and/or systemic GH/IGF-1 axis deficiency. Interestingly, the skeletal development defects observed in TRα1$^{+/m}$β$^{+/-}$ mice, which present normal serum levels of TH, are overcome in adulthood (Tinnikov et al., 2002), suggesting that other factors compensate for the TRα defect (Tinnikov et al., 2002). These animals (TRα1$^{+/m}$β$^{+/-}$) present reduced bone resorption surfaces and osteoclast number, which are normalized in TRα1$^{+/m}$β$^{-/-}$ mice (that present 10-fold increase in serum T4) or after T3 treatment, suggesting that TRα1 controls bone mass by regulating osteoclast bone resorption. This is supported by an increased number of osteoclasts in TRα1$^{+/+}$β$^{-/-}$ mice (Bassett, Nordstrom, et al., 2007). Accordingly, transient T3 treatment of juvenile TRα1$^{+/+}$β$^{+/-}$ mice significantly increased adult endosteal and trabecular resorption areas in 53% and 18%, respectively. Unexpectedly, T3 treatment of TRα1$^{+/m}$β$^{+/-}$ resulted in drastic increases in adult endosteal and trabecular resorption surfaces, 300% and 71%, respectively. Consequently, transient T3 treatment of juveniles resulted in a 63% decrease in trabecular bone volume in TRα1$^{+/m}$β$^{+/-}$ mice and in an insignificant decrease (12%) in TRα1$^{+/+}$β$^{-/-}$ mice. Surprisingly again, T3 treatment of juveniles also resulted in a remarkable reduction of trabecular bone mineralization in adult TRα1$^{+/m}$β$^{+/-}$ mice, with smaller loss in TRα1$^{+/+}$β$^{+/-}$ mice (Bassett, Nordstrom, et al., 2007). These unpredictable findings show that the skeleton of TRα1$^{+/m}$β$^{+/-}$ mice is more sensitive to T3 treatment than the skeleton of TRα1$^{+/+}$β$^{+/-}$ mice during development, which strongly suggest that TRβ is responsible for this responsiveness to T3, mediating bone remodeling. These findings are also supported by the study of Monfoulet et al. (2011) that showed that the negative effects of TH excess on bone remodeling are conserved in TRα1$^{0/0}$ mice but abolished in TRβ$^{-/-}$ mice, indicating an important role of TRβ in mediating bone remodeling.

Treatment of TRα1$^{PV/+}$ mice with a supraphysiological dose of T4, between 4 and 20 weeks of age, had no beneficial effect on the shorter tibial length nor on the increased growth plate thickness of these animals, suggesting no effect on linear bone growth and EO (Bassett et al., 2014). Nevertheless, the long-term T4 supplementation in TRα1$^{PV/+}$ mice promoted, in their tibias, a gradual increase in cortical thickness and diameter, a paradoxical increase in bone mineral content, a further reduction in osteoclast surface and number (reduced in TRα1$^{PV/+}$ vs WT mice), and gradual increases in bone strength and stiffness (Bassett et al., 2014). These unexpected responses were probably mediated by TRβ. Finally, another dominant-negative TRα mutant mice (TRα1$^{P398H/+}$), due to a point mutation in THRA gene (P398H), do not manifest dwarfism. These TRα1$^{P398H/+}$ animals are normal in size for the first 2–3 months, and then the males show increased body weight, whereas female mice are still normal in size (Liu, Schultz, & Brent, 2003). These different phenotypes reinforce the hypothesis that TRβ may exert some compensation for the transcriptional downregulation exerted by TRα mutants. Thus, and surprisingly, these studies with mouse models of RTHα reveal sensitivity of their skeletons to TH, suggesting that TRβ may be responsible for some of these responses. On the other hand, it is possible that the doses of TH administered to these RTHα mouse models were high enough to heterogeneously overcome the reduced ligand-binding affinity and dominant-negative activity of the TRα mutants, depending on the cell type, tissue type, diverse TR isoform expression, and differential expression of coactivator and corepressors, promoting, therefore, unexpected responses.

In conclusion, this extensive series of elegant studies in genetically modified mice has shown that: (i) TH interacts with the GH/IGF-1 axis to regulate skeletal development but also has direct effects, mediated by TRs, and independent of the GH/IGF-1 axis; (ii) TH promotes direct anabolic effects on bone tissue during skeletal development; (iii) TRα is the major TR isoform to mediate TH actions on skeletal development in mice; (iv) the TRα aporeceptor activity is more detrimental to the skeleton than the lack of TRα, which reveals the relevance of TRα aporeceptor activity to skeletal development; and, finally, (v) TRβ partially mediates TH actions on skeletal development, or at least partially compensate for the lack of TRα.

7.2 TSHR Mutant Mice

Global TSHR-null (TSHR$^{-/-}$) mice present disorganized thyroid follicles, reduced serum levels of T3 and T4, high circulating TSH, reduced body growth, and high bone turnover osteoporosis with focal bone sclerosis (Abe et al., 2003). In addition, these animals require thyroid extracts treatment, to render them euthyroid, which allows survival and corrects the hypothyroid abnormalities, including reduced growth, but does not correct the decreased bone mass. In contrast, heterozygote (TSHR$^{-/+}$) mice present normal thyroid follicles, are euthyroid, and display typical growth pattern and normal circulating TH and TSH levels. Regardless of the normal thyroid function, the TSHR haploinsufficient mice display a generalized reduced BMD, indicating that the osteopenic skeleton was due to the

50% decrease in TSHR concentration, which suggests a direct and antiresorptive action of TSH on bone remodeling (Fig. 2; Abe et al., 2003). Therefore, these findings have challenged the classical interpretation that the osteopenic effects of thyrotoxicosis are caused by actions of supraphysiological T3 acting on TR isoforms. Accordingly, exogenous administration of low doses of TSH (that do not increase TH serum levels) to aged rats just after ovariectomy (OVX) prevented the resulting and typical progressive bone loss (Sampath et al., 2007). Likewise, intermittent administration of rhTSH promoted powerful antiresorptive actions in OVX mice, also avoiding bone loss (Sun et al., 2008). In addition, T4-induced thyrotoxicosis promoted more severe bone loss and resorption in $TSHR^{-/-}$ mice than in WT mice (Baliram et al., 2012), reinforcing that reduction or absence of TSH signaling contributes to bone loss.

In line with this hypothesis, *in vitro* osteoclastogenesis was shown to be decreased by rhTSH treatment; and the resorptive activity of osteoclasts, quantitated as the number of pits per slice of dentin (pit assay), was significantly increased in $TSHR^{-/-}$ mice-derived cells as compared with WT cells. Expectedly, rhTSH inhibited bone resorption (in pit assays) by WT osteoclasts, but not by $TSHR^{-/-}$ osteoclasts. This antiresorptive effect of TSH was mediated by low TNFα (Fig. 2), since its levels were higher in the supernatant of BM cells derived from $TSHR^{-/-}$ mice, and anti-TNFα neutralizing antibody added to the cell cultures blocked the increase in osteoclastogenesis (Abe et al., 2003). Suitably, the increased osteoclastogenesis was abrogate in BM cultures derived from compound $TSHR^{+/-}/TNF\alpha^{+/-}$ mice and even more in $TSHR^{-/-}/TNF\alpha^{-/-}$ mice (Hase et al., 2006). The TSH-mediated suppression of osteoclast formation was more profound in cells that overexpress TSHR; and the overexpression of a ligand-independent constitutively active TSHR (caTSHR) abolished osteoclast formation even in the absence of TSH. Altogether, these results established that TNFα is a critical cytokine mediating the antiresorptive effects of TSH on the skeleton (Hase et al., 2006).

Interestingly, the TSH administration to rats after 7 months of OVX increased BMD and both trabecular and cortical bone volume; and improved trabecular microarchitecture and mechanical strength (Sampath et al., 2007). Intermittent TSH administration was also able to promote bone mass gain in OVX mice (Sun et al., 2008), suggesting that TSH exerts anabolic effects in addition to antiresorptive effects on bone (Fig. 2). In fact, addition of TSH to osteoblast cultures increased production of ALP activity by rat osteosarcoma-derived UMR-106 cells, rat calvaria-derived primary osteoblasts, and MC3T3-E1 preosteoblasts; and increased OCN mRNA expression and mineral deposition in rat calvaria-derived osteoblast cultures (Sampath et al., 2007). Similar effects of TSH were seen in ES cells, which were found to express TSHR. TSH promoted osteoblast differentiation of ES cells and induced the expression of OPG (Baliram et al., 2011; Ma, Morshed, Latif, Zaidi, & Davies, 2011), contributing to the antiresorptive action of TSH. In addition, rhTSH induced the expression the coreceptor Frz and Wnt-5a, which both constitute the noncanonical arm of the Wnt signaling pathway (Baliram et al., 2011). These *in vitro* studies show direct and positive actions of TSH on osteoblast differentiation and suggests that the noncanonical Wnt signaling pathway mediates the TSH-induced osteoblastogenesis. In contrast, Abe et al. (2003) had shown that TSH decreased osteoblast differentiation of BM-derived MSCs by attenuating the expression of LRP-5, a Wnt coreceptor; and Flk-1 (kinase insert domain protein receptor), a VEGF receptor. This TSH-induced decrease of LRP-5 expression was also observed in ES cells (Baliram et al., 2011), but the contradictory effects of TSH on osteoblast differentiation indicate that this effect needs to be further clarified.

More recently, a novel and functional form of the TSH-β subunit, designated TSH-βv, was identified in BM cells (Vincent et al., 2009). Baliram et al. (2013) found that fresh BM-derived macrophages (MØs) preferentially produced TSH-βv and, when cocultured with CHO cells that overexpress TSHR, generated intracellular cAMP, a phenomenon not seen in control CHO cells, confirming the TSH variant bioactivity. Moreover, cocultures of MØs and osteoblasts induced osteoblastogenesis (Fig. 2), a phenomenon markedly reduced by TSH-β antibody, proposing direct interaction between MØs and osteoblasts, and suggesting a local and paracrine modulation of osteoblast differentiation by TSH-βv.

Contrasting all these findings, Bassett et al. (2008) did not find the effect of TSH on the differentiation or function of osteoblasts or osteoclasts *in vitro*. Besides, in their hands, treatment of primary osteoblasts and osteoclasts with TSH or a TSHR-stimulating antibody failed to induce a cAMP response. In addition, they compared the skeletal phenotype of two mouse models of congenital hypothyroidism with opposing conditions regarding TSH action, $Pax8^{-/-}$ mice and hyt/hyt mice. As previously described in this review, $Pax8^{-/-}$ mice have no follicular cells and, therefore, do not produce TH; consequently, these animals display a 1900-fold increase in serum TSH, with functional TSHR (Mansouri et al., 1998). Hyt/hyt mice harbor a TSHR loss-of-function mutation, resulting in nonfunctional TSHR, with subsequent thyroid hypoplasia, severe hypothyroidism, and a 2300 elevation in serum TSH (Gu et al., 1995). The study of Bassett et al. (2008) showed that both mouse models displayed very similar skeletal phenotypes, characterized by severe growth and ossification delay, decreased cortical bone mass, reduced trabecular bone remodeling, and lower bone mineralization, which was also observed in young adult animals. In addition, TH-replacement therapy leads to rapid and equal catch-up in both mouse models. Therefore, the very similar skeletal phenotypes and equal skeletal responses to TH in these animals, with opposing conditions regarding TSH actions, suggest that the skeletal abnormalities of congenital hypothyroidism are independent of TSH and that the HPT axis regulates skeletal development and bone turnover via T3 actions, but not via TSH. Accordingly, in adult rats, high doses of the TRβ-selective analog GC-1 decreased serum levels of TSH as much as equimolar doses of T3, but differently of T3, had no osteopenic effects (Freitas et al., 2003), suggesting a predominant role of TH over TSH on bone mass.

More recently, however, Endo and Kobayashi (2013), in a more detailed analysis, confirmed the lower body length in hyt/hyt mice during juvenile life, but observed that adult hyt/hyt mice display the same body length as WT mice, despite the fact that their serum T4 is very low through life. In addition, they compared the skeletal phenotypes of WT thyroidectomized (THYx) mice and hyt/hyt mice, which display hypothyroidism, similar serum levels of calcium and GH, but opposing situations regarding TSH actions. They found that hyt/hyt mice (without TSH/TSHR action) grow faster than THYx mice, and that hyt/hyt mice under TH reposition also grow faster than WT mice. They also observed shorter tibias and narrower PZ and HZ in the growth plates of THYx mice when compared with hyt/hyt mice. Moreover, they demonstrated that growth plate and various cartilage tissues of mouse express TSHR. *In vitro* studies with chondrocytes from femoral head cartilage demonstrated high-affinity TSH-binding activity and a TSH dose-dependent increase in cAMP production in these cells, supporting that TSH expressed in chondrocytes is functional. In addition, gene expression of Sox9 and Col II (which are chondrocyte differentiation markers) was 52% and 60% lower, respectively, in the growth plate of THYx mice than in hyt/hyt mice. Altogether, these findings suggest that TSH has direct and negative effects in growth plate chondrocytes, resulting in epiphyseal dysgenesis and delayed linear bone growth (Fig. 2). However, further studies are necessary to reveal the effects of TSH in growth plate chondrocytes.

In conclusion, these findings reveal a direct antiresorptive action of TSH in the bone tissue, inhibiting osteoclastogenesis and osteoclastic activity. In addition, they show stimulatory effects of TSH in osteoblasts. More recent data suggest a negative role of high serum TSH in skeletal development. Considering the predominant effects of TH on bone, TH, but not TSH, is the main responsible for the skeletal development defects observed in conditions of TH deficiency or excess. On the other hand, it is likely that TSH contributes to the pathogenesis of growth retardation, especially to the epiphyseal dysgenesis, observed in hypothyroidism.

7.3 Deiodinase Mutant Mice

Mice with an inherited D1 deficiency, C3H/HeJ (C3H) animals, and mice with general gene deletion of DIO1 ($D1^{-/-}$) display normal serum T3 and normal bone growth (Berry et al., 1993; Schneider et al., 2006), which is consistent with the lack of D1 activity in the skeleton (Gouveia et al., 2005; Williams et al., 2008). In addition, these findings reveal that D1 is not essential for life and normal body growth. D2-deficient mice ($D2^{-/-}$) show pituitary resistance to T4, a 2-fold increase in TSH, and a 1.4-fold increase in serum T4 with normal serum T3 (Galton et al., 2007). These animals present a moderate reduction in the thickness of all growth plate zones (RZ, PZ, and HZ), but show normal linear growth and EO (Bassett et al., 2010), and unaltered AC (Waung, Bassett, & Williams, 2015), which is also consistent with the lack of D2 activity in chondrocytes of mice (Williams et al., 2008). On the other hand, bones from adult $D2^{-/-}$ mice present increased mineralization, are brittle, and have reduced toughness, suggesting an increase in fracture risk (Bassett et al., 2010). It seems clear that this bone phenotype is not due to abnormal osteogenesis and growth, since bone development is normal in $D2^{-/-}$ mice. In addition, these animals display normal bone resorption, which is coherent with the lack of D2 activity in osteoclast (Williams et al., 2008), and normal osteoblast number, but reduced bone formation rate, suggesting reduced osteoblastic activity (Bassett et al., 2010). These findings indicate a restricted cellular hypothyroidism in osteoblasts of D2 KO mice and reveal the physiological importance of D2 activity for these cells. Growth was also normal in $C3H/D2^{-/-}$ mice (Christoffolete et al., 2007), and the double general deletion of D1 and D2 (D1D2 KO) has barely modified the phenotype of $D2^{-/-}$ mice, which reinforces the irrelevance of D1 to the skeleton (Bassett et al., 2010). Interestingly, transgenic rats overexpressing D2 in a chondrocyte-specific manner present normal growth and an apparent normal general physiology, but show advanced AC destruction in an OA model (Nagase et al., 2013). This is consistent with microarray analyses, showing that D2 is highly upregulated (15-fold) in AC of OA patients (Nagase et al., 2013), with increased amount of D2-positive cell in hip AC cartilage of OA patients (Bos et al., 2012), and with associations of common single nucleotide polymorphisms in D2 gene with OA (Bos, Slagboom, & Meulenbelt, 2008; Kerkhof et al., 2010). These findings suggest that D2 activity may be involved in the pathogenesis of OA pathogenesis, which requires further analyses to be confirmed. Mice lacking general D3 expression and activity ($D3^{-/-}$) show impaired clearance of T3 and markedly elevated serum T3 early in life. From P15 to adulthood, D3 KO mice display central and peripheral hypothyroidism, with low serum levels of T4 and T3, and a modest or no increase in serum TSH. In addition, hypothalamic T3 content is decreased, while TRH expression is elevated. Therefore, these animals show important regulatory defects at all levels of the HPT axis, resulting in perinatal thyrotoxicosis, followed by a moderate hypothyroidism in adulthood (Hernandez, Martinez, Fiering, Galton, & St Germain, 2006; Hernandez et al., 2007). These animals display partial perinatal mortality, impaired fertility, and a marked reduction in body weight and body length at weaning that persists in adulthood (Hernandez et al., 2007). Given the defects in the HPT axis followed by altered thyroid status in $D3^{-/-}$ mice, the importance of D3 to skeletal development cannot be inferred. Further evaluation of these animals and the development of tissue- or cell-specific D3 gene inactivation will be necessary to determine the role of D3 in the skeleton.

In conclusion, D1 and D2 activities are not necessary for skeletal development. D2 is important to maintain a local euthyroidism in osteoblasts, keeping its normal bone formation activity. D3 is relevant for skeletal development, but, at this point, it is unknown if the skeletal development defects observed in the absence of D3 expression occur because of the lack of D3 activity locally in the skeleton, or are consequence of the altered TH status and disrupted HPT axis.

7.4 TH Transporters Mutant Mice

The CNS development is only partially affected in MCT8-null (MCT8 KO) mice, but the circulating iodothyronines and TSH levels closely resemble those of humans with MCT8 mutations: low serum T4 and rT3, and high serum T3 and TSH (Liao et al., 2011; Trajkovic et al., 2007). There is evidence that D1 contributes to the increased serum T3 in MCT8 deficiency, whereas D2 mainly functions locally, converting T4 to T3 to compensate the distinct cellular TH depletion in MCT8 KO mice (Liao et al., 2011). Apparently, skeletal development of MCT8 KO mice is normal, since their body weight and body length curves are not different from those of WT animals (Liao et al., 2011; Trajkovic et al., 2007), suggesting that this transporter is not essential for skeletal development. Adding D1 deficiency to that of MCT8 (MCT8D1 KO) corrects serum TH abnormalities, normalizes brain T3 content, and leads to a minimal decrease in body length (1.9% and 2.3% vs WT) (Liao et al., 2011). The combination of MCT8 deficiency with that of D2 (MCT8D2 KO) maintained the serum TH abnormalities but increased serum TSH levels in 27-fold, suggesting an important impact on the HPT axis. Interestingly, MCT8D2 KO mice present a slightly greater reduction in body length (2.5%–4.3%) than MCT8D1 KO (Liao et al., 2011). The combination of MCT8, D1, and D2 deficiencies (MCT8D1D2 KO) markedly increased serum TH levels, promoted a further increase in TSH levels (100-fold vs WT and 6-fold vs MCT8D2 KO), produced milder brain defects, and led to an 8% decrease in body length at early age (28–42 days), with a slight catch-up growth thereafter, resulting in only 3%–4% lower length than WT mice. Body weight was significantly lower in MCT8D1 KO (11%–15%), MCT8D2 KO (16%–20%), and MCT8D1D2 KO (19%–25%) relative to WT (Liao et al., 2011). Since D1- and D2-deficient mice develop normally (discussed in Section 7.3), the increase in TSH may partially explain the reduction in the body length observed in MCT8D2 KO and MCT8D1D2 KO mice. As discussed in Section 7.2, there is evidence that TSH has a negative effect on skeletal linear growth. MCT10-null mice (MCT10 KO) grow and reproduce normally, show no gross phenotype and no obvious neurological defect, in addition to normal serum TH levels and normal TH metabolism (Mariotta et al., 2012; Muller et al., 2014). Mice with combined MCT10 and MCT8 deficiency (MCT8/MCT10 KO) revealed normal serum T4, with elevated serum T3 and TSH levels. MCT8/MCT10 KO pups develop normally and do not show any signs of growth retardation, indicating that both transporters are not important for body growth (Muller et al., 2014). OATP1C1-null (OATP1C1 KO) mice present normal serum T3 and T4 concentrations, and unaltered renal and hepatic D1 expression levels, as well as normal hypothalamic TRH and pituitary TSH mRNA levels. These animals are born with the expected frequency, develop without any overt neurological abnormalities, and are not growth retarded (Mayerl, Visser, Darras, Horn, & Heuer, 2012), which is consistent with the lack of OATP1C1 mRNA expression in the skeleton (discussed in Section 4.4). Mice deficient for both MCT8 and OATP1C1 (MCT8/OATP1C1 KO) exhibited alterations in peripheral TH homeostasis that were like those observed in MCT8KO mice; however, there was a higher increase in serum TSH levels. In addition, unlike the single mutant MCT8 KO and OATP1C1 KO animals, MCT8/OATP1C1 KO mice displayed CNS-specific deficits that have also been observed in patients with MCT8 mutations, underscoring the importance of TH transporters for proper brain development in mice (Mayerl et al., 2014). MCT8/OATP1C1 KO mice reproduced normally were born with the expected Mendelian frequency and were indistinguishable from WT mice during the first 2 weeks of postnatal development; however, starting from P17, these animals revealed significantly reduced body weight. The growth reduction persisted after weaning and resulted in a 15%–20% decrease in the body weight of adult MCT8/OATP1C1 KO mice compared with WT mice. In line with a normal growth curve, GH mRNA expression by ISH did not reveal any differences at P6 and P12. However, at P21 (i.e., around the time when MCT8/OATP1C1 KO growth reduction became evident), pituitary GH expression as well as hepatic IGF-1 expression was reduced in MCT8/OATP1C1 KO mice. Moreover, a decay in expression of hypothalamic growth hormone-releasing hormone (GHRH) was observed, suggesting that decreased pituitary GH expression is reflective of reduced GHRH stimulation. Therefore, the body weight reduction presented by MCT8/OATP1C1 KO is probably due to alteration in the GH/IGF-1 axis, in addition to possible negative effects of elevated TSH in skeletal development (Mayerl et al., 2014).

In conclusion, these mutant mouse models reveal that MCT8, MCT10, and OATP1C1 are not essential for skeletal growth. It seems that each one can compensate for the lack of the others. The combination of MCT8 KO with D1 KO, D2 KO, or both D1 and D2 KOs slightly impaired body growth, which may be the result of a local interplay between MCT8 and D2 activities, or the result of the diverse circulating iodothyronines and TSH levels that result from these combined KOs. The notable reduction in body growth observed in MCT8/OATP1C1 seems to be due to an alteration in the GH/IGF-1 axis. Nonetheless, all these studies just evaluated body weight and/or body length, whereas important skeletal development processes and features were not explored, such as EO, IO, and growth plate morphology, impairing further thoughts regarding the role of these TH transporters in skeletal development.

8. THE SKELETAL PHENOTYPE AND DEVELOPMENT IN PATIENTS WITH RTH

RTH is characterized by TRs reduced affinity for TH, with a consequent TR failure to release corepressors and to recruit coactivators, resulting in a persistent repression of gene transcription (Brent, 2012; Ortiga-Carvalho, Sidhaye, & Wondisford, 2014). This condition is caused by mutations in the THRB and THRA genes, located in the LBD. The mutant receptors exert variable dominant-negative effects on the functional TR isoforms, resulting in inconstant phenotypes and adding complexity to correlate genotype with phenotype (Ortiga-Carvalho et al., 2014). RTH due to THRB mutations (RTHβ) have been reported since 1989 (Sakurai et al., 1989), and more than 170 mutations have been described (Cardoso, Maciel, & Paula, 2014; Dumitrescu & Refetoff, 2013). RTHβ is characterized by elevated serum levels of T4 and usually T3, with normal or increased TSH (nonsuppressed TSH), reflecting central (hypothalamic–pituitary) refractoriness to TH (Ortiga-Carvalho et al., 2014). Clinical phenotype is a mix of tissue hypothyroidism and tissue thyrotoxicosis, which reflects, respectively, the resistance of tissues predominantly expressing TRβ isoforms and the overstimulation of tissues mainly expressing TRα1 isoform, such as bone. RTHβ has been shown to be mediated by heterozygous mutations in THRB, and rarely by homozygous mutations, which results in more severe phenotypes than those due to heterozygous mutations (Ono et al., 1991; Ortiga-Carvalho et al., 2014). The most common manifestations are goiter, sinus tachycardia, raised energy expenditure, attention-deficit hyperactivity disorder, and short stature (Onigata & Szinnai, 2014). RTH due to THRA mutations (RTHα) was first identified in 2012 (Bochukova et al., 2012) and, since then, around 9 mutations in 15 patients have been identified (Tang, Yu, & Lian, 2016). For now, the disorder is mediated by heterozygous mutations affecting either TRα1 alone or both TRα1 and TRα2, with no apparent extra phenotype attributable to defective TRα2 (Moran & Chatterjee, 2015). In addition, the mutant receptors have been shown to exert dominant-negative activity against both TRα1 and TRβ (Tinnikov et al., 2002; Vlaeminck-Guillem, Espiard, Flamant, & Wemeau, 2015). As expected, RTHα patients present TH resistance in tissues with predominance of TRα expression, including the CNS, heart, and bone. The clinical manifestations observed so far exhibit hypothyroid features in these TRα-predominant tissues: mental retardation of different degrees, short stature (due to growth retardation), dysmorphic syndrome (affecting face and limbs), chronic constipation, and bradycardia (Onigata & Szinnai, 2014; Tang et al., 2016). These patients usually present a quite normal TH axis: normal TSH (low in some cases), high-normal FT3, low-normal FT4, with an abnormally low FT4:FT3 ratio, and a variable reduced rT3, which is probably explained by reduced D3 activity and/or increased T4 to T3 conversion, which remains to be clarified (Tang et al., 2016). Although the common clinical aspects of RTHα patients are emerging, diversity in their phenotypes is becoming apparent.

It is noteworthy that in RTHα, apparent clinical signs of hypothyroidism are paradoxically associated with quite normal laboratory tests for the thyroid axis, which makes its diagnosis challenging. On the other hand, patients with RTHβ usually present minimal symptoms of thyrotoxicosis, whereas laboratory tests indicate hyperthyroidism with nonsuppressed TSH. Therefore, patients with RTHβ and RTHα have different clinical phenotypes, including the skeletal phenotypes, which will be discussed below.

8.1 RTHα Patients

Child RTHα patients showed delayed bone age and dysmorphic syndrome, which combines disproportionate short stature, with decreased subischial inferior limb and a normal sitting height (normal trunk height); macrocephalia, probably due to delayed IO of the sutures; and other multiple facial anomalies, such as hypertelorism of the eyeballs, a short and wide nose, micrognathism, and delayed tooth eruption, in addition to a short neck. The feet and hands may be enlarged and hip dislocation and coxa valga have also been reported (Vlaeminck-Guillem et al., 2015). In severe situations, ovoid-shaped vertebrae, clavicular and/or rib agenesis may be present (Espiard et al., 2015; Tang et al., 2016). X-ray analyses have shown excessively serpiginous cranial sutures (wormian bone appearance), delayed ossification of the fontanels, epiphyseal dysgenesis (femoral heads), and delayed bone age (Vlaeminck-Guillem et al., 2015), revealing a delay in both EO and IO. It is clear, therefore, that the decreased linear growth occurs due to growth retardation, which, in some patients, can be severe, approaching dwarfism (Espiard et al., 2015; Vlaeminck-Guillem et al., 2015). These skeletal development-related manifestations substantiate the predominant role of TRα in the skeleton of mice and are consistent with the skeletal phenotype of most of all the RTHα mouse models reported so far (discussed in Section 7.1), which present retarded EO and IO, associated with delayed longitudinal bone growth.

It is noteworthy that child and adolescent (but not adult) RTHα patients may present low or normal serum IGF-1 levels (Bochukova et al., 2012; Moran et al., 2013; van Mullem et al., 2013, 2012), with slightly altered-normal GH stimulation test (Bochukova et al., 2012; van Mullem et al., 2013). T4 treatment has been shown to increase serum IGF-1 in RTHα children (Bochukova et al., 2012; van Mullem et al., 2013), whereas cessation of T4 treatment leads to a decrease in serum IGF-1 levels (van Mullem et al., 2013). Serum levels of IGF-1 were not determined in juvenile mouse models of RTHα, but normal pituitary GH mRNA expression along with low IGF-1, IGF-1R, and GHR mRNA expressions was detected in the growth plate of $TR\alpha1^{PV/+}$ mice (Kaneshige et al., 2001; O'Shea et al., 2005), suggesting that the severe skeletal consequences of the PV mutation (discussed in Sections 6.1 and 7.1) result from a dysregulated local GH/IGF-1 signaling in the growth plate (O'Shea et al., 2005). On the other hand, $TR\alpha1^{+/m}$ mice, which harbor a milder $TR\alpha1^{R384C}$ mutation than $TR\alpha1^{PV/+}$, present a transiently delayed ossification and growth retardation, with decreased pituitary GH mRNA (Tinnikov et al., 2002). Interestingly, the additional deletion of the THRB gene in $TR\alpha1^{+/m}$ ($TR\alpha1^{+/m}\beta^{-/-}$) mice caused a 10-fold increase in serum T4, which restored GH mRNA expression and the growth retardation (Bassett, Nordstrom, et al., 2007; Tinnikov et al., 2002). Thus, the skeletal development delay observed in RTHα may be partially explained by a local and/or systemic GH/IGF-1 axis deficiency. However, GH treatment has not improved linear growth in an RTHα child (van Mullem et al., 2013).

Surprisingly, initiation of T4 treatment in early childhood (between 3 months and 3 years of age) improved body growth and bone maturation of four RTHα patients (Moran et al., 2014, 2013). Remarkably, in an RTHα girl who had been treated with T4 since age 5 months, advanced bone age was observed at age 3.2 years, which prompted T4 discontinuation. At age 10 years, T4 treatment was restarted and height velocity improved (Moran et al., 2013), suggesting sensitivity of the developing RTHα skeleton to T4. Accordingly, transient treatment of juvenile $TR\alpha1^{+/m}$ ($TR\alpha1^{R384C}$) mice with T3 did ameliorate their skeletal abnormalities (Bassett, Nordstrom, et al., 2007). In addition, treatment of $TR\alpha1^{PV/+}$ mice with a supraphysiological dose of T4 between 4 and 20 weeks of age had no effect on the growth plate thickness nor on the linear growth of their tibias, but promoted a gradual increase in cortical thickness and diameter, a paradoxical increase in bone mineral content, a reduction in bone resorption (reduced in TRα1$^{PV/+}$ vs WT mice), and gradual increases in bone strength and stiffness (Bassett et al., 2014). Thus, and surprisingly, these studies with mouse models of RTHα reveal sensitivity of their developing skeletons to TH, suggesting that TRβ may have a role or exhibit compensatory actions in the developing skeleton. Interestingly, adult RTHα patients seem to present normal BMD but also exhibit macrocephaly with thickening of the skull vault (cranial hyperostosis), especially in the occipital, and enlargement of the frontal bone, in addition to cortical bone thickening of the long bones (appendicular hyperostosis) (Moran et al., 2014, 2013). These adult findings are coherent with the reduced bone resorption observed in adult TRα1$^{+/m}$ mice (Bassett, Nordstrom, et al., 2007). Remarkably, the reduced bone resorption in TRα1$^{+/m}$ mice was normalized or inverted by TH treatment or by the additional deletion of TRβ (TRα1$^{+/m}$β$^{-/-}$), which promotes a 10-fold increase in serum T4 (Bassett, Nordstrom, et al., 2007). These findings suggest that the increases in serum TH levels had overcome the dominant-negative effect of TRα1$^{+/m}$ on WT TRα (Bassett, Nordstrom, et al., 2007). Accordingly, T4 therapy progressively increased bone turnover markers in adult RTHα patients, indicating enhanced osteoblastic and osteoclastic activities (Moran et al., 2014, 2013). The responsiveness of adult and developing RTHα skeleton to TH could also be the result of an abolished dominant-negative effect of mutant TRα on functional TRs. Interestingly, *in vitro* functional analysis of the E403X and F397fs406X TRα proteins showed that the dominant-negative effect of these mutant receptors on WT TRα was resistant to high levels of T3 (Bochukova et al., 2012; van Mullem et al., 2013). However, high doses of T3 were able to partially overcome the dominant-negative effect of F397fs406X TRα on TRβ1 (van Mullem et al., 2013). Considering these *in vitro* functional studies, it is possible that some effects of TH on the skeleton of RTHα patients are mediated by TRβ.

In conclusion, RTHα patients present important delayed skeletal development, leading to short stature that may be severe approaching dwarfism. The GH/IGF-1 axis seems to be slightly altered in some patients, but GH therapy has not shown beneficial effects. Preliminary experiences with T4 supplementation therapy in RTHα patients show that their skeletons are responsive to TH, during development and adulthood, and suggest that this therapy can be beneficial for skeletal development in RTHα patients, when administered in childhood. However, clinical effects are inconclusive, since there is evidence of no effects or lack of valuable effects of TH therapy on skeletal development as well (Bochukova et al., 2012). More evident benefits have been shown for constipation, bradycardia, energy levels, and carpal tunnel syndrome in adults and/or children (Moran et al., 2014, 2013; Moran & Chatterjee, 2015). Nevertheless, the effectiveness of T4 therapy to skeletal development and to other physiological processes and clinical manifestations in RTHα patients needs to be clarified. In addition, given the skeleton responsiveness of RTHα skeleton to TH, a possible role of TRβ in mediating TH actions in the skeleton should be considered, as well as possible detrimental consequences of TH therapy to bone development and maintenance of RTHα patients.

8.2 RTHβ Patients

Different skeletal phenotypes have been reported in RTHβ patients, especially during development. Among the most common clinical manifestations (goiter, sinus tachycardia, raised energy expenditure, attention-deficit hyperactivity disorder, and short stature), short stature has been the less frequent in RTHβ patients with heterozygous mutations (Brucker-Davis et al., 1995; Refetoff et al., 1993; Takeda, Balzano, Sakurai, DeGroot, & Refetoff, 1991), but was shown to be present in all subjects of four families with homozygous mutations (Ferrara et al., 2012). In RTHβ due to heterozygous mutations of THRB, short stature was detected in 8%–26% of patients, and delayed bone age was identified in a proportion of 29%–47% (Brucker-Davis et al., 1995; Cardoso, Maciel, et al., 2014; Weiss & Refetoff, 1996), whereas advanced bone age was documented only in two families (among 49 families) (Weiss & Refetoff, 1996). Besides, in children homozygous for mutant THRB gene, delayed bone age was a constant (Ferrara et al., 2012; Ono et al., 1991; Usala et al., 1991), suggesting the predominance of a hypothyroid skeleton in RTHβ patients. Stippled epiphyses, similar to those described in hypothyroid children, are another skeletal abnormality eventually detected in RTHβ patients (Weiss & Refetoff, 1996). These findings are inconsistent with the phenotype of skeletal hyperthyroidism displayed by mouse models of RTHβ (discussed in Section 7.2), characterized by short stature due to advanced bone maturation (O'Shea et al., 2003). It is noteworthy that, although infrequently, other skeletal abnormalities suggestive of skeletal hyperthyroidism have also been detected in RTHβ children, such as craniosynostosis, and dysmorphic features (birdlike facies, pigeon breast, winged scapulae, and short fourth metacarpals) (Refetoff et al., 1993; Takeda,

Sakurai, DeGroot, & Refetoff, 1992; Usala et al., 1991). In addition, increased serum levels of total calcium were detected in children (Cardoso, de Paula, & Maciel, 2014). In adults, RTHβ has been associated with decreased bone mass and increased serum levels of total calcium as well (Cardoso, de Paula, et al., 2014). There is evidence that differing mutant-to-normal TRβ ratios may partially explain the variety and severity of phenotypes. In addition, it was shown that the ratios of normal and mutant TRs may be age and growth related, which could account for the reported attenuation of phenotypic symptoms of RTHβ with age (Mixson et al., 1993). Thus, the amount of TRβ mutant, exerting dominant-negative effects on both functional TRβ and TRα, undoubtedly influences the severity of the phenotype (Ferrara et al., 2012).

In conclusion, most of the bone development defects detected so far in RTHβ patients suggest delayed and not advanced bone growth and maturation, which is inconsistent with the findings from the RTHβ mouse models, that reveal a thyrotoxic skeleton and advanced skeletal development. This discrepancy is probably explained by a putative less predominant role of TRα in the skeleton of humans as compared with a crucial role of TRα in the skeleton of mice. At this point, it is not known whether the skeletal development defects occurring in RTHβ patients are primarily due to (1) the dominant-negative effects of TRβ mutants on functional TRα1 and/or TRβ, or due to (2) a TRβ dysfunction per se, or due to (3) a combination of both circumstances.

9. CONCLUDING REMARKS

1. TH is essential for postnatal skeletal development from late fetal life, having key roles in EO, IO, and longitudinal bone growth, which are directly and indirectly regulated by TH. The presence of all the factors necessary for direct actions of TH in skeletal cells suggests that direct actions of TH are crucial for skeletal development, which are supported by *in vitro* and *in vivo* studies.
2. TRα1, TRα2, and TRβ1 are expressed in the whole skeleton and in its mains components: bone, cartilage, and bone marrow tissues. TRα1 is the predominant TR isoform in the bone tissue and in osteoblasts of rodents, which suggests a predominant role of TRα1 in mediating TH actions on skeletal development and maintenance. A large body of studies in genetically modified mice confirm this hypothesis and reveal that TRα1 aporeceptor activity is more detrimental to skeleton

than its absence in skeletal cells. On the other hand, these studies show evidence that TRβ may partially compensate for the lack of TRα1 in skeletal development of rodents.

3. RHTα patients present short stature, due to retarded skeletal development, which may be severe approaching dwarfism. This skeletal phenotype is consistent with findings of skeletal hypothyroidism in mouse models of RTHα and substantiates the predominant role of TRα for skeletal development. However, preliminary experiences with T4 supplementation therapy in RTHα patients showed beneficial effects on body growth and maturation, which suggests that TRβ partially mediates TH actions in the human skeleton. Short stature is detected in 8%–26% of RTHβ patients with heterozygous mutations, but is present in all RTHβ patients with homozygous mutations. In contrast to mouse models of RTHβ that present a thyrotoxic skeleton, 29%–47% of the heterozygous RTHβ patients present delayed bone age, whereas advance bone age was documented only in two families, among 49 families. In addition, delayed bone age was identified in all children with homozygous THRβ mutations. Other signs of skeletal thyrotoxicosis (craniosynostoses and skeletal dysmorphic features) were found infrequently in RTHβ patients. Altogether, these findings corroborate the predominant role of TRα on skeletal development but suggest that TRβ partially mediates TH actions on skeletal development of humans, revealing that TRα/TRβ relevance to skeleton may be specie-specific. This is supported by *in vitro* studies that showed higher expression of TRα vs TRβ in osteoblast-like cells of mice than in osteoblastic cells of rats, in addition to a lower responsiveness of mouse osteoblastic cells to the TRβ-selective analog GC-1, which had effects similar to those of T3 in rat cells (Beber et al., 2009).
4. Functional THSR has been described in osteoblasts, osteoclasts, and chondrocytes. A series of studies have shown that TSH acts directly in the skeleton to inhibit osteoclast-mediated bone resorption and to induce bone formation. In addition, studies in mice expressing nonfunctional TSHR suggest that TSH has negative effects on skeletal development; however, the effects of TSH in chondrocytes were not investigated yet. Evidence suggests that the effects of TH deficiency and excess on the skeleton predominate over the effects of TSH excess and deficiency, respectively. However, it likely that TSH excess contributes to the delayed linear bone growth and maturation observed in hypothyroidism.

5. D1 is not expressed in the skeleton, D2 activity was detected only in osteoblasts, and D3 activity was shown in growth plate chondrocytes, osteoblasts, and osteoclasts. D3 is the only deiodinase relevant for skeletal development. Evidence suggests that D3 is important to avoid premature exposure of skeletal tissues to TH during development.
6. Several TH transporters have been identified in the skeleton. Studies in transgenic mice suggest redundancy among MCT8, MCT10, or OATP1C1 regarding the skeletal development, and that none of them is essential for this process.
7. TH directly increases osteoblastic differentiation and activity, increasing bone formation and turnover. There is no doubt that TH acts directly in the skeleton to stimulates bone resorption, but it is unclear if TH acts directly in osteoclast precursors to stimulate osteoclastogenesis, and in mature osteoclasts to stimulate their bone resorptive activity. Assumption is that TH induces osteoclastogenesis and osteoclastic activity mainly indirectly, acting in osteoblastic cells. On the other hand, direct actions of TH on chondrocytes are evident. These cells are highly responsive to TH, which is essential for inducing terminal differentiation.
8. TH is crucial for EO, acting directly in the chondrocytes to promote chondrocyte terminal differentiation. TH is also crucial for normal linear bone growth, acting directly in the growth plate to induce the columnar organization of the PCs, and to induce the differentiation of PCs into HCs, which are key steps of bone elongation.
9. TH interacts with the GH/IGF-1 axis to promote skeletal development. This interaction is complex and occurs at multiple levels and in both directions. TH regulation of linear bone growth and EO involves both GH/IGF-1-dependent and -independent mechanisms. During postnatal development, TH is the main regulator of EO and depends on GH/IGF-1 to induce full linear bone growth.
10. Evidence suggests that TH interacts with the Ihh/PTHrP-R loop to control the pace of chondrocyte differentiation in the growth plate. On the other hand, TH induces chondrocyte differentiation by activating the Wnt-4/β-catenin canonical pathway, to induce expression of Runx-2, which plays essential roles during chondrocyte hypertrophy, activating transcription of several chondrocyte hypertrophy-associated genes. This TH/Wnt-4/β-catenin/Runx2 pathway involves TH–IGF-1/IGF-1R–leptin interplays. TH stimulates expression of IGF-1/IGF-1R, which, in turn, acts in the Wnt-4/β-catenin pathway to

partially mediate TH actions on chondrocyte proliferation and differentiation. TH also stimulates leptin/Ob-R expression in chondrocytes, which has positive effects on IGF-1/IGF-1R and Wnt-4/β-catenin pathway, in addition to increase TRα expression in chondrocytes. In parallel, leptin antagonizes the suppressive effects of PPARδ on T3-mediated chondrocyte hypertrophy. Another pathway through which T3 promotes growth plate chondrocyte terminal differentiation is inducing BMP4 and inhibiting Noggin expression.

11. Evidence suggests that TH induces IO acting in osteoblast precursors, to induce osteoblastogenesis, in a mechanism that involves the FGF2/FGFR1/MAPK signaling pathway. TH modulates the FGF/FGFR3 signaling pathways in chondrocytes, which probably mediates TH actions in these cells; however, at this point, the consequences of this modulation to linear bone growth and EO are unknown.
12. TH regulates the ECM composition of the growth plate, including the synthesis and degradation of HSPGs, which modulate FGF and Ihh signaling pathways, raising the hypothesis that TH modulates signaling pathways in the growth plate by modifying its cartilage matrix composition.

10. FINAL CONCLUSIONS AND FUTURE DIRECTIONS

There is no doubt that TH is a key factor for postnatal skeletal development, acting directly on skeletal cells to modulate the skeletal ossification and the longitudinal bone growth. In the past 20 years, great advances have been made on understanding the mechanisms through which TH regulates the skeletal development. However, the precise mechanisms of TH action on skeletal development are far from being fully understood. It seems that TH acts directly in skeletal cells modulating the expression of diverse local factors that compose different and interconnected key-skeletal development signaling pathways. The challenges for the next years are the identification of these TH-regulated factors and signaling pathways, the understanding on how TH modulates them, and, finally, the understanding of the meaning of the modulation of each one of these key factors and signaling pathways to skeletal development.

ACKNOWLEDGMENTS

The author acknowledges the support of the Fundação de Amparo a Pesquisa do Estado de São Paulo (FAPESP), São Paulo, SP, Brazil (C.H.A.G., Grant No. 2015/10671-5). M.M.-R.

and B.N.P. are recipients of fellowships from FAPESP (Grant Nos. 2014/19020-4 and 2015/12554-6, respectively). The authors apologize for the omission of many references owing to space limitations.

REFERENCES

Abdalla, S. M., & Bianco, A. C. (2014). Defending plasma T3 is a biological priority. *Clinical Endocrinology*, *81*(5), 633–641.

Abe, E., Marians, R. C., Yu, W., Wu, X. B., Ando, T., Li, Y., et al. (2003). TSH is a negative regulator of skeletal remodeling. *Cell*, *115*(2), 151–162.

Abe, S., Namba, N., Abe, M., Fujiwara, M., Aikawa, T., Kogo, M., et al. (2012). Monocarboxylate transporter 10 functions as a thyroid hormone transporter in chondrocytes. *Endocrinology*, *153*(8), 4049–4058.

Abel, E. D., Boers, M. E., Pazos-Moura, C., Moura, E., Kaulbach, H., Zakaria, M., et al. (1999). Divergent roles for thyroid hormone receptor beta isoforms in the endocrine axis and auditory system. *The Journal of Clinical Investigation*, *104*(3), 291–300.

Abu, E. O., Bord, S., Horner, A., Chatterjee, V. K., & Compston, J. E. (1997). The expression of thyroid hormone receptors in human bone. *Bone*, *21*(2), 137–142.

Abu, E. O., Horner, A., Teti, A., Chatterjee, V. K., & Compston, J. E. (2000). The localization of thyroid hormone receptor mRNAs in human bone. *Thyroid*, *10*(4), 287–293.

Ahmed, S. F., & Farquharson, C. (2010). The effect of GH and IGF1 on linear growth and skeletal development and their modulation by SOCS proteins. *The Journal of Endocrinology*, *206*(3), 249–259.

Akita, S., Hirano, A., & Fujii, T. (1996). Identification of IGF-I in the calvarial suture of young rats: Histochemical analysis of the cranial sagittal sutures in a hyperthyroid rat model. *Plastic and Reconstructive Surgery*, *97*(1), 1–12.

Akiyama, H., Lyons, J. P., Mori-Akiyama, Y., Yang, X., Zhang, R., Zhang, Z., et al. (2004). Interactions between Sox9 and beta-catenin control chondrocyte differentiation. *Genes & Development*, *18*(9), 1072–1087.

Allain, T. J., Chambers, T. J., Flanagan, A. M., & McGregor, A. M. (1992). Tri-iodothyronine stimulates rat osteoclastic bone resorption by an indirect effect. *The Journal of Endocrinology*, *133*(3), 327–331.

Allain, T. J., & McGregor, A. M. (1993). Thyroid hormones and bone. *The Journal of Endocrinology*, *139*(1), 9–18.

Allain, T. J., Thomas, M. R., McGregor, A. M., & Salisbury, J. R. (1995). A histomorphometric study of bone changes in thyroid dysfunction in rats. *Bone*, *16*(5), 505–509.

Allain, T. J., Yen, P. M., Flanagan, A. M., & McGregor, A. M. (1996). The isoform-specific expression of the tri-iodothyronine receptor in osteoblasts and osteoclasts. *European Journal of Clinical Investigation*, *26*(5), 418–425.

Apriletti, J. W., Ribeiro, R. C., Wagner, R. L., Feng, W., Webb, P., Kushner, P. J., et al. (1998). Molecular and structural biology of thyroid hormone receptors. *Clinical and Experimental Pharmacology & Physiology. Supplement*, *25*, S2–11.

Aranda, A., & Pascual, A. (2001). Nuclear hormone receptors and gene expression. *Physiological Reviews*, *81*(3), 1269–1304.

Asling, C. W., Simpson, M. E., Li, C. H., & Evans, H. M. (1954). The effects of chronic administration of thyroxin to hypophysectomized rats on their skeletal growth, maturation and response to growth hormone. *The Anatomical Record*, *119*(1), 101–117.

Baliram, R., Chow, A., Huber, A. K., Collier, L., Ali, M. R., Morshed, S. A., et al. (2013). Thyroid and bone: Macrophage-derived TSH-beta splice variant increases murine osteoblastogenesis. *Endocrinology*, *154*(12), 4919–4926.

Baliram, R., Latif, R., Berkowitz, J., Frid, S., Colaianni, G., Sun, L., et al. (2011). Thyroid-stimulating hormone induces a Wnt-dependent, feed-forward loop for osteoblastogenesis in embryonic stem cell cultures. *Proceedings of the National Academy of Sciences of the United States of America*, *108*(39), 16277–16282.

Baliram, R., Sun, L., Cao, J., Li, J., Latif, R., Huber, A. K., et al. (2012). Hyperthyroid-associated osteoporosis is exacerbated by the loss of TSH signaling. *The Journal of Clinical Investigation*, *122*(10), 3737–3741.

Ballock, R., Mita, B. C., Zhou, X., Chen, D. H., & Mink, L. M. (1999). Expression of thyroid hormone receptor isoforms in rat growth plate cartilage in vivo. *Journal of Bone and Mineral Research*, *14*(9), 1550–1556.

Ballock, R. T., & O'Keefe, R. J. (2003). The biology of the growth plate. *The Journal of Bone and Joint Surgery. American Volume*, *85A*(4), 715–726.

Ballock, R. T., & Reddi, A. H. (1994). Thyroxine is the serum factor that regulates morphogenesis of columnar cartilage from isolated chondrocytes in chemically defined medium. *The Journal of Cell Biology*, *126*(5), 1311–1318.

Ballock, R. T., Zhou, X., Mink, L. M., Chen, D. H., Mita, B. C., & Stewart, M. C. (2000). Expression of cyclin-dependent kinase inhibitors in epiphyseal chondrocytes induced to terminally differentiate with thyroid hormone. *Endocrinology*, *141*(12), 4552–4557.

Barnard, J. C., Williams, A. J., Rabier, B., Chassande, O., Samarut, J., Cheng, S. Y., et al. (2005). Thyroid hormones regulate fibroblast growth factor receptor signaling during chondrogenesis. *Endocrinology*, *146*(12), 5568–5580.

Bassett, J. H., Boyde, A., Howell, P. G., Bassett, R. H., Galliford, T. M., Archanco, M., et al. (2010). Optimal bone strength and mineralization requires the type 2 iodothyronine deiodinase in osteoblasts. *Proceedings of the National Academy of Sciences of the United States of America*, *107*(16), 7604–7609.

Bassett, J. H., Boyde, A., Zikmund, T., Evans, H., Croucher, P. I., Zhu, X., et al. (2014). Thyroid hormone receptor alpha mutation causes a severe and thyroxine-resistant skeletal dysplasia in female mice. *Endocrinology*, *155*(9), 3699–3712.

Bassett, J. H., Nordstrom, K., Boyde, A., Howell, P. G., Kelly, S., Vennstrom, B., et al. (2007). Thyroid status during skeletal development determines adult bone structure and mineralization. *Molecular Endocrinology*, *21*(8), 1893–1904.

Bassett, J. H., O'Shea, P. J., Chassande, O., Samarut, J., Cheng, S. Y., Vennstrom, B., et al. (2006). Analysis of skeletal phenotypes in thyroid hormone receptor mutant mice. *Scanning*, *28*(2), 91–93.

Bassett, J. H., O'Shea, P. J., Sriskantharajah, S., Rabier, B., Boyde, A., Howell, P. G., et al. (2007). Thyroid hormone excess rather than thyrotropin deficiency induces osteoporosis in hyperthyroidism. *Molecular Endocrinology*, *21*(5), 1095–1107.

Bassett, J. H., Swinhoe, R., Chassande, O., Samarut, J., & Williams, G. R. (2006). Thyroid hormone regulates heparan sulfate proteoglycan expression in the growth plate. *Endocrinology*, *147*(1), 295–305.

Bassett, J. H., & Williams, G. R. (2009). The skeletal phenotypes of TRalpha and TRbeta mutant mice. *Journal of Molecular Endocrinology*, *42*(4), 269–282.

Bassett, J. H., & Williams, G. R. (2016). Role of thyroid hormones in skeletal development and bone maintenance. *Endocrine Reviews*, *37*(2), 135–187.

Bassett, J. H., Williams, A. J., Murphy, E., Boyde, A., Howell, P. G., Swinhoe, R., et al. (2008). A lack of thyroid hormones rather than excess thyrotropin causes abnormal skeletal development in hypothyroidism. *Molecular Endocrinology*, *22*(2), 501–512.

Beber, E. H., Capelo, L. P., Fonseca, T. L., Costa, C. C., Lotfi, C. F., Scanlan, T. S., et al. (2009). The thyroid hormone receptor (TR) beta-selective agonist GC-1 inhibits proliferation but induces differentiation and TR beta mRNA expression in mouse and rat osteoblast-like cells. *Calcified Tissue International*, *84*(4), 324–333.

Berry, M. J., Grieco, D., Taylor, B. A., Maia, A. L., Kieffer, J. D., Beamer, W., et al. (1993). Physiological and genetic analyses of inbred mouse strains with a type I iodothyronine 5′ deiodinase deficiency. *The Journal of Clinical Investigation*, *92*(3), 1517–1528.
Bianco, A. C., Salvatore, D., Gereben, B., Berry, M. J., & Larsen, P. R. (2002). Biochemistry, cellular and molecular biology, and physiological roles of the iodothyronine selenodeiodinases. *Endocrine Reviews*, *23*(1), 38–89.
Bochukova, E., Schoenmakers, N., Agostini, M., Schoenmakers, E., Rajanayagam, O., Keogh, J. M., et al. (2012). A mutation in the thyroid hormone receptor alpha gene. *The New England Journal of Medicine*, *366*(3), 243–249.
Bookout, A. L., Jeong, Y., Downes, M., Yu, R. T., Evans, R. M., & Mangelsdorf, D. J. (2006). Anatomical profiling of nuclear receptor expression reveals a hierarchical transcriptional network. *Cell*, *126*(4), 789–799.
Bordji, K., Grillasca, J. P., Gouze, J. N., Magdalou, J., Schohn, H., Keller, J. M., et al. (2000). Evidence for the presence of peroxisome proliferator-activated receptor (PPAR) alpha and gamma and retinoid Z receptor in cartilage. PPARgamma activation modulates the effects of interleukin-1beta on rat chondrocytes. *The Journal of Biological Chemistry*, *275*(16), 12243–12250.
Bos, S. D., Bovee, J. V., Duijnisveld, B. J., Raine, E. V., van Dalen, W. J., Ramos, Y. F., et al. (2012). Increased type II deiodinase protein in OA-affected cartilage and allelic imbalance of OA risk polymorphism rs225014 at DIO2 in human OA joint tissues. *Annals of the Rheumatic Diseases*, *71*(7), 1254–1258.
Bos, S. D., Slagboom, P. E., & Meulenbelt, I. (2008). New insights into osteoarthritis: Early developmental features of an ageing-related disease. *Current Opinion in Rheumatology*, *20*(5), 553–559.
Brent, G. A. (2012). Mechanisms of thyroid hormone action. *The Journal of Clinical Investigation*, *122*(9), 3035–3043.
Britto, J. M., Fenton, A. J., Holloway, W. R., & Nicholson, G. C. (1994). Osteoblasts mediate thyroid hormone stimulation of osteoclastic bone resorption. *Endocrinology*, *134*(1), 169–176.
Brucker-Davis, F., Skarulis, M. C., Grace, M. B., Benichou, J., Hauser, P., Wiggs, E., et al. (1995). Genetic and clinical features of 42 kindreds with resistance to thyroid hormone. The National Institutes of Health Prospective Study. *Annals of Internal Medicine*, *123*(8), 572–583.
Burgess, T. L., Qian, Y., Kaufman, S., Ring, B. D., Van, G., Capparelli, C., et al. (1999). The ligand for osteoprotegerin (OPGL) directly activates mature osteoclasts. *The Journal of Cell Biology*, *145*(3), 527–538.
Burstein, S., Grumbach, M. M., Kaplan, S. L., & Li, C. H. (1978). Immunoreactivity and receptor binding of mixed recombinants of human growth hormone and chorionic somatomammotropin. *Proceedings of the National Academy of Sciences of the United States of America*, *75*(11), 5391–5394.
Busse, M., Feta, A., Presto, J., Wilen, M., Gronning, M., Kjellen, L., et al. (2007). Contribution of EXT1, EXT2, and EXTL3 to heparan sulfate chain elongation. *The Journal of Biological Chemistry*, *282*(45), 32802–32810.
Butler, A. A., & Le Roith, D. (2001). Control of growth by the somatropic axis: Growth hormone and the insulin-like growth factors have related and independent roles. *Annual Review of Physiology*, *63*, 141–164.
Cadigan, K. M., & Nusse, R. (1997). Wnt signaling: A common theme in animal development. *Genes & Development*, *11*(24), 3286–3305.
Campbell, P. G., & Andress, D. L. (1997). Insulin-like growth factor (IGF)-binding protein-5-(201-218) region regulates hydroxyapatite and IGF-I binding. *The American Journal of Physiology*, *273*(5 Pt. 1), E1005–E1013.

Capelo, L. P., Beber, E. H., Fonseca, T. L., & Gouveia, C. H. (2009). The monocarboxylate transporter 8 and L-type amino acid transporters 1 and 2 are expressed in mouse skeletons and in osteoblastic MC3T3-E1 cells. *Thyroid*, *19*(2), 171–180.

Capelo, L. P., Beber, E. H., Huang, S. A., Zorn, T. M., Bianco, A. C., & Gouveia, C. H. (2008). Deiodinase-mediated thyroid hormone inactivation minimizes thyroid hormone signaling in the early development of fetal skeleton. *Bone*, *43*(5), 921–930.

Cardoso, L. F., de Paula, F. J., & Maciel, L. M. (2014). Resistance to thyroid hormone due to mutations in the THRB gene impairs bone mass and affects calcium and phosphorus homeostasis. *Bone*, *67*, 222–227.

Cardoso, L. F., Maciel, L. M., & Paula, F. J. (2014). The multiple effects of thyroid disorders on bone and mineral metabolism. *Arquivos Brasileiros de Endocrinologia e Metabologia*, *58*(5), 452–463.

Carlsson, B., Billig, H., Rymo, L., & Isaksson, O. G. (1990). Expression of the growth hormone-binding protein messenger RNA in the liver and extrahepatic tissues in the rat: Co-expression with the growth hormone receptor. *Molecular and Cellular Endocrinology*, *73*(1), R1–R6.

Carrascosa, A., Ferrandez, M. A., Audi, L., & Ballabriga, A. (1992). Effects of triiodothyronine (T3) and identification of specific nuclear T3-binding sites in cultured human fetal epiphyseal chondrocytes. *The Journal of Clinical Endocrinology and Metabolism*, *75*(1), 140–144.

Chiellini, G., Apriletti, J. W., Yoshihara, H. A., Baxter, J. D., Ribeiro, R. C., & Scanlan, T. S. (1998). A high-affinity subtype-selective agonist ligand for the thyroid hormone receptor. *Chemistry & Biology*, *5*(6), 299–306.

Christoffolete, M. A., Arrojo e Drigo, R., Gazoni, F., Tente, S. M., Goncalves, V., Amorim, B. S., et al. (2007). Mice with impaired extrathyroidal thyroxine to 3,5,3′-triiodothyronine conversion maintain normal serum 3,5,3′-triiodothyronine concentrations. *Endocrinology*, *148*(3), 954–960.

Ciarmatori, S., Kiepe, D., Haarmann, A., Huegel, U., & Tonshoff, B. (2007). Signaling mechanisms leading to regulation of proliferation and differentiation of the mesenchymal chondrogenic cell line RCJ3.1C5.18 in response to IGF-I. *Journal of Molecular Endocrinology*, *38*(4), 493–508.

Colvin, J. S., Bohne, B. A., Harding, G. W., McEwen, D. G., & Ornitz, D. M. (1996). Skeletal overgrowth and deafness in mice lacking fibroblast growth factor receptor 3. *Nature Genetics*, *12*(4), 390–397.

Costa-e-Sousa, R. H., & Hollenberg, A. N. (2012). Minireview: The neural regulation of the hypothalamic-pituitary-thyroid axis. *Endocrinology*, *153*(9), 4128–4135.

Coumoul, X., & Deng, C. X. (2003). Roles of FGF receptors in mammalian development and congenital diseases. *Birth Defects Research. Part C, Embryo Today*, *69*(4), 286–304.

Cray, J. J., Jr., Khaksarfard, K., Weinberg, S. M., Elsalanty, M., & Yu, J. C. (2013). Effects of thyroxine exposure on osteogenesis in mouse calvarial pre-osteoblasts. *PLoS One*, *8*(7), e69067.

Cruz Grecco Teixeira, M. B., Martins, G. M., Miranda-Rodrigues, M., De Araujo, I. F., Oliveira, R., Brum, P. C., et al. (2016). Lack of alpha2C-adrenoceptor results in contrasting phenotypes of long bones and vertebra and prevents the thyrotoxicosis-induced osteopenia. *PLoS One*, *11*(1), e0146795.

Daughaday, W. H., & Reeder, C. (1966). Synchronous activation of DNA synthesis in hypophysectomized rat cartilage by growth hormone. *The Journal of Laboratory and Clinical Medicine*, *68*(3), 357–368.

Davies, T., Marians, R., & Latif, R. (2002). The TSH receptor reveals itself. *The Journal of Clinical Investigation*, *110*(2), 161–164.

Delezoide, A. L., Benoist-Lasselin, C., Legeai-Mallet, L., Le Merrer, M., Munnich, A., Vekemans, M., et al. (1998). Spatio-temporal expression of FGFR 1, 2 and 3 genes during human embryo-fetal ossification. *Mechanisms of Development*, 77(1), 19–30.

Dentice, M., Bandyopadhyay, A., Gereben, B., Callebaut, I., Christoffolete, M. A., Kim, B. W., et al. (2005). The Hedgehog-inducible ubiquitin ligase subunit WSB-1 modulates thyroid hormone activation and PTHrP secretion in the developing growth plate. *Nature Cell Biology*, *7*(7), 698–705.

Dentice, M., Luongo, C., Huang, S., Ambrosio, R., Elefante, A., Mirebeau-Prunier, D., et al. (2007). Sonic hedgehog-induced type 3 deiodinase blocks thyroid hormone action enhancing proliferation of normal and malignant keratinocytes. *Proceedings of the National Academy of Sciences of the United States of America*, *104*(36), 14466–14471.

Desjardin, C., Charles, C., Benoist-Lasselin, C., Riviere, J., Gilles, M., Chassande, O., et al. (2014). Chondrocytes play a major role in the stimulation of bone growth by thyroid hormone. *Endocrinology*, *155*(8), 3123–3135.

Di Rocco, F., Biosse Duplan, M., Heuze, Y., Kaci, N., Komla-Ebri, D., Munnich, A., et al. (2014). FGFR3 mutation causes abnormal membranous ossification in achondroplasia. *Human Molecular Genetics*, *23*, 2914–2925.

Dong, Y., Drissi, H., Chen, M., Chen, D., Zuscik, M. J., Schwarz, E. M., et al. (2005). Wnt-mediated regulation of chondrocyte maturation: Modulation by TGF-beta. *Journal of Cellular Biochemistry*, *95*(5), 1057–1068.

Dong, Y. F., Soung do, Y., Schwarz, E. M., O'Keefe, R. J., & Drissi, H. (2006). Wnt induction of chondrocyte hypertrophy through the Runx2 transcription factor. *Journal of Cellular Physiology*, *208*(1), 77–86.

Dumitrescu, A. M., Liao, X. H., Best, T. B., Brockmann, K., & Refetoff, S. (2004). A novel syndrome combining thyroid and neurological abnormalities is associated with mutations in a monocarboxylate transporter gene. *American Journal of Human Genetics*, *74*(1), 168–175.

Dumitrescu, A. M., Liao, X. H., Weiss, R. E., Millen, K., & Refetoff, S. (2006). Tissue-specific thyroid hormone deprivation and excess in monocarboxylate transporter (mct) 8-deficient mice. *Endocrinology*, *147*(9), 4036–4043.

Dumitrescu, A. M., & Refetoff, S. (2013). The syndromes of reduced sensitivity to thyroid hormone. *Biochimica et Biophysica Acta*, *1830*(7), 3987–4003.

Egrise, D., Martin, D., Neve, P., Verhas, M., & Schoutens, A. (1990). Effects and interactions of 17 beta-estradiol, T3 and 1,25(OH)2D3 on cultured osteoblasts from mature rats. *Bone and Mineral*, *11*(3), 273–283.

Endo, T., & Kobayashi, T. (2013). Excess TSH causes abnormal skeletal development in young mice with hypothyroidism via suppressive effects on the growth plate. *American Journal of Physiology. Endocrinology and Metabolism*, *305*(5), E660–E666.

Engler, D., & Burger, A. G. (1984). The deiodination of the iodothyronines and of their derivatives in man. *Endocrine Reviews*, *5*(2), 151–184.

Eriksen, E. F., Mosekilde, L., & Melsen, F. (1986). Kinetics of trabecular bone resorption and formation in hypothyroidism: Evidence for a positive balance per remodeling cycle. *Bone*, *7*(2), 101–108.

Ernst, M., & Froesch, E. R. (1987). Triiodothyronine stimulates proliferation of osteoblast-like cells in serum-free culture. *FEBS Letters*, *220*(1), 163–166.

Espiard, S., Savagner, F., Flamant, F., Vlaeminck-Guillem, V., Guyot, R., Munier, M., et al. (2015). A novel mutation in THRA gene associated with an atypical phenotype of resistance to thyroid hormone. *The Journal of Clinical Endocrinology and Metabolism*, *100*(8), 2841–2848.

Fekkes, D., Hennemann, G., & Visser, T. J. (1982). Evidence for a single enzyme in rat liver catalysing the deiodination of the tyrosyl and the phenolic ring of iodothyronines. *The Biochemical Journal*, *201*(3), 673–676.

Fernandez-Pernas, P., Fafian-Labora, J., Lesende-Rodriguez, I., Mateos, J., De la Fuente, A., Fuentes, I., et al. (2016). 3,3′,5-Triiodo-L-thyronine increases in vitro chondrogenesis of mesenchymal stem cells from human umbilical cord stroma through SRC2. *Journal of Cellular Biochemistry*, *117*(9), 2097–2108.

Ferrara, A. M., Onigata, K., Ercan, O., Woodhead, H., Weiss, R. E., & Refetoff, S. (2012). Homozygous thyroid hormone receptor beta-gene mutations in resistance to thyroid hormone: Three new cases and review of the literature. *The Journal of Clinical Endocrinology and Metabolism*, *97*(4), 1328–1336.

Filanti, C., Dickson, G. R., Di Martino, D., Ulivi, V., Sanguineti, C., Romano, P., et al. (2000). The expression of metalloproteinase-2, -9, and -14 and of tissue inhibitors-1 and -2 is developmentally modulated during osteogenesis in vitro, the mature osteoblastic phenotype expressing metalloproteinase-14. *Journal of Bone and Mineral Research*, *15*(11), 2154–2168.

Filmus, J., & Selleck, S. B. (2001). Glypicans: Proteoglycans with a surprise. *The Journal of Clinical Investigation*, *108*(4), 497–501.

Firth, S. M., & Baxter, R. C. (2002). Cellular actions of the insulin-like growth factor binding proteins. *Endocrine Reviews*, *23*(6), 824–854.

Flamant, F., Poguet, A. L., Plateroti, M., Chassande, O., Gauthier, K., Streichenberger, N., et al. (2002). Congenital hypothyroid Pax8(−/−) mutant mice can be rescued by inactivating the TRalpha gene. *Molecular Endocrinology*, *16*(1), 24–32.

Forrest, D., Hanebuth, E., Smeyne, R. J., Everds, N., Stewart, C. L., Wehner, J. M., et al. (1996). Recessive resistance to thyroid hormone in mice lacking thyroid hormone receptor beta: Evidence for tissue-specific modulation of receptor function. *The EMBO Journal*, *15*(12), 3006–3015.

Forsberg, E., & Kjellen, L. (2001). Heparan sulfate: Lessons from knockout mice. *The Journal of Clinical Investigation*, *108*(2), 175–180.

Fraichard, A., Chassande, O., Plateroti, M., Roux, J. P., Trouillas, J., Dehay, C., et al. (1997). The T3R alpha gene encoding a thyroid hormone receptor is essential for post-natal development and thyroid hormone production. *The EMBO Journal*, *16*(14), 4412–4420.

Freedman, L. P. (1992). Anatomy of the steroid receptor zinc finger region. *Endocrine Reviews*, *13*(2), 129–145.

Freitas, F. R., Capelo, L. P., O'Shea, P. J., Jorgetti, V., Moriscot, A. S., Scanlan, T. S., et al. (2005). The thyroid hormone receptor beta-specific agonist GC-1 selectively affects the bone development of hypothyroid rats. *Journal of Bone and Mineral Research*, *20*(2), 294–304.

Freitas, F. R., Moriscot, A. S., Jorgetti, V., Soares, A. G., Passarelli, M., Scanlan, T. S., et al. (2003). Spared bone mass in rats treated with thyroid hormone receptor TR beta-selective compound GC-1. *American Journal of Physiology. Endocrinology and Metabolism*, *285*(5), E1135–E1141.

Friesema, E. C., Jansen, J., Milici, C., & Visser, T. J. (2005). Thyroid hormone transporters. *Vitamins and Hormones*, *70*, 137–167.

Fruhbeck, G. (2006). Intracellular signalling pathways activated by leptin. *The Biochemical Journal*, *393*(Pt 1), 7–20.

Galton, V. A., Wood, E. T., St Germain, E. A., Withrow, C. A., Aldrich, G., St Germain, G. M., et al. (2007). Thyroid hormone homeostasis and action in the type 2 deiodinase-deficient rodent brain during development. *Endocrinology*, *148*(7), 3080–3088.

Gat-Yablonski, G., & Phillip, M. (2008). Leptin and regulation of linear growth. *Current Opinion in Clinical Nutrition and Metabolic Care*, *11*(3), 303–308.

Gauthier, K., Chassande, O., Plateroti, M., Roux, J. P., Legrand, C., Pain, B., et al. (1999). Different functions for the thyroid hormone receptors TRalpha and TRbeta in the control of thyroid hormone production and post-natal development. *The EMBO Journal*, *18*(3), 623–631.

Gauthier, K., Plateroti, M., Harvey, C. B., Williams, G. R., Weiss, R. E., Refetoff, S., et al. (2001). Genetic analysis reveals different functions for the products of the thyroid hormone receptor alpha locus. *Molecular and Cellular Biology*, *21*(14), 4748–4760.

Gevers, E. F., van der Eerden, B. C., Karperien, M., Raap, A. K., Robinson, I. C., & Wit, J. M. (2002). Localization and regulation of the growth hormone receptor and growth hormone-binding protein in the rat growth plate. *Journal of Bone and Mineral Research*, *17*(8), 1408–1419.

Giustina, A., & Veldhuis, J. D. (1998). Pathophysiology of the neuroregulation of growth hormone secretion in experimental animals and the human. *Endocrine Reviews*, *19*(6), 717–797.

Glantschnig, H., Varga, F., & Klaushofer, K. (1996). Thyroid hormone and retinoic acid induce the synthesis of insulin-like growth factor-binding protein-4 in mouse osteoblastic cells. *Endocrinology*, *137*(1), 281–286.

Glass, C. K., Franco, R., Weinberger, C., Albert, V. R., Evans, R. M., & Rosenfeld, M. G. (1987). A c-erb-A binding site in rat growth hormone gene mediates trans-activation by thyroid hormone. *Nature*, *329*(6141), 738–741.

Gothe, S., Wang, Z., Ng, L., Kindblom, J. M., Barros, A. C., Ohlsson, C., et al. (1999). Mice devoid of all known thyroid hormone receptors are viable but exhibit disorders of the pituitary-thyroid axis, growth, and bone maturation. *Genes & Development*, *13*(10), 1329–1341.

Gouveia, C. H., Christoffolete, M. A., Zaitune, C. R., Dora, J. M., Harney, J. W., Maia, A. L., et al. (2005). Type 2 iodothyronine selenodeiodinase is expressed throughout the mouse skeleton and in the MC3T3-E1 mouse osteoblastic cell line during differentiation. *Endocrinology*, *146*(1), 195–200.

Gouveia, C. H., Schultz, J. J., Bianco, A. C., & Brent, G. A. (2001). Thyroid hormone stimulation of osteocalcin gene expression in ROS 17/2.8 cells is mediated by transcriptional and post-transcriptional mechanisms. *The Journal of Endocrinology*, *170*(3), 667–675.

Green, H., Morikawa, M., & Nixon, T. (1985). A dual effector theory of growth-hormone action. *Differentiation*, *29*(3), 195–198.

Grimsrud, C. D., Romano, P. R., D'Souza, M., Puzas, J. E., Reynolds, P. R., Rosier, R. N., et al. (1999). BMP-6 is an autocrine stimulator of chondrocyte differentiation. *Journal of Bone and Mineral Research*, *14*(4), 475–482.

Gruber, R., Czerwenka, K., Wolf, F., Ho, G. M., Willheim, M., & Peterlik, M. (1999). Expression of the vitamin D receptor, of estrogen and thyroid hormone receptor alpha- and beta-isoforms, and of the androgen receptor in cultures of native mouse bone marrow and of stromal/osteoblastic cells. *Bone*, *24*(5), 465–473.

Gu, W. X., Du, G. G., Kopp, P., Rentoumis, A., Albanese, C., Kohn, L. D., et al. (1995). The thyrotropin (TSH) receptor transmembrane domain mutation (Pro556-Leu) in the hypothyroid hyt/hyt mouse results in plasma membrane targeting but defective TSH binding. *Endocrinology*, *136*(7), 3146–3153.

Habuchi, H., Miyake, G., Nogami, K., Kuroiwa, A., Matsuda, Y., Kusche-Gullberg, M., et al. (2003). Biosynthesis of heparan sulphate with diverse structures and functions: Two alternatively spliced forms of human heparan sulphate 6-O-sulphotransferase-2 having different expression patterns and properties. *The Biochemical Journal*, *371*(Pt 1), 131–142.

Habuchi, H., Tanaka, M., Habuchi, O., Yoshida, K., Suzuki, H., Ban, K., et al. (2000). The occurrence of three isoforms of heparan sulfate 6-O-sulfotransferase having different specificities for hexuronic acid adjacent to the targeted N-sulfoglucosamine. *The Journal of Biological Chemistry*, *275*(4), 2859–2868.

Halestrap, A. P. (2012). The monocarboxylate transporter family—Structure and functional characterization. *IUBMB Life*, *64*(1), 1–9.

Hall, B. K., & Miyake, T. (2000). All for one and one for all: Condensations and the initiation of skeletal development. *BioEssays*, *22*(2), 138–147.

Han, C., Belenkaya, T. Y., Wang, B., & Lin, X. (2004). Drosophila glypicans control the cell-to-cell movement of Hedgehog by a dynamin-independent process. *Development*, *131*(3), 601–611.

Harakawa, S., Yamashita, S., Tobinaga, T., Matsuo, K., Hirayu, H., Izumi, M., et al. (1990). In vivo regulation of hepatic insulin-like growth factor-1 messenger ribonucleic acids with thyroid hormone. *Endocrinologia Japonica, 37*(2), 205–211.

Hase, H., Ando, T., Eldeiry, L., Brebene, A., Peng, Y., Liu, L., et al. (2006). TNFalpha mediates the skeletal effects of thyroid-stimulating hormone. *Proceedings of the National Academy of Sciences of the United States of America, 103*(34), 12849–12854.

Hashmi, S. S., Canfield, M. A., Marengo, L., Moffitt, K. B., Belmont, J. W., Freedenberg, D., et al. (2012). The association between neonatal thyroxine and craniosynostosis, Texas, 2004-2007. *Birth Defects Research. Part A, Clinical and Molecular Teratology, 94*(12), 1004–1009.

Hay, E., Faucheu, C., Suc-Royer, I., Touitou, R., Stiot, V., Vayssiere, B., et al. (2005). Interaction between LRP5 and Frat1 mediates the activation of the Wnt canonical pathway. *The Journal of Biological Chemistry, 280*(14), 13616–13623.

Hernandez, A., Martinez, M. E., Fiering, S., Galton, V. A., & St Germain, D. (2006). Type 3 deiodinase is critical for the maturation and function of the thyroid axis. *The Journal of Clinical Investigation, 116*(2), 476–484.

Hernandez, A., Martinez, M. E., Liao, X. H., Van Sande, J., Refetoff, S., Galton, V. A., et al. (2007). Type 3 deiodinase deficiency results in functional abnormalities at multiple levels of the thyroid axis. *Endocrinology, 148*(12), 5680–5687.

Hervas, F., Morreale de Escobar, G., & Escobar Del Rey, F. (1975). Rapid effects of single small doses of L-thyroxine and triiodo-L-thyronine on growth hormone, as studied in the rat by radioimmunoassay. *Endocrinology, 97*(1), 91–101.

Heuer, H., Maier, M. K., Iden, S., Mittag, J., Friesema, E. C., Visser, T. J., et al. (2005). The monocarboxylate transporter 8 linked to human psychomotor retardation is highly expressed in thyroid hormone-sensitive neuron populations. *Endocrinology, 146*(4), 1701–1706.

Heyerman, W., & Weiner, D. (1984). Slipped epiphysis associated with hypothyroidism. *Journal of Pediatric Orthopedics, 4*(5), 569–573.

Hilton, M. J., Tu, X., & Long, F. (2007). Tamoxifen-inducible gene deletion reveals a distinct cell type associated with trabecular bone, and direct regulation of PTHrP expression and chondrocyte morphology by Ihh in growth region cartilage. *Developmental Biology, 308*(1), 93–105.

Hirano, A., Akita, S., & Fujii, T. (1995). Craniofacial deformities associated with juvenile hyperthyroidism. *The Cleft Palate-Craniofacial Journal, 32*(4), 328–333.

Hoffmann, O., Klaushofer, K., Koller, K., Peterlik, M., Mavreas, T., & Stern, P. (1986). Indomethacin inhibits thrombin-, but not thyroxin-stimulated resorption of fetal rat limb bones. *Prostaglandins, 31*(4), 601–608.

Holly, S. S. K., Martim, J., Chappel, J., Teitelbaum, S. L., Cao, X., & Ross, F. P. (1993). Thyroid hormone enhances expression of beta2 and beta5 integrin mRNA by osteoclast precursors. *Journal of Bone and Mineral Research, 8*(Suppl), S136.

Holt, I., Davie, M. W., Braidman, I. P., & Marshall, M. J. (1994). Prostaglandin E2 stimulates the production of interleukin-6 by neonatal mouse parietal bones. *Bone and Mineral, 25*(1), 47–57.

Horton, W. A. (2003). Skeletal development: Insights from targeting the mouse genome. *The Lancet, 362*(9383), 560–569.

Hoskins, W. E., & Asling, C. W. (1977). Influence of growth hormone and thyroxine on endochondral osteogenesis in the mandibular condyle and proximal tibial epiphysis. *Journal of Dental Research, 56*(5), 509–517.

Houseknecht, K. L., Baile, C. A., Matteri, R. L., & Spurlock, M. E. (1998). The biology of leptin: A review. *Journal of Animal Science, 76*(5), 1405–1420.

Howie, R. N., Durham, E. L., Black, L., Bennfors, G., Parsons, T. E., Elsalanty, M. E., et al. (2016). Effects of in utero thyroxine exposure on murine cranial suture growth. *PLoS One, 11*(12), e0167805.

Hsu, H., Lacey, D. L., Dunstan, C. R., Solovyev, I., Colombero, A., Timms, E., et al. (1999). Tumor necrosis factor receptor family member RANK mediates osteoclast differentiation and activation induced by osteoprotegerin ligand. *Proceedings of the National Academy of Sciences of the United States of America*, *96*(7), 3540–3545.

Hu, X., & Lazar, M. A. (2000). Transcriptional repression by nuclear hormone receptors. *Trends in Endocrinology and Metabolism*, *11*(1), 6–10.

Huang, B. K., Golden, L. A., Tarjan, G., Madison, L. D., & Stern, P. H. (2000). Insulin-like growth factor I production is essential for anabolic effects of thyroid hormone in osteoblasts. *Journal of Bone and Mineral Research*, *15*(2), 188–197.

Hung, I. H., Yu, K., Lavine, K. J., & Ornitz, D. M. (2007). FGF9 regulates early hypertrophic chondrocyte differentiation and skeletal vascularization in the developing stylopod. *Developmental Biology*, *307*(2), 300–313.

Inoue, M., Tawata, M., Yokomori, N., Endo, T., & Onaya, T. (1998). Expression of thyrotropin receptor on clonal osteoblast-like rat osteosarcoma cells. *Thyroid*, *8*(11), 1059–1064.

Ishikawa, Y., Genge, B. R., Wuthier, R. E., & Wu, L. N. (1998). Thyroid hormone inhibits growth and stimulates terminal differentiation of epiphyseal growth plate chondrocytes. *Journal of Bone and Mineral Research*, *13*(9), 1398–1411.

Jacob, A. L., Smith, C., Partanen, J., & Ornitz, D. M. (2006). Fibroblast growth factor receptor 1 signaling in the osteo-chondrogenic cell lineage regulates sequential steps of osteoblast maturation. *Developmental Biology*, *296*(2), 315–328.

Jeong, J., & McMahon, A. P. (2002). Cholesterol modification of Hedgehog family proteins. *The Journal of Clinical Investigation*, *110*(5), 591–596.

Kamegai, J., Tamura, H., Ishii, S., Sugihara, H., & Wakabayashi, I. (2001). Thyroid hormones regulate pituitary growth hormone secretagogue receptor gene expression. *Journal of Neuroendocrinology*, *13*(3), 275–278.

Kanatani, M., Sugimoto, T., Sowa, H., Kobayashi, T., Kanzawa, M., & Chihara, K. (2004). Thyroid hormone stimulates osteoclast differentiation by a mechanism independent of RANKL-RANK interaction. *Journal of Cellular Physiology*, *201*(1), 17–25.

Kaneshige, M., Kaneshige, K., Zhu, X., Dace, A., Garrett, L., Carter, T. A., et al. (2000). Mice with a targeted mutation in the thyroid hormone beta receptor gene exhibit impaired growth and resistance to thyroid hormone. *Proceedings of the National Academy of Sciences of the United States of America*, *97*(24), 13209–13214.

Kaneshige, M., Suzuki, H., Kaneshige, K., Cheng, J., Wimbrow, H., Barlow, C., et al. (2001). A targeted dominant negative mutation of the thyroid hormone alpha 1 receptor causes increased mortality, infertility, and dwarfism in mice. *Proceedings of the National Academy of Sciences of the United States of America*, *98*(26), 15095–15100.

Karl, A., Olbrich, N., Pfeifer, C., Berner, A., Zellner, J., Kujat, R., et al. (2014). Thyroid hormone-induced hypertrophy in mesenchymal stem cell chondrogenesis is mediated by bone morphogenetic protein-4. *Tissue Engineering. Part A*, *20*(1–2), 178–188.

Karolak, M. R., Yang, X., & Elefteriou, F. (2015). FGFR1 signaling in hypertrophic chondrocytes is attenuated by the Ras-GAP neurofibromin during endochondral bone formation. *Human Molecular Genetics*, *24*(9), 2552–2564.

Kasono, K., Sato, K., Han, D. C., Fujii, Y., Tsushima, T., & Shizume, K. (1988). Stimulation of alkaline phosphatase activity by thyroid hormone in mouse osteoblast-like cells (MC3T3-E1): A possible mechanism of hyperalkaline phosphatasia in hyperthyroidism. *Bone and Mineral*, *4*(4), 355–363.

Kassem, M., Mosekilde, L., & Eriksen, E. F. (1993). Effects of triiodothyronine on DNA synthesis and differentiation markers of normal human osteoblast-like cells in vitro. *Biochemistry and Molecular Biology International*, *30*(4), 779–788.

Katz, D., & Lazar, M. A. (1993). Dominant negative activity of an endogenous thyroid hormone receptor variant (alpha 2) is due to competition for binding sites on target genes. *The Journal of Biological Chemistry*, *268*(28), 20904–20910.

Kerkhof, H. J., Lories, R. J., Meulenbelt, I., Jonsdottir, I., Valdes, A. M., Arp, P., et al. (2010). A genome-wide association study identifies an osteoarthritis susceptibility locus on chromosome 7q22. *Arthritis and Rheumatism, 62*(2), 499–510.

Kiepe, D., Ciarmatori, S., Hoeflich, A., Wolf, E., & Tonshoff, B. (2005). Insulin-like growth factor (IGF)-I stimulates cell proliferation and induces IGF binding protein (IGFBP)-3 and IGFBP-5 gene expression in cultured growth plate chondrocytes via distinct signaling pathways. *Endocrinology, 146*(7), 3096–3104.

Kindblom, J. M., Gothe, S., Forrest, D., Tornell, J., Vennstrom, B., & Ohlsson, C. (2001). GH substitution reverses the growth phenotype but not the defective ossification in thyroid hormone receptor alpha 1 −/−beta −/− mice. *The Journal of Endocrinology, 171*(1), 15–22.

Kindblom, J. M., Roboson, H., Stevens, D. A., Siebler, T., Gothe, S., Tornell, J., et al. (1999). Regulation of the Ihh/PTHrP negative loop in growth plate of mice lacking all known thyroid hormone receptors. *Journal of Bone and Mineral Research, 14*(Suppl. 1), S140.

King, R. A., Smith, R. M., Meller, D. J., Dahlenburg, G. W., & Lineham, J. D. (1988). Effect of growth hormone on growth and myelination in the neonatal hypothyroid rat. *The Journal of Endocrinology, 119*(1), 117–125.

Kishida, Y., Hirao, M., Tamai, N., Nampei, A., Fujimoto, T., Nakase, T., et al. (2005). Leptin regulates chondrocyte differentiation and matrix maturation during endochondral ossification. *Bone, 37*(5), 607–621.

Klaushofer, K., Hoffmann, O., Gleispach, H., Leis, H. J., Czerwenka, E., Koller, K., et al. (1989). Bone-resorbing activity of thyroid hormones is related to prostaglandin production in cultured neonatal mouse calvaria. *Journal of Bone and Mineral Research, 4*(3), 305–312.

Kobayashi, T., Soegiarto, D. W., Yang, Y., Lanske, B., Schipani, E., McMahon, A. P., et al. (2005). Indian hedgehog stimulates periarticular chondrocyte differentiation to regulate growth plate length independently of PTHrP. *The Journal of Clinical Investigation, 115*(7), 1734–1742.

Koenig, R. J., Brent, G. A., Warne, R. L., Larsen, P. R., & Moore, D. D. (1987). Thyroid hormone receptor binds to a site in the rat growth hormone promoter required for induction by thyroid hormone. *Proceedings of the National Academy of Sciences of the United States of America, 84*(16), 5670–5674.

Kosher, R. A., Kulyk, W. M., & Gay, S. W. (1986). Collagen gene expression during limb cartilage differentiation. *The Journal of Cell Biology, 102*(4), 1151–1156.

Koziel, L., Kunath, M., Kelly, O. G., & Vortkamp, A. (2004). Ext1-dependent heparan sulfate regulates the range of Ihh signaling during endochondral ossification. *Developmental Cell, 6*(6), 801–813.

Krishnan, V., Bryant, H. U., & Macdougald, O. A. (2006). Regulation of bone mass by Wnt signaling. *The Journal of Clinical Investigation, 116*(5), 1202–1209.

Kusano, K., Miyaura, C., Inada, M., Tamura, T., Ito, A., Nagase, H., et al. (1998). Regulation of matrix metalloproteinases (MMP-2, -3, -9, and -13) by interleukin-1 and interleukin-6 in mouse calvaria: Association of MMP induction with bone resorption. *Endocrinology, 139*(3), 1338–1345.

Lanske, B., Amling, M., Neff, L., Guiducci, J., Baron, R., & Kronenberg, H. M. (1999). Ablation of the PTHrP gene or the PTH/PTHrP receptor gene leads to distinct abnormalities in bone development. *The Journal of Clinical Investigation, 104*(4), 399–407.

Lassova, L., Niu, Z., Golden, E. B., Cohen, A. J., & Adams, S. L. (2009). Thyroid hormone treatment of cultured chondrocytes mimics in vivo stimulation of collagen X mRNA by increasing BMP 4 expression. *Journal of Cellular Physiology, 219*(3), 595–605.

Lazarenko, O. P., Rzonca, S. O., Hogue, W. R., Swain, F. L., Suva, L. J., & Lecka-Czernik, B. (2007). Rosiglitazone induces decreases in bone mass and strength that are reminiscent of aged bone. *Endocrinology, 148*(6), 2669–2680.

Leboy, P. S., Sullivan, T. A., Nooreyazdan, M., & Venezian, R. A. (1997). Rapid chondrocyte maturation by serum-free culture with BMP-2 and ascorbic acid. *Journal of Cellular Biochemistry*, *66*(3), 394–403.

LeBron, B. A., Pekary, A. E., Mirell, C., Hahn, T. J., & Hershman, J. M. (1989). Thyroid hormone 5′-deiodinase activity, nuclear binding, and effects on mitogenesis in UMR-106 osteoblastic osteosarcoma cells. *Journal of Bone and Mineral Research*, *4*(2), 173–178.

Lefebvre, V., Huang, W., Harley, V. R., Goodfellow, P. N., & de Crombrugghe, B. (1997). SOX9 is a potent activator of the chondrocyte-specific enhancer of the pro alpha1(II) collagen gene. *Molecular and Cellular Biology*, *17*(4), 2336–2346.

Levy, L., Wei, Y., Labalette, C., Wu, Y., Renard, C. A., Buendia, M. A., et al. (2004). Acetylation of beta-catenin by p300 regulates beta-catenin-Tcf4 interaction. *Molecular and Cellular Biology*, *24*(8), 3404–3414.

Lewinson, D., Harel, Z., Shenzer, P., Silbermann, M., & Hochberg, Z. (1989). Effect of thyroid hormone and growth hormone on recovery from hypothyroidism of epiphyseal growth plate cartilage and its adjacent bone. *Endocrinology*, *124*(2), 937–945.

Li, J., & Dong, S. (2016). The signaling pathways involved in chondrocyte differentiation and hypertrophic differentiation. *Stem Cells International*, 2016, 12, Article ID 2470351. https://dx.doi.org/10.1155/2016/2470351.

Li, X., Udagawa, N., Itoh, K., Suda, K., Murase, Y., Nishihara, T., et al. (2002). p38 MAPK-mediated signals are required for inducing osteoclast differentiation but not for osteoclast function. *Endocrinology*, *143*(8), 3105–3113.

Liao, X. H., Di Cosmo, C., Dumitrescu, A. M., Hernandez, A., Van Sande, J., St Germain, D. L., et al. (2011). Distinct roles of deiodinases on the phenotype of Mct8 defect: A comparison of eight different mouse genotypes. *Endocrinology*, *152*(3), 1180–1191.

Lindahl, A., Isgaard, J., Carlsson, L., & Isaksson, O. G. (1987). Differential effects of growth hormone and insulin-like growth factor I on colony formation of epiphyseal chondrocytes in suspension culture in rats of different ages. *Endocrinology*, *121*(3), 1061–1069.

Liu, J. P., Baker, J., Perkins, A. S., Robertson, E. J., & Efstratiadis, A. (1993). Mice carrying null mutations of the genes encoding insulin-like growth factor I (Igf-1) and type 1 IGF receptor (Igf1r). *Cell*, *75*(1), 59–72.

Liu, Y. Y., Schultz, J. J., & Brent, G. A. (2003). A thyroid hormone receptor alpha gene mutation (P398H) is associated with visceral adiposity and impaired catecholamine-stimulated lipolysis in mice. *The Journal of Biological Chemistry*, *278*(40), 38913–38920.

Loder, R. T., Wittenberg, B., & DeSilva, G. (1995). Slipped capital femoral epiphysis associated with endocrine disorders. *Journal of Pediatric Orthopedics*, *15*(3), 349–356.

Long, F., Chung, U. I., Ohba, S., McMahon, J., Kronenberg, H. M., & McMahon, A. P. (2004). Ihh signaling is directly required for the osteoblast lineage in the endochondral skeleton. *Development*, *131*(6), 1309–1318.

Long, F., & Ornitz, D. M. (2013). Development of the endochondral skeleton. *Cold Spring Harbor Perspectives in Biology*, *5*(1), a008334.

Long, F., Zhang, X. M., Karp, S., Yang, Y., & McMahon, A. P. (2001). Genetic manipulation of hedgehog signaling in the endochondral skeleton reveals a direct role in the regulation of chondrocyte proliferation. *Development*, *128*(24), 5099–5108.

Lotz, M., Blanco, F. J., von Kempis, J., Dudler, J., Maier, R., Villiger, P. M., et al. (1995). Cytokine regulation of chondrocyte functions. *The Journal of Rheumatology. Supplement*, *43*, 104–108.

Ma, R., Morshed, S., Latif, R., Zaidi, M., & Davies, T. F. (2011). The influence of thyroid-stimulating hormone and thyroid-stimulating hormone receptor antibodies on osteoclastogenesis. *Thyroid*, *21*(8), 897–906.

Mackie, E. J., Ahmed, Y. A., Tatarczuch, L., Chen, K. S., & Mirams, M. (2008). Endochondral ossification: How cartilage is converted into bone in the developing skeleton. *The International Journal of Biochemistry & Cell Biology*, *40*(1), 46–62.

Mackie, E. J., Tatarczuch, L., & Mirams, M. (2011). The skeleton: A multi-functional complex organ: The growth plate chondrocyte and endochondral ossification. *The Journal of Endocrinology*, *211*(2), 109–121.
Makihira, S., Yan, W., Murakami, H., Furukawa, M., Kawai, T., Nikawa, H., et al. (2003). Thyroid hormone enhances aggrecanase-2/ADAM-TS5 expression and proteoglycan degradation in growth plate cartilage. *Endocrinology*, *144*(6), 2480–2488.
Mansouri, A., Chowdhury, K., & Gruss, P. (1998). Follicular cells of the thyroid gland require Pax8 gene function. *Nature Genetics*, *19*(1), 87–90.
Mao, B., Wu, W., Davidson, G., Marhold, J., Li, M., Mechler, B. M., et al. (2002). Kremen proteins are Dickkopf receptors that regulate Wnt/beta-catenin signalling. *Nature*, *417*(6889), 664–667.
Maor, G., Rochwerger, M., Segev, Y., & Phillip, M. (2002). Leptin acts as a growth factor on the chondrocytes of skeletal growth centers. *Journal of Bone and Mineral Research*, *17*(6), 1034–1043.
Maranduba, C. M., Friesema, E. C., Kok, F., Kester, M. H., Jansen, J., Sertie, A. L., et al. (2006). Decreased cellular uptake and metabolism in Allan-Herndon-Dudley syndrome (AHDS) due to a novel mutation in the MCT8 thyroid hormone transporter. *Journal of Medical Genetics*, *43*(5), 457–460.
Mariotta, L., Ramadan, T., Singer, D., Guetg, A., Herzog, B., Stoeger, C., et al. (2012). T-type amino acid transporter TAT1 (Slc16a10) is essential for extracellular aromatic amino acid homeostasis control. *The Journal of Physiology*, *590*(24), 6413–6424.
Matsuo, K., Owens, J. M., Tonko, M., Elliott, C., Chambers, T. J., & Wagner, E. F. (2000). Fosl1 is a transcriptional target of c-Fos during osteoclast differentiation. *Nature Genetics*, *24*(2), 184–187.
Matsushita, T., Wilcox, W. R., Chan, Y. Y., Kawanami, A., Bukulmez, H., Balmes, G., et al. (2009). FGFR3 promotes synchondrosis closure and fusion of ossification centers through the MAPK pathway. *Human Molecular Genetics*, *18*, 227–240.
Mayerl, S., Muller, J., Bauer, R., Richert, S., Kassmann, C. M., Darras, V. M., et al. (2014). Transporters MCT8 and OATP1C1 maintain murine brain thyroid hormone homeostasis. *The Journal of Clinical Investigation*, *124*(5), 1987–1999.
Mayerl, S., Visser, T. J., Darras, V. M., Horn, S., & Heuer, H. (2012). Impact of Oatp1c1 deficiency on thyroid hormone metabolism and action in the mouse brain. *Endocrinology*, *153*(3), 1528–1537.
Melmed, S., & Yamashita, S. (1986). Insulin-like growth factor-I action on hypothyroid rat pituitary cells: Suppression of triiodothyronine-induced growth hormone secretion and messenger ribonucleic acid levels. *Endocrinology*, *118*(4), 1483–1490.
Menking, M., Wiebel, J., Schmid, W. U., Schmidt, W. T., Ebel, K. D., & Ritter, R. (1972). Premature craniosynostosis associated with hyperthyroidism in 4 children with reference to 5 further cases in the literature. *Monatsschrift für Kinderheilkunde*, *120*(3), 106–110.
Milne, M., Kang, M. I., Cardona, G., Quail, J. M., Braverman, L. E., Chin, W. W., et al. (1999). Expression of multiple thyroid hormone receptor isoforms in rat femoral and vertebral bone and in bone marrow osteogenic cultures. *Journal of Cellular Biochemistry*, *74*(4), 684–693.
Milne, M., Kang, M. I., Quail, J. M., & Baran, D. T. (1998). Thyroid hormone excess increases insulin-like growth factor I transcripts in bone marrow cell cultures: Divergent effects on vertebral and femoral cell cultures. *Endocrinology*, *139*(5), 2527–2534.
Miura, T., Perlyn, C. A., Kinboshi, M., Ogihara, N., Kobayashi-Miura, M., Morriss-Kay, G. M., et al. (2009). Mechanism of skull suture maintenance and interdigitation. *Journal of Anatomy*, *215*(6), 642–655.
Miura, M., Tanaka, K., Komatsu, Y., Suda, M., Yasoda, A., Sakuma, Y., et al. (2002a). A novel interaction between thyroid hormones and 1,25(OH)(2)D(3) in osteoclast formation. *Biochemical and Biophysical Research Communications*, *291*(4), 987–994.

Miura, M., Tanaka, K., Komatsu, Y., Suda, M., Yasoda, A., Sakuma, Y., et al. (2002b). Thyroid hormones promote chondrocyte differentiation in mouse ATDC5 cells and stimulate endochondral ossification in fetal mouse tibias through iodothyronine deiodinases in the growth plate. *Journal of Bone and Mineral Research*, *17*(3), 443–454.

Mixson, A. J., Hauser, P., Tennyson, G., Renault, J. C., Bodenner, D. L., & Weintraub, B. D. (1993). Differential expression of mutant and normal beta T3 receptor alleles in kindreds with generalized resistance to thyroid hormone. *The Journal of Clinical Investigation*, *91*(5), 2296–2300.

Mohan, S., Nakao, Y., Honda, Y., Landale, E., Leser, U., Dony, C., et al. (1995). Studies on the mechanisms by which insulin-like growth factor (IGF) binding protein-4 (IGFBP-4) and IGFBP-5 modulate IGF actions in bone cells. *The Journal of Biological Chemistry*, *270*(35), 20424–20431.

Monfoulet, L. E., Rabier, B., Dacquin, R., Anginot, A., Photsavang, J., Jurdic, P., et al. (2011). Thyroid hormone receptor beta mediates thyroid hormone effects on bone remodeling and bone mass. *Journal of Bone and Mineral Research*, *26*(9), 2036–2044.

Moran, C., Agostini, M., Visser, W. E., Schoenmakers, E., Schoenmakers, N., Offiah, A. C., et al. (2014). Resistance to thyroid hormone caused by a mutation in thyroid hormone receptor (TR)alpha1 and TRalpha2: Clinical, biochemical, and genetic analyses of three related patients. *The Lancet Diabetes and Endocrinology*, *2*(8), 619–626.

Moran, C., & Chatterjee, K. (2015). Resistance to thyroid hormone due to defective thyroid receptor alpha. *Best Practice & Research. Clinical Endocrinology & Metabolism*, *29*(4), 647–657.

Moran, C., Schoenmakers, N., Agostini, M., Schoenmakers, E., Offiah, A., Kydd, A., et al. (2013). An adult female with resistance to thyroid hormone mediated by defective thyroid hormone receptor alpha. *The Journal of Clinical Endocrinology and Metabolism*, *98*(11), 4254–4261.

Morimura, T., Tsunekawa, K., Kasahara, T., Seki, K., Ogiwara, T., Mori, M., et al. (2005). Expression of type 2 iodothyronine deiodinase in human osteoblast is stimulated by thyrotropin. *Endocrinology*, *146*(4), 2077–2084.

Morte, B., Manzano, J., Scanlan, T., Vennstrom, B., & Bernal, J. (2002). Deletion of the thyroid hormone receptor alpha 1 prevents the structural alterations of the cerebellum induced by hypothyroidism. *Proceedings of the National Academy of Sciences of the United States of America*, *99*(6), 3985–3989.

Mosekilde, L., & Melsen, F. (1978a). Effect of antithyroid treatment on calcium-phosphorus metabolism in hyperthyroidism. II: Bone histomorphometry. *Acta Endocrinologica*, *87*(4), 751–758.

Mosekilde, L., & Melsen, F. (1978b). A tetracycline-based histomorphometric evaluation of bone resorption and bone turnover in hyperthyroidism and hyperparathyroidism. *Acta Medica Scandinavica*, *204*(1–2), 97–102.

Mosekilde, L., Melsen, F., Bagger, J. P., Myhre-Jensen, O., & Schwartz Sorensen, N. (1977). Bone changes in hyperthyroidism: Interrelationships between bone morphometry, thyroid function and calcium-phosphorus metabolism. *Acta Endocrinologica*, *85*(3), 515–525.

Mugniery, E., Dacquin, R., Marty, C., Benoist-Lasselin, C., de Vernejoul, M. C., Jurdic, P., et al. (2012). An activating Fgfr3 mutation affects trabecular bone formation via a paracrine mechanism during growth. *Human Molecular Genetics*, *21*(11), 2503–2513.

Muller, J., Mayerl, S., Visser, T. J., Darras, V. M., Boelen, A., Frappart, L., et al. (2014). Tissue-specific alterations in thyroid hormone homeostasis in combined Mct10 and Mct8 deficiency. *Endocrinology*, *155*(1), 315–325.

Mundy, G. R., Shapiro, J. L., Bandelin, J. G., Canalis, E. M., & Raisz, L. G. (1976). Direct stimulation of bone resorption by thyroid hormones. *The Journal of Clinical Investigation*, *58*(3), 529–534.

Murakami, S., Balmes, G., McKinney, S., Zhang, Z., Givol, D., & de Crombrugghe, B. (2004). Constitutive activation of MEK1 in chondrocytes causes Stat1-independent achondroplasia-like dwarfism and rescues the Fgfr3-deficient mouse phenotype. *Genes & Development, 18*(3), 290–305.

Murray, S. S., Brochmann Murray, E. J., Wang, J. C., & Duarte, M. E. (2016). The history and histology of bone morphogenetic protein. *Histology and Histopathology, 31*(7), 721–732.

Murray, P. G., & Clayton, P. E. (2013). Endocrine control of growth. *American Journal of Medical Genetics. Part C, Seminars in Medical Genetics, 163C*, 76–85.

Nagase, H., Nagasawa, Y., Tachida, Y., Sakakibara, S., Okutsu, J., Suematsu, N., et al. (2013). Deiodinase 2 upregulation demonstrated in osteoarthritis patients cartilage causes cartilage destruction in tissue-specific transgenic rats. *Osteoarthritis and Cartilage, 21*(3), 514–523.

Ng, L. J., Wheatley, S., Muscat, G. E., Conway-Campbell, J., Bowles, J., Wright, E., et al. (1997). SOX9 binds DNA, activates transcription, and coexpresses with type II collagen during chondrogenesis in the mouse. *Developmental Biology, 183*(1), 108–121.

Ohishi, K., Ishida, H., Nagata, T., Yamauchi, N., Tsurumi, C., Nishikawa, S., et al. (1994). Thyroid hormone suppresses the differentiation of osteoprogenitor cells to osteoblasts, but enhances functional activities of mature osteoblasts in cultured rat calvaria cells. *Journal of Cellular Physiology, 161*(3), 544–552.

Ohlsson, C., Bengtsson, B. A., Isaksson, O. G., Andreassen, T. T., & Slootweg, M. C. (1998). Growth hormone and bone. *Endocrine Reviews, 19*(1), 55–79.

Ohlsson, C., Nilsson, A., Isaksson, O., Bentham, J., & Lindahl, A. (1992). Effects of triiodothyronine and insulin-like growth factor-I (IGF-I) on alkaline phosphatase activity, [3H]thymidine incorporation and IGF-I receptor mRNA in cultured rat epiphyseal chondrocytes. *The Journal of Endocrinology, 135*(1), 115–123.

Okubo, Y., & Reddi, A. H. (2003). Thyroxine downregulates Sox9 and promotes chondrocyte hypertrophy. *Biochemical and Biophysical Research Communications, 306*(1), 186–190.

Onigata, K., & Szinnai, G. (2014). Resistance to thyroid hormone. *Endocrine Development, 26*, 118–129.

Ono, S., Schwartz, I. D., Mueller, O. T., Root, A. W., Usala, S. J., & Bercu, B. B. (1991). Homozygosity for a dominant negative thyroid hormone receptor gene responsible for generalized resistance to thyroid hormone. *The Journal of Clinical Endocrinology and Metabolism, 73*(5), 990–994.

Opitz, R., Lutz, I., Nguyen, N. H., Scanlan, T. S., & Kloas, W. (2006). Analysis of thyroid hormone receptor betaA mRNA expression in *Xenopus laevis* tadpoles as a means to detect agonism and antagonism of thyroid hormone action. *Toxicology and Applied Pharmacology, 212*(1), 1–13.

Ornitz, D. M., & Itoh, N. (2015). The fibroblast growth factor signaling pathway. *Wiley Interdisciplinary Reviews: Developmental Biology, 4*(3), 215–266.

Ornitz, D. M., & Marie, P. J. (2002). FGF signaling pathways in endochondral and intramembranous bone development and human genetic disease. *Genes & Development, 16*(12), 1446–1465.

Ornitz, D. M., & Marie, P. J. (2015). Fibroblast growth factor signaling in skeletal development and disease. *Genes & Development, 29*(14), 1463–1486.

Ortiga-Carvalho, T. M., Chiamolera, M. I., Pazos-Moura, C. C., & Wondisford, F. E. (2016). Hypothalamus-pituitary-thyroid Axis. *Comprehensive Physiology, 6*(3), 1387–1428.

Ortiga-Carvalho, T. M., Sidhaye, A. R., & Wondisford, F. E. (2014). Thyroid hormone receptors and resistance to thyroid hormone disorders. *Nature Reviews. Endocrinology, 10*(10), 582–591.

O'Shea, P. J., Bassett, J. H., Sriskantharajah, S., Ying, H., Cheng, S. Y., & Williams, G. R. (2005). Contrasting skeletal phenotypes in mice with an identical mutation targeted to thyroid hormone receptor alpha1 or beta. *Molecular Endocrinology, 19*(12), 3045–3059.

O'Shea, P. J., Harvey, C. B., Suzuki, H., Kaneshige, M., Kaneshige, K., Cheng, S. Y., et al. (2003). A thyrotoxic skeletal phenotype of advanced bone formation in mice with resistance to thyroid hormone. *Molecular Endocrinology*, *17*(7), 1410–1424.

Owens, J. M., Matsuo, K., Nicholson, G. C., Wagner, E. F., & Chambers, T. J. (1999). Fra-1 potentiates osteoclastic differentiation in osteoclast-macrophage precursor cell lines. *Journal of Cellular Physiology*, *179*(2), 170–178.

Pasca di Magliano, M., Di Lauro, R., & Zannini, M. (2000). Pax8 has a key role in thyroid cell differentiation. *Proceedings of the National Academy of Sciences of the United States of America*, *97*(24), 13144–13149.

Pellizas, C. G., Coleoni, A. H., Costamagna, M. E., Di Fulvio, M., & Masini-Repiso, A. M. (1998). Insulin-like growth factor I reduces thyroid hormone receptors in the rat liver. Evidence for a feed-back loop regulating the peripheral thyroid hormone action. *The Journal of Endocrinology*, *158*(1), 87–95.

Pepene, C. E., Kasperk, C. H., Pfeilschifter, J., Borcsok, I., Gozariu, L., Ziegler, R., et al. (2001). Effects of triiodothyronine on the insulin-like growth factor system in primary human osteoblastic cells in vitro. *Bone*, *29*(6), 540–546.

Pereira, R. C., Jorgetti, V., & Canalis, E. (1999). Triiodothyronine induces collagenase-3 and gelatinase B expression in murine osteoblasts. *The American Journal of Physiology*, *277*(3 Pt 1), E496–E504.

Phornphutkul, C., Wu, K. Y., Yang, X., Chen, Q., & Gruppuso, P. A. (2004). Insulin-like growth factor-I signaling is modified during chondrocyte differentiation. *The Journal of Endocrinology*, *183*(3), 477–486.

Qiu, J., Ma, X. L., Wang, X., Chen, H., & Huang, B. R. (2012). Insulin-like growth factor binding protein-6 interacts with the thyroid hormone receptor alpha1 and modulates the thyroid hormone-response in osteoblastic differentiation. *Molecular and Cellular Biochemistry*, *361*(1–2), 197–208.

Quignodon, L., Vincent, S., Winter, H., Samarut, J., & Flamant, F. (2007). A point mutation in the activation function 2 domain of thyroid hormone receptor alpha1 expressed after CRE-mediated recombination partially recapitulates hypothyroidism. *Molecular Endocrinology*, *21*(10), 2350–2360.

Rabier, B., Williams, A. J., Mallein-Gerin, F., Williams, G. R., & Chassande, O. (2006). Thyroid hormone-stimulated differentiation of primary rib chondrocytes in vitro requires thyroid hormone receptor beta. *The Journal of Endocrinology*, *191*(1), 221–228.

Radetti, G., Zavallone, A., Gentili, L., Beck-Peccoz, P., & Bona, G. (2002). Foetal and neonatal thyroid disorders. *Minerva Pediatrica*, *54*(5), 383–400.

Rasmussen, S. A., Yazdy, M. M., Carmichael, S. L., Jamieson, D. J., Canfield, M. A., & Honein, M. A. (2007). Maternal thyroid disease as a risk factor for craniosynostosis. *Obstetrics and Gynecology*, *110*(2 Pt 1), 369–377.

Ray, R. D., Asling, C. W., Walker, D. G., Simpson, M. E., Li, C. H., & Evans, H. M. (1954). Growth and differentiation of the skeleton in thyroidectomized-hypophysectomized rats treated with thyroxin, growth hormone, and combination. *The Journal of Bone and Joint Surgery. American Volume*, *36-A*(1), 94–103.

Reddy, P. A., Harinarayan, C. V., Sachan, A., Suresh, V., & Rajagopal, G. (2012). Bone disease in thyrotoxicosis. *The Indian Journal of Medical Research*, *135*, 277–286.

Refetoff, S., Weiss, R. E., & Usala, S. J. (1993). The syndromes of resistance to thyroid hormone. *Endocrine Reviews*, *14*(3), 348–399.

Ribeiro, M. O., Carvalho, S. D., Schultz, J. J., Chiellini, G., Scanlan, T. S., Bianco, A. C., et al. (2001). Thyroid hormone—Sympathetic interaction and adaptive thermogenesis are thyroid hormone receptor isoform—Specific. *The Journal of Clinical Investigation*, *108*(1), 97–105.

Rice, D. P., & Rice, R. (2008). Locate, condense, differentiate, grow and confront: Developmental mechanisms controlling intramembranous bone and suture formation and function. *Frontiers of Oral Biology*, *12*, 22–40.

Rice, D. P., Rice, R., & Thesleff, I. (2003). Fgfr mRNA isoforms in craniofacial bone development. *Bone, 33*(1), 14–27.

Riggs, W., Jr., Wilroy, R. S., Jr., & Etteldorf, J. N. (1972). Neonatal hyperthyroidism with accelerated skeletal maturation, craniosynostosis, and brachydactyly. *Radiology, 105*(3), 621–625.

Rivkees, S. A., Bode, H. H., & Crawford, J. D. (1988). Long-term growth in juvenile acquired hypothyroidism: The failure to achieve normal adult stature. *The New England Journal of Medicine, 318*(10), 599–602.

Rizzoli, R., Poser, J., & Burgi, U. (1986). Nuclear thyroid hormone receptors in cultured bone cells. *Metabolism, 35*(1), 71–74.

Robson, H., Siebler, T., Stevens, D. A., Shalet, S. M., & Williams, G. R. (2000). Thyroid hormone acts directly on growth plate chondrocytes to promote hypertrophic differentiation and inhibit clonal expansion and cell proliferation. *Endocrinology, 141*(10), 3887–3897.

Ross, D. S. (1994). Hyperthyroidism, thyroid hormone therapy, and bone. *Thyroid, 4*(3), 319–326.

Sakurai, A., Miyamoto, T., & DeGroot, L. J. (1992). Cloning and characterization of the human thyroid hormone receptor beta 1 gene promoter. *Biochemical and Biophysical Research Communications, 185*(1), 78–84.

Sakurai, A., Takeda, K., Ain, K., Ceccarelli, P., Nakai, A., Seino, S., et al. (1989). Generalized resistance to thyroid hormone associated with a mutation in the ligand-binding domain of the human thyroid hormone receptor beta. *Proceedings of the National Academy of Sciences of the United States of America, 86*(22), 8977–8981.

Salto, C., Kindblom, J. M., Johansson, C., Wang, Z., Gullberg, H., Nordstrom, K., et al. (2001). Ablation of TRalpha2 and a concomitant overexpression of alpha1 yields a mixed hypo- and hyperthyroid phenotype in mice. *Molecular Endocrinology, 15*(12), 2115–2128.

Sampath, T. K., Simic, P., Sendak, R., Draca, N., Bowe, A. E., O'Brien, S., et al. (2007). Thyroid-stimulating hormone restores bone volume, microarchitecture, and strength in aged ovariectomized rats. *Journal of Bone and Mineral Research, 22*(6), 849–859.

Sato, K., Han, D. C., Fujii, Y., Tsushima, T., & Shizume, K. (1987). Thyroid hormone stimulates alkaline phosphatase activity in cultured rat osteoblastic cells (ROS 17/2.8) through 3,5,3′-triiodo-L-thyronine nuclear receptors. *Endocrinology, 120*(5), 1873–1881.

Schiller, C., Gruber, R., Ho, G. M., Redlich, K., Gober, H. J., Katzgraber, F., et al. (1998). Interaction of triiodothyronine with 1alpha,25-dihydroxyvitamin D3 on interleukin-6-dependent osteoclast-like cell formation in mouse bone marrow cell cultures. *Bone, 22*(4), 341–346.

Schlessinger, J., Plotnikov, A. N., Ibrahimi, O. A., Eliseenkova, A. V., Yeh, B. K., Yayon, A., et al. (2000). Crystal structure of a ternary FGF-FGFR-heparin complex reveals a dual role for heparin in FGFR binding and dimerization. *Molecular Cell, 6*(3), 743–750.

Schmid, T. M., & Linsenmayer, T. F. (1985). Immunohistochemical localization of short chain cartilage collagen (type X) in avian tissues. *The Journal of Cell Biology, 100*(2), 598–605.

Schmid, C., Zwimpfer, C., Brandle, M., Krayenbuhl, P. A., Zapf, J., & Wiesli, P. (2006). Effect of thyroxine replacement on serum IGF-I, IGFBP-3 and the acid-labile subunit in patients with hypothyroidism and hypopituitarism. *Clinical Endocrinology, 65*(6), 706–711.

Schneider, M. J., Fiering, S. N., Thai, B., Wu, S. Y., St Germain, E., Parlow, A. F., et al. (2006). Targeted disruption of the type 1 selenodeiodinase gene (Dio1) results in marked changes in thyroid hormone economy in mice. *Endocrinology, 147*(1), 580–589.

Sgariglia, F., Candela, M. E., Huegel, J., Jacenko, O., Koyama, E., Yamaguchi, Y., et al. (2013). Epiphyseal abnormalities, trabecular bone loss and articular chondrocyte hypertrophy develop in the long bones of postnatal Ext1-deficient mice. *Bone, 57*(1), 220–231.

Shao, Y. Y., Wang, L., & Ballock, R. T. (2006). Thyroid hormone and the growth plate. *Reviews in Endocrine & Metabolic Disorders*, 7(4), 265–271.

Shao, Y. Y., Wang, L., Hicks, D. G., Tarr, S., & Ballock, R. T. (2005). Expression and activation of peroxisome proliferator-activated receptors in growth plate chondrocytes. *Journal of Orthopaedic Research*, *23*(5), 1139–1145.

Shim, K. S. (2015). Pubertal growth and epiphyseal fusion. *Annals of Pediatric Endocrinology & Metabolism*, *20*(1), 8–12.

Shimokawa, K., Kimura-Yoshida, C., Nagai, N., Mukai, K., Matsubara, K., Watanabe, H., et al. (2011). Cell surface heparan sulfate chains regulate local reception of FGF signaling in the mouse embryo. *Developmental Cell*, *21*(2), 257–272.

Sims, N. A., Clement-Lacroix, P., Da Ponte, F., Bouali, Y., Binart, N., Moriggl, R., et al. (2000). Bone homeostasis in growth hormone receptor-null mice is restored by IGF-I but independent of Stat5. *The Journal of Clinical Investigation*, *106*(9), 1095–1103.

Sjogren, K., Bohlooly, Y. M., Olsson, B., Coschigano, K., Tornell, J., Mohan, S., et al. (2000). Disproportional skeletal growth and markedly decreased bone mineral content in growth hormone receptor −/− mice. *Biochemical and Biophysical Research Communications*, *267*(2), 603–608.

Smeets, T., & van Buul-Offers, S. (1983). The influence of growth hormone, somatomedins, prolactin and thyroxine on the morphology of the proximal tibial epiphysis and growth plate of Snell dwarf mice. *Growth*, *47*(2), 160–173.

Stevens, D. A., Harvey, C. B., Scott, A. J., O'Shea, P. J., Barnard, J. C., Williams, A. J., et al. (2003). Thyroid hormone activates fibroblast growth factor receptor-1 in bone. *Molecular Endocrinology*, *17*(9), 1751–1766.

Stevens, D. A., Hasserjian, R. P., Robson, H., Siebler, T., Shalet, S. M., & Williams, G. R. (2000). Thyroid hormones regulate hypertrophic chondrocyte differentiation and expression of parathyroid hormone-related peptide and its receptor during endochondral bone formation. *Journal of Bone and Mineral Research*, *15*(12), 2431–2442.

Stratikopoulos, E., Szabolcs, M., Dragatsis, I., Klinakis, A., & Efstratiadis, A. (2008). The hormonal action of IGF1 in postnatal mouse growth. *Proceedings of the National Academy of Sciences of the United States of America*, *105*(49), 19378–19383.

Sun, L., Vukicevic, S., Baliram, R., Yang, G., Sendak, R., McPherson, J., et al. (2008). Intermittent recombinant TSH injections prevent ovariectomy-induced bone loss. *Proceedings of the National Academy of Sciences of the United States of America*, *105*(11), 4289–4294.

Suzuki, F. (1992). Effects of various growth factors on a chondrocyte differentiation model. *Advances in Experimental Medicine and Biology*, *324*, 101–106.

Takayanagi, H., Kim, S., Matsuo, K., Suzuki, H., Suzuki, T., Sato, K., et al. (2002). RANKL maintains bone homeostasis through c-Fos-dependent induction of interferon-beta. *Nature*, *416*(6882), 744–749.

Takeda, K., Balzano, S., Sakurai, A., DeGroot, L. J., & Refetoff, S. (1991). Screening of nineteen unrelated families with generalized resistance to thyroid hormone for known point mutations in the thyroid hormone receptor beta gene and the detection of a new mutation. *The Journal of Clinical Investigation*, *87*(2), 496–502.

Takeda, K., Sakurai, A., DeGroot, L. J., & Refetoff, S. (1992). Recessive inheritance of thyroid hormone resistance caused by complete deletion of the protein-coding region of the thyroid hormone receptor-beta gene. *The Journal of Clinical Endocrinology and Metabolism*, *74*(1), 49–55.

Tang, Y., Yu, M., & Lian, X. (2016). Resistance to thyroid hormone alpha, revelation of basic study to clinical consequences. *Journal of Pediatric Endocrinology & Metabolism*, *29*(5), 511–522.

Thorngren, K. G., & Hansson, L. I. (1973). Effect of thyroxine and growth hormone on longitudinal bone growth in the hypophysectomized rat. *Acta Endocrinologica*, *74*(1), 24–40.

Tinnikov, A., Nordstrom, K., Thoren, P., Kindblom, J. M., Malin, S., Rozell, B., et al. (2002). Retardation of post-natal development caused by a negatively acting thyroid hormone receptor alpha1. *The EMBO Journal, 21*(19), 5079–5087.

Trajkovic, M., Visser, T. J., Mittag, J., Horn, S., Lukas, J., Darras, V. M., et al. (2007). Abnormal thyroid hormone metabolism in mice lacking the monocarboxylate transporter 8. *The Journal of Clinical Investigation, 117*(3), 627–635.

Trost, S. U., Swanson, E., Gloss, B., Wang-Iverson, D. B., Zhang, H., Volodarsky, T., et al. (2000). The thyroid hormone receptor-beta-selective agonist GC-1 differentially affects plasma lipids and cardiac activity. *Endocrinology, 141*(9), 3057–3064.

Tsumaki, N., Nakase, T., Miyaji, T., Kakiuchi, M., Kimura, T., Ochi, T., et al. (2002). Bone morphogenetic protein signals are required for cartilage formation and differently regulate joint development during skeletogenesis. *Journal of Bone and Mineral Research, 17*(5), 898–906.

Twigg, S. R., Healy, C., Babbs, C., Sharpe, J. A., Wood, W. G., Sharpe, P. T., et al. (2009). Skeletal analysis of the Fgfr3(P244R) mouse, a genetic model for the Muenke craniosynostosis syndrome. *Developmental Dynamics, 238*, 331–342.

Underwood, L. E., & Van Wyk, J. J. (1992). Normal and aberrant growth. In J. D. Wilson & D. W. Foster (Eds.), *Williams textbook of endocrinology* (pp. 1079–1138). Philadelphia, PA: WB Saunders Company.

Usala, S. J., Menke, J. B., Watson, T. L., Wondisford, F. E., Weintraub, B. D., Berard, J., et al. (1991). A homozygous deletion in the c-erbA beta thyroid hormone receptor gene in a patient with generalized thyroid hormone resistance: Isolation and characterization of the mutant receptor. *Molecular Endocrinology, 5*(3), 327–335.

van Mullem, A. A., Chrysis, D., Eythimiadou, A., Chroni, E., Tsatsoulis, A., de Rijke, Y. B., et al. (2013). Clinical phenotype of a new type of thyroid hormone resistance caused by a mutation of the TRalpha1 receptor: Consequences of LT4 treatment. *The Journal of Clinical Endocrinology and Metabolism, 98*(7), 3029–3038.

van Mullem, A., van Heerebeek, R., Chrysis, D., Visser, E., Medici, M., Andrikoula, M., et al. (2012). Clinical phenotype and mutant TRalpha1. *The New England Journal of Medicine, 366*(15), 1451–1453.

Van Vliet, G., Larroque, B., Bubuteishvili, L., Supernant, K., & Leger, J. (2003). Sex-specific impact of congenital hypothyroidism due to thyroid dysgenesis on skeletal maturation in term newborns. *The Journal of Clinical Endocrinology and Metabolism, 88*(5), 2009–2013.

Varga, F., Rumpler, M., & Klaushofer, K. (1994). Thyroid hormones increase insulin-like growth factor mRNA levels in the clonal osteoblastic cell line MC3T3-E1. *FEBS Letters, 345*(1), 67–70.

Varga, F., Rumpler, M., Luegmayr, E., Fratzl-Zelman, N., Glantschnig, H., & Klaushofer, K. (1997). Triiodothyronine, a regulator of osteoblastic differentiation: Depression of histone H4, attenuation of c-fos/c-jun, and induction of osteocalcin expression. *Calcified Tissue International, 61*(5), 404–411.

Varga, F., Spitzer, S., Rumpler, M., & Klaushofer, K. (2003). 1,25-Dihydroxyvitamin D3 inhibits thyroid hormone-induced osteocalcin expression in mouse osteoblast-like cells via a thyroid hormone response element. *Journal of Molecular Endocrinology, 30*(1), 49–57.

Vincent, B. H., Montufar-Solis, D., Teng, B. B., Amendt, B. A., Schaefer, J., & Klein, J. R. (2009). Bone marrow cells produce a novel TSHbeta splice variant that is upregulated in the thyroid following systemic virus infection. *Genes and Immunity, 10*(1), 18–26.

Vlaeminck-Guillem, V., Espiard, S., Flamant, F., & Wemeau, J. L. (2015). TRalpha receptor mutations extend the spectrum of syndromes of reduced sensitivity to thyroid hormone. *Presse Médicale, 44*(11), 1103–1112.

Vortkamp, A., Lee, K., Lanske, B., Segre, G. V., Kronenberg, H. M., & Tabin, C. J. (1996). Regulation of rate of cartilage differentiation by Indian hedgehog and PTH-related protein. *Science, 273*(5275), 613–622.

Wang, Y., Cheng, Z., Elalieh, H. Z., Nakamura, E., Nguyen, M. T., Mackem, S., et al. (2011). IGF-1R signaling in chondrocytes modulates growth plate development by interacting with the PTHrP/Ihh pathway. *Journal of Bone and Mineral Research, 26*(7), 1437–1446.

Wang, L., Shao, Y. Y., & Ballock, R. T. (2005). Peroxisome proliferator activated receptor-gamma (PPARgamma) represses thyroid hormone signaling in growth plate chondrocytes. *Bone, 37*(3), 305–312.

Wang, L., Shao, Y. Y., & Ballock, R. T. (2006). Peroxisome proliferator-activated receptor-gamma promotes adipogenic changes in growth plate chondrocytes in vitro. *PPAR Research, 2006*, 67297.

Wang, L., Shao, Y. Y., & Ballock, R. T. (2007). Thyroid hormone interacts with the Wnt/beta-catenin signaling pathway in the terminal differentiation of growth plate chondrocytes. *Journal of Bone and Mineral Research, 22*(12), 1988–1995.

Wang, L., Shao, Y. Y., & Ballock, R. T. (2009). Carboxypeptidase Z (CPZ) links thyroid hormone and Wnt signaling pathways in growth plate chondrocytes. *Journal of Bone and Mineral Research, 24*(2), 265–273.

Wang, L., Shao, Y. Y., & Ballock, R. T. (2010). Thyroid hormone-mediated growth and differentiation of growth plate chondrocytes involves IGF-1 modulation of beta-catenin signaling. *Journal of Bone and Mineral Research, 25*(5), 1138–1146.

Wang, L., Shao, Y. Y., & Ballock, R. T. (2011). Leptin synergizes with thyroid hormone signaling in promoting growth plate chondrocyte proliferation and terminal differentiation in vitro. *Bone, 48*(5), 1022–1027.

Wang, L., Shao, Y. Y., & Ballock, R. T. (2012). Leptin antagonizes peroxisome proliferator-activated receptor-gamma signaling in growth plate chondrocytes. *PPAR Research, 2012*, 756198.

Wang, L., Waltenberger, B., Pferschy-Wenzig, E. M., Blunder, M., Liu, X., Malainer, C., et al. (2014). Natural product agonists of peroxisome proliferator-activated receptor gamma (PPARgamma): A review. *Biochemical Pharmacology, 92*(1), 73–89.

Waung, J. A., Bassett, J. H., & Williams, G. R. (2015). Adult mice lacking the type 2 iodothyronine deiodinase have increased subchondral bone but normal articular cartilage. *Thyroid, 25*(3), 269–277.

Weiss, R. E., & Refetoff, S. (1996). Effect of thyroid hormone on growth. Lessons from the syndrome of resistance to thyroid hormone. *Endocrinology and Metabolism Clinics of North America, 25*(3), 719–730.

Wells, D., King, J. D., Roe, T. F., & Kaufman, F. R. (1993). Review of slipped capital femoral epiphysis associated with endocrine disease. *Journal of Pediatric Orthopedics, 13*(5), 610–614.

White, K. E., Cabral, J. M., Davis, S. I., Fishburn, T., Evans, W. E., Ichikawa, S., et al. (2005). Mutations that cause osteoglophonic dysplasia define novel roles for FGFR1 in bone elongation. *American Journal of Human Genetics, 76*(2), 361–367.

Wikstrom, L., Johansson, C., Salto, C., Barlow, C., Campos Barros, A., Baas, F., et al. (1998). Abnormal heart rate and body temperature in mice lacking thyroid hormone receptor alpha 1. *The EMBO Journal, 17*(2), 455–461.

Wilkie, A. O., & Morriss-Kay, G. M. (2001). Genetics of craniofacial development and malformation. *Nature Reviews. Genetics, 2*(6), 458–468.

Williams, G. R. (2013). Thyroid hormone actions in cartilage and bone. *European Thyroid Journal, 2*(1), 3–13.

Williams, G. R., Bland, R., & Sheppard, M. C. (1994). Characterization of thyroid hormone (T3) receptors in three osteosarcoma cell lines of distinct osteoblast phenotype: Interactions among T3, vitamin D3, and retinoid signaling. *Endocrinology, 135*(6), 2375–2385.

Williams, A. J., O'Shea, P. J., & Williams, G. R. (2007). Complex interactions between thyroid hormone and fibroblast growth factor signalling. *Current Opinion in Endocrinology, Diabetes, and Obesity*, *14*(5), 410–415.
Williams, A. J., Robson, H., Kester, M. H., van Leeuwen, J. P., Shalet, S. M., Visser, T. J., et al. (2008). Iodothyronine deiodinase enzyme activities in bone. *Bone*, *43*(1), 126–134.
Willson, T. M., Brown, P. J., Sternbach, D. D., & Henke, B. R. (2000). The PPARs: From orphan receptors to drug discovery. *Journal of Medicinal Chemistry*, *43*(4), 527–550.
Wu, Y., Sun, H., Basta-Pljakic, J., Cardoso, L., Kennedy, O. D., Jasper, H., et al. (2013). Serum IGF-1 is insufficient to restore skeletal size in the total absence of the growth hormone receptor. *Journal of Bone and Mineral Research*, *28*(7), 1575–1586.
Xing, W., Aghajanian, P., Goodluck, H., Kesavan, C., Cheng, S., Pourteymoor, S., et al. (2016). Thyroid hormone receptor-beta1 signaling is critically involved in regulating secondary ossification via promoting transcription of the Ihh gene in the epiphysis. *American Journal of Physiology. Endocrinology and Metabolism*, *310*(10), E846–E854.
Xing, W., Cheng, S., Wergedal, J., & Mohan, S. (2014). Epiphyseal chondrocyte secondary ossification centers require thyroid hormone activation of Indian hedgehog and osterix signaling. *Journal of Bone and Mineral Research*, *29*(10), 2262–2275.
Xing, W., Govoni, K. E., Donahue, L. R., Kesavan, C., Wergedal, J., Long, C., et al. (2012). Genetic evidence that thyroid hormone is indispensable for prepubertal insulin-like growth factor-I expression and bone acquisition in mice. *Journal of Bone and Mineral Research*, *27*(5), 1067–1079.
Yakar, S., Courtland, H. W., & Clemmons, D. (2010). IGF-1 and bone: New discoveries from mouse models. *Journal of Bone and Mineral Research*, *25*(12), 2543–2552.
Yasoda, A., Komatsu, Y., Chusho, H., Miyazawa, T., Ozasa, A., Miura, M., et al. (2004). Overexpression of CNP in chondrocytes rescues achondroplasia through a MAPK-dependent pathway. *Nature Medicine*, *10*(1), 80–86.
Yu, K., Xu, J., Liu, Z., Sosic, D., Shao, J., Olson, E. N., et al. (2003). Conditional inactivation of FGF receptor 2 reveals an essential role for FGF signaling in the regulation of osteoblast function and bone growth. *Development*, *130*(13), 3063–3074.
Zhao, Q., Eberspaecher, H., Lefebvre, V., & De Crombrugghe, B. (1997). Parallel expression of Sox9 and Col2a1 in cells undergoing chondrogenesis. *Developmental Dynamics*, *209*(4), 377–386.
Zhu, X. G., Yu, C. L., McPhie, P., Wong, R., & Cheng, S. Y. (1996). Understanding the molecular mechanism of dominant negative action of mutant thyroid hormone beta 1-receptors: The important role of the wild-type/mutant receptor heterodimer. *Endocrinology*, *137*(2), 712–721.
Zumbrunn, J., & Trueb, B. (1996). Primary structure of a putative serine protease specific for IGF-binding proteins. *FEBS Letters*, *398*(2–3), 187–192.

CHAPTER SEVENTEEN

Thyroid Hormone Role and Economy in the Developing Testis

Arturo Hernandez[1]
Center for Molecular Medicine, Maine Medical Center Research Institute, Scarborough, ME, United States
[1]Corresponding author: e-mail address: hernaa@mmc.org

Contents

1. Introduction 474
2. Thyroid Hormone Status and the Developing Testis: Studies on Animal Models of Developmental Hypo- and Hyperthyroidism 475
   2.1 Effects on Sertoli Cells 475
   2.2 Effects on Testis Size and Cytoarchitecture 477
   2.3 Effects on Spermatogenesis 478
   2.4 Effects on Leydig Cells, Steroidogenesis, and the HPG Axis 480
3. Endogenous Determinants of TH Action in the Developing Testis 482
   3.1 Ontogeny of Serum TH Levels and Milestones in Testis Development 483
   3.2 Thyroid Hormone Receptors: Role of Thra 484
   3.3 Thyroid Hormone Transporters 487
   3.4 Type 1 and Type 2 Deiodinases 488
   3.5 Type 3 Deiodinase 489
   3.6 Summary 491
4. Conclusion 491
Acknowledgments 492
References 492

Abstract

Thyroid hormones (TH) exhibit pleiotropic regulatory effects on growth, development, and metabolism, and it is becoming increasingly apparent that the developing testis is an important target for them. Testicular development is highly dependent on TH status. Both hypo- and hyperthyroidism affect testis size and the proliferation and differentiation of Sertoli, Leydig, and germ cells, with consequences for steroidogenesis, spermatogenesis, and male fertility. These observations suggest that an appropriate content of TH and by implication TH action in the testis, whether the result of systemic hormonal levels or regulatory mechanisms at the local level, is critical for normal testicular and reproductive function. The available evidence indicates the presence in the developing testis of a number of transporters, deiodinases and receptors that could play a role in the timely delivery of TH action on testicular cells. These include the thyroid hormone receptor

Vitamins and Hormones, Volume 106
ISSN 0083-6729
http://dx.doi.org/10.1016/bs.vh.2017.06.005

alpha (THRA), the MCT8 transporter, the TH-activating deiodinase DIO2, and the TH-inactivating deiodinase DIO3, all of which appear to modulate testicular TH economy and testis outcomes.

ABBREVIATIONS

AMH antimullerian hormone
DIO1, DIO2, DIO3 deiodinase 1, 2, and 3
FSH follicle-stimulating hormone
LH luteinizing hormone
MCT8 monocarboxylate transporter 8
OATP1C1 organic anion transport peptide 1C1
T3 3,5,3′-triiodothyronine
T4 thyroxine
TH thyroid hormones
THRA and THRB thyroid hormone receptor alpha and beta
TRH thyrotropin-releasing hormone

1. INTRODUCTION

The thyroid hormones (TH) thyroxine (T4) and 3,5,3′-triiodothyronine (T3) exert profound effects in developing and adult tissues. Their regulation of brain development, heart physiology, tissue metabolism, and energy balance has been profusely documented. However, much less is known about their biological effects on the male reproductive system. However, evidence that TH play a critical role in testis development and the function of the hypothalamic–pituitary–gonadal (HPG) axis is accumulating. In adult humans and rodents, alterations in TH status may cause testicular abnormalities (Hudson & Edwards, 1992; Jannini, Ulisse, & D'Armiento, 1995), influence testis steroidogenesis and androgen physiology (Maran, 2003), and impair spermatogenesis (Wagner, Wajner, & Maia, 2008), ultimately leading to a reduction in male fertility.

Despite the influence of TH status on adult testis function and male fertility, it is the regulation of TH action in the developing testis that appears of critical functional significance. A large body of research shows that transient, abnormal levels of TH during particular developmental stages lead to molecular, cellular and functional defects in the testis and the regulation of the HPG axis (Castaneda Cortes, Langlois, & Fernandino, 2014; Gao, Lee, & Cheng, 2014; Wagner et al., 2008). These defects may not necessarily be corrected by exogenous interventions to modify TH status, and may ultimately result in irreversibly impaired reproductive function in males.

The levels of TH in the circulation are an important factor in determining tissue TH action. However, recent research has shown that the degree of TH action in a specific tissue is not always dependent on this factor alone. The data have revealed the existence of several cellular factors that together determine the appropriate intracellular TH content for a given cell (Astapova & Hollenberg, 2013; Bianco, 2011; Forrest & Visser, 2013; St Germain, Galton, & Hernandez, 2009; Visser, 2013). This can be significantly different from that predicted by serum TH levels. In the case of the developing testis, past observations reveal an essential role for systemic TH status in the testis cell maturation and function, but many questions about the importance of endogenous determinants of testicular TH action remain to be answered.

This chapter focuses on the importance of TH economy on the neonatal and prepubertal testis. First, we review the biological effects observed in experimental animal models of hypo- and hyperthyroidism. Then, in the context of the available literature, we evaluate the relative importance of the factors that define local TH availability and action in testicular tissue during development. Although much is yet to be learned, the current data increasingly depict a model of locally controlled TH action that is tailored to particular developmental stages and specific testicular cell types.

2. THYROID HORMONE STATUS AND THE DEVELOPING TESTIS: STUDIES ON ANIMAL MODELS OF DEVELOPMENTAL HYPO- AND HYPERTHYROIDISM

Neonatal and prepubertal alterations in TH status lead to abnormalities in testis development and function. Investigators have used mostly rodent models to manipulate TH levels surgically or pharmacologically, and the subsequent testicular analysis in these models has revealed a panoply of molecular, cellular, hormonal, and organ abnormalities that are summarized later and in Table 1.

2.1 Effects on Sertoli Cells

The effects of hypothyroidism on the testis have been extensively studied in thyrodectomized prepubertal rats and in rodent neonates treated with antithyroid drugs. Extensive and consistent data demonstrate a critical effect on Sertoli cells (SCs). Thus, transient neonatal hypothyroidism promotes an

Table 1 Summary of Testicular Effects of Developmental Hypo- and Hyperthyroidism

Parameter	Hypothyroidism	Hyperthyroidism
Testis size	Increase	Decrease
SC proliferation	Prolonged	Premature arrest
SC differentiation	Delay	Stimulated
LC proliferation	Increased[a]	Increased[b]
LC differentiation	Reduced[a]	Stimulated[b]
Spermatogonia proliferation	Prolonged	Premature arrest
Spermatogonia differentiation	Delay	Stimulated

[a]Interstitial cells.
[b]Immature LC.
LC, Leydig cell; *SC*, Sertoli cell.

extended period of proliferation of SCs, delaying their differentiation (Bunick, Kirby, Hess, & Cooke, 1994; Francavilla et al., 1991; Maran, Ravisankar, et al., 1999; Oluwole, Bartlewski, & Hahnel, 2013; Palmero, de Marchis, Gallo, & Fugassa, 1989; Sakai, Yamashina, & Furudate, 2004; Van Haaster, De Jong, Docter, & De Rooij, 1992). This delay leads to an increase in SC number in the adult testis. However, prolonged hypothyroidism from birth to adulthood may ultimately lead to a lower number of SCs (Maran, Sivakumar, et al., 1999).

In contrast, transient neonatal hyperthyroidism, usually achieved by the administration of supraphysiological or pharmacological doses of TH, arrests SC proliferation and triggers their premature differentiation (Palmero, Prati, Bolla, & Fugassa, 1995; van Haaster, de Jong, Docter, & de Rooij, 1993). The effects of T3 on promoting the exit of SCs from their cell cycle and stimulating and maintaining their differentiation program have been thoroughly confirmed in primary cultures of neonatal SCs (Boulogne, Habert, & Levacher, 2003; Buzzard, Wreford, & Morrison, 2003; Cooke, Zhao, & Bunick, 1994; Palmero et al., 1995).

Consistent with their extended proliferation and differentiation delay, the expression of dividing SC marker antimullerian hormone (*Amh*) was prolonged in neonatal hypothyroidism (Mendis-Handagama & Ariyaratne, 2008), while the expression markers of SC maturation were delayed, as was that of androgen-binding protein and inhibin B (Bunick et al., 1994; Maran, Sivakumar, et al., 1999).

The effects of TH status on the proliferation and differentiation of SCs suggest the existence of molecular targets of T3 with cell cycle-regulatory roles. Using cultures of SCs and also animal models, investigators have currently identified several of these targets, including the PI3K signaling pathway (Sun et al., 2015), the cell cycle inhibitors p27Kip1 and p21Cip1 (Holsberger, Buchold, et al., 2005; Holsberger, Jirawatnotai, Kiyokawa, & Cooke, 2003), and the oncogenes JunD and c-myc (Fumel et al., 2012).

In addition, studies in neonatal SCs in culture have identified several T3 target genes of relevance to the physiology of these cells. Notably, T3 influences the levels of androgen signaling in SCs at different levels by (i) increasing the relative expression ratio of androgen and estrogen receptors (Arambepola, Bunick, & Cooke, 1998; Cardone, Angelini, Esposito, Comitato, & Varriale, 2000; Panno et al., 1994; Panno, Sisci, Salerno, Lanzino, Mauro, et al., 1996; Panno, Sisci, Salerno, Lanzino, Pezzi, et al., 1996; Sisci et al., 1997), (ii) inhibiting the expression of aromatase (Ando et al., 2001; Catalano et al., 2003; Panno et al., 1994, 1995; Pezzi et al., 2001; Ulisse et al., 1994), and (iii) decreasing the expression of androgen-binding protein (Fugassa, Palmero, & Gallo, 1987; Rao, Liang, Chakraborti, & Feng, 2003). These observations indicate that the effects of T3 on SC maturation include increasing their sensitivity to androgens.

Other molecular targets of T3 in differentiating SCs in culture of potential significance to cell phenotypes include the glucose transporter GLUT1 (Carosa et al., 2005; Ulisse, Jannini, Pepe, De Matteis, & D'Armiento, 1992) and the insulin-like growth factor-binding protein-4 (Bardi, Bottazzi, Demori, & Palmero, 1999).

2.2 Effects on Testis Size and Cytoarchitecture

Testis weight is markedly influence by developmental thyroid status. Since SC number is the largest determinant of adult testis size (Cooke et al., 1994; Saxena, Yoshinaga, Tanii, & Toshimori, 2002), the longer proliferative period of SCs caused by transient neonatal hypothyroidism ultimately leads to testicular enlargement in rodents (Kimura & Furudate, 1996; Kirby et al., 1992; Kobayashi, Kubota, Hojo, & Miyagawa, 2014; Umezu, Kagabu, Jiang, Niimura, & Sato, 2004) and other animal models (Kirby, Mankar, Hardesty, & Kreider, 1996). This is usually associated with a reduction in the outer and luminal diameter of seminiferous tubules and the area occupied by the seminiferous epithelium (Lagu, Bhavsar, Sharma, & Ramachandran, 2005; Oncu et al., 2004). Interestingly, the testis enlargement caused by hypothyroidism is reversed if the animals are raised at a high ambient temperature (Lagu et al., 2005). In contrast to the effects of hypothyroidism, TH administration during neonatal life leads to a reduction in testis size (Bakke, Lawrence, & Wilber, 1974; Majdic et al., 1998; Majeed, Injidi, & Forbes, 1984), expanded seminiferous tubule area, and accelerated formation of the tubular lumen (van Haaster et al., 1993).

The cytoskeleton and cell–cell interactions also play a role in regulating SC differentiation. SCs have a critical role in the establishment of the blood–testis barrier, which provides the necessary environment for the progression of spermatogenesis (Pelletier, 2011). In this context, T3 regulates the expression and localization of connexin 43 (also called Gja1) (Marchlewska, Kula, Walczak-Jedrzejowska, Oszukowska, Filipiak, et al., 2011; St-Pierre, Dufresne, Rooney, & Cyr, 2003), an essential protein for SC gap junctions (Cyr, 2011), and also for the control of SC proliferation (Gilleron et al., 2006; Marchlewska et al., 2013; Sridharan et al., 2007). Furthermore, T3 regulates the dynamics of the intermediate filament vimentin (Zamoner et al., 2005, 2007), the expression of basement membrane components (Ulisse et al., 1998), and the expression of the neural cell adhesion molecule (NCAM), which affects the attachment of gonocytes and SCs *in vitro* (Laslett, Li, Jester, & Orth, 2000). These observations suggest that the effects of TH on SC proliferation arrest and maturation may be partially mediated by proteins with cell-contact and cytoskeletal functions, and associated with the establishment of the blood–testis barrier during SC maturation.

Finally, T3 also regulates the expression of insulin-like growth factor 1 (IGF1) in immature SCs in culture (Palmero et al., 1990), suggesting that T3 increases the levels of IGF1 signaling in the testis. Recent observations indicate that genetic deletion of both insulin and IGF1 receptors specifically in SCs cells leads to a reduction in SC proliferation and testis size (Pitetti et al., 2013). Since hyperthyroidism leads to reduced testis size, these two observations appear contradictory and do not support the concept that the effects of T3 on SC proliferation and testis size are mediated by IGF1 signaling. It is possible that IGF1 signaling mostly affects testis growth during the prepubertal period, and not in early neonatal life, but additional research is needed to clarify this potential mechanism.

2.3 Effects on Spermatogenesis

Spermatogenesis is a germ cell differentiation process that is initiated approximately at postnatal day 5 in rodents by the transition of gonocytes to spermatogonia and culminates at puberty with the appearance of mature spermatids and spermiogenesis (Picut et al., 2015). This transition proceeds through numerous stages of differentiation some of which are more prominent at particular stages of development. Spermatogenesis requires the normal function of SCs (Pelletier, 2011) and the appropriate hormonal milieu controlled by the HPG axis (see below). Thus, given the SCs abnormalities observed in models of altered TH status, it is not surprising that they are associated with deficits in spermatogenesis. In this regard, neonatal hypothyroidism leads to impaired spermatogenesis (Kirby et al., 1996; Tousson, Ali, Ibrahim, & Mansour, 2011), mostly due to a delay in spermatogonia differentiation (Simorangkir, Wreford, & De Kretser, 1997). A developmental delay in the appearance of spermatids and spermatocytes has been also observed in a genetic model of congenital hypothyroidism (Sakai et al., 2004). The failure of germ cell differentiation to proceed at an adequate pace may lead later in life to a larger reservoir of undifferentiated spermatogonia (Auharek & de Franca, 2010), an increased number of spermatocytes and spermatids (Maran, Sivakumar, et al., 1999), and enhanced sperm production (Kirby et al., 1996). In this regard, AKT1 signaling has been implicated in the extended proliferation of undifferentiated germ cells in the face of hypothyroidism, as *Akt1*$^{-/-}$ mice made hypothyroid do not exhibit an increase in sperm production. This effect seems specific to the germ line, as other consequences of hypothyroidism are still observed in these mice, including delayed lumen formation in smaller seminiferous tubules and decreased *Amh* expression (Santos-Ahmed et al., 2011).

The developmental delay in spermatogenesis caused by a hypothyroid state has been associated with permanently reduced levels of follicle-stimulating hormone (FSH) (Kirby et al., 1997; Maran & Aruldhas, 2002; Van Haaster et al., 1992), raising the possibility that both low FSH and the immaturity of SCs contribute to the deficits in spermatogenesis caused by neonatal hypothyroidism.

Similarly to the findings in SCs, the action of T3 reduces germ cell proliferation (Boulogne et al., 2003) and promotes the initiation of spermatogenesis (Rijntjes, Wientjes, Swarts, de Rooij, & Teerds, 2008; Souhila, Zohra, Kamel, & Hadj-Bekkouche, 2013). Using comparable experimental designs, investigators have reported that T3 treatment of neonatal rat leads to increase gonocyte differentiation and number of spermatogonia, as well as apoptosis of germ cells (Faraone-Mennella et al., 2009; Marchlewska, Kula, Walczak-Jedrzejowska, Oszukowska, Orkisz, et al., 2011).

Thus, hypothyroidism and hyperthyroidism seem to affect both SCs and germ cells in a similar fashion. Interestingly, as occurs in SCs (Maran,

Sivakumar, et al., 1999), if the induction of hypothyroidism is not transient but extended into adulthood, investigators have noted not an increase but a reduction in germ cells (Maran, Sivakumar, et al., 1999), probably due to increased apoptosis after maturation (Sakai et al., 2004). These observations strongly suggest that an adequate level of TH action must be appropriately timed for normal spermatogenesis.

Investigators have also shown that TH can also affect lipid metabolism (Aruldhas, Valivullah, Srinivasan, & Govindarajulu, 1986; Goswani, Kar, & Chowdhury, 1963), regulate lipogenic enzymes (Aruldhas, Valivullah, & Govindarajulu, 1983, 1984), and influence oxidative stress (Palmero et al., 1994; Sahoo, Roy, Bhanja, & Chainy, 2008; Zamoner et al., 2007). The latter observation could be especially relevant for spermatogenesis since increased oxidative stress stimulate the expression of testicular orphan receptor 4 (Li et al., 2008). A lack of this receptor in mice disrupts the late meiotic prophase and subsequent meiotic divisions during spermatogenesis (Mu et al., 2004), raising the possibility that TH may affect this facet of spermatogenesis by influencing oxidative stress and testicular orphan receptor 4 signaling.

2.4 Effects on Leydig Cells, Steroidogenesis, and the HPG Axis

Additional findings reveal a role for TH in the development and function of Leydig cells (LCs), which are essential for androgen synthesis and the subsequent hormonal support of spermatogenesis and the establishment and physiology of the HPG axis. Although Teerds, de Rooij, de Jong, and van Haaster (1998) do not observed a significant effect, a large majority of studies find that neonatal hypothyroidism increases the proliferation of mesenchymal cells destined to become LC progenitors (Cristovao, Bisi, Mendonca, Bianco, & Bloise, 2002; Hardy et al., 1996), reducing the number of those cells that achieve complete adult-like LC differentiation (Ariyaratne, Mendis-Handagama, & Mason, 2000; Kobayashi et al., 2014; Mendis-Handagama, Ariyaratne, Teunissen van Manen, & Haupt, 1998; Rijntjes, Swarts, Anand-Ivell, & Teerds, 2009; Siril Ariyaratne, Ian Mason, & Mendis-Handagama, 2000; Teerds et al., 1998). These findings support the notion that TH play an important role in triggering the onset of mesenchymal precursor cell differentiation into Leydig progenitor cells, and their subsequent proliferation and their differentiation into mature LCs (Mendis-Handagama & Ariyaratne, 2001, 2004a). This notion is further supported by observations in models of neonatal T3 administration, in which T3 increases the proliferation and differentiation of LCs (Rijntjes et al., 2008; Teerds et al., 1998) and partially counteracts the inhibitory effect of AMH on prepubertal LC differentiation (Ariyaratne & Mendis-Handagama, 2012).

The effects of an altered TH status on SC differentiation appear associated with hormonal changes of relevance to the HPG axis. Neonatal hypothyroidism leads to deficits in serum progesterone, testosterone, and prolactin (Kimura & Furudate, 1996; Maran, Arunakaran, Jeyaraj, et al., 2000). Prepubertal hypothyroidism also leads to deficient levels of prolactin, luteinizing hormone (LH), and testosterone (Lagu et al., 2005; Valle, Oliveira-Filho, Romaldini, & Lara, 1985). Thyrodectomized immature rats manifest deficits in LC development and impaired spermatogenesis (Chowdhury & Arora, 1984). Interestingly, these effects of hypothyroidism on the late development of the HPG axis can be reversed by T4 treatment (Chowdhury, Gautam, & Chatterjee, 1984). Although reduced TH levels may affect the HPG centrally (Chandrasekhar, D'Occhio, Holland, & Setchell, 1985), at least some of these effects may be the direct consequence of a reduced TH action in the testis, as suggested by decreased expression of FSH receptor and androgen-binding protein in SCs and reduced expression of the LH receptor in LCs (Rao et al., 2003). On the other hand, neonatal hyperthyroidism leads to alterations in FSH and testosterone pulses (Fallah-Rad, Connor, & Del Vecchio, 2001), and to a significant increase in serum inhibin levels (van Haaster et al., 1993).

The disruption of hormonal physiology related to the HPG axis as a result of altered TH status is probably partially due to T3 effects of LC steroidogenesis. Neonatal hypothyroidism causes deficiencies in the testicular expression of steroidogenic transcription factors and enzymes including SF-1, StAR, CYP11A1, and 3beta-HSD (Sarkar & Singh, 2017). Studies on *in vitro* LC models demonstrate that T3 stimulates steroidogenesis (Mendis-Handagama & Siril Ariyaratne, 2005) and modulate the LH induction of testosterone and estrogen production (Maran, Arunakaran, & Aruldhas, 2000). *In vitro* analysis of testicular slices from heat-acclimated rats indicates that the steroidogenic response to T3 is altered by heat acclimation (Kurowicka, Chrusciel, Zmijewska, & Kotwica, 2016), suggesting a contribution of impaired TH action to the deficits in spermatogenesis associated with elevated temperature.

The potential role of thyrotropin-releasing hormone (TRH) deserves a special mention. TRH is critical for the regulation of TH physiology; suppression of its release by TH contributes to the regulation of the HPT axis.

However, both TRH and its receptor are expressed in the testis, specifically in LCs (Feng, Gu, Kim, Carnell, & Wilber, 1993; Jiang et al., 2011; Mendis-Handagama & Ariyaratne, 2004b; Mendis-Handagama & Siril Ariyaratne, 2005; Wilber & Xu, 1998). Since it is unlikely that testicular TRH exerts a systemic influence on the HPT axis, these observations suggest a potential autocrine role for TRH in LCs. Since the TRH gene is suppressed by T3, it is possible that TRH signaling may have opposite effects to those of T3 in LCs. This is supported by the observation that TRH, contrary to the effect of T3, inhibits LH-induced testosterone secretion (Wilber & Xu, 1998). Additional work is required to delineate the role of TRH signaling in the developing testis.

Expression levels of adiponectin (Caminos et al., 2008) and insulin-like peptide 3 (Mendis-Handagama, Ariyaratne, Mrkonjich, & Ivell, 2007) are two molecular endpoints that are regulated by TH specifically in LCs, although the significance of these findings for testis function is yet unclear.

In summary, the work briefly reviewed earlier indicates that both an excess or a deficit of TH in early neonatal or prepubertal age markedly influence testis size, spermatogenesis, and the timely proliferation and differentiation of SCs, LCs, and germ cells. This influence is often associated with altered steroidogenesis and hormonal disruptions in the HPG axis.

3. ENDOGENOUS DETERMINANTS OF TH ACTION IN THE DEVELOPING TESTIS

Thus, based on observations in *in vivo* and *in vitro* models of altered TH status, it is evident that appropriate levels of TH are essential for normal development of the testis. An important factor on tissue TH action is the circulating level of TH, which is primarily determined by secretion from the thyroid gland and regulated by the HPT axis. The most abundant circulating hormone is T4, which functions largely as a prohormone; T3 is the active hormone due to its much higher affinity for the nuclear receptor (Forrest & Visser, 2013). Upon binding to its receptor, which is a transcription factor with DNA-binding capacity, T3 modulates transcription of specific genes and thereby produces its biological effects.

In addition, there are a number of molecular factors specific to particular tissues or cell types that modulate tissue T3 availability to its receptor. TH are transported into (and out of) target cells by transporters exhibiting different affinities for each of the hormones. These transporters mostly belong to the monocarboxylate (MCTs) and the organic anion transporting polypeptide

(OATPs) families of cell membrane transporters (Heuer & Visser, 2009). Once inside the cell, T4 can be activated by conversion to T3, a process that is catalyzed by the type 1 and the type 2 deiodinases (DIO1 and DIO2, also called D1 and D2) (St Germain et al., 2009). Locally generated T3 by these enzymes may also exit the cell, affecting neighboring cells and contributing to the pool of circulating T3. Alternatively, both T4 and T3 can be converted locally into inactive metabolites by the type 3 deiodinase (DIO3, also D3), thus preventing their biological action (Hernandez, 2005). These mechanisms signify that TH signaling in a target tissue not only depends on circulating levels of TH but also on the specific complement of transporters, deiodinases, and receptors that are present in a particular cell or tissue. Hence, a given level of TH in the serum does not necessarily translate into a proportional level of TH signaling in a particular tissue. Later, we review the published literature concerning the relative importance of endogenous factors controlling TH action in the developing testis.

3.1 Ontogeny of Serum TH Levels and Milestones in Testis Development

The primary determinant of TH action in the developing testis is the level of TH in the circulation. Serum TH levels change dramatically during development in human and rodents. In early fetal development, rodent TH levels are very low compared to those present in the mother or an adult. They start to increase slowly and steadily in late gestation, after fetal thyroid gland function is initiated; then they continue to increase during neonatal life, reach a modest peak 2 weeks after birth, and slowly decrease reaching normal adult levels around puberty, when the HPT axis is fully established (Dussault & Labrie, 1975) (Fig. 1). The ontogeny of serum TH levels indicates that in normal conditions, the testis is exposed to low TH levels in early development, and high, adult-like levels after the first 2 weeks of life. This developmental profile of low-to-high TH action fits rather well with cellular events in testicular development and with the observations from the models of developmental hypo- and hyperthyroidism reviewed earlier. Immature SCs and spermatogonia mostly proliferate in the rodent for the first 2–3 weeks of life (Picut et al., 2015), and their differentiation coincides with peaking levels of TH, since TH action is needed for the maturation of the three main testicular cell types. Thus, as reviewed earlier, excessive TH action inappropriately early leads to premature cell differentiation, while extended neonatal hypothyroidism delays cell differentiation.

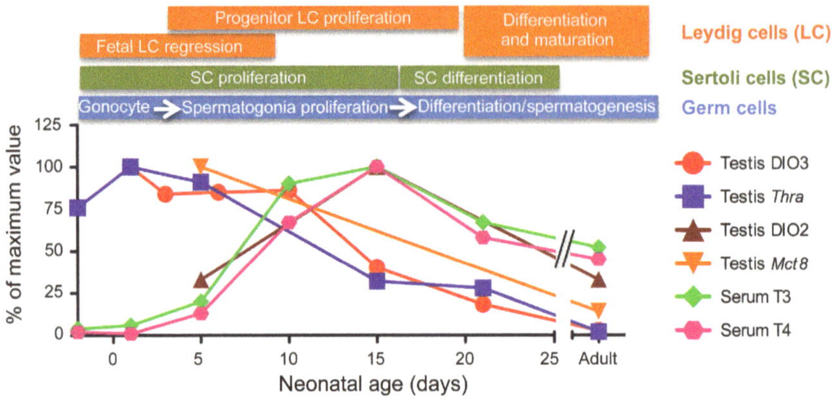

Fig. 1 Stages in testicular cell development and ontogeny of determinants of TH action. The newborn testis appears primed for TH action, based on the high expression of the transporter *Mct8* and the receptor *Thra*. The high expression at this time of the TH-inactivating deiodinase DIO3 seems to extend the period of low TH availability for the testis, despite the rising serum levels of TH. This occurs in synchrony with a period of intense proliferation of testicular cells. Approximately 2 weeks after birth, the marked decrease in DIO3 expression, higher expression of activating deiodinase DIO2, and peaking TH levels suggest a sharp increase in TH action in the testis, which coincides with major shifts toward differentiation for most testicular cells. *The graph is based on data adapted from Bates, JM., St Germain, D. L., & Galton, V. A. (1999). Expression profiles of the three iodothyronine deiodinases, D1, D2, and D3, in the developing rat.* Endocrinology, 140*(2), 844–851; Jannini, E. A., Dolci, S., Ulisse, S., & Nikodem, V. M. (1994). Developmental regulation of the thyroid hormone receptor alpha 1 mRNA expression in the rat testis.* Molecular Endocrinology, 8*(1), 89–96; Martinez, M. E., Karaczyn, A., Stohn, J. P., Donnelly, W. T., Croteau, W., Peeters, R. P., et al. (2016). The Type 3 deiodinase is a critical determinant of appropriate thyroid hormone action in the developing testis.* Endocrinology, 157*(3), 1276–1288; Picut, C. A., Remick, A. K., de Rijk, E. P., Simons, M. L., Stump, D. G., & Parker, G. A. (2015). Postnatal development of the testis in the rat: Morphologic study and correlation of morphology to neuroendocrine parameters.* Toxicologic Pathology, 43*(3), 326–342.*

3.2 Thyroid Hormone Receptors: Role of Thra

A substantial amount of research has identified the presence of TH receptors in the developing testis. Initially, investigators noted low capacity, high affinity-binding sites for T3 in the testis (DeGroot, Hill, & Rue, 1976; Oppenheimer, Schwartz, & Surks, 1974), as well as immunoreactivity for the putative TH receptor c-erb-A (Tagami et al., 1990), suggesting the presence of TH receptors in this tissue. Of the two genes encoding for TH receptors (*Thra* and *Thrb*) (Lazar & Chin, 1990), *Thra* is the one predominantly expressed in the rodent testis both during development and in adulthood (Canale et al., 2001; Jannini et al., 1990). Similarly, in the human testis, Jannini et al. observed no detectable expression of *Thrb* by in situ hybridization at different ages, while they noted a significant level of *Thra* expression (Jannini et al., 2000).

Testicular *Thra* expression exhibits a developmental profile, with peak expression during early neonatal age and significantly diminished expression later in life and in adulthood (Arambepola et al., 1998; Buzzard, Morrison, O'Bryan, Song, & Wreford, 2000; Jannini, Carosa, Rucci, Screponi, & D'Armiento, 1999; Jannini, Dolci, Ulisse, & Nikodem, 1994; Jannini et al., 1990) (Fig. 1). Furthermore, *Thra* expression is upregulated or prolonged in neonatal hypothyroidism (Bunick et al., 1994; Palmero, De Marco, Trucchi, Prati, & Fugassa, 1991; Palmero, Prati, De Marco, Trucchi, & Fugassa, 1993; Rao et al., 2003; Sarkar & Singh, 2017). This regulation enhances sensitivity to TH action in the developing testis.

Studies in testicular tissue and cell culture have produced additional insight about the cell types in the testis that are targeted by TH. TH-binding sites have been located in developing SCs (Jannini et al., 1990; Palmero, Benahmed, Morera, Trucchi, & Fugassa, 1992; Palmero, Maggiani, & Fugassa, 1988), where *Thra* mRNA appears to be predominantly expressed (Buzzard et al., 2000; Jannini et al., 1994, 2000). In addition, *Thra* is also expressed in developing germ cells, from spermatogonia to spermatocyte stages and also in a subset of unidentified interstitial cells (Buzzard et al., 2000), some of which may later commit to the LC lineage. However, no THRA expression has been detected in germ or interstitial cells of the human testis (Jannini et al., 2000). *Thra* expression in SCs and germ cells render these cells a target for TH, which is consistent with their abnormal outcomes in developmental models of hypo- and hyperthyroidism.

Additional studies have demonstrated the functional role of *Thra* in mediating the effects of TH in the developing testis. In a model of neonatal T3 administration, genetic inactivation of *Thra* fully rescues the testicular abnormalities associated with neonatal hyperthyroidism, including reduced testis size, premature SC proliferation, enlarged size of seminiferous tubules and premature lumen formation, and elevated FSH (Holsberger & Cooke, 2005; Holsberger, Kiesewetter, & Cooke, 2005). Important additional insight about the role of THRA in SCs was provided by the work of Fumel et al. Global and SC-specific genetic introduction of a dominant-negative *Thra* leads to testis enlargement (Table 2) and extended SC proliferation, the latter partially mediated by a Cdk4/c-myc-dependent pathway (Fumel et al., 2012). Similar findings are obtained when either an *Amh* or

Table 2 Adult Testis Size Changes in Mice With Genetic Alterations in Components of TH Signaling

Genetic Modification	Effects on Testis Size	Functional Defect
Thra$^{0/0}$	↑	No thyroid receptor alpha signaling
Thra (p43)$^{-/-}$	↑	No thyroid receptor alpha p43 variant signaling
Amh-Cre/ ThraAMI	↑	Dominant-negative thyroid receptor alpha in Sertoli cells
ARO-Cre/ ThraAMI	↑	Dominant-negative thyroid receptor alpha in Sertoli and Leydig cells
Mct8$^{-/-}$	↑	No MCT8-mediated transport of thyroid hormones
Dio3$^{-/-}$/*Thra*$^{-/-}$	=	No thyroid hormone clearance by DIO3 and no thyroid receptor alpha signaling
Dio2$^{-/-}$	=	No T4 to T3 conversion by DIO2
Dio3$^{-/-}$/*Dio2*$^{-/-}$	↓↓	No thyroid hormone clearance by DIO3 and no T4 to T3 conversion by DIO2
Dio3$^{-/-}$/*Mct8*$^{-/-}$	↓↓	No thyroid hormone clearance by DIO3 and no transport by MCT8
Dio3$^{-/-}$	↓↓↓	No thyroid hormone clearance by DIO3

Based on data from Holsberger, D. R., Buchold, G. M., Leal, M. C., Kiesewetter, S. E., O'Brien, D. A., Hess, R. A., et al. (2005). Cell-cycle inhibitors p27Kip1 and p21Cip1 regulate murine Sertoli cell proliferation. *Biology of Reproduction, 72*(6), 1429–1436; Fumel, B., Guerquin, M. J., Livera, G., Staub, C., Magistrini, M., Gauthier, C., et al. (2012). Thyroid hormone limits postnatal Sertoli cell proliferation *in vivo* by activation of its alpha1 isoform receptor (TRalpha1) present in these cells and by regulation of Cdk4/JunD/c-myc mRNA levels in mice. *Biology of Reproduction, 87*(1), 16, 11–19; Fumel, B., Roy, S., Fouchecourt, S., Livera, G., Parent, A. S., Casas, F., et al. (2013). Depletion of the p43 mitochondrial T3 receptor increases Sertoli cell proliferation in mice. *PLoS One, 8*(9), e74015; Fumel, B., Froment, P., Holzenberger, M., Livera, G., Monget, P., & Fouchecourt, S. (2015). Expression of dominant-negative thyroid hormone receptor alpha1 in Leydig and Sertoli cells demonstrates no additional defect compared with expression in Sertoli cells only. *PLoS One, 10*(3), e0119392; Martinez, M. E., Karaczyn, A., Stohn, J. P., Donnelly, W. T., Croteau, W., Peeters, R. P., et al. (2016). The Type 3 deiodinase is a critical determinant of appropriate thyroid hormone action in the developing testis. *Endocrinology, 157*(3), 1276–1288.

an aromatase-specific promoter (Fumel et al., 2015) drives the expression of the dominant-negative *Thra*. These observations are largely recapitulated using a mouse knockout model of p43 (Table 2), a truncated form of *Thra* expressed in the mitochondrial matrix (Fumel et al., 2013), suggesting an important role for this *Thra* isoform in mediating TH action in developing SCs. Gene expression profiling by RNA sequencing in the testis of postnatal day 3 mice lacking *Thra* in SCs provides an initial compilation of SC target genes (Chatonnet, Livera, Fumel, Fouch, & Flamant, 2014). This work reveals changes in the expression of genes that are enriched in germ cells, suggesting that TH signaling in SC may influence germ cells at this early developmental stage.

Although LCs outcomes are affected by alterations in the TH status during development, more limited evidence is currently available on significant TH receptor expression and action in LCs. T3 binding by isolated LCs has been reported in association with increased androgen release (Jana & Bhattacharya, 1994), and *in vitro* assays demonstrate regulation of steroidogenic gene promoters by *Thra* (Park, Kim, Lee, & Lee, 2014). Recent observations in mice with a dominant-negative *Thra* isoform expressed in LCs did not modify the testis phenotype caused by expression of that transgene in SCs only, suggesting that *Thra* may not mediate TH action in SCs. Although the data support the idea that LCs are a direct target of TH, additional work is needed to define which receptors are involved at developmental and adult stages.

The marked predominance of THRA over THRB observed in the testis at different developmental stages and the observations in transgenic mouse models deficient in THRA function suggest that most testicular actions of TH are mediated by THRA. However, *Thrb* expression has been detected, together with that of *Thra*, in adult spermatocytes (Vanacker, Bonnelye, Delmarre, & Laudet, 1998) and prepubertal testis (Rao et al., 2003). Also, a rare THRB isoform is relatively abundant vs other THRB isoforms in the human testis (Tagami et al., 2010). Thus, although uncertain, a role for THRB in mediating TH action in specific testicular cells types or during particular developmental stages is possible.

3.3 Thyroid Hormone Transporters

TH action requires that T3 and/or T4 are transported into the cell by specific transporters. Limited information is available about the set of transporters that are expressed in specific testicular cell types at different developmental stages. Current data indicate that both MCT and OATP types of transporters are present in the testis (Hagenbuch & Meier, 2003; Martinez et al., 2016). OATP1C1 (also *Slc21a14*) is expressed in the testis (Hagenbuch, 2007) and has been identified at least in LCs (Pizzagalli et al., 2002). Its functional significance remains undefined.

The MCT8 (also *Slc16a2*) transporter, which exhibits the highest affinity for T3 (Heuer & Visser, 2009), is strongly expressed in the chicken

(Bourgeois et al., 2016) and mouse testis (Martinez et al., 2016). Mouse MCT8 expression is much higher in neonatal than in adult testis, and the genetic loss of MCT8 results in significant testicular enlargement (Martinez et al., 2016) (Table 2). This phenotype is similar to that observed in mouse models of neonatal hypothyroidism or dominant-negative THRA function, suggesting that the lack of MCT8-mediated transport is rendering hypothyroid some testicular cells, which may include SCs, extending their proliferation and ultimately affecting testis size. Surprisingly, the neonatal testes of MCT8-deficient mice show elevated expression of the T3-dependent gene *hairless* (Martinez et al., 2016). This may due to the unexplained, transient and modest, perinatal serum hyperthyroidism recently reported in these animals and noted in other tissues (Ferrara et al., 2013). These observations suggest that MCT8 is not the exclusive mediator of TH transport in the neonatal testis.

3.4 Type 1 and Type 2 Deiodinases

The conversion of T4 to the more biologically active T3 can be achieved by deiodinases DIO1 and DIO2 (Bianco, Salvatore, Gereben, Berry, & Larsen, 2002). In rodents, significant DIO2 activity is present in the developing testis from postnatal day 5 until adulthood (Bates, St Germain, & Galton, 1999; Wagner et al., 2003), and in the adult, it is located in germ cells at late stages of differentiation (Wajner et al., 2007). Interestingly, a limited ontogenic profile of DIO2 expression in the testis reveals that highest level of DIO2 activity occurs around 3 weeks of age (Bates et al., 1999) (Fig. 1). The timing of maximal DIO2 activity found in this study coincides with the peaking of TH in the serum and also with the end of the infantile period characterized by high proliferation rate of SCs and spermatogonia (Picut et al., 2015). This suggests that DIO2 action may work concurrently with rising levels of circulating TH to trigger the active differentiation of both cell types.

The testes of mice genetically devoid of DIO2 activity do not show abnormal size in adult age, but at neonatal age, they exhibit increased T3-dependent gene expression (Martinez et al., 2016). This is probably due to a transient neonatal elevation of serum T3 (Ng et al., 2004), the origin of which has not been investigated. Thus, although testis size and reproduction appear normal in DIO2-deficient mice, it is possible that their testicular TH economy is disrupted at some developmental stages. The testicular cell type(s) that express DIO2 in early development remain unknown.

Modest DIO1 activity has also been detected in the developing testis (Bates et al., 1999), but its functional significance is unknown, and the cells expressing the enzyme have not been identified.

3.5 Type 3 Deiodinase

DIO3 (also D3) can convert T4 and T3 into the inactive metabolites 3,3′,5′-triiodothyronine (reverse T3) and 3,3′-diiodothyronine, respectively. Thus, DIO3 action leads to a reduction in TH availability and signaling (Hernandez, 2005). DIO3 expression is high in the developing testis and exhibits a marked developmental profile in the rat (Bates et al., 1999) and mouse (Martinez et al., 2016), with the highest expression during the first 2 weeks of life (Fig. 1). It is not known which cell type(s) are expressing DIO3 at this age. Mice lacking a functional DIO3 (*Dio3*$^{-/-}$ mice) exhibit severe impairment in TH clearance and experience thyrotoxicosis during fetal and early neonatal life, characterized by an elevated serum T3 level, which is associated with low serum T4 level due to T3 suppression of the HPT axis (Hernandez, Martinez, Fiering, Galton, & St Germain, 2006) (Fig. 2). This T3 excess results later in life in similar but generally more severe abnormalities than those observed in experimental models of neonatal hyperthyroidism. These include stunted testis growth and markedly reduced testis size in adulthood (Fig. 2 and Table 2), decreased neonatal proliferation of testicular cells, increased testicular expression of T3-dependent genes and reduced *Amh* expression during neonatal life, larger seminiferous tubules with premature lumen formation, reduced spermatogenesis and sperm production, increased testicular expression of inhibin, elevated levels of FSH and LH, reduced levels of testosterone, and very reduced fertility (Martinez et al., 2016). Interestingly, DIO3 expression in the normal testis remains high during early neonatal life, at the same time that serum T3 is increasing, and the levels of *Thra* expression are highest (Fig. 1). Taken together, these findings demonstrate that a very low testicular level of TH action during neonatal (and possibly fetal) life is an integral component of the developmental program of the testis and the HPG axis and that DIO3 plays an essential role in it. This role possibly involves extending the period of limited T3 availability in the testis, allowing the proliferation of testicular cells to proceed in a context of high TH receptor content and increasing levels of serum T3.

The severe testicular phenotype of *Dio3*$^{-/-}$ mice provides a useful baseline on which to use genetic manipulation to assess the role of other

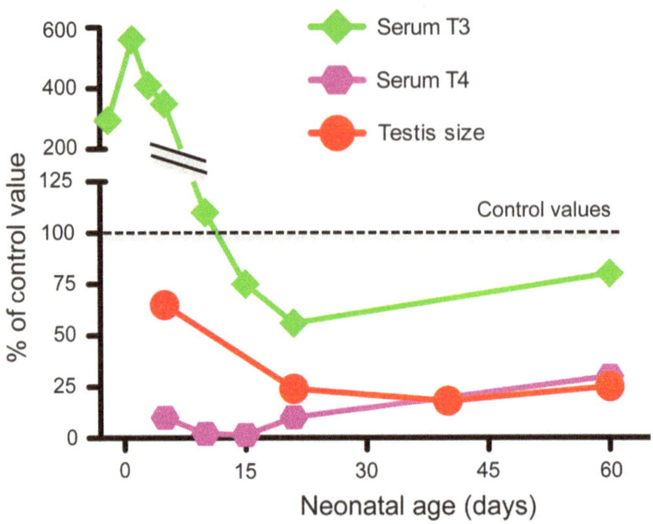

Fig. 2 Ontogeny of testis size and serum TH parameters in *Dio3*$^{-/-}$ mice relative to control values (*dotted line*). *Dio3*$^{-/-}$ experience severe thyrotoxicosis in fetal and early neonatal life, leading to an irreversible reduction of testis size. *Graph is based on data adapted from Martinez, M. E., Karaczyn, A., Stohn, J. P., Donnelly, W. T., Croteau, W., Peeters, R. P., et al. (2016). The Type 3 deiodinase is a critical determinant of appropriate thyroid hormone action in the developing testis.* Endocrinology, 157*(3), 1276–1288.*

determinants of testicular TH action. A lack of THRA completely corrects the testicular phenotype of DIO3-deficient mice (Martinez et al., 2016) (Table 2), further demonstrating the importance of THRA in mediating TH action in the neonatal testis. Since SC number is the larger contributor to adult testis size, this finding is also consistent with a critical role of this receptor in arresting the proliferation of SCs.

Genetic inactivation of the MCT8 transporter partially corrects the reduced testis size of *Dio3*$^{-/-}$ mice (Table 2), as well as the neonatal *Amh* and T3-dependent gene expression in the neonatal testis (Martinez et al., 2016). Since *Amh* expression is specific to SCs, these findings suggest that testicular MCT8 is located at least in SCs, although this requires confirmation. The fact that the testis size correction is only partial indicates that additional TH transporters are also mediating the excessive T3 action that occurs in the neonatal *Dio3*$^{-/-}$ testis.

Likewise, genetic inactivation of *Dio2* partially normalizes the size of the *Dio3*$^{-/-}$ testes (Table 2) and their neonatal *Amh* and T3-dependent gene expression (Martinez et al., 2016). This raises the possibility that part of the T3 excess targeting the testis in *Dio3*$^{-/-}$ mice originates from T4 by conversion by the DIO2, either systemically or in specific cells within testicular tissue.

Loss of MCT8 does not correct the elevated expression of testicular inhibin and pituitary *Fshb* (FSH beta) that is characteristic of *Dio3*$^{-/-}$ mice, while DIO2 deficiency does (Martinez et al., 2016). As these studies use mouse models of global gene deficiencies, these divergent observations are possibly due to differential effects of MCT8 and DIO2 in other components of the HPG axis, including the pituitary and the hypothalamus, or to their different cellular localizations within the developing testis.

3.6 Summary

Among the endogenous determinants of TH action, THRA plays critical role in mediating TH signaling in the developing testis and most of its biological effects. Similarly, MCT8-mediated transport of TH and T3 generation from T4 by the action of DIO2 seems to contribute to increase the T3 that is available in testicular cells for receptor binding. TH inactivation by DIO3 during early neonatal development is critical to prevent premature proliferation arrest in testicular cells, aberrant testis outcomes, and impaired fertility. The ontogeny of the testicular expression of *Thra* and *Mct8* places the newborn testis in a position of high T3 sensitivity. During this time, in the context of raising serum levels of TH, the maintenance of high DIO3 expression prevents excessive TH action, allowing testicular cells for an extended period of proliferation that would otherwise be arrested. The decrease in DIO3 activity and the increase in DIO2 activity later in neonatal life, together with the rise in serum levels of TH appear to be timely coordinated to stimulate the differentiation of most testicular cells.

4. CONCLUSION

Work from the last three decades leaves no doubt that the developing testis is an important target for TH. Adequate TH action at particular developmental stages is necessary for normal testicular cell development, overall testis function, hormonal regulation of the HPG axis, and normal fertility. Some local determinants of cell TH action, including THRA, MCT8, DIO2, and DIO3, are emerging as critical players in the regulation of TH economy and signaling in the developing testis, markedly influencing testicular and hormonal outcomes. However, additional work is needed to provide a complete understanding of the cellular and developmental basis by which the different TH transporters, deiodinases, and receptors interplay to timely deliver the level of TH action that is required by each type of testicular cell.

ACKNOWLEDGMENTS

The author thanks V.A. Galton for critical review of the manuscript and M.E. Martinez for assistance. This work was supported by Grants DK095908 and MH096050 from the National Institutes of Health.

REFERENCES

Ando, S., Sirianni, R., Forastieri, P., Casaburi, I., Lanzino, M., Rago, V., et al. (2001). Aromatase expression in prepubertal Sertoli cells: Effect of thyroid hormone. *Molecular and Cellular Endocrinology, 178*(1–2), 11–21.

Arambepola, N. K., Bunick, D., & Cooke, P. S. (1998). Thyroid hormone effects on androgen receptor messenger RNA expression in rat Sertoli and peritubular cells. *The Journal of Endocrinology, 156*(1), 43–50.

Ariyaratne, H. B., & Mendis-Handagama, S. M. (2012). Thyroid hormone and anti-Mullerian hormone (AMH) on Leydig cell differentiation: Studies using C57BL/6 mice and AMH over expressing mice. *Histology and Histopathology, 27*(3), 365–376.

Ariyaratne, H. B., Mendis-Handagama, S. M., & Mason, J. I. (2000). Effects of triiodothyronine on testicular interstitial cells and androgen secretory capacity of the prepubertal rat. *Biology of Reproduction, 63*(2), 493–502.

Aruldhas, M. M., Valivullah, H. M., & Govindarajulu, P. (1983). Effect of thyroidectomy on testicular enzymes of the pyruvate/malate cycle involved in lipogenesis. *Biochimica et Biophysica Acta, 755*(1), 90–94.

Aruldhas, M. M., Valivullah, H. M., & Govindarajulu, P. (1984). Effect of thyroxine-induced hyperthyroidism on some testicular enzymes of the pyruvate/malate cycle. *Biochimica et Biophysica Acta, 797*(2), 143–146.

Aruldhas, M. M., Valivullah, H. M., Srinivasan, N., & Govindarajulu, P. (1986). Role of thyroid on testicular lipids in prepubertal, pubertal and adult rats. I. Hyperthyroidism. *Biochimica et Biophysica Acta, 881*(3), 462–469.

Astapova, I., & Hollenberg, A. N. (2013). The in vivo role of nuclear receptor corepressors in thyroid hormone action. *Biochimica et Biophysica Acta, 1830*(7), 3876–3881.

Auharek, S. A., & de Franca, L. R. (2010). Postnatal testis development, Sertoli cell proliferation and number of different spermatogonial types in C57BL/6J mice made transiently hypo- and hyperthyroidic during the neonatal period. *Journal of Anatomy, 216*(5), 577–588.

Bakke, J. L., Lawrence, N., & Wilber, J. F. (1974). The late effects of neonatal hyperthyroidism upon the hypothalamic-pituitary-thyroid axis in the rat. *Endocrinology, 95*(2), 406–411.

Bardi, G., Bottazzi, C., Demori, I., & Palmero, S. (1999). Thyroid hormone and retinoic acid induce the synthesis of insulin-like growth factor-binding protein-4 in prepubertal pig sertoli cells. *European Journal of Endocrinology, 141*(6), 637–643.

Bates, J. M., St Germain, D. L., & Galton, V. A. (1999). Expression profiles of the three iodothyronine deiodinases, D1, D2, and D3, in the developing rat. *Endocrinology, 140*(2), 844–851.

Bianco, A. C. (2011). Minireview: Cracking the metabolic code for thyroid hormone signaling. *Endocrinology, 152*(9), 3306–3311.

Bianco, A. C., Salvatore, D., Gereben, B., Berry, M. J., & Larsen, P. R. (2002). Biochemistry, cellular and molecular biology, and physiological roles of the iodothyronine selenodeiodinases. *Endocrine Reviews*, *23*(1), 38–89.

Boulogne, B., Habert, R., & Levacher, C. (2003). Regulation of the proliferation of cocultured gonocytes and Sertoli cells by retinoids, triiodothyronine, and intracellular signaling factors: Differences between fetal and neonatal cells. *Molecular Reproduction and Development*, *65*(2), 194–203.

Bourgeois, N. M., Van Herck, S. L., Vancamp, P., Delbaere, J., Zevenbergen, C., Kersseboom, S., et al. (2016). Characterization of chicken thyroid hormone transporters. *Endocrinology*, *157*(6), 2560–2574.

Bunick, D., Kirby, J., Hess, R. A., & Cooke, P. S. (1994). Developmental expression of testis messenger ribonucleic acids in the rat following propylthiouracil-induced neonatal hypothyroidism. *Biology of Reproduction*, *51*(4), 706–713.

Buzzard, J. J., Morrison, J. R., O'Bryan, M. K., Song, Q., & Wreford, N. G. (2000). Developmental expression of thyroid hormone receptors in the rat testis. *Biology of Reproduction*, *62*(3), 664–669.

Buzzard, J. J., Wreford, N. G., & Morrison, J. R. (2003). Thyroid hormone, retinoic acid, and testosterone suppress proliferation and induce markers of differentiation in cultured rat Sertoli cells. *Endocrinology*, *144*(9), 3722–3731.

Caminos, J. E., Nogueiras, R., Gaytan, F., Pineda, R., Gonzalez, C. R., Barreiro, M. L., et al. (2008). Novel expression and direct effects of adiponectin in the rat testis. *Endocrinology*, *149*(7), 3390–3402.

Canale, D., Agostini, M., Giorgilli, G., Caglieresi, C., Scartabelli, G., Nardini, V., et al. (2001). Thyroid hormone receptors in neonatal, prepubertal, and adult rat testis. *Journal of Andrology*, *22*(2), 284–288.

Cardone, A., Angelini, F., Esposito, T., Comitato, R., & Varriale, B. (2000). The expression of androgen receptor messenger RNA is regulated by tri-iodothyronine in lizard testis. *The Journal of Steroid Biochemistry and Molecular Biology*, *72*(3–4), 133–141.

Carosa, E., Radico, C., Giansante, N., Rossi, S., D'Adamo, F., Di Stasi, S. M., et al. (2005). Ontogenetic profile and thyroid hormone regulation of type-1 and type-8 glucose transporters in rat Sertoli cells. *International Journal of Andrology*, *28*(2), 99–106.

Castaneda Cortes, D. C., Langlois, V. S., & Fernandino, J. I. (2014). Crossover of the hypothalamic pituitary-adrenal/interrenal, -thyroid, and -gonadal axes in testicular development. *Frontiers in Endocrinology*, *5*, 139.

Catalano, S., Pezzi, V., Chimento, A., Giordano, C., Carpino, A., Young, M., et al. (2003). Triiodothyronine decreases the activity of the proximal promoter (PII) of the aromatase gene in the mouse Sertoli cell line, TM4. *Molecular Endocrinology*, *17*(5), 923–934.

Chandrasekhar, Y., D'Occhio, M. J., Holland, M. K., & Setchell, B. P. (1985). Activity of the hypothalamo-pituitary axis and testicular development in prepubertal ram lambs with induced hypothyroidism or hyperthyroidism. *Endocrinology*, *117*(4), 1645–1651.

Chatonnet, F., Livera, G., Fumel, B., Fouch, E. S., & Flamant, F. (2014). Direct and indirect consequences on gene expression of a thyroid hormone receptor alpha 1 mutation restricted to Sertoli cells. *Molecular Reproduction and Development*, *81*(12), 1159–1166.

Chowdhury, A. R., & Arora, U. (1984). Role of thyroid in testicular development of immature rat. *Archives of Andrology*, *12*(1), 49–51.

Chowdhury, A. R., Gautam, A. K., & Chatterjee, B. B. (1984). Thyroid-testis interrelationship during the development and sexual maturity of the rat. *Archives of Andrology*, *13*(2–3), 233–239.

Cooke, P. S., Zhao, Y. D., & Bunick, D. (1994). Triiodothyronine inhibits proliferation and stimulates differentiation of cultured neonatal Sertoli cells: Possible mechanism for increased adult testis weight and sperm production induced by neonatal goitrogen treatment. *Biology of Reproduction*, *51*(5), 1000–1005.

Cristovao, F. C., Bisi, H., Mendonca, B. B., Bianco, A. C., & Bloise, W. (2002). Severe and mild neonatal hypothyroidism mediate opposite effects on Leydig cells of rats. *Thyroid, 12*(1), 13–18.
Cyr, D. G. (2011). Connexins and pannexins: Coordinating cellular communication in the testis and epididymis. *Spermatogenesis, 1*(4), 325–338.
DeGroot, L. J., Hill, L., & Rue, P. (1976). Binding of nuclear triiodothyronine (T3) binding protein-T3 complex to chromatin. *Endocrinology, 99*(6), 1605–1611.
Dussault, J. H., & Labrie, F. (1975). Development of the hypothalamic-pituitary-thyroid axis in the neonatal rat. *Endocrinology, 97*, 1321–1324.
Fallah-Rad, A. H., Connor, M. L., & Del Vecchio, R. P. (2001). Effect of transient early hyperthyroidism on onset of puberty in Suffolk ram lambs. *Reproduction, 121*(4), 639–646.
Faraone-Mennella, M. R., Ferone, A., Marino, L., Cardone, A., Comitato, R., Venditti, P., et al. (2009). Poly(ADP-ribosyl)ation of proteins and germ cell development in hyperthyroid rat testes. *Molecular and Cellular Biochemistry, 323*(1–2), 119–129.
Feng, P., Gu, J., Kim, U. J., Carnell, N. E., & Wilber, J. F. (1993). Identification, localization and developmental studies of rat prepro thyrotropin-releasing hormone mRNA in the testis. *Neuropeptides, 24*(2), 63–69.
Ferrara, A. M., Liao, X. H., Gil-Ibanez, P., Marcinkowski, T., Bernal, J., Weiss, R. E., et al. (2013). Changes in thyroid status during perinatal development of MCT8-deficient male mice. *Endocrinology, 154*(7), 2533–2541.
Forrest, D., & Visser, T. J. (2013). Thyroid hormone signaling. *Biochimica et Biophysica Acta, 1830*(7), 3859.
Francavilla, S., Cordeschi, G., Properzi, G., Di Cicco, L., Jannini, E. A., Palmero, S., et al. (1991). Effect of thyroid hormone on the pre- and post-natal development of the rat testis. *The Journal of Endocrinology, 129*(1), 35–42.
Fugassa, E., Palmero, S., & Gallo, G. (1987). Triiodothyronine decreases the production of androgen binding protein by rat Sertoli cells. *Biochemical and Biophysical Research Communications, 143*(1), 241–247.
Fumel, B., Froment, P., Holzenberger, M., Livera, G., Monget, P., & Fouchecourt, S. (2015). Expression of dominant-negative thyroid hormone receptor alpha1 in Leydig and Sertoli cells demonstrates no additional defect compared with expression in Sertoli cells only. *PloS One, 10*(3), e0119392.
Fumel, B., Guerquin, M. J., Livera, G., Staub, C., Magistrini, M., Gauthier, C., et al. (2012). Thyroid hormone limits postnatal Sertoli cell proliferation in vivo by activation of its alpha1 isoform receptor (TRalpha1) present in these cells and by regulation of Cdk4/JunD/c-myc mRNA levels in mice. *Biology of Reproduction, 87*(1), 16, 11–19.
Fumel, B., Roy, S., Fouchecourt, S., Livera, G., Parent, A. S., Casas, F., et al. (2013). Depletion of the p43 mitochondrial T3 receptor increases Sertoli cell proliferation in mice. *PloS One, 8*(9), e74015.
Gao, Y., Lee, W. M., & Cheng, C. Y. (2014). Thyroid hormone function in the rat testis. *Frontiers in Endocrinology, 5*, 188.
Gilleron, J., Nebout, M., Scarabelli, L., Senegas-Balas, F., Palmero, S., Segretain, D., et al. (2006). A potential novel mechanism involving connexin 43 gap junction for control of Sertoli cell proliferation by thyroid hormones. *Journal of Cellular Physiology, 209*(1), 153–161.
Goswani, A., Kar, A. B., & Chowdhury, S. R. (1963). Effect of 3:5:3′-triiodothyronine on testicular lipid pattern in prepuberal rhesus monkeys. *Annals of Biochemistry and Experimental Medicine, 23*, 325–328.
Hagenbuch, B. (2007). Cellular entry of thyroid hormones by organic anion transporting polypeptides. *Best Practice & Research. Clinical Endocrinology & Metabolism, 21*(2), 209–221.

Hagenbuch, B., & Meier, P. J. (2003). The superfamily of organic anion transporting polypeptides. *Biochimica et Biophysica Acta, 1609*(1), 1–18.
Hardy, M. P., Sharma, R. S., Arambepola, N. K., Sottas, C. M., Russell, L. D., Bunick, D., et al. (1996). Increased proliferation of Leydig cells induced by neonatal hypothyroidism in the rat. *Journal of Andrology, 17*(3), 231–238.
Hernandez, A. (2005). Structure and function of the type 3 deiodinase gene. *Thyroid, 15*(8), 865–874.
Hernandez, A., Martinez, M. E., Fiering, S., Galton, V. A., & St Germain, D. (2006). Type 3 deiodinase is critical for the maturation and function of the thyroid axis. *Journal of Clinical Investigation, 116*(2), 476–484.
Heuer, H., & Visser, T. J. (2009). Minireview: Pathophysiological importance of thyroid hormone transporters. *Endocrinology, 150*(3), 1078–1083.
Holsberger, D. R., Buchold, G. M., Leal, M. C., Kiesewetter, S. E., O'Brien, D. A., Hess, R. A., et al. (2005). Cell-cycle inhibitors p27Kip1 and p21Cip1 regulate murine Sertoli cell proliferation. *Biology of Reproduction, 72*(6), 1429–1436.
Holsberger, D. R., & Cooke, P. S. (2005). Understanding the role of thyroid hormone in Sertoli cell development: A mechanistic hypothesis. *Cell and Tissue Research, 322*(1), 133–140.
Holsberger, D. R., Jirawatnotai, S., Kiyokawa, H., & Cooke, P. S. (2003). Thyroid hormone regulates the cell cycle inhibitor p27Kip1 in postnatal murine Sertoli cells. *Endocrinology, 144*(9), 3732–3738.
Holsberger, D. R., Kiesewetter, S. E., & Cooke, P. S. (2005). Regulation of neonatal Sertoli cell development by thyroid hormone receptor alpha1. *Biology of Reproduction, 73*(3), 396–403.
Hudson, R. W., & Edwards, A. L. (1992). Testicular function in hyperthyroidism. *Journal of Andrology, 13*(2), 117–124.
Jana, N. R., & Bhattacharya, S. (1994). Binding of thyroid hormone to the goat testicular Leydig cell induces the generation of a proteinaceous factor which stimulates androgen release. *The Journal of Endocrinology, 143*(3), 549–556.
Jannini, E. A., Carosa, E., Rucci, N., Screponi, E., & D'Armiento, M. (1999). Ontogeny and regulation of variant thyroid hormone receptor isoforms in developing rat testis. *Journal of Endocrinological Investigation, 22*(11), 843–848.
Jannini, E. A., Crescenzi, A., Rucci, N., Screponi, E., Carosa, E., de Matteis, A., et al. (2000). Ontogenetic pattern of thyroid hormone receptor expression in the human testis. *The Journal of Clinical Endocrinology and Metabolism, 85*(9), 3453–3457.
Jannini, E. A., Dolci, S., Ulisse, S., & Nikodem, V. M. (1994). Developmental regulation of the thyroid hormone receptor alpha 1 mRNA expression in the rat testis. *Molecular Endocrinology, 8*(1), 89–96.
Jannini, E. A., Olivieri, M., Francavilla, S., Gulino, A., Ziparo, E., & D'Armiento, M. (1990). Ontogenesis of the nuclear 3,5,3′-triiodothyronine receptor in the rat testis. *Endocrinology, 126*(5), 2521–2526.
Jannini, E. A., Ulisse, S., & D'Armiento, M. (1995). Thyroid hormone and male gonadal function. *Endocrine Reviews, 16*(4), 443–459.
Jiang, X., Cai, Z., Zhao, X., Zhang, L., Chen, Z., Wang, Y., et al. (2011). Mapping, cDNA cloning and tissue expression of the porcine thyrotropin-releasing hormone receptor gene. *Animal Biotechnology, 22*(1), 30–36.
Kimura, T., & Furudate, S. (1996). Pituitary GH and prolactin deficiency and testis enlargement in hypothyroid rats caused by goitrogen methimazole. *Experimental Animals, 45*(4), 369–375.
Kirby, J. D., Arambepola, N., Porkka-Heiskanen, T., Kirby, Y. K., Rhoads, M. L., Nitta, H., et al. (1997). Neonatal hypothyroidism permanently alters follicle-stimulating hormone and luteinizing hormone production in the male rat. *Endocrinology, 138*(7), 2713–2721.

Kirby, J. D., Jetton, A. E., Cooke, P. S., Hess, R. A., Bunick, D., Ackland, J. F., et al. (1992). Developmental hormonal profiles accompanying the neonatal hypothyroidism-induced increase in adult testicular size and sperm production in the rat. *Endocrinology*, *131*(2), 559–565.
Kirby, J. D., Mankar, M. V., Hardesty, D., & Kreider, D. L. (1996). Effects of transient prepubertal 6-N-propyl-2-thiouracil treatment on testis development and function in the domestic fowl. *Biology of Reproduction*, *55*(4), 910–916.
Kobayashi, K., Kubota, H., Hojo, R., & Miyagawa, M. (2014). Dose-dependent effects of perinatal hypothyroidism on postnatal testicular development in rat offspring. *The Journal of Toxicological Sciences*, *39*(6), 867–874.
Kurowicka, B., Chrusciel, M., Zmijewska, A., & Kotwica, G. (2016). Effect of neonatal or adult heat acclimation on plasma fT3 level, testicular thyroid receptors expression in male rats and testicular steroidogenesis in vitro in response to triiodothyronine treatment. *Polish Journal of Veterinary Sciences*, *19*(2), 379–386.
Lagu, S. K., Bhavsar, N. G., Sharma, R. K., & Ramachandran, A. V. (2005). Neonatal hypothyroidism-induced changes in rat testis size, dependence on temperature. *Neuro Endocrinology Letters*, *26*(6), 780–788.
Laslett, A. L., Li, L. H., Jester, W. F., Jr., & Orth, J. M. (2000). Thyroid hormone downregulates neural cell adhesion molecule expression and affects attachment of gonocytes in Sertoli cell-gonocyte cocultures. *Endocrinology*, *141*(5), 1633–1641.
Lazar, M. A., & Chin, W. W. (1990). Nuclear thyroid hormone receptors. *The Journal of Clinical Investigation*, *86*, 1777–1782.
Li, G., Lee, Y. F., Liu, S., Cai, Y., Xie, S., Liu, N. C., et al. (2008). Oxidative stress stimulates testicular orphan receptor 4 through forkhead transcription factor forkhead box O3a. *Endocrinology*, *149*(7), 3490–3499.
Majdic, G., Snoj, T., Horvat, A., Mrkun, J., Kosec, M., & Cestnik, V. (1998). Higher thyroid hormone levels in neonatal life result in reduced testis volume in postpubertal bulls. *International Journal of Andrology*, *21*(6), 352–357.
Majeed, S. K., Injidi, M. H., & Forbes, J. M. (1984). Effects of thyroxine on the reproductive organs and growth of young chickens. *Experientia*, *40*(3), 281–283.
Maran, R. R. (2003). Thyroid hormones: Their role in testicular steroidogenesis. *Archives of Andrology*, *49*(5), 375–388.
Maran, R. R., & Aruldhas, M. M. (2002). Adverse effects of neonatal hypothyroidism on Wistar rat spermatogenesis. *Endocrine Research*, *28*(3), 141–154.
Maran, R. R., Arunakaran, J., & Aruldhas, M. M. (2000). T3 directly stimulates basal and modulates LH induced testosterone and oestradiol production by rat Leydig cells in vitro. *Endocrine Journal*, *47*(4), 417–428.
Maran, R. R., Arunakaran, J., Jeyaraj, D. A., Ravichandran, K., Ravisankar, B., & Aruldhas, M. M. (2000). Transient neonatal hypothyroidism alters plasma and testicular sex steroid concentration in puberal rats. *Endocrine Research*, *26*(3), 411–429.
Maran, R. R., Ravisankar, B., Ravichandran, K., Valli, G., Arunakaran, J., & Aruldhas, M. M. (1999). Impact of neonatal onset hypothyroidism on Sertoli cell number, plasma and testicular interstitial fluid androgen binding protein concentration. *Endocrine Research*, *25*(3–4), 307–322.
Maran, R. R., Sivakumar, R., Arunakaran, J., Ravisankar, B., Ravichandran, K., Sidharthan, V., et al. (1999). Duration-dependent effect of transient neonatal hypothyroidism on Sertoli and germ cell number, and plasma and testicular interstitial fluid androgen binding protein concentration. *Endocrine Research*, *25*(3–4), 323–340.
Marchlewska, K., Kula, K., Walczak-Jedrzejowska, R., Kula, W., Oszukowska, E., Filipiak, E., et al. (2013). Maturational changes in connexin 43 expression in the seminiferous tubules may depend on thyroid hormone action. *Archives of Medical Science*, *9*(1), 139–145.

Marchlewska, K., Kula, K., Walczak-Jedrzejowska, R., Oszukowska, E., Filipiak, E., & Slowikowska-Hilczer, J. (2011a). Role of FSH and triiodothyronine in Sertoli cell development expressed by formation of connexin 43-based gap junctions. *Journal of Experimental Zoology. Part A, Ecological Genetics and Physiology, 315*(6), 329–336.
Marchlewska, K., Kula, K., Walczak-Jedrzejowska, R., Oszukowska, E., Orkisz, S., & Slowikowska-Hilczer, J. (2011b). Triiodothyronine modulates initiation of spermatogenesis in rats depending on treatment timing and blood level of the hormone. *Molecular and Cellular Endocrinology, 341*(1–2), 25–34.
Martinez, M. E., Karaczyn, A., Stohn, J. P., Donnelly, W. T., Croteau, W., Peeters, R. P., et al. (2016). The type 3 deiodinase is a critical determinant of appropriate thyroid hormone action in the developing testis. *Endocrinology, 157*(3), 1276–1288.
Mendis-Handagama, S. M., & Ariyaratne, H. B. (2001). Differentiation of the adult Leydig cell population in the postnatal testis. *Biology of Reproduction, 65*(3), 660–671.
Mendis-Handagama, C., & Ariyaratne, S. (2004a). Prolonged and transient neonatal hypothyroidism on Leydig cell differentiation in the postnatal rat testis. *Archives of Andrology, 50*(5), 347–357.
Mendis-Handagama, S. M., & Ariyaratne, H. B. (2004b). Effects of thyroid hormones on Leydig cells in the postnatal testis. *Histology and Histopathology, 19*(3), 985–997.
Mendis-Handagama, S. M., & Ariyaratne, H. B. (2008). Effects of hypothyroidism on anti-mullerian hormone expression in the prepubertal rat testis. *Histology and Histopathology, 23*(2), 151–156.
Mendis-Handagama, S. M., Ariyaratne, H. B., Mrkonjich, L., & Ivell, R. (2007). Expression of insulin-like peptide 3 in the postnatal rat Leydig cell lineage: Timing and effects of triiodothyronine-treatment. *Reproduction, 133*(2), 479–485.
Mendis-Handagama, S. M., Ariyaratne, H. B., Teunissen van Manen, K. R., & Haupt, R. L. (1998). Differentiation of adult Leydig cells in the neonatal rat testis is arrested by hypothyroidism. *Biology of Reproduction, 59*(2), 351–357.
Mendis-Handagama, S. M., & Siril Ariyaratne, H. B. (2005). Leydig cells, thyroid hormones and steroidogenesis. *Indian Journal of Experimental Biology, 43*(11), 939–962.
Mu, X., Lee, Y. F., Liu, N. C., Chen, Y. T., Kim, E., Shyr, C. R., et al. (2004). Targeted inactivation of testicular nuclear orphan receptor 4 delays and disrupts late meiotic prophase and subsequent meiotic divisions of spermatogenesis. *Molecular and Cellular Biology, 24*(13), 5887–5899.
Ng, L., Goodyear, R. J., Woods, C. A., Schneider, M. J., Diamond, E., Richardson, G. P., et al. (2004). Hearing loss and retarded cochlear development in mice lacking type 2 iodothyronine deiodinase. *Proceedings of the National Academy of Sciences of the United States of America, 101*, 3474–3479.
Oluwole, O. A., Bartlewski, P. M., & Hahnel, A. (2013). Relationships of serum thyroid hormones and follicle-stimulating hormone concentrations to Sertoli cell differentiation during the first wave of spermatogenesis in euthyroid ram lambs. *Reproductive Biology, 13*(2), 150–160.
Oncu, M., Kavakli, D., Gokcimen, A., Gulle, K., Orhan, H., & Karaoz, E. (2004). Investigation on the histopathological effects of thyroidectomy on the seminiferous tubules of immature and adult rats. *Urologia Internationalis, 73*(1), 59–64.
Oppenheimer, J. H., Schwartz, H. L., & Surks, M. I. (1974). Tissue differences in the concentration of triiodothyronine nuclear binding sites in the rat: Liver, kidney, pituitary, heart, brain, spleen, and testis. *Endocrinology, 95*(3), 897–903.
Palmero, S., Benahmed, M., Morera, A. M., Trucchi, P., & Fugassa, E. (1992). Identification of nuclear tri-iodothyronine receptors in Sertoli cells from immature piglet testes. *Journal of Molecular Endocrinology, 9*(1), 55–59.
Palmero, S., de Marchis, M., Gallo, G., & Fugassa, E. (1989). Thyroid hormone affects the development of Sertoli cell function in the rat. *The Journal of Endocrinology, 123*(1), 105–111.

Palmero, S., De Marco, P., Trucchi, P., Prati, M., & Fugassa, E. (1991). Nuclear triiodothyronine receptors in the testis of the growing rat are regulated by the thyroid status. *Bollettino della Società Italiana di Biologia Sperimentale*, *67*(5), 501–508.

Palmero, S., Maggiani, S., & Fugassa, E. (1988). Nuclear triiodothyronine receptors in rat Sertoli cells. *Molecular and Cellular Endocrinology*, *58*(2–3), 253–256.

Palmero, S., Prati, M., Barreca, A., Minuto, F., Giordano, G., & Fugassa, E. (1990). Thyroid hormone stimulates the production of insulin-like growth factor I (IGF-I) by immature rat Sertoli cells. *Molecular and Cellular Endocrinology*, *68*(1), 61–65.

Palmero, S., Prati, M., Bolla, F., & Fugassa, E. (1995). Tri-iodothyronine directly affects rat Sertoli cell proliferation and differentiation. *The Journal of Endocrinology*, *145*(2), 355–362.

Palmero, S., Prati, M., De Marco, P., Trucchi, P., & Fugassa, E. (1993). Thyroidal regulation of nuclear tri-iodothyronine receptors in the developing rat testis. *The Journal of Endocrinology*, *136*(2), 277–282.

Palmero, S., Trucchi, P., Prati, M., Fugassa, E., Lanni, A., & Goglia, F. (1994). Effect of thyroid status on the oxidative capacity of Sertoli cells isolated from immature rat testis. *European Journal of Endocrinology*, *130*(3), 308–312.

Panno, M. L., Beraldi, E., Pezzi, V., Salerno, M., De Luca, G., Lanzino, M., et al. (1994). Influence of thyroid hormone on androgen metabolism in peripuberal rat Sertoli cells. *The Journal of Endocrinology*, *140*(3), 349–355.

Panno, M. L., Salerno, M., Lanzino, M., De Luca, G., Maggiolini, M., Straface, S. V., et al. (1995). Follow-up study on the effects of thyroid hormone administration on androgen metabolism of peripubertal rat Sertoli cells. *European Journal of Endocrinology*, *132*(2), 236–241.

Panno, M. L., Sisci, D., Salerno, M., Lanzino, M., Mauro, L., Morrone, E. G., et al. (1996a). Effect of triiodothyronine administration on estrogen receptor contents in peripuberal Sertoli cells. *European Journal of Endocrinology*, *134*(5), 633–638.

Panno, M. L., Sisci, D., Salerno, M., Lanzino, M., Pezzi, V., Morrone, E. G., et al. (1996b). Thyroid hormone modulates androgen and oestrogen receptor content in the Sertoli cells of peripubertal rats. *The Journal of Endocrinology*, *148*(1), 43–50.

Park, E., Kim, Y., Lee, H. J., & Lee, K. (2014). Differential regulation of steroidogenic enzyme genes by TR alpha signaling in testicular Leydig cells. *Molecular Endocrinology*, *28*(6), 822–833.

Pelletier, R. M. (2011). The blood-testis barrier: The junctional permeability, the proteins and the lipids. *Progress in Histochemistry and Cytochemistry*, *46*(2), 49–127.

Pezzi, V., Panno, M. L., Sirianni, R., Forastieri, P., Casaburi, I., Lanzino, M., et al. (2001). Effects of tri-iodothyronine on alternative splicing events in the coding region of cytochrome P450 aromatase in immature rat Sertoli cells. *The Journal of Endocrinology*, *170*(2), 381–393.

Picut, C. A., Remick, A. K., de Rijk, E. P., Simons, M. L., Stump, D. G., & Parker, G. A. (2015). Postnatal development of the testis in the rat: Morphologic study and correlation of morphology to neuroendocrine parameters. *Toxicologic Pathology*, *43*(3), 326–342.

Pitetti, J. L., Calvel, P., Zimmermann, C., Conne, B., Papaioannou, M. D., Aubry, F., et al. (2013). An essential role for insulin and IGF1 receptors in regulating Sertoli cell proliferation, testis size, and FSH action in mice. *Molecular Endocrinology*, *27*(5), 814–827.

Pizzagalli, F., Hagenbuch, B., Stieger, B., Klenk, U., Folkers, G., & Meier, P. J. (2002). Identification of a novel human organic anion transporting polypeptide as a high affinity thyroxine transporter. *Molecular Endocrinology*, *16*(10), 2283–2296.

Rao, J. N., Liang, J. Y., Chakraborti, P., & Feng, P. (2003). Effect of thyroid hormone on the development and gene expression of hormone receptors in rat testes in vivo. *Journal of Endocrinological Investigation*, *26*(5), 435–443.

Rijntjes, E., Swarts, H. J., Anand-Ivell, R., & Teerds, K. J. (2009). Prenatal induced chronic dietary hypothyroidism delays but does not block adult-type Leydig cell development. *American Journal of Physiology. Endocrinology and Metabolism*, *296*(2), E305–E314.

Rijntjes, E., Wientjes, A. T., Swarts, H. J., de Rooij, D. G., & Teerds, K. J. (2008). Dietary-induced hyperthyroidism marginally affects neonatal testicular development. *Journal of Andrology*, *29*(6), 643–653.

Sahoo, D. K., Roy, A., Bhanja, S., & Chainy, G. B. (2008). Hypothyroidism impairs antioxidant defence system and testicular physiology during development and maturation. *General and Comparative Endocrinology*, *156*(1), 63–70.

Sakai, Y., Yamashina, S., & Furudate, S. (2004). Developmental delay and unstable state of the testes in the rdw rat with congenital hypothyroidism. *Development, Growth & Differentiation*, *46*(4), 327–334.

Santos-Ahmed, J., Brown, C., Smith, S. D., Weston, P., Rasoulpour, T., Gilbert, M. E., et al. (2011). Akt1 protects against germ cell apoptosis in the postnatal mouse testis following lactational exposure to 6-N-propylthiouracil. *Reproductive Toxicology*, *31*(1), 17–25.

Sarkar, D., & Singh, S. K. (2017). Effect of neonatal hypothyroidism on prepubertal mouse testis in relation to thyroid hormone receptor alpha 1 (THRalpha1). *General and Comparative Endocrinology*, (in press). https://doi.org/10.1016/j.ygcen.2016.08.001.

Saxena, D. K., Yoshinaga, K., Tanii, I., & Toshimori, K. (2002). Are germ cell factors essential in the testicular enlargement after neonatal hypothyroidism recovery? A study using W/Wv mutant mice model. *International Journal of Andrology*, *25*(1), 11–18.

Simorangkir, D. R., Wreford, N. G., & De Kretser, D. M. (1997). Impaired germ cell development in the testes of immature rats with neonatal hypothyroidism. *Journal of Andrology*, *18*(2), 186–193.

Siril Ariyaratne, H. B., Ian Mason, J., & Mendis-Handagama, S. M. (2000). Effects of thyroid and luteinizing hormones on the onset of precursor cell differentiation into Leydig progenitor cells in the prepubertal rat testis. *Biology of Reproduction*, *63*(3), 898–904.

Sisci, D., Panno, M. L., Salerno, M., Maggiolini, M., Pezzi, V., Morrone, E. G., et al. (1997). A time course study on the "in vitro" effects of T3 and testosterone on androgen and estrogen receptors in peripuberal primary rat Sertoli cells. *Experimental and Clinical Endocrinology & Diabetes*, *105*(4), 218–224.

Souhila, D. S., Zohra, H. S., Kamel, A., & Hadj-Bekkouche, F. (2013). Effects of thyroxine treatment during lactation on the testicular function of rats across different ages. *Folia Histochemica et Cytobiologica*, *51*(2), 107–114.

Sridharan, S., Simon, L., Meling, D. D., Cyr, D. G., Gutstein, D. E., Fishman, G. I., et al. (2007). Proliferation of adult Sertoli cells following conditional knockout of the gap junctional protein GJA1 (connexin 43) in mice. *Biology of Reproduction*, *76*(5), 804–812.

St Germain, D., Galton, V. A., & Hernandez, A. (2009). Defining the roles of iodothyronine deiodinases: Current concepts and challenges. *Endocrinology*, *150*(3), 1097–1107.

St-Pierre, N., Dufresne, J., Rooney, A. A., & Cyr, D. G. (2003). Neonatal hypothyroidism alters the localization of gap junctional protein connexin 43 in the testis and messenger RNA levels in the epididymis of the rat. *Biology of Reproduction*, *68*(4), 1232–1240.

Sun, Y., Yang, W., Luo, H., Wang, X., Chen, Z., Zhang, J., et al. (2015). Thyroid hormone inhibits the proliferation of piglet Sertoli cell via PI3K signaling pathway. *Theriogenology*, *83*(1), 86–94.

Tagami, T., Nakamura, H., Sasaki, S., Mori, T., Yoshioka, H., Yoshida, H., et al. (1990). Immunohistochemical localization of nuclear 3,5,3′-triiodothyronine receptor proteins in rat tissues studied with antiserum against C-ERB A/T3 receptor. *Endocrinology*, *127*(4), 1727–1734.

Tagami, T., Yamamoto, H., Moriyama, K., Sawai, K., Usui, T., Shimatsu, A., et al. (2010). Identification of a novel human thyroid hormone receptor beta isoform as a transcriptional modulator. *Biochemical and Biophysical Research Communications*, *396*(4), 983–988.

Teerds, K. J., de Rooij, D. G., de Jong, F. H., & van Haaster, L. H. (1998). Development of the adult-type Leydig cell population in the rat is affected by neonatal thyroid hormone levels. *Biology of Reproduction*, *59*(2), 344–350.

Tousson, E., Ali, E. M., Ibrahim, W., & Mansour, M. A. (2011). Proliferating cell nuclear antigen as a molecular biomarker for spermatogenesis in PTU-induced hypothyroidism of rats. *Reproductive Sciences*, *18*(7), 679–686.

Ulisse, S., Jannini, E. A., Carosa, E., Piersanti, D., Graziano, F. M., & D'Armiento, M. (1994). Inhibition of aromatase activity in rat Sertoli cells by thyroid hormone. *The Journal of Endocrinology*, *140*(3), 431–436.

Ulisse, S., Jannini, E. A., Pepe, M., De Matteis, S., & D'Armiento, M. (1992). Thyroid hormone stimulates glucose transport and GLUT1 mRNA in rat Sertoli cells. *Molecular and Cellular Endocrinology*, *87*(1–3), 131–137.

Ulisse, S., Rucci, N., Piersanti, D., Carosa, E., Graziano, F. M., Pavan, A., et al. (1998). Regulation by thyroid hormone of the expression of basement membrane components in rat prepubertal Sertoli cells. *Endocrinology*, *139*(2), 741–747.

Umezu, M., Kagabu, S., Jiang, J. Y., Niimura, S., & Sato, E. (2004). Developmental hormonal profiles in rdw rats with congenital hypothyroidism accompanying increased testicular size and infertility in adulthood. *The Journal of Reproduction and Development*, *50*(6), 675–684.

Valle, L. B., Oliveira-Filho, R. M., Romaldini, J. H., & Lara, P. F. (1985). Pituitary-testicular axis abnormalities in immature male hypothyroid rats. *Journal of Steroid Biochemistry*, *23*(3), 253–257.

Van Haaster, L. H., De Jong, F. H., Docter, R., & De Rooij, D. G. (1992). The effect of hypothyroidism on Sertoli cell proliferation and differentiation and hormone levels during testicular development in the rat. *Endocrinology*, *131*(3), 1574–1576.

van Haaster, L. H., de Jong, F. H., Docter, R., & de Rooij, D. G. (1993). High neonatal triiodothyronine levels reduce the period of Sertoli cell proliferation and accelerate tubular lumen formation in the rat testis, and increase serum inhibin levels. *Endocrinology*, *133*(2), 755–760.

Vanacker, J. M., Bonnelye, E., Delmarre, C., & Laudet, V. (1998). Activation of the thyroid hormone receptor alpha gene promoter by the orphan nuclear receptor ERR alpha. *Oncogene*, *17*(19), 2429–2435.

Visser, T. J. (2013). Thyroid hormone transporters and resistance. *Endocrine Development*, *24*, 1–10.

Wagner, M. S., Morimoto, R., Dora, J. M., Benneman, A., Pavan, R., & Maia, A. L. (2003). Hypothyroidism induces type 2 iodothyronine deiodinase expression in mouse heart and testis. *Journal of Molecular Endocrinology*, *31*(3), 541–550.

Wagner, M. S., Wajner, S. M., & Maia, A. L. (2008). The role of thyroid hormone in testicular development and function. *The Journal of Endocrinology*, *199*(3), 351–365.

Wajner, S. M., dos Santos Wagner, M., Melo, R. C., Parreira, G. G., Chiarini-Garcia, H., Bianco, A. C., et al. (2007). Type 2 iodothyronine deiodinase is highly expressed in germ cells of adult rat testis. *The Journal of Endocrinology*, *194*(1), 47–54.

Wilber, J. F., & Xu, A. H. (1998). The thyrotropin-releasing hormone gene 1998: Cloning, characterization, and transcriptional regulation in the central nervous system, heart, and testis. *Thyroid*, *8*(10), 897–901.

Zamoner, A., Barreto, K. P., Filho, D. W., Sell, F., Woehl, V. M., Guma, F. C., et al. (2007). Hyperthyroidism in the developing rat testis is associated with oxidative stress and hyperphosphorylated vimentin accumulation. *Molecular and Cellular Endocrinology*, *267*(1–2), 116–126.

Zamoner, A., Corbelini, P. F., Funchal, C., Menegaz, D., Silva, F. R., & Pessoa-Pureur, R. (2005). Involvement of calcium-dependent mechanisms in T3-induced phosphorylation of vimentin of immature rat testis. *Life Sciences*, 77(26), 3321–3335.

CHAPTER EIGHTEEN

Signaling Pathways in Thyroid Cancer

Shan Jin[1], Yun-Tian Yang, Wuyuntu Bao
Affiliated Hospital of Inner Mongolia Medical University, Hohhot, China
[1]Corresponding author: e-mail address: jinshangood@163.com

Contents

1. Introduction 501
2. Thyroid Cancer-Related Signaling Pathways 502
3. Residual Tumor-Related Signaling Pathways 504
4. "Reclaim Therapy"-Related Signaling Pathways 505
5. Improvement in Anticancer Ability and Signaling Pathways 507
6. Thyroid Cancer-Targeted Therapeutic Drugs 508
7. Conclusion and Future Directions 510
Acknowledgments 511
References 511

Abstract

The morbidity of thyroid cancer is increasing gradually year by year, showing an increasing tendency in nationality, sex, age, tumor size, and tumor staging. The changes of thyroid cell genes, signaling pathways, and related molecular dysfunction promote the occurrence, development, invasion, and metastasis of thyroid cancer. Surgical operation, radioiodine, and endocrinotherapy models can achieve a better prognosis for most patients with thyroid cancer. Although targeted therapeutic drugs bring possible therapeutic opportunities for refractory thyroid cancer, there is a great gap between their predictive value and their actual efficacy. Currently, there is still no completely effective drug for the treatment. Based on the signaling pathways, the "reclaim therapy" for residual tumor and systemic intervention aims to increase anticancer ability and to encourage new directions and thoughts in the treatment of refractory thyroid cancer.

1. INTRODUCTION

Thyroid cancer is the most common malignant tumor in the endocrine system, and its rate of morbidity is increasing gradually year by year. In South Korea, thyroid cancer is the leading type among all cancers, and between 2008 and 2010, the overall morbidity was 64.1 per 100,000

Vitamins and Hormones, Volume 106
ISSN 0083-6729
http://dx.doi.org/10.1016/bs.vh.2017.05.002

individuals: female morbidity was 107.3, male morbidity was 21.1, and the average annual growth was 24.2% (Ahn et al., 2016; Ahn, Kim, & Welch, 2014). In China, in 2013 the overall morbidity was 2.07 per 100,000 individuals: female morbidity was 3.44, male morbidity was 0.92, and the average annual growth was 4% (Cong et al., 2016). In 2012, about 56,460 patients were newly diagnosed with thyroid cancer in the United States, resulting in 1780 deaths. Compared with data from 1989, the incident numbers had increased 4.99-fold by 2012. It has been predicted that by 2014, there may be 62,980 newly diagnosed cases and 1890 deaths (Pellegriti, Frasca, Regalbuto, Squatrito, & Vigneri, 2013; Siegel et al., 2016). Thyroid cancer now has becoming the fastest increasing cancer among all the malignancies (Keinan-Boker & Silverman, 2016; Meza & Chang, 2015). Surgical operation, radioiodine, and endocrinotherapy models were used to obtain a better prognosis in most of the patients with differentiated thyroid carcinoma (DTC). Medullary thyroid carcinoma (MTC) is mainly treated by surgery, but the occurrence of local and distant metastasis is very high, and treatment strategies are limited. The traditional therapeutic model has no effectiveness on anaplastic thyroid cancer (ATC), so therapy is limited, with poor prognosis. During therapy or the natural course of the disease, 2%–5% of the DTC lose their differentiation phenotype, and radioiodine and thyroid-stimulating hormone (TSH) suppressive therapy cannot achieve results showing good efficacy (Rivera et al., 2008). About 10% of patients with thyroid cancer were diagnosed with distant metastasis, which was the main reason for their mortality rate (Sáez, 2013). The 5-year survival rate of metastatic refractory DTC was 66% (Nixon et al., 2012), while the 10-year survival rate was only 10% (Durante et al., 2006). Surgical operation, radioiodine, and endocrine suppression therapy demonstrated poor therapeutic effects on the above patients; however, targeted molecular treatment possibly offers opportunities. As for therapy for refractory thyroid cancer, excluding the earlier therapies, differentiating and inducing residual tumor, making it "reclaimed," as well as immune-inducing bodies and increasing anticancer ability, may possess the same important effects (Jin, Borkhuu, Bao, & Yang, 2016).

2. THYROID CANCER-RELATED SIGNALING PATHWAYS

The occurrence, development, invasion, and metastasis of thyroid cancer involve various cellular signaling pathways and dysfunction of related molecules. The signaling pathways include the Src signaling pathway,

JAK–STAT signaling pathway, MAPK signaling pathway, PI3K/AKT signaling pathway, NF-κB signaling pathway, TSHR signaling pathway, Wnt-β-catenin signaling pathway, Notch signaling pathway, RASSF1–MST1–FOXO3 signaling pathway, C-met signaling pathway, Sonic hedgehog signaling pathway, etc. Among these, MAPK and PI3K/AKT were closely related to occurrences of DTC. Src family kinase is related to epithelial–mesenchymal transition (Kim et al., 2012). The activation of the Src signaling pathway can activate other signaling pathways such as MAPK, PI3K, FAK, STAT, etc. to influence growth, invasion, and metastasis of thyroid cancer (Chan et al., 2012; Henderson et al., 2014; Kim et al., 2012), all of which play pivotal roles in the signal transduction pathway of occurrence and development of thyroid cancer. The MAPK signaling pathway plays an important role in the occurrence of papillary thyroid carcinoma (PTC), and its activation is mainly related to BRAF genetic mutation (Leonardi et al., 2012; Vidal et al., 2013). Xing analyzed 29 clinical research studies about $BRAF^{V600E}$ genetic mutation in thyroid tumors. The results showed that 44% (810/1856) of PTC patients and 24% (23/94) of ATC patients had $BRAF^{V600E}$ genetic mutation. Furthermore, 165 patients with follicular thyroid carcinoma (FTC), 65 patients with MTC, and 542 patients with benign thyroid tumors showed no $BRAF^{V600E}$ gene mutations, suggesting that the $BRAF^{V600E}$ gene mutation was closely related to the occurrence of PTC (Xing, 2007). The mutation of $BRAF^{V600E}$ was related to many pathological indexes such as large tumor size of PTC, polycentric occurrence, thyroid extension and lymphatic metastasis, etc. (Li, Lee, Schneider, & Zeiger, 2012; Lim et al., 2013). PI3K/AKT signaling pathway activation can activate Wnt-β-catenin, HIF1α, FOXO3, and NF-κB signaling pathways (Krejci et al., 2012; Miao & Zhao, 2016; Nozhat & Hedayati, 2016; Weidinger, Krause, Mueller, Klagge, & Fuhrer, 2011). Akt, which is activated by P13K, activates or inhibits its downstream target proteins (such as Bad, caspase9, forkhead, Par-4, p2l, mTOR, etc.) by phosphorylation. It is an important antiapoptotic adjustment factor that promotes cell survival via various pathways. RAS mutation is the second most common genetic mutation, followed by BRAF, which mainly activates PI3K/AKT pathway. It was found that 93% of FTC and 96% of ATC underlie the protein-coding gene mutation of the PI3K/AKT signaling pathway (Liu et al., 2008). Xing believed that the activation of the MAPK pathway promoted the occurrence of PTC. Based on this, the activation of the PI3K/AKT pathway promoted the dedifferentiation focus and developed into ATC. The activation of the PI3K/AKT pathway promoted the occurrence of FTC, and based on this, the PI3K/AKT pathway further activated and/or accompanied with the activation of MAPK pathway, which promoted the transformation from focus into ATC (Xing, 2009). In addition, the C-met signal transduction pathway is related to thyroid cancer growth, infiltration, and lymphatic metastasis (Choe, Yun, Nam, & Kim, 2012; Koo et al., 2014). The Sonic hedgehog signaling pathway is widely expressed in thyroid cancer and is related to tumor occurrence, staging, and lymphatic metastasis (Xu et al., 2012). No matter how each signaling pathway works, it is very hard to regulate the occurrence, invasion, and metastasis of thyroid cancer with activation of just one signaling pathway. Each signaling transduction pathway plays its role in the form of a network. With a relationship that involves interaction, mutual coordination, and mutual promotion, they form a complicated set of networks and achieve whole cell regulation. The imbalance of the network promotes the occurrence, development, invasion, and metastasis of thyroid cancer.

3. RESIDUAL TUMOR-RELATED SIGNALING PATHWAYS

Tumors in the new generation possess eight characteristics: everlasting proliferation signal, escape from growth-resistant genes, frustration of cell death, infinite lifetime, vasculogenesis, activation of infiltration and metastasis, reconstitution of energy metabolism, and avoidance of immune damage. The instability of a genome composes the characteristics of genes, and the instability of genes leads to diversification of genes (Hanahan & Weinberg, 2011). The relevant genome of tumor metastasis shows dynamic changes, and all therapies that are intended to destroy the tumor directly (such as surgery, radiotherapy, chemotherapy, molecular-targeted therapy, etc.) may cause an increase in metastatic potential of residual tumor and are accompanied by genome changes. Sohn et al. studied 17 cases of DTC patients with primary tumor and paired distant metastasis focus, and 3 unpaired samples from 2 patients with metastatic tumors. It was found that 8 of the 10 PTC patients had BRAF mutation, and NRAS mutation was detected in 1 patient with FTC. TERT promoter mutation was found in 8 of the 10 PTC patients; this was often associated with BRAF mutation (BRAF mutation was found in 7 of the 8 patients with TERT promoter mutation). In FTC, the most common mutation is NRAS (4 cases in 9 patients), followed by HRAS (2 cases) and KRAS (1 case). In 7 FTC patients with positive RAS, 6 of them had TERT promoter mutation. Somatic mutations, such as BRAF and RAS, showed high integration without any difference in the primary and matching metastatic focus. BRAF and RAS mutations had a higher mutation frequency in the matched distant metastatic focus (compared with primary focus), among which the mutation rate of BRAF was 35% vs 25% ($P=0.04$) and the RAS mutation rate was 40% vs 34% ($P=0.002$). TERT promoter mutation was also highly consistent in the matched samples, and was in concordance rate of up to 93% (Sohn et al., 2016). The activation of the MAPK pathway is mainly related to BRAF genetic mutation, suggesting that activation of the MAPK pathway was increased in residual tumors. RAS mutation activated the PI3K/AKT pathway, which indicated that the activation of the PI3K/AKT pathway and/or MAPK pathway promoted the focus to transform ATC. Similarly, thyroid cancer with TERT promoter mutation showed stronger invasiveness (Landa et al., 2013). This suggested that although there is a close genetic linkage between the tumor cells in the primary focus and metastatic focus, heterogeneity within the tumor in the metastatic focus exists, and the metastatic potential of a residual tumor is increased. This demonstrates the high complexity and diversity in the evolutionary process of a malignant tumor, and results in huge difficulties in terms of therapy. As for the primary focus, even if there is heterogeneity within the tumor, the focus can be successfully removed by operative treatment. However, the metastatic focus (residual tumor) cannot be removed completely, and the removal of the residual tumor depends on systemic therapy. All these residual tumor signaling pathways and the changes in the molecular level are not only the basis for invasion and metastasis of thyroid cancer, but also possible targets for therapy in the future.

4. "RECLAIM THERAPY"-RELATED SIGNALING PATHWAYS

As for iodine refractory thyroid cancer, using drugs to induce and redifferentiate thyroid cancer, and further increasing the iodine uptake rate to make the tumor "reclaim," is a therapy that is worth trying. Some therapies, such as peroxisome proliferator-activated receptor (PPAR) agonists, retinoic acids, histone deacetylase inhibitors, DNA methyltransferase inhibitors, and some other targeted therapeutic drugs, showed a "reclamation" which made a little progress in treating iodine refractory thyroid cancer. In a Phase II clinical trial, applying the PPAR agonist, rosiglitazone, to treat DTC demonstrated an increased iodine uptake rate in 40% (4/10) of the patients (Kebebew et al., 2006). In another clinical trial, the increased rate was observed in 30% (7/23) of patients (Tepmongkol, Keelawat, Honsawek, & Ruangvejvorachai, 2008). After binding with the PPAR agonist, the receptor PPAR-γ forms a heterodimer with RXR, which in turn combines with the target gene promoter region PPPE. It then activates the transcription of the target gene, promotes expression of the target gene PTEN, and this inhibits the PI3K signaling pathway, which is abnormally activated in thyroid cancer cells (Kapiteijn et al., 2012). Sodium iodide symporter (NIS) mRNA levels in thyroid cancer were significantly lower than those in benign nodule and normal thyroid tissue (Wang et al., 2011). The upregulation of NIS expression promoted iodine transport into the cytomembrane and increased the efficacy of radioiodine, and the expression was regulated by the TSHR signaling pathway (Lakshmanan et al., 2015). Clinical and *in vitro* experiments proved that the application of retinoic acid can induce redifferentiation of thyroid cancer cells, increase the expression of NIS and uptake of radioiodine, and improve the efficacy of radioiodine in thyroid cancer (Kogai, Taki, & Brent, 2006; Lan et al., 2016). A histone deacetylase inhibitor, Trichostatin A, was applied to intervene in three thyroid cancer cell lines (TPC-1, XTC-1, and FTC-133). NIS expression in these cell lines increased significantly, which made the tumor cells redifferentiate and recover the uptake of ^{131}I. The histone deacetylase inhibitor improved the NIS gene expression at the posttranslational modification level (Zarnegar et al., 2002). In another research, another histone deacetylase inhibitor, panobinostat, increased the expression level of NIS, and also improved the cytotoxicity of ^{131}I (Pugliese et al., 2013). In experiments using transgenic mice with thyroid cancer, targeted drugs were applied to induce and differentiate thyroid cancer with the drug-induced BRAFV^{600E} mutation. After giving an MEK inhibitor, PD0325901, and a BRAF inhibitor, PLX4720, sensitivity to radioiodine therapy was regained in the tumor (Chakravarty et al., 2011). This is because the targets that act upon MAPK and PI3K/AKT pathways, such as BRAF, MEK, and AKT, increase the expression level of thyroid genes and iodine uptake rate of cells, and the effect can also be improved by TSH (Hou, Bojdani, & Xing, 2010). In a clinical trial using a combination therapy of Selumetinib and ^{131}I to evaluate 20 refractory thyroid cancer patients, iodine uptake was increased in 12 patients. The ^{124}I PET/CT scanning results showed that the iodine uptake in 8 patients was more than 20 Gy, and after receiving ^{131}I therapy, patients showed a therapeutic response with imaging evidence. In 20 patients, the iodine uptake rates along with NRAS mutation were all increased (Ho et al., 2013). Above all, the MAPK signaling pathway and

PI3K/AKT signaling pathway played a very important role in the "reclaim therapy" of residual tumors.

5. IMPROVEMENT IN ANTICANCER ABILITY AND SIGNALING PATHWAYS

Gatenby published a paper regarding the change of strategic ideas of antitumor therapy in the journal *Nature*. "Patients and politicians anxiously await and increasingly demand a 'cure' for cancer. But trying to control the disease may prove a better plan than striving to cure it," says Gatenby (2009). Cancer is a systemic disease, and controlling the tumor without destroying it is obviously not comprehensive. We believe that in the case of refractory thyroid cancer, except for therapies aimed at eliminating the tumor, differentiation and induction of a residual tumor, making it reclaim, and immune induction on the body, which increases the antitumor ability, are more important. Hanahan also believed that in the past 40-year anticancer war, we did not succeed while targeted therapies were not radical and persistent. The resistance of cancer was generated due to repeated attacks, so novel multidimensional views were needed to fight the cancer war (Hanahan, 2014). Komarova thought that cancer was a moving target, showing dynamic change, and thus, precision medicine based on genetic targets was limited (Komarova, 2015). It has been proven that the invasion and metastasis of tumor cells was mediated by neurotransmitters. Various neurotransmitters are expressed in tumor cells, which further supported the theory that psychological–social factors are related to tumor development (Entschladen, Drell, Lang, Joseph, & Zaenker, 2004). Through a breast cancer mice model that was stress induced, Sloan et al. found that potential breast cancer metastasis was increased 30-fold, and the effect was mediated by the β-adrenergic signaling pathway. The β-antagonist, propranolol, was applied to treat stress-induced animals, and the drug reversed the tumor metastasis induced by stress. The application of a CSF-1 receptor kinase inhibitor, GW2580, also inhibited the influence of stress on distant metastasis (Sloan et al., 2010). Ondicova and Mravec thought that the effect of the nervous system in the onset of cancer was through the body fluids and neural pathway, which passed the information of cancer cells into the brain, and the brain regulated tumor growth by the neuroendocrine–immune system (Ondicova & Mravec, 2010). Psychological–social factors not only regulate the healthy human genome but also regulate the cancer genome. The sympathetic nervous system promotes cancer metastasis (Cole, 2013).

A β-endorphin neuron transplanted into a rat hypothalamus (sympathetic nerve center) could excite the vagus, which could inhibit the sympathetic nerve, alleviate inflammatory response and epithelial–mesenchymal transition in tumor tissues, and decrease the metastatic rate of breast cancer cells (Sarkar et al., 2011). Under chronic psychological stress conditions, the signals from the locus coeruleus and hypothalamus ventral median nuclei can cause disorder in the functioning of the hypothalamic–pituitary–adrenal and sympathetic nerves. This in turn leads to the release of increased neurotransmitters, such as catecholamines, mainly expressed as increased levels of epinephrine and noradrenaline, and decreased levels of dopamine. Epinephrine and noradrenaline can promote proliferation, and the adrenal gland can promote invasion and metastasis of tumor cells, while the effects of dopamine are just the opposite (Lutgendorf, Sood, & Antoni, 2010). In contrast, an enriched environment can significantly upregulate the level of brain-derived neurotrophic factor in the hypothalamus, which inhibits leptin expression, increases adiponectin, and inhibits tumor growth by the β-adrenergic receptor signaling pathway (Cao et al., 2010; Li et al., 2015). In addition, exercise, music, etc. can also change the hypothalamus–pituitary–adrenal axis and neurological function, further inhibiting the development of tumor (Archie, Bruera, & Cohen, 2013; Nyhuis, Masini, Sasse, Day, & Campeau, 2010). All these points suggest that the biobehavioral signaling pathway related to social–psychology–nerve–immune factors plays an important role in the occurrence and development of cancer. Although direct evidence regarding the influence of social–psychology–nerve–immune factors on refractory thyroid cancer is still lacking, the author believes that the biobehavioral signaling pathway can effectively regulate by increasing the systemic anticancer ability and intervene in the development of refractory thyroid cancer.

6. THYROID CANCER-TARGETED THERAPEUTIC DRUGS

The exploration of signaling pathways and molecular changes in thyroid cancer promote the development of targeted therapeutic drugs. So far, Sorafenib, Lenvatinib, Vandetanib, and Cabozantinib have been approved as targeted therapies for thyroid cancer by the United States Food and Drug Authority (FDA). Sorafenib is a kind of tyrosine kinase inhibitor (TKI), and the targets were VEGFR 1–3, PDGFR, c-KIT, BRAF, RET/PTC, etc. It is worth mentioning that a Phase III clinical trial of Sorafenib has been performed by Brose et al. (2014). In 417 patients with locally advanced or metastatic radioiodine refractory DTC in a DECISION clinical trial conducted in United States, Europe, and Asia, there was a median age of 63 years and patients with distant metastasis (96%), papillary type (66%), and Eastern Cooperative Oncology Group performance status of 0 (61%). Patients did not receive any chemotherapy or targeted therapy. All the patients showed disease development in the 14 months before enrollment. After randomization, the patients were given either Sorafenib 400 mg or placebo twice daily. Sorafenib did not induce a complete response. There were 12.2% and 0.5% of patients in the Sorafenib group and placebo group, respectively, who showed a partial response. The disease was stable for more than 6 months in 42% of the patients, and the disease control rates of the two groups were 54% and 34%, respectively. Above all, 73% of the patients in Sorafenib group showed tumor reduction in different degrees, while the placebo group showed only 27% tumor reduction. Based on these results, Sorafenib was approved by FDA for the treatment of radioiodine refractory DTC in terminal stages. Lenvatinib is also a multitarget TKI, and the targets were VEGFR1–3, FGFR1–4, PDGFR, RET, c-KIT, etc. Schlumberger et al. (2015) performed a Phase III clinical trial which included 392 patients with relapse or metastasis DTC failed by iodine therapy in a SELECT clinical trial. Patients randomly received either Lenvatinib 24 mg once daily or placebo therapy orally. Results showed that Lenvatinib significantly improved the median progression-free survival rate (PFS) (18.3 months vs 3.6 months, $P < 0.001$), and the benefit ran through all the subgroups, including subtypes that had previously received TKI therapy and low differentiation and columnar cells subtypes. In the Lenvatinib group, 64.8% of the patients experienced tumor remission, while in the placebo group this figure was only 1.5%. Based on the efficacy of significant improvement in median PFS and tumor remission, FDA approved the drug Lenvatinib for the treatment of DTC that has been failed by radioiodine therapy. Vandetanib is a kind of powerful small TKI molecule that is administrated orally. It can selectively act on RET, and competitively inhibit VEGF receptor 2 and slightly inhibit EGFR. Wells et al. reported a randomized, double-blind, Phase III clinical trial (ZETA) on Vandetanib in the treatment of locally advanced or distant metastatic MTC. The results showed that the median follow-up period was 24 months; the median PFS in the therapy group (231 cases) was significantly longer than that in the placebo group (100 cases) (30.5 months vs 19.3 months, HR, 0.46; 95% CI 0.31–0.69, $P < 0.001$). This is because more than half of the patients were still alive when the data were analyzed, and hence the survival rates could not be compared. However, the subjective efficiency, disease control rate, and biochemical index in the therapy group were improved significantly compared to those in the placebo group. Therefore, the drug was approved by the FDA to treat unresectable, locally advanced, or metastatic MTC in adults (Wells et al., 2012). Cabozantinib is a multitarget TKI, and this selectively inhibits RET, VEGF receptor 2, and MET. Elisei et al. reported a Phase III randomized clinical trial (EXAM) on Cabozantinib therapy for locally advanced or distant metastatic MTC on 330 patients. The main outcome of the trial was median progression-free survival. The secondary outcomes were response rates of tumor therapy, overall survival rates, and safety. The results showed that patients in the therapy group were administered 140 mg Cabozantinib every day. Results showed a median progression-free survival of 11.2 months, which was longer compared to the placebo group (4.0 months) (HR, 0.28; 95% CI 0.19–0.40; $P < 0.001$). The response rate in the therapy group was 28%, and 1-year progression-free survival of therapy group was 47.3%, compared to 7.2% in the placebo group. Therefore, the drug has been approved by FDA in the treatment of MTC with unresectable and symptomatic local development and distant metastasis (Elisei et al., 2013). Currently, many targeted therapeutic drugs have entered Phase I and/or Phase II clinical trials (Jin et al., 2016). According to the molecular biological theory as well as the results of current preclinical experiments of targeted drugs, these targeted drugs should be efficacious. However, there is a large gap between predictive value and practical efficacy, and the results of clinical trials of targeted drugs for thyroid cancer are still not satisfactory. Hence, more and more drug experiments that can help provide more options and with better efficacy for targeted therapies in the treatment of thyroid cancer are warranted.

7. CONCLUSION AND FUTURE DIRECTIONS

The changes of thyroid cell genes, signaling pathways, and related dysfunction of molecules promoted the occurrence, development, invasion, and metastasis of thyroid cancer. Although targeted therapeutic drugs bring possible therapeutic opportunities for refractory thyroid cancer, there is a great gap between predictive value and actual efficacy. Currently, there is still no completely effective drug in the treatment of refractory thyroid cancer. However, based on the signaling pathways, the "reclaim therapy" for residual tumor and systemic intervention aim at increasing anticancer's ability to encourage new directions and thoughts in the therapies for refractory thyroid cancer.

ACKNOWLEDGMENTS

This work is supported by the National Natural Science Foundation of China (Grant No. 81460157) and Program for Young Talents of Science and Technology in the Universities of Inner Mongolia Autonomous Region (Grant No. NJYT-14-B18).

REFERENCES

Ahn, H. S., Kim, H. J., Kim, K. H., Lee, Y. S., Han, S. J., Kim, Y., et al. (2016). Thyroid cancer screening in South Korea increases detection of papillary cancers with no impact on other subtypes or thyroid cancer mortality. *Thyroid*, *26*, 1535–1540. http://dx.doi.org/10.1089/thy.2016.0075. 27627550.

Ahn, H. S., Kim, H. J., & Welch, H. G. (2014). Korea's thyroid-cancer "epidemic"—Screening and overdiagnosis. *The New England Journal of Medicine*, *371*, 1765–1767. http://dx.doi.org/10.1056/NEJMp1409841. 25372084.

Archie, P., Bruera, E., & Cohen, L. (2013). Music-based interventions in palliative cancer care: A review of quantitative studies and neurobiological literature. *Support Care Cancer*, *21*, 2609–2624. http://dx.doi.org/10.1007/s00520-013-1841-4. 23715815.

Brose, M. S., Nutting, C. M., Jarzab, B., Elisei, R., Siena, S., Bastholt, L., et al. (2014). Sorafenib in radioactive iodine-refractory, locally advanced or metastatic differentiated thyroid cancer: A randomised, double-blind, phase 3 trial. *Lancet*, *384*, 319–328. http://dx.doi.org/10.1016/S0140-6736(14)60421-9. 24768112.

Cao, L., Liu, X., Lin, E. J., Wang, C., Choi, E. Y., Riban, V., et al. (2010). Environmental and genetic activation of a brain-adipocyte BDNF/leptin axis causes cancer remission and inhibition. *Cell*, *142*, 52–64. http://dx.doi.org/10.1016/j.cell.2010.05.029. 20603014.

Chakravarty, D., Santos, E., Ryder, M., Knauf, J. A., Liao, X. H., West, B. L., et al. (2011). Small-molecule MAPK inhibitors restore radioiodine incorporation in mouse thyroid cancers with conditional BRAF activation. *The Journal of Clinical Investigation*, *121*, 4700–4711. http://dx.doi.org/10.1172/JCI46382. 22105174.

Chan, C. M., Jing, X., Pike, L. A., Zhou, Q., Lim, D. J., Sams, S. B., et al. (2012). Targeted inhibition of Src kinase with dasatinib blocks thyroid cancer growth and metastasis. *Clinical Cancer Research*, *18*, 3580–3591. http://dx.doi.org/10.1158/1078-0432.CCR-11-3359. 22586301.

Choe, J. Y., Yun, J. Y., Nam, S. J., & Kim, J. E. (2012). Expression of c-Met is different along the location and associated with lymph node metastasis of head and neck carcinoma. *Korean Journal of Pathology*, *46*, 515–522. http://dx.doi.org/10.4132/KoreanJPathol.2012.46.6.515. 23323101.

Cole, S. W. (2013). Nervous system regulation of the cancer genome. *Brain, Behavior, and Immunity*, *Suppl*, S10–S18. http://dx.doi.org/10.1016/j.bbi.2012.11.008. 23207104.

Cong, S., Fang, L. W., Bao, H. L., Feng, Y. J., Wang, N., Yin, P., et al. (2016). Disease burden of thyroid cancer in the Chinese population, in 1990 and 2013. *Zhonghua Liu Xing Bing Xue Za Zhi*, *37*, 773–777. http://dx.doi.org/10.3760/cma.j.issn.0254-6450.2016.06.006. 27346100.

Durante, C., Haddy, N., Baudin, E., Leboulleux, S., Hartl, D., Travagli, J. P., et al. (2006). Long-term outcome of 444 patients with distant metastases from papillary and follicular thyroid carcinoma: Benefits and limits of radioiodine therapy. *The Journal of Clinical Endocrinology and Metabolism*, *91*, 2892–2899. http://dx.doi.org/10.1210/jc.2005-2838. 16684830.

Elisei, R., Schlumberger, M. J., Müller, S. P., Schöffski, P., Brose, M. S., Shah, M. H., et al. (2013). Cabozantinib in progressive medullary thyroid cancer. *Journal of Clinical Oncology*, *31*, 3639–3646. http://dx.doi.org/10.1200/JCO.2012.48.4659. 24002501.

Entschladen, F., Drell, T. L., 4th., Lang, K., Joseph, J., & Zaenker, K. S. (2004). Tumour-cell migration, invasion, and metastasis: Navigation by neurotransmitters. *The Lancet. Oncology, 5*, 254–258. http://dx.doi.org/10.1016/S1470-2045(04)01431-7. 15050959.

Gatenby, R. A. (2009). A change of strategy in the war on cancer. *Nature, 459*, 508–509. http://dx.doi.org/10.1038/459508a. 19478766.

Hanahan, D. (2014). Rethinking the war on cancer. *Lancet, 383*, 558–563. http://dx.doi.org/10.1016/S0140-6736(13)62226-6. 24351321.

Hanahan, D., & Weinberg, R. A. (2011). Hallmarks of cancer: The next generation. *Cell, 144*, 646–674. http://dx.doi.org/10.1016/j.cell.2011.02.013. 21376230.

Henderson, Y. C., Toro-Serra, R., Chen, Y., Ryu, J., Frederick, M. J., Zhou, G., et al. (2014). Src inhibitors in suppression of papillary thyroid carcinoma growth. *Head & Neck, 36*, 375–384. http://dx.doi.org/10.1002/hed.23316. 23729178.

Ho, A. L., Grewal, R. K., Leboeuf, R., Sherman, E. J., Pfister, D. G., Deandreis, D., et al. (2013). Selumetinib-enhanced radioiodine uptake in advanced thyroid cancer. *The New England Journal of Medicine, 368*, 623–632. http://dx.doi.org/10.1056/NEJMoa1209288. 23406027.

Hou, P., Bojdani, E., & Xing, M. (2010). Induction of thyroid gene expression and radioiodine uptake in thyroid cancer cells by targeting major signaling pathways. *The Journal of Clinical Endocrinology and Metabolism, 95*, 820–828. http://dx.doi.org/10.1210/jc.2009-1888. 20008023.

Jin, S., Borkhuu, O., Bao, W., & Yang, Y. T. (2016). Signaling pathways in thyroid cancer and their therapeutic implications. *Journal of Clinical Medicine Research, 8*, 284–296. http://dx.doi.org/10.14740/jocmr2480w. 26985248.

Kapiteijn, E., Schneider, T. C., Morreau, H., Gelderblom, H., Nortier, J. W., & Smit, J. W. (2012). New treatment modalities in advanced thyroid cancer. *Annals of Oncology, 23*, 10–18. http://dx.doi.org/10.1093/annonc/mdr117. 21471561.

Kebebew, E., Peng, M., Reiff, E., Treseler, P., Woeber, K. A., Clark, O. H., et al. (2006). A phase II trial of rosiglitazone in patients with thyroglobulin-positive and radioiodine-negative differentiated thyroid cancer. *Surgery, 140*, 960–966. discussion 966–967. http://dx.doi.org/10.1016/j.surg.2006.07.038 17188145.

Keinan-Boker, L., & Silverman, B. G. (2016). Trends of thyroid cancer in Israel: 1980–2012. *Rambam Maimonides Medical Journal*, 7(1), e0001. http://dx.doi.org/10.5041/RMMJ.1022826886958.

Kim, W. G., Guigon, C. J., Fozzatti, L., Park, J. W., Lu, C., Willingham, M. C., et al. (2012). SKI-606, an Src inhibitor, reduces tumor growth, invasion, and distant metastasis in a mouse model of thyroid cancer. *Clinical Cancer Research, 18*, 1281–1290. http://dx.doi.org/10.1158/1078-0432.CCR-11-2892. 22271876.

Kogai, T., Taki, K., & Brent, G. A. (2006). Enhancement of sodium/iodide symporter expression in thyroid and breast cancer. *Endocrine-Related Cancer, 13*, 797–826. http://dx.doi.org/10.1677/erc.1.01143. 16954431.

Komarova, N. L. (2015). Cancer: A moving target. *Nature, 525*, 198–199. http://dx.doi.org/10.1038/nature15210. 26308890.

Koo, B. S., Kim, J. M., Seo, S. T., Yoon, Y. H., Kwon, K. R., Kim, S. H., et al. (2014). Upregulation of HGF and c-MET is associated with subclinical central lymph node metastasis in papillary thyroid microcarcinoma. *Annals of Surgical Oncology, 21*, 2310–2317. http://dx.doi.org/10.1245/s10434-014-3553-5. 24562934.

Krejci, P., Aklian, A., Kaucka, M., Sevcikova, E., Prochazkova, J., Masek, J. K., et al. (2012). Receptor tyrosine kinases activate canonical WNT/β-catenin signaling via MAP kinase/LRP6 pathway and direct β-catenin phosphorylation. *PLoS One*7, e35826. http://dx.doi.org/10.1371/journal.pone.0035826 22558232.

Lakshmanan, A., Scarberry, D., Green, J. A., Zhang, X., Selmi-Ruby, S., & Jhiang, S. M. (2015). Modulation of thyroidal radioiodide uptake by oncological pipeline inhibitors and Apigenin. *Oncotarget, 6,* 31792–31804. http://dx.doi.org/10.18632/oncotarget.5172. 26397139.

Lan, L., Deng, W., Chen, H. L., Huo, L. L., Deng, L. L., Zhang, G. Y., et al. (2016). All-trans retinoic acid improves iodine uptake of thyroid cancer cells via repressing transcriptional activity of β-catenin. *Zhonghua Yi Xue Za Zhi, 96,* 553–558. 26902198.

Landa, I., Ganly, I., Chan, T. A., Mitsutake, N., Matsuse, M., Ibrahimpasic, T., et al. (2013). Frequent somatic TERT promoter mutations in thyroid cancer: Higher prevalence in advanced forms of the disease. *The Journal of Clinical Endocrinology and Metabolism, 98,* E1562–E1566. http://dx.doi.org/10.1210/jc.2013-2383. 23833040.

Leonardi, G. C., Candido, S., Carbone, M., Raiti, F., Colaianni, V., Garozzo, S., et al. (2012). BRAF mutations in papillary thyroid carcinoma and emerging targeted therapies (review). *Molecular Medicine Reports, 6,* 687–694. http://dx.doi.org/10.3892/mmr.2012.1016. 22858857.

Li, G., Gan, Y., Fan, Y., Wu, Y., Lin, H., Song, Y., et al. (2015). Enriched environment inhibits mouse pancreatic cancer growth and down-regulates the expression of mitochondria-related genes in cancer cells. *Scientific Reports, 5,* 7856. http://dx.doi.org/10.1038/srep07856. 25598223.

Li, C., Lee, K. C., Schneider, E. B., & Zeiger, M. A. (2012). BRAF V600E mutation and its association with clinicopathological features of papillary thyroid cancer: A meta-analysis. *The Journal of Clinical Endocrinology and Metabolism, 97,* 4559–4570. http://dx.doi.org/10.1210/jc.2012-2104. 23055546.

Lim, J. Y., Hong, S. W., Lee, Y. S., Kim, B. W., Park, C. S., Chang, H. S., et al. (2013). Clinicopathologic implications of the BRAF(V600E) mutation in papillary thyroid cancer: A subgroup analysis of 3130 cases in a single center. *Thyroid, 23,* 1423–1430. http://dx.doi.org/10.1089/thy.2013.0036. 23496275.

Liu, Z., Hou, P., Ji, M., Guan, H., Studeman, K., Jensen, K., et al. (2008). Highly prevalent genetic alterations in receptor tyrosine kinases and phosphatidylinositol 3-kinase/akt and mitogen-activated protein kinase pathways in anaplastic and follicular thyroid cancers. *The Journal of Clinical Endocrinology and Metabolism, 93,* 3106–3116. http://dx.doi.org/10.1210/jc.2008-0273. 18492751.

Lutgendorf, S. K., Sood, A. K., & Antoni, M. H. (2010). Host factors and cancer progression: Biobehavioral signaling pathways and interventions. *Journal of Clinical Oncology: Official Journal of the American Society of Clinical Oncology, 28,* 4094–4099. http://dx.doi.org/10.1200/JCO.2009.26.9357. 20644093.

Meza, R., & Chang, J. T. (2015). Multistage carcinogenesis and the incidence of thyroid cancer in the US by sex, race, stage and histology. *BMC Public Health, 15,* 789. 26282269. http://dx.doi.org/10.1186/s12889-015-2108-4.

Miao, X., & Zhao, Y. (2016). ST6GalNAcII mediates tumor invasion through PI3K/Akt/NF-κB signaling pathway in follicular thyroid carcinoma. *Oncology Reports, 35,* 2131–2140. 26820593. http://dx.doi.org/10.3892/or.2016.4590.

Nixon, I. J., Whitcher, M. M., Palmer, F. L., Tuttle, R. M., Shaha, A. R., Shah, J. P., et al. (2012). The impact of distant metastases at presentation on prognosis in patients with differentiated carcinoma of the thyroid gland. *Thyroid, 22,* 884–889. http://dx.doi.org/10.1089/thy.2011.0535. 22827579.

Nozhat, Z., & Hedayati, M. (2016). PI3K/AKT pathway and its mediators in thyroid carcinomas. *Molecular Diagnosis & Therapy, 20,* 13–26. http://dx.doi.org/10.1007/s40291-015-0175-y. 26597586.

Nyhuis, T. J., Masini, C. V., Sasse, S. K., Day, H. E., & Campeau, S. (2010). Physical activity, but not environmental complexity, facilitates HPA axis response habituation to repeated audiogenic stress despite neurotrophin mRNA regulation in both conditions. *Brain Research, 1362,* 68–77. http://dx.doi.org/10.1016/j.brainres.2010.09.038.20851112.

Ondicova, K., & Mravec, B. (2010). Role of nervous system in cancer aetiopathogenesis. *The Lancet Oncology*, *11*, 596–601. http://dx.doi.org/10.1016/S1470-2045(09)70337-7. 20522385.

Pellegriti, G., Frasca, F., Regalbuto, C., Squatrito, S., & Vigneri, R. (2013). Worldwide increasing incidence of thyroid cancer: Update on epidemiology and risk factors. *Journal of Cancer Epidemiology*, *2013*, 965212. http://dx.doi.org/10.1155/2013/965212. 23737785.

Pugliese, M., Fortunati, N., Germano, A., Asioli, S., Marano, F., Palestini, N., et al. (2013). Histone deacetylase inhibition affects sodium iodide symporter expression and induces 131I cytotoxicity in anaplastic thyroid cancer cells. *Thyroid*, *23*, 838–846. http://dx.doi.org/10.1089/thy.2012.0359. 23531031.

Rivera, M., Ghossein, R. A., Schoder, H., Gomez, D., Larson, S. M., & Tuttle, R. M. (2008). Histopathologic characterization of radioactive iodine-refractory fluorodeoxyglucose-positron emission tomography-positive thyroid carcinoma. *Cancer*, *113*, 48–56. 18484584. http://dx.doi.org/10.1002/cncr.23515.

Sáez, J. M. (2013). Treatment directed to signalling molecules in patients with advanced differentiated thyroid cancer. *Anti-Cancer Agents in Medicinal Chemistry*, *13*, 483–49522583421.

Sarkar, D. K., Zhang, C., Murugan, S., Dokur, M., Boyadjieva, N. I., Ortigüela, M., et al. (2011). Transplantation of β-endorphin neurons into the hypothalamus promotes immune function and restricts the growth and metastasis of mammary carcinoma. *Cancer Research*, *71*, 6282–6291. http://dx.doi.org/10.1158/0008-5472.CAN-11-1610. 21835894.

Schlumberger, M., Tahara, M., Wirth, L. J., Robinson, B., Brose, M. S., Elisei, R., et al. (2015). Lenvatinib versus placebo in radioiodine-refractory thyroid cancer. *The New England Journal of Medicine*, *372*, 621–630. http://dx.doi.org/10.1056/NEJMoa1406470. 25671254.

Siegel, R., DeSantis, C., Virgo, K., Stein, K., Mariotto, A., Smith, T., et al. (2016). Cancer treatment and survivorship statistics, 2012. *CA: A Cancer Journal for Clinicians*, *62*, 220–241. http://dx.doi.org/10.3322/caac.21149. 22700443.

Sloan, E. K., Priceman, S. J., Cox, B. F., Yu, S., Pimentel, M. A., Tangkanangnukul, V., et al. (2010). The sympathetic nervous system induces a metastatic switch in primary breast cancer. *Cancer Research*, *70*, 7042–7052. http://dx.doi.org/10.1158/0008-5472.CAN-10-0522. 20823155.

Sohn, S. Y., Park, W. Y., Shin, H. T., Bae, J. S., Ki, C. S., Oh, Y. L., et al. (2016). Highly concordant key genetic alterations in primary tumors and matched distant metastases in differentiated thyroid cancer. *Thyroid*, *26*, 672–682. http://dx.doi.org/10.1089/thy.2015.0527. 26971368.

Tepmongkol, S., Keelawat, S., Honsawek, S., & Ruangvejvorachai, P. (2008). Rosiglitazone effect on radioiodine uptake in thyroid carcinoma patients with high thyroglobulin but negative total body scan: A correlation with the expression of peroxisome proliferator-activated receptor-gamma. *Thyroid*, *18*, 697–704. 18630997. http://dx.doi.org/10.1089/thy.2008.0056.

Vidal, A. P., Andrade, B. M., Vaisman, F., Cazarin, J., Pinto, L. F., Breitenbach, M. M., et al. (2013). AMP-activated protein kinase signaling is upregulated in papillary thyroid cancer. *European Journal of Endocrinology*, *169*, 521–528. http://dx.doi.org/10.1530/EJE-13-0284. 23904275.

Wang, Z. F., Liu, Q. J., Liao, S. Q., Yang, R., Ge, T., He, X., et al. (2011). Expression and correlation of sodium/iodide symporter and thyroid stimulating hormone receptor in human thyroid carcinoma. *Tumori*, *97*, 540–546. http://dx.doi.org/10.1700/950.10410. 21989446.

Weidinger, C., Krause, K., Mueller, K., Klagge, A., & Fuhrer, D. (2011). FOXO3 is inhibited by oncogenic PI3K/Akt signaling but can be reactivated by the NSAID sulindac sulfide. *The Journal of Clinical Endocrinology and Metabolism, 96*, E1361–E1371. http://dx.doi.org/10.1210/jc.2010-2453. 21752881.

Wells, S. A., Jr., Robinson, B. G., Gagel, R. F., Dralle, H., Fagin, J. A., Santoro, M., et al. (2012). Vandetanib in patients with locally advanced or metastatic medullary thyroid cancer: A randomized, double-blind phase III trial. *Journal of Clinical Oncology, 30*, 134–141. http://dx.doi.org/10.1200/JCO.2011.35.5040. 22025146.

Xing, M. (2007). BRAF mutation in papillary thyroid cancer: Pathogenic role, molecular bases, and clinical implications. *Endocrine Reviews, 28*, 742–762. 17940185. http://dx.doi.org/10.1210/er.2007-0007.

Xing, M. (2009). Identifying genetic alterations in poorly differentiated thyroid cancer: A rewarding pursuit. *The Journal of Clinical Endocrinology and Metabolism, 94*, 4661–4664. 19959752. http://dx.doi.org/10.1210/jc.2009-2147.

Xu, X., Ding, H., Rao, G., Arora, S., Saclarides, C. P., Esparaz, J., et al. (2012). Activation of the Sonic Hedgehog pathway in thyroid neoplasms and its potential role in tumor cell proliferation. *Endocrine-Related Cancer, 19*, 167–179. http://dx.doi.org/10.1530/ERC-11-0305. 22241722.

Zarnegar, R., Brunaud, L., Kanauchi, H., Wong, M., Fung, M., Ginzinger, D., et al. (2002). Increasing the effectiveness of radioactive iodine therapy in the treatment of thyroid cancer using Trichostatin A, a histone deacetylase inhibitor. *Surgery, 132*, 984–990. discussion 990. http://dx.doi.org/10.1067/msy.2002.128690 12490845.

CHAPTER NINETEEN

Roles of Thyroid Transcription Factor 1 in Lung Cancer Biology

Cody A. Phelps, Shao-Chiang Lai, David Mu[1]
Leroy T. Canoles Jr. Cancer Research Center, Eastern Virginia Medical School, Norfolk, VA, United States
[1]Corresponding author: e-mail address: mud@evms.edu

Contents

1. Introduction 518
2. Genetic Alterations of TTF-1 in Cancer 518
3. Dynamics of Gene Content of a Core Amplicon 519
4. TTF-1$^-$ Lung Squamous Cell Carcinomas and TTF-1$^-$ Lung ADs 520
5. TTF-1 and Small Cell Lung Cancer 521
6. Pro- and Antioncogenic Activities of TTF-1 522
   6.1 Oncogenic Activities of TTF-1 in Lung Cancer 522
   6.2 Oncogenic Activities of TTF-1 in Other Cancers 523
   6.3 Antioncogenic Activities of TTF-1 524
7. The Link Between TTF-1 and Lipid/Cholesterol Metabolism in the Lung 530
8. The Connection Between TTF-1 and Secretome 532
9. Conclusion and Future Directions 533
Acknowledgment 533
References 535

Abstract

Thyroid transcription factor 1 (TTF-1 or NKX2-1) is a transcription factor of fundamental importance in driving lung maturation and morphogenesis. In the last decade, scientists began to appreciate the functional roles of TTF-1 in lung tumorigenesis. This movement was triggered by the discoveries of genetic alterations of *TTF-1* in the form of gene amplification in lung cancer. Many downstream target genes of TTF-1 relevant to the lung cancer biology of TTF-1 have been documented. One of the most surprising findings was that TTF-1 may exhibit either pro- or antitumorigenic activities, an outcome with the complexity exceeding the original anticipation purely based on the fact that *TTF-1* undergoes gene amplification in lung cancer. In the coming decade, we believe, we will witness additional surprises as the research exploring the cancer roles of TTF-1 progresses.

Vitamins and Hormones, Volume 106
ISSN 0083-6729
http://dx.doi.org/10.1016/bs.vh.2017.05.007

1. INTRODUCTION

Thyroid transcription factor 1 (TTF-1 or known as NKX2-1 for NK2 homeobox 1) has long been known as a critical development regulator of multiple organs and tissues including the lung (Kimura et al., 1996; Lazzaro, Price, de Felice, & Di Lauro, 1991; Minoo et al., 1995). While the developmental biology of TTF-1 was under active investigation, it became evident that TTF-1 immunopositivity may be used as a marker of lung adenocarcinomas (ADs) for diagnosis by pathologists (Goldstein & Thomas, 2001; Ordonez, 2000a; Travis et al., 2011). However, the genetic evidence that *TTF-1* may be functionally connected to lung tumorigenesis was not obtained until 2007–08 when four laboratories, including ours, independently reported the gene amplification event at the 14q13.3 locus harboring *TTF-1* in lung cancer (Kendall et al., 2007; Kwei et al., 2008; Tanaka et al., 2007; Weir et al., 2007). Since then, there has been a wealth of knowledge uncovered for the biological and mechanistic roles of TTF-1 in lung and other cancer types. Readers are referred to the two 2013 review articles for a synopsis of TTF-1-relevant cancer biology (Mu, 2013; Yamaguchi, Hosono, Yanagisawa, & Takahashi, 2013). In this chapter, we discuss the more recent advances in the cancer research of TTF-1 with an emphasis on lung cancer.

2. GENETIC ALTERATIONS OF TTF-1 IN CANCER

The *TTF-1* gene amplification documented by the initial four studies strongly suggested that *TTF-1* is a gain-of-function oncogene (Kendall et al., 2007; Kwei et al., 2008; Tanaka et al., 2007; Weir et al., 2007). Intriguingly, the core amplicon detected by each study varied in size and gene content. Our 2013 review has thoroughly contrasted the *TTF-1* gene amplification findings among these four studies (Mu, 2013). In 2014, The Cancer Genome Atlas (TCGA) consortium published a genomic analysis of 230 previously untreated lung ADs (Cancer Genome Atlas Research, 2014). Out of the 230 lung ADs, 143 ADs (62%) harbored known activating mutations in known driver oncogenes. These 143 ADs were collectively termed "oncogene positive." TCGA used the GISTIC method (Beroukhim et al., 2007) to analyze the DNA copy number alterations of either the entire 230 ADs or the 143 oncogene-positive ADs. In both sample sets, the 14q13.3 *TTF-1* amplicon was identified as the most recurrent and focal gene amplification event, but the core amplicon size varied from 169 kb in size with two genes (*TTF-1* and *SFTA3*) over the entire 230 ADs to 653 kb in size with four genes (*TTF-1*, *NKX2–8*, *MBIP*, and *SFTA3*) over the 143 oncogene-positive ADs. While the 14q13.3 *TTF-1* locus is unmistakably recurrently amplified in lung ADs, it has been known that this locus can suffer from microdeletion, encompassing *TTF-1* and *PAX9*, in congenital conditions (Gentile et al., 2016; Hayashi, Yagi, Morisaki, & Inazawa, 2015). The DNA copy number losses at 14q3.3 do not appear to be limited in congenital noncancer disorders. Indeed using a microarray-based tool, both DNA copy number gains and losses were observed in lung cancer, with amplification occurring in ADs and deletion arising in squamous and adenosquamous carcinomas (Harris et al., 2011). Loss-of-function frameshift and nonsense mutations of *TTF-1* have been found in a fraction of mucinous lung ADs (Hwang et al., 2016). The subject of TTF-1 and mucinous lung ADs will be discussed further later.

3. DYNAMICS OF GENE CONTENT OF A CORE AMPLICON

Functional elements of the human genome continue to evolve with time (Consortium, 2012). Consequently, the gene content of a core amplicon is destined to alter with time. Although *TTF-1* was a common target gene of the core amplicons discovered back in 2007–08 (Kendall et al., 2007; Kwei et al., 2008; Tanaka et al., 2007; Weir et al., 2007), based on the current genome annotation the core amplicon at 14q13.3, regardless of the sources of discovery, must contain these two additional genes—surfactant associated 3 (*SFTA3*) and NKX2-1 antisense RNA 1 (*NKX2-1-AS1*)—due to their proximity to *TTF-1*—*SFTA3* being only 3 kb away from *TTF-1* and *NKX2-1-AS1* overlapping with *TTF-1*. When we first discussed the concept of *SFTA3* being a coamplified gene of the 14q13.3 amplicon in 2013 (Mu, 2013), little biology of *SFTA3* was known. Subsequently, the murine ortholog of the human *SFTA3* gene, *Nanci* (Nkx2.1-associated noncoding intergenic RNA), was shown to act upstream of *Ttf-1* and downstream from Wnt signaling to regulate lung endoderm gene expression (Herriges et al., 2014). Knocking down *SFTA3* in human cells also resulted in decreased *TTF-1* expression (Herriges et al., 2014), suggesting a functional conservation between *Nanci* and *SFTA3* in spite of a low sequence conservation between the two genes (26%). While Herriges et al. provided data demonstrating that *Nanci* and *SFTA3* are long noncoding RNA, human *SFTA3* has also been annotated as a novel lung surfactant protein gene (Diler et al., 2014; Schicht et al., 2014) and the production of bacteria-derived recombinant SFTA3 protein has been reported (Schroder, Sollfrank, Paulsen, Brauer, & Schicht, 2016). Considering the well-documented activities of *TTF-1*/*Ttf-1* in tumorigenesis, it is tempting to speculate that *SFTA3* commands a functional role in human lung tumorigenesis. Future work will answer this important puzzle.

The second newcomer gene to the 14q13.3 amplicon is *NKX2-1-AS1*, which belongs to the class of natural antisense transcripts (NATs) (Balbin et al., 2015). The RNA expression of *TTF-1* and *NKX2-1-AS1* is strongly correlated, suggesting a cis-NAT regulation in lung cancer. Knockdown of *NKX2-1-AS1* decreased *TTF-1* expression level. Likewise, knockdown of *NKX2-1* with siRNAs also reduced expression of *NKX2-1-AS1*, establishing a reciprocal regulation of this cis-NAT pair (Balbin et al., 2015). Importantly, knockdown of *NKX2-1-AS1* impaired cell proliferation of the lung cell line NCI-H441, which is highly amplified at the 14q13.3 locus. This observation implicates a putative functional dependence on *NKX2-1-AS1* by lung AD cells with the *TTF-1* amplicon. Another recent study reported *NKX2-1-AS1* upregulation in lung ADs (Zhao, Luo, & Jiao, 2014), providing evidence in line with the idea that *NKX2-1-AS1* may be an active participant orchestrating lung tumorigenesis. It should be a priority to further dissect the mechanism of how *SFTA3* and *NKX2-1-AS1* are intertwined with lung cancer progression in either TTF-1$^+$ or TTF-1$^-$ background.

4. TTF-1$^-$ LUNG SQUAMOUS CELL CARCINOMAS AND TTF-1$^-$ LUNG ADS

After lung ADs, lung squamous cell carcinomas (SQCCs) are the most frequent lung cancer subtype within the nonsmall cell lung cancer (NSCLC) spectrum (Travis et al., 2011). But unlike lung ADs, lung SQCCs are characterized by the absence of TTF-1 expression (Pelosi et al., 2011; Rekhtman, Ang, Sima, Travis, & Moreira, 2011). Interestingly, oncogenic *EGFR*/*KRAS* mutations are also absent in lung SQCCs (Rekhtman et al., 2012). On the contrary, TTF-1 immunopositivity is associated with lung ADs and TTF-1-positive ADs are enriched with *EGFR* mutations (Chung et al., 2012; Shanzhi, Yiping, Ling, Jianming, & Qiang, 2014). Given the opposite expression profiles of TTF-1 between SQCCs and ADs, it has been suggested that the negative TTF-1 immunostaining, which has a high negative predictive value for the absence of *EGFR* mutations, may be taken as a surrogate marker for the choice of starting with chemotherapy to manage lung SQCCs in a clinical urgent setting (Vincenten et al., 2012). A persisting enigma regarding TTF-1 in lung SQCCs is that the amplification of the 14q13.3 *TTF-1* locus has been reported repeatedly in the literature (Kwei et al., 2008; Perner et al., 2009; Tang et al., 2011). TCGA has also detected the *TTF-1* amplicon in lung SQCCs (Cancer Genome Atlas Research Network, 2012). These observations are seemingly inconsistent with the well-accepted notion that the lack of TTF-1 expression is a molecular hallmark of lung SQCCs. We have offered hypotheses elsewhere (Mu, 2013) to rationalize this conflicting observation. Briefly, the first hypothesis is that the other coamplified genes in the *TTF-1* amplicon may be the functional target(s) of the gene amplification occurring in SQCCs. The two "newcomer" genes (*SFTA3* and *NKX2-1-AS1*) to the 14q13.3 amplicon are candidates to fulfill such a role. The second hypothesis is that the 14q13.3 amplicon detected in lung SQCCs is a passenger genetic alteration without actual functional consequences. More studies are needed to resolve this fundamental issue of the TTF-1-related cancer biology. In view of the poor prognosis faced by patients who have TTF-1$^-$ lung ADs (Barlesi et al., 2005; Barletta et al., 2009; Berghmans et al., 2006; Perner et al., 2009), Cardnell et al. conducted proteomic profiling to uncover that TTF-1$^-$ lung ADs harbor increased expression of DNA repair proteins and suppressed PI3K/mTOR signaling (Cardnell et al., 2015), exposing putative vulnerabilities of TTF-1$^-$ lung ADs for drug discovery.

5. TTF-1 AND SMALL CELL LUNG CANCER

Approximately 15%–17% of all diagnosed lung cancers are small cell lung cancer (SCLC) (Kahnert, Kauffmann-Guerrero, & Huber, 2016). A common molecular feature between SCLCs and ADs is that both lung cancer subtypes are positive for TTF-1 expression. In fact, TTF-1 immunopositivity is higher in SCLC (>90%) than in ADs (70%) (Di Loreto et al., 1997; Fabbro et al., 1996; Folpe et al., 1999). Although TTF-1 is not a marker specific for SCLCs, the prevalence of TTF-1 immunopositivity in SCLCs suggests that TTF-1 may be used to distinguish SCLCs from some nonpulmonary small cell carcinomas (Ordonez, 2000b). However, a significant fraction of extrapulmonary small cell carcinomas are positive for TTF-1 (Agoff et al., 2000; Travis, 2012), casting doubts on the utility of TTF-1 as a marker to distinguish SCLCs from extrapulmonary small cell carcinomas. Unlike lung ADs, SCLCs do not appear to harbor genetic alterations of *TTF-1* (George et al., 2015; McFadden et al., 2014; Peifer et al., 2012; Rudin et al., 2012). Therefore, this apparent question remains open: Is TTF-1 in any way functionally relevant to the biology of SCLC?

6. PRO- AND ANTIONCOGENIC ACTIVITIES OF TTF-1

6.1 Oncogenic Activities of TTF-1 in Lung Cancer

Following the study showing *MITF* (microphthalmia-associated transcription factor) as a lineage survival oncogene in melanomas (Garraway et al., 2005), *TTF-1* was proposed to be a candidate lineage lung oncogene (Garraway & Sellers, 2006). Thus, perhaps not surprisingly, as soon as the discovery of a multigenic 14q13.3 amplicon-containing *TTF-1* was first reported in 2007, *TTF-1* received the most attention from researchers. The exclusive occurrence of the 14q13.3 amplicon in NSCLC implicates that the amplification targets a lung lineage gene. At that time, *TTF-1* was already a well-known lung lineage master regulator (Maeda, Dave, & Whitsett, 2007; Minoo, 2000; Warburton et al., 2000). Logically, the spotlight was focused on *TTF-1*. In line with the notion that *TTF-1* may act as a lung lineage oncogene, *TTF-1* knockdown reduced growth, and/or viability of lung cancer cells containing the *TTF-1* amplicon (Kwei et al., 2008; Tanaka et al., 2007; Weir et al., 2007). To further dissect, the oncogenic mechanism of TTF-1, the Takahashi group demonstrated that TTF-1 directly induced the expression of the receptor tyrosine kinase-like orphan receptor 1 (ROR1) (Yamaguchi et al., 2012). The *ROR1* upregulation by TTF-1 stimulated the AKT pathway, pushing the cell into a more prosurvival state. This is complemented by the reduction of the proapoptotic, p38 signaling pathway. How the latter occurs mechanistically was later elucidated to be mediated by ROR1-induced inhibition of proapoptotic apoptosis signal-regulating kinase 1 (ASK1) in a kinase-dependent manner (Ida et al., 2016). ASK1 was shown to mediate the p38-stimulated apoptotic program. Knockdown of *ASK1* would reverse the growth inhibition brought about by *ROR1* knockdown. Unexpectedly, ROR1 was also shown to act independently of its kinase activity and interacts with EGFR and sustains the EGFR-ERBB3 signaling, further stimulating the AKT pathway. More recently, it was shown that ROR1 had yet an additional role, independent of its kinase functions, as a scaffolding protein for caveolin-1 (CAV1) and cavin-1. This partnership resulted in sustained caveolae formation and prosurvival signaling from multiple RTKs converging on AKT. This sustained AKT signaling is the result of ROR1 preventing lysosomal degradation of CAV1 (Yamaguchi et al., 2016).

In addition to *ROR1*, the *LMO3* (LIM-only protein 3) oncogene is also a direct transcriptional target of TTF-1 (Watanabe et al., 2013). This activity appears to be cooperatively performed with Forkhead box protein A1 (FOXA1) which also resides in chromosome 14 (14q21.1), about 1.1 Mb away from *TTF-1*. Interestingly, *FOXA1* gene amplification in lung ADs has been observed (Lin et al., 2002), but the TCGA study did not score the *FOXA1* amplicon as significant in lung ADs (Cancer Genome Atlas Research, 2014). Outside lung cancer, oncogenic point mutations of *FOXA1* were reported in castration-resistant prostate cancer (Grasso et al., 2012). The prooncogenic role of *LMO3* may be dependent on its ability to repress the tumor suppressor *TP53* and inhibit the transcription of its target genes (Larsen et al., 2010). Moreover, LMO3 interacts with the neuronal-specific basic helix-loop-helix transcription factor (HEN2) and acts as an oncogene in neuroblastoma (Aoyama et al., 2005). EGFR has also been documented to be a downstream mediator in *TTF-1*-amplified NSCLC and forms a negative feedback loop with *TTF-1* as *EGFR* knockdown led to *TTF-1* upregulation (Clarke et al., 2015). Undoubtedly, future studies will uncover additional oncogenic mediators of *TTF-1*.

In a study to characterize the molecular heterogeneity of early-stage, chemorefractory *KRAS*-mutant lung ADs, three robust subsets of tumors were detected (Skoulidis et al., 2015). These three subgroups of *KRAS*-mutant lung ADs are enriched with different co-occurring genetic events: (i) the KL subgroup (*STK11/LKB1*), (ii) the KP subgroup (*TP53*), and (iii) the KC subgroup (*CDKN2A/B* inactivation with low expression of *TTF-1*). In the tumors of the KC group, the PI3K pathway appeared depressed and 6 of the 15 KC lung ADs were classified as invasive mucinous TTF-1^{-} ADs. This histologic subtype was not observed in the KL and KP subgroups, consistent with the known concept that TTF-1 represses mucin expression (Maeda et al., 2011).

6.2 Oncogenic Activities of TTF-1 in Other Cancers

TTF-1 immunoreactivity is used by anatomic pathologists as a marker of carcinomas of lung or thyroid origin. But positive immunostaining of TTF-1 has been infrequently, but reproducibly, detected in other primary cancers originating from colon, ovary, endometrium, breast, and endocervix (Graham, Williams, & Salter, 2006; Kubba et al., 2008; Penman, Downie, &

Roberts, 2006; Siami et al., 2007; Zhang, Gao, Pasha, Litzky, & Livolsi, 2009). In breast cancer, an epidemiologic study found that TTF-1 expression is significantly associated with grade III tumor and triple negative breast cancer (Klingen et al., 2013). In endometrioid ADs of the uterine corpus, patients with TTF-1 immunoreactivity had a statistically significant worse prognosis with poorer disease-specific survival (Ervine, Leung, Gilks, & McCluggage, 2014). This negative clinical correlation is in contrast with the fact that positive TTF-1 expression in lung ADs is correlated with better patients' survival in multiple studies (Barlesi et al., 2005; Barletta et al., 2009; Berghmans et al., 2006; Cardnell et al., 2015; Perner et al., 2009). Functionally, it is not clear whether the ectopically expressed TTF-1 is an active player modulating the biology of these nonpulmonary TTF-1$^+$ tumors. Based on the two functional analyses of ectopically expressed TTF-1 in blood cancers, we theorize that ectopically expressed TTF-1 is likely an active participant of primary extrapulmonary solid tumors that are TTF-1$^+$. The first study uncovered oncogenic activities of ectopically expressed TTF-1 in T cell acute lymphoblastic leukemias (T-ALLs) (Homminga et al., 2011). Genomic rearrangements resulted in the expression of TTF-1 in T-ALLs; *TTF-1* transformed cells synergistically when combined with either *RAS* or *MYC*, two oncogenes frequently activated in T-ALLs (Kawamura et al., 1999; Palomero et al., 2006; Weng et al., 2004). The second study identified ectopic expression of *TTF-1* in diffuse large B-cell lymphomas (DLBCLs) as a result of aberrant chromatin modifications (Nagel et al., 2013). A key consequence of ectopic TTF-1 in this case is the direct transactivation HEY1 which then inhibits B-cell differentiation. More fascinatingly, expression analysis of DLBCL patient samples detected aberrant activity of TTF-1 in 5% of the total samples, suggesting the existence of a previous unrecognized (TTF-1$^+$) molecular subgroup of DLBCLs (Nagel et al., 2013).

6.3 Antioncogenic Activities of TTF-1

While many studies have demonstrated prooncogenic roles of *TTF-1*, as implied by *TTF-1* gene amplification, there exists literature suggesting antioncogenic activities of TTF-1. In a study published in 2004, Ttf-1 expression of early lung tumorigenesis was analyzed in an AJBL6 TGFβ heterozygous mouse model (Kang et al., 2004). They found that after treatment with the carcinogen ethyl carbamate, there was a reduction in *Ttf-1* mRNA expression in lung adenomas compared to normal tissues. This decrease was even stronger in ADs compared to normal tissues; this was also in agreement with immunohistochemical staining they performed. In another 2009 study examining TTF-1 and its effect on epithelial–mesenchymal transition (EMT) induced by TGFβ (Saito et al., 2009), it was found that TTF-1 can antagonize the TGFβ-induced EMT and can downregulate Snail and Slug, two transcription factors known to promote EMT (Polyak & Weinberg, 2009; Shih & Yang, 2011). The study went on to show that ectopic expression of TTF-1 reduced the invasion and motile properties of cancer cells. Likewise, knockdown of *TTF-1* in TTF-1-expressing cells resulted in an increase in cellular motility and invasion. Furthermore, a reduction in tumor growth and blood vessel area was detected in the tumor xenographs derived from Lewis lung carcinoma cells stably expressing *TTF-1* (Saito et al., 2009). All these studies are consistent with a tumor-suppressive role of TTF-1/Ttf-1.

The KP mouse system is a popular lung cancer animal model (DuPage, Dooley, & Jacks, 2009), which has been used to study the functional impacts of *Ttf-1* on lung tumorigenesis. This model couples the oncogenic activation of $Kras^{G12D}$ with the loss of *p53* function and has been used in several studies to elucidate *in vivo* functions of Ttf-1. In 2011, Winslow et al. showed that Ttf-1 can confer an antimetastatic phenotype to its host cell by repressing the expression of *Hmga2*, a potent prometastasis gene (Fusco & Fedele, 2007; Morishita et al., 2013), in the KP mouse model (Winslow et al., 2011). Further probing of their model system showed that nonmetastatic and metastatic primary lung ADs in the KP mice are $Ttf\text{-}1^{+}/Hmga2^{-}$ and $Ttf\text{-}1^{-}/Hmga2^{+}$, respectively. As to how *Ttf-1* represses *Hmga2*, it was not resolved by Winslow et al. MicroRNAs (miRNAs), such as let-7s, are a possible means for Ttf-1 to repress *Hmga2* because let-7s miRNAs have been shown to target *HMGA2/Hmga2* (Lee & Dutta, 2007; Mayr, Hemann, & Bartel, 2007; Park et al., 2007). But Winslow et al. ruled out a potential involvement of let-7 miRNAs in mediating the regulation of Hmga2 by Ttf-1 in view of the unaltered expression levels of let-7s in their study. This puzzle was resolved by a 2013 study showing that TTF-1/Ttf-1 upregulates the expression of miR-33a which targets *HMGA2/Hmga2* (Rice et al., 2013). This occurs by TTF-1 transactivating the promoter of *SREBF2*, the host gene of miR-33a (Horie et al., 2010; Marquart, Allen, Ory, & Baldan, 2010; Najafi-Shoushtari et al., 2010; Rayner et al., 2010). Importantly, this activity is conserved from mouse to human cells. Recently, Li et al. discovered an alternative mechanism of *Hmga2* repression by Ttf-1 (Li et al., 2015). Loss of Foxa2 and Cdx2 synergizes with loss of Ttf-1 to fully activate the metastatic program, and silencing of these three transcription factors is sufficient to account for a significant fraction of the gene expression differences between the nonmetastatic and metastatic states in lung ADs, including upregulated expression of the invadopodia component $Tks5_{long}$, the metastasis promoter Hmga2, and the EMT mediator Snail. Shortly after the study by Winslow et al. (2011), the Takahashi group reported that TTF-1 directly transactivates myosin-binding protein H (*MYBPH*) (Hosono et al., 2012). They showed that MYBPH binds to Rho kinase 1 (ROCK1), inhibiting the phosphorylation of myosin regulatory light chain, which resulted in a disruption in the actomyosin organization of the cell and a reduction of invasion and metastasis *in vivo*. Regarding the timing of *Ttf-1* downregulation along the progression of murine lung ADs, Caswell et al. determined that *Ttf-1* downregulation occurs prior to tumor cell dissemination from the primary tumor (Caswell et al., 2014), suggesting that dissemination is a hindrance for lung AD metastasis and that *Ttf-1* downregulation may alleviate this bottleneck step.

Pulmonary surfactant is a complex mixture of lipids and proteins, which is found in the fluid lining the alveolar surface of the lungs. Surfactant prevents alveolar collapse at low lung volume and preserves bronchiolar patency during normal and forced respiration (biophysical functions) (Bernhard, 2016). About 2%–5% of surfactant consists of surfactant proteins (SP) SP-A, -B, -C, and -D, named according to the order of detection, and TTF-1 is known to regulate the expression of these surfactant proteins (Bohinski, Di Lauro, & Whitsett, 1994; Dave, Childs, & Whitsett, 2004; Kelly, Bachurski, Burhans, & Glasser, 1996). Studies focusing on SP-D and SP-A demonstrated that both factors can suppress lung cancer progression—the former downregulates epidermal growth factor (EGF) signaling (Hasegawa et al., 2015), whereas the latter shifts the polarization of tumor-associated macrophages toward the M1 antitumorigenic phenotype (Mitsuhashi et al., 2013). These observations establish surfactant proteins as candidate mediators of the antioncogenic function of TTF-1.

TTF-1 has also been shown to regulate the factors within cell junctions. With the adherens junctions, TTF-1 induces the expression of a key adherens junction factor (E-cadherin) through the promoter of *CDH1* (Saito et al., 2009). In addition to regulating E-cadherin, TTF-1 was shown to be a modulator of tight junction (TJ) proteins, i.e., Occludin and Claudin-1. Both TJ factors are positively regulated by and are direct transcriptional targets of TTF-1 (Runkle et al., 2012). Furthermore, Claudin-18, another TJ factor, was shown to be upregulated by TTF-1 (Niimi et al., 2001).

These junctional factors (E-cadherin, Occludin, and Claudin-1) are important modulators of EMT, as their loss destabilizes their respective junctions and is an early step in the progression to a mesenchymal state (Lamouille, Xu, & Derynck, 2014; Martin & Jiang, 2009; Runkle & Mu, 2013). Collectively, the regulation of cellular phenotypes toward metastasis by TTF-1 appears to be multifactorial due to its influence on a myriad of factors that can impact cellular behavior relevant to metastasis.

In addition to its regulation of mediators that confer an antimetastatic phenotype, TTF-1 has also been shown to repress a mucinous phenotype in lung cancer. In human lung ADs, there is a subtype named mucinous lung ADs which comprise 5%–10% of human lung ADs. This mucin^{+} lung AD subtype typically lacks TTF-1 expression (Hata et al., 2010; Kunii et al., 2011); and frame-shift and nonsense mutations of *TTF-1* have been found in a fraction of mucinous lung ADs (Hwang et al., 2016), shedding light on the mechanism of negative TTF-1 expression in mucinous lung ADs. Maeda et al. revealed that haploinsufficiency of *Ttf-1* in *Kras*^{G12D} mice induced mucin expression and adopted properties of mucinous ADs, as indicated by an increase in the expression of gel-forming polymeric Muc5ac and Muc5b (Maeda et al., 2012). By investigating the promoter of *Muc5ac*, they found that Ttf-1 associated with the same DNA-binding motif recognized by AP-1, an oncogene frequently activated in cancer (Eferl & Wagner, 2003; Heintz, Janssen-Heininger, & Mossman, 2010; Shen et al., 2008). Their results paint a model in which TTF-1 binds to the AP-1-binding sites in the *MUC5AC* promoter and inhibits the AP-1 transactivation of *MUC5AC*. In an independent study, a reduction of the *Ttf-1* gene dosage enhanced lung tumorigenesis in a KRas-driven model and induced a loss of pulmonary identity accompanied by a conversion to gastric lineage, implicating that Ttf-1 prevents lung ADs from gastric differentiation (Snyder et al., 2013). The role of mucins in lung cancer has been investigated as aberrant mucin expression is associated with tumorigenesis (Kufe, 2009). Lakshmanan et al. reported that MUC5AC enhances the migratory properties of lung cancer cells using a Boyden chamber assay (Lakshmanan et al., 2016). This increase in cellular migration was the result of MUC5AC interaction with integrin β4 and subsequent activation of focal adhesion kinase. In this study, they also found that MUC5AC confers resistance to the lung cancer chemotherapeutic cisplatin, a finding reminiscent of how mucins modulate cellular responses to other chemotherapies (Bafna, Kaur, Momi, & Batra, 2009; Ham et al., 2016; Kharbanda, Rajabi, Jin, Raina, & Kufe, 2013; Leteurtre et al., 2004). It is important to note that

Maeda et al. also found that overexpression of TTF-1 enhances cellular sensitivity to cisplatin. However, it is unknown if the regulation of MUC5AC by TTF-1 plays a role in cisplatin sensitivity. Furthermore, the mucin proteins have also been implicated in a conferral of a more mesenchymal phenotype (Ponnusamy et al., 2013). An intriguing realization is that many of the cellular phenotypic characteristics associated with *MUC5AC* knockdown are similar to those found in TTF-1-overexpressing cells. These behaviors include an increase in chemotherapy sensitivity (Lakshmanan et al., 2016; Maeda et al., 2012), a reduction in cell motility (Hosono et al., 2012; Lakshmanan et al., 2016), and an adoption of a more epithelial phenotype (Ponnusamy et al., 2013). It is likely that the regulation of MUC5AC by TTF-1 mediates the cellular phenotypes conferred by TTF-1; however, if and the degree to which the suppression of mucin proteins mediating the TTF-1 activities remain to be fully elucidated.

The tumor suppressor, *TP53*, has been shown to be positively regulated by TTF-1 (Chen, Wu, Cheng, Chen, & Lee, 2015; Tsai et al., 2014). Tsai et al. furthered their study by focusing on the TTF-1/p53 axis and the regulation of the expression of the LKB1 tumor suppressor (Tsai et al., 2014). Their studies suggest that TTF-1 upregulated LKB1 through expression of wild-type p53. However, when p53 is mutated, TTF-1 would reduce the expression of LKB1 through a reduction in SP1-mediated transcription of *LKB1*. Unexpectedly, in human lung ADs, TTF-1 expression is negatively correlated with p53 staining (Yatabe, Mitsudomi, & Takahashi, 2002). But another study failed to detect a statistically significant correlation between p53 expression and TTF-1 status among TTF-1^+ NSCLCs (Myong, 2003). The relationship between TTF-1 and p53 is indeed intriguing. When the endogenous *Ttf-1* gene was suppressed using the CRISPR technology in a $Kras^{G12D/+}$; $p53^{fl/fl}$ mouse model, the tumor burden of the animals increased (Sanchez-Rivera et al., 2014), in agreement with other reports (Maeda et al., 2012; Snyder et al., 2013; Winslow et al., 2011). However, in the study of Sanchez-Rivera et al., when the endogenous *Ttf-1* was suppressed in the $Kras^{G12D}$ mouse model with wt-*p53*, tumor burden remained relatively light and unchanged. Their results show that *Ttf-1* knockdown led to a more dedifferentiated mucinous phenotype.

As discussed earlier, TTF-1 has been shown to modulate both pro- and antitumorigenic functions. This dichotomous behavior of TTF-1 is best manifested when comparing the role of Ttf-1 in the tumor progression between *Egfr*- and *Ras*-driven murine lung cancer. As described earlier, Ttf-1 has been shown to inhibit tumorigenesis in *Ras*-driven mouse models of lung cancer (Maeda et al., 2012; Snyder et al., 2013; Winslow et al., 2011). Intriguingly, in an *Egfr*^{L858R}-driven mouse model of lung cancer, *Ttf-1* haploinsufficiency decreased tumor number and volume *in vivo* compared to their *Ttf-1*$^{+/+}$ littermates (Maeda et al., 2012). This study provided the *in vivo* evidence for the dual roles that TTF-1 plays in lung cancer, highlighting the significance of genetic background in manifesting TTF-1 activities.

Overall, TTF-1 proves to be a multitasking protein in cancer (Fig. 1). While it is a driver gene in the amplified 14q13.3 region in lung cancer as the survival and proliferation of the amplified lung cancer cells depend on *TTF-1* (Kwei et al., 2008; Tanaka et al., 2007; Weir et al., 2007), it also allows cells to adopt antimetastatic properties. Furthermore, the influence of TTF-1 on growth and tumorigenesis appears to be contingent on other driver mutations. While TTF-1 expression generally is correlated with a better clinical outcome in lung ADs (Barlesi et al., 2005; Barletta et al., 2009; Berghmans et al., 2006; Cardnell et al., 2015; Perner et al., 2009), TTF-1 DNA copy number amplification (Barletta et al., 2009) and coactivation of the TTF-1 and NKX2–8 signaling pathways (Hsu et al., 2009) are negatively correlated with better clinical outcome in NSCLC patients. The research discussed earlier is in line with *TTF-1* being a context-dependent cancer gene with either pro- or antitumorigenic functions, a feature shared by other well-validated cancer genes such as *MYC* (Hoffman & Liebermann, 2008; Nilsson & Cleveland, 2003; Uribesalgo, Benitah, & Di Croce, 2012), *Notch* (Roy, Pear, & Aster, 2007), *SOX4* (Vervoort, van Boxtel, & Coffer, 2013), and *AKT1* (Liu et al., 2006).

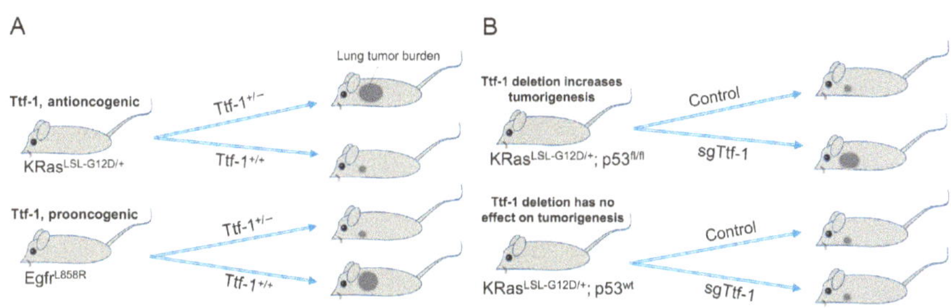

Fig. 1 Mouse lung cancer models manifesting either pro- or antitumorigenic activities of Ttf-1. (A) A *Ttf-1* gene dosage reduction enhances lung tumorigenesis in a *KRas*-driven lung AD model (Maeda et al., 2012; Snyder et al., 2013), but the opposite effect was observed in an *Egfr*-driven model (Maeda et al., 2012). (B) The antitumorigenic function of Ttf-1 requires mutant *Trp53* in a *KRas*-driven mouse lung cancer system (Sanchez-Rivera et al., 2014). *sgTtf-1*, single guide RNA-targeting Ttf-1.

7. THE LINK BETWEEN TTF-1 AND LIPID/CHOLESTEROL METABOLISM IN THE LUNG

Pulmonary alveolar membrane is the main structure where gas exchange occurs with blood in the lung. The alveolar wall consists of three types of cells, type I cells, type II cells, and macrophages. Pulmonary surfactant is produced and secreted by type II cells, and the major function of surfactant is to reduce surface tension of liquid in the lung in order to increase capability of alveolar membrane to exchange gas. Pulmonary surfactant is made up by lipid (80%–90%), surfactant-associated proteins, and cholesterol (Bernhard, 2016). Phospholipids, such as phosphatidylcholine and phosphatidylglycerol, are major components in pulmonary surfactant (Wright & Clements, 1987). The role of phosphatidylglycerol is to increase interaction between phospholipids and surfactant apoproteins (King & MacBeth, 1981). Out of the four types of pulmonary surfactant proteins (Han & Mallampalli, 2015), SP-A and SP-D belong to a protein family named collectins and may play a role in innate and adaptive immunity in the lung (Crouch, 1998; Crouch, Hartshorn, & Ofek, 2000; Crouch & Wright, 2001). SP-B and SP-C proteins are highly hydrophobic and both proteins are stored and secreted along with phospholipids by type II cells to produce surfactant (Schurch, Ospina, Cruz, & Perez-Gil, 2010). The function of SP-B is to facilitate surfactants to decrease surface tension (Whitsett, Nogee, Weaver, & Horowitz, 1995). The gene of SP-C is highly conserved among species, but its function in the lung is still not clear (Perez-Gil & Weaver, 2010).

Even though both lipid/cholesterol metabolism and TTF-1 have critical roles in modulating the function of surfactants through regulating various constituents of surfactants, the link between lipid/cholesterol metabolism and TTF-1 is poorly understood. In fact, there are only a few scientific studies implicating transcriptional regulation of lipid/cholesterol metabolism factors by TTF-1. Sterol regulatory element-binding proteins (SREBPs) are master transcriptional regulators supervising genes involved in lipid/cholesterol metabolism (Brown & Goldstein, 1997; Horton, Goldstein, & Brown, 2002). There are three members of SREBPs—SREBP1a and SREBP1c are encoded by *SREBF1*, whereas SREBP2 is encoded by *SREBF2*. The function and mechanism of SREBPs in lipid/cholesterol metabolism have been well studied in hepatocytes. In general, SREBP1a and 1c regulate genes involved in fatty acid synthesis and SREBP2 regulates genes of cholesterol biosynthesis (Horton et al., 2002). SREBP cleavage-activating protein (SCAP) is a sterol sensing protein that proteolytically activates SREBPs on ER membrane to produce mature/shortened SREBPs (Hua, Sakai, Brown, & Goldstein, 1996). The mechanism of SREBPs activation mediated by SCAP/INSIG pathway is very important to lung lipogenesis and maturation during mouse embryonic development (Bridges et al., 2014). Studies have shown that TTF-1 is one of the lung-specific transcriptional factors that regulate *ABCA3*, a lipid transporter which is critical to the function of pulmonary surfactants (Kaminski, Piehler, & Wenzel, 2006). TTF-1 transactivates *ABCA3* gene by directly interacting with two *cis*-regulatory regions in the *ABCA3* promoter (Besnard, Xu, & Whitsett, 2007). In the same study, overexpressing SREBP1c, but not SREBP1a and SREBP2, increased *ABCA3* promoter activity. Moreover, increased SREBP1c expression enhanced *ABCA3* mRNA and protein level during mouse lung development. This suggests a collaborative effort among TTF-1, SREBP1c, and ABCA3 in maintaining lipid and surfactant homeostasis in the lung (Besnard et al., 2007). Moreover, ZEB1 overexpression competes with endogenous TTF-1 in DNA binding, leading to a suppression of the expression of SP-A, LPCAT1, and ABCA3. The latter two proteins are involved in the supply of surfactant phosphatidylcholine in the type II pulmonary cells (Benlhabib, Guo, Pierce, & Mendelson, 2015). There is another study linking TTF-1 to the regulation of a cholesterol metabolism regulator—miR-33a (Rottiers & Naar, 2012) by demonstrating that TTF-1 can upregulate miR-33a by transactivating its host gene *SREBF2* (Rice et al., 2013). SREBP2 regulates most if not all of the genes mediating the mevalonate cholesterol biosynthesis pathway, including the HMG-CoA reductase to maintain cellular cholesterol homeostasis; whereas miR-33a directly represses ABCA1, a cholesterol efflux transporter (Kaminski et al., 2006), to assist cholesterol homeostasis (Najafi-Shoushtari et al., 2010; Rayner et al., 2010). The expression regulation of *SREBF2* and miR-33a by TTF-1 provides a hint that TTF-1 could functionally impact cholesterol metabolism locally in the lung. Moreover, ABCA1 is known to exhibit antitumorigenic activities (Smith & Land, 2012) and TTF-1 may impact ABCA1 activity via its control of miR-33a. These data raise a hypothesis that ABCA1 is a component of the TTF-1 signaling pathway. More studies are required to sharpen our understanding of a putative TTF-1/ABCA1 connection, and how TTF-1 is involved in lipid/cholesterol metabolism during not only lung development but also NSCLC progression.

8. THE CONNECTION BETWEEN TTF-1 AND SECRETOME

As chromatin immunoprecipitation experiments clearly indicate that TTF-1 can regulate the expression of hundreds of genes (Clarke et al., 2015; Maeda et al., 2012; Snyder et al., 2013; Tagne et al., 2012), perhaps unsurprisingly many of the genes under the regulation of TTF-1 may produce cell-secreted proteins. In 2011, the plasma protein profiles of murine lung cancer models were analyzed using a mass spectrometry approach and found to harbor a plasma protein signature for Ttf-1 (Taguchi et al., 2011). In this study, the circulating surfactant proteins (Sftpb and Sftpd), known targets of Ttf-1, were upregulated in the plasma. Additionally, the cholesterol transfer protein *Npc2* was also increased in the lung cancer plasma connected with Ttf-1^+ tumors.

In addition to surfactant proteins, club cell secretory protein (CCSP), another TTF-1 target gene (Whitsett & Glasser, 1998; Zhou, Lim, Costa, & Whitsett, 1996), can also function as an immunomodulator of the inflammatory response in an immunosuppressive role (Broeckaert & Bernard, 2000; Wang, Rosenberger, Bao, Stark, & Harrod, 2003; Zhang, Whitsett, & Stripp, 1997) and inhibit the production and activity of IFN-γ (Dierynck, Bernard, Roels, & De Ley, 1995). Conversely, IFN-γ and TNF-α have been shown to upregulate CCSP expression (Yao, Ikezono, et al., 1998; Yao, Levine, Cowan, Logun, & Shelhamer, 1998). Intriguingly, CCSP serum levels were inversely correlated with lung cancer mortality using data obtained from a prospective cohort study of participants in the Tucson Epidemiological Study of Airway Obstructive Disease (TESAOD) (Guerra, Vasquez, Spangenberg, Halonen, & Martinez, 2013).

Wood et al. shed more light on the secreted proteome under TTF-1 regulation by demonstrating that TTF-1 expression resulted in an alteration of the abundances of secreted angiogenic proteins present in the cell culture media (Wood et al., 2016). Their findings indicate that TTF-1 directly transactivates vascular endothelial growth factor (*VEGF*), leading to an increase of VEGF in the conditioned media (CM) of TTF-1-expressing cells. Paradoxically, the VEGF-enriched CM derived from TTF-1^+ cells did not exhibit a proangiogenic activity per the *in vitro* endothelial tube formation assay using primary human umbilical vein endothelial cells (HUVECs). Using an antibody array targeting angiogenesis factors, they discovered that the VEGF-enriched CM of TTF-1^+ cells also contains a higher level of granulocyte-macrophage colony-stimulating factor (GM-CSF) which would induce HUVECs to secrete soluble VEGF receptor 1 (sVEGFR1). sVEGFR1 is known to act as a VEGF sponge (Wu et al., 2010) and interferes with the angiogenic activity of the CM of TTF-1$^+$ cells (Wood et al., 2016). Consistently, a reduction in the blood vessel density has been observed in TTF-1$^+$ tumor xenographs (Saito et al., 2009). Previous work has also shown that TTF-1 directly represses the expression of MUC5AC (Maeda et al., 2011), a secreted mucin protein with known roles in tumorigenesis. In addition to its already discussed roles in exacerbating cancer as discussed earlier, MUC5AC also serves as a protective barrier to lung infection (Ehre et al., 2012). This is in line with the notion that TTF-1 regulates various secreted immunomodulatory factors such as GM-CSF, SP-A, and CCSP. Wood et al. further extended their study to exosomes by showing that the TTF-1-dependent upregulation of VEGF is also present in the exosomes secreted into the CM (Wood et al., 2016). This is the first evidence that TTF-1 alters an exosomal protein cargo load. It will be interesting to conduct proteomic studies to quantify the exact impact of TTF-1 globally on the entire secreted proteomes of the exosome and nonexosomal fractions of lung cancer secretome.

9. CONCLUSION AND FUTURE DIRECTIONS

In this chapter, we review recent findings relevant to the lung cancer biology of TTF-1. From the bountiful data discussed, it is evident that we have witnessed a robust advancement in the understanding of how TTF-1 is intertwined with lung cancer progression at the molecular level. Table 1 lists a selection of downstream functional mediators of TTF-1 in either oncogenic- or tumor-suppressive direction. In Table 2, we provide a list of questions touching on an array of TTF-1 characteristics discussed in this chapter for future research surrounding the roles of TTF-1 in cancer biology. It is our hope that this chapter would help the readers understand the status quo of TTF-1 biology in the realm of lung cancer; and we look forward to learning novel and surprising discoveries related to TTF-1 in the future.

ACKNOWLEDGMENT

The Mu laboratory was supported in part by NIH (CA198327). The content is solely the responsibility of the authors and does not necessarily represent the official views of the National Institutes of Health.

Table 1 Mediators of Both Pro- and Antioncogenic Activities of TTF-1

Mediators	**Biological Effects**	**Sources**
Oncogenic mediators		
EGFR	Cell proliferation	Clarke et al. (2015)
LMO3	Inhibits transcription of p53 target genes	Watanabe et al. (2013)
ROR1	Reduces p38 apoptotic signaling	Yamaguchi et al. (2012)
	Sustains Akt survival signaling	Yamaguchi et al. (2012)
	Sustains caveolae	Yamaguchi et al. (2016)
Antioncogenic mediators		
TGF-beta	Enhances mesenchymal phenotype	Saito et al. (2009)
miR-33a	Targets HMGA2	Rice et al. (2013)
HMGA2	Alters chromatins	Winslow et al. (2011)
MYBPH	Binds to ROCK1	Hosono et al. (2012)
	Inhibits phosphorylation of RLC	
	Reduces motility and metastasis	
E-cadherin	Reduces metastasis	Saito et al. (2009)
Occludin	Increases tight junctions	Runkle et al. (2012)
Claudin-1	Increases tight junctions	Runkle et al. (2012)
MUC5AC	Reduction in mucinous phenotype	Maeda et al. (2012) and Snyder et al. (2013)
p53	Upregulates LKB1	Tsai et al. (2014)
SP-A	Increasing antitumor microphages	Mitsuhashi et al. (2013)
SP-D	Inhibits binding of EGF to EGFR	Hasegawa et al. (2015)

Table 2 Questions for Future Research of TTF-1 Biology

1	Are *SFTA3* and/or *NKX2–1-AS1* involved in lung cancer progression in either TTF-1$^+$ or TTF-1$^-$ background?
2	Are there novel and druggable vulnerabilities of TTF-1$^+$ and TTF-1$^-$ lung adenocarcinomas?
3	Is TTF-1 functionally relevant to tumorigenesis of small cell lung cancer?
4	What functional consequences are afforded, if any, by the ectopically expressed TTF-1 in extrapulmonary, primary solid tumors?
5	As TTF-1 is not amenable to small-molecule drug discovery, are there druggable target proteins of TTF-1, which could serve as a surrogate drug discovery target?
6	What is the exact extent of the interactions between TTF-1 and long noncoding RNAs?
7	Do gel-forming polymeric mucins mediate TTF-1 activities?
8	Does TTF-1 regulate the local cholesterol metabolism in the lung?
9	Do secreted exosomes mediate the biology of TTF-1?
10	How does TTF-1 influence the tumor immunomicroenvironment through the TTF-1$^+$ tumor cell-secreted immunomodulators?

REFERENCES

Agoff, S. N., Lamps, L. W., Philip, A. T., Amin, M. B., Schmidt, R. A., True, L. D., et al. (2000). Thyroid transcription factor-1 is expressed in extrapulmonary small cell carcinomas but not in other extrapulmonary neuroendocrine tumors. *Modern Pathology*, *13*, 238–242.

Aoyama, M., Ozaki, T., Inuzuka, H., Tomotsune, D., Hirato, J., Okamoto, Y., et al. (2005). LMO3 interacts with neuronal transcription factor, HEN2, and acts as an oncogene in neuroblastoma. *Cancer Research*, *65*, 4587–4597.

Bafna, S., Kaur, S., Momi, N., & Batra, S. K. (2009). Pancreatic cancer cells resistance to gemcitabine: The role of MUC4 mucin. *British Journal of Cancer*, *101*, 1155–1161.

Balbin, O. A., Malik, R., Dhanasekaran, S. M., Prensner, J. R., Cao, X., Wu, Y. M., et al. (2015). The landscape of antisense gene expression in human cancers. *Genome Research*, *25*, 1068–1079.

Barlesi, F., Pinot, D., Legoffic, A., Doddoli, C., Chetaille, B., Torre, J. P., et al. (2005). Positive thyroid transcription factor 1 staining strongly correlates with survival of patients with adenocarcinoma of the lung. *British Journal of Cancer*, *93*, 450–452.

Barletta, J. A., Perner, S., Iafrate, A. J., Yeap, B. Y., Weir, B. A., Johnson, L. A., et al. (2009). Clinical significance of TTF-1 protein expression and TTF-1 gene amplification in lung adenocarcinoma. *Journal of Cellular and Molecular Medicine*, *13*, 1977–1986.

Benlhabib, H., Guo, W., Pierce, B. M., & Mendelson, C. R. (2015). The miR-200 family and its targets regulate type II cell differentiation in human fetal lung. *The Journal of Biological Chemistry*, *290*, 22409–22422.

Berghmans, T., Paesmans, M., Mascaux, C., Martin, B., Meert, A. P., Haller, A., et al. (2006). Thyroid transcription factor 1—A new prognostic factor in lung cancer: A meta-analysis. *Annals of Oncology*, *17*, 1673–1676.

Bernhard, W. (2016). Lung surfactant: Function and composition in the context of development and respiratory physiology. *Annals of Anatomy*, *208*, 146–150.

Beroukhim, R., Getz, G., Nghiemphu, L., Barretina, J., Hsueh, T., Linhart, D., et al. (2007). Assessing the significance of chromosomal aberrations in cancer: Methodology and application to glioma. *Proceedings of the National Academy of Sciences of the United States of America*, *104*, 20007–20012.

Besnard, V., Xu, Y., & Whitsett, J. A. (2007). Sterol response element binding protein and thyroid transcription factor-1 (Nkx2.1) regulate Abca3 gene expression. *American Journal of Physiology Lung Cellular and Molecular Physiology*, *293*, L1395–L1405.

Bohinski, R. J., Di Lauro, R., & Whitsett, J. A. (1994). The lung-specific surfactant protein B gene promoter is a target for thyroid transcription factor 1 and hepatocyte nuclear factor 3, indicating common factors for organ-specific gene expression along the foregut axis. *Molecular and Cellular Biology*, *14*, 5671–5681.

Bridges, J. P., Schehr, A., Wang, Y., Huo, L., Besnard, V., Ikegami, M., et al. (2014). Epithelial SCAP/INSIG/SREBP signaling regulates multiple biological processes during perinatal lung maturation. *PLoS One*, *9*, e91376.

Broeckaert, F., & Bernard, A. (2000). Clara cell secretory protein (CC16): Characteristics and perspectives as lung peripheral biomarker. *Clinical and Experimental Allergy*, *30*, 469–475.

Brown, M. S., & Goldstein, J. L. (1997). The SREBP pathway: Regulation of cholesterol metabolism by proteolysis of a membrane-bound transcription factor. *Cell*, *89*, 331–340.

Cancer Genome Atlas Research Network. (2012). Comprehensive genomic characterization of squamous cell lung cancers. *Nature*, *489*, 519–525.

Cancer Genome Atlas Research Network. (2014). Comprehensive molecular profiling of lung adenocarcinoma. *Nature*, *511*, 543–550.

Cardnell, R. J., Behrens, C., Diao, L., Fan, Y., Tang, X., Tong, P., et al. (2015). An integrated molecular analysis of lung adenocarcinomas identifies potential therapeutic targets among TTF1-negative tumors, including DNA repair proteins and Nrf2. *Clinical Cancer Research*, *21*, 3480–3491.

Caswell, D. R., Chuang, C. H., Yang, D., Chiou, S. H., Cheemalavagu, S., Kim-Kiselak, C., et al. (2014). Obligate progression precedes lung adenocarcinoma dissemination. *Cancer Discovery*, *4*, 781–789.

Chen, P. M., Wu, T. C., Cheng, Y. W., Chen, C. Y., & Lee, H. (2015). NKX2-1-mediated p53 expression modulates lung adenocarcinoma progression via modulating IKKbeta/NF-kappaB activation. *Oncotarget*, *6*, 14274–14289.

Chung, K. P., Huang, Y. T., Chang, Y. L., Yu, C. J., Yang, C. H., Chang, Y. C., et al. (2012). Clinical significance of thyroid transcription factor-1 in advanced lung adenocarcinoma under epidermal growth factor receptor tyrosine kinase inhibitor treatment. *Chest*, *141*, 420–428.

Clarke, N., Biscocho, J., Kwei, K. A., Davidson, J. M., Sridhar, S., Gong, X., et al. (2015). Integrative genomics implicates EGFR as a downstream mediator in NKX2-1 amplified non-small cell lung cancer. *PLoS One*, *10*, e0142061.

Consortium, E. P. (2012). An integrated encyclopedia of DNA elements in the human genome. *Nature*, *489*, 57–74.

Crouch, E. C. (1998). Collectins and pulmonary host defense. *American Journal of Respiratory Cell and Molecular Biology*, *19*, 177–201.

Crouch, E., Hartshorn, K., & Ofek, I. (2000). Collectins and pulmonary innate immunity. *Immunological Reviews*, *173*, 52–65.

Crouch, E., & Wright, J. R. (2001). Surfactant proteins a and d and pulmonary host defense. *Annual Review of Physiology*, *63*, 521–554.

Dave, V., Childs, T., & Whitsett, J. A. (2004). Nuclear factor of activated T cells regulates transcription of the surfactant protein D gene (Sftpd) via direct interaction with thyroid transcription factor-1 in lung epithelial cells. *The Journal of Biological Chemistry, 279*, 34578–34588.

Di Loreto, C., Di Lauro, V., Puglisi, F., Damante, G., Fabbro, D., & Beltrami, C. A. (1997). Immunocytochemical expression of tissue specific transcription factor-1 in lung carcinoma. *Journal of Clinical Pathology, 50*, 30–32.

Dierynck, I., Bernard, A., Roels, H., & De Ley, M. (1995). Potent inhibition of both human interferon-gamma production and biologic activity by the Clara cell protein CC16. *American Journal of Respiratory Cell and Molecular Biology, 12*, 205–210.

Diler, E., Schicht, M., Rabung, A., Tschernig, T., Meier, C., Rausch, F., et al. (2014). The novel surfactant protein SP-H enhances the phagocytosis efficiency of macrophage-like cell lines U937 and MH-S. *BMC Research Notes, 7*, 851.

DuPage, M., Dooley, A. L., & Jacks, T. (2009). Conditional mouse lung cancer models using adenoviral or lentiviral delivery of Cre recombinase. *Nature Protocols, 4*, 1064–1072.

Eferl, R., & Wagner, E. F. (2003). AP-1: A double-edged sword in tumorigenesis. *Nature Reviews. Cancer, 3*, 859–868.

Ehre, C., Worthington, E. N., Liesman, R. M., Grubb, B. R., Barbier, D., O'Neal, W. K., et al. (2012). Overexpressing mouse model demonstrates the protective role of Muc5ac in the lungs. *Proceedings of the National Academy of Sciences of the United States of America, 109*, 16528–16533.

Ervine, A., Leung, S., Gilks, C. B., & McCluggage, W. G. (2014). Thyroid transcription factor-1 (TTF-1) immunoreactivity is an adverse prognostic factor in endometrioid adenocarcinoma of the uterine corpus. *Histopathology, 64*, 840–846.

Fabbro, D., Di Loreto, C., Stamerra, O., Beltrami, C. A., Lonigro, R., & Damante, G. (1996). TTF-1 gene expression in human lung tumours. *European Journal of Cancer, 32A*, 512–517.

Folpe, A. L., Gown, A. M., Lamps, L. W., Garcia, R., Dail, D. H., Zarbo, R. J., et al. (1999). Thyroid transcription factor-1: Immunohistochemical evaluation in pulmonary neuroendocrine tumors. *Modern Pathology, 12*, 5–8.

Fusco, A., & Fedele, M. (2007). Roles of HMGA proteins in cancer. *Nature Reviews. Cancer, 7*, 899–910.

Garraway, L. A., & Sellers, W. R. (2006). Lineage dependency and lineage-survival oncogenes in human cancer. *Nature Reviews. Cancer, 6*, 593–602.

Garraway, L. A., Widlund, H. R., Rubin, M. A., Getz, G., Berger, A. J., Ramaswamy, S., et al. (2005). Integrative genomic analyses identify MITF as a lineage survival oncogene amplified in malignant melanoma. *436*, 117–122.

Gentile, M., De Mattia, D., Pansini, A., Schettini, F., Buonadonna, A. L., Capozza, M., et al. (2016). 14q13 distal microdeletion encompassing NKX2-1 and PAX9: Patient report and refinement of the associated phenotype. *American Journal of Medical Genetics Part A, 170*, 1884–1888.

George, J., Lim, J. S., Jang, S. J., Cun, Y., Ozretic, L., Kong, G., et al. (2015). Comprehensive genomic profiles of small cell lung cancer. *Nature, 524*, 47–53.

Goldstein, N. S., & Thomas, M. (2001). Mucinous and nonmucinous bronchioloalveolar adenocarcinomas have distinct staining patterns with thyroid transcription factor and cytokeratin 20 antibodies. *American Journal of Clinical Pathology, 116*, 319–325.

Graham, A. D., Williams, A. R., & Salter, D. M. (2006). TTF-1 expression in primary ovarian epithelial neoplasia. *Histopathology, 48*, 764–765.

Grasso, C. S., Wu, Y. M., Robinson, D. R., Cao, X., Dhanasekaran, S. M., Khan, A. P., et al. (2012). The mutational landscape of lethal castration-resistant prostate cancer. *Nature, 487*, 239–243.

Guerra, S., Vasquez, M. M., Spangenberg, A., Halonen, M., & Martinez, F. D. (2013). Serum concentrations of club cell secretory protein (Clara) and cancer mortality in adults:

A population-based, prospective cohort study. *The Lancet Respiratory Medicine*, *1*, 779–785.
Ham, S. Y., Kwon, T., Bak, Y., Yu, J. H., Hong, J., Lee, S. K., et al. (2016). Mucin 1-mediated chemo-resistance in lung cancer cells. *Oncogene*, *5*, e185.
Han, S., & Mallampalli, R. K. (2015). The role of surfactant in lung disease and host defense against pulmonary infections. *Annals of the American Thoracic Society*, *12*, 765–774.
Harris, T., Pan, Q., Sironi, J., Lutz, D., Tian, J., Sapkar, J., et al. (2011). Both gene amplification and allelic loss occur at 14q13.3 in lung cancer. *Clinical Cancer Research*, *17*, 690–699.
Hasegawa, Y., Takahashi, M., Ariki, S., Asakawa, D., Tajiri, M., Wada, Y., et al. (2015). Surfactant protein D suppresses lung cancer progression by downregulation of epidermal growth factor signaling. *Oncogene*, *34*, 838–845.
Hata, A., Katakami, N., Fujita, S., Kaji, R., Imai, Y., Takahashi, Y., et al. (2010). Frequency of EGFR and KRAS mutations in Japanese patients with lung adenocarcinoma with features of the mucinous subtype of bronchioloalveolar carcinoma. *Journal of Thoracic Oncology: Official Publication of the International Association for the Study of Lung Cancer*, *5*, 1197–1200.
Hayashi, S., Yagi, M., Morisaki, I., & Inazawa, J. (2015). Identical deletion at 14q13.3 including PAX9 and NKX2-1 in siblings from mosaicism of unaffected parent. *Journal of Human Genetics*, *60*, 203–206.
Heintz, N. H., Janssen-Heininger, Y. M., & Mossman, B. T. (2010). Asbestos, lung cancers, and mesotheliomas: From molecular approaches to targeting tumor survival pathways. *American Journal of Respiratory Cell and Molecular Biology*, *42*, 133–139.
Herriges, M. J., Swarr, D. T., Morley, M. P., Rathi, K. S., Peng, T., Stewart, K. M., et al. (2014). Long noncoding RNAs are spatially correlated with transcription factors and regulate lung development. *Genes & Development*, *28*, 1363–1379.
Hoffman, B., & Liebermann, D. A. (2008). Apoptotic signaling by c-MYC. *Oncogene*, *27*, 6462–6472.
Homminga, I., Pieters, R., Langerak, A. W., de Rooi, J. J., Stubbs, A., Verstegen, M., et al. (2011). Integrated transcript and genome analyses reveal NKX2-1 and MEF2C as potential oncogenes in T cell acute lymphoblastic leukemia. *Cancer Cell*, *19*, 484–497.
Horie, T., Ono, K., Horiguchi, M., Nishi, H., Nakamura, T., Nagao, K., et al. (2010). MicroRNA-33 encoded by an intron of sterol regulatory element-binding protein 2 (Srebp2) regulates HDL in vivo. *Proceedings of the National Academy of Sciences of the United States of America*, *107*, 17321–17326.
Horton, J. D., Goldstein, J. L., & Brown, M. S. (2002). SREBPs: Activators of the complete program of cholesterol and fatty acid synthesis in the liver. *The Journal of Clinical Investigation*, *109*, 1125–1131.
Hosono, Y., Yamaguchi, T., Mizutani, E., Yanagisawa, K., Arima, C., Tomida, S., et al. (2012). MYBPH, a transcriptional target of TTF-1, inhibits ROCK1, and reduces cell motility and metastasis. *The EMBO Journal*, *31*, 481–493.
Hsu, D. S., Acharya, C. R., Balakumaran, B. S., Riedel, R. F., Kim, M. K., Stevenson, M., et al. (2009). Characterizing the developmental pathways TTF-1, NKX2-8, and PAX9 in lung cancer. *Proceedings of the National Academy of Sciences of the United States of America*, *106*, 5312–5317.
Hua, X., Sakai, J., Brown, M. S., & Goldstein, J. L. (1996). Regulated cleavage of sterol regulatory element binding proteins requires sequences on both sides of the endoplasmic reticulum membrane. *The Journal of Biological Chemistry*, *271*, 10379–10384.
Hwang, D. H., Sholl, L. M., Rojas-Rudilla, V., Hall, D. L., Shivdasani, P., Garcia, E. P., et al. (2016). KRAS and NKX2-1 mutations in invasive mucinous adenocarcinoma of the lung. *Journal of Thoracic Oncology: Official Publication of the International Association for the Study of Lung Cancer*, *11*, 496–503.
Ida, L., Yamaguchi, T., Yanagisawa, K., Kajino, T., Shimada, Y., Suzuki, M., et al. (2016). Receptor tyrosine kinase-like orphan receptor 1, a target of NKX2-1/TTF-1

lineage-survival oncogene, inhibits apoptosis signal-regulating kinase 1-mediated pro-apoptotic signaling in lung adenocarcinoma. *Cancer Science, 107*, 155–161.
Kahnert, K., Kauffmann-Guerrero, D., & Huber, R. M. (2016). SCLC-state of the art and what does the future have in store? *Clinical Lung Cancer, 17*, 325–333.
Kaminski, W. E., Piehler, A., & Wenzel, J. J. (2006). ABC A-subfamily transporters: Structure, function and disease. *Biochimica et Biophysica Acta, 1762*, 510–524.
Kang, Y., Hebron, H., Ozbun, L., Mariano, J., Minoo, P., & Jakowlew, S. B. (2004). Nkx2.1 transcription factor in lung cells and a transforming growth factor-β1 heterozygous mouse model of lung carcinogenesis. *Molecular Carcinogenesis, 40*, 212–231.
Kawamura, M., Ohnishi, H., Guo, S. X., Sheng, X. M., Minegishi, M., Hanada, R., et al. (1999). Alterations of the p53, p21, p16, p15 and RAS genes in childhood T-cell acute lymphoblastic leukemia. *Leukemia Research, 23*, 115–126.
Kelly, S. E., Bachurski, C. J., Burhans, M. S., & Glasser, S. W. (1996). Transcription of the lung-specific surfactant protein C gene is mediated by thyroid transcription factor 1. *The Journal of Biological Chemistry, 271*, 6881–6888.
Kendall, J., Liu, Q., Bakleh, A., Krasnitz, A., Nguyen, K. C., Lakshmi, B., et al. (2007). Oncogenic cooperation and coamplification of developmental transcription factor genes in lung cancer. *Proceedings of the National Academy of Sciences of the United States of America, 104*, 16663–16668.
Kharbanda, A., Rajabi, H., Jin, C., Raina, D., & Kufe, D. (2013). Oncogenic MUC1-C promotes tamoxifen resistance in human breast cancer. *Molecular Cancer Research, 11*, 714–723.
Kimura, S., Hara, Y., Pineau, T., Fernandez-Salguero, P., Fox, C. H., Ward, J. M., et al. (1996). The T/ebp null mouse: Thyroid-specific enhancer-binding protein is essential for the organogenesis of the thyroid, lung, ventral forebrain, and pituitary. *Genes & Development, 10*, 60–69.
King, R. J., & MacBeth, M. C. (1981). Interaction of the lipid and protein components of pulmonary surfactant. Role of phosphatidylglycerol and calcium. *Biochimica et Biophysica Acta, 647*, 159–168.
Klingen, T. A., Chen, Y., Suhrke, P., Stefansson, I. M., Gundersen, M. D., & Akslen, L. A. (2013). Expression of thyroid transcription factor-1 is associated with a basal-like phenotype in breast carcinomas. *Diagnostic Pathology, 8*, 80.
Kubba, L. A., McCluggage, W. G., Liu, J., Malpica, A., Euscher, E. D., Silva, E. G., et al. (2008). Thyroid transcription factor-1 expression in ovarian epithelial neoplasms. *Modern Pathology, 21*, 485–490.
Kufe, D. W. (2009). Mucins in cancer: Function, prognosis and therapy. *Nature Reviews. Cancer, 9*, 874–885.
Kunii, R., Jiang, S., Hasegawa, G., Yamamoto, T., Umezu, H., Watanabe, T., et al. (2011). The predominant expression of hepatocyte nuclear factor 4alpha (HNF4alpha) in thyroid transcription factor-1 (TTF-1)-negative pulmonary adenocarcinoma. *Histopathology, 58*, 467–476.
Kwei, K. A., Kim, Y. H., Girard, L., Kao, J., Pacyna-Gengelbach, M., Salari, K., et al. (2008). Genomic profiling identifies TITF1 as a lineage-specific oncogene amplified in lung cancer. *Oncogene, 27*, 3635–3640.
Lakshmanan, I., Rachagani, S., Hauke, R., Krishn, S. R., Paknikar, S., Seshacharyulu, P., et al. (2016). MUC5AC interactions with integrin beta4 enhances the migration of lung cancer cells through FAK signaling. *Oncogene, 35*, 4112–4121.
Lamouille, S., Xu, J., & Derynck, R. (2014). Molecular mechanisms of epithelial-mesenchymal transition. *Nature Reviews. Molecular Cell Biology, 15*, 178–196.
Larsen, S., Yokochi, T., Isogai, E., Nakamura, Y., Ozaki, T., & Nakagawara, A. (2010). LMO3 interacts with p53 and inhibits its transcriptional activity. *Biochemical and Biophysical Research Communications, 392*, 252–257.

Lazzaro, D., Price, M., de Felice, M., & Di Lauro, R. (1991). The transcription factor TTF-1 is expressed at the onset of thyroid and lung morphogenesis and in restricted regions of the foetal brain. *Development*, *113*, 1093–1104.

Lee, Y. S., & Dutta, A. (2007). The tumor suppressor microRNA let-7 represses the HMGA2 oncogene. *Genes & Development*, *21*, 1025–1030.

Leteurtre, E., Gouyer, V., Rousseau, K., Moreau, O., Barbat, A., Swallow, D., et al. (2004). Differential mucin expression in colon carcinoma HT-29 clones with variable resistance to 5-fluorouracil and methotrexate. *Biology of the Cell*, *96*, 145–151.

Li, C. M., Gocheva, V., Oudin, M. J., Bhutkar, A., Wang, S. Y., Date, S. R., et al. (2015). Foxa2 and Cdx2 cooperate with Nkx2-1 to inhibit lung adenocarcinoma metastasis. *Genes & Development*, *29*, 1850–1862.

Lin, L., Miller, C. T., Contreras, J. I., Prescott, M. S., Dagenais, S. L., Wu, R., et al. (2002). The hepatocyte nuclear factor 3 alpha gene, HNF3{alpha} (FOXA1), on chromosome band 14q13 is amplified and overexpressed in esophageal and lung adenocarcinomas. *Cancer Research*, *62*, 5273–5279.

Liu, H., Radisky, D. C., Nelson, C. M., Zhang, H., Fata, J. E., Roth, R. A., et al. (2006). Mechanism of Akt1 inhibition of breast cancer cell invasion reveals a protumorigenic role for TSC2. *Proceedings of the National Academy of Sciences of the United States of America*, *103*, 4134–4139.

Maeda, Y., Chen, G., Xu, Y., Haitchi, H. M., Du, L., Keiser, A. R., et al. (2011). Airway epithelial transcription factor NK2 homeobox 1 inhibits mucous cell metaplasia and Th2 inflammation. *American Journal of Respiratory and Critical Care Medicine*, *184*, 421–429.

Maeda, Y., Dave, V., & Whitsett, J. A. (2007). Transcriptional control of lung morphogenesis. *Physiological Reviews*, *87*, 219–244.

Maeda, Y., Tsuchiya, T., Hao, H., Tompkins, D. H., Xu, Y., Mucenski, M. L., et al. (2012). Kras(G12D) and Nkx2-1 haploinsufficiency induce mucinous adenocarcinoma of the lung. *The Journal of Clinical Investigation*, *122*, 4388–4400.

Marquart, T. J., Allen, R. M., Ory, D. S., & Baldan, A. (2010). miR-33 links SREBP-2 induction to repression of sterol transporters. *Proceedings of the National Academy of Sciences of the United States of America*, *107*, 12228–12232.

Martin, T. A., & Jiang, W. G. (2009). Loss of tight junction barrier function and its role in cancer metastasis. *Biochimica et Biophysica Acta*, *1788*, 872–891.

Mayr, C., Hemann, M. T., & Bartel, D. P. (2007). Disrupting the pairing between let-7 and Hmga2 enhances oncogenic transformation. *Science*, *315*, 1576–1579.

McFadden, D. G., Papagiannakopoulos, T., Taylor-Weiner, A., Stewart, C., Carter, S. L., Cibulskis, K., et al. (2014). Genetic and clonal dissection of murine small cell lung carcinoma progression by genome sequencing. *Cell*, *156*, 1298–1311.

Minoo, P. (2000). Transcriptional regulation of lung development: Emergence of specificity. *Respiratory Research*, *1*, 109–115.

Minoo, P., Hamdan, H., Bu, D., Warburton, D., Stepanik, P., & deLemos, R. (1995). TTF-1 regulates lung epithelial morphogenesis. *Developmental Biology*, *172*, 694–698.

Mitsuhashi, A., Goto, H., Kuramoto, T., Tabata, S., Yukishige, S., Abe, S., et al. (2013). Surfactant protein A suppresses lung cancer progression by regulating the polarization of tumor-associated macrophages. *The American Journal of Pathology*, *182*, 1843–1853.

Morishita, A., Zaidi, M. R., Mitoro, A., Sankarasharma, D., Szabolcs, M., Okada, Y., et al. (2013). HMGA2 is a driver of tumor metastasis. *Cancer Research*, *73*, 4289–4299.

Mu, D. (2013). The complexity of thyroid transcription factor 1 with both pro- and anti-oncogenic activities. *The Journal of Biological Chemistry*, *288*, 24992–25000.

Myong, N. H. (2003). Thyroid transcription factor-1 (TTF-1) expression in human lung carcinomas: Its prognostic implication and relationship with wxpressions of p53 and Ki-67 proteins. *Journal of Korean Medical Science*, *18*, 494–500.

Nagel, S., Ehrentraut, S., Tomasch, J., Quentmeier, H., Meyer, C., Kaufmann, M., et al. (2013). Ectopic expression of homeobox gene NKX2-1 in diffuse large B-cell lymphoma is mediated by aberrant chromatin modifications. *PLoS One, 8,* e61447.
Najafi-Shoushtari, S. H., Kristo, F., Li, Y., Shioda, T., Cohen, D. E., Gerszten, R. E., et al. (2010). MicroRNA-33 and the SREBP host genes cooperate to control cholesterol homeostasis. *Science, 328,* 1566–1569.
Niimi, T., Nagashima, K., Ward, J. M., Minoo, P., Zimonjic, D. B., Popescu, N. C., et al. (2001). Claudin-18, a novel downstream target gene for the T/EBP/NKX2.1 homeodomain transcription factor, encodes lung- and stomach-specific isoforms through alternative splicing. *Molecular and Cellular Biology, 21,* 7380–7390.
Nilsson, J. A., & Cleveland, J. L. (2003). Myc pathways provoking cell suicide and cancer. *Oncogene, 22,* 9007–9021.
Ordonez, N. G. (2000a). Thyroid transcription factor-1 is a marker of lung and thyroid carcinomas. *Advances in Anatomic Pathology, 7,* 123–127.
Ordonez, N. G. (2000b). Value of thyroid transcription factor-1 immunostaining in distinguishing small cell lung carcinomas from other small cell carcinomas. *The American Journal of Surgical Pathology, 24,* 1217–1223.
Palomero, T., Lim, W. K., Odom, D. T., Sulis, M. L., Real, P. J., Margolin, A., et al. (2006). NOTCH1 directly regulates c-MYC and activates a feed-forward-loop transcriptional network promoting leukemic cell growth. *Proceedings of the National Academy of Sciences of the United States of America, 103,* 18261–18266.
Park, S.-M., Shell, S., Radjabi, A. R., Schickel, R., Feig, C., Boyerinas, B., et al. (2007). Let-7 prevents early cancer progression by suppressing expression of the embryonic gene HMGA2. *Cell Cycle, 6,* 2585–2590.
Peifer, M., Fernandez-Cuesta, L., Sos, M. L., George, J., Seidel, D., Kasper, L. H., et al. (2012). Integrative genome analyses identify key somatic driver mutations of small-cell lung cancer. *Nature Genetics, 44,* 1104–1110.
Pelosi, G., Rossi, G., Bianchi, F., Maisonneuve, P., Galetta, D., Sonzogni, A., et al. (2011). Immunohistochemistry by means of widely agreed-upon markers (cytokeratins 5/6 and 7, p63, thyroid transcription factor-1, and vimentin) on small biopsies of non-small cell lung cancer effectively parallels the corresponding profiling and eventual diagnoses on surgical specimens. *Journal of Thoracic Oncology: Official Publication of the International Association for the Study of Lung Cancer, 6,* 1039–1049.
Penman, D., Downie, I., & Roberts, F. (2006). Positive immunostaining for thyroid transcription factor-1 in primary and metastatic colonic adenocarcinoma: A note of caution. *Journal of Clinical Pathology, 59,* 663–664.
Perez-Gil, J., & Weaver, T. E. (2010). Pulmonary surfactant pathophysiology: Current models and open questions. *Physiology, 25,* 132–141.
Perner, S., Wagner, P. L., Soltermann, A., LaFargue, C., Tischler, V., Weir, B. A., et al. (2009). TTF1 expression in non-small cell lung carcinoma: Association with TTF1 gene amplification and improved survival. *The Journal of Pathology, 217,* 65–72.
Polyak, K., & Weinberg, R. A. (2009). Transitions between epithelial and mesenchymal states: Acquisition of malignant and stem cell traits. *Nature Reviews. Cancer, 9,* 265–273.
Ponnusamy, M. P., Seshacharyulu, P., Lakshmanan, I., Vaz, A. P., Chugh, S., & Batra, S. K. (2013). Emerging role of mucins in epithelial to mesenchymal transition. *Current Cancer Drug Targets, 13,* 945–956.
Rayner, K. J., Suarez, Y., Davalos, A., Parathath, S., Fitzgerald, M. L., Tamehiro, N., et al. (2010). MiR-33 contributes to the regulation of cholesterol homeostasis. *Science, 328,* 1570–1573.
Rekhtman, N., Ang, D. C., Sima, C. S., Travis, W. D., & Moreira, A. L. (2011). Immunohistochemical algorithm for differentiation of lung adenocarcinoma and squamous cell

carcinoma based on large series of whole-tissue sections with validation in small specimens. *Modern Pathology, 24*, 1348–1359.
Rekhtman, N., Paik, P. K., Arcila, M. E., Tafe, L. J., Oxnard, G. R., Moreira, A. L., et al. (2012). Clarifying the spectrum of driver oncogene mutations in biomarker-verified squamous carcinoma of lung: Lack of EGFR/KRAS and presence of PIK3CA/AKT1 mutations. *Clinical Cancer Research, 18*, 1167–1176.
Rice, S. J., Lai, S. C., Wood, L. W., Helsley, K. R., Runkle, E. A., Winslow, M. M., et al. (2013). MicroRNA-33a mediates the regulation of high mobility group AT-hook 2 gene (HMGA2) by thyroid transcription factor 1 (TTF-1/NKX2-1). *The Journal of Biological Chemistry, 288*, 16348–16360.
Rottiers, V., & Naar, A. M. (2012). MicroRNAs in metabolism and metabolic disorders. *Nature Reviews. Molecular Cell Biology, 13*, 239–250.
Roy, M., Pear, W. S., & Aster, J. C. (2007). The multifaceted role of notch in cancer. *Current Opinion in Genetics & Development, 17*, 52–59.
Rudin, C. M., Durinck, S., Stawiski, E. W., Poirier, J. T., Modrusan, Z., Shames, D. S., et al. (2012). Comprehensive genomic analysis identifies SOX2 as a frequently amplified gene in small-cell lung cancer. *Nature Genetics, 44*, 1111–1116.
Runkle, E. A., & Mu, D. (2013). Tight junction proteins: From barrier to tumorigenesis. *Cancer Letters, 337*(1), 41–48.
Runkle, E. A., Rice, S. J., Qi, J., Masser, D., Antonetti, D. A., Winslow, M. M., et al. (2012). Occludin is a direct target of thyroid transcription factor-1 (TTF-1/NKX2-1). *The Journal of Biological Chemistry, 287*, 28790–28801.
Saito, R. A., Watabe, T., Horiguchi, K., Kohyama, T., Saitoh, M., Nagase, T., et al. (2009). Thyroid transcription factor-1 inhibits transforming growth factor-beta-mediated epithelial-to-mesenchymal transition in lung adenocarcinoma cells. *Cancer Research, 69*, 2783–2791.
Sanchez-Rivera, F. J., Papagiannakopoulos, T., Romero, R., Tammela, T., Bauer, M. R., Bhutkar, A., et al. (2014). Rapid modelling of cooperating genetic events in cancer through somatic genome editing. *Nature, 516*, 428–431.
Schicht, M., Rausch, F., Finotto, S., Mathews, M., Mattil, A., Schubert, M., et al. (2014). SFTA3, a novel protein of the lung: Three-dimensional structure, characterisation and immune activation. *The European Respiratory Journal, 44*, 447–456.
Schroder, H., Sollfrank, L., Paulsen, F., Brauer, L., & Schicht, M. (2016). Recombinant expression of surfactant protein H (SFTA3) in *Escherichia coli*. *Annals of Anatomy - Anatomischer Anzeiger, 208*, 129–134.
Schurch, D., Ospina, O. L., Cruz, A., & Perez-Gil, J. (2010). Combined and independent action of proteins SP-B and SP-C in the surface behavior and mechanical stability of pulmonary surfactant films. *Biophysical Journal, 99*, 3290–3299.
Shanzhi, W., Yiping, H., Ling, H., Jianming, Z., & Qiang, L. (2014). The relationship between TTF-1 expression and EGFR mutations in lung adenocarcinomas. *PLoS One, 9*, e95479.
Shen, Q., Uray, I. P., Li, Y., Krisko, T. I., Strecker, T. E., Kim, H. T., et al. (2008). The AP-1 transcription factor regulates breast cancer cell growth via cyclins and E2F factors. *Oncogene, 27*, 366–377.
Shih, J. Y., & Yang, P. C. (2011). The EMT regulator slug and lung carcinogenesis. *Carcinogenesis, 32*, 1299–1304.
Siami, K., McCluggage, W. G., Ordonez, N. G., Euscher, E. D., Malpica, A., Sneige, N., et al. (2007). Thyroid transcription factor-1 expression in endometrial and endocervical adenocarcinomas. *The American Journal of Surgical Pathology, 31*, 1759–1763.
Skoulidis, F., Byers, L. A., Diao, L., Papadimitrakopoulou, V. A., Tong, P., Izzo, J., et al. (2015). Co-occurring genomic alterations define major subsets of KRAS-mutant lung adenocarcinoma with distinct biology, immune profiles, and therapeutic vulnerabilities. *Cancer Discovery, 5*, 860–877.

Smith, B., & Land, H. (2012). Anticancer activity of the cholesterol exporter ABCA1 gene. *Cell Reports*, *2*, 580–590.
Snyder, E. L., Watanabe, H., Magendantz, M., Hoersch, S., Chen, T. A., Wang, D. G., et al. (2013). Nkx2-1 represses a latent gastric differentiation program in lung adenocarcinoma. *Molecular Cell*, *50*, 185–199.
Tagne, J. B., Gupta, S., Gower, A. C., Shen, S. S., Varma, S., Lakshminarayanan, M., et al. (2012). Genome-wide analyses of Nkx2-1 binding to transcriptional target genes uncover novel regulatory patterns conserved in lung development and tumors. *PloS One*, *7*, e29907.
Taguchi, A., Politi, K., Pitteri, S. J., Lockwood, W. W., Faca, V. M., Kelly-Spratt, K., et al. (2011). Lung cancer signatures in plasma based on proteome profiling of mouse tumor models. *Cancer Cell*, *20*, 289–299.
Tanaka, H., Yanagisawa, K., Shinjo, K., Taguchi, A., Maeno, K., Tomida, S., et al. (2007). Lineage-specific dependency of lung adenocarcinomas on the lung development regulator TTF-1. *Cancer Research*, *67*, 6007–6011.
Tang, X., Kadara, H., Behrens, C., Liu, D. D., Xiao, Y., Rice, D., et al. (2011). Abnormalities of the TITF-1 lineage-specific oncogene in NSCLC: Implications in lung cancer pathogenesis and prognosis. *Clinical Cancer Research*, *17*, 2434–2443.
Travis, W. D. (2012). Update on small cell carcinoma and its differentiation from squamous cell carcinoma and other non-small cell carcinomas. *Modern Pathology*, *25*(Suppl. 1), S18–S30.
Travis, W. D., Brambilla, E., Noguchi, M., Nicholson, A. G., Geisinger, K. R., Yatabe, Y., et al. (2011). International association for the study of lung cancer/american thoracic society/european respiratory society international multidisciplinary classification of lung adenocarcinoma. *Journal of Thoracic Oncology: Official Publication of the International Association for the Study of Lung Cancer*, *6*, 244–285.
Tsai, L. H., Chen, P. M., Cheng, Y. W., Chen, C. Y., Sheu, G. T., Wu, T. C., et al. (2014). LKB1 loss by alteration of the NKX2-1/p53 pathway promotes tumor malignancy and predicts poor survival and relapse in lung adenocarcinomas. *Oncogene*, *33*, 3851–3860.
Uribesalgo, I., Benitah, S. A., & Di Croce, L. (2012). From oncogene to tumor suppressor: The dual role of Myc in leukemia. *Cell Cycle*, *11*, 1757–1764.
Vervoort, S. J., van Boxtel, R., & Coffer, P. J. (2013). The role of SRY-related HMG box transcription factor 4 (SOX4) in tumorigenesis and metastasis: Friend or foe? *Oncogene*, *32*, 3397–3409.
Vincenten, J., Smit, E. F., Vos, W., Grunberg, K., Postmus, P. E., Heideman, D. A., et al. (2012). Negative NKX2-1 (TTF-1) as temporary surrogate marker for treatment selection during EGFR-mutation analysis in patients with non-small-cell lung cancer. *Journal of Thoracic Oncology: Official Publication of the International Association for the Study of Lung Cancer*, *7*, 1522–1527.
Wang, S. Z., Rosenberger, C. L., Bao, Y. X., Stark, J. M., & Harrod, K. S. (2003). Clara cell secretory protein modulates lung inflammatory and immune responses to respiratory syncytial virus infection. *Journal of Immunology*, *171*, 1051–1060.
Warburton, D., Schwarz, M., Tefft, D., Flores-Delgado, G., Anderson, K. D., & Cardoso, W. V. (2000). The molecular basis of lung morphogenesis. *Mechanisms of Development*, *92*, 55–81.
Watanabe, H., Francis, J. M., Woo, M. S., Etemad, B., Lin, W., Fries, D. F., et al. (2013). Integrated cistromic and expression analysis of amplified NKX2-1 in lung adenocarcinoma identifies LMO3 as a functional transcriptional target. *Genes & Development*, *27*, 197–210.
Weir, B. A., Woo, M. S., Getz, G., Perner, S., Ding, L., Beroukhim, R., et al. (2007). Characterizing the cancer genome in lung adenocarcinoma. *Nature*, *450*, 893–898.
Weng, A. P., Ferrando, A. A., Lee, W., Morris, J. P. t., Silverman, L. B., Sanchez-Irizarry, C., et al. (2004). Activating mutations of NOTCH1 in human T cell acute lymphoblastic leukemia. *Science*, *306*, 269–271.

Whitsett, J. A., & Glasser, S. W. (1998). Regulation of surfactant protein gene transcription. *Biochimica et Biophysica Acta*, *1408*, 303–311.

Whitsett, J. A., Nogee, L. M., Weaver, T. E., & Horowitz, A. D. (1995). Human surfactant protein B: Structure, function, regulation, and genetic disease. *Physiological Reviews*, *75*, 749–757.

Winslow, M. M., Dayton, T. L., Verhaak, R. G., Kim-Kiselak, C., Snyder, E. L., Feldser, D. M., et al. (2011). Suppression of lung adenocarcinoma progression by Nkx2-1. *Nature*, *473*, 101–104.

Wood, L. W., Cox, N. I., Phelps, C. A., Lai, S. C., Poddar, A., Talbot, C., Jr., et al. (2016). Thyroid transcription factor 1 reprograms angiogenic activities of secretome. *Scientific Reports*, *6*, 19857.

Wright, J. R., & Clements, J. A. (1987). Metabolism and turnover of lung surfactant. *The American Review of Respiratory Disease*, *136*, 426–444.

Wu, F. T., Stefanini, M. O., Mac Gabhann, F., Kontos, C. D., Annex, B. H., & Popel, A. S. (2010). A systems biology perspective on sVEGFR1: Its biological function, pathogenic role and therapeutic use. *Journal of Cellular and Molecular Medicine*, *14*, 528–552.

Yamaguchi, T., Hosono, Y., Yanagisawa, K., & Takahashi, T. (2013). NKX2-1/TTF-1: An enigmatic oncogene that functions as a double-edged sword for cancer cell survival and progression. *Cancer Cell*, *23*, 718–723.

Yamaguchi, T., Lu, C., Ida, L., Yanagisawa, K., Usukura, J., Cheng, J., et al. (2016). ROR1 sustains caveolae and survival signalling as a scaffold of cavin-1 and caveolin-1. *Nature Communications*, *7*, 10060.

Yamaguchi, T., Yanagisawa, K., Sugiyama, R., Hosono, Y., Shimada, Y., Arima, C., et al. (2012). NKX2-1/TITF1/TTF-1-induced ROR1 is required to sustain EGFR survival signaling in lung adenocarcinoma. *Cancer Cell*, *21*, 348–361.

Yao, X. L., Ikezono, T., Cowan, M., Logun, C., Angus, C. W., & Shelhamer, J. H. (1998a). Interferon-gamma stimulates human Clara cell secretory protein production by human airway epithelial cells. *The American Journal of Physiology*, *274*, L864–L869.

Yao, X. L., Levine, S. J., Cowan, M. J., Logun, C., & Shelhamer, J. H. (1998b). Tumor necrosis factor-alpha stimulates human Clara cell secretory protein production by human airway epithelial cells. *American Journal of Respiratory Cell and Molecular Biology*, *19*, 629–635.

Yatabe, Y., Mitsudomi, T., & Takahashi, T. (2002). TTF-1 expression in pulmonary adenocarcinomas. *The American Journal of Surgical Pathology*, *26*, 767–773.

Zhang, P. J., Gao, H. G., Pasha, T. L., Litzky, L., & Livolsi, V. A. (2009). TTF-1 expression in ovarian and uterine epithelial neoplasia and its potential significance, an immunohistochemical assessment with multiple monoclonal antibodies and different secondary detection systems. *International Journal of Gynecological Pathology*, *28*, 10–18.

Zhang, L., Whitsett, J. A., & Stripp, B. R. (1997). Regulation of Clara cell secretory protein gene transcription by thyroid transcription factor-1. *Biochimica et Biophysica Acta*, *1350*, 359–367.

Zhao, W., Luo, J., & Jiao, S. (2014). Comprehensive characterization of cancer subtype associated long non-coding RNAs and their clinical implications. *Scientific Reports*, *4*, 6591.

Zhou, L., Lim, L., Costa, R. H., & Whitsett, J. A. (1996). Thyroid transcription factor-1, hepatocyte nuclear factor-3beta, surfactant protein B, C, and Clara cell secretory protein in developing mouse lung. *The Journal of Histochemistry and Cytochemistry*, *44*, 1183–1193.

FSC
www.fsc.org
MIX
Papier | Fördert gute Waldnutzung
FSC® C083411

Kolibri 360 GmbH
Ferdinand-Jühlke-Straße 7
99095 Erfurt, Deutschland
produktsicherheit@kolibri360.de